Design, Manufacturing and Mechatronics

Proceedings of the International Conference on Design, Manufacturing and Mechatronics (ICDMM2016)

Design, Manufacturing and Mechatronics

Proceedings of the International Conference on Design, Manufacturing and Mechatronics (ICDMM2016)

The 3rd Annual International Conference on Design, Manufacturing and Mechatronics (ICDMM2016)

Wuhan, China, 13 – 15 May 2016

Editor

A Mehran Shahhosseini
Indiana State University, USA

World Scientific

NEW JERSEY • LONDON • SINGAPORE • BEIJING • SHANGHAI • HONG KONG • TAIPEI • CHENNAI • TOKYO

Published by

World Scientific Publishing Co. Pte. Ltd.
5 Toh Tuck Link, Singapore 596224
USA office: 27 Warren Street, Suite 401-402, Hackensack, NJ 07601
UK office: 57 Shelton Street, Covent Garden, London WC2H 9HE

Library of Congress Cataloging-in-Publication Data
Names: International Conference on Design, Manufacturing and Mechatronics (3rd : 2016 : Wuhan, China), author. | Shahhosseini, A. Mehran, editor.
Title: Design, manufacturing and mechatronics : proceedings of the International Conference on Design, Manufacturing and Mechatronics (ICDMM2016) / editor, A. Mehran Shahhosseini, Indiana State University, USA.
Description: [Hackensack] New Jersey : World Scientific, [2017] | 3rd annual international conference held in Wuhan, China in 2016.
Identifiers: LCCN 2016055345| ISBN 9789813208315 (hardcover : alk. paper) | ISBN 9813208317 (hardcover : alk. paper)
Subjects: LCSH: Engineering design--Congresses. | Manufacturing processes--Congresses. | Mechatronics--Congresses. | Electrical engineering--Congresses.
Classification: LCC TA174 .I4717 2016 | DDC 621--dc23
LC record available at https://lccn.loc.gov/2016055345

British Library Cataloguing-in-Publication Data
A catalogue record for this book is available from the British Library.

Printed in Singapore

Preface

The 3rd Annual International Conference on Design, Manufacturing and Mechatronics (ICDMM2016) was successfully held in Wuhan, China in 2016.

The ICDMM2016 covers a wide range of fundamental studies, technical innovations and industrial applications in industry design, manufacturing and mechatronics. The ICDMM2016 programs consists of 4 keynote speeches, 96 oral and poster presentations. We were pleased to have more than 80 participants from China, South Korea, Taiwan, Japan, Malaysia, and Saudi Arabia. However, finally, only 83 articles were selected after peer review to be included in this proceeding.

Finally, we would like to specifically thank all members of keynote speakers, local organizing committee, international scientific committee, as well as all members and conference volunteers for their hard work and contributions to the success of ICDMM2016.

Prof. Ali "Mehran" Shahhosseini
Guest Editor
Department of Applied Engineering and Technology Management
Indiana State University
Terre Haute, Indiana 47809
USA

Committees

Technical Program Committee

Dr. Wei Wu, Hong Kong Scholars and Wuhan University, China
Dr. A. Mehran Shahhosseini, Indiana State University, USA.
Prof. Wen-Tsai Sung, National Chin-Yi University of Technology Department of Electrical Engineering, Taiwan
Ph.D. Sead Spuzic, The University of South Australia, Australia
Associate Prof. Ray-Hwa Wong, Department of Mechanical Eng., Hwa-Hsia University of Technology, Taiwan, R.O.C., Taiwan
Prof. Hung-Chun Chien, Jinwen University of Science and Technology, Taiwan
Ph.D. Jan Ziobro, Technical Institute of the Jan Grodek State Vocational Academy (PWSZ) in Sanok, Poland
Lecturer, Dr. Mak Kai-Long, The Hong Kong Polytechnic University, Hong Kong
Prof. Chia-Pao Chang, National Chin-Yi University of Technology/ Department of Industrial and Management, Taiwan
Ph.D. Imran Memon, Zhejiang University, China
Prof. Junwei Cao, Institute of Information Technology, Tsinghua University, China
Prof. Dr. Xinggao Liu, Zhejiang University, China
Prof. Yingmei Chen, Southeast University, China
Ph.D. Wensheng Cao, Jimei University, Xiamen, China
Prof. Xiaorong Luo, University of Electronic Science and Technology, China

Prof. Jian Wang, Huazhong University of Science and Technology, China
Ph.D. Zhongsheng Chen, National University of Defense Technology, China
Ph.D. Hui Yang, Beijing University of Posts and Telecommunications, China
Ph.D. Yidong Tan, Tsinghua University, China
Ph.D. Liang Hu, Department of Mechanical Engineering, Zhejiang University, China
Ph.D. Ray Y. Zhong, The University of Hong Kong, China
Dr. Bingjie Zhang, Beijing Technology and Business University, China
Dr. Weizun Zeng, Guangdong University of Technology, China
Dr. Jie Dai, Guangdong University of Technology, China
Dr. Xinshu Feng, Chengdu University, China
Dr. Weiyue Zhen, Jinan University, China
Dr. Ningli Kong, Hebei University, China
Dr. Jianwei Zhou, Shandong Polytechnic University, China
Dr. Xinyi Yu, Chengdu University, China
Dr. Junfeng Guo, Harbin Institute of Technology, China
Dr. Hui Liu, Hebei University, China
Dr. Qianqian Ye, Southwest University, China
Dr. Wufu Sheng, Hebei University, China

Keynote Speakers

Keynote Speaker I. Prof. Wen-Tsai Sung
Affiliation: National Chin-Yi University of Technology
Department of Electrical Engineering
Introduction: Wen-Tsai Sung is working with the Department of Electrical Engineering, National Chin-Yi University of Technology as a professor and Vice-Dean of Academic Affairs. He received a PhD and MS degree from the Department of Electrical Engineering, National Central University, Taiwan in 2007 and 2000. He has won the 2009 JMBE Best Annual Excellent Paper Award and the dragon thesis award that sponsor is Acer Foundation. His research interests include Wireless Sensors Network, Data Fusion, System Biology, System on Chip, Computer-Aided Design for Learning, Bioinformatics, and Biomedical Engineering. He has published a number of international journal and conferences article related to these areas. Currently, he is the chief of Wireless Sensors Networks Laboratory. At present, he serves as the Editor-in-Chief in three international journals: International Journal of Communications (IJC), Communications in Information Science and Management Engineering (CISME) and Journal of Vibration Analysis, Measurement, and Control (JVAMC), he also serves as the other international journals in Associate-Editor and Guest Editor (IET Systems Biology).

Keynote Speaker II. Prof. Ren-Jie Ji

Affiliation: College of Mechanical and Electronic Engineering, China University of Petroleum, Qingdao, Shandong

Introduction: Prof. Ren-Jie Ji received his Ph.D from Mechanical and Electronic Engineering in 2011. He is now associate professor in China University of Petroleum, Qingdao and as a Hong Kong Scholar in the Hong Kong Polytechnic University, Hong Kong. He has published more than 80 refereed journal papers and has presented more than 10 international conferences papers and is the member of Sinking EDM Branch of Chinese Non-Traditional Machining Society, Chinese Mechanical Engineering Society and Hong Kong association of scholars. For the commercialization of the results of his research work, 17 patents have also been granted and 10 patents have been filed. His research interests include advanced manufacturing technology, electrical discharge machining of difficult-to-machine materials and micro and precision machining of sand control screen, etc.

Keynote Speaker III. Prof. Feng Gao

Affiliation: School of Electrical Engineering, Chongqing University

Introduction: Prof. Feng Gao received his Ph.D from Tsinghua University in 2007. He is now research fellow of School of Electrical Engineering, Chongqing University and Vice Chairman of Research Center of Vehicle E/E and Intelligence. On February, 2013, Prof. Feng Gao was introduced to Chongqing University through One-hundred talent program and now is mainly engaged in vehicle dynamic control, ADAS (Advanced Driver Assistance Systems) and electromagnetic compatibility in automobile and other related topics. He has published 39 academic papers most of which has been indexed by SCI or EI and applied 11 patents for invention 6 of which has been authorized. He has won the Best

Paper Award of China Automotive Engineering Institute for 2 times and the third prize of improvement award, Chongqing Changan Technology.

Keynote Speaker IV. Prof. Yi-Dong Tan

Affiliation: Dept. Precision Instrument, Tsinghua University, Beijing, China

Introduction: Prof Yi-Dong Tan is currently an associate professor in Department of Precision Instruments, Tsinghua University, Beijing, China and also a visiting associate professor in Electrical Engineering, California Institute of Technology, USA. He received Ph.D in Optical Engineering from Department of Precision Instruments & Mechanology, Tsinghua University in 2008. Prof Yi-Dong Tan is mainly engaged in laser measuring technology of Precision Manufacturing, especially the laser feedback effect and applied research of precision measurement. More than 60 SCI papers have been published by Prof. Yi-Dong Tan, including International Top Journals 'Applied Physics Letters', 'Optics Letters', 'Optics Express'. He has applied 27 patents for invention and 8 patents have been authorized. He acts as reviewers in the following journal 'Applied Optics', 'Journal of Physics B, Journal of Physics D', 'Applied Physics B', 'Journal of Optics', 'Optics Express', 'Optics Letters'. As so far, he has won the Gold Prize in Jin Guofan Prize for excellent youth of China Instrument Society Scholarships and first prize in Electronic Information Science and Technology Award of Chinese Institute of Electronics and so on.

Table of Contents

Chapter 1

Automation and Mechatronics

Topography image segmentation based on multiphase level set method

Min-Rong Zhao
Natural Science College,
Air Force Engineering University,
Xi'an 710000, China
E-mail:zmr698963@qq.com

Xi-Wen Zhang
Information and Navigation College,
Air Force Engineering University,
Xi'an 710000, China

For the complexity of uneven and multiple target landform, this paper used the polyphase level set image segmentation to reconstruction 3d landform successfully. In order to improve the sliced effect and sliced speed of complex landform image, the RSF active contour evolution mode, improved Chan-Vese mode and polyphase level set mode combined to complete the segmentation of several landform images, and lay the foundation for 3d landform reconstruction based on image.

Keywords: Level set mode, image segmentation, landform reconstruction.

1. Introduction

The foundation for 3d landform reconstruction based on image is image segmentation the effect of the segmentation directly affect the accuracy of 3d reconstruction model. Identify the corresponding area of interested object that is image segmentation is the crucial step for reconstruction 3d object. Image segmentation plays an important role in the actual system such as in medical image processing, in order to rebuild the structure of an organ, the segmentation of the human CT scan image is needed[1]. In the remote sensing satellite image processing, in order to research the distribution of military target, the segmentation of the landform image is needed, the reconstruction of position model is based on landform image. Image segmentation is to separate different areas of the image has a special meaning and make the consistency of mutually disjoint areas, it is a classical problem of image processing and analysis.

2. A Level Set Method to Solve M-S Segmentation Model

Digital image has developed from 2d to 3d and from traditional gray level image to color image. At the same time, the traditional image processing techniques cannot meet the high demand, The shape change process cannot described in parameterized curve bight or hook face because of the landform complicated and the topological change of silhouette of key component is complexity and randomly

The image processing techniques based on partial differential equation excel in this respect, such as the traditional image edge detection method is difficult to realize 3d image segmentation, however , the mobile partition method of contour models based on the level set method is easier to do these. This way can handle the topological change very well, and divide the image which including many targets and extract the outline of targets[1,2], as shown in the Figure 1.

However , this way has a large calculation amount, a long sliced time, and even more serious for complex image, therefore, a new implementation model which increased the sliced speed of complex image in the way without requiring an initialization of signed distance function by introducing a new energy term was put forward, and the Chan-Vese mode was improved, so the iterations of segmentation was reduced, the sliced time was shortened, and the sliced speed was improved[3, 4], as shown in the Figure 2.

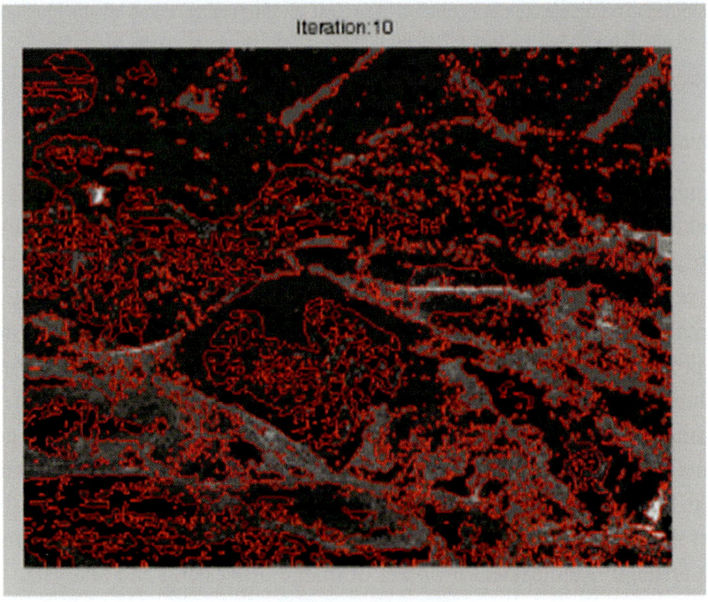

(a)300 times interated

Fig. 1. The sliced landform image based on C-V mode.

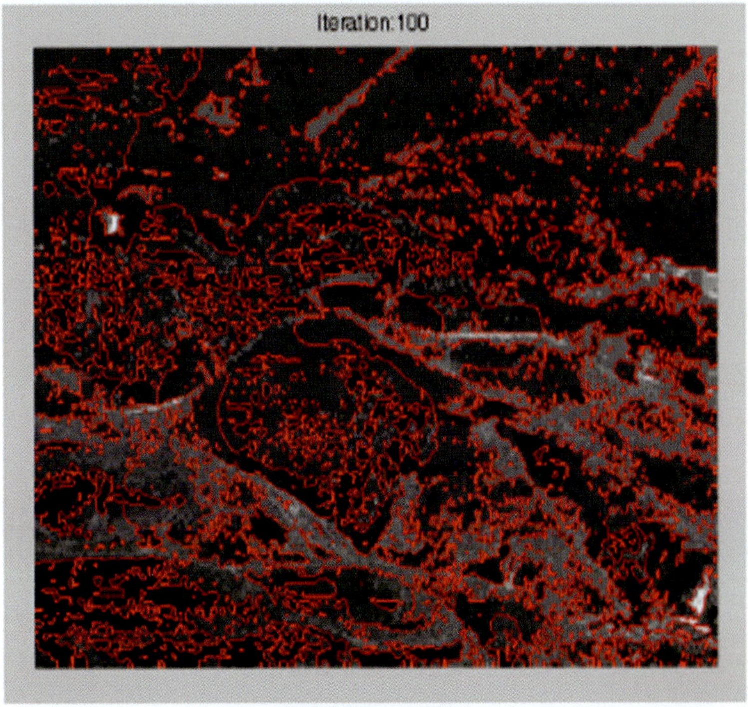

(b)500 times interated

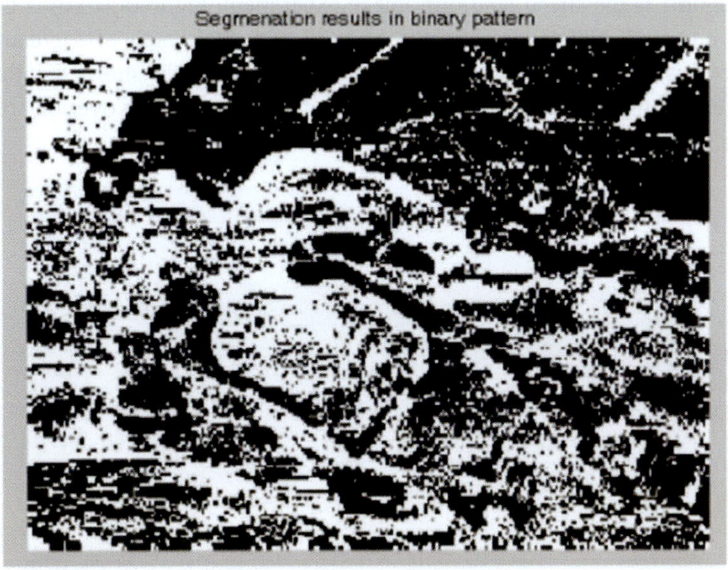

(c)The grey-scale map of segmentation

Fig. 1. (*Continued*)

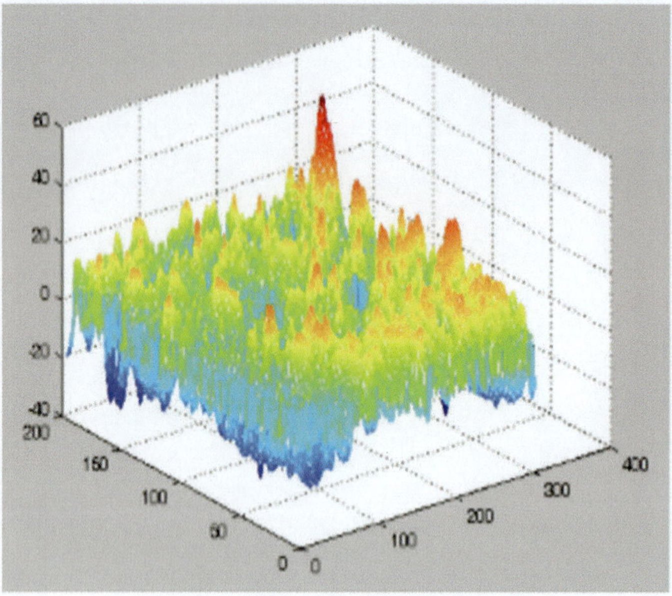

(d)Level set function

Fig. 1. (*Continued*)

Figure1 is the sliced landform image experiment. Figure(a)(b)(c) is the 300 times interated, 500 times interated and the grey-scale map of segmentation. Figure(d) is the level set mode after stabilization. The results shows that the C-V mode can be used to slice landform image, but the sliced time is overlong or even several hours.

(a) Original image

Fig. 2. The sliced landform image based on improved C-V mode.

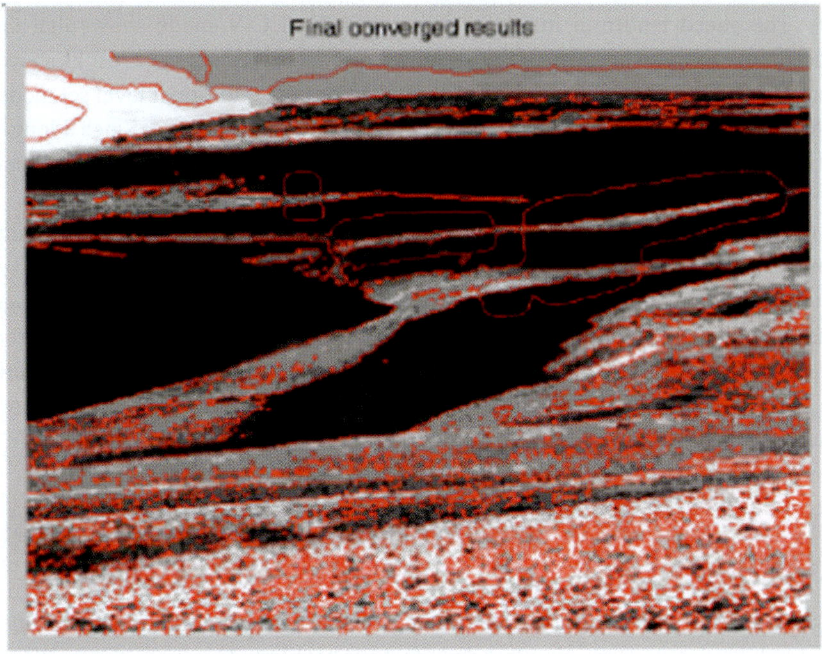

(b) The split graph based on C-V mode

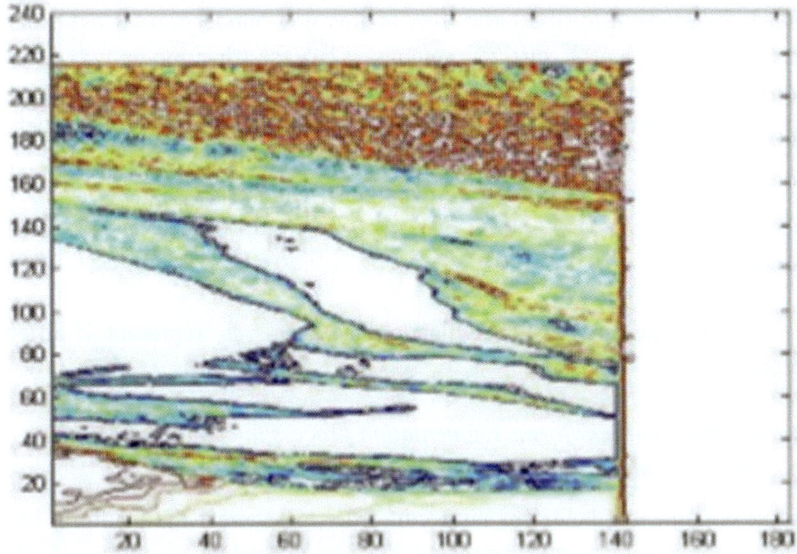

(c) The split graph based on improved mode

Fig. 2. (*Continued*)

The sliced landform image based on improved C-V mode shows that the sliced time reduced because of the initialization was leaved out. The results shows that the choice of Δt in improved mode can larger than traditional level set mode. The evolution of level set can remain stable as long as μ·Δt <1/4s. Since a longer time step can be chosen, the iterations of segmentation is reduced, the sliced time is shortened and the sliced speed is improved.

All the results of the experiment show that the modified method not only improved the sliced speed but also improved the segmentation effect and stability. But the problem existing in reality is that the uneven strength of landform image result in difficult object extraction. In order to correct the segmentation method and achieve more desirable segmentation results, all kinds of model is put forward to solve the unavailability of C-V mode in uneven image management and achieve good results.

3. Landform Image Segmentation based on Polyphase Level Set Mode

3.1. *Polyphase level set*

The image is divided into two part in the C-V mode and modified mode and the research begins with the simplest dichotomous problem. However, the landform image consists of multiple target regions and deals principally with multi-objects image. In 2000, Chan and Vese provided a polyphase level set mode based on the Chan-Vese mode[5, 6], the basic idea of this new mode is used n Level Set functions to represent 2^n different regions, it is Vese-Chan mode: two different region distinguished from ϕ can be expressed by a Level Set function, $\{\phi > 0\}$ and $\{\phi < 0\}$; Two Level Set functions can distinguish four parts, such as $\{\phi_1 > 0, \phi_2 > 0\}$, $\{\phi_1 > 0, \phi_2 < 0\}$, $\{\phi_1 < 0, \phi_2 > 0\}$, and $\{\phi_1 < 0, \phi_2 < 0\}$. Three Level Set functions ϕ_1, ϕ_2, ϕ_3 can distinguish eight parts, such as $\{\phi_1 > 0, \phi_2 > 0, \phi_3 > 0\}$, $\{\phi_1 > 0, \phi_2 > 0, \phi_3 < 0\}$, $\{\phi_1 > 0, \phi_2 < 0, \phi_3 > 0\}$, $\{\phi_1 > 0, \phi_2 < 0, \phi_3 < 0\}$, $\{\phi_1 < 0, \phi_2 > 0, \phi_3 > 0\}$, $\{\phi_1 > 0, \phi_2 > 0, \phi_3 < 0\}$, $\{\phi_1 < 0, \phi_2, 0, \phi_3 > 0\}$ and $\{\phi_1 < 0, \phi_2 < 0, \phi_3 < 0\}$.

By this token, n Level Set functions $\phi_1, \phi_2, \cdots, \phi_n$ represent 2^n different regions.

3.2. *The regional landforms segmentation process in comouflage region based on polyphase level set mode*

For the complicated landform image, in order to segment the landform image better, a more ideal way must be finded. The idea is combine the virtues of Chan-Vese mode and RSF mode, consider both global and local gray level information and modeling based on polyphase level set method to segment the complicated landform image. Because of adopting level set method to realize active contour evolution, the algorithm can be used to several targets segmentation and expanded to multichannel image segmentation such as color image without modification. Realizing steps are:

(1) *Constructed energy function*

The segmentation energy mode of dichotomous problem in level set is:

$$F(C,c_1,c_2)=\lambda_1 \int_{insed(C)} |u_0(x,y)-c_1|^2 dxdy+\lambda_2 \int_{outside(C)} |u_0(x,y)-c_2|^2 dxdy$$
$$+\mu L(C)+vS(insideC) \tag{1}$$

The 2^n positions segmentation energy function based on the four positions segmentation energy function is[5]:

$$\begin{aligned} F_4(C,\phi) = &\int_\Omega |\mu_0 - C_{11}|^2 H(\phi_1)H(\phi_2)d_xd_y + \int_\Omega |\mu_0 - C_{10}|^2 H(\phi_1)(1-H(\phi_2))d_xd_y \\ &+\int_\Omega |\mu_0 - C_{01}|^2 (1-H(\phi_1))H(\phi_2)d_xd_y \\ &+\int_\Omega |\mu_0 - C_{00}|^2 (1-H(\phi_1))(1-H(\phi_2))d_xd_y \\ &+v\int_\Omega |\nabla H(\phi_1)| + v\int_\Omega |\nabla H(\phi_2)| \end{aligned} \tag{2}$$

(2) *The polyphase level set segmentation*

The same as the previous level set function, calculating $C_{i,j-k}$ separately and introducing time parameters. Taking the derivative of two Level Set functions in four positions segmentation energy function, the evolution equation is received:

$$\frac{\partial \varphi_1}{\partial t} = \delta(\varphi_1)\begin{bmatrix} vdiv[\frac{\nabla \varphi_1}{|\nabla \varphi_1|}]-\left((\mu_0-c_{11})^2-(\mu_0-c_{01})^2\right)H(\varphi_2) \\ -\left((\mu_0-c_{10})^2-(\mu_0-c_{00})^2(1-H(\varphi_2))\right) \end{bmatrix} \tag{3}$$

$$\frac{\partial \varphi_2}{\partial t} = \delta(\varphi_2)\begin{bmatrix} vdiv[\frac{\nabla \varphi_2}{|\nabla \varphi_{21}|}]-\left((\mu_0-c_{11})^2-(\mu_0-c_{01})^2\right)H(\varphi_1) \\ -\left((\mu_0-c_{01})^2-(\mu_0-c_{00})^2(1-H(\varphi_1))\right) \end{bmatrix} \tag{4}$$

The evolution curve $\{\phi_1=0\}\cup\{\phi_2=0\}$ is given according to the evolution equation.

The 2^n positions segmentation energy function denote by n Level Set functions is:

$$F_{2^n}(c,\phi)=\sum_{1\le I\le 2^n}\int(\mu_0-C_{ij-k})^2 p_i p_j \quad p_k d_x d_y+\sum_{1\le I\le N} v\int|\nabla H(\phi_i)|$$

$$i,j,k=0 \text{ or } 1 \tag{5}$$

When $i,j,k=0$, $p_i=1-H(\phi_i),\ p_j=1-H(\phi_j),\ ,p_k=1-H(\phi_k)$

When $i,j,k=1$, $p_i=H(\phi_i),\ p_j=H(\phi_j),\ ,p_k=H(\phi_k)$

The $C_{i,j-k}$ represent the average gray-value of 2^n regions, the evolution result of multi-curve is:

(3) *Solution*

Although the polyphase segmentation can get several different regions, but sensitive to initial placement, the segmentation result is widely different. However, if the number of objects $2^{n-1}<N<2^n$, the segmentation result will appear a number of the same part, the Level Set functions does not utilize fully and the sliced time is increased. Jeon, Alexander and Pizzi come up with unsupervised hierarchical segmentation, and use only one Level Set functions in every segmentation step. Divide image into two parts, and choose the next part to segment until the original image segmentation finished

(4) *Experiment*

Figure 3 and Figure 4 is the segmentation result of mountain topography image and yuan area topography image used above steps.

1)The segmentation image of Anze area

(a) The original topographic map

Fig. 3. The segmentation result of mountain topography.

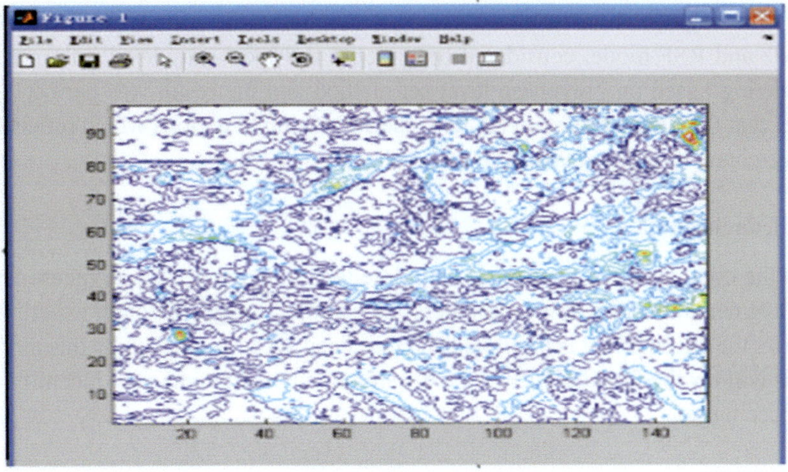

(b) The segmentation result

Fig. 3. (*Continued*)

2) The segmentation image of yuan area

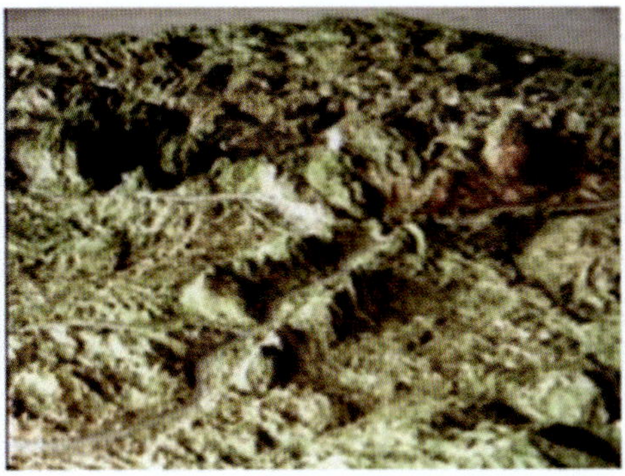

(a) The original topographic map

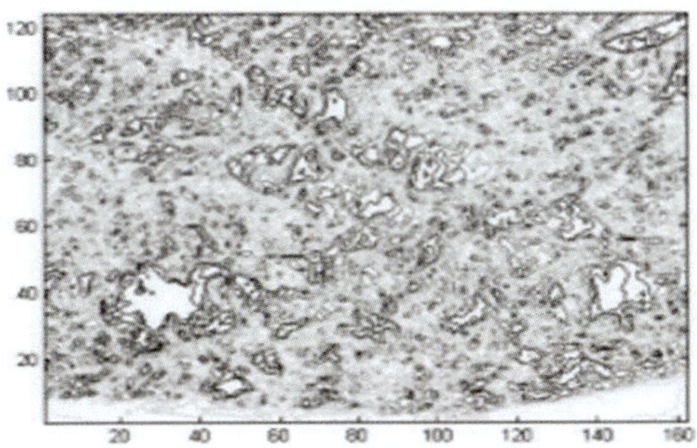

(b) The segmentation result

Fig. 4. The segmentation result of yuan topography.

The segmentation results shows that combined the virtues of Chan-Vese mode and RSF mode, considered both global and local gray level information, modeling based on polyphase level set method and the result was perfect. The goal that divide the target image into a series of simple regions which reflect the basic characteristics of landform was realized.

4. Conclusion

For the complexity of uneven and multiple target landform, the polyphase level set segmentation was used to segment the complex image. Through experiment prove, the sliced effect is improved and the sliced speed is quicken by this model. This will lay the foundation for 3d reconstruction based on image to indentify the surface target.

References

1. Ilya P., Alan S.W., Hamid K. Image segmentation and edge enhancement with stabilized inverse diffusion equations [J]. IEEE Transaction on Image Processing, 2000, 9(2):256~266.
2. Li C., Kao C, Gore J., et al. Implicit active contours driven by local binary fitting energy [J]. IEEE Conference Computer Vision and Pattern Recognition, 2007.
3. 2010 sea Min., University of Science and Technology of China segmentation studies (Master) based on regional-based level set method.
4. Wang Xiaofeng level set method and its application in image segmentation. [D]. Chinese Scientific and Technical University, 2009.
5. Zhao Minrong study [D] based on topography image of camouflage exterior design methods. Air Force Engineering University, 2010.
6. Aubert G. and Kornprobst P. 2002. Mathematical Problems in Image Processing: Partial Differential Equations and the Calculus of Variations, Springer, New York.

Research of logistics park location based on regional economic development

Zhan-Zhong Wang* and Hong-Yu Shi

School of Transportation, PhD Supervisor, Professor,
Department of Transportation Planning and Management ,
Jilin University, P.O. Box 130025, Changchun , Renmin Road NO. 5988;
**Email: wangzz@jlu.edu.cn, Tel:13904316393*

Hong-Yu Shi

School of Transportation, Current Master Student,
Department of Transportation Planning and Management ,
Jilin University, P.O. Box 130025, Changchun ,
Renmin Road NO.5988;
Email: 18744015521@163.com, Tel:18744015521

In this paper, taking the 2015-2030 development plan of logistics park in Hunchun as an example, through the number and distribution of logistics enterprises in the location area, using the MATLAB software to program cellular genetic algorithm, according to the actual situation to set parameters, so that a reasonable logistics park location model can be built. In this model, the strong, weak logistics enterprises and the surrounding environment can be seen as a natural ecosystem, following the expansion-destruction strategy, and long-term dynamic development of logistics enterprise can be achieved. At the end, the result is accord with the S-Curve, which is follow the nature laws, combining enterprise cluster effect and actual development plan, and choose a reasonable position, for providing theoretical basis for the actual logistics park location problem.

Keywords: Logistics park; MATL AB; cellular genetic algorithm; S-Curve.

1. Introduction

The traditional logistics park location method is not correspond with the reality, which is based on the quantitative factors, such as cost, path, freight and so on. Those models also ignore the qualitative factors, such as governmental control, industrial competition, service level and so on. Then, the Fuzzy Evaluation, Analytic Hierarchy Process, Dynamic Dual Diamond Evaluation method are produced. These algorithms can combine quantitative and qualitative analysis to establish a model, so that they are more suitable for the actual situation, at the same time, they also provide some good references for the location selection, Because of the complexity, dynamics combination of qualitative and quantitative factors, now some scholars have been studying location ,using Cellular Automation. Gang Zhu has proposed the logistics system location model based on the Cellular Automation, with the minimum freight turnover as the calculation principle to do a simulation [1][2]. On the basis of the Zhu Gang's model, Zheng Wang has made some improvements and added the competition factors[3][4]. Taking the service radius of logistics to finish the simulation analysis. In this paper, the strong and weak logistics enterprises and the surrounding environment can be seen as a natural ecosystem, following the expansion-destruction strategy, and long-term dynamic development of logistics enterprise can be achieved.

2. Model

2.1. *Cellular genetic algorithm*

Cellular Genetic Algorithm is a new calculation tool, which is the combination Genetic Algorithm and Cellular Automation. At present, the application of the algorithm mainly has two direction[5][6]. One direction is that the abundant evolving rules of Cellular Automation replace some simple genetic manipulation in traditional Genetic Algorithm. At the meantime, a single cellular can be seen as a gene in a chromosome. The other direction is that on the basis of the complex system theory, realize the operation of Genetic Algorithm with the neighborhood structure of Cellular Automation. In the direction, a single cell is compared to individual in the population.

The Cellular Automation includes the cellular, the state space of cellular, the neighbors and the local rules. These 4 parts can be represented in a function [7].

$$A = (L_d, S, N, f) \tag{1}$$

Where A is the Cellular Genetic Algorithm system, and Ld is State Space of cellulars, at the meantime d is the number of dimension CGA lattice ($L \times M$) that represents the location area, where L is the length of the area, and M is the width of the area. S is a finite and discrete set of states, and N is the number of cellulars, and f represents the criteria, which the updating of the model follows.

2.2. *Analysis of logistics park location*

The establishment of logistics park is an important measure to integrate logistics resources, improve the utilization of logistics resource and meet the needs of the logistics operation and supply capacity. According to the theory of spatial gravitation in the economic system, there is a mutual attraction between the logistics park its surrounding service area. The former can attractive the latter settled, at the same time, the latter can decide the former to site. In the same region, one logistics enterprise's size, economy level, competitiveness are differ from the others, hence the jungle rules still applicable among the enterprises for survival and development [8].

In this paper, the logistics enterprise should be divided into strong and weak two types. The strong ones can occupy the location of the weak ones or the blank with a certain probability, while the weak ones only can replace the blank with a certain probability. Of course, all of the enterprises will be destroyed with a certain probability, so that there can appear the blank. The phenomenon likes a living have a process of reproduction and death in the nature. Every enterprise (every cellular) represents a certain economy level, and the total economy level vary as the cellulars' number and layout vary. To be honest, only consider the factors that are complicated and in agreement with the order of nature, we can draw a right conclusion.

2.3. *Modeling*

According to the evolution process of nature, the existence of inter-specific competition have promoted the development of different populations in the ecological circle. In this paper, the logistics enterprises should be devoted into strong and weak two types. Follow the expansion-destruction strategy, and a large number of industrial enterprises and related supporting industries in the space begin to gather. At the end, the phenomenon with a strong and competitive advantage is going to be formed.

Consider the CGA model in a two-dimensional CA lattice $(L \times M)$ that represents the location area, where L is the length of the area, and the M is the width of the area. Each site(i,j), where $1 \le i \le L, 1 \le j \le M$, can be empty or used. The S(i,j) denotes the state of site(i,j), and empty S(i,j)=0, the strong enterprise S(i,j)=1,the weak enterprise S(i,j)=2 and special S(i,j)=3. The enterprises change their states according to the rules of CGA model.

The criteria of changing states are as follows.

C1 For simplicity, the cellulars marked 1, namely S(i,j)t=1, is going to reproduce with a reproduction rate c1 in the other eight adjacent cellulars randomly.

$$S(m,n)^{t+1} = \begin{cases} 1 \ \ S(m,n)^t = 0,1,2 \ \ (m \neq i \ or \ n \neq j) \\ 0 \ \ other \ \ situations \end{cases} \tag{2}$$

Where S(m,n)t is filial generation of S(i,j)t. If S(m,n) in the time step t is 0 or 1 or 2,the next time(the time step t+1) will be turn into 1 at a certain probability.

$$Da(L \times M) > 1 - c1 \tag{3}$$

Where Da is random probability matrix $(L \times M)$ .The other situations S(m,n)t will turn into 0.

C2 Using parallel dynamics, the cellular marked with 2, namely S(i,j)t=2,

$$S(m,n)^{t+1} = \begin{cases} 2 & S(m,n)^t = 0,2 \ \ (m \neq i \ or \ n \neq j) \\ 0 & other \ situations \end{cases} \tag{4}$$

Where if S(m,n) in the time step t is 0 or 2,the next time step t+1 will turn into 2 at a certain probability.

$$Da(L \times M) > 1 - c2 \tag{5}$$

Where c2 is the reproduction rate of the cellular marked 2.The other situations S(m,n)t will turn into 0.

C3 the cellulars marked with 1 or 2 the next time step t+1 will perish at a certain mortality rate.

$$Dr(L \times M) > 1 - m(i) \tag{6}$$

Where Dr is a random probability matrix $(L \times M)$, m(i)(m(1), m(2)) represents the two mortality rates.

C4 The level of strong enterprise economic development can be presented by the number 5 while the weak ones are presented by number 2.

$$QN^t = N_1^t * 5 + N_2^t * 2 \tag{7}$$

Where QNt is the sum of the logistics enterprise economic strength at the time step t, N1t ,N2t represents the number of the strong and weak ones, respectively.

3. Example

In this paper, we take the logistics park location in Hunchun as an example. To do such numerical analyses, we have considered a two-dimensional lattice ($L \times M$) with periodic boundaries. Here L=300 is the length of the location, and M=300 is the width. In this case, the simulation begins with particles distributed according to the actual location around the lattice. As shown in Fig.1.Hear N1=212, and N2=3994.The number of cellulars marked 0 is 5135, and the rest marked 3 is 80659.QN1=13042.

Using the MATLAB to program, suppose the c=0.1, m=0.3, and follow the expansion-destruction strategy. After 50 generation, N150=6543, N250=0, as shown in Fig.2.

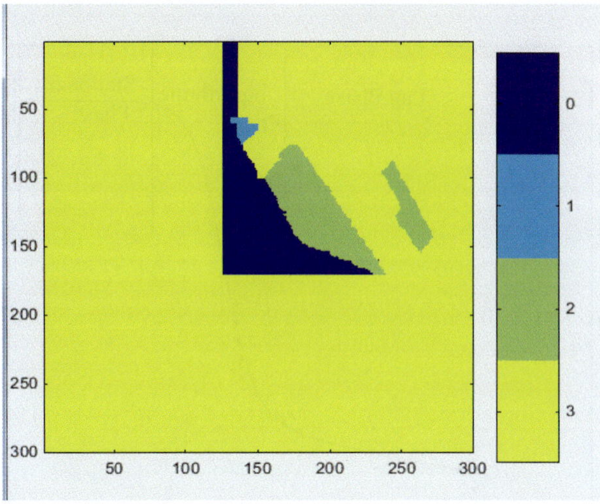

Fig. 1. The layout of the location area, which is made up for 90000 cellulars.

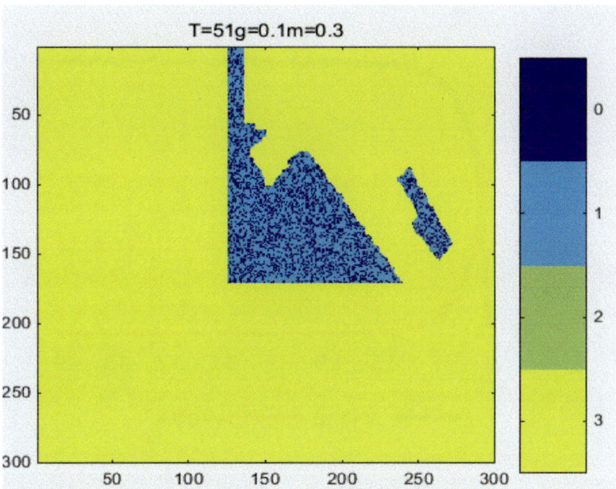

Fig. 2. The layout after 50 generations.

Actually, the second generation N_1^2=3840, reaching 1/2 K value (3320), has entered the logarithmic phase of S curve, as show in Fig. 3, Fig. 4.

$$y = 2331.7 * \ln(x) + 1513.9 \tag{8}$$

Where y is N_1^t, and t represents the number of generation. To the 22 generation, when N2=0, and N1 is going to be stabilized.

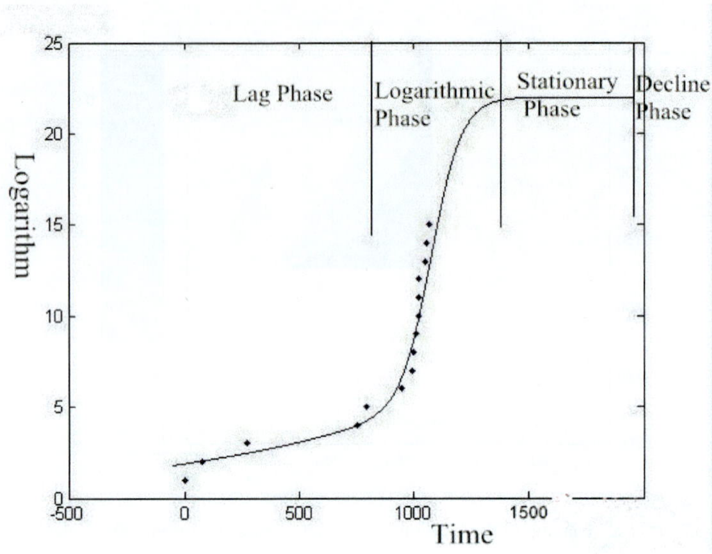

Fig. 3. S-curve.

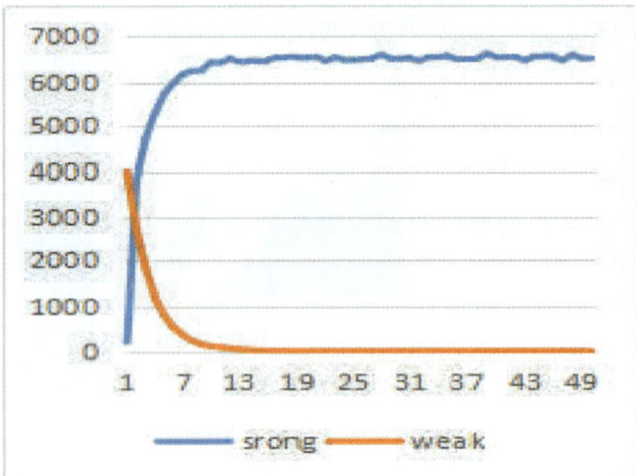

Fig. 4. Number changing.

The changing trend of this CGA model is in agreement with the law of nature that a living is going to growth under the comfortable conditions. There for, removing governmental guidance and human control factors, the strong ones can wipe out the weak ones, and occupy the corresponding blank space, so that can achieve a stable, harmonious ecological circle. For further more development, the strong enterprise use modern industrial cluster theory to accomplish an ideal result, which will make full use of resources, and low input but get a high output. Hence, we should choose a location, which the strong ones concentrate to. Throughout to the map, a reasonable position should be selected at the red circle mark, as shown in Fig.5.According to the actual location and the corresponding location principles, the final position at the red circle mark, as shown in Fig.6.

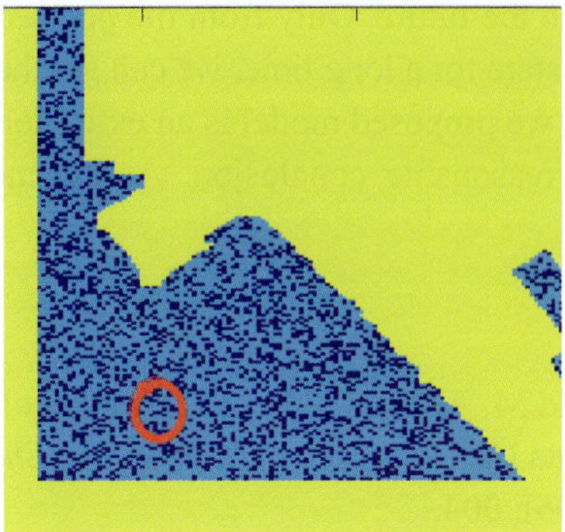

Fig. 5. The location of CGA model.

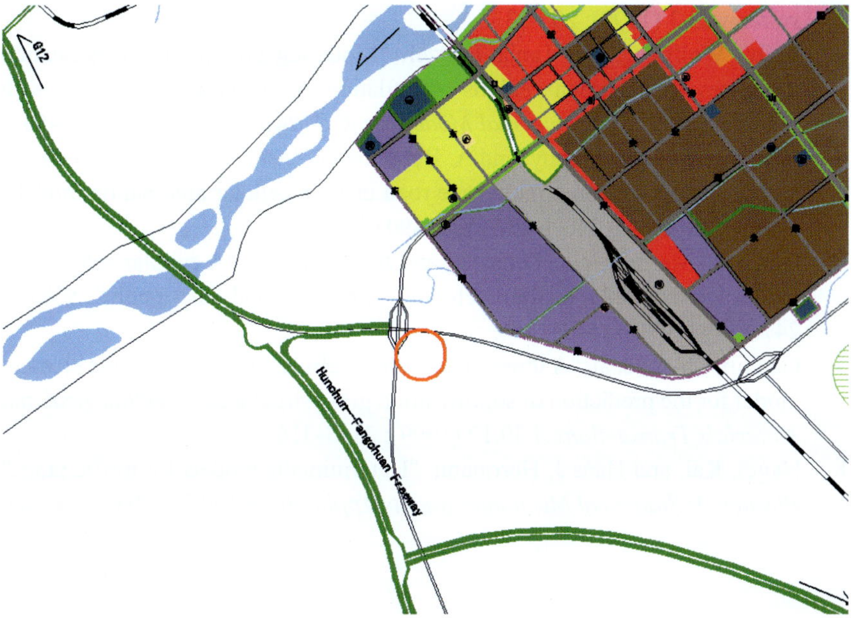

Fig. 6. The actual Location finally.

4. Conclusion

At present, the logistics park location consider some static economic indicators, such as the lowest freight turnover and so on, and less consider the development of logistics enterprise in the future. Only from the perspective of enterprise itself organization, dynamic state for a long time, we can get the answer to the location question. In this paper, we proposed model is an extension to the CGA model for location, and draw a reasonable conclusion. At the meantime, it provides a scientific theory basis.

References

1. Gang, Z. H. U., and M. A. Liang. "Location model of logistics system based on cellular automata [J]." *Journal of University of Shanghai for Science and Technology* 1 (2006): 004.
2. Couclelis, Helen. "From cellular automata to urban models: new principles for model development and implementation." *environment and Planning B* 24 (1997): 165-174.
3. Batty, Michael, Yichun Xie, and Zhanli Sun. "Modeling urban dynamics through GIS-based cellular automata." *Computers, environment and urban systems* 23.3 (1999): 205-233.
4. Loibl, Wolfgang, and Tanja Toetzer. "Modeling growth and densification processes in suburban regions—simulation of landscape transition with spatial agents." *Environmental Modelling & Software* 18.6 (2003): 553-563.
5. Jetto, K., H. Ez-Zahraouy, and A. Benyoussef. "An investigation of merging and diverging cars on a multi-lane road using a cellular automation model." *Chinese Physics B* 21.11 (2012): 118901.
6. Gerhardt, Martin, Heike Schuster, and John J. Tyson. "A cellular automation model of excitable media including curvature and dispersion." *Science* 247.4950 (1990): 1563-1566.
7. Gandin, Ch-A., et al. "A three-dimensional cellular automation-finite element model for the prediction of solidification grain structures." *Metallurgical and Materials Transactions A* 30.12 (1999): 3153-3165.
8. Nagel, Kai, and Hans J. Herrmann. "Deterministic models for traffic jams." *Physica A: Statistical Mechanics and its Applications* 199.2 (1993): 254-269.

Reinforcement learning method based on radial basis function neural network

Yun-Ming Du,
College of Information and Electronic Technology,
Jiamusi University, Jiamusi 154007, China
E-mail: duyunming@126.com

Bing-Bing Yan*, Yong-Cheng Jiang and Qi-Sheng Liu
College of Mechanical Engineering,
Jiamusi University, Jiamusi 154007, China
**E-mail: yanbingbing@126.com*

In complex and continuous environment, the state space of reinforcement learning system faces the dimension disaster problem. This paper puts forward the Q learning method based on radial basis function neural network. The method adopts the dynamic clustering algorithm and the least mean square algorithm to realize the hidden layer parameter learning and the output layer weights learning respectively, which can quickly and steadily approximate the state space and Q function and make the learning system has good generalization ability. The experimental results show that the proposed method can achieve continuous state control tasks in the most optimal behavior, which further verifies the effectiveness of the mentioned method.

Keywords: Reinforcement learning; continuous space; radial basis function network.

1. Introduction

Reinforcement learning, as an important method of machine learning, mainly adopts trial-and-error and utilizes the reward and punishment of environment feedback to adjust or improve its behavior and finally obtain the optimal strategy[1]. As reinforcement learning has low requirements for the prior knowledge of environment and can carry out online learning in the real-time environment, this method has be noticed by many researchers[2].

The basic task of reinforcement learning is to use parameterized function to approximate the mapping from the state to the action, which can be determined by the state value function or state-action function. Classic reinforcement learning is based on the lookup table method, which describes the value function

* Corresponding author.

in the small-scale discrete state and action space. However, the state space is usually a continuous and infinite process in actual system and the expression of high dimensional state space by a lookup table takes up a lot of memory. The starting point for solving the problem is to use function mapping method naturally. As neural network can approximate a function with arbitrary precision and has strong adaptability and generalization ability, combining neural network with reinforcement learning and using neural network to fit the Q function can obtain better results.

2. Reinforcement Learning

Reinforcement learning is refers to the mapping learning from environment state to action, which makes the system action obtain the cumulative reward from environment. The optimal strategy can be found by trial-and-error. In the reinforcement learning system, the environment is a discrete Markov process with finite states. The decision-making can be realized by the quadruples <*S, A, T, R*>, in which *S* and *A* denote a limited environment state set and action set, and *T* and *R* represent the state transfer function and the reward function respectively.

The reinforcement learning system accepts the environment state input $s \in S$, and then the system outputs the corresponding action $a \in A$ according to the internal reasoning mechanism. Under the action *a*, the system environment changes to a new state *s'*. The change of environment state is determined by the state transfer function *T*. When the system accepts the new environment state input, it also gets the instantaneous reward and punishment feedback *r* form the effect of environment on system synchronously. For the reinforcement learning system, the goal is to learn the action strategies π that makes the system obtain the maximum accumulated value of environment reward. In other words, when the system selects a certain action, Eq. (1) has a maximum value.

$$V^{\pi}(s_t) = r_t + \gamma r_{t+1} + \gamma^2 r_{t+2} + \quad . \tag{1}$$

Q learning is a kind of reinforcement learning widely used[3], which does not directly apply the value function in Eq. (1), but use a similar Q function as expressed in Eq. (2).

$$\mathrm{Q}(s_t, a_t) \leftarrow r_t + \gamma V(s_{t+1}) \, . \tag{2}$$

where a_t represents the action selected from the set *A* at time t, and γ denotes the discount factor.

As the system is designed to make the total reward value maximum, substituting $\max_{a \in A} Q(s_{t+1}, a)$ for $V(s_{t+1})$ we can rewrite Eq. (2) as

$$Q(s_t, a_t) \leftarrow r_t + \gamma \max_{a \in A} Q(s_{t+1}, a). \tag{3}$$

At time t, the agent chooses an action a according to the current state, and then uses Eq. (4) to update the Q value.

$$Q(s_t, a_t) \leftarrow Q(s_t, a_t) + \alpha \left[r_t + \gamma \max_{a \in A} Q(s_{t+1}, b) - Q(s_t, a_t) \right]. \tag{4}$$

where b represents the action selected in the state t+1 and α denotes the learning rate.

The running steps of Q learning take the following form:

(1) initialize Q (s, a) in a random manner,

(2) repeat the following steps,

  (a) obtain the current state of the agent,

  (b) select an action and perform it,

  (c) update the state and acquire a reward at the same time,

  (d) update the Q value according to Eq. (4).

3. Radial Basis Function Neural Network

Reinforcement learning is generally applied in the discrete state information, and stores the corresponding state-action value in the memory in the form of table. It not only takes up a lot of memory space, but also slows the learning convergence speed. When the state information is continuous, the above method will not be achieved. Based on the above reasons, this paper presents the Q learning based on neural network, which uses neural network to approximate the Q function. It is effective to overcome the deficiency in using a table to store Q values.

3.1. *Basic structure*

Radial basis function (RBF) neural network is a kind of feedforward network with three unidirectional propagation layers, which is composed of the input layer, hidden layer and output layer[4]. Network structure used in this paper is a multiinput-multioutput mode, as shown in Fig. 1.

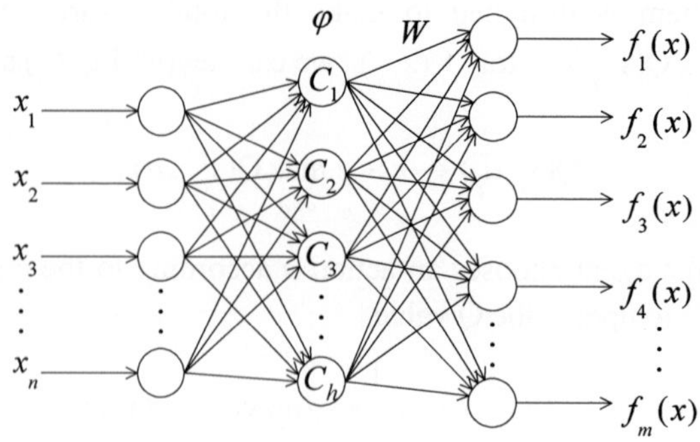

Fig. 1. RBF neural network structure.

where X=[x_1, x_2,..., x_n]^T and $F(x)$=[$f_1(x)$, $f_2(x)$,..., $f_m(x)$]^T denote the input vector and the network output vector, C_i=[c_{i1}, c_{i2},..., c_{in}]represents the ith hidden node location, φ=[$\varphi(x,c_1)$, $\varphi(x,c_2)$,..., $\varphi(x,c_h)$]and W are the the output matrix of hidden layer and the weight matrix of network output respectively.

RBF network utilizes the space generated by the radial basis function to approximate a nonlinear function[5]. Several basis functions are discussed in the literature[6], where the most widely used function is Gaussian function.

$$\varphi(x,u_i) = \exp\left(-\frac{(\|X-u_i\|)^2}{2\sigma^2}\right) \quad \text{for } i = 1,2, \quad ,k . \tag{5}$$

where u_i represents a center of basis function, and σ denotes the width of hidden layer that determines the shape of basis function.

The output of output layer is equal to the linear weighted sum of hidden layer. The mth node output value of the output layer is

$$f_m(x) = \sum_{k=1}^{K} \varphi(x,u_k) W_{km} = \sum_{k=1}^{K} \exp\left(-\frac{(\|X-u_k\|)^2}{2\sigma^2}\right) W_{km} . \tag{6}$$

where W_{km} represents the connection weights between the kth node of hidden layer and the mth node of output layer.

3.2. *RBF parameter learning*

RBF neural network learning is divided into two stages: the hidden layer parameters learning and the weights learning form hidden layer to output layer.

3.2.1. *Hidden layer parameters learning*

In this learning stage, the dynamic clustering algorithm is used to find the cluster centers of all input samples, which are set as the centers of radial basis function. Specific operation is as follows:

(1) initialize all the clustering centers C_j (j=1,…,N),

(2) group all samples X according to the nearest clustering center, in other words, when the condition $\| X^P - C_i \| = \min \| X^P - C_j \|$ for $1 \le i \le P, 1 \le j \le N$ is meet the sample X^P is classified as the cluster θ_i,

(3) calculate the average of samples in θ_i, and readjust the clustering center,

(4) repeat step (2) and (3) until the distribution of the cluster centers is no longer change, and then assigne all the cluster centers to different basis function units as the centers of RBF,

(5) determine the width of base function.

3.2.2. *Output layer weights learning*

The learning task of this stage is using the error function and the least mean square algorithm to update the weights from the hidden layer to the output layer. The error of each output neuron is first obtained. The error equation is

$$e_{error} = d_m - f_m(x) . \tag{7}$$

where d_m denotes desired output and $f_m(x)$ represents RBF network output.

After calculating the error, we can update the weights

$$w_{mi} = w_{mi} + \eta e_{error} \varphi(x, u_i) . \tag{8}$$

where η denotes the learning rate, and i represents the ith node of hidden layer.

4. Experiment Design and Simulation

Mountain car is a classic reinforcement learning problem with continuous state space and discrete action space, as shown in Fig. 2.

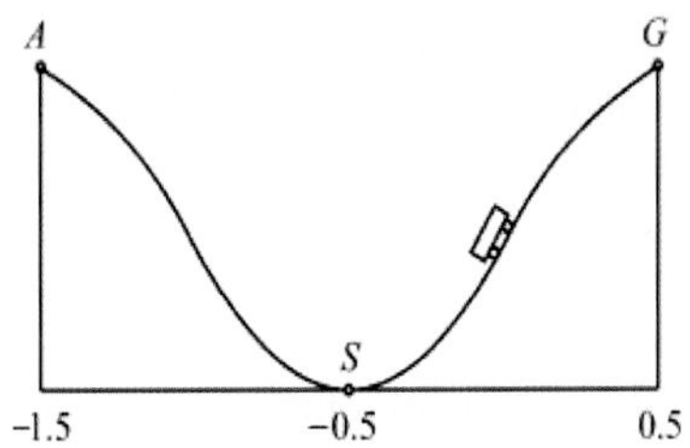

Fig. 2. Sketch map of mountain car.

In Fig. 2 the curve represents a valley terrain, *A* and *G* denote the highest point of left and right end, and *S* is the valley floor. The task of mountain car is to start from the valley floor *S* and as soon as possible move to the highest point *G*. The system has a two-dimensional continuous state space, which is expressed in the horizontal displacement $x \in [-1.5, 0.5]$ and the horizontal movement speed $v \in [-0.07, 0.07]$.

The dynamic equation of simulation system of mountain car is

$$\begin{cases} x_{t+1} = \text{bound}[x_t + v_{t+1}] \\ v_{t+1} = \text{bound}[v_t + 0.001u - g\cos(3x_t)] \end{cases}. \tag{9}$$

where u represents a control action of the car, which has three discrete values ($+1$, 0, -1) denoting the acceleration, uniform speed and deceleration respectively. $g=0.025$ is a gravitational constant, and *bound[.]* denotes a bounded function. In addition, the reward function of the car movement is set as

$$r(t) = \begin{cases} -1, & x<0.5 \\ 0, & x\geq 0.5 \end{cases}. \tag{10}$$

The proposed method is applied to the simulation of mountain car and the greedy algorithm (ε-greedy) is used in exploring behavior strategy. When the car arrives at the point *G* or the time step is more than 1000, the system accomplishs a learning experiment, and then reinitializes the system state and starts the next learning. Experimental parameters are as follows: learning rate $\alpha=0.5$, discount factor $\gamma=0.95$, greedy probability $\varepsilon=0.1$. The initial states of mountain car are $x=-0.5$ and $v=0$. The simulation results are shown in Fig. 3, which are the mountain car self-rescue process curves after 200 learnings.

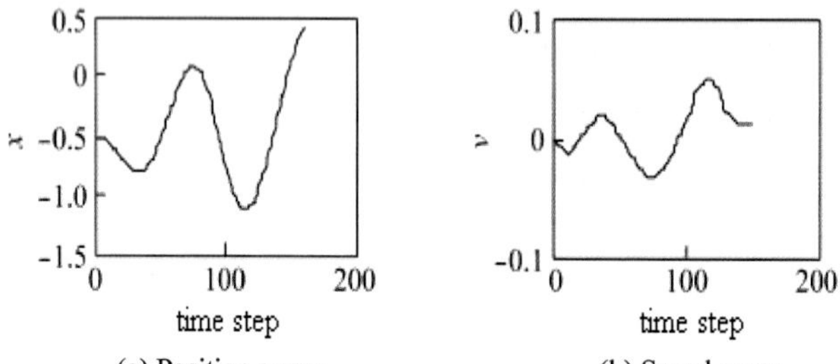

(a) Position curve (b) Speed curve

Fig. 3. Q learning control based on RBF neural network of mountain car.

As shown in Fig. 3, after using the Q learning based on RBF neural network, the mountain car can realize self-help task by the optimal control action. It also fully proves that the proposed reinforcement learning algorithm can effectively deal with continuous state control problems.

5. Conclusion

This paper uses the RBF network to realize the mapping from state space to high dimensional feature space, and builds the Q value function approximation model through linear combination of Gaussian basis function. The model effectively improves the expression ability by using the kernel function mechanism, and also has the simplicity of the linear approximation model. This can solve the disastrous disturbance problem caused by the model approximation of reinforcement learning, therefore be used in continuous space reinforcement learning.

Considering the self-learning and online learning characteristics of reinforcement learning, it needs to be further studied how to design an effective approximation model to better adapt and handle online reinforcement learning problems.

Acknowledgments

This research was partially supported by by Natural Science Foundation of Heilongjiang Province (E201254) and the Key Scientific Research Project of Jiamusi University (12Z1201519).

References

1. R.S. Sutton and A.G. Barto, *Reinforcement Learning: An Introduction.* (MIT Press, Cambridge, MA, 1998).
2. L.P. Kaelbling, M.L. Littman and A.W. Moore, *Reinforcement Learning: A Survey*. Machine learning, Vol. 3, 1988, pp. 9–44.
3. C. Watkins and P. Dayan, *Q-Learning*. Machine learning, Vol. 8, 1992, pp. 279-292.
4. J. Moody and C. Darken, *Learning with Localized Receptive Fields*, in *Proceedings of the 1988 Connectionist Models Summer School*, eds. D. Touretsky, G. Hinton, and T. Sejnowski. (Morgan-Kaufmann, 1988).
5. L. Jun, Learning Reactive Behaviors with Constructive Neural Networkin Mobile Robotics, PhD thesis, Orebro University, (Orebro, Sweden, 2006).
6. A. R. Webb and S. Shannon, *Shape-Adaptive Radial Basis Functions*. IEEE Transactions on Neural Networks, Vol. 9, 1998, pp. 1155-1166.

The design of ship course sliding mode controller based on RBF neural network

Guan-Shan Hu* and Hai-Rong Xiao

Shandong Jiaotong University,
Jinan,, 250357, China
**hugshan@126.com*

In this paper, an adaptive intelligence sliding mode method was studied to improve the maneuverability and safety of ship motion control. In the design process, a ship nonlinear model was established, the control law of SMC and the structure of RBF neural network were proposed. By adopting an online learning algorithm, the parameters of the RBF neural network were optimized. In the intelligence controller, sliding model variable structure was utilized to solve the parameter perturbation and environment disturbances in ship sailing, and RBF was utilized to enhance the controller performance and to reduce vibration caused by the SMC. At the end of the paper, course keeping simulations were done to test the applicability of the designed controller. The results show that the intelligence controller can reduce the vibration trouble in SMC, and achieve high stability and fairly strong robustness.

Keywords: Sliding mode control; RBF neural network; ship course.

1. Introduction

Due to the complex dynamic nonlinearity, time varying aspects characteristics and uncertainties of ship motion, the design of ship course controller is a complicated problem[1]. Since the application of PID control law to ship navigation control in 1920's, autopilots installed on ships usually use the algorithm of the PID controller[2]. But the PID algorithm too dependent the accurate mathematical models so can not to achieve high performance control.

The sliding mode variable structure control has merits of high response and the invariability to systemic parameters and external disturbance, further-more, it's algorithm is simple and is easy to be realized engineering. So the sliding mode controller(SMC) gets a major advance in solving the nonlinear system[3]. But the SMC has some problem in using. The first is the chattering phenomenon which is caused by high-frequency switching. Next, SMC is easy to be influenced by measurement noise.

Recent advances in computing technology have prompted the so-called "Intelligent Control" theory. Neural networks represent the most popular of such intelligent control algorithm that can be used in order to provide the control systems with the ability to learn and adapt to different operating conditions. In 1985, Powell proposed radial basis function (RBF) method. A combination of RBF and neural network formed the RBF neural network (RBFNN). It has good nonlinear mapping ability and it can overcome the shortcomings of SMC[4]. The paper studied how to combine sliding mode control with RBFNN to reduced the existing buffeting and enhance the performance of sliding mode control, and applied the design method in ship motion control.

2. Modeling of Ship Motion

The course control of ship is essentially achieved by controlling the rudder. An automatic steering measures the ship's course to automatically maintain a ship sailing in a given course. As shown in Fig. 1.

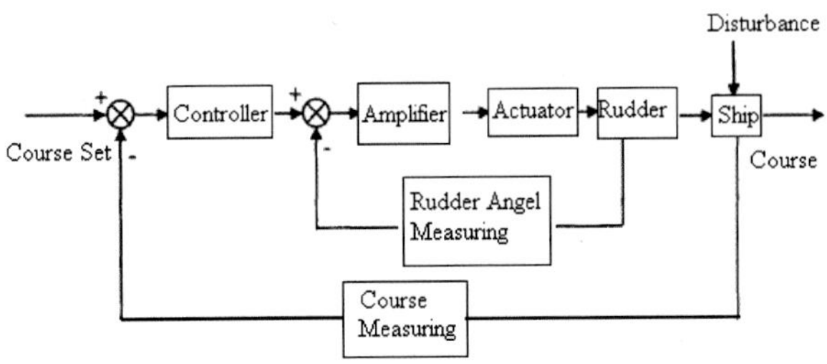

Fig. 1. The structure of ship automatic steering.

For the control of ship motion, there are several mathematical models describing the dynamic characteristics of the ship, such as Nomoto model, Bech equation, and so on. When making the ship course control of ship, the Nomoto model is generally adopt as equation (1).

$$(T_1 + T_2)\varphi + \varphi = K\delta \tag{1}$$

Where, T1, T2 and K are associated parameters relation with the ship's structure, speed and other factors. δ is the rudder angle. Because the Nomoto model is unable to explain unstable and nonlinear phenomena of ship heading, this paper adopts the Norbbin equation describe the motion of the ship model. The model also developed the consideration of external disturbance and ship system parameter perturbation as shown equation (2).

$$T\varphi + \varphi + a\varphi^3 = K\delta + \Delta\omega \tag{2}$$

In the expression, T and K are associated parameters relation, $\Delta\omega$ is external disturbance and ship system parameter perturbation.

The control input of ship is made from the servo rudder machine. The feature formula of rudder servo system can be described as:

$$T_E\delta + \delta = K_E\delta_E \tag{3}$$

Where, T_E and K_E are rudder time constant, δ is command rudder angel of controller, δ_E is the actual rudder angel.

3. The Structure of SMC Base on RBF Neural Network

The intelligent adaptive ship course control system is shown in Fig. 2.

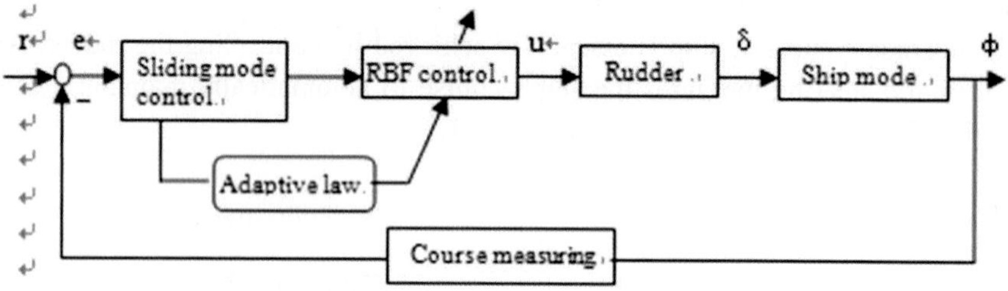

Fig. 2. The intelligent adaptive ship course control system.

By using switching function of sliding mode control as input of RBF network, variable structure control based on RBF is achieved. The switching function of the SMC is as a factor to adjust the parameters of RBF. The output of RBF is send to the rudder to control the ship course. The sliding model variable structure was used to solve the parameter perturbation and environment disturbances in ship sailing, and RBF to reduce chattering caused by the SMC and to optimize the controller performance.

3.1. *The design of sliding mode variable structure control*

We can obtain the dynamical equation in the state-space form of ship from equations (2) and (3).

$$\begin{cases} x_1 = x_2 \\ x_2 = -\dfrac{1}{T_0}(x_2 + a{x_2}^3) + gx_3 + \Delta w \\ x_3 = -\dfrac{1}{T_E}x_3 + \dfrac{K_E}{T_E}u \end{cases} \tag{4}$$

Among equation (3), $x_1 = \varphi$, $x_2 = \varphi$, $x_3 = \delta$, $u = \delta_E$. The tracking error is defined as $e(t) = r(t) - x_1$, $e(t) = r(t) - x_2$.

Inputs of SMC are $e(t)$ and $e(t)$. Define a sliding surface in the space of the error state as the function of SMC: $s(t) = ce(t) + e(t)$. Where, c is a coefficients of s(t).

3.2. *The structure of RBF neural network*

It has been proved that RBFNN can approach any continuous function at any accuracy. The RBF network mode is shown in Fig. 3.

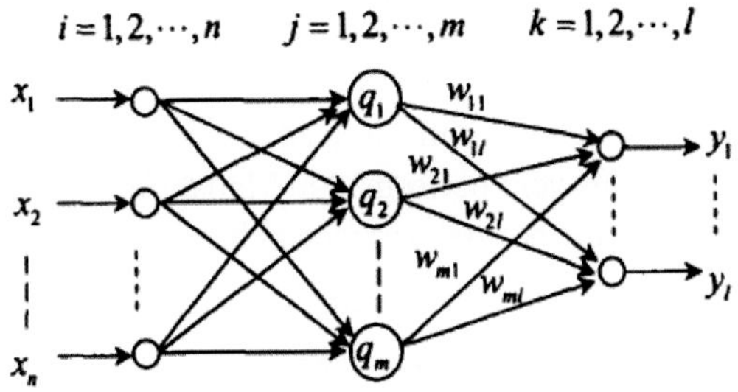

Fig. 3. Structure of an RBF neural network.

The first layer is the premise of the network input. The nodes directly connect with the input vector of the various components. In the second layer, each node represents a linguistic variable value. It is a Gaussian function.

$$h_j = \exp\left(-\frac{\left\|X - C_j\right\|^2}{2b_j^2}\right), j = 1,2,...m \tag{5}$$

The out layer realizes clarity for the calculation.

$$u = \omega_1 h_1 + \omega_2 h_2 + ... + \omega_m h_m \tag{6}$$

Here, we set structure of the RBF is 1 input, 6 hidden layers and 1 output. Use the sliding mode to adjust the parameters of RBF controller.

Parameters learning include the study of premise and conclusions parameters. In this paper, use the SMC to optimize the parameters.

The control destination of RBF is to make $s(t) \cdot s(t) \to 0$, so we set the weight adjust target is $E = s(t) \cdot s(t)$ to obtain the parameters learning algorithm of RBF. By adopting learning algorithm as equations (7) to (9), the parameters of the RBF neural network were optimized online.

$$d\omega_j = -\eta \frac{\partial E}{\partial \omega_j(t)} = -\eta \frac{\partial s(t)s(t)}{\partial \omega_j(t)} = -\eta \frac{\partial s(t)s(t)}{\partial u(t)} \frac{\partial u(t)}{\partial w_j(t)} \tag{7}$$

$$dc_j = -\eta \frac{\partial E}{\partial c_j(t)} = -\eta \frac{\partial s(t)s(t)}{\partial c_j(t)} = -\eta \frac{\partial s(t)s(t)}{\partial u(t)} \frac{\partial u(t)}{\partial c_j(t)} \tag{8}$$

$$db_j = -\eta \frac{\partial E}{\partial b_j(t)} = -\eta \frac{\partial s(t)s(t)}{\partial b_j(t)} = -\eta \frac{\partial s(t)s(t)}{\partial u(t)} \frac{\partial u(t)}{\partial b_j(t)} \tag{9}$$

In above formula, $\frac{\partial s(t)s(t)}{\partial u(t)} = -bs(t)$, $u = \sum_{j=1}^{m} \omega_j \exp\left(-\frac{\left\|s - c_j\right\|^2}{2b_j^2}\right)$.

4. Simulation Result

In order to evaluate the quality of the derived algorithms of the design SMC based on RBF, simulation tests were performed in Matlab/Simulink.

In our experiment, the normal parameters in equation (3), T_0=161s, k_0=0.25, T_E =2.5, a_0=30 and K_E =1 by calculating one ship model. The disturbances are white noise on second-order low-pass filter formed by model spectrum of wave.

Moreover, to compare the obtained results, additional tests are performed on only slide model controller with the function of SMC: $s(t) = ce(t) + e(t)$.

We adopt the same initial conditions and related parameters in the simulation of the SMC based on RBF and the only SMC controller. The simulation results of ship course and the actual rudder angel δ_E are shown from Fig. 4 to Fig. 5, where ship course is set 10^0.

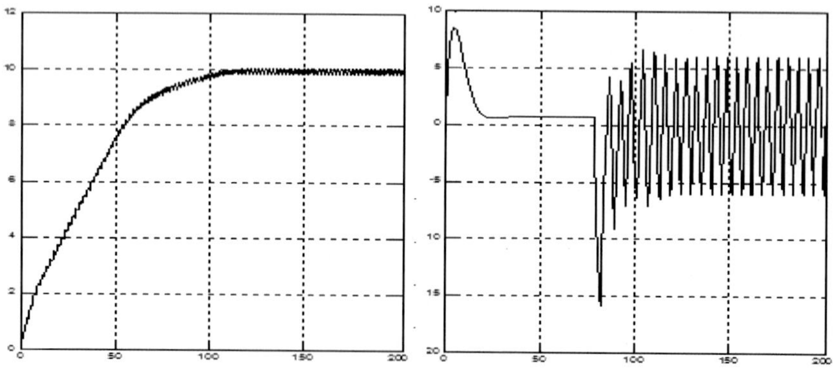

Fig. 4. The output of ship course and rudder angel of only SMC.

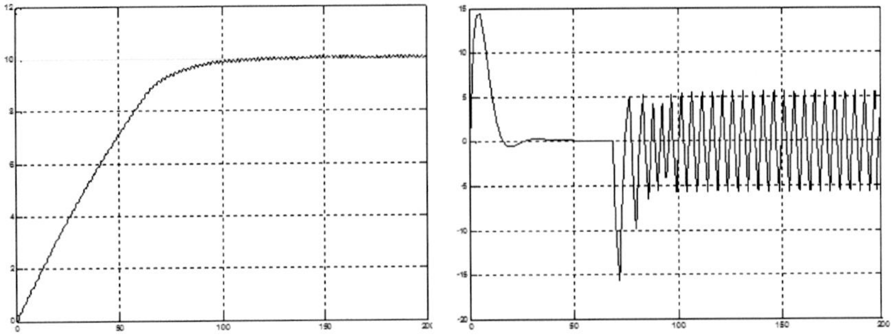

Fig. 5. The output of ship course and rudder angel of SMC base on RBF.

In the simulation, the Gauss function parameters of RBF were randomly generated and were optimized online according the equations (7) to (9). The learning rate η in the equations (7) to (9) was set 3.6, the coefficient C of the sliding mode surface was set 35.

The simulation results showed that the designed SMC controller base on RBF has much better dynamic characteristic and steady state performance. Even in the presence of external disturbances or parameter perturbation, the course control has more stability, the time response process is more quickly, and the control rudder angle is reasonable and fewer frequency shake than the only SMC. The designed controller adopt optimization algorithm to enhance itself learning ability.

5. Conclusions

In the paper, based on nonlinear mathematic model of ship, a nonlinear design scheme which combines SMC and RBFNN was put forward for ship motion control. The numerical simulations illustrate that the design controller has better characteristic than using sliding mode control only, has good stability and more high robust performance when there are parameter perturbations or external disturbances. The further work we would study is how to optimize others parameters such as the learning rate and the coefficient of SMC in the controller and how to apply others intelligence methods in ship motion control.

Acknowledgments

This work was supported by the traffic science and technology innovation projects of Shandong Province Traffic Department (2012A-03) and the applied basic research project of Transport Ministry (2014329817130).

References

1. B.Z. Jia, G. Ren and H.P. Zhang, *An Approach of Signal Detecting on Ship Control Test-bed,* Proceedings of 6th International Symposium on Test and Measurement, vol. 4, pp. 3407-3410, 2005.
2. J.L. Du, C. Guo and X. K. Zhang, *Simulation studies on adaptive nonlinear control of ship motion course,* Journal of System Simulation, vol. 17, pp. 1445-1449, 2005.
3. R.X. Bu and Z.J. Liu, *Iterative sliding mode based increment feedback control and its application on ship autopilot,* Journal of Harbin Engineering University, vol. 28, pp. 268-274, 2007.
4. L. Qin, M. Yang and Z.C. Wang, *Attitude sliding mode controller based on RBF network,* Aerospace Control, vol. 25, pp. 41-45, 2007.

Speed up big integer multiplication in the Block Wiedemann on graphics processing unit

Peng-Bo Wu[†], Jing-Fei Jiang and Yang Zhao

School of Computer Science, National University of Defense Technology, Changsha, 410000, China

[†]E-mail: pengbo026@163.com

The big integer multiplication play a vital role in fields of cryptography, biological information, genetic engineering and so on. To accelerate it has become the one of the main objective of these fields' algorithms research. In this paper, we choose to use static array to store big integer and design the parallel algorithms of big integer multiplication to improve the computational efficiency based on the Block Wiedemann algorithms. Using 32 threads on Graphics Processing Unit to carry out multiplication result for 1024-bit integer and 32-bit integer. And this implementation achieves the speedup of 7.44 times over the big integer multiplication implementation on a single CPU.

Keywords: Big number multiplication; the block wiedemann; multi-core; parallel; CUDA.

1. Introduction

The Block Wiedemann (BW) algorithm[1] is used to solve Bw = 0[2], where B is the input matrix and w is the solution for solving linear equations represented by B. The steps for solving the iterative method of the BW algorithm are as follows [3][4]:

Step 1: The matrix sequence is computed, as in Eq. (1).

$$a^i = \left(x^T B^i y\right)^T, 0 \le i \le \frac{N}{m} + \frac{N}{n} + o(1) \tag{1}$$

where x and y are randomly chosen binary matrices of size $m \times N$ and $N \times n$, respectively.

Step 2: Berlekamp–Massey algorithm is used to compute a generating polynomial of the matrix sequence a^i from step 1 in the form, as in Eq. (2).

$$F(X) = \sum_{i=1}^{\frac{N}{n}+o(1)} C_i \bullet X^i \tag{2}$$

where C_i is $n \times n$, which is a binary matrix.

Step 3: The sequence of matrices S of size $N \times n$ is calculated such as in Eq. (3).

$$S_i = B^i \bullet y \bullet C_i, 0 \leq i \leq \frac{N}{m} + \frac{N}{n} + o(1) \tag{3}$$

With high probability, $B \bullet \sum S_i = 0$ for which $\sum S_i$ is output as a solution.

This paper focuses on solving the matrix sequence in the first step of BW. The main operating process is as in Eq. (4).

$$\begin{array}{cc} By & xBy = Z_1 \\ B \bullet By = B^2 y & xB^2 y = Z_2 \\ \vdots & \vdots \\ B \bullet By = B^2 y & xB^2 y = Z_2 \end{array} \tag{4}$$

The key operation of the matrix-vector multiplication in BW algorithm is defined as $B \times y$, where B is a large-scale matrix whose element is a small number that can usually be represented in machine-word size, such as 32 bits. A large vector with big number elements is present in the field of p[5][6], where p is a large prime number that must be expressed in a long word, such as 1024 bits. This operation can be iterated n times. And it includes a large number of multiplication of big integer. So we can know that big integer multiplication has a widely used algorithm in the BW[7].

As for big integer multiplication, Jie Chen et al.[8] implement efficient GCD (greatest common divisor) computation for it on Xeon Phi coprocessor in 2014 and achieve a substantial speedup over a GCD implementation on a single CPU core. And Cheng Chang et al.[9] achieve vectorized big integer operations for cryptosystems on the inter MIC architecture and make their RSA implementation achieve a good speedup compared to that in OpenSSL in 2015. Thus, the parallelization of this multiplication can improve the performance of the BW algorithm.

In this paper, we present our design and implementation of parallel big integer multiplication on Graphics Processing Unit (GPU). Then we compare its performance with its implementation on single CPU. Our main contributions are summarized as follows:

- We design the parallel algorithm of big integer multiplication on GPU, which achieve using 32 threads to carry out multiplication result for 1024-bit integer and 32-bit integer.
- We improve the multiplication's throughput and achieve higher speedup by using latency hiding for data transfer.

The paper is organized as follows: The calculation method is described and its complexity is studied in Section 2. Latency hiding is discussed in Section 3. The experimental result is shown in Section 4. The conclusion and future work is presented in Section 5.

2. Algorithm and Implementation

The value of bignum is not within the scope of the number of computers that can be expressed. Thus, an array can only be used to represent a bignum, which means that an array that has 32 elements of integer (32 bits) needs to be used to express a bignum (1024 bits). An array is supposed to be adopted to represent the result (33×32 bits) and intermediate result (32×64 bits) of this multiplication. The carry problem between the elements also needs to be considered because of this condition.

2.1. *CPU serial algorithm*

This part introduces the multiplication of bignum and integer algorithm processes by using CPU. For example, $C[i] = A[i] \times B.$

The program on the CPU is a serial execution. First, a cycle of 32 times is needed to obtain the intermediate results. The program must then carry and achieve the intermediate results in the cycle 31 times to obtain the final result.

The algorithm is as follows：

CPU Serial algorithm for 1024_bits * 32_bits

```
Input: A[i]: array of represent a large prime number
       B: a small number (32_bits)
Output:  Ms = A×B
    for i=0 to 31 do
        C[i] = A[i]×B
    end for
    M [31] =C [31] & 0x00000000FFFFFFFF
    for i=30 to 0 do
        M [i] =C [i] & 0x00000000FFFFFFFF+
        (C [i+1]>>32) &0x00000000FFFFFFFF+M[i+1]
                       &0x00000000FFFFFFFF
        Ms [i+2] =M[i+1]&0x00000000FFFFFFFF
    end for
    C [0] =C [0] &0x00000000FFFFFFFF+M [0]
    Ms [1] =C [0] &0x00000000FFFFFFFF
    Ms [0] =(C [0]>>32) &0x00000000FFFFFFFF
```

2.2. *GPU Parallel algorithm*

Multiple threads can be used to make an instruction in parallel [10] . Thus, some operations, which do not have a relation, can be executed in parallel.

Step 1: Intermediate results are calculated by using multiple threads. The values of array M are computed with 32 threads in parallel (Fig. 1).

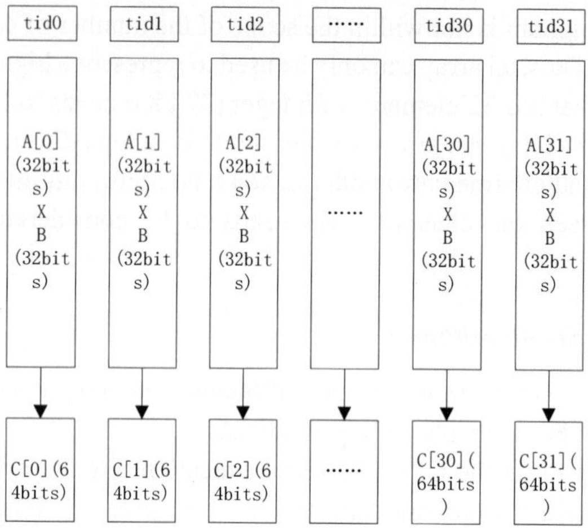

Fig. 1. Calculation of intermediate results by using multiple threads.

Step 2: Shifting and addition is performed by using multiple threads.

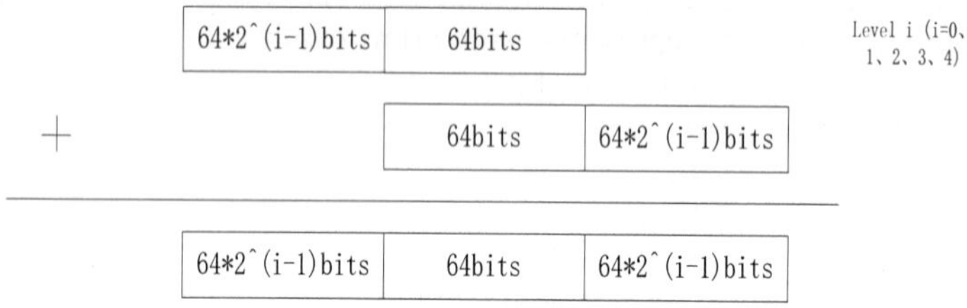

Fig. 2. Addition in parallel with 5 levels.

The present paper uses five levels of shifting and adding operations to calculate the array C [33] (Fig. 2). In this process, the number of threads used per level is $32/2^{i+1}$ (level i).

2.3. *Analysis of two algorithms*

The execution time of multiplication (32 bits×32 bits) is assumed to be T_m, and the time of addition (64 bit+64 bits) is supposed to be T_a. The quantitative analysis for the execution of two algorithms will be shown in the next paragraphs.

A total of 32 multiplications and 63 additions are known to exist in a traditional CPU implementation, whereas a GPU implementation uses 1 multiplication and 20 additions.

Thus,

$$T_{cpu} = 32T_m + 63T_a \tag{5}$$

$$T_{gpu} = T_m + 20T_a \tag{6}$$

Obviously, $T_{gpu} < T_{cpu}$

The execution time of GPU is concluded to be less than that of the CPU in the calculating section.

3. Latency Hiding for Data Transfer

As for the GPU parallel [11] algorithm, data needs to be transferred from the CPU to the GPU before computing and moving data back to the CPU after calculation. Therefore, the effect of the data transfer should be considered [12].

If the amount of data is assumed to be N, the amount of data that can be computed by per kernel is M, the time of transferring data per M from CPU to GPU is $T_{H\to D}$, the execution time of kernel is T_{kernel}, and the time of transferring data per M from GPU to CPU is $T_{D\to H}$, then the execution time of GPU is

$$T_{gpu} = K\times\left(T_{H\to D} + T_{\text{kernel}} + T_{D\to H}\right) \tag{7}$$

without latency hiding for data transfer.

The number of running kernel times is

$$K = \lceil N/M \rceil \tag{8}$$

As a result, three schemes are designed to reduce the effect of data transfer on execution time.

The three schemes are elaborated as follows:

Scheme A: Two streams are used to hide the time of data transfer [Fig. 3(a)]. As a result, the transmission time, which can be hidden, is

$$(K/2)\times T_{H\to D} + (K-1)\times T_{D\to H} \tag{9}$$

Thus,

$$T_A = (K/2)\times T_{H\to D} + T_{D\to H} + K\times T_{\text{kernel}} \tag{10}$$

Scheme B: First, according to the total amount of global memory (G is assumed), a considerable amount of data (G) are transferred into the GPU (transmission time is T_{HGD}). The kernel ($\lceil G/M \rceil$ times) is then processed for computation. Finally, data (G) are transferred back to GPU (transmission time is T_{GDH}). In addition, this cycle must be repeated $\lceil N/G \rceil$ times [Fig. 3(b)].

Therefore, the hidden transmission time is

$$\lceil N/G \rceil \times (\lceil G/M \rceil \times T_{H\to D} - T_{HGD} + \lceil G/M \rceil \times T_{D\to H} - T_{DGH}) \tag{11}$$

Hence,

$$T_B = \lceil N/G \rceil \times (T_{HGD} + T_{DGH}) + K \times T_{\mathrm{kernel}} \tag{12}$$

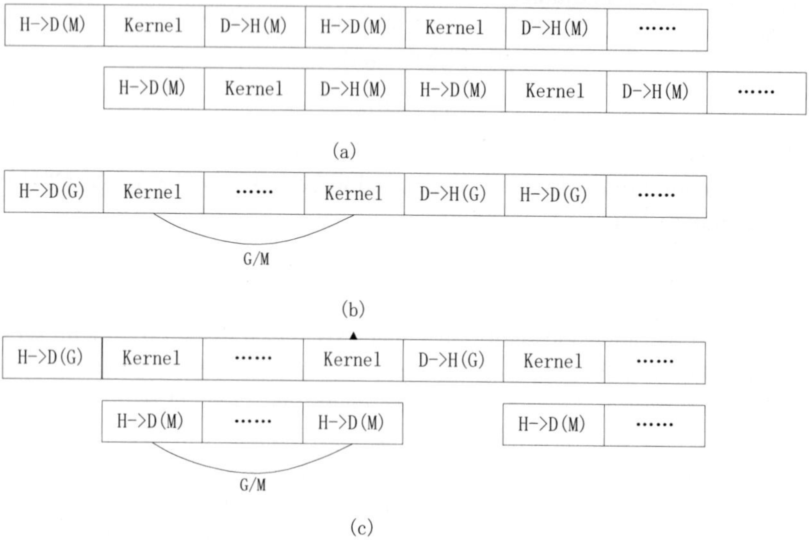

Fig. 3. Scheme design of latency hiding.

Scheme C: The first cycle is the same as scheme B. The difference is that the other stream is adopted to transfer the data of M and run the kernel simultaneously [Fig. 3(c)].

The hidden transmission time is

$$\left(\left\lceil \frac{G}{M} \right\rceil \times T_{H\to D} - T_{HGD}\right) + \left(\left\lceil \frac{N}{G} \right\rceil - 1\right) \times \left\lceil \frac{G}{M} \right\rceil \times T_{H\to D} + \left\lceil \frac{N}{G} \right\rceil \times \left(\left\lceil \frac{G}{M} \right\rceil \times T_{D\to H} - T_{DGH}\right) \tag{13}$$

Thus,

$$T_C = T_{HGD} + (\lceil N/G \rceil \times T_{DGH}) + K \times T_{\mathrm{kernel}} \tag{14}$$

Obviously, $T_C < T_B$;

In general,

$$T_{HGD} = (\lceil G/M \rceil \times T_{H\to D})/2 \tag{15}$$

In the same way,

$$T_{DGH} = (\lceil G/M \rceil \times T_{D\to H})/2 \tag{16}$$

Hence,

$$T_C = \left(\lceil G/M \rceil \times T_{H\to D}\right)/2 + (K/2)\times T_{D\to H} + K\times T_{\text{kernel}} \tag{17}$$

So,

$$T_A < T_C$$

As discussed in the preceding paragraphs, the execution speed of scheme A is the fastest, whereas scheme B is theoretically the slowest.

4. Experiment Result

4.1. *Introduction of experiment platform and determining the parameters*

The processors that are used in this experiment are Inter (R) Xeon (R) CPU E5-2650 (CPU) and NVIDIA TESLA k20c (GPU).

K20c has 13 multiprocessors. The maximum number of threads per multiprocessor is 2048. In the calculation, K20c uses 32 threads to compute for 1024*32. Thus, the number of 1024*32, which K20c can calculate every time, is 832 (26*32); thus, M=832. It then needs to store 215 MB (A [i], B [i], C [i] included) in the global memory of the GPU, while the amount of data that needs to be calculated is M. The total global memory of K20c is 4096 MB (G). Hence, the amount of data that can be stored once in the global memory is 18M (G=18M).

4.2 *Result analysis*

(1) The amount of data is within M; thus, it obtains the curve of graph (Fig. 4).

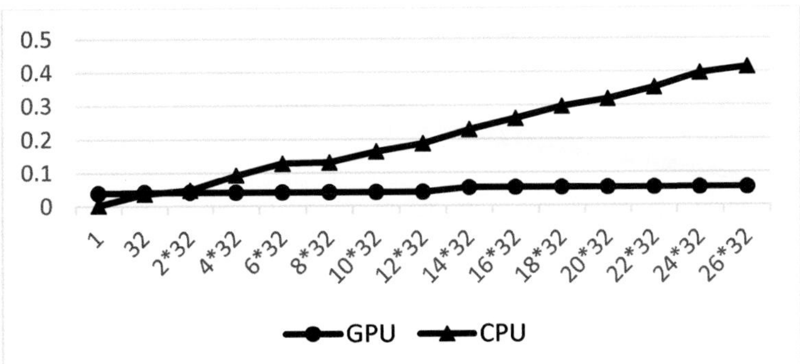

Fig.4.The x-axis represents the amount of data/1024*32 bits, y-axis represents the execution time of program/ms. This shows that the execution time comparison between GPU and CPU.

The curve of the execution time of GPU is less fluctuant than that of CPU, which means that the execution time is not reduced with the use of GPU when the amount of data is small. However, the execution time is decreased when the amount of data increases. When the amount is 256 KB, the time of GPU is less than that of CPU.

The result for the transmission time is not positive (Fig. 5).

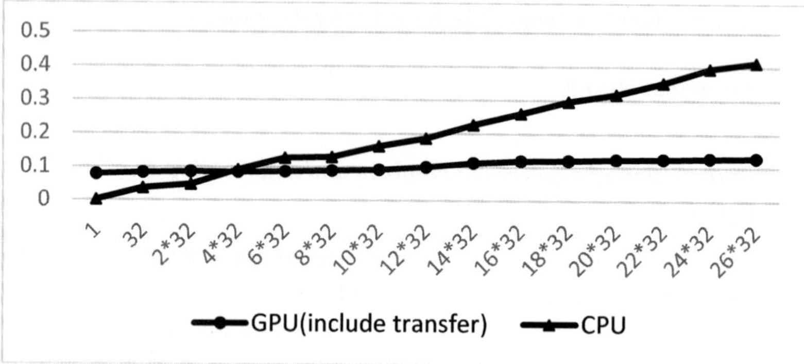

Fig. 5. The x-axis represents the amount of data/1024*32 bits, y-axis represents the execution time of program/ms. This shows that the execution time comparison between GPU (include tranfer) and CPU.

According to Figure 5, the GPU requires a large amount of data (512 KB) after adding the transmission time to make the execution time equal to the CPU time. The speedup is also simultaneously smaller.

(2) When the amount of data is more than M, three schemes are used to hide the transmission time. The result is shown in Fig. 6.

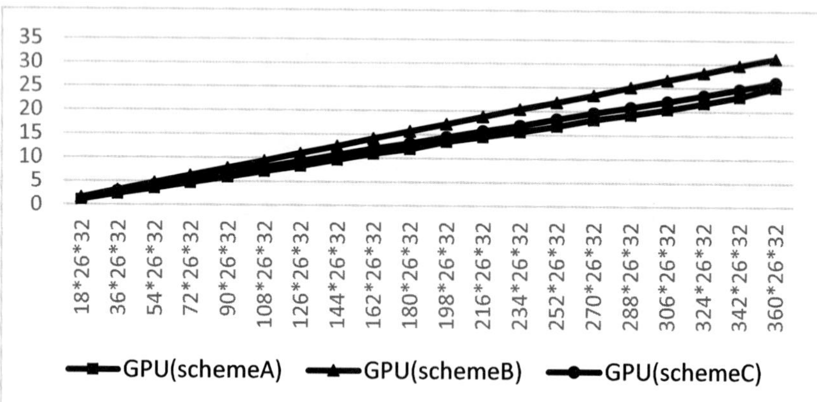

Fig. 6. The x-axis represents the amount of data/1024*32 bits, y-axis represents the execution time of program/ms. This shows that the execution time comparison of three schemes.

The result proves that the theoretical calculations are correct. The execution speed of scheme A is the fastest, whereas that of scheme B is the slowest.

In the end, scheme A is selected for comparison with the execution time of GPU (Fig. 7).

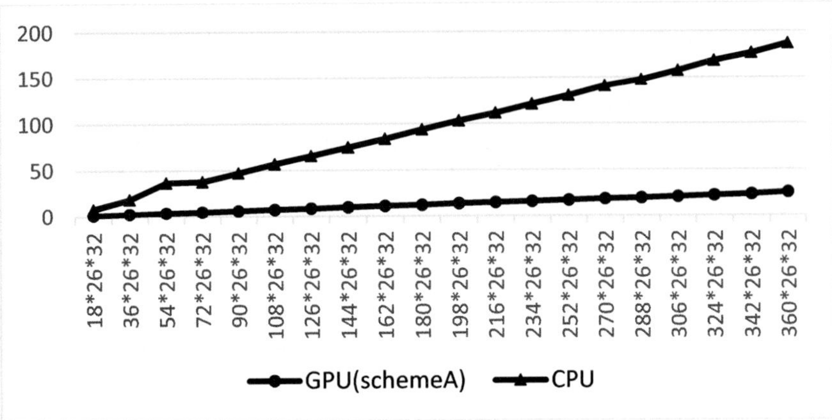

Fig. 7. The x-axis represents the amount of data/1024*32 bits, y-axis represents the execution time of program/ms. This shows that the execution time comparison between GPU (scheme A) and CPU.

The graph shows that the execution time of GPU is less than that of CPU when the amount of data is more than M. The speedup is greater after latency hiding.

5. Conclusion and Future Work

Acceleration achieved by using a single GPU is not obvious with a small amount of data; sometimes, the running speed, which uses a single GPU, is even slower. However, a substantial speedup can be achieved with an increased amount of data. The speedup is 3.2 when the stream is not used in this paper. The speed can be improved significantly if the algorithm is optimized further (e.g., with hidden transmission) to achieve a speedup 7.44 times faster than that of a single CPU. And we will port BW algorithm on GPU based on this speedup and checkout its performance in the future work.

Acknowledgments

This work is funded by National Science Foundation of China (number 61303070). Dr. Jingfei Jiang was an academic visitor at University of Manchester. We acknowledge TianHe-1A supercomputing system service.

References

1. Li J, Zou P, Wu Q. Parallel Algorithms for Solving Large Sparse Linear Equations. In: 2nd International Symposium on Computer, Communication, Control and Automation. Atlantis Press, 2013
2. Vialla B. Block Wiedemann algorithm on multicore architectures. In:ACM Communications in Computer Algebra, 2014, 47(3/4): 102-103
3. Yan Zhong. " Parallel menthod and preconditioning studies for large sparse liner system" PHD thesis, school of computer science, NUDT.
4. Giorgi P, Lebreton R. Online order basis algorithm and its impact on the block Wiedemann algorithm. In: Proceedings of the 39th International Symposium on Symbolic and Algebraic Computation. ACM, 2014: 202-209.
5. Erich Kaltofen, B. David Saunders. On Wiedemann's Method of Solving Sparse Linear Systems. In: National Science Foundation. No. CCR•90-06077 and No. CDA-88-05910.
6. G.Villard. Further Analysis of Coppersmith's Block Wiedemann Algorithm for the Solution of Sparse Linear Systems.
7. Harrison G, Johnson J, Saunders B D. Probabilistic analysis of Wiedemann's algorithm for minimal polynomial computation.In: ACM Communications in Computer Algebra, 2014, 47(3/4): 118-119.
8. Chen J, Watson W, Chen M F. Efficient GCD Computation for Big Integers on Xeon Phi Coprocessor[C]//Networking, Architecture, and Storage (NAS), 2014 9th IEEE International Conference on. IEEE, 2014: 113-117.
9. Chang C, Yao S, Yu D. Vectorized Big Integer Operations for Cryptosystems on the Intel MIC Architecture[C]//2015 IEEE 22nd International Conference on High Performance Computing (HiPC). IEEE, 2015: 194-203.
10. Nickolls J, Dally WJ. The GPU computing era. IEEE Micro 2010; 30:56–69.
11. Nickolls J, Buck I, Garland M, Skadron K.: Scalable parallel programming with CUDA. Queue2008; 6:40-53. DOI:10.1145/ 1365490.1365500.
12. NVIDIA. CUDA C Programming Guide, 2011. Available from: http://developer.download.nvidia.com/compute/DevZone/docs/html/C/doc/CUDA_C_Programming_Guide.pdf [Accessed on June 21, 2012].

A research in image transmission based on fractal compression

Tian-Ming Yu, Jian-Hua Yang[†], Wei Lu and Guo-Liang Feng

Institute of Advanced Control Technology,
Dalian University of Technology, Dalian, 116024, China
[†]*E-mail: yqh0711@foxmail.com*

For the characteristics of many monitory points, high cost of data transmission and sending scene pictures periodically in the Coal Bed Methane mining site, fractal theory is used in the compression of images. The compression rate of fractal is high, but it takes too much time to encode, so an improved compact genetic algorithm is used in this paper to improve the efficiency of encode. Experimental result shows that, the size of compressed image is smaller than before, so that it can be transmitted more quickly. And the image has minimum distortion after uncompressing, does not affect the correct judgment of site conditions.

Keywords: Fractal algorithm; compression ; MCGA.

1. Introduction

Due to the high risk of CBM monitoring is an indispensable part of it. But the places of CBM mining are more remote, and the signal collecting devices are not gathered in one region, so that the video capture system with cable is unfit for this situation. The characters of it are its collection points are large number, geographically dispersed and far away from city, in this case, the video capture system with wireless of mobile network is usable. Although GPRS and 3G are wireless, and without the limits of geographical positions, the cost of video traffic is a problem should be considered. Depending on test experience, if the front camera was CIF, PAL, 25 frames per second, the video data is more than 100MB per hour, according to the current mobile flow rate calculation, that is very uneconomic.

50 monitoring devices are used in this project, if displayers show video pictures from each monitoring devices, it will be a good deal of work for security personnel, and it is easy to have poor judgment. Long time of video traffic with traditional way of monitoring will be a huge cost. Actually, the video signals are not necessary to be sent all the time, the frames can be sent while event occurs,

when there is no event, frames will be sent to a specified server regular in order to review. In view of the case above, initiative monitor mode is adopted. DSP video processing device is installed on each monitoring camera to process monitoring pictures intelligently, in case of emergency, such as fire or presence of unauthorized persons, the relative device will give an alarm and send the picture back. The designed intelligent monitoring system needs to send pictures regularly in normal times. It is significant to compress images before they are sent, because it can effectively save the mobile network traffic[1]. Due to the images are sent at some interval, the time takes by compression is confined little. Fractal is cited in the paper, it has high compression ratio feature, but it takes long time to code and decode, so genetic algorithm is used to improve the image matching speed of fractal, meet the requirement of high compression ratio and high fidelity transmission.

The structure of monitoring system is shown in Fig. 1, each monitory point is geographically dispersed, chooses GPRS or 3G to send pictures based on the situation of mobile network. The front camera is developed by our research group. The scene was constantly monitored at supervisory end, the monitoring data was stored in the front end SD card, and the pictures are sent to the specified sever periodically to backup. When abnormal situation occurs, the system will give the alarm, and the front-end monitoring turns into real-time video mode.

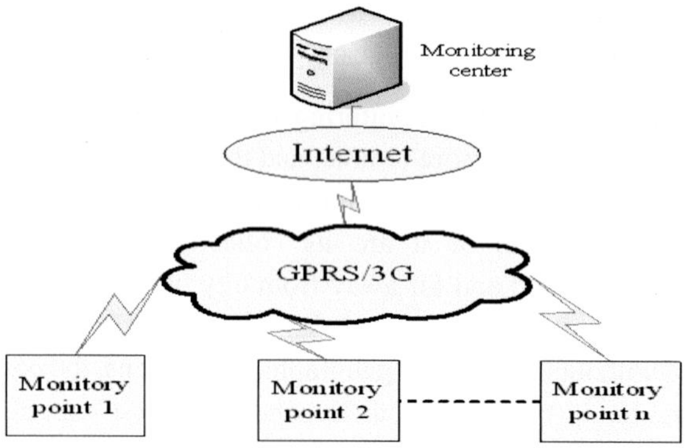

Fig. 1. The structure of monitoring system.

2. Method of fractal compression and Compact Genetic Algorithm

2.1. *Fractal compression*

The important feature of fractal graphics compression technology is greatly compressing image data according to the structure symmetry and similarity of images, having a very high compression ratio[2-4]. Compressed data can recover source images with no distortion reduction after iterative transformation decoding, it also can random enlarge images, resolution-independent. Rebuilt images have high signal-to-noise ratio and better restored detail[5-6]. The advantage of fractal compression technology makes it to be an important direction in the image compression field[7-8]. The theoretical bases of fractal compression technology are IFS (Iterated Function System) and Banach fixed-point theorem. An IFS has complete metric space (X,d), mapping due to this space $w_n : X \to X$, and mapping-factor S_n, $n = 1, 2 \cdots N$. The system can be expressed as: $\{X; w_n \mid n = 1, 2, \cdots N\}$, compressibility factor $s = \max \{s_n \mid n = 1, 2, \cdots N\}$. IFS with high compression ratio can use fewer coefficients to represent an image. Iteration method which is often used segments image into small parts, then search their corresponding iteration function system from IFS libraries.

The encoding method of fractal image is as follows: segment image into domain R and domain D, D is twice as large as R. Shrinkage sampling domain D, make it shrink to the size of domain R, then rotation and symmetry transformation are used to get code pool. While encoding, the minimum variance of R and D is used to get fractal code. And the fractal code is iterated to restore image while decoding.

When fractal is used in image compression, data calculation of compression is extremely large, and the speed is slow, how to match search effectively is the main problem of the fractal method. Genetic algorithm as a kind of bionics random search algorithm, has a strong search capability, and doesn't need a lot of prior knowledge and specific mathematical models, so it is suitable for solving nonlinear problem[9-11].

2.2. *Compact Genetic Algorithm*

CGA (Compact Genetic Algorithm)[12] uses a probability vector to represent a population, this can reduce the storage cost. Each element value of PV represents the probability of its corresponding binary being 1. Initial values of PV are set to 0.5, through this competition, each element of PV finally increases to 1 or reduces to 0, PV is the optimal chromosome at this time. Implementation step of CGA as is shown in Fig. 2. In traditional binary genetic algorithm, the length of each individual is L bit, the number of population is N, and the storage space is NL bit. Individual of CGA only save the elements of PV whose probability is 1 or 0, and the memory space of each element of PV is 16L, this save the storage space effectively. This has very important implication for image analysis and processing by using less memory microprocessor.

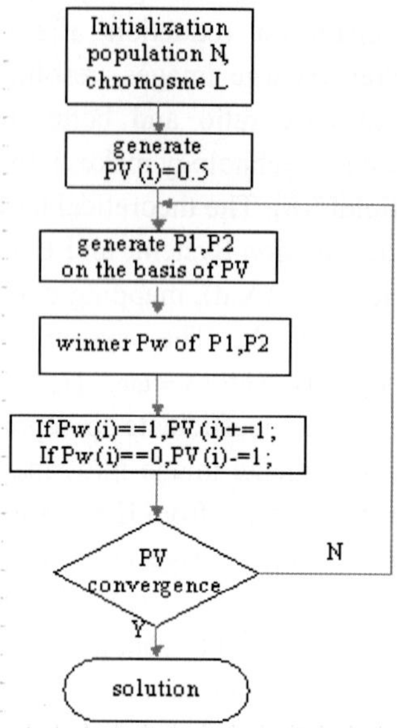

Fig. 2. Algorithm flow of CGA.

3. Mutated-Based Compact Genetic Algorithm

CGA embodies selection and crossover but lacks variation, it is easy to have local optimal results. MCGA (Mutated-Based Compact Genetic Algorithms) adopts a improved compact genetic algorithm, which mutates chromosome bit by bit, and reserves the most optimal one. It combines with matching error calculation of fractal image compression, constructs the fitness function. It is used for fractal compression matching to improve the matching efficiency. On the base of CGA, production P1 and P2 are mutated respectively, and the winners of variation are reserved to the next round of operation.

Encoding process of fractal algorithm takes longer, MCGA is used in fractal coding with the goal of constructing fitness function to achieve fast fractal encoding process and improve the speed of fractal algorithm. The objective function of fractal coding is the minimum-variance of D and R, fitness function can take the reciprocal of the variance. The fitness function of the structure is:

$$f = \frac{1}{E(R, D)^2} \tag{1}$$

where $E(R,D)=\min\left(\min_{s,o}\|R\text{-}sD-oI\|\right)$, $E(R,D)$ denotes the error of transformation mapping of D and R. s, o respectively denote scaling and translation of sub block gray.

Fractal image compression process based on mutated-based compact genetic algorithm is shown as Fig. 3, Fig. 4 is the subroutine of CGA.

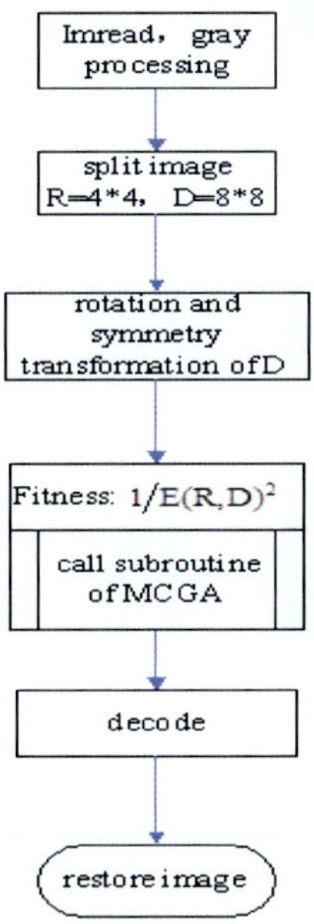

Fig. 3. Fractal image compression process based on MCGA.

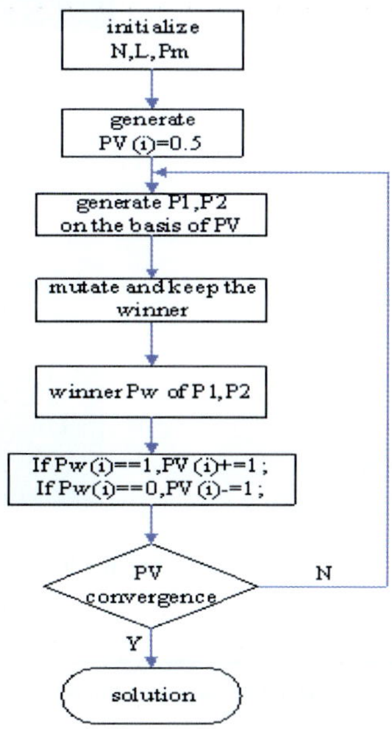

Fig. 4. MCGA algorithm.

4. Experiments

The resolution of test image is 640×480, the monitoring image of the scene as is shown in Fig. 5, PC is used in the experiment, the decoding image which is processed by MCGA fractal compression is shown in Fig. 6. The performance table based on MCGA fractal image compression and GA fractal compression as shown in Table 1, encoding time-consuming(s), PSNR and compression ratio are taken as index.

Fig. 5. Original image.

Fig. 6. MCGA fractal compressing decoding image.

Table 1

Compression method	Encoding time(S)	Compression ratio	PSNR(dB)
GA	60.31	5.48	29.23
MCGA	45.15	5.33	29.56

The experimental results show that on the base of MCGA fractal compression, the change of compression ratio and SNR were very little, but the encoding time reduced 1/4, that is a great advantage in practical application. In monitoring field, the surveillance image is sent every few minutes. Fractal encoding time based on MCGA is less than one minute. this method can meet the requirement of the actual monitoring system. Moreover, MCGA can greatly reduce the storage cost, for internal resources strain of the single chip microprocessor, the system hardware is not necessary to extend additional corresponding memory chips, so it reduces the difficulty of designing hardware system.

5. Conclusions

Due to the necessary of high compression ratio and timed upload image on the scene monitoring, the fractal image compression method with high compression ratio is adopted to compress image. In order to solve shortcoming of encoding time consuming when fractal image is compressing, CGA which can save storage space is used in match searching during encoding process. Mutation method is introduced to prevent genetic algorithm converging to local optimum. The experiment results express that the MCGA fractal image compression method can meet the initial demands of on-site supervision. But it still needs further analysis and research on how to speed up encoding.

Acknowledgments

This work is supported by the National Science and Technology Major Project of the Ministry of Science and Technology of China "12th Five-Year Plan" (No.2011ZX05039).

References

1. Dhawan S. A review of image compression and comparison of its algorithms[J]. International Journal of electronics & Communication technology, 2011.
2. Zhao E, Liu D. Fractal image compression methods: A review[C]. Information Technology and Applications, International Conference on (2005).
3. Wang H, Wang M, Hintz T, et al. VSA-based fractal image compression[J]. Journal of WSCG. 2005.
4. Poobal S, Ravindran G. The performance of fractal image compression on different imaging modalities using objective quality measures[J]. International Journal of Engineering Science and Technology (IJEST), 2011.
5. Wang M, Lai C H. A hybrid fractal video compression method[J]. Computers & Mathematics with Applications, 2005.
6. Kaur G, Hitashi G S. Performance Evaluation of Image Quality Based on Fractal Image Compression[J]. International Journal of Computers & Technology, 2012.
7. Wang H, Wang M, Hintz T, et al. Fractal image compression on a pseudo Spiral Architecture[C]//Proceedings of the Twenty-eighth Australasian

conference on Computer Science-Volume 38. Australian Computer Society, Inc., 2005.

8. Koli NA, Ali M S. A survey on fractal image compression key issues[J]. Information Technology Journal, 2008.
9. Zheng Y, Liu G, Niu X. An improved fractal image compression approach by using iterated function system and genetic algorithm[J]. Computers & Mathematics with Applications, 2006.
10. Xi L, Zhang L. A study of fractal image compression based on an improved genetic algorithm[J]. International Journal of Nonlinear Science, 2007.
11. Chakrapani Y, Rajan K S. Genetic algorithm applied to fractal image compression[J]. ARPN Journal of Engineering and applied Sciences, 2009.
12. Rastegar R, Hariri A. A step forward in studying the compact genetic algorithm[J]. Evolutionary computation, 2006.

An iterative list scheduling algorithm for static task scheduling problem on multiprocessor systems

Yu-Kun Song
School of Electronic Science & Applied Physics, Hefei University of Technology, HFUT,
Hefei, China
songyukun@hfut.edu.cn

Jun Yang
School of Electronic Science & Applied Physics, Hefei University of Technology, HFUT,
Hefei, China
yangjun@163.com

The performance of the Simple List Scheduling algorithm is unsatisfactory. In order to get the better schedule solution, this paper designs a novel task scheduling algorithm which is named as the Iterative List Scheduling algorithm. Compared with the Simple List Scheduling algorithm, the Iterative List Scheduling algorithm enlarges the search space of graph topological order to get smaller schedule length. Experiments with different scale of macroblock and different communication cost are provided to compare the performance of the Iterative List Scheduling and that of the Simple List Scheduling algorithm. Comparison result indicates that, the Iterative List Scheduling algorithm can schedule tasks more effectively than the Simple List Scheduling algorithm, especially under the circumstance of high communication cost. The maximum accelerate ratio reaches 102.8%

Keywords: Iterative list scheduling; graph; multiprocessor.

1. Introduction

The task scheduling problem is NPC and it still needs further research in order to give full paly to the advantage of the parallel computing system [1]. Scholars have proposed many algorithms for this problem [2] [3] [4] [5].

The Simple List Scheduling (SLS) algorithm is a frequently used algorithm for scheduling static tasks on a fully connected homogenous multiprocessor system [6]. While, the SLS algorithm can't get favorable schedule solution generally. In order to get smaller schedule length, this paper designs a novel one called the Iterative List Scheduling (ILS) algorithm.

The remainder of this paper is organized as following. The next section gives a description of the task scheduling problem and the SLS algorithm. Section 3 will put forward the ILS algorithm. Section 4 makes a comparison between the performance of the SLS algorithm and that of the ILS algorithm. Section 5 will make a conclusion for this paper.

2. Background

According to [7], a directed acyclic graph (DAG) G = (V, E, ω, c) as shown in Fig. 1 can denote a parallel program.

The time cost that all tasks accomplish is defined as schedule length (SL). Many algorithms have been raised to get an acceptable schedule length for the static task scheduling problem. Algorithm 1 shows the procedure of the SLS algorithm [6].

Take graph G1 as an example, the BL (Bottom Level) value of each task consists of the bl_list.

$$bl\_list = \{160,138,36,78,9,9,10,15,3,9,65,2\}$$

The SLS algorithm gets the TPL (Task Priority List) by sorting all tasks in the descending order of their BL value.

$$TPL = \{0,1,3,10,2,7,6,4,5,9,8,11\}$$

The SLS schedule solution of G1 is shown in Fig. 2 which denotes the schedule length of the SLS algorithm is equal to 98.

Generally, the SLS algorithm can't give a satisfactory schedule solution because only the TPL is taken into consideration for LS ( ) (List Scheduling) and all other graph topological orders (GTOs) are neglected. In other words, the schedule solution of the SLS algorithm just is one of the schedule solutions of the LS ( ). A better schedule result can be found by taking more GTOs into consideration. Based on that assumption, the next section will present the implementation of the Iterative List Scheduling (ILS) algorithm.

Algorithm 1: $SL\_SLS = SLS\,(G, P)$

Input: graph G, the set of processors P
Output: schedule length SL_SLS

- Part 1: $TasksSort(G)$

for each task $v_i \in V$
    if v_i is an exit-node
        $bl(v_i) = \omega(v_i)$
    else
        $bl(v_i) = max\ \{\ bl(v_j) + c(v_i, v_j)$

$+ \omega(v_i) \}, v_j \in V$

5) end for

6) sort all tasks into the list *TPL* in the descending order of their BL value.

7) set graph topological order $GTO = TPL$.

● Part 2: *LS* (*G*, *P*, *GTO*)
 // LS: List Scheduling

8) for each task $v_i \in GTO$

9) schedule v_i on the optimum processor $p_i \in P$ which offers the earliest start time

10) end for

11) compute
 $SL\_SLS = max \{ ts(v_j) + \omega(v_i) \}, v_j \in V$
 // $ts(v_i)$ denotes the start time of v_i

12) return *SL_SLS*

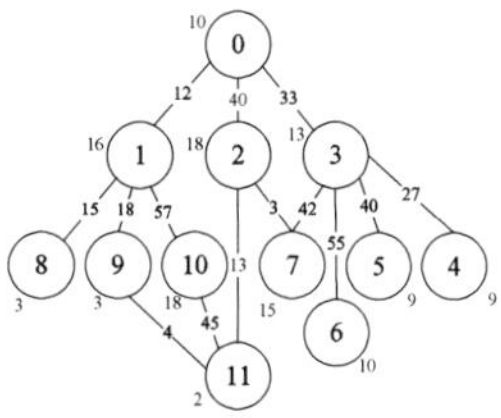

Fig. 1. Graph G1.

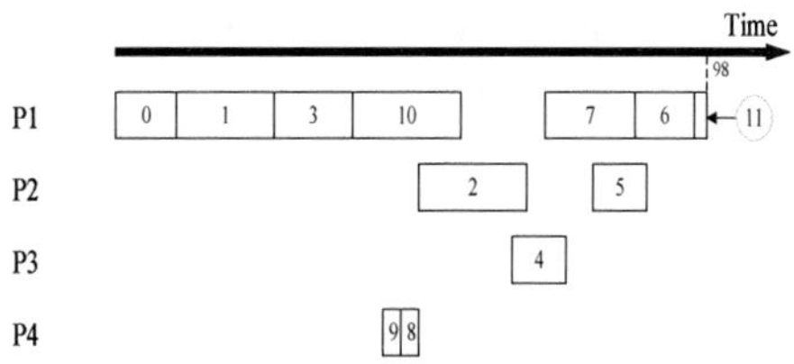

Fig. 2. The SLS schedule solution of G1.

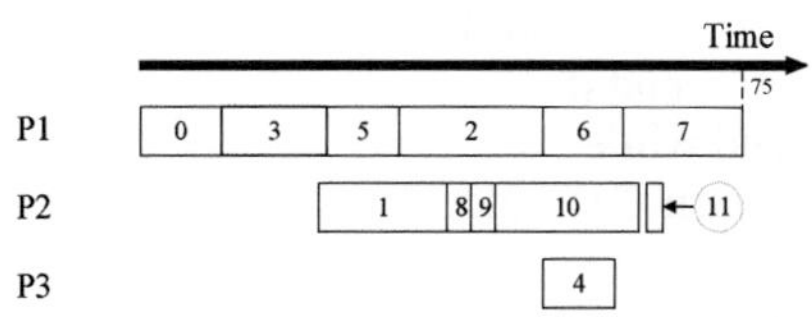

Fig. 3. An ILS schedule solution of G1.

3. The Proposed Algorithm

According to Algorithm 2, the ILS algorithm selects the TPL and LS (G, P, TPL) as the initial schedule solution. Then, it optimizes the solution by executing LS ( ) repeatedly. There is no doubt that the schedule length of the ILS algorithm is no larger than the SLS algorithm. In other words, Equation (1) is right for any graph which indicates that the ILS algorithm has higher performance.

$$\text{SL\_ILS} \leqslant \text{SL\_SLS} \tag{1}$$

```
Algorithm 2: (GTO, SL) = ILS (G, P)
1)  set times limit for loop: N = |V| ^2
2)  initialize n = 0
3)  for each task v ∈ V
4)      compute bl (v)
5)  end for
6)  get the TPL
7)  initialize GTO: GTO = TPL
8)  initialize SL: SL = LS (G, P, GTO)
9)  while n < N
10)     get two random numbers r and
        s in the range of [0, |V| - 1]
11)     generate GTO_swap by exchanging
        the position of the rth task and the
        sth task in GTO
12)     if GTO_swap is a topological order
        of G
13)         n = n + 1
14)         SL_swap =
                LS (G, P, GTO_swap)
15)         if SL_swap < SL
16)             GTO = GTO_swap
17)             SL = SL _swap
18)         end if
19)     end if
20) end while
21) return GTO and SL
```

Fig. 3 shows an ILS schedule solution of G1. As we can see from Fig. 3, SL_ILS = 75. While, according to Fig. 2, SL_SLS = 98. The schedule solution of the ILS algorithm is much better than that of the SLS algorithm. In order to compare the performance of these two algorithms, this paper defines the accelerate ration as following.

$$\text{Acc (SLS, ILS)} = (\text{SL\_SLS} / \text{SL\_ILS} - 1) \cdot 100\% \tag{2}$$

According to (1) and (2), the following expression is right for any graph.

$$\text{Acc (SLS, ILS)} \geqslant 0 \tag{3}$$

Take G1 for instance,

$$\text{Acc (SLS, ILS)} = 30.67\% \tag{4}$$

4. Experiments

In order to compare the performance of these two algorithms, this paper adopts the C language to implement the SLS and ILS algorithm. This paper selects four types of graph as sample graph: Input, Output, Meanvalue and FFT [8]. The number of tasks for each type of graph is 31, 31, 31 and 39. For each type, 1000 samples are generated with CCR = 1/9, 1/3, 1, 3 and 9. CCR denotes the ratio of average communication cost to average computation cost. The SLS and ILS algorithm will schedule all these graphs and we will make a comparison between the SLs of these two algorithms. The result is shown in Fig. 4, where the Acc denotes the mean value of the Acc (SLS, ILS).

According to Fig. 4, the ILS algorithm performs remarkable better than the SLS algorithm, especially under the circumstance of high communication cost. The larger the value of CCR, the larger the Acc will be. The maximum value of Acc (SLS, ILS) reaches 1.028 when CCR = 9. From the side, it reflects that the performance of the SLS algorithm has great promotion space and the ILS algorithm is a great improved version.

5. Conclusion

Based on the deep analysis on the SLS algorithm, this paper gives an assumption that taking more graph-topological-orders into consideration will lead to better schedule solution. According to this assumption, the ILS algorithm which outperforms the SLS algorithm has been presented. The proposed algorithm works evidently better if the communication cost is higher.

Acknowledgment

This work is supported by the National Natural Science Foundation of China (grant NSFC- 61179036).

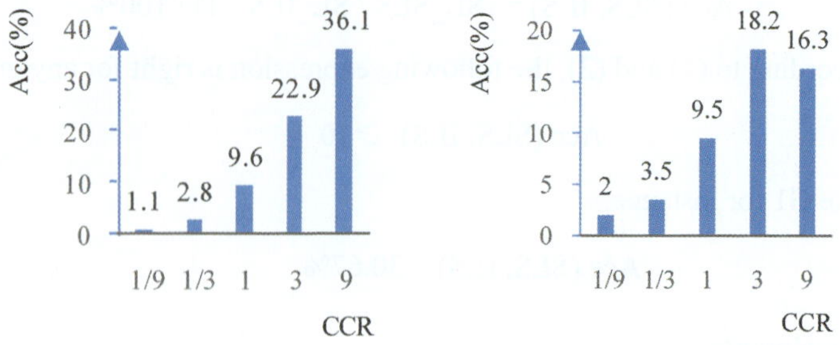

(a) Input (b) Output

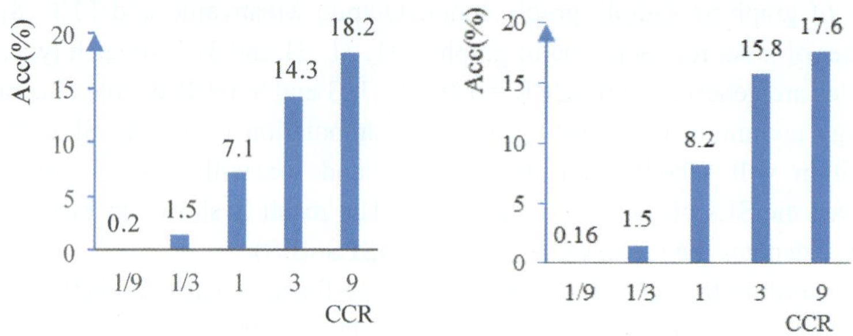

(c) Meanvalue (d) FFT

Fig. 4. Acc with different CCRs and type.

References

1. M.R. Gary, D.S. Johnson. Computers and Intractability: A Guide to the Theory of NP-Completeness, USA: Bell Telephone Laboratories, Inc., 1979, pp. 238-239.
2. J.J. Hwang, Y.C. Chow, F.D. Anger, and C.Y. Lee. "Scheduling Precedence Graphs in Systems with Interprocessor Communication Times," SIAM J. Computing, vol. 2, April. 1989, pp. 244-257.
3. HB Prajapati, VA Shah. "Scheduling in Grid Computing Environment," Fourth International Conference on Advanced Computing & Communication Technologies, IEEE Press, Feb. 2014, pp. 315-324, doi: 10.1109/ACCT.2014.32.
4. Yan Kang, Ying Lin. "A recursive algorithm for scheduling of tasks in a heterogeneous distributed environment," Fourth International Conference

on Biomedical Engineering and Informatics, IEEE Press, April. 2011, pp. 2099-2103, doi: 10.1109/BMEI.2011.6098753.

5. N. Saranya, R. C. Hansdah. "Dynamic Partitioning Based Scheduling of Real-Time Tasks in Multicore Processors," IEEE 18th International Symposium on Real-Time Distributed Computing, IEEE Press, Apr. 2015, pp. 190-197, doi: 10.1109/ISORC.2015.23.
6. Oliver Sinnen. Task schedule for parallel system, New Jersey: John Wiley & Sons, Inc., 2007, pp. 108-113.
7. MU Peng-Cheng, NEZAN Jean-Francois, RAULET Mickael, COUSIN Jean-Gabriel. "Advanced list scheduling heuristic for task scheduling with communication contention for parallel embedded systems", Scientia Sinica Informationis, vol. 41, 2011, pp. 349-364.
8. Yu-Kwong Kwok, Member, IEEE Computer Society and Ishfaq Ahmad, Member, IEEE Computer Society. "Dynamic Critic Path Scheduling: An Effective Technique for Allocation Task Graphs to Multiprocessors", IEEE Transactions on Parallel and Distributed Systems, vol. 7, 1996, pp. 506-521.

A design of denture grinding CAM system based on the theory of digital copying

Li-Cheng Fan, Lai-Feng Zhang* and Si-Qi Liu
School of Mechanical and Electric Engineering , Soochow University, Suzhou, 215006, China
**E-mail: sudalaifeng@yeah.net*

All-ceramic dentures grinding is a typical sculptured surface machining, this paper presented a digital copying machining method for sculptured surface machining. First, for the existing temporary metal dentures, probe with the same shape of grinding bur was used to achieve the contact measuring of temporary dentures according to the planned measurement path, which could obtain the grinding bur path data directly. In addition, a five-axis machining measurement system was designed to realize parallel measurement and grinding with double tools. Finally, experiments were conducted on incisor and molar to measure the tool path and implement grinding test, and the all-ceramic dentures with high-quality surface shape were possessed. The results verified that the digital copying machining method raised for sculptured surface machining was feasible

Keywords: Digital copying; contact measuring; tool path; denture grinding.

1. Introduction

With the reverse engineering and CAD / CAM technology introduced into the field of prosthodontics, bio-ceramic with better biocompatible and non-toxic effect has been a major denture material, which has caused great changes in denture materials. First of all, a specialized scanners and dental CAD / CAM system were used to achieve the denture CAD model. After that, a dedicated CAM module of denture repair system was applied to generate the grinding bur path which could implement the forming grinding of bio-ceramic blocks and produce the perfect all-ceramic dentures [1-3]. This denture production technology has changed the conventional manual manufacturing method fundamentally and it is a major technological breakthrough in prosthodontics.

General production process of all-ceramic dentures is: Firstly, obtaining the scanning point cloud data from the plaster model which sampled from the patient's mouth by the dedicated scanner. Secondly, using the special oral CAD software to join these point clouds together automatically and generated a teeth 3D model. And then, according to the denture CAD model and the tool shape, the CAM module of the system generated the tool path of processing in accordance with the planned grinding path. Finally, installing ceramic blanks and grinding all-ceramic dentures[4-5].

Currently, there are two kinds of dental bio-ceramic mainly used: one type is called the zirconia ceramic, which could be used to repair the oral cavity of patients until it is grinding-formed and glazed at high temperature. The disadvantage of this denture is that it would shrink after sintering at high temperature, influencing its final shape and precision. The other type is the glazed glass-ceramic, which can be directly applied to patients after grinding-forming. The grinding shape is also the final shape with high precision, yet the material is expensive, and the high hardness which could lead to greater loss of grinding bur. At present, the development of dedicated CAM systems according to the two kinds of denture materials also have different characteristics, the processing system applying to denture zirconia ceramic material used a single tool usually, which could achieve four or five axes. Firstly, the external crown of the denture was processed, and then turned the blank 180 degrees, and the internal crown was processed, see Figure 1.This CAM system was less efficient, but the ceramic blanks could be made into round cakes[6], besides, the blank could processed more dentures with high material utilization. The Sirona company have developed a CEREC-III grinding system with 5-aixes for glass-ceramic materials, which used double tools to grind the internal and external crown of the denture in parallel[7], but it is expensive and high-cost.

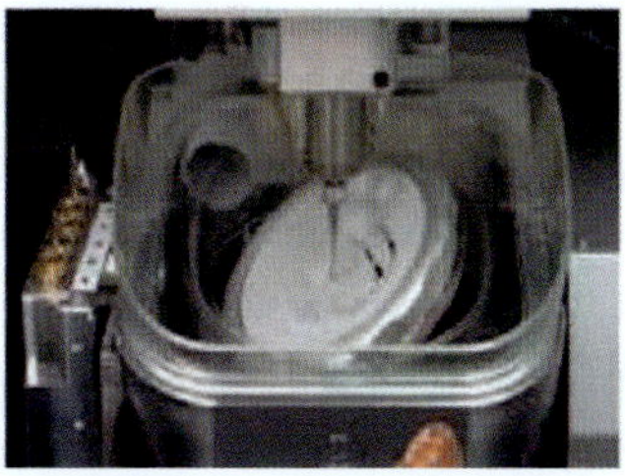

Fig. 1. The DECLAM system.

A digital copying processing method for grinding dentures was presented in this paper. It could get grinding bur path withoutCAD modeling and tool path calculation module, and then developed a 5-axis CAM denture grinding system according to this principle. This system with the measurementfunctioncould obtain the tool path of denture grinding directly. Finally, experiments were carried out on measuring and grinding, machined all-ceramic dentures. The experiment results were presented in the end of this paper. And the results indicated that the CAM system introduced in this paper was feasible.

2. Principle of Digital Copying Measurement

The basic method of digital copying principle is that a probe which shape is exactly the same to grinding bur was used to scan the surface of a temporary metal denture with desired shapeaccording to the plannedtool path of surface machining. The measurement data is the path data of the tool grinding motion due to the identical shape of probe and tool. Using this data to drive and control the movement of grinding bur could grind out an all-ceramic denture with the same shape to temporary denture on the ceramic material. The advantage of digital copying machining principle is: it is not necessary to develop expensive and dedicated denture CAD/CAM software, and the temporary dentures could be obtained by engraving of dental technicians, low cost.

2.1. *The planning of measuring and grinding path*

For sculptured surface of denture, whether the measuring on metal denture or the grinding on ceramic block, the planned path of measuring and grinding are fully identical due to the same shape of measuring probe and grinding bur, so only the planning method of measuring path is described here. Without a doubt, the method is also applicable to the subsequent all-ceramic dentures grinding. The appearance of denture is mainly made of internal crown and external crown, see Figure 2, and it is a typical complex sculptured surface. If using a single probe, not only needing multi-axis, but also the measuring efficiency is low. Therefore, a planning path method which used two probes to measure each layer at the same time was proposed in this paper. In the measurement process of each layer the two probes moved in two directions relatively, measuring the internal and external crown of the denture separately and recording the trajectory data in real time, see Figure 3. When the measuring in the same level was over, metal denture increased one pitch, then the two probes continued to measure the trajectory data of the next layer, so continue, as a result the whole denture was measured.

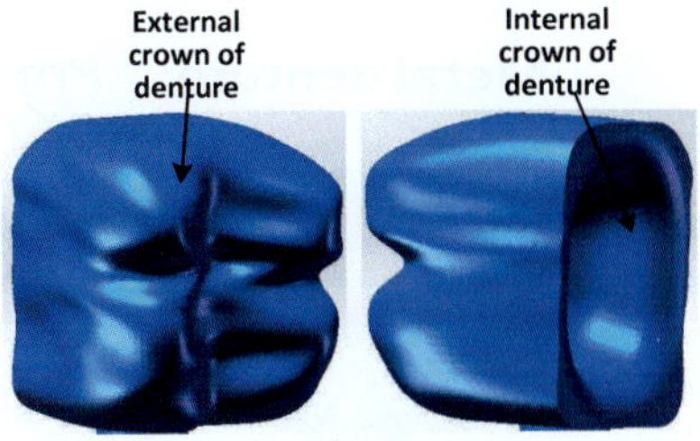

Fig. 2. The denture model.

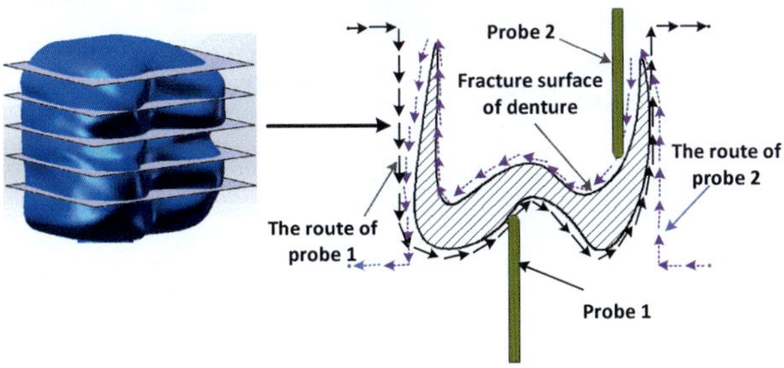

Fig. 3. The planning measuring path of molar.

2.2. *The design of measuring contact sensor*

The most critical issue was to check the contact status of probe with denture in real time when measuring. There were two jobs to be done at the moment when the probe and the denture contacted, the one was to decide the next direction of probe, the other one was to record the position data of probe. Consequently, the requirement to touch sensor sensitivity was very high, and a method for detecting electrical signals was presented in this article. A design of measuring contact sensor was shown in Figure 4. Two detection circuits were made of temporary metal denture, metal denture mounting shaft, measuring probe and probe mounting spindle, and the temporary metal denture and metal denture mounting shaft was a common circuit among them. The left and right probe moved by the planned measuring path.When the left measuring probe and the metal denture contacted, the system formed a left closed loop, left circuit conducting, and when the right measuring probe and the metal denture contacted, the system formed a right closed loop, right circuit conducting. The controller could judge whether the probe and denture metal contacted when it detected the current signal. Finally, the system could obtain the instantaneous probe tip point trajectory data from the feedback of motion axis position information.

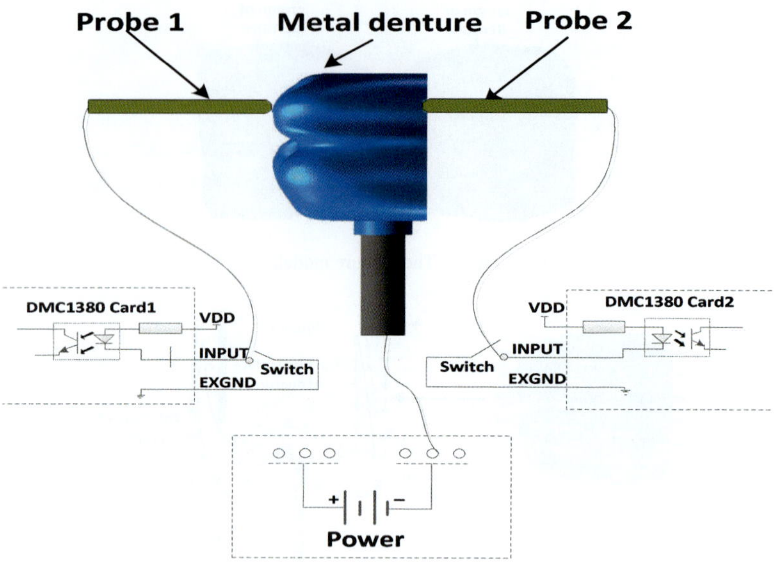

Fig. 4. The design of measuring contact sensor.

3. Design of System Structure

Denture measuring and grinding system was a measuring instrument and a NC machine tool, the specific layout of the structure as shown in Figure 5. The system was composed of the left X axis, the left Y axis, the right X axis, the right Y axis and the Z axis, five motion axes totally. The left X axis, the left Y axis and the right X axis, the right Y axis were arranged horizontally and symmetrically among them, and grinding bur or measuring probe was mounted on the Y-axis. The system could realize the plane movement of probe or tool with the linkage of each side of the X axis and Y axis. The Z axis was the mounting axis of the ceramic block and metal denture, and the movement direction of the Z axis was vertical. The main reason why the power of the tool rotation was driven by the pneumatic turbine instead of the motor was that the probe, the pneumatic turbine and the Y axis were made of metal materials with conductive property, and these components could be combined into a circuit, while the motor could not achieve the conductive function because of insulation.

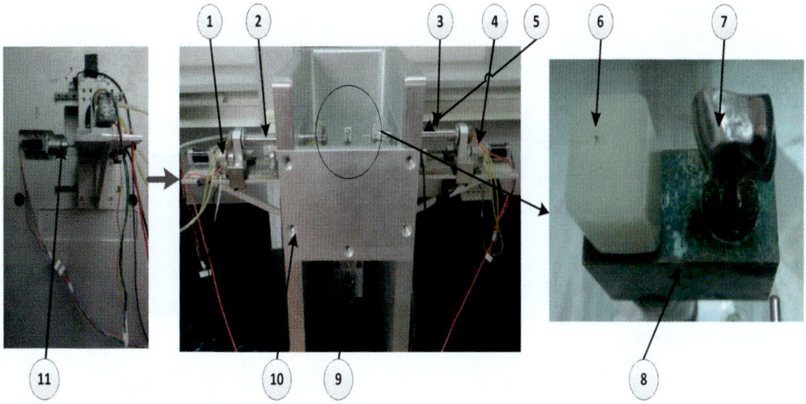

1-Left X axis, 2-Left main axle, 3-Right main axle, 4-Right X axis, 5-Right Y axis, 6- Ceramic block, 7- Temporary mental dentures, 8-Work bench, 9-Z axis, 10- Lathe bed, 11-Left Y axis

Fig. 5. The specific layout of the structure.

4. Experiment on Measuring and Grinding

In order to verify the feasibility of the digital copying theory of processing dentures, in this paper, for two typical teeth: incisor and molar, the measuring and grinding experiments were carried out according to the system of denture measuring and grinding. The specific experimental data and results as shown in Table 1, Table 2 and Figure 6, Figure 8, Figure 9.

Fig. 6. The scene photographs of measuring.

Table 1. Measurement experimental parameters.

Tooth	Probe	Speed of probe	The interval	Times
Incisor	Φ1.6mm probe	2.4mm/s	0.05mm	30minutes
Molar	Φ1.6mm probe	2.4mm/s	0.05mm	35minutes

The tool path was simulated by MATLAB according to the measuring cutter location coordinate data and the results were shown in Figure 7.

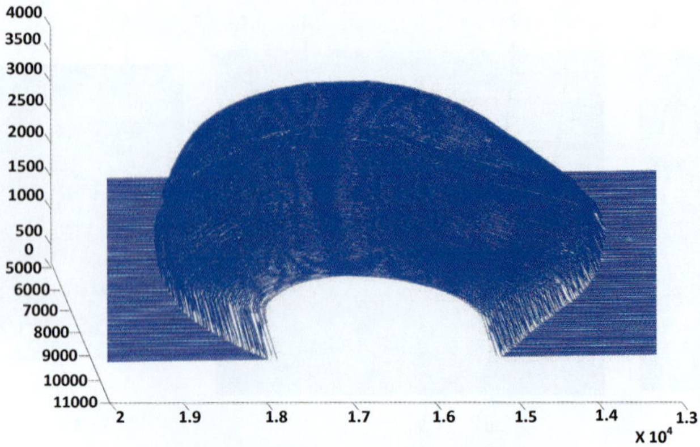

(a)The tool path of incisor external crown

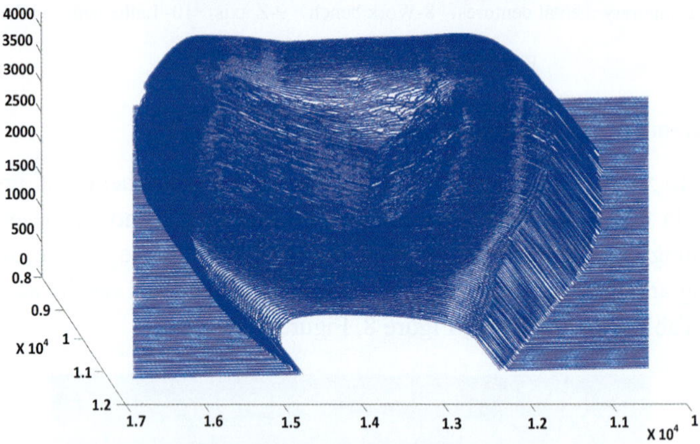

(b)The tool path of incisor internalcrown

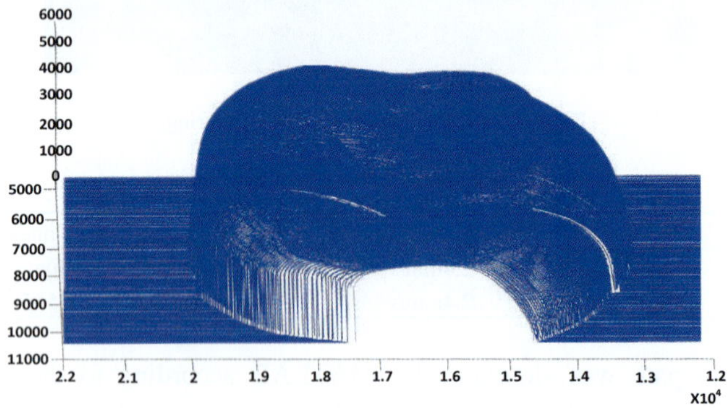

(c) The tool path of molar externalcrown

Fig. 7. The tool path of denture grinding.

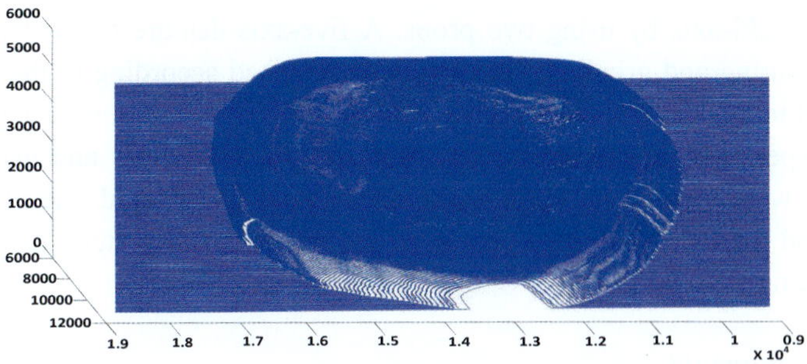

(d) The tool path of molar internalcrown

Fig. 7. (*Continued*)

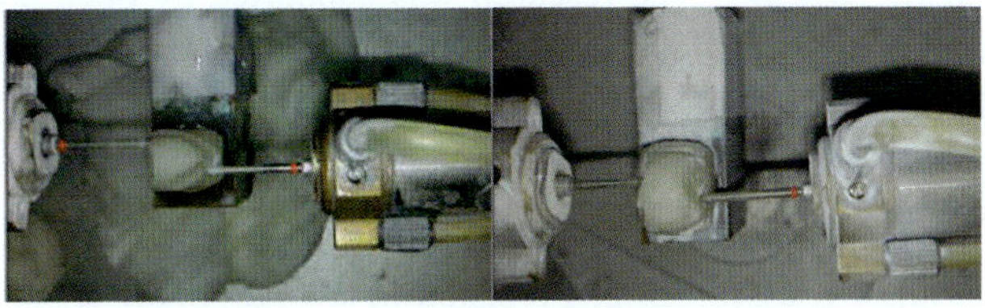

Fig. 8.The scene photographs of grinding.

Table 2. Grinding experimental parameters.

Tooth	Tool	Blank	Speed of tool	Times
Incisor	Φ1.6mm grinding bur	CEREC Blocs ceramics	1.6mm/s	60 minutes
Molar	Φ1.6mm grinding bur	CEREC Blocs ceramics	1.6mm/s	75 minutes

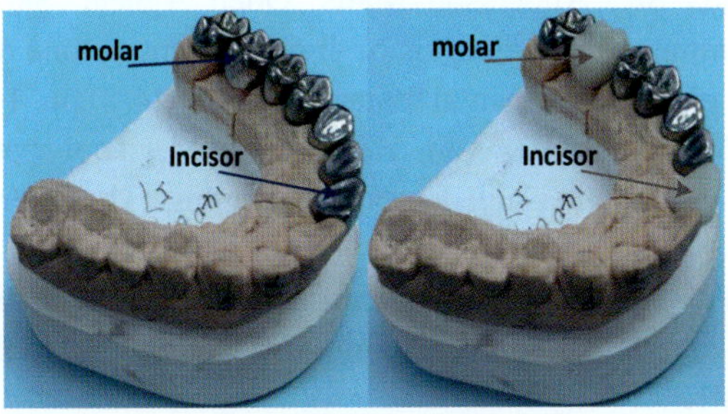

Fig. 9. Mental and all-ceramic denture.

5. Conclusions

This paper proposed a principle of digital copying machining, which could obtain the coordinate data of grinding bur directly. The method of planning measuring path and the design of measuring contact sensor of the system were introduced in this article, which was high measuring efficiency and without the need for multi-axis by using two probs. A five-axis denture processing system with measuring and grinding functions was designed according to the principle presented in this paper, then the layout of the structure were introduced. Finally, as the experiment of measuring on temporary metal denture and grinding on ceramics were conducted, getting one all-ceramic incisor and one all-ceramic molar, and the feasibility of digital copying principle and system of measuring and grinding were verified.

Acknowledgments

The research was supported by Natural Science Foundation of Jiangsu Province of China, No BK2011299. The research was supported by open project of the State Key Laboratory of Digital Manufacturing Equipment and Technology of China, No DMETKF2010005

References

1. Coli Pierluigi, Karlsson Stig, Precision of a CAD/CAM technique for the production of zirconium dioxide copings. International Journal of Prosthodontics. 2004, 17(5), pp. 577-80.
2. Takashi MIYAZAKI, Yasuhiro HOTTA, Jun KUNII, Soichi KURIYAMA and Yukimichi TAMAKI, A review of dental CAD/CAM: current status and future perspectives from 20 years of experience. Dental Materials Journal. 2009, 28(1), pp. 44-56.
3. Persson Anna SK, Andersson Matts, Odén Agneta, Sandborgh-Englund Gunilla, Computer aided analysis of digitized dental stone replicas by dental CAD/CAM technology. Dental materials. 2008, 24(8),pp. 1123-30.
4. Boogert RM, Kals HJJ, van Houten FJAM, Tool paths and cutting technology in computer-aided process planning. The International Journal of Advanced Manufacturing Technology, 1996, 11(1), pp. 86-97.
5. M.Gaspar, F.Weichert, Integrated construction and simulation of tool paths for milling dental crowns and bridges. Computer-Aided Design 2013, 45:1170-1181.
6. Fenfen Tu, Development of Small-Scale Rapid Dental Processing Machine and Research on its Control Performance[D]. Nanjing University of Aeronautics and Astronautics(Nanjing, China, 2013).
7. Allen K L, Schenkel A B, Estafan D, An Overview of the CEREC 3D CAD/CAM System. General Dentistry, 2004, 52(3), pp. 234-235.

Geometric error analysis for casting/forging blades by an optical inspection device

Rui-rui Song[1,a], Lai-xin Yu[1,b], He Xie[1,2,c], Chang-tao Pang[3,d] and Wen-long Li[1,2,†]

[1]*State Key Laboratory of Digital Manufacturing Equipment and Technology, Huazhong University of Science & Technology, Wuhan, P. R. China*

[2]*Guangdong HUST Industrial Technology Research Institute, Dongguan, P. R. China*

[3]*Aviation Key Laboratory of Science & Technology on Precision Manufacturing, Beijing, 100007, P.R. China*

[a]*260834471@qq.com*

[b]*365375332@qq.com*

[c]*xiehe@mail.hust.edu.cn*

[d]*13611309735@163.com*

[†]*wlli@mail.hust.edu.cn*

Optical inspection of blades is a meaningful work in automation manufacturing. In this paper, a laser inspection device is designed for inspecting the parameters of casting/forging blades. For the manufacturing and installation errors, an effective way is given to improve the inspection accuracy. First, a mathematical model of the device is established to find out the main parameters affecting the inspection accuracy. Second, the influence of parameters on the inspection accuracy can be calculated by the blade CAD model. Third, the maximum deflection of the translational and rotational axis is acquired to adjust the deflection angle using shims and a laser tracker. Furthermore, experiments show that the inspection accuracy is 0.023mm, which can meet the inspection requirements very well.

Keywords: Parameters of blades; maximum deflection; turbine blade inspection.

1. Introduction

Blades play a vital role in aero-engine, nuclear turbine and propeller blades. They have complicated internal structure and high precision, and even work in harsh environment with high temperature and pressure. So the blades tend to experience geometric deformation for clamping and cutting forces, especially in thin-walled and easy-deformed structure[1, 2]. Therefore, it has always been a challenge for turbine blade inspection in a precision and efficient way.

At present, blade inspection can be divided into two ways: contact measurement (standard measuring instrument, CMM) and non-contact

[†] Corresponding author.

measurement (laser scanner, CT). Most Chinese enterprises used the standard measuring instrument and observed the size of the light-leaking gap between the standard measuring instrument and blade[3]; Kang et al.[4] used CMM to detect a blade and automatically calculated the geometry size; Makem et al.[5] developed a virtual casting blade detection system, including blade measuring method, efficient localization algorithm and software, and the effectiveness is verified by means of turbine blade inspection. Gao et al.[6] launched an important work for on-line inspection, where a GOM-ATOS equipment is used to scan the blade. Recently, the Breuckmann company developed a 3D inspection system using adapted structured light projector[7]. The GE company developed an inspection machine to measure blades for comparison with CAD data[8]. Contact measurement methods are inefficient and cannot guarantee the accuracy of measurement. And the traditional non-contact measurement methods are affected by the data scale and noise. Therefore, the accuracy and efficiency of the turbine blade inspection, refining point cloud processing method are difficult problems to be solved.

This paper designs an optical blade inspection device named iBlade Scan Ⅰ for casting/forging blades, which meet the required inspection accuracy. The blade section can be located accurately on the rotational stage by a clamp. With the movement of translational and rotational axis, there are manufacturing and installing errors brought into the device. In order to make both errors in required range, it is necessary to identify the biggest change of blade section parameters, such as radius of leading/trailing edge and maximum length. Then, we can find out the maximum deflection angle of this section and the overall inspection error is less than 0.03mm. In the following, a mathematical model of the device is established in Section 2, which is used to find out the main parameters affecting the accuracy. Following that, the maximum deflection of the translational and rotational axis can be obtained by a blade model in Section 3. Then, experiments and conclusion are given in Section 4 and Section 5, respectively.

2. Mathematical Model of Measuring Device

The model of the optical blade inspection device is shown in Figure 1. The sensor can move linearly and the blade installed by a clamp can rotate in 360 degrees. The developed software iCloud3D of motion planning, data collection, data processing and generating report is programmed using C++.

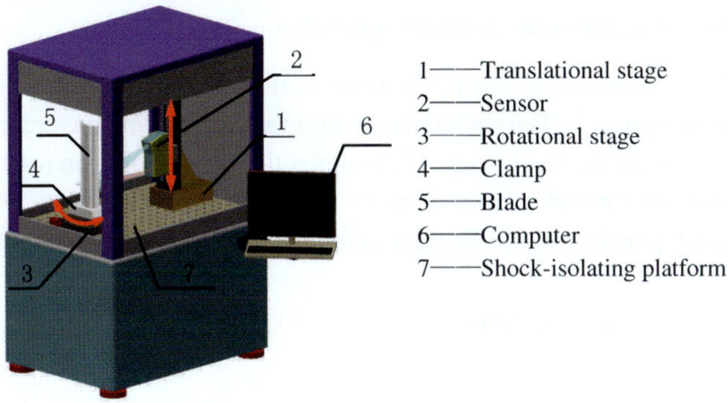

Fig. 1. The optical blade inspection device.

2.1. *The error model with transition deflection*

Because of the manufacturing and installation errors of transitional stage, transitional axis has a deflection. The deflection causes that the laser plane is not vertical with measured blade section and shock-isolating platform. The deviation of the laser plane can be expressed as

$$\Delta s_1 = h - h \cdot \cos\alpha + d \cdot \tan\alpha \tag{1}$$

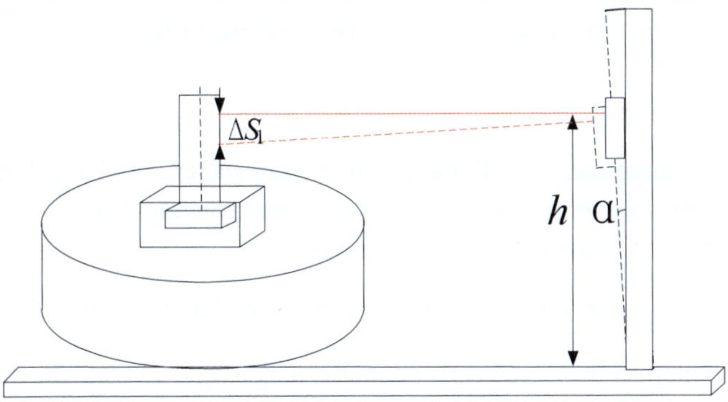

Fig. 2. The inspection device with transitional stage deflection.

Where symbol α and h are the deviation angle of the transitional stage and the height of the measured blade section, respectively. And the symbol d is a fixed value which represents the distance between transitional and rotational axes. The solid and dashed lines represent that there is an error of the device or not in Figure 2.

2.2. *The error model with rotation deflection*

The similar to transitional stage, the rotational stage, clamp, and blade all have deflections in Figure 3. The deflections cause that the blade measurement section is not vertical with the laser plane. If the deflection angle between rotational stage and platform, clamp and rotational stage, blade and clamp are $\beta_1, \beta_2, \beta_3$. And the height of each part is r_1, r_2, r_3 . The deviation of the measured blade section is

$$\Delta s_2 = r_1 \cdot \tan \beta_1 + r_2 \cdot \tan \beta_2 + r_3 \cdot \tan \beta_3 \tag{2}$$

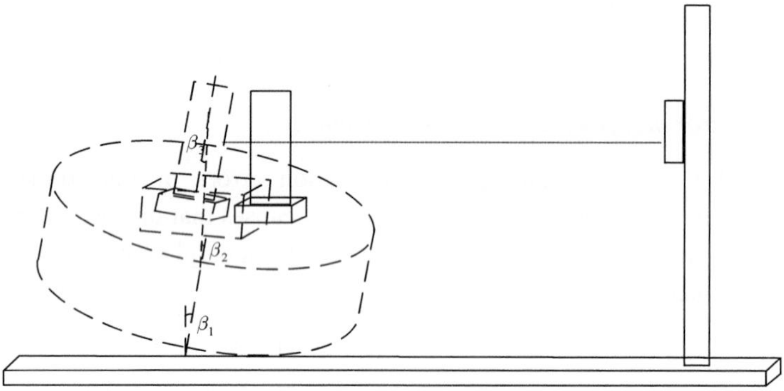

Fig. 3. The inspection device with rotational stage deflection.

2.3. *The combined error model of inspection device*

During the measurement, transitional and rotational stages both have deflections. So there is always an error between transitional and rotational stage. Then, laser plane and measured blade section have a deviation

$$\Delta s = (h - h \cdot \cos\alpha + d \cdot \tan\alpha) + (r_1 \cdot \tan \beta_1 + r_2 \cdot \tan \beta_2 + r_3 \cdot \tan \beta_3) \tag{3}$$

The deviation Δs in Eq. (3) varies with the height h, r_1, r_2, r_3 and the deflection angles $\beta_1, \beta_2, \beta_3$. This paper mainly analyzes the influence of the height and the deflection angles on the inspection accuracy.

3. Calculation of Blade Model

All variables affected the deviation can be seen as the synthesis of three directions, including X, Y and Z, so the error of the blade inspection is a result of these three directions. According to the Taylor formula

$$
\begin{aligned}
f_i(h+\Delta h,\theta_x+\Delta\theta_x,\theta_y+\Delta\theta_y) &= f_i(h,\theta_x,\theta_y) \\
&+\frac{1}{1!}(\Delta h\cdot\frac{\partial}{\partial h}+\Delta\theta_x\cdot\frac{\partial}{\partial\theta_x}+\Delta\theta_y\cdot\frac{\partial}{\partial\theta_y})f_i(h,\theta_x,\theta_y)+\ldots \\
&\ldots+\frac{1}{n!}\left(\Delta h\cdot\frac{\partial}{\partial h}+\Delta\theta_x\cdot\frac{\partial}{\partial\theta_x}+\Delta\theta_y\cdot\frac{\partial}{\partial\theta_y}\right)^n f_i(h,\theta_x,\theta_y)\ \ (i=1,2,3)
\end{aligned} \tag{4}
$$

The three variables h,θ_x,θ_y represent the height of the blade section, X-axis angle and Y-axis angle, respectively. And the three variables $\Delta h,\Delta\theta_x,\Delta\theta_y$ represent the Z-axis distance of the blade section, X-axis deflection angle and Y-axis deflection angle, respectively. The equation $f_i(h,\theta_x,\theta_y)$ $(i\quad 1,2,3)$ represents the blade parameters (leading edge radius, trailing edge radius, maximal thickness) vary with the height of the blade section and the deflection angle of each axis.

3.1. *The blade model without deflection*

The blade sections and the point clouds of this blade can be acquired from the design CAD model in figure 4, then in figure 5 we can extract the parameters of this blade section, such as leading edge radius, trailing edge radius and maximal thickness[9,10].

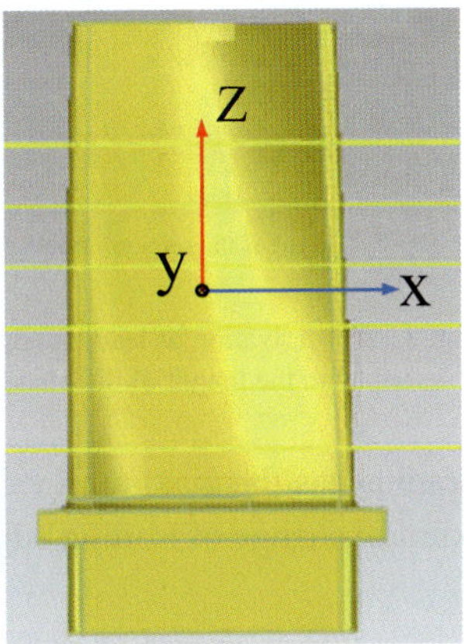

Fig. 4. The blade sections CAD model.

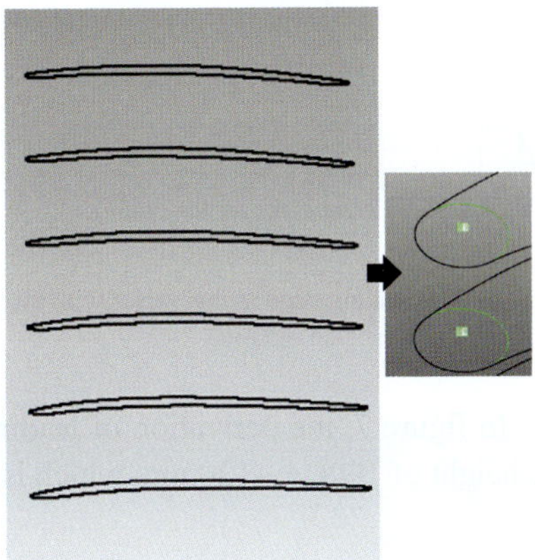

Fig. 5. The point clouds of blade.

Leading/Trailing edge radius: The points can be selected from the leading/trailing edge, and the circle can be fitted by least square method.

Maximal thickness: The diameter of the largest inscribed circle about the blade section.

3.1.1. *The error model of the leading edge radius*

Putting the parameters data into MATLAB, the curve represents that the leading edge radius $f_1(h,0,0)$ varies with the height of the blade section h in figure 6, and the curve expressions can be given with Gaussian fitting

$$f_1(h,0,0)=5.028\cdot e^{-\left(\frac{h-164.1}{500.6}\right)^2}+0.6382\cdot e^{-\left(\frac{h+31.53}{76.59}\right)^2} \tag{5}$$

After derivation, the expression is

$$\frac{\partial f_1}{\partial h}=-0.02\cdot\left(\frac{h-164.1}{500.6}\right)\cdot e^{-\left(\frac{h-164.1}{500.6}\right)^2}-0.0167\cdot\left(\frac{h+31.53}{76.59}\right)\cdot e^{-\left(\frac{h+31.53}{76.59}\right)^2} \tag{6}$$

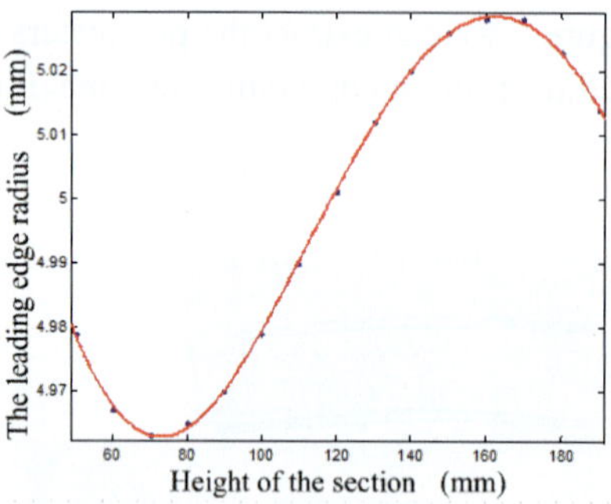

Fig. 6. The leading edge radius varies with the height of the blade section.

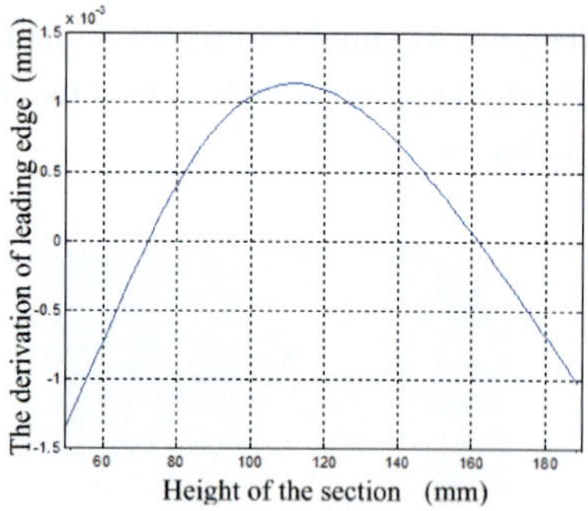

Fig. 7. The derivation of leading edge radius varies with the height of the blade section.

In figure 7, the derivation of leading edge radius has the maximum value at the height of 100mm~120mm, which is the maximum change of leading edge.

3.1.2. *The error model of the trailing edge radius*

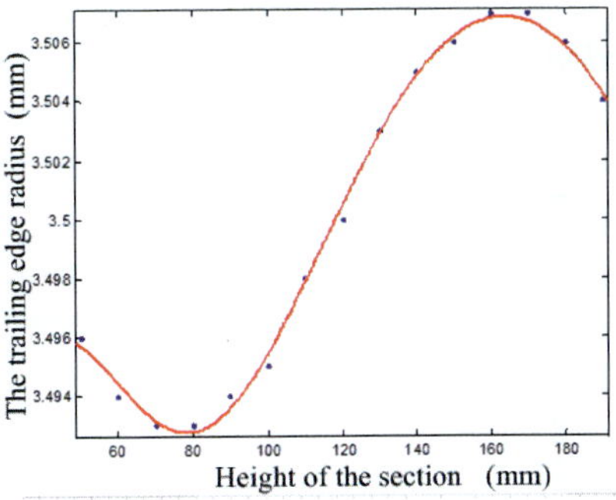

Fig. 8. The trailing edge radius varies with the height of the blade section.

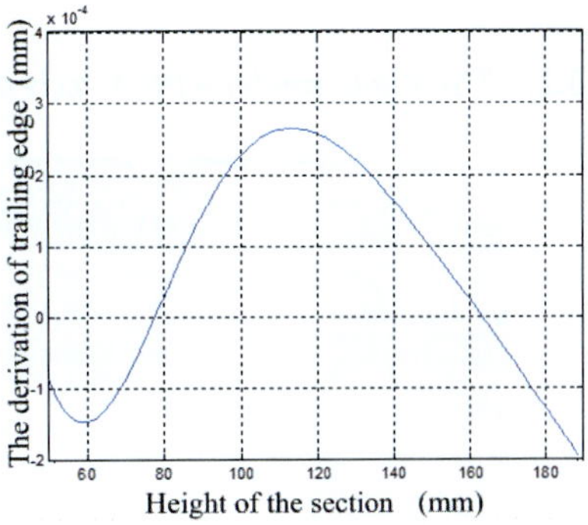

Fig. 9. The derivation of trailing edge radius varies with the height of the blade section.

Same as above, trailing edge radius and derivation of trailing edge radius curves can be acquired in figure 8 and figure 9.

In figure 9, the derivation of trailing edge radius has the maximum value at the height of 100mm~120mm, which is the maximum change of trailing edge.

3.1.3. *The error model of the maximal thickness*

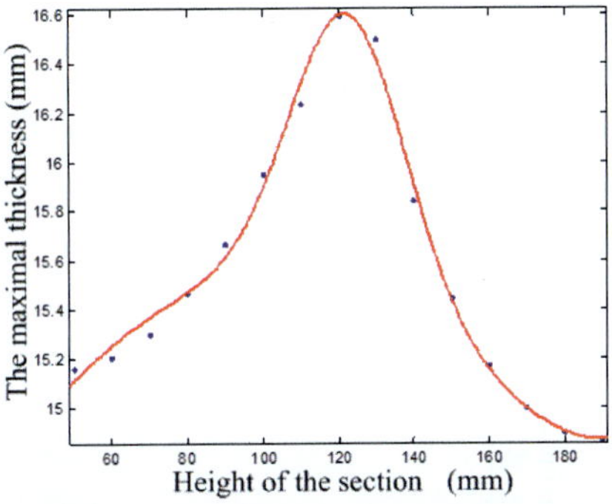

Fig. 10. The maximal thickness varies with the height of the blade section.

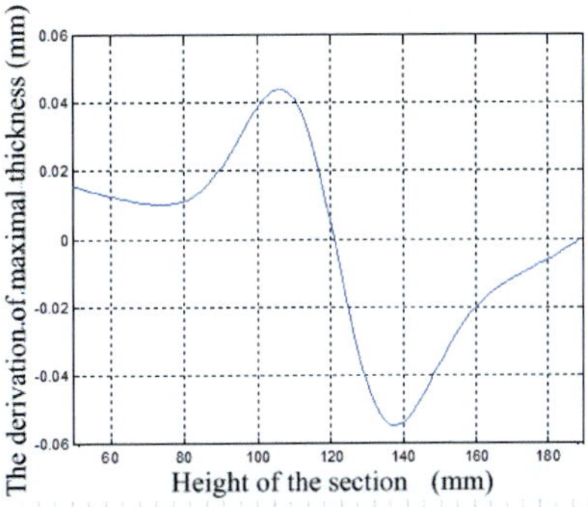

Fig. 11. The derivation of maximal thickness varies with the height of the blade section.

Similarly, the maximum change of maximal thickness is at the height of 95mm~105mm (Figs. 10 and 11).

All things considered, the height of 100mm is regarded as the maximum change of all parameters. So it is necessary to find out the X and Y axes maximum deflection angles at the height of 100mm. The angles are its maximum allowable values of the whole blade.

3.2. *The blade model with X-axis deflection*

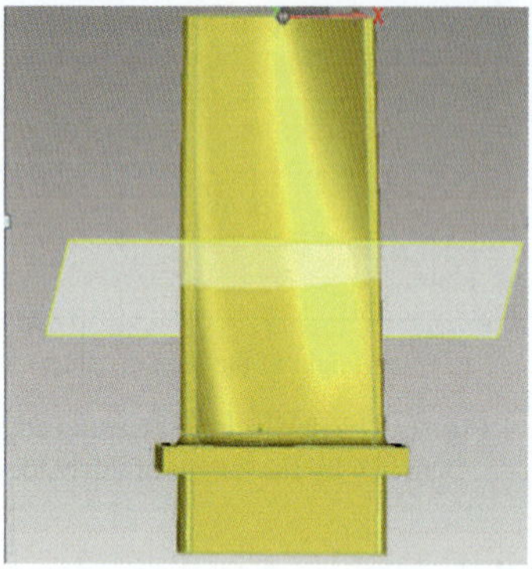

Fig. 12. The blade model with X-axis deflection.

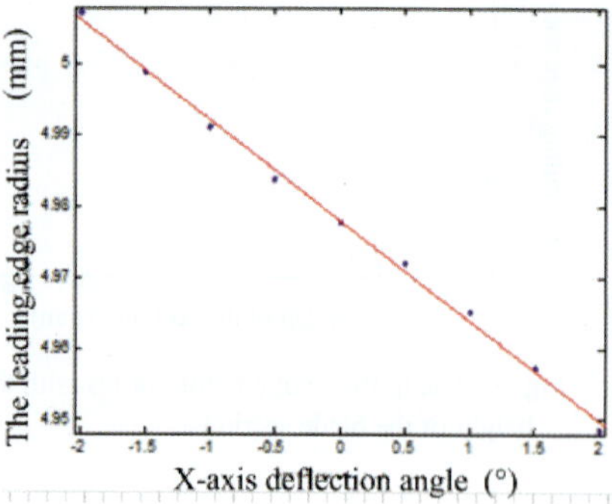

Fig. 13. The variation of leading edge radius.

If the blade has a X-axis deflection, the blade section at the height of 100mm can be required in Fig. 12. Referring to section 3.1, the variation curves of blade parameters (Figs. 13~15) and their expressions can be acquired.

The following expression represents that the leading edge radius varies with X-axis deflection angle.

$$f_1(100,\theta_x,0) = -0.0141\cdot\theta_x + 4.978 \tag{7}$$

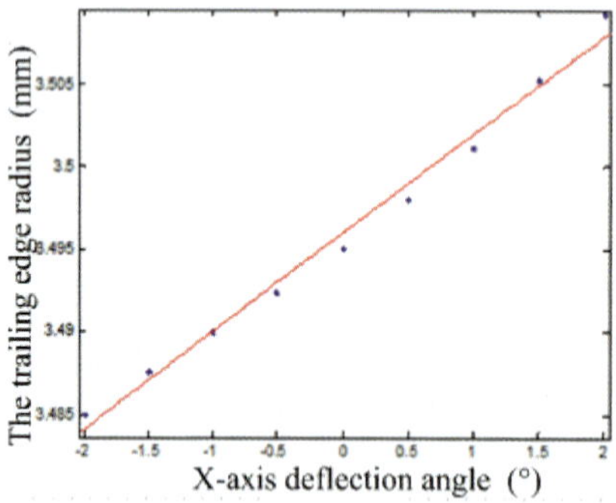

Fig. 14. The variation of trailing edge radius.

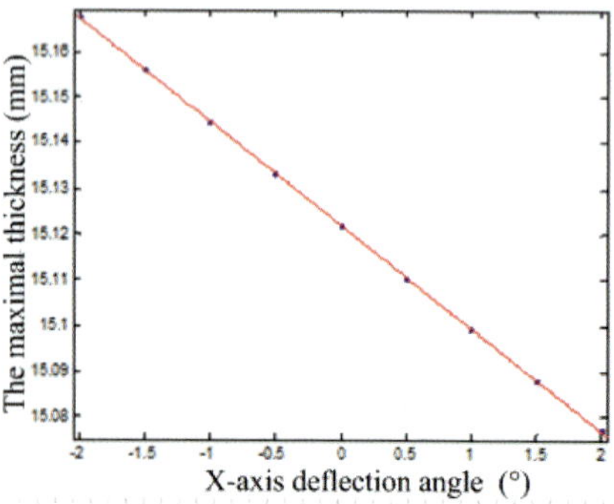

Fig. 15. The variation of maximal thickness.

The Eq. (8) and Eq. (9) represent that trailing edge radius and maximal thickness vary with X-axis deflection angle.

$$f_2(100,\theta_x,0)=0.005918\cdot\theta_x+3.496 \tag{8}$$

$$f_3(100,\theta_x,0)=-0.02265\cdot\theta_x+15.12 \tag{9}$$

3.3. *The blade model with Y-axis deflection*

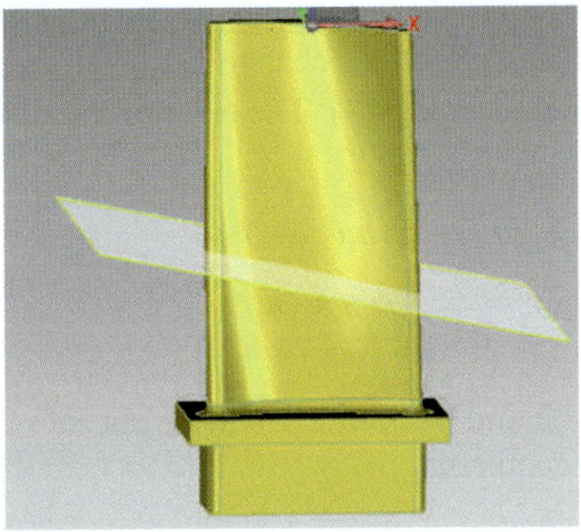

Fig. 16. The blade model with Y-axis deflection.

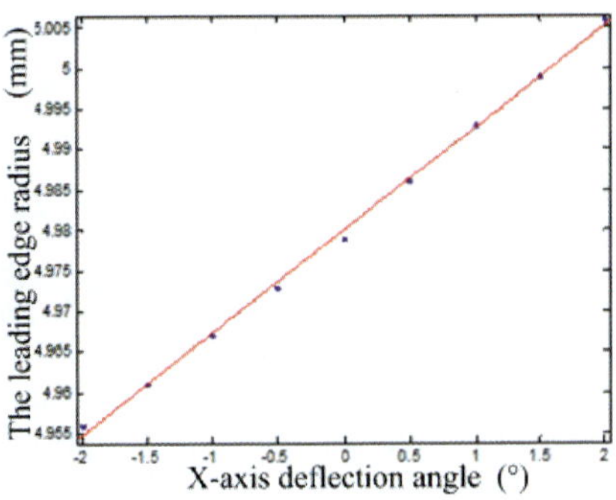

Fig. 17. The variation of leading edge radius.

If the blade has a Y-axis deflection, the blade section at the height of 100mm will be received as Fig. 16. Referring to Sections 3.1 and 3.2, the variation of blade parameters and their expressions can be acquired in figure 17~19.

The following expression represents that the leading edge radius varies with Y-axis deflection angle.

$$f_1(100,0,\theta_y)=0.01263\cdot\theta_y+4.98 \tag{10}$$

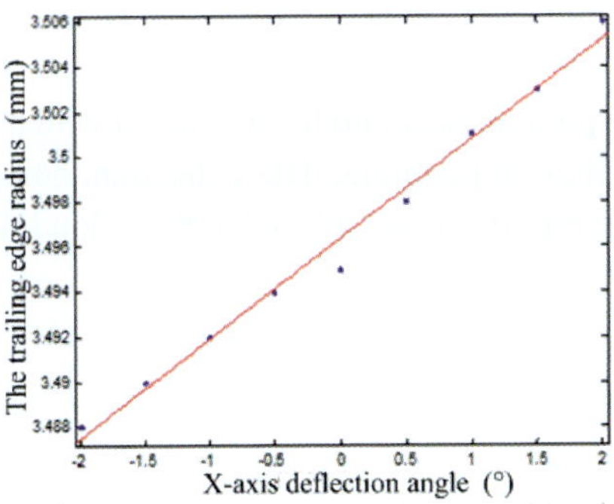

Fig. 18. The variation of trailing edge radius.

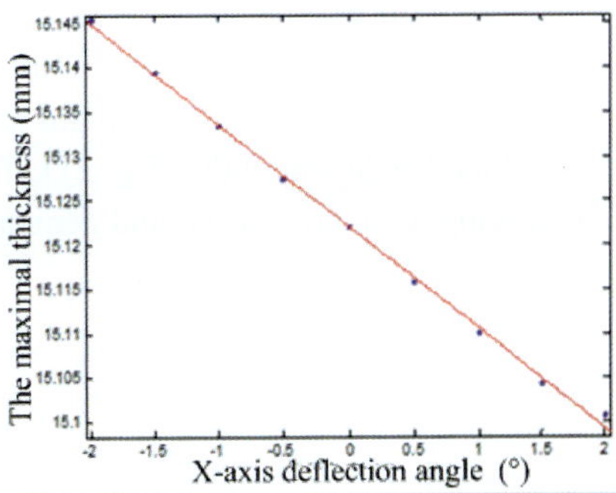

Fig. 19. The variation of maximal thickness.

The Eq. (11) and Eq. (12) represent that trailing edge radius and maximal thickness vary with Y-axis deflection angle.

$$f_2(100,0,\theta_y) = 0.004433 \cdot \theta_y + 3.496 \tag{11}$$

$$f_3(100,0,\theta_y) = -0.01144 \cdot \theta_y + 15.12 \tag{12}$$

3.4. *The error of the blade inspection*

All deflections should be considered together during the measurement. From Section 3.1 to Section 3.3, we know the three deflections change in a small range as above, so higher derivatives can be ignored in Eq. (4). The error of the blade parameters can be acquired as Eq. (13):

$$\begin{aligned} \Delta &= f_i(h+\Delta h, \theta_x + \Delta\theta_x, \theta_y + \Delta\theta_y) - f_i(h, \theta_x, \theta_y) \\ &\approx \left(\Delta h \cdot \frac{\partial}{\partial h} + \Delta\theta_x \cdot \frac{\partial}{\partial \theta_x} + \Delta\theta_y \cdot \frac{\partial}{\partial \theta_y} \right) f(h, \theta_x, \theta_y) \end{aligned} \tag{13}$$

All the data of Section 3.1, Section 3.2 and Section 3.3 will be put into the Eq. (13) together. Then the results can be acquired:

$$\begin{cases} |h| \le 0.802 \\ |\theta_x| \le 0.42 \\ |\theta_y| \le 0.83 \end{cases} \tag{14}$$

Eq. (14) shows that the blade deflection cannot exceed 0.42° in the X-axis direction, 0.83° in the Y-axis direction, 0.802mm in the Z-axis direction.

4. Experiment

If the deflection angle of transitional axis and rotational axis is bigger than the maximum during the measurement, the shims and the laser tracker can be used to adjust the deflection angle. The point clouds of this blade section can be extracted by the device (Fig. 20). We can put these point clouds into the iCloud3D which is developed independently (Fig. 21). The motion planning, data collection, noises removing, data processing and generating report can be finished in the iCloud3D.

Fig. 20. Scan and acquire point cloud by iBlade Scan I .

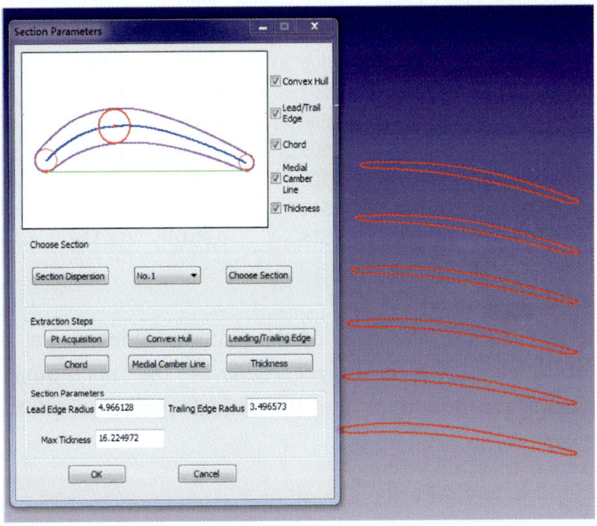

Fig. 21. Extract the parameters by iCloud3D.

According to the requirements of the factories, six sections are selected to inspect and extract the parameters. If there is no deflection, the variation of blade parameters along with the height of the blade section is shown in Table 1.

Table 1. If there is no deflection, the blade parameters vary with the height of the blade section (mm).

Height	Leading edge radius	Trailing edge radius	Maximal thickness
60	4.966	3.497	16.225
80	4.965	3.501	15.624
100	4.982	3.502	15.184
120	4.997	3.505	14.955
140	4.991	3.508	14.950
160	5.038	3.511	15.124

If the deflections reach maximum angle, the blade parameters vary with the height of the blade section is shown in Table 2.

Table 2. If there are maximum deflections, the blade parameters vary with the height of the blade section (mm).

Height	Leading edge radius		Trailing edge radius		Maximal thickness	
	Measured value	Error	Measured value	Error	Measured value	Error
60	4.973	0.007	3.503	0.006	16.212	-0.013
80	4.972	0.007	3.506	0.005	15.621	-0.003
100	4.989	0.007	3.507	0.005	15.161	-0.023
120	4.999	0.002	3.509	0.004	14.954	-0.001
140	4.993	0.002	3.512	0.004	14.957	0.007
160	5.034	-0.004	3.515	0.004	15.139	0.015

From the above table, the error of the maximal thickness at the height of 100mm is -0.023mm, which is regarded as the maximum error. The experimental result is less than the theoretical result, to meet the design requirements very well.

5. Conclusion

This article figures out and testifies that the error of the device is smaller than the allowable error, and it is concluded that the blade deflection cannot exceed 0.42° in the X-axis direction, 0.83° in the Y-axis direction, 0.802mm in the Z-axis direction. The precision of the blade inspection can reach 0.023mm and it can inspect a whole blade in 90s. So the accuracy and efficiency of the turbine blade inspection is high, which meets the requirement of blade inspection very well. The inspection device can be gradually applied into the margin detection and deformation correction of nuclear power forged blade.

Acknowledgments

This work is supported by the National Natural Science Foundation of China (No. 51475187), the National Basic Research Program of China (No. 2015CB057304), the National Key Projects (No. 2014ZX04001051-05), and the Guangdong Innovative Research Team Program (Grant No. 2011G006).

References

1. W.L. Li, L.P. Zhou and S.J. Yan, A case study of blade inspection based on optical scanning method, *International Journal of Production Research.* **53**, 2165-2178 (2015).
2. H. Xie *et al.*, Hand-eye calibration and its accuracy analysis in robotic grinding, in *IEEE Int. Conf. Automation Science and Engineering (CASE)*, (Gothenburg, Sweden, 2015).
3. J. Li, Comparison of measurement methods of aero-engine blade, *China New Technologies and Products.* **09**, 6 (2015).
4. M. Kang and J.W. Xu, Measurement of profile error of Monolithic impeller's blade by using CMM, *Tool Engineering.* **08**, 55-57 (2002).
5. J.E. Makem *et al.*, A virtual inspection framework for precision manufacturing of aerofoil components, *Computer-Aided Design.* **9**, 858-874 (2012).
6. J. Gao, N. Gindy and X. Chen, An automated GD&T inspection system based on non-contact 3D digitization, *International Journal of Production Research.* **44**, 117-134. (2006).
7. Hsu T H *et al.*, On the development of airfoil section inspection and analysis technique, *The International Journal of Advanced Manufacturing Technology.* **30**, 129-140 (2006).
8. T. Chen *et al.*, Notice of Retraction Application of optical inspection and metrology in quality control for aircraft components, in *2nd Int. Conf. Computer Engineering and Technology (ICCET)*, (Chengdu, China, 2010).
9. W.L. Li *et al.*, Section Curve Reconstruction and Mean-Camber Curve Extraction of a Point-Sampled Blade Surface, *PloS one.* **9**, e115471 (2014).
10. Ohtake Y *et al.*, Mesh regularization and adaptive smoothing, *Computer-Aided Design.* **33**, 789-800 (2001).

Research on NURBS path interpolation method with controllable error

Liang-Ji Chen and Hui-Ying Li†

School of Mechatronics Engineering, Zhengzhou University of Aeronautics, Zhengzhou, Henan Province, China

†E-mail: chenjiaxiclj@163.com

Non-Uniform Rational B-Spline(NURBS) curves are mostly used to generate arbitrary contour of a complex shape in the CAD field. A new method with NURBS technology is presented and realized in a location and federate controlling system. In the method, the tool paths and the cutter locations(CL) were represented into NURBS curves based on the same knot vector. The above CL data was firstly calculated and then transformed to five motion commands of five axes of machine through the real-time post-procession algorithm. The acceleration/deceleration controlling method is also presented to avoid the impact of machine. The proposed 5-axis spline interpolation method is realized and the experimental result of machining shows that the method is valid.

Keywords: NURBS curve; location and federate controlling; cutter location data.

1. Introduction

Till now, most of complex products such as aeronautical turbine blades, impellors, dies, molds etc. are machined on 5-axis Computer Numerical Controlled (CNC) machines. The core of the machines is the contour controlling system which decides machining efficiency and accuracy of a part. However, the current 5-axis CNC system mostly provide only line interpolators, that is, tool motion along straight lines is supported. To machine a surface, the tool paths, which are also known as the cutter location (CL) paths, are typically approximated with piecewise linear segments by CAD/CAM systems. The approximately machining method has its own drawbacks in advancing efficiency and accuracy of a machined part. To overcome the disadvantages, it must be taken into account that the 5-axis control system has intelligent ability to generate the tool paths of the machined surfaces.

†Corresponding author.

Non-Uniform Rational B-Spline (NURBS), as the only standard of data-exchange of product sanctioned by ISO, has been mostly used in almost all CAD/CAM systems to describe work-piece's surface. Based on the reality, the CNC system should be provided with the ability to directly generate 5-axis NURBS curves or surfaces in real-time.

Some NURBS interpolation methods had been proposed by several investigators[2-6]. Cheng et al.[4] had proposed a real-time NURBS curve motion command generator for CNC machines. Zhiming et al.[6] developed a NURBS curve interpolator for CNC machining based on the geometric properties of the tool path. However, most of them had concentrated their attentions on 3-axis ball-end machining, but for 5-axis NURBS surface interpolator, little has been done.

When developing and realizing a 5-axis CNC system with function of NURBS interpolation, related study must be carried out. In this paper, a real-time 5-axis NURBS curve interpolation method is given and also its' format of G code.

2. NURBS Expression of Freeform Surface

In the Cartesian coordinate system (o-xyz), a free surface can be expressed as the following parameter form

$$\boldsymbol{S}(u,v)=[x(u,v),y(u,v),z(u,v)]^{\mathrm{T}}=x(u,v)\boldsymbol{i}+y(u,v)\boldsymbol{j}+z(u,v)\boldsymbol{k}\qquad 0\leq u,v\leq 1 \quad (1)$$

Where ***i***, ***j***, ***k*** respectively is the unit vector of the direction of *x*, *y*, *z*, *u* and v are the parameters.

If the free curved surface in the equation (1) is expressed with NURBS[5]

$$\boldsymbol{S}(u,v)=\frac{\sum_{s=0}^{n}\sum_{t=0}^{m}B_{s,p}(u)B_{t,q}(v)\omega_{s,t}\boldsymbol{T}_{s,t}}{\sum_{s=0}^{n}\sum_{t=0}^{m}B_{s,p}(u)B_{t,q}(v)\omega_{s,t}}\qquad 0\leq u,v\leq 1 \quad (2)$$

where $\boldsymbol{T}_{s,t}$ is control point; $\omega_{i,j}$ is the corresponding weight factor of $\boldsymbol{T}_{s,t}$; (n+1) is the number of control point of u direction and (m+1) is the number of control point of v direction; $B_{s,p}(u)$ is the B-spline basis function of u direction and $B_{t,q}(v)$ is the B-spline basis function of v direction, p and q are the times.

The calculation of $B_{s,p}(u)$ and $B_{t,q}(v)$ can be consulted in literature [5].

In the Eq.(2), if $v = v_0$ and $u = u_0$, it can respectively obtain parametric curve of u direction and v direction on the $\boldsymbol{S}(u,v)$ surface. $\boldsymbol{S}(u_0,v_0)$ which is a point on the NURBS curved surface is the intersection point of the curve of $\boldsymbol{S}(u,v_0)$ and $\boldsymbol{S}(u_0,v)$, and it is shown as Figure 1.

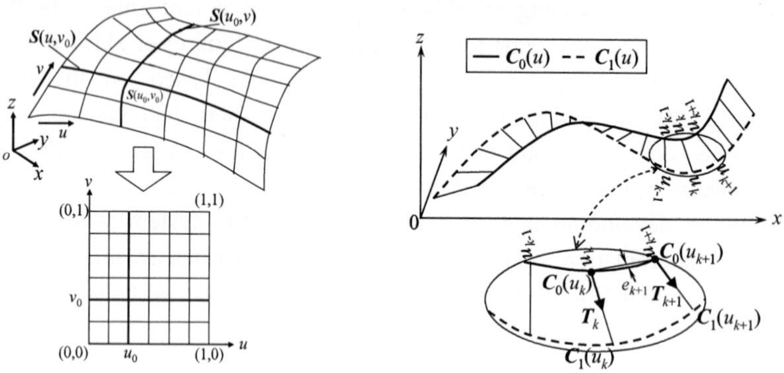

Fig. 1. NURBS surface.

3. Location controlling method

The task of real-time NURBS spline interpolation is to calculate next interpolation period's CL data that including tool-tip location coordination and tool-axis orientation and then transfer the CL data into machine tool's motion commands such as X,Y,Z,A,C through post-procession.

3.1. *Calculation of CL data*

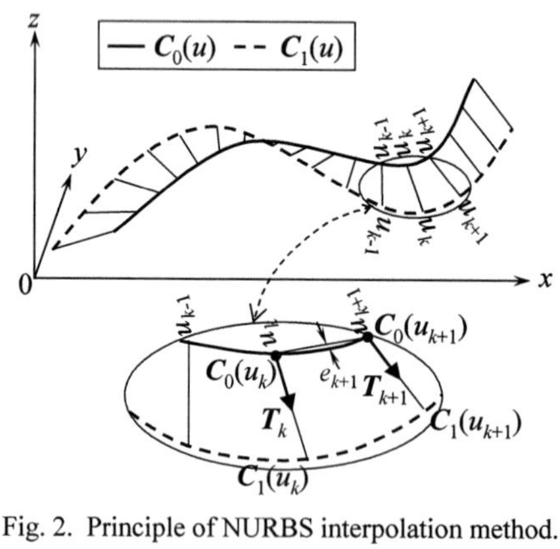

Fig. 2. Principle of NURBS interpolation method.

As shown in Figure 2, $C_0(u)$ and $C_1(u)$ is the cutter center point spline and the orientation spline, respectively. The two splines are constructed in the same knot vector. Assumed that parameter u is function of time t, that is, u=u(t). By using Taylor's expansion of the parameter u with respect to time t to obtain the first order approximation interpolation algorithm, the first order approximation up to the first derivatives is

$$u_{k+1}=u_k+T_s\frac{\mathrm{d}u}{\mathrm{d}t}\Big|_{t=t_k} \tag{3}$$

where T_s is interpolation period, u_k and u_{k+1} are corresponding parameters of current and next time t_k and t_{k+1}.

The federate of tool-tip point along the curve $\boldsymbol{C}_0(u)$ is defined by

$$\boldsymbol{V}(u)=\frac{\mathrm{d}\boldsymbol{C}_0(u)}{\mathrm{d}t}=\left(\frac{\mathrm{d}\boldsymbol{C}_0(u)}{\mathrm{d}u}\right)\left(\frac{\mathrm{d}u}{\mathrm{d}t}\right) \tag{4}$$

Since the curve speed

$$V(u_k)=\left\|\frac{\mathrm{d}\boldsymbol{C}_0(u)}{\mathrm{d}t}\right\|_{u=u_k} \tag{5}$$

The first derivative of u with t is obtained as

$$\frac{\mathrm{d}u}{\mathrm{d}t}\Big|_{t=t_k}=\frac{V(u_k)}{\left\|\frac{\mathrm{d}\boldsymbol{C}_0(u)}{\mathrm{d}u}\right\|_{u=u_k}} \tag{6}$$

Therefore, the first-order interpolation algorithm is obtained by substituting Eq. (6) into Eq. (5), Eq. (5) can be processed as follows:

$$u_{k+1}=u_k+\frac{T_sV(u_k)}{\left\|\frac{\mathrm{d}\boldsymbol{C}_0(u)}{\mathrm{d}u}\right\|_{u=u_k}} \tag{7}$$

The first and second derivative of $\boldsymbol{C}_0(u)$ with u is obtained as

$$\frac{\mathrm{d}\boldsymbol{C}_0(u)}{\mathrm{d}u}=\frac{\sum_{i=0}^{n}N'_{i,p}(u)W_i\boldsymbol{P}_i}{\sum_{i=0}^{n}N_{i,p}(u)W_i}-\frac{\sum_{i=0}^{n}N'_{i,p}(u)W_i\sum_{i=0}^{n}N_{i,p}(u)W_i\boldsymbol{P}_i}{\left(\sum_{i=0}^{n}N_{i,p}(u)W_i\right)^2} \tag{8}$$

where the general algorithm for 1st order derivative of $N_{i,p}(u)$ is

$$N'_{i,p}(u)=p\left[\frac{N_{i,p-1}(u)}{u_{i+p-1}-u_i}-\frac{N_{i+1,p-1}(u)}{u_{i+p}-u_{i+1}}\right] \tag{9}$$

$\boldsymbol{C}_0(u_{k+1})$ and $\boldsymbol{C}_1(u_{k+1})$ are obtained by substituting the calculated u_{k+1} into $\boldsymbol{C}_0(u)$ and $\boldsymbol{C}_1(u)$. Suppose the tool-axis unit vector is $\boldsymbol{T}_{k+1}$, so

$$\boldsymbol{T}_{k+1}=\frac{C_1(u_{k+1})-C_0(u_{k+1})}{\left\|C_1(u_{k+1})-C_0(u_{k+1})\right\|} \tag{10}$$

3.2. *Federate controlling method*

In this paper, it assumed that the federate controlling law is linear variable with acceleration *a* and deceleration -*a* and the destination controlling federate is *v* as shown in Figure 3.

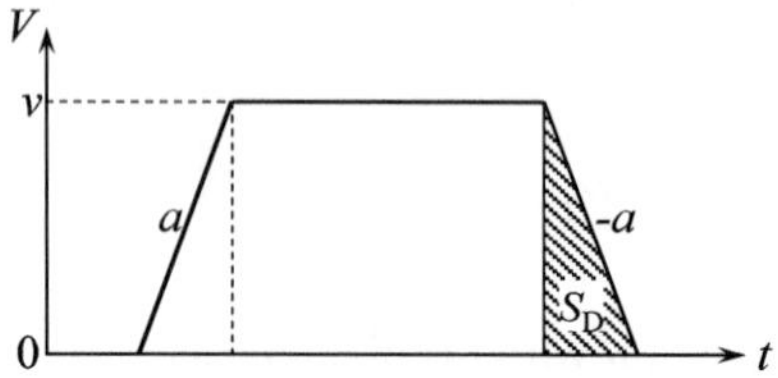

Fig. 3. Model of linear feedrate controlling.

For the acceleration process, the following method can be used

$$V(u_k)=at_k$$

For the acceleration process, the key problem to be solved is evaluation of deceleration point. In Figure 3, S_{D}, distance of deceleration, can be calculated as follows

$$S_{\mathrm{D}}=\frac{v^2}{2a} \tag{11}$$

The distance of deceleration, which is also the length between deceleration point and destination point along the spline curve, can be calculated as follows

$$\int_{u_{\mathrm{D}}}^{1}\left\|\frac{d\boldsymbol{C}_0(u)}{du}\right\|du=S_{\mathrm{D}} \tag{12}$$

where u_{D} is corresponding parameter of the deceleration point. The following Newton-Rapson method is used to calculate u_{D}.

If

$$f(u)=\int_{u}^{1}\left\|\frac{d\boldsymbol{C}_0(u)}{du}\right\|du-S_{\mathrm{D}} \tag{13}$$

we have

$$f(u_{\mathrm{D}})=0$$

Assumed the initial value of u_{D} is 1, that is, $f(u_0) = -S_{\mathrm{D}}$, then we have the following equation

$$u_{i+1} = u_i - f(u_i) / f'(u_i) \qquad i \geq 0 \tag{14}$$

where i is integer. Eq.(14) can be equivalent to the following formula

$$u_{i+1} = u_i + \frac{f(u_i)}{\left\| \dfrac{dC_0(u_i)}{du} \right\|} \qquad i \geq 0 \tag{15}$$

$f(u_i)$ in Eqs.(14) and (15) can be obtained using Simpson integral method.

4. Conclusions

Thoroughly analyzing the defects of the existing 5-axis linear interpolation method used in the machining of sculptured surface, 5-axis NURBS interpolation method with location and federate controlling is presented and realized in a real-time CNC system. Different from the traditional machining method of the 5-axis linear interpolation, the proposed method can directly obtain a curve on a free-form surface. Consequently, the number of the NC code for the same machining path can be greatly decreased and machining efficiency and accuracy is also improved.

Acknowledgments

This work is supported by grant 51275485 of the National Natural Science Foundation of China and grant 13HASTIT036 of the program for Science & Technology Innovation Talents in Universities of Henan Province.

References

1. L. Piegl, On NURBS: a survey, *IEEE Computer Graphics & Application* **11**, 55 (1991).
2. M. Tikhon, NURBS interpolator for constant material removal rate in open NC machine tools, *International Journal of Machine Tools & Manufacture* **44**, 237(2004).
3. B. Bahr, X. Xiao, and K. Krishnan, A real-time scheme of cubic parametric curve interpolations for CNC systems, *Computers in Industry* **45**, 309 (2001).

4. M. Cheng, M. Tsai, and J. Kuo, Real-time NURBS command generators for CNC servo controllers, *International Journal of Machine Tools & Manufacture* **42**, 801 (2002).
5. R. E. Sorace, V. S. Reinhardt, and S. A. Vaughn, High-speed digital-to-RF converter, U.S. Patent 5 668 842, Sep. 16, (1997).
6. X. Zhiming, C. Jincheng, and F. Zhengjin, Performance evaluation of a real-time interpolation algorithm for NURBS curves, *International Journal of Advanced Manufacturing Technology* **20**, 270 (2002).
7. G. Qiyi, R. Zhang, and B. Greenway, Development and implementation of a NURBS curve motion interpolator, *Robotics and Computer-Integrated Manufacturing* **14**, 27 (1998).
8. M. Tsai, C. Cheng, and M. Cheng, A real-time NURBS surface interpolator for precision three-axis CNC machining, *International Journal of Machine Tools & Manufacture* **43**, 1217 (2003).

Semi-automatic assembly method of wire-supported micro-target with various parts

Juan Zhang†, Wen-Rong Wu, Guo Li and Kai Du

Research Center of Laser Fusion, China Academy of Engineering Physics, Mianyang, Sichuan, China

†*E-mail: lgdzhangjuan@163.com*

For assembly of wire-supported micro-target which has four different micro-scale parts, a semi-automated assembly process which includes artificial operation, semi-automation control, automatic alignment control and bonding process is proposed. To control micro-parts alignment automatically, position detection method of two micro-parts is proposed. The method is based on image feature extraction and information fusion of two microscopic vision systems. A tracking method based on prior knowledge of movement is introduced to track micro-part in real-time in assembly process. Visual servo control strategy is designed to align micro-parts automatically in 3-D Cartesian space. The micro-parts are connected by bonding process. Experiment results verified the effectiveness of proposed methods.

Keywords: Complex assembly; assembly control; micro-assembly; ICF.

1. Introduction

Inertial Confinement Fusion (ICF) is a major method to achieve nuclear fusion. Many countries invest lots of money and human in it. Micro-target assembly is one important condition to achieve ICF research. However, the scale of micro-target is commonly in mm range, and is composed of different micro-parts. As the scale of micro-part is commonly in the range of μm, it's difficult to assemble different micro-parts together.

Many research results have showed that micro-assembly robot is an effect way to assemble micro-parts[1-3]. The University of Toronto in Canada developed an MEMS micro-robot[4]. The MEMS micro-robot includes micromanipulator and operating platform. The micromanipulator has two degree of freedom (DOF) of rotation, while the operating platform has three DOF of translation and one DOF of rotation. In the guide of micro-vision, the MEMS micro-robot can assemble two micro-parts together. Research results in process design, micro-gripping technology, movement control technology based on MEMS micro-robot are listed in paper [5], [6], [7], respectively.

The FEMTO-ST Institute in France developed a five-DOF micro-assembly robot, and studied on calibration method of microscope vision [8], auto-focusing technology [9], visual servo technology [10]. The five-DOF micro-assembly robot can assemble 3-D MEMS, while position alignment error is less than 1.4μm, and pose alignment error is less than 0.5 degree [11, 12].

Shanghai Jiao Tong University in China produced an micro-assembly system with mobile robot [13], the volume of mobile robot is less than $1cm^3$. In the guide of micro-vision, mobile robot can be controlled to assemble pole and hole, the diameter of pole is 0.35mm and the diameter of hole is 0.6mm.

We can see that there are many research achievements in the field of micro-assembly robot, and the achievements are focused on structural design of robot, process design, micro-gripping technology, and assembly process control. However, as the structure of micro-parts and assembly process in the achievements are usually simple, the achievements normally don't apply to assemble complex 3-D micro-target.

Wire-supported micro-target is a type of micro-target in ICF, and it's composed of different micro-parts. While the micro-parts have different size, structure, and material science, it's not easy to put them together in high precision. To solve the problem, in the paper, an assembly robot is designed, and a semi-automation assembly method is studied to assemble wire-supported micro-target in high precision.

2. Micro-Assembly Robot and Assembly Process

2.1. *Micro-assembly robot*

To assemble complex 3-D micro-target, a semi-automation micro-assembly robot is designed. The structure is shown in Figure 1. The robot system includes two microscopic vision systems, which are named 1st, and 2nd microscopic vision system, respectively. The two vision systems are arranged orthogonal to each other in space, and both of them have one degree of freedom (DOF) to focus on the object. In addition, the system includes two manipulators which are named 1st and 2nd manipulator, respectively, and a console. The two manipulators and console all have three translational degree of freedoms (DOFs), and two rotational DOFs such as yawing and pitching.

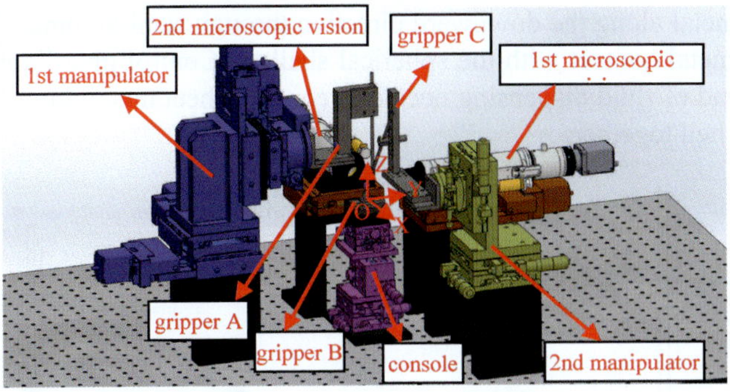

Fig. 1. The structure of micro-assembly robot system.

The 1st manipulator is connected with gripper A to gripe micro-part A, while the console is connected with gripper B to gripe micro-part B. In the guide of microscopic vision systems, the 1st manipulator operates micro-part A to alignment and assembly with micro-part B.

The 2st manipulator is connected with gripper C to gripe fluid dispensing needle. As the two micro-parts are assembled, the fluid dispensing needle is controlled movement to the dispensing position. Later, in the effect of external pressure, the glue liquid flow out from the end of fluid dispensing needle to connect micro-part A and B together.

2.2. *Assembly process design*

Wire-supported micro-target is composed of four micro-parts, as shown in Figure 2(c). The micro-parts are target stand, sheet metal, glass microsphere, spherical shell, respectively. The diameter of glass microsphere is 200μm, and the diameter of spherical shell is 800μm. Based on structure analysis of wire-supported micro-target and semi-automation assembly robot system, a semi-automation assembly process is designed.

The flow sheet of assembly process is shown in Figure 2. Firstly, hang the sheet metal at the top of the target stand through artificial way, as shown in Figure 1(a). Secondly, in the guide of microscopic vision system, assemble the target stand and the glass microsphere trough semi-automated way, as shown in Figure 2(b). The process is divided into two steps: 1) Smear glue at the end of the target stand through artificial way; 2) Align the target stand and glass microsphere through automated way. The combination of target stand and glass microsphere is named target combination. Thirdly, assemble the target combination and the spherical shell trough semi-automated way, as shown in Figure 2(c). The process can be divided into several steps: 1) Align the target combination and the spherical shell in X-axis and Y-axis through automated way; 2) Control the target combination movement in Z-axis direction until sphere centers of glass microsphere and spherical shell are coincidence. 3) Poke the sheet metal along the downward direction through semi-automated way until the sheet metal contact with the spherical shell. 4) Control glue liquid flow out from the end of fluid dispensing needle to connect sheet metal, target stand, and spherical shell together.

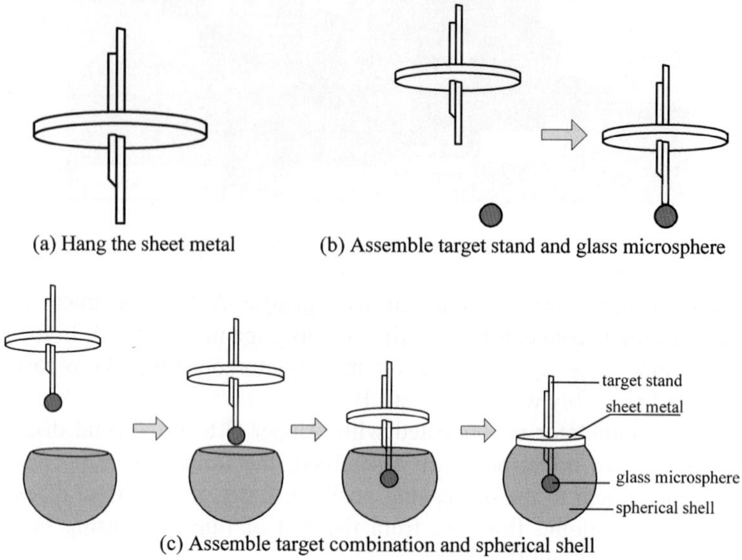

Fig. 2. The flow sheet of assembly process.

3. Micro-Target Detection and Tracking

3.1. *Relative position detection*

To align the micro-parts through automated way, relative position of two micro-parts in Cartesian space are detected. As noted in section 2.1, the micro-assembly robot has two microscopic vision systems which are arranged orthogonal to each other. The 1st microscopic vision system can detect relative position of two micro-parts in X-axis direction and Z-axis direction in image, while the 2nd microscopic vision system can detect relative position in Y-axis direction and Z-axis direction in image. According to the data information from two directions, the relative position of two micro-parts in 3-D Cartesian space can be calculated, as in Eq. (1).

$$\begin{cases} P_X = k_{cx} I_{px} \\ P_Y = k_{cy} I_{py} \\ P_Z = r_1 k_{cz1} I_{pz1} + r_2 k_{cz2} I_{pz2} \end{cases} \tag{1}$$

where P_X, P_Y, and P_Z are the relative positions in X, Y, and Z directions in Cartesian space. I_{px} and I_{py} are the relative positions in X-axis and Y-axis directions in image. I_{pz1} detected in 1st microscopic vision system and I_{pz2} detected in 2nd microscopic vision system are the relative positions in Z-axis in image. k_{cx}, k_{cy}, k_{cz1}, and k_{cz2} are the scales from image space to Cartesian space, where k_{cx}, k_{cz1} are scales in X-axis and Y-axis directions of 1st microscopic vision system, k_{cy} and k_{cz2} are scales in X-axis and Y-axis directions of 2nd microscopic vision system. r_1 and r_2 are weighting coefficients, and in the range of 0~1. The sum of r_1 and r_2 is 1.

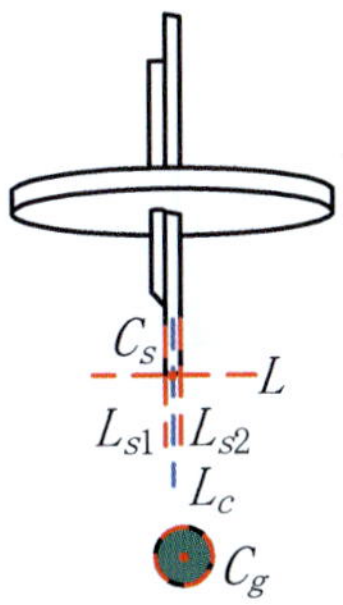

(a) Image feature extraction I

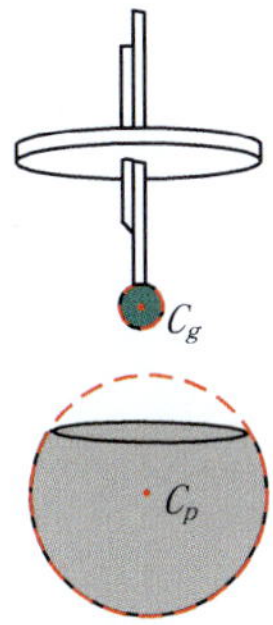

(b) Image feature extraction II

Fig. 3. Image feature extraction in assembly process.

Image features of two micro-parts in two microscopic vision systems are extracted to calculate relative position of two micro-parts in image. In the assembly process of target stand and glass microsphere, as shown in Figure 3(a), the vertical edge line L_{s1}, L_{s2} and the horizontal edge line L of target stand are extracted, while the sphere center C_g of glass microsphere are extracted. With the linear equation L_{s1} and L_{s2}, the center line L_c of target stand can be calculated. The center point C_s of the end-effector can be calculated by Line L_c and L. At last, the relative position in two directions in image can be calculated by point C_s and C_g.

In the assembly process of target combination and spherical shell, as shown in Figure 3(b), the sphere center C_g of glass microsphere and the sphere center C_p of spherical shell are extracted, the relative position in two directions in image can be calculated by point C_g and C_p. Line and ellipse fitting methods proposed in [14] are used to detect edge line and sphere center.

3.2. *Micro-target tracking method*

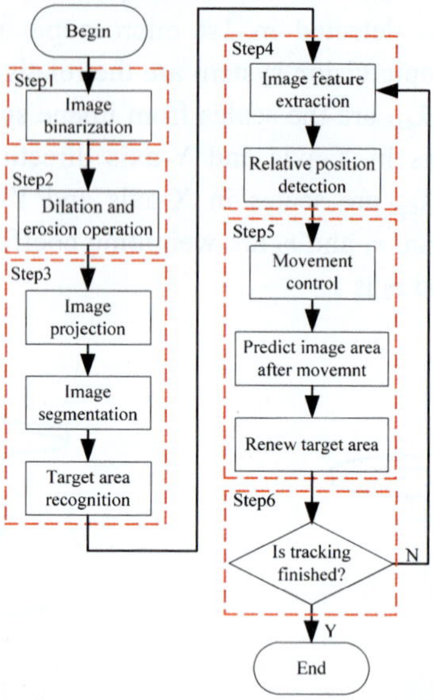

Fig. 4. The flow diagram of target detection and tracking.

Object tracking is necessary for automatic micro-assembly, the flow chart of micro-part detection and tracking method is shown in Figure 4, the process can be divided into six steps:

1) Get binary image through binarization method.

2) Remove image noise in binary image by dilation and erosion operation.

3) Target area recognition. Firstly, make the binary image projects respectively on the X-axis and Y-axis, and then get two 1-D data. Secondly, analyze the two 1-D data to realize image segmentation. Lastly, recognize target area by combining the prior knowledge of micro-parts and image segmentation results.

4) Get image features of two micro-parts and calculate the relative position in 3-D Cartesian space. The methods have introduced in section 3.1.

5) Get target area after movement. Firstly, control micro-part movement in 3-D space. Secondly, according to motion information, predict target area after movement, and expand target area with a certain factor. Lastly, according to image feature extraction method introduced in section 3.1, get the target area after movement.

6) Step 4) and 5) are constantly repeated until the tracking process in finished.

4. Visual Servo Control Strategy

Visual servo control method is used to align two micro-parts automatically. The block diagram of alignment control is shown in Figure 5, where Δu_{x1} and Δu_{y1} are image position errors of two micro-parts in X-axis and Y-axis directions of 1st microscopic vision, Δu_{x2} and Δu_{y2} are image position errors in X-axis and Y-axis directions of 2nd microscopic vision. Δp_x and Δp_{z1} are the relative positions of micro-parts in X-axis and Z-axis directions of Cartesian space detected by 1st microscopic vision, while Δp_y and Δp_{z2} are the relative positions in Y-axis and Z-axis directions detected by 2nd microscopic vision. Δp_z is the calculated relative position in Z-axis direction. Δp_x, Δp_y, Δp_z can be calculated by formula (1). E_x, E_y, and E_z are thresholds used to judge whether relative positions are in the required ranges. e_x, e_y, and e_z are calculated errors. Proportional controller is used to improve alignment efficiency.

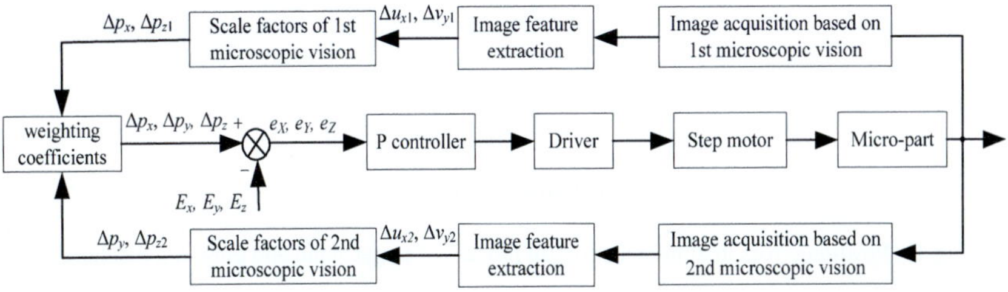

Fig. 5. The block diagram of alignment control.

5. Experimental Results

In the guide of two microscopic vision systems, the target stand and glass microsphere are aligned automatically. Figure 6 shows images in the alignment process. We can see from Figure 6(a) that the center point of target stand end and the sphere center of glass microsphere are extracted. Figure 6(b) shows the status after alignment, the glue at the end of target stand connect the two micro-parts together. We can also see from Figure 6 that in the alignment process, the end of target stand is tracked.

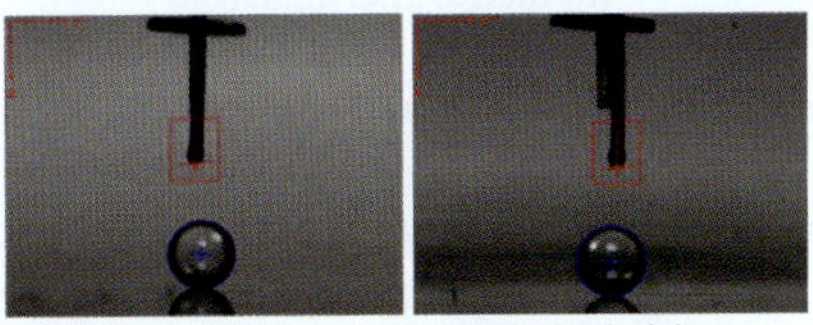

Image in 1st vision Image in 2nd vision

(a) Images in alignment process

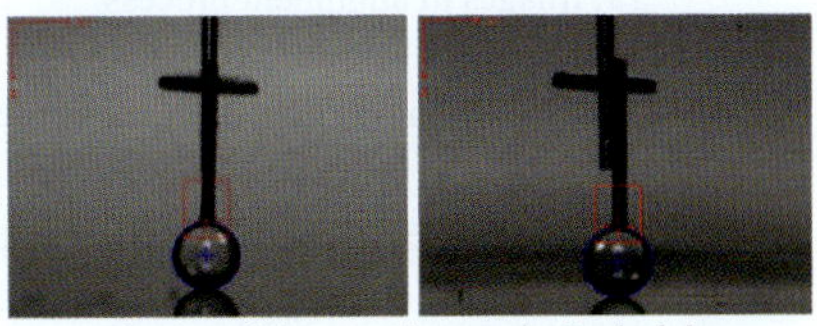

Image in 1st vision Image in 2nd vision

(b) Images after alignment

Fig. 6. Images in position alignment process of target stand and glass microsphere.

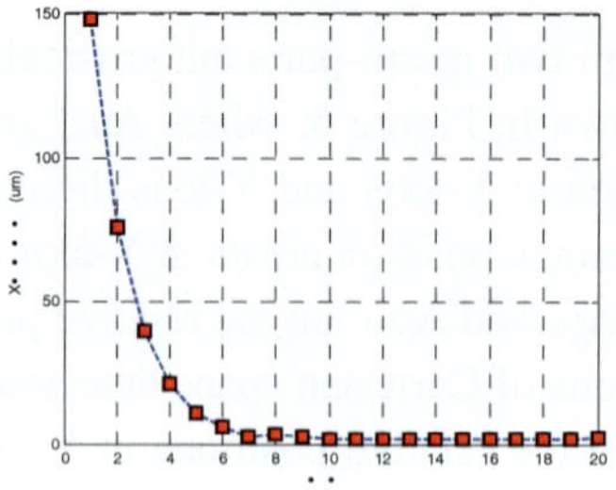

(a) Error in X-axis

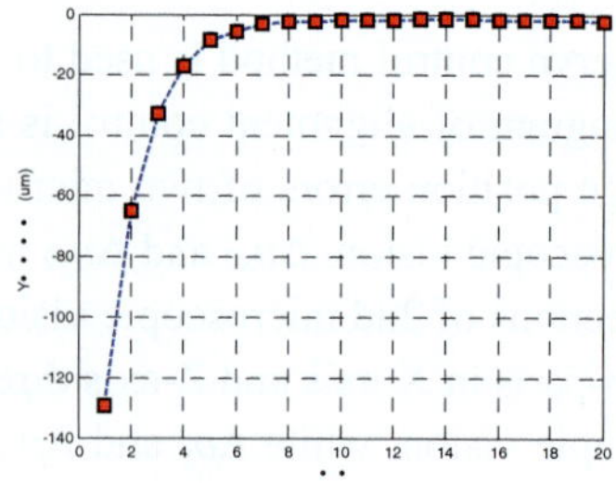

(b) Error in Y-axis

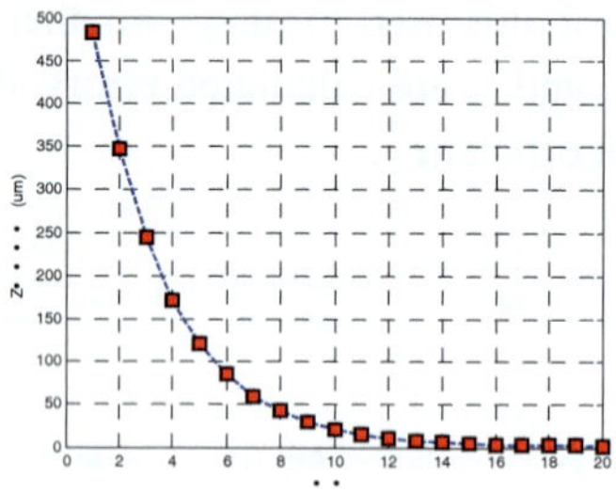

(c) Error in Z-axis

Fig. 7. Position errors in alignment control process of target stand and glass microsphere.

Figure 7 shows position errors in alignment process, Figure 7(a), (b), (c) is the position error in X-axis, Y-axis, and Z-axis direction, respectively. We can see that after several control cycles, position error in X-axis, Y-axis, and Z-axis direction are all in the given range. E_x, E_y, and E_z referred in Section 4 are all set to 3μm.

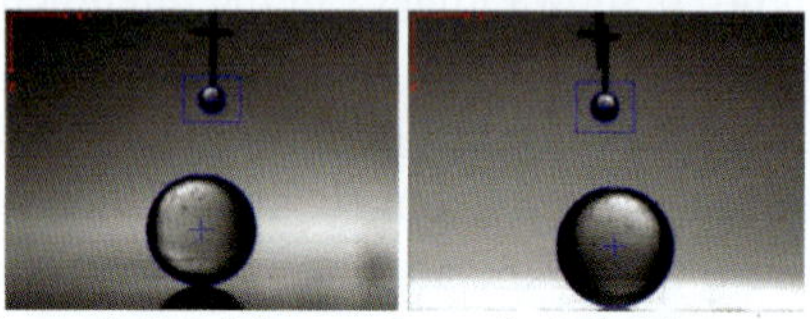

Image in 1st vision   Image in 2nd vision

(a) Images in alignment process

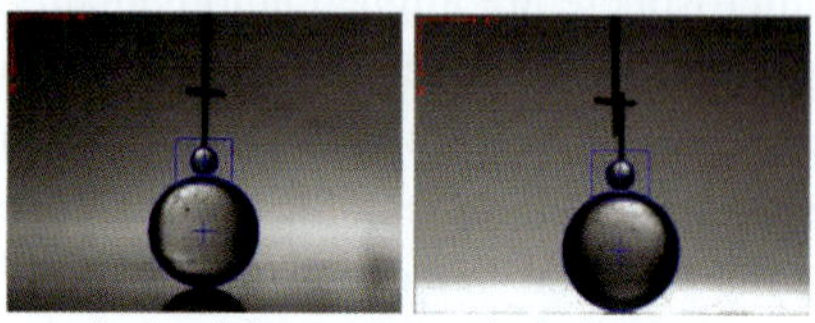

Image in 1st vision   Image in 2nd vision

(b) Images after alignment

Fig. 8. Images in position alignment process of target combination and spherical shell.

Figure 8 shows images in the alignment process of target combination and spherical shell. We can see from Figure 8(a) that sphere centers of glass microsphere and spherical shell are extracted. Figure 8(b) shows the status after alignment, the glass microsphere is alignment with spherical shell in X-axis and Y-axis direction, respectively, and on the top of spherical shell. We can also see from Figure 8 that in the alignment process, the glass microsphere is tracked.

Figure 9 shows position errors in alignment process, Figure 9(a), (b), (c) is the position error in X-axis, Y-axis, and Z-axis direction, respectively. We can also see that after several control cycles, position error in X-axis, Y-axis, and Z-axis direction are all in the given range. E_x and E_y are set to 3μm, while E_z is set to 410μm.

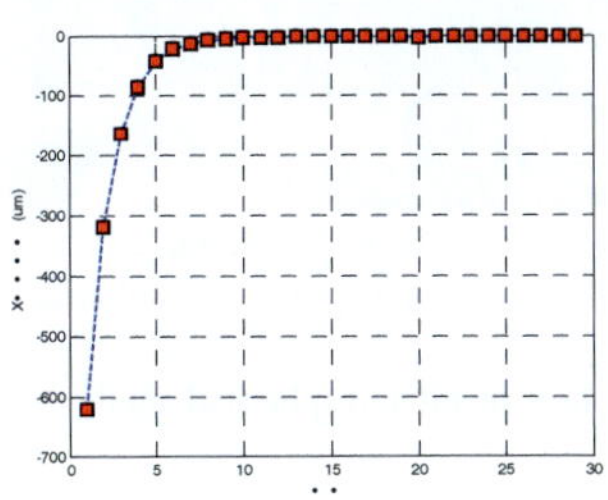

(a) Error in X-axis

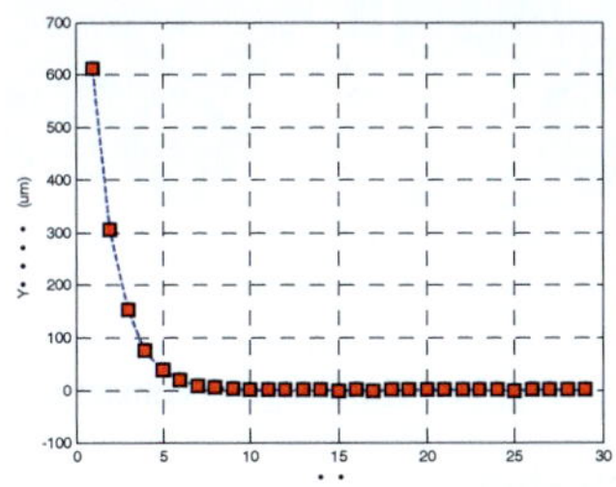

(b) Error in Y-axis

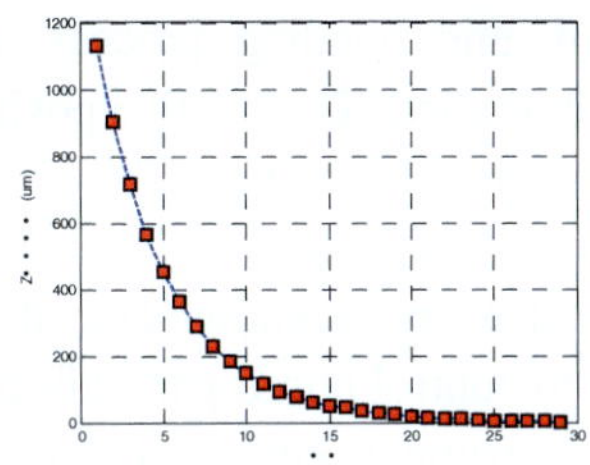

(c) Error in Z-axis

Fig. 9. Position errors in alignment control process of target combination and spherical shell.

After alignment of target combination and spherical shell, the target combination is controlled movement in Z-axis direction until the sphere centers of glass microsphere and spherical shell are coincidence, as shown in Figure 10. The sheet metal is poked along the downward direction by one fluid dispensing needle until the sheet metal contact with spherical shell, as shown in Figure 11. In the effect of external pressure, glue liquid flow out from the end of fluid dispensing needle to connect sheet metal, target stand, and spherical shell together.

Image in 1st vision Image in 2nd vision Image in 1st vision Image in 2nd vision
(a) Images in control process (b) Images after control process

Fig. 10. The image in the process of target combination movement control.

Image in 1st vision Image in 2nd vision Image in 1st vision Image in 2nd vision
(a) Poke the sheet metal (b) Sheet metal contact with spherical shell

Fig. 11. The image in contact process of sheet metal and spherical shell.

6. Conclusion

The semi-automation assembly method of wire-supported micro-parts is studied in the paper. It includes artificial operation, semi-automation operation, automatic assembly control, and bonding process. An automatic alignment method of two micro-parts is studied. The relative position of micro-parts in 3-D Cartesian space is calculated in real-time based on image feature extraction and image information fusion of two microscopic vision systems. The moving target in assembly process is based on prior knowledge of movement. Visual servo control strategy is designed to control micro-parts alignment automatically. Glue fluid flow out of fluid dispensing needle in the effect of extra pressure to connect micro-parts together. The experimental results show effectiveness of proposed methods.

Acknowledgment

This work is supported by Development Fund of CAEP (2014B0403066).

References

1. Q. S. Xu. Adaptive Discrete-Time Sliding Mode Impedance Control of a Piezoelectric Microgripper[J]. *IEEE Transactions on Robotics*, 2013, 29(3): 663-673.
2. X. Ye, J. Gao, Z. J. Zhang, et al. Robotic Microassembly for Meso-scale Application[J]. *Industrial Robot*, 2015, 42(2): 142-148.

3. X. Ye, J. Gao, Z. J. Zhang, et al. An Improved Vision Calibration Method for Coaxial Alignment Microassembly. *Assembly Automation*, 2014, 34(3): 237-243.
4. L. Ren, L. D. Wang, J. K. Mills, et al. 3-D Automatic Microassembly by Vision-Based Control[C]. *IEEE International Conference on Intelligent Robots and Systems*, 2007, pp: 297-302.
5. L. Ren, L. D. Wang, J. K. Mills, et al. Vision-Based 2-D Automatic Micrograsping Using Coarse-to-Fine Grasping Strategy[J]. *IEEE Transactions on Industrial Electronics*, 2008, 55(9): 3324-3331.
6. L. D. Wang, J. K. Mills, and W. L. Cleghorn. Automatic Microassembly Using Visual Servo Control[J]. *IEEE Transactions on Electronics Packaging Manufacturing*, 2008, 31(4): 316-325.
7. L. D. Wang, L. Ren, J. K. Mills, et al. Automated 3-D Micrograsping Tasks Performed by Vision-Based Control[J]. *IEEE Transactions on Automation Science and Engineering*, 2010, 7(3): 417-426.
8. B. Tamadazte, S. Dembele and N. L. Piat. A Multiscale Calibration of a Photon Video Microscope for Visual Servo Control: Application to Micromanipulation[J]. *IEEE International Workshop on Robotic and Sensors Environments*, 2008, pp: 29-34.
9. G. Duceux, B. Tamadazte, N. L. Piat, et al. Autofocusing-Based Visual Servoing: Application to MEMS Micromanipulation[J]. *International Symposium on Optomechatronic Technologies*, 2010, pp: 1-6.
10. B. Tamadazte, N. L. Piat, and E. Marchand. A Direct Visual Servoing Scheme for Automatic Nanopositioning[J]. *IEEE Transactions on Mechatronics*, 2012, 17(4): 728-736.
11. B. Tamadazte, N. L. Piat, and S. Dembele. Robotic Micromanipulation and Microassembly Using Monoview and Multiscale Visual Servoing[J]. *IEEE Transactions on Mechatronics*, 2011, 16(2): 277-287.
12. B. Tamadazte, N. L. Piat, and S. Dembele. Robust Trajectory Tracking and Visual Servoing Schemes for MEMS Manipulation[C]. *IEEE International Conference on Advanced Intelligent Mechatronics*, 2010, pp: 860-865.
13. J. P. Chen, Z. B. Li, X. N. Tang. Omni-directional Mobile Microrobot and its Micro Assembly System[J]. *China Mechanical Engineering*, 2005, 16(14): 1223-1225 (in Chinese).
14. J. Zhang, Z. T. Zhang, D. Xu, W. S. Zhang. Aligning Micro-gripper to Ring Object in High Precision with Microscope Vision[C]. *IEEE International Conference on Computer Science and Automation Engineering*, 2011, vol. 3, pp: 123-127.

Parameters optimization in laser texturing assisted with electrochemical machining

Hua Zhang†, Xin-Dong Jiang and Yan Lu
School of Mechanical Engineering, Nantong University,
Nantong 226019, China
†*E-mail: ntuzhh@ntu.edu.cn*

Laser texturing assisted with electrochemical machining (LT-ECM) is combined the high efficiency of laser machining and the high surface quality of electrochemical machining. The light-trapping structures are mainly formed by laser action and the slag produced in laser texturing is removed by electrochemical machining. In order to obtain the better textured quality, parameters optimization for LT-ECM has been carried out in this paper. Three important parameters such as pulse laser energy and applied electrochemical have been optimized to improve the optical property of mc-Si. The textured surface reflectance (TSR) and the slag removal rate (SRR) have been used as the evaluation indicators to evaluate the textured results on mc-Si by LT-ECM. The experiments were performed on 30×30mm^2 mc-Si wafer with pulsed Nd:YAG laser at second harmonic wavelength and NaOH solution. The experimental results showed that pulse laser energy has the most significant influence on TSR. Applied electrochemical voltage has significant influence on SRR. Optimized parameters for LT-ECM are: pulse laser energy is about 30mJ; electrochemical applied voltage is about 15V.

Keywords: Laser texturing; electrochemical machining; parameters optimization; reflectance; surface quality.

1. Introduction

Reduction of optical losses in multicrystalline silicon solar cells by texturing surface is one of the most important issues of modern silicon photo-voltaic. Anisotropic chemical etching, widely used for single crystalline silicon, is less efficient when it is used for mc-Si. An ideal isotropic texturing process forms hemispherical shaped light trapping structures in the mc-Si surface. The smallest fraction of mc-Si is achieved by a hexagonal arrangement of the tubs, which result in the so-called honeycomb texture. Various techniques have been reported in the literature to fabricate honeycomb textures in silicon using: lithography [1], reactive ion etching [2], ink jet masking [3] and direct laser texturing [4]. Direct laser texturing has the advantages of noncontact processing and high efficiency. However, there are some inherent defects such as slag and spatter during laser texturing on mc-Si. These defects must be removed after laser texturing.

A new hybrid method of laser texturing assisted with electrochemical machining (LT-ECM) has been developed. LT-ECM combines two different sources of energy simultaneously: energy of photons and energy of ions. The main aim is to obtain better textured results with high efficiency and good surface quality. The defects produced in laser texturing can be removed effectively by electrochemical machining. In this research, three important parameters such as pulse laser energy, applied electrochemical voltage and temperature of electrolyte have been optimized to improve the textured quality. The textured surface reflectance (TSR) and the slag removal rate (SRR) have been used as the evaluation indicators to evaluate the textured results on mc-Si.

2. Experimental Procedures

2.1. *Experimental apparatus*

Fig. 1 shows a LT-ECM apparatus system. It consists of three systems: Laser system, Electrolyte system and Power supply system. As shown in Fig.1, the workpiece is mc-Si wafer. It is fixed in the middle of the cell and the cell is divided to be two chambers. The mc-Si wafer becomes to be the only current path. Under the back side of the mc-Si, a platinum plate being connected to the positive power supply, is the an-ode. Under the front side of the mc-Si, a platinum plate being connected to the negative power supply, is the cathode. Laser action and electrochemical reaction are composited on the surface of mc-Si. The array light trapping structures are formed on the surface of mc-Si.

2.2. *Experimental parameters*

The used samples are 156×156mm2 mc-Si wafers commercially available by REC Corp. The thickness is about 230μm and the resistivity is about 2Ω•cm. The used workpiece is 20×20mm2 mc-Si wafer which is cut by the used samples. The used green laser is Nd:YAG laser at second harmonic wavelength. The used electrolyte is NaOH solution. LT-ECM main processing parameters include pulse laser energy (Ep) and applied electrochemical voltage (Ve).These parameters and levels are listed in Table 1. Every experiment with one group parameters is repeated five times. All experimental data is the average value of the five experiments.

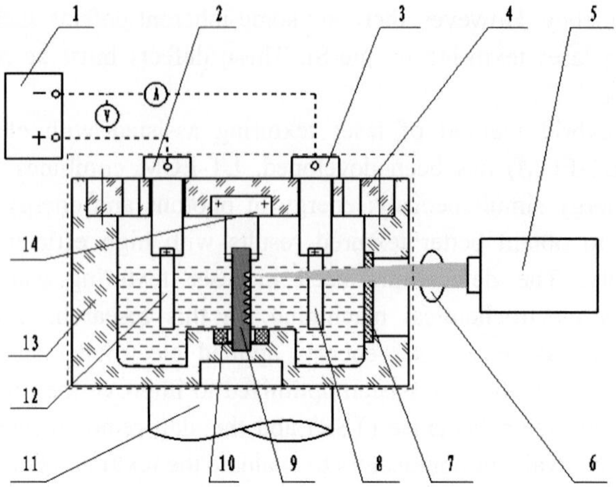

Fig. 1. Schematic diagram of LT-ECM for texturization.
(1) DC power supply system; (2)Anode base; (3) Cathode base (4) Electrolyzer cover; (5) Laser system; (6) Focus lens; (7) Optical lens;(8) Cathode; (9) Workpiece; (10)seals; (11) Table; (12)Anode; (13) Electrolyzer; (14) Fixture.

Table 1. Processing parameters and levels in LT-ECM.

Processing parameters	Unit	Level			
E_p	mJ	10	20	30	40
V_e	V	10	15	20	25

2.3. *Evaluation indicators for experiments*

In this research, two evaluation indicators are used to evaluate LT-ECM experimental results. The first evaluation indicator is textured surface reflectance (TSR) and slag remove rate (SRR). TSR was tested on the samples with different parameters by UV-Vis Spectrophotometer. The second evaluation indicator is slag removal rate (SRR). SRR is defined as follows Eq.1:

$$SRR(\%) = \frac{T_{\text{LT-ECM}}}{T_{\text{DLT}}} \times 100\% \tag{1}$$

Where, T_{DLT} is the thickness of the slag around light trapping structure machined by direct laser texturing; $T_{\text{LT-ECM}}$ is thickness of the thickness of the slag of light trapping structure machined by LT-ECM.

A scanning electronic microscopy (JSM- 5610LV/VANTAGE) was used to measure the thickness of slag of the experiments by direct laser texturing compared with LT-ECM.

3. Results and Discussion

3.1. *Effects of parameters on TSR*

Fig. 2 shows the relationship between the different pulse laser energy and TSR. The in-crease of laser pulse energy results in corresponding decrease of TSR. The reason for the substantial increase in TSR can be attributed to the higher laser energy results in increase of the depth of light trapping structure on the textured surface. With the laser pulse energy increase, the angle of two-sides of light trapping structure is decreased. Therefore, the incident light will have more reflection on the textured mc-Si surface by LT-ECM with higher pulse energy. However, the thickness of mc-Si sample is only about 230μm, the depth of light trapping structure should not be more than 80μm. When the depth of light trapping structure is more than 80μm, the electrical properties of mc-Si is poor for using to form solar cell. In this experiment, when pulse laser energy is 40mJ, the average depth of light trapping structure is 90μm. So, the optimized pulse laser energy is 30mJ in these experiments.

The relationship between electrochemical voltage and TSR is shown in Fig. 3. An increase in electrochemical voltage results in increases TSR. It is can be explained by the entrance of light trapping structure is enhanced by electrochemical reaction, and then the angle of two-sides of light trapping structure is increased. With the increased angle, the reflection of incident will be reduced and then the reflectance is less.

Compared with Fig. 2 and Fig. 3, the effect on TSR by pulse laser energy is stronger than the effect of electrochemical voltage. It is proved that the materials are mainly removed by laser action during LT-ECM. The electrochemical reaction is used to remove the slag produced in laser texturing on line.

The above experimental results are indicated that textured surface reflectance of mc-Si by LT-ECM varies significantly with pulse laser en-ergy and electrochemical voltage. The dominant factor effect on TSR is pulse laser energy. The more pulse laser energy, TSR of mc-Si can be less. However, the more pulse laser energy, the depth of light trapping structure is too deep to not be effectively for electrical properties.

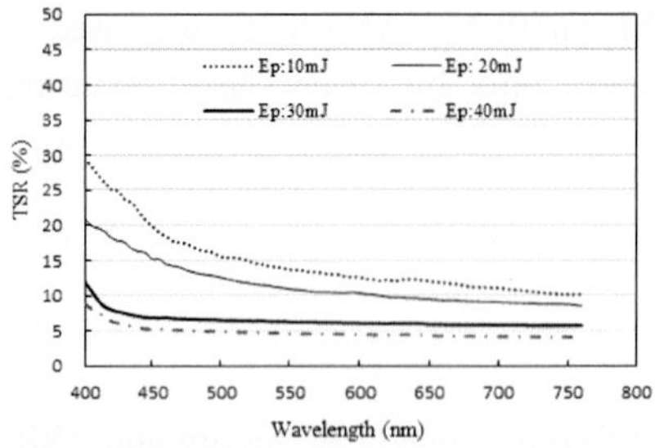

Fig. 2. Effect of pulse laser energy on TSR.

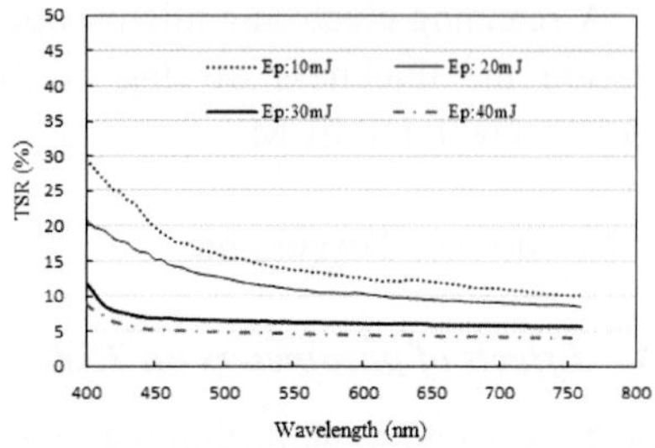

Fig. 3. Effect of electrochemical voltage on TSR.

3.2. *Effects of parameters on SRR*

The effect of laser pulse energy on the slag remove rate is shown in Table 2. The other parameters are V_e:15V; T:60℃. With the increase of pulse laser energy, the slag remove rate is decrease. The reason is that the increase of pulse laser energy results in more slag around the entrance of light trapping structure. All of the SRR are more than 80%.

The effect of electrochemical voltage on the slag remove rate is shown in Table 3. It is obvious that the increase of voltage leads to remove slag effectively.

Table 2. Effect of pulse laser energy on SRR.

Ep (mJ)	*SRR (%)*
10	95.3
20	92.2
30	87.8
40	83.4

Table 3. Effect of electrochemical voltage on SRR.

V_e (V)	*SRR (%)*
10	84.9
15	92.2
20	95.6
25	97.2

3.3. *Optimized results with LT-ECM*

Based on the optimization parameters experiments for LT-ECM, effective texturing by LT-ECM on mc-Si wafer was carried out. Figs. 4 and 5 compare the entrance of light trapping structure machined by direct laser texturing and LT-ECM. It is shown that slag is much more pronounced on the entrance and wall of pits in direct laser texturing. These slag like some damage to mc-Si and are poor to mc-Si being formed solar cell. Fig. 5 shows that the slags have been effectively removed with LT-ECM. Compared with direct laser texturing and LT-ECM, added the anode electrochemical action is significantly to remove the slag. The optimization of parameters for LT-ECM is successful.

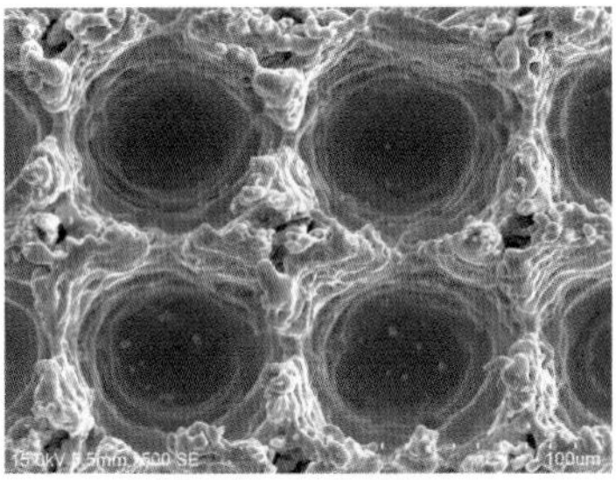

Fig. 4. Entrance of light trapping structure by direct laser texturing on mc-Si.

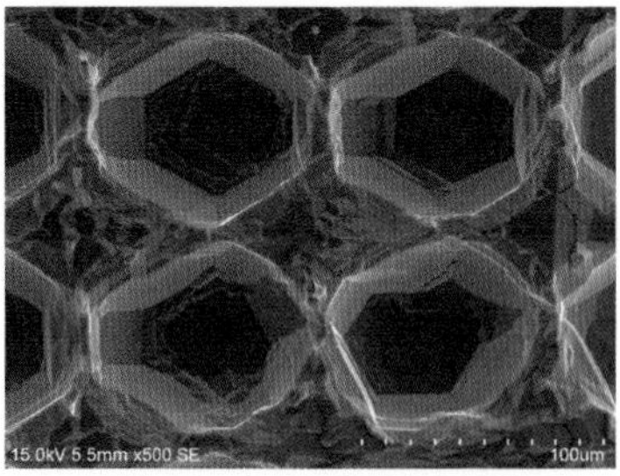

Fig. 5. Entrance of light trapping structure by LT-ECM on mc-Si.

4. Conclusions

1) During LT-ECM, the materials are removed by laser action, electrochemical action and single wet etching by NaOH solution; the slag can be effectively removed by latter two actions.

2) Laser energy is dominant parameter for textured reflectance. Optimized pulse laser energy is 30mJ. Slags around the entrance of light trap-ping structure are removed mainly by electro-chemical reaction. Slags around the wall of light trapping structure are removed mainly by single wet etching. Optimized electrochemical voltage is 15V.

Acknowledgment

This work is supported by the National Natural Science Foundation of China under Grant No. 51205212, Natural Science Fund of Jiangsu Province of China under Grant No.BK2012233, Applied Basic Research Project of Nantong City under Grant No.GY12015019 and Qing Lan Project of Jiangsu.

References

1. M.J. Stocks, A.J. Carr, and A. W. Blakers. Texturing of polycrystalline silicon. *Solar Energy Materials & Solar Cells*, 40, 33 (1996).
2. O. Schultz. Multicrystalline silicon solar cells exceeding 20% efficiency. *Progress in Photovoltaics Research & Applications*, 12, 553 (2004).
3. J. Nievendick and J. Specht. Formation of a honeycomb texture for multicrystalline silicon solar cells using an ink jetted mask. *Physica Status Solidi*, 6,7 (2011), 6(1):7-9.
4. H. Morikawa. Processes for over 18.5% high-efficiency multi-crystalline silicon solar cell. *Current Applied Phys*ics, 10, 5210 (2010).

Influences of three-dimensional printing to product innovation design thinking

Ling Liu* and Bo-Cong Lin
The Electromechanical Engineering School,
Beijing Institute of Graphic Communication,
Beijing, China
**E-mail: miji@vip.sina.com*

Based on the three-dimensional printing technology, to explore the creative products custom design approach, custom design paper products market demand and the meaning and importance in terms of resource conservation; explore the custom design and product combining the advantages of 3D printing, 3D printing by advocate to customize the product design, so as to achieve to meet market demand, the purpose of saving resources

Keywords: Three-dimensional printing; product design; innovative thinking.

1. Introduction

Along with the enrichment of physical goods and the development of technology, more and more customers demand individuation service. Enterprises establish their customization system one by one to satisfy customers and win the competition. In the design process, more attention has been paid to user's individuation factors. In other words, the increasing demand of individuation brings new challenges to traditional product design. Facing with new customer demand, it is possible to realize product custom design through three-dimensional printing. So far, researches on custom design using three-dimensional printing have gained extensive attention. Researching on creative products custom design technology has profound theoretical significance and practical value of the establishment of custom design platform which meets customers' demand.

2. Characteristics of Products Innovation Design Thinking

According to the once shaping principle, 3D printer can manufacture complex structures which are hard to process or even cannot be processed under traditional methods. This pushes the limitations of traditional manufacturing. 3D printer shortens the distance between ideas and products by manufacturing three-dimensional objects as long as the designers have ideas and turn them into three-dimensional data model. Yan Yongnian, professor and doctoral supervisor of Tsinghua University, mentioned that the biggest effect of three-dimensional printer is to train and inspire young men's intelligence when he gave a report in the lecture hall of Zhongtian institute of technology.

To meet the customers' changing demand, product manufacturers need to transform their old pattern which production promotes sales into a new way which requires manufacturers to arrange production according to customer requirement. Realizing product innovation design and personalized customization can help manufacturers reduce product cost, improve product quality, shorten time to market and provide quality service. So far, customization pattern, which means implementing individual mass production to meet customer's individualized needs, has been applied widely. Facing with new customer demand, it is possible to realize product custom design through three-dimensional printing. This custom method is an interactive, authentic, seasonable method which enables customers to see custom outcome immediately and get satisfied product. Customers can also participate in the design at conceptual design phase.

3. Influences of Three-dimensional Printing to Product Innovation Design Thinking

Throughout the entire history, design has been playing an important role that makes life more efficient, advanced and interesting by applying technical innovation. Sometimes it plays the same role of art. Applying digital technology into producing also belongs to this traditional. For the same reason, 3D print has become a controversial topic in the design community. Some people regard it as an innovative technology as important as printing press and steam engine, while others consider that is too exaggerated. Hence it is the perfect time to evaluate the influence of three-dimensional printing to design and its future development. Designers take more and more advantages of the progress of digital technology to manufacture products with complex geometric structure and elusive shape. As a result, an aesthetic pursuit of this kind of structure which as if comes from other worlds will become more and more common. Designers no longer pursue to control the final form of their product, they prefer to design a method which enables others personalize the production. The reasons come from aspects of style as well as function. Amputees can build a complete suitable artificial lime using 3D print and arthritics can adjust kitchen wares' hand shank for an easy use.

3D printer can not only build new objects but copy existing ones in case of lost and damage as well. Therefore, designers must consider the interchangeability of individual part when they design original items. Meanwhile, the growing requirement of maintenance brings a new challenge to the design of a new business model.

Firstly, conceptual model: at early stage of design flow, using 3D print technology structure models can examine the construction, shape and functions of an object. Any defects can be modified at the first time. After this, if it needs, the process of construction, examination and modification can be done again and again until the best conceptual model has been designed. Translate two dimensional design drawing into real objects can surely accelerate the development process of a product and reduce costs. In addition, three-dimensional objects can help designers make better decisions faster by presenting design preferably.

Secondly, function prototypes: designers can prove the rationality of their designs through manufacturing function prototypes. At the same time, they can test the performance and evaluate engineering strictly using three-dimensional objects. 3D print technology can shorten the time to market and maximize product performance.

Thirdly, finished products: 3D print technology has been a new fashion within the industry. The application fields are various including aviation enterprises, medical equipment manufacturers and small cars manufacturers. Using 3D print technology to replace traditional manufacturing process can save time and also reduce costs. What's more, 3D print technology breaks the limitations of traditional manufacturing process and enables designers to modify their design whenever there is a need. In this way, enterprises can exploit new markets of customized services or small amount application apparatus.

4. Three-dimensional Print Technology Promotes the Future Development of Product Innovation Design

4.1. *Shorten the design period using 3D print technology*

Digital production can not only change the relation between designers and users but also improve the cooperation between designers and producers. Ultimately, 3D printing equipment may spread every town, where local residents can acquire custom products or maintain existing products by themselves, just like their ancestors ask blacksmiths for forging products. Without question, enterprises which have technical experience have begun their preparation.

If this imagination comes true, designers no longer need to find producers who appreciate their designs. What they need to do is to make a prototype of their products and then sell them to consumers like entrepreneurs.

Producing products with complex shapes rapidly give a solution to the crazy ideas of designers. On the surface, 3D print can bring some changes temporarily. But if we connect it with some modern words like internet, cloud technology, open-source sharing

And big data, we can see a whole new connotation of the possibilities provided by 3D print. At this moment, 3D print becomes a key technology to subversion manufacturing industry. In aspect of production pattern, 3D can bring us changes of customization. Because of the affiliation dissociation between internet and manufacturing industry, clients are willing to design the product with manufacturers, which is feasible. Product manufacturing turns from patterned mass production into customization low-volume production, leading to some outcomes including the elimination of traditional industrial production line, the simplification of supply chain and even changes of world manufacturing economic distribution.

The longer a design cycle of one product is, the longer it takes to market, which means less potential profit of enterprise. Because of the importance to shorten time to market, enterprises have to shorten decision making time during the conceptual phase of a design. These decisions will decide most cost factors including material selection, manufacturing technology and designed life. 3D print technology can optimize design process and maximize potential profit for enterprises. There are many difficulties between an excellent idea and a successful product. When it comes to estimate an idea and decide whether it is worth to be invested, 3D print technology can help enterprises by shortening evaluation time and improving efficiency.

Design process which experienced prototype iterative optimization repeatedly can lower the risks of failure products effectively. Because that 3D printer can manufacture models which have excellent properties and can stand rigorous tests, it is much easier for designers to design. Models manufactured by 3D printer can help designers and engineers understand everything about potential products and decrease leftover problems.

4.2. *Customize products using 3D print technology*

The greatest advantage of 3D print is to expand the imaginary space of designers and develop their imagination and creativity. Future equipment manufacturing industry will focus on personalized customization. 3D printer does not manufacture objects through incision or molds as traditional manufacturing machines. From physical objects by laying one by one expand the range of number concept in aspect of physics. 3D printer can manufacture cultural and creative products which demand accurate internal depressions or interlock shapes in physical world, which could develop huge design space. 3D print technology enables consumers customize products based on their requirements. Mass production will turn into mass customization. Current mass production method pursues to gain scale benefit through minimizing costs, detailing labor division and normalizing operation procedure. It cannot meet the customers' diversified demand any longer. The basic strategy of customization is to strengthen the standardization of inside structure of products in the manufacturing process and to increase the diversity of external structure which can be perceived by customers.

3D print will change our life tremendously in the future. This change, actually, has been taken its place for a long time. This revolution includes the invention of computers and the prosperity of internet. As said by Chris Anderson in *Makers: New Industrial Revolution*: In the process of manufacturing industry having a digital revolution, physical objects have become different designs on the screen which can be shared on line in forms of documents. In the past several decades, factories and design studios experienced this transformation. Today, it happens on the clients' computers as well as in their basements. From retail industry to publishing industry, once an industry shift to digital, it will go through profound revolution. The biggest change is who is manufacturing rather than how to do it. As long as the process can be done by computer, everyone can participate in it. This is the revolution which manufacturing industry is experiencing.

5. Conclusion

No matter what revolution occurs to the manufacturing industry because of the combination between internet and 3D print technology, the essence of industrial design remains unchanged. It is to solve and coordinate the relationship between people and objects. Products bring enjoyments to people not only in physical level but also spiritual level. No matter using traditional artisan craftsmanship or 3D print technology, the essence of design is to bring people a beautiful life. This is the fundamental which would not change forever.

3D print technology provides a cost-effectiveness way to accomplish repeatedly design iteration. It can obtain immediate feedback information of product design on the critical initial phase of development process. 3D product technology can modify creative products immediately, which can help to reduce costs and shorten time to market. To enterprises which use 3D print technology in their design process, this gives them competitive advantages. Along with the decline of price, 3D printing equipment will enjoy a larger market share. The advantages of 3D printing equipment including speediness, precision and low-cost can help creative products enterprises shorten product manufacturing period and make them powerful in competition.

References

1. Wang YG, Wang X, Three-dimensional Printing [M]. Beijing. Wuhan: Huazhong University of Science and Technology Press, 2013.
2. Brian Evans, Analysis 3D Printer: Science and Art of 3D Printer [M]. Beijing, China Machine Press, 2013.
3. Wu HY, 3D Print: Three-dimensional Intelligent Digital Manufacturing [M], Electronic Industry Press, 2014.
4. Chris Anderson, Makers: New Industrial Revolution [M]. Beijing, China Citic Press, 2012.

Design of a 2-DOF compliant parallel mechanism with excellent decoupling performance

Yao Jiang*, Tie-Min Li and Li-Ping Wang
Department of Mechanical Engineering, Tsinghua University,
Beijing, 100084, China
E-mail: jiangyao11@mails.tsinghua.edu.cn

This paper discusses the design of a decoupled 2-DOF compliant parallel mechanism (CPM). The decoupling performances of the CPM are guaranteed by the structure design and geometry parameters optimization both. The structure design of the CPM is firstly discussed in detail. A totally output decoupling character is realized through a symmetric design of the CPM, and the input coupling is reduced with resorting to a flexure-based decoupler. Then the specific definition and analytical expression of the input coupling degree (ICD) of the CPM are given. And analytical models of other performances, including the input and output stiffness, motion stroke, and natural frequency, are also derived. The geometry parameters of the CPM are optimized to minimize the ICD and guarantee the other key performance indicators at the same time. The ICD of the optimized CPM can be as low as 0.279%. Finally, a prototype of the 2-DOF CPM is constructed. Experimental tests of the workspace, natural frequency, input and output coupling, and circular tracking are conducted.

Keywords: Compliant parallel mechanism; design; decoupling performance.

1. Introduction

With the development of science and technology, manipulators with ultrahigh precision and resolution are urgently needed and play more and more important roles in applications such as the atomic force microscope, wafer alignment in microlithography, and microsurgical operation[1-3]. Mechanisms with conventional joints cannot meet the requirements due to the inevitable backlash, friction, and wear will lead to coarse accuracy. By contrast, compliant mechanism, composed of rigid links and flexure hinges, transmit motions solely through the deformations of the materials and can therefore provide high accuracy micro-scale motion.

Compliant parallel mechanism (CPM) has the advantages of high stiffness, large load capacity, low inertia, and high natural frequency over its serial counterpart[4]. Therefore, it has already aroused scholar's great attention and numerous CPMs with various types of motion have been developed[5-8]. Despite its merits, there are specific challenges in the design and development of the CPM, mainly include the realization of output and input coupling[9-11]. The output coupling will lead to a complicated kinematic model and additional kinematic calibration should be performed. The input coupling will bring additional loads on the actuators and therefore cause unwanted deformations and even clearance between the actuators and the mechanism. Piezoelectric actuator is commonly adopted in the compliant mechanism, which is made of multiple piezoelectric layer glued together. The pulling and bending forces caused by the input coupling may bring damage to this kind of actuator. Therefore, the input coupling should be reduced in the development of the CPM. There are several literatures investigating on the design of the decoupled CPMs. S. Awtar[9] firstly proposed the design requirements on the decoupling performances for the XY CPM. L. Lai[12] presented a novel decoupled 2-DOF compliant stage employing a mirror symmetric structure and distributed-compliance compound flexure modules to reduce the output and input coupling. Y. Qin[13] utilized statically indeterminate symmetric structures to attenuate the cross-axis coupling of the XY compliant stage. Y. Li[14,15] investigated on the design and development of a decoupled flexure-based compliant stage with displacement amplifier. The output decoupling was achieved with resorting to a compound parallelogram flexure, and a double compound parallelogram flexure and the amplifier were acted as the decoupler to ensure the stage with an excellent input decoupling character.

Instead of providing a symmetric method for guiding the design of a CPM with excellent decoupling performance, the CPMs are directly presented. Considering the 2-DOF CPM is widely utilized, it is necessary to investigate the detailed design and development process according to the specific application. Therefore, this main purpose of this paper is to present a symmetric design process for a 2-DOF decoupled CPM, including its structure design and geometric parameter optimization. The organization of this paper is introduced as follows. The requirement on the kinematic of the CPM for realizing a totally output decoupling character is firstly discussed. A flexure-based decoupler is introduced and its structure design is detailed discussed to reduce the input coupling of the CPM. Then the geometric parameters of the CPM are optimized to minimize its input coupling degree while also guarantee the workspace, output stiffness and natural frequency. According to the optimized geometric parameters, a prototype of the 2-DOF CPM is fabricated. Several experimental tests, including the workspace, natural frequency, output and input coupling, and circular tracking, are performed to verify the effectiveness of the design method.

2. Structure Design

In this section, the structure design of the 2-DOF CPM will be discussed according to the requirements of the decoupling performance.

2.1. *Structure design considering the output decoupling performance*

The output decoupling character of the 2-DOF CPM requires that the translational motions of the terminal platform along the *x*- and *y*-axes are independent. Therefore, its output stiffness should satisfy the form as

$$\boldsymbol{K} = \begin{bmatrix} K_{x-Fx} & 0 & 0 \\ 0 & K_{y-Fy} & 0 \\ 0 & 0 & K_{\alpha-Mz} \end{bmatrix}. \tag{1}$$

In order to determine a satisfied kinematic structure type easily, the CPM is divided into two parts, as shown in Fig. 1. The local coordinates of each part can be established in two ways, i.e., mirror symmetry and rotational symmetry.

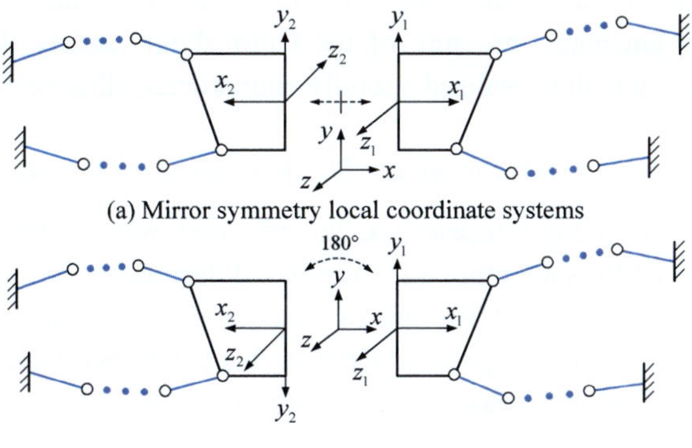

(a) Mirror symmetry local coordinate systems

(b) Rotation symmetry local coordinate systems.

Fig. 1. Division of the 2-DOF CPM.

The stiffness matrices of each subsystems shown in Fig. 1(a) under their corresponding local coordinate systems can be expressed as

$$\boldsymbol{K}_i^m = \begin{bmatrix} K_{i-Fx}^m & K_{ix-Fy}^m & K_{ix-Mz}^m \\ K_{iy-Fx}^m & K_{iy-Fy}^m & K_{iy-Mz}^m \\ K_{i\alpha-Fx}^m & K_{i\alpha-Fy}^m & K_{i\alpha-Mz}^m \end{bmatrix} (i = 1, 2). \tag{2}$$

The output stiffness matrix under the globe coordinate frame can be calculated through the summation of the stiffness of each subsystem as

$$\boldsymbol{K}^m = \sum_{i=1}^{2} \left(\boldsymbol{T}_{if}^m\right)^{-1} \boldsymbol{K}_i^m \left(\boldsymbol{T}_{id}^m\right)^{-1}, \tag{3}$$

where $\boldsymbol{T}_{1f}^m = \boldsymbol{T}_{1d}^m = \operatorname{diag}([1, 1, 1])$, and $\boldsymbol{T}_{2f}^m = \boldsymbol{T}_{2d}^m = \operatorname{diag}([-1, -1, 1])$.

When the kinematic structure of the CPM is mirror symmetry, the local stiffness of each subsystem is identical. Then Eq. (3) can be simplified as

$$\boldsymbol{K}^m = \begin{bmatrix} 2K_{1x-Fx}^m & 0 & 2K_{1x-Mz}^m \\ 0 & 2K_{1y-Fy}^m & 0 \\ 2K_{1\alpha-Fx}^m & 0 & 2K_{1\alpha-Mz}^m \end{bmatrix}. \tag{4}$$

Eq. (4) shows that the translational motion along the y-axis will not affect the other two axes.

Similarly, when the kinematic structure of the CPM is rotation symmetry, the rotation motion about the z-axis has no influence on the other two axes. Therefore, the kinematic structure of the CPM should satisfy the mirror and rotation symmetry both to realized a totally output decoupling performance.

2.2. *Structure design considering the input coupling performance*

According to the present researches, an effective way to reduce the input coupling of the CPM is adopting a flexure-based decoupler between the actuator and the terminal platform. There are two common types of the decouplers, as shown in Fig. 2. The first type is adopted in this paper, because of its compact structure, reduced moving mass and high natural frequency over the second type.

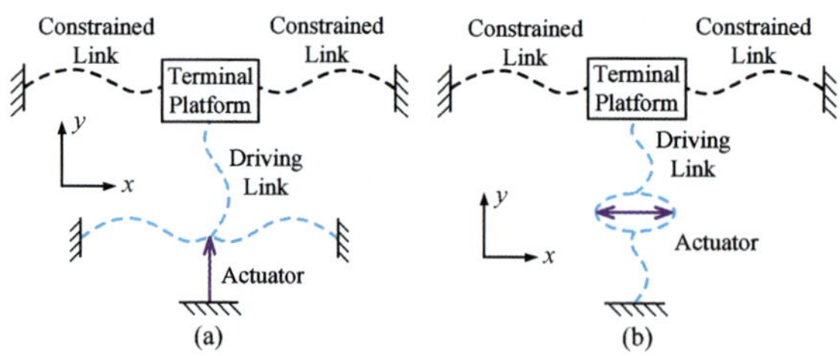

Fig. 2. Kinematic structure of the decoupler.

The stiffness model of the decoupler is shown in Fig. 3. The input decoupling character of the 2-DOF CPM requires that the motion of one actuator will not affect the other. However, it is impossible in the 2-DOF CPM due to its parallel structure. According to Fig. 3, the relationship between the displacement$\boldsymbol{\delta}_o$ of terminal platform and the deformation$\boldsymbol{\delta}_a$ at the actuation point A_1 as

$$\boldsymbol{\delta}_a = \left(\boldsymbol{K}_1 + \boldsymbol{K}_2\right)^{-1} \boldsymbol{K}_2 \boldsymbol{\delta}_o, \tag{5}$$

where are the $\boldsymbol{K}_1$ and $\boldsymbol{K}_2$ are the stiffness matrices of parts I and II of the decoupler, respectively.

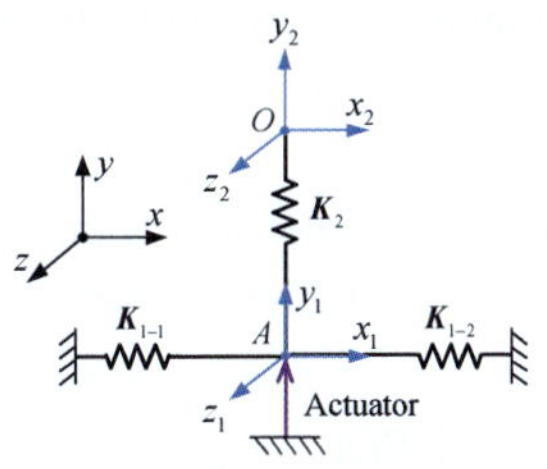

Fig. 3. Stiffness model of the flexure-based decoupler.

The excellent input decoupling character requires that the bending and rotation deformations at the actuation point are small enough when the terminal platform is actuated by the other actuator. According to Eq. (5), the stiffness requirements of the decoupler are: the bending stiffness along the x-axis and the rotation stiffness of part II should be much smaller than those of part I.

Considering the actuator of the CPM is transmitted through the decoupler, the axial stiffness along the y-axis of part I should not be large, otherwise the motion stroke of the actuator will losing too much due to its deformation. On the contrary, the axial stiffness of part II should be large to reduce the motion lost between the actuator and the terminal platform.

The detailed structure design of the decoupler will be discussed according to its stiffness requirement. Figure 4 shows the kinematic structure of part I, which adopts a parallel type and single leaf spring is selected to construct each limb due to its advantages of compact size, light weight, and easy manufacture.

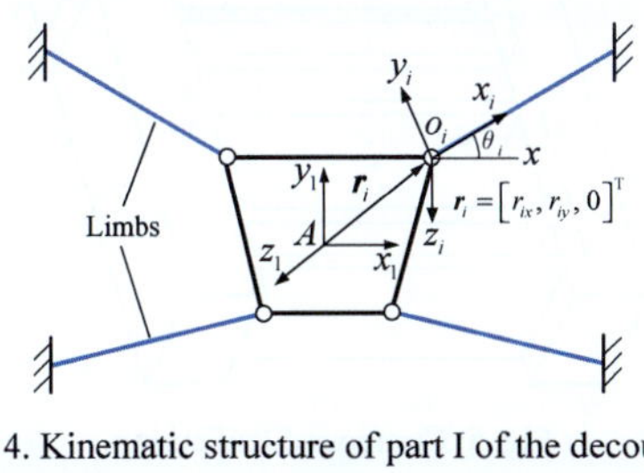

Fig. 4. Kinematic structure of part I of the decoupler.

The stiffness of part I of the decoupler is the summation of the stiffness of each limb. Therefore, the influences of the geometric parameters of the limbs on the stiffness of part I are firstly studied. The stiffness of the ith limb can be expressed in coordinate frame $\{A\text{-}x_1y_1z_1\}$ as

$$\boldsymbol{k}_i^o = \boldsymbol{T}_{if}^{-1}\boldsymbol{k}_i^l\boldsymbol{T}_{id}^{-1}, \tag{6}$$

Where $\boldsymbol{k}_i^l$ is the is the stiffness matrix of the ith limb expressed in its local coordinate frame.

According to Eq. (6), the angle θ_i of the leaf spring of part I should be zero to satisfy its stiffness requirement, i.e., the limbs of part I should be parallel to the x-axis.

Then the influences of the geometric parameters of the leaf sprig on the stiffness of part I are studied. The stiffness matrix of a single leaf spring can be expressed as

$$\boldsymbol{k}_L = \begin{bmatrix} \frac{EDt}{L} & 0 & 0 \\ 0 & \frac{EDt^3}{L^3} & \frac{EDt^3}{2L^2} \\ 0 & \frac{EDt^3}{2L^2} & \frac{EDt^3}{3L} \end{bmatrix}, \tag{7}$$

where E is elastic modulus, and L, t, and D are the length, thick and height of the leaf spring, respectively.

According to the stiffness requirement of part I and Eq. (6), the bending stiffness of the single leaf spring should be small while the axial and rotation stiffness should be large. However, it is impossible to reduce the bending stiffness of the single leaf spring while the axial and rotation stiffness remain unchanged or increased just through adjusting its geometrical parameter. To conquer this problem, a compound hinge composed of several identical parallel leaf spring is chosen, as shown in Fig. 5.

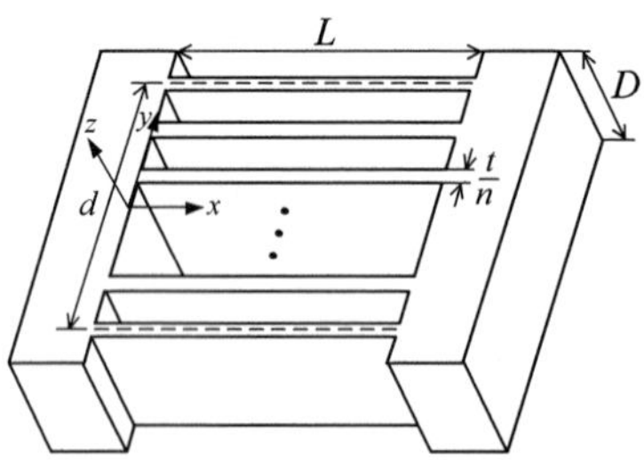

Fig. 5. Compound leaf spring.

The length, height and total mass of the compound hinge are identical to the single leaf spring, and its stiffness matrix can be expressed as

$$\boldsymbol{K}_n = \begin{bmatrix} k_{1x-Fx} & 0 & 0 \\ 0 & (1/n)^2 \cdot k_{1y-Fy} & (1/n)^2 \cdot k_{1y-Mz} \\ 0 & (1/n)^2 \cdot k_{1\alpha-Fy} & \left((n+1)/(n-1)\cdot\gamma^2 + (1/n)^2\right)\cdot k_{1\alpha-Mz} \end{bmatrix}, \qquad (8)$$

Where $\gamma = d/2t$, and $k_{1m\text{-}n}$ is the stiffness component of the single leaf spring.

Equation (8) indicates that the axial stiffness of the compound hinge is always equal to the single leaf spring, while the bending stiffness is always smaller. In addition, the rotation stiffness of the compound hinge will be larger than the single leaf spring when $\gamma > 1$. Therefore, the compound hinge will meet the design requirement of part I well.

The stiffness requirement of part II of the decoupler is contrary to part I, therefore, the limb of part II should be parallel to y-axis according to Eq. (6). The leaf spring utilized in part II should meet the requirements as: the axial stiffness should be large while the bending and rotation stiffness should be small. Unfortunately, no matter the single leaf spring and the compound hinge cannot meet the stiffness requirement. Considering the terminal platform of the 2-DOF CPM only has the translational motion, minimizing the bending stiffness and shear stiffness $k_{a\text{-}Fy}$ can already ensure the decoupler with an excellent input decoupling character according to (5). The bending and shear stiffness of the compound hinge are all decreased sharply as the number n grows up. From the above discussions, the compound hinge is still the best choice for constructing the limbs of part II of the decoupler.

Without loss of generality, the number n of the compound hinges of part I and II are chosen to be two and three respectively in this paper. The detailed structure of the 2-DOF CPM and its key geometry parameters are given in Fig. 6

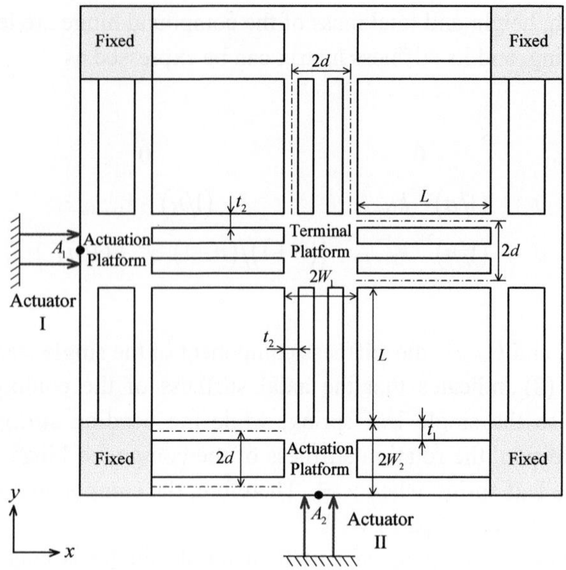

Fig. 6. Detailed structure of the 2-DOF CPM.

3. Analytical Modeling

The detailed structure of the 2-DOF CPM with excellent decoupling characters has already introduced in Section 2. Its geometric parameters are expected to be optimized to obtain an ideal mechanism. Except for the decoupling characters, the concerned performance indicators include the input and output stiffness, motion stroke, and natural frequency. Their analytical models should be established before the geometrical parameters optimization.

3.1. *Input and output stiffness*

The stiffness model of the 2-DOF CPM and the globe coordinate and local coordinates of each parts are given in Fig. 7.

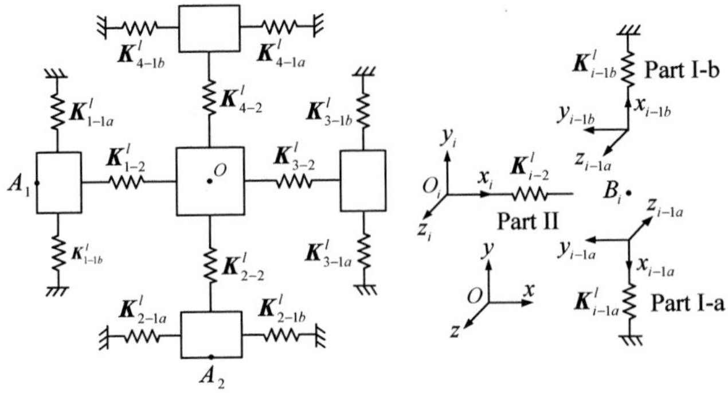

Fig. 7. Stiffness model and coordinate frames of the 2-DOF CPM.

In this section, we utilize $\boldsymbol{K}_{i\text{-}m}^{\,l}$ and ${}^{P}\boldsymbol{K}_{i\text{-}m}$ representing the stiffness of part $m$of the ith chain expressed in its local coordinate frame and the coordinate frame $\{P\}$. The stiffness of the same part expressed in different coordinate frames can be transformed through the following expression as

$$ {}^{Q}\boldsymbol{K} = \boldsymbol{T}_f^{-1}\,{}^{P}\boldsymbol{K}\,\boldsymbol{T}_d^{-1}, \tag{9} $$

where $\boldsymbol{T}_f$ and $\boldsymbol{T}_d$ are the force and displacement transformation matrices between the coordinates $\{P\}$ and $\{Q\}$.

The stiffness of the kinematic chain of the 2-DOF CPM under the coordinate system $\{O_i\}$ can be calculated as

$$ {}^{O_i}\boldsymbol{K}_i = \left({}^{O_i}\boldsymbol{K}_{i-1a} + {}^{O_i}\boldsymbol{K}_{i-1b}\right)^{-1} + {}^{O_i}\boldsymbol{K}_{i-2}^{-1}. \tag{10} $$

The output stiffness of the CPM can be obtained through the summation of the four identical chains as

$$ \boldsymbol{K}_{output} = \sum_{i=1}^{4} {}^{O}\boldsymbol{T}_i^{-1}\left({}^{O_i}\boldsymbol{K}_i\right){}^{O}\boldsymbol{T}_i^{-\mathrm{T}}. \tag{11} $$

The input stiffness model of the CPM along the x-axis is depicted in Fig. 8, where the stiffness matrix ${}^{O}\boldsymbol{K}_{T1}$ in Fig. 9 is the total stiffness of three chains from the second to the fourth chain and can be expressed as

$$ {}^{O}\boldsymbol{K}_{T1} = \sum_{i=2}^{4} {}^{O}\boldsymbol{T}_i^{-1}\left({}^{O_i}\boldsymbol{K}_i\right){}^{O}\boldsymbol{T}_i^{-\mathrm{T}}. \tag{12} $$

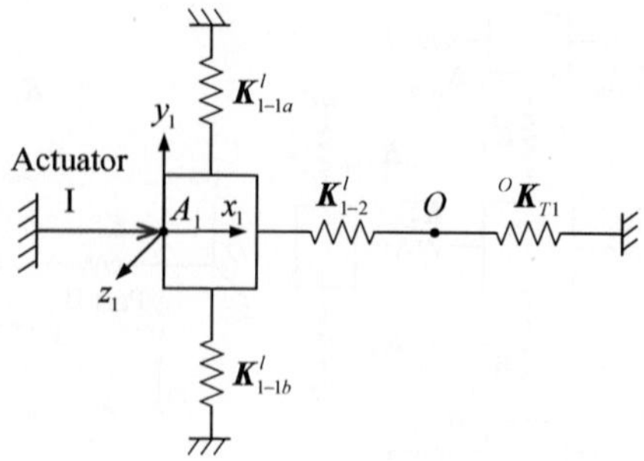

Fig. 8. Input stiffness model of the CPM in the x-axis.

According to Fig. 8, the input stiffness of the CPM can be obtained as

$$\boldsymbol{K}_{Input} = {}^{A_1}\boldsymbol{K}_{1-1a} + {}^{A_1}\boldsymbol{K}_{1-1b} + \left({}^{A_1}\boldsymbol{K}_{1-2}^{-1} + {}^{A_1}\boldsymbol{T}^{\mathrm{T}} \left({}^{O}\boldsymbol{K}_{T1}^{-1} \right) {}^{A_1}\boldsymbol{T} \right)^{-1}. \tag{13}$$

3.2. *Motion stroke*

To determine the motion stroke of the CPM, the deformation of the actuator and the motion lost between the actuator and the terminal platform should be taken into consideration. Considering the compliance of the actuator, the relationship between the actual and nominal displacement output of the actuator can be obtained as

$$\delta_{a1} = \frac{k_n}{k_n + K_{x-Fx}^{In}} \delta_{n1}, \tag{14}$$

where $\boldsymbol{\delta}_{a1}$ and $\boldsymbol{\delta}_{n1}$ are the actual and nominal displacement output of the actuator, k_n is the axial stiffness of the actuator, and K_{x-Fx}^{In} is the input axial stiffness of the CPM along the x-axis.

Then the efficiency of the motion transferred from the actuator to the terminal platform is calculated. According to Fig. 8, the relationship between the displacement output of the terminal platform and the deformation at actuation point A_1 can be obtained with resorting to the equilibrium condition of point O as

$$\boldsymbol{\delta}_o = \left[\left({}^{O}\boldsymbol{K}_{1-2} + {}^{O}\boldsymbol{K}_{T1} \right)^{-1} {}^{O}\boldsymbol{K}_{1-2} \right] \boldsymbol{\delta}_{a1}. \tag{15}$$

Therefore, the motion stroke of the CPM along the x-axis can be obtained as

$$\delta_{ox} = \frac{\left[\left({}^{O}\boldsymbol{K}_{1-2} + {}^{O}\boldsymbol{K}_{T1} \right)^{-1} {}^{O}\boldsymbol{K}_{1-2}\boldsymbol{e}_1 \right]^{\mathrm{T}} \boldsymbol{e}_1}{1 + K_{x-Fx}^{In} / k_n} \delta_n. \tag{16}$$

The motion stroke of the 2-DOF along the y-axisis identical toδ_{ox}.

3.3. *Input coupling degree*

The symmetric structure of the 2-DOF will ensure it with a totally output decoupling character. Therefore, only the analytical model of the input coupling degree (ICD) is established. The relationship between the deformations at the two actuation points should be obtained to determine the ICD of the CPM.The stiffness model of the CPM along the y-axis is shown in Fig. 9.

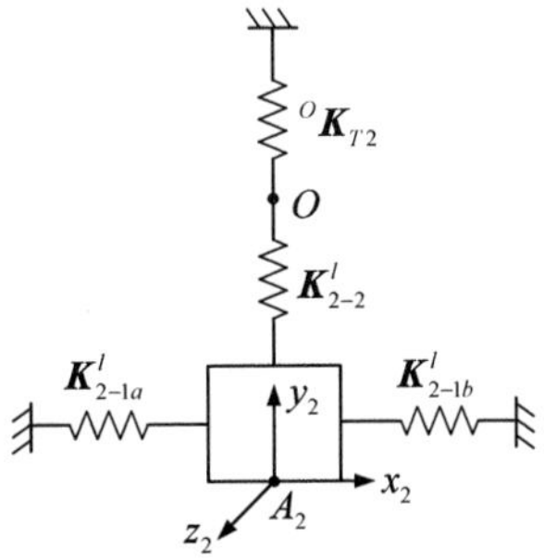

Fig. 9. Stiffness model of the CPM along the y-axis.

Similarly, the deformation at actuation point A_2 caused by the motion of the terminal platform can be obtained utilizing the equilibrium condition of point A_2 as

$$\boldsymbol{\delta}_{a2} = \left[\left({}^{A_2}\boldsymbol{K}_{2-1a} + {}^{A_2}\boldsymbol{K}_{2-1b} + {}^{A_2}\boldsymbol{K}_{2-2}\right)^{-1} {}^{A_2}\boldsymbol{K}_{2-2}\right]\boldsymbol{\delta}_o. \tag{17}$$

Combing Eqs. (15) and (17), the relationship between the deformations at the two actuation points when actuator I is actuated can be obtained as

$$\boldsymbol{\delta}_{a2} = \left({}^{A_2}\boldsymbol{K}_{2-1a} + {}^{A_2}\boldsymbol{K}_{2-1b} + {}^{A_2}\boldsymbol{K}_{2-2}\right)^{-1} {}^{A_2}\boldsymbol{K}_{2-2}\left({}^{O}\boldsymbol{K}_{1-2} + {}^{O}\boldsymbol{K}_{T1}\right)^{-1} {}^{O}\boldsymbol{K}_{1-2}\boldsymbol{\delta}_{a1}. \tag{18}$$

The analytical model of the ICD of the 2-DOF CPM can be defined as

$$ICD = \left[\left({}^{A_2}\boldsymbol{K}_{2-1a} + {}^{A_2}\boldsymbol{K}_{2-1b} + {}^{A_2}\boldsymbol{K}_{2-2}\right)^{-1} {}^{A_2}\boldsymbol{K}_{2-2}\left({}^{O}\boldsymbol{K}_{1-2} + {}^{O}\boldsymbol{K}_{T1}\right)^{-1} {}^{O}\boldsymbol{K}_{1-2}\boldsymbol{e}_1\right]^{\mathrm{T}} \boldsymbol{e}_1. \tag{19}$$

3.4. *Natural frequency*

For the 2-DOF CPM, its first two natural frequencies are identical and can be estimated by the Rayleigh method which treats the CPM as an equivalent single degree-of-freedom system. And the natural frequency can be obtained as

$$f = \frac{1}{2\pi}\sqrt{\frac{K_{eq}}{M_{eq}}}. \tag{20}$$

Where K_{eq} and M_{eq} are the equivalent stiffness and mass of the system.

4. Geometric Parameters Optimization

In order to further guarantee the performances of the CPM, it is necessary to optimize its geometry parameters. The CPM is fabricated from a single block with a specific thickness of 15mm, and the independent geometry parameters to be optimized are L, t_1, t_2, and d. The minimization of the ICD of the CPM is selected as the optimization objective, and the specifications of the key performances of the 2-DOF will utilized as constrains which are described as follows.

(1) Workspace of the CPM. The maximum nominal motion stroke of the piezoelectric actuator adopted in this CPM is 60μm, and its static axial stiffness is about 15N/μm. We expect the workspace of the CPM is larger than 47.5μm×47.5μm, and (16) can be utilized to determine the satisfied size ranges of the geometry parameters.

(2) Output rotation stiffness. The rotation deformation about the z-axis cannot be compensated by the actuators, thus will affect the motion precision of the CPM. Therefore, the output rotation stiffness is expected to be large enough, and it is chosen to be larger than 3.5×10^4Nm/rad.

(3) Natural frequency. The first and second natural frequencies of the 2-DOF CPM are identical and correspond to the translational vibration modes along the two axes. Therefore, they should be as high as possible to guarantee the CPM with a fast response and robustness against the disturbances. Here, we expect the natural frequency of the CPM to be higher than 800Hz.

(4) Geometry parameters range: The ranges of the geometry parameters are given as 50mm $\leq L \leq$ 75mm, 1mm $\leq t_1 \leq$ 5mm, 1mm $\leq t_2 \leq$ 5mm, 5mm $\leq d \leq$ 10mm.

The optimization process is performed with the "fmincon" function in MATLAB. The initial values of the geometry parameters are given as: L=60mm, t_1=3mm, t_2=3mm, and d=7.5mm, and the optimized dimensions are: L=50mm, t_1=3.56mm, t_2=2.57mm, and d=10mm. The ICD of the optimized mechanism is as low as 0.279%, which indicates the 2-DOF CPM possesses an excellent input decoupling character. The other performances are also calculated by the analytical models and given as follows:

(1) Workspace: 47.54μm×47.54μm;

(2) Output rotation stiffness: 3.50×10^4Nm/rad;

(3) Natural frequency: f_1=f_2=876Hz.

The analytical results show that all the key performance indicators of the optimized CPM all satisfy the design requirements.

5. Experimental Tests

A prototype of the 2-DOF CPM is fabricated, as shown in Fig. 10. Several experimental tests on the performances of the CPM are carried out to demonstrate the effectiveness of the design method.

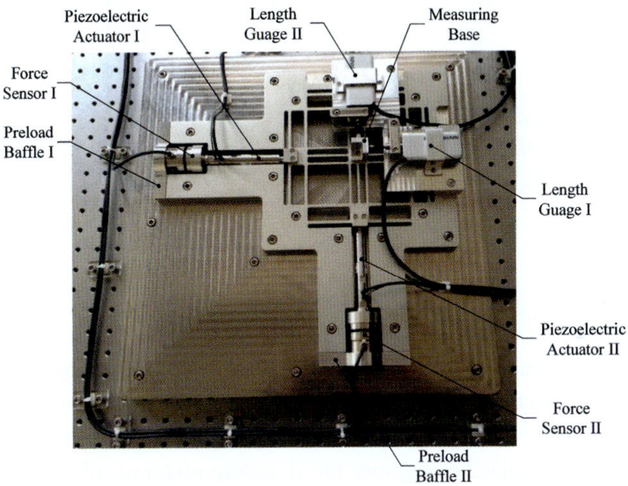

Fig. 10. Prototype of the 2-DOF CPM.

The main body of the CPM is fabricated from a single block of Al-7075, where its features are machined utilizing the milling process. Two piezoelectric actuators (P-843.40 from PI) are adopted and they are driven by a modular piezo controller & driver (E-621 from PI). Two force sensors (U93-1kN from HBM) are mounted at the ends of each actuator to detect the actuation forces. The analog voltage outputs of the force sensors are amplified by a high dynamic signal processor (MX410 from HBM). The displacements of the terminal platform are measured by two length gauges (MT-1281 from Heidenhain) installed vertically. Preloads are applied along the two axes of the stage to prevent the piezoelectric actuators from suffering pulling forces and the preloading forces are set to about 300N.

Firstly, the workspace of the CPM is measured by running a square trajectory in the open-loop control. The experimental results and the estimated workspace are given in Fig. 11. The measured workspace is 45.21μm×45.63μm, which is about 4% smaller than the estimated result. The decreased motions of the CPM mainly result from the deformations of the pretensioners and the contact interface between the actuator and mechanism.

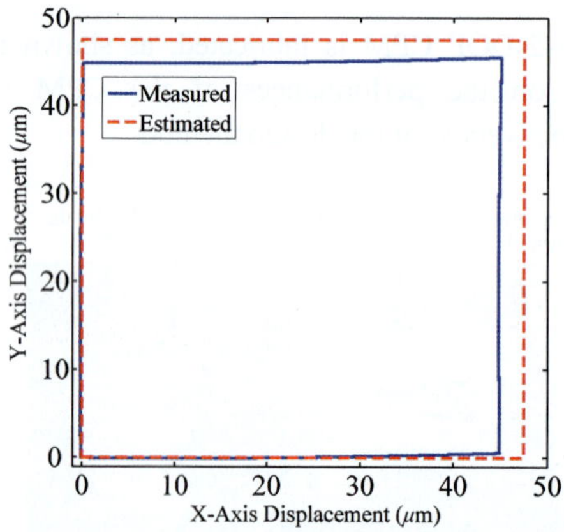

Fig. 11. Workspace of the CPM.

Then the natural frequency of the CPM is tested through the hammering method with the actuators dismantled. A modal hammer is utilized to apply impulse forces at the actuation points along the two axes, respectively. The length gauges are utilized to capture the dynamic response of the stage and the spectral analyses on the experimental results are given in Fig. 12. The first two natural frequencies of the CPM can be obtained as 793Hz and 799Hz. They are supposed to be equal and the slight difference may lie in the machining errors. Besides, the experimental results are approximately 9.5% smaller compared to the analytical values. It should be the analytical calculation errors of the equivalent stiffness or mass of the CPM. If we replace the analytical equivalent stiffness with the value simulated by the FEM software, the first and second natural frequencies can be calculated as 805Hz, which is very close to the experimental results.

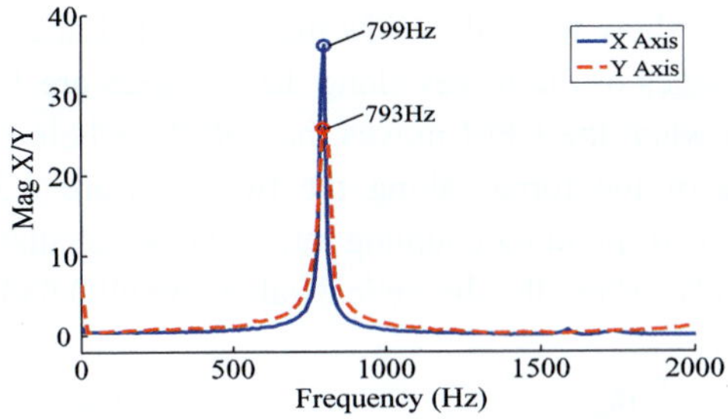

Fig. 12. Natural frequency tests of the CPM.

The output coupling is evaluated by actuating one axis and measuring the displacement output of the terminal platform along the other axis. The maximum output motion ranges of the CPM along the two axes are both set as 40μm.The results of the output coupling evaluation are shown in Fig. 13. The results show that the maximum coupling displacement is0.24μm when the *x*-axis is actuated, corresponding to a coupling ratio of 0.60%. Similarly, the coupling ratio can be calculated as 0.15% when the *y*-axis is actuated. The cross-axis motions may arise from the manufacturing errors of the dimensions of the CPM and the flatness of the measuring base.

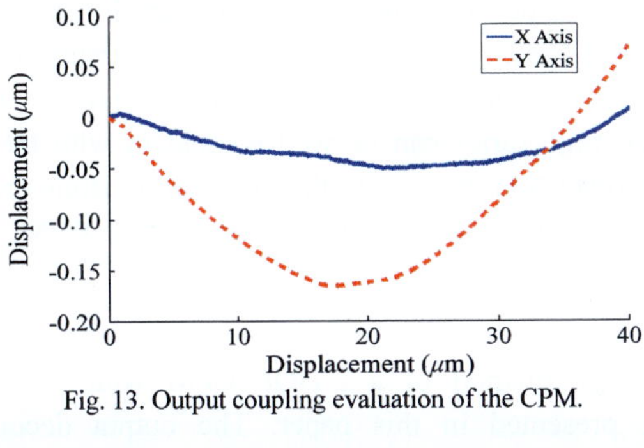

Fig. 13. Output coupling evaluation of the CPM.

The deformations at the actuation points should be measured to evaluate the ICD of the CPM, while this cannot be carried out in this prototype. Therefore, we propose utilizing the ratio of the axial deformations at the two actuation points to reflect the input coupling character. Considering the relation of the axial deformation and the force detected by the force sensor conforms to a linear relationship, the changes in the axial forces along the two axes can be measured to evaluate the input coupling performance of the CPM. The force changes along the two axes when the*x*-axis or *y*-axis is actuated are tested and shown in

Fig. 14. The results show when the CPM moves through the whole range along the *x*-axis, the changes of the forces along the two axes are 172.6N and 0.41N, respectively. And when the CPM moves through the whole range along the *y*-axis, the changes of the forces along the two axes are 0.39N and 175.4N, respectively. The corresponding coupling ratios can be calculated as 0.238% and 0.223%, respectively, where the theoretical values are all 0.084%.

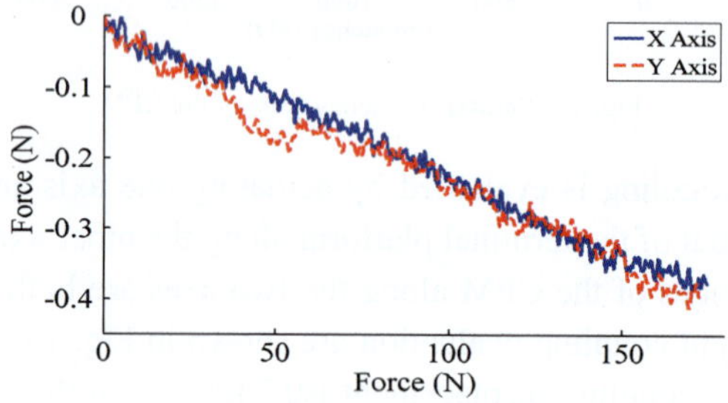

Fig. 14. Input coupling evaluation of the CPM.

The circular tracking performance of the CPM is also tested as well. In order to verify its decoupling performance adequately, the two actuators are controlled independently without any additional kinematic calibration and compensation. The Proportion-Integration algorithm is utilized to realize the closed-loop control. The experimental results of the circular contouring with a radius of 5μm at four different frequencies from 0.5-5Hz are shown in Fig. 15. Theactual circular trajectories can be well coincident with the ideal trajectory without elliptical distortion. It owes to the identical dynamic characteristics and excellent decoupling performances of the two axes of the CPM.

6. Conclusion

A systematic design method of a 2-DOF CPM with excellent decoupling performances is presented in this paper. The output decoupling character requires the CPM possesses a symmetry structure, and the input coupling is reduced through introducing a flexure-based decoupler. And the geometry parameters of the CPM are optimized to reduce the ICD of the CPM while also guarantee its motion stroke, output stiffness, and natural frequency. A prototype is developed according to the optimized result with a measured workspace of 45.21μm×45.63μm, and the first and second natural frequencies are 793Hz and 799Hz. The output coupling ratios of the two axes are 0.60% and 0.15%, respectively. And the input coupling ratios evaluated by detecting the axial force

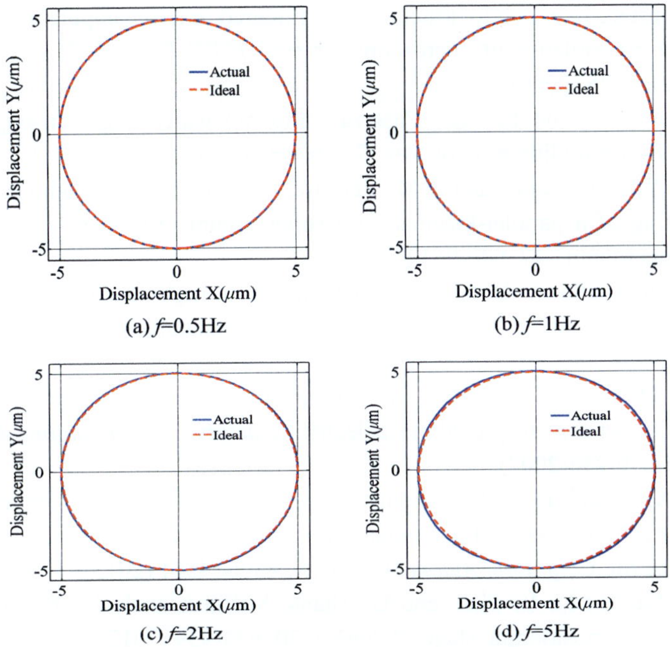

Fig. 15. Circular tracking performances test of the CPM.

variations along the two axes are 0.238% and 0.223%, respectively. Besides, high tracking accuracy is obtained in the circular contouring tests with the two actuators controlled independently. The experimental results all demonstrate the excellent decoupling performances of the CPM, and thereby prove the effectiveness of the design method of the decoupled 2-DOF CPM.

Acknowledgments

This work was supported in part by National Natural Science Foundation of China (Grant No. 51275260), China National Funds for for Distinguished Young Scientists (Grant No. 51225503), and Tsinghua University Initiative Scientific Research Program (No. 2014z22068).

References

1. G. Schitter, P.J. Thurner, and P.K. Hansma, Design and input-shaping control of a novel scanner for high-speed atomic force microscopy, *Mechatronics*, 18, 282-288 (2006).

2. A.A. Ramadan, T. Takubo, Y. Mae, et al., Development process of a chopstick-like hybrid-structure two-fingered micromanipulator hand for 3-D manipulation of microscopic objects, *IEEE Trans. Ind. Electron.*, 56, 1121-1135 (2009).
3. K.B. Choi, and J.J. Lee, Passive compliant wafer stage for single- step nano-imprint lithography, *Rev. Sci. Instrum.*, 76, 075106 (2005).
4. Q. Yao, J. Dong, and P.M. Ferreira, Design, analysis, fabrication and testing of a parallel-kinematic micropositioning XY stage, Int. J. Mach. Tools Manuf., 47, 946-961 (2007).
5. S. Poilt and J. Dong, Development of a high-bandwidth XY nano-positioning stage for high-rate micro-nanomanufacturing, *IEEE Trans. Mechatronics*, 16, 724-733 (2011).
6. W. Dong, L. Sun, and Z. Du, Design of a precision compliant parallel positioner driven by dual piezoelectric actuators, *Sens. Actuators, A: Phys.*, 135,250-256(2007).
7. S. Xiao, and Y. Li, Development of a large working range flexure- based 3-DOF micro-parallel manipulator driven by Electromagnetic Actuators, in *IEEE Int. Conf. Rob. Autom.*, Germany, 4506- 4511 (2013).
8. X. Jia, J. Liu, Y. Tian, and D. Zhang, Stiffness analysis of a compliant precision positioning stage, *Robotica*, 30, 925-939 (2012).
9. S. Awtar and A. H. Slocum, Design of parallel kinematic XY flexure mechanisms, in *Proc. Int. Des. Eng. Technol. Conf. &Comput. Info. Eng. Conf.*, 7, 89-99 (2005).
10. S. Awtar and A. H. Slocum, Constraint-based design of parallel kinematic XY flexure mechanisms, *J. Mech. Des.*, 129, 816-830 (2007).
11. Y. Li and Q. Xu, Design of a new decoupled XY flexure parallel kinematic manipulator with actuator isolation, in *Proc. IEEE/RSJ Int. Conf. Intell. Rob. Sys.*, 470-475 (2008).
12. L. Lai, G. Gu, and L. Zhu, Design and control of a decoupled two degree of freedom translational parallel micro-positioning stage, *Rev. Sci. Instrum.*, 83, 045105 (2012).
13. Y. Qin, B. Shirinzadeh, Y. Tian, D. Zhang, and U. Bhagat, Design and computational optimization of a decoupled 2-DOF monolithic mechanism, *IEEE Trans. Mechatronics*, 19, 872-881(2014).
14. Y. Li and Q. Xu, Design and analysis of a totally decoupled flexure-based XY parallel micromanipulator, *IEEE Trans. Rob.*, 25, 645-657 (2009).
15. Y. Li and Q. Xu, Development and assessment of a novel decoupled XY parallel micropositioning platform, *IEEE Trans. Mechatronics*, 15, 125-135, (2010).

Precision assembly oriented method to determine the evaluation parameters for surface form error

Shao-Wei Guo, Hong-Li Fan, Fa-Ping Zhang*, Ti-Guang Zhang and Ge Wang

School of Mechanical Engineering, Beijing Institute of Technology,
Beijing, State ZIP/Zone, China
**E-mail: 947337867@qq.com*

Surface form error of a part has a significant impact on the assembly precision of the entire mechanical system. The traditional form error evaluation methods, based on the principles of a minimum tolerance, are inadequate to quantitatively describe the vital influence of the distribution of the surface form error on the assembly precision . This paper proposes a method to determine the potential evaluation parameters of surface form error for assembly precision. Firstly, non-Gaussian surface construction is used to simulate machining surface, and wavelet analysis with the selected base function is used to obtain the surface form error, and the candidate parameters characterizing the surface form error distribution are selected. Then the virtual assembly of the filtered surface with ideal smooth surface is conducted. hence the ideal surface spatial location is calculated. The relationship between the surface parameters and the second surface spatial location is established by correlation analysis, which can determine the form error evaluation parameters by a certain criteria. Thus provides a scientific basis for optimizing the assembly process to improve assembly precision. Finally a case is used to verify the feasibility proposed method.

Keywords: Assembly precision; surface form error; evaluation parameters; related analysis.

1. Introduction

In the assembly of precision mechanical systems, parts surface form error is one of the important factors affecting the relative position accuracy of components in the whole of system[1]. The present surface form error evaluation method, is mainly concerned about whether machining parts is qualified [2-3]. However, from the assembly point of view, as the same mounting surface form error, due to their different shapes error distribution, it will affect the accuracy of the assembly. As shown in Figure 1. Domestic [4-6] first proposed by entropy to characterize the form error evaluation method for assembly, but the quantitative description methods about surface parameters and assembly precision has not been established. to solve these problems, this paper proposes a method to determine the surface form error estimation parameters for assembly precision.

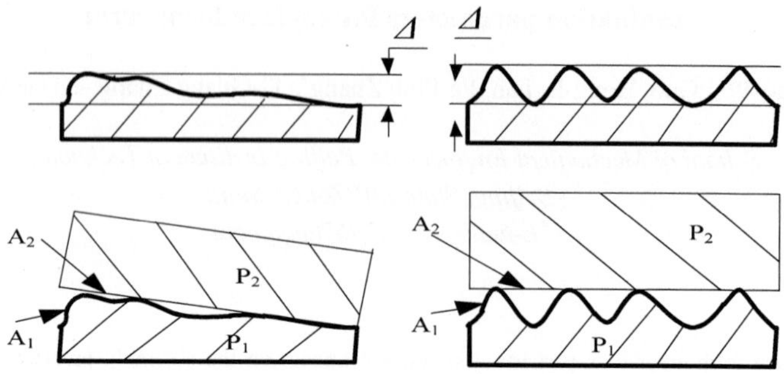

Fig. 1. Assembly errors caused by different distributions of the same shape errors.

2. Methods to Determine the Form Error Evaluation Parameter for Precision Assembly

This paper provide methods to form the cutting surface by a non-Gaussian plane simulation method, Based on that, Wavelet analysis to obtain surface form error, And extract the parameters characterizing the surface shape of the error distribution; the surface after filtering through search contact points simulate assembly, determine the relationship between the surface parameters and part spatial orientation after assembly distribution parameters by correlation analysis. The proposed form error assembly evaluation parameter determination method shown in Figure 2.

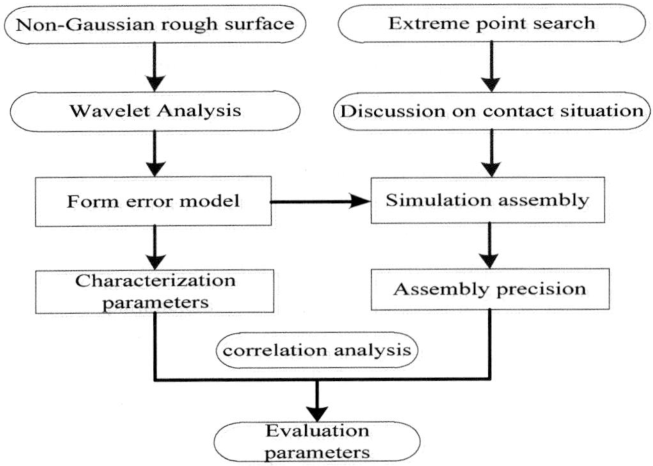

Fig. 2. Overall framework map.

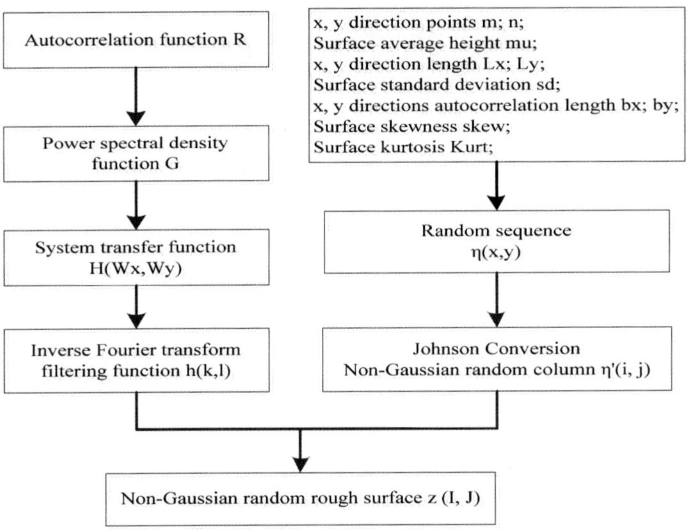

Fig. 3. Flow chart of simulation of rough surface.

2.1. *Analog and generation of machined surface*

Studies have shown that the general processing methods such as turning, milling, grinding the resulting plane are in line with non-Gaussian surface characteristics. Non-Gaussian rough plane [7-8] can be generated by computers. The flow chart of rough plan simulation is shown in Figure 3.

In order to isolate the form error profile, the wavelet analysis is used to separate the feature information of the surface topography of the parts from the non-Gaussian plane. E.g., input parameters are as follows: m=64; n=64; mu=1.6; Lx=4; Ly=4; sd=1; bx=5; by=5; skew=0.25; kurt=3,You can generate a non-Gaussian plane shown in Figure 4, and obtain a form error plane, shown in Figure 5, through the wavelet analysis.

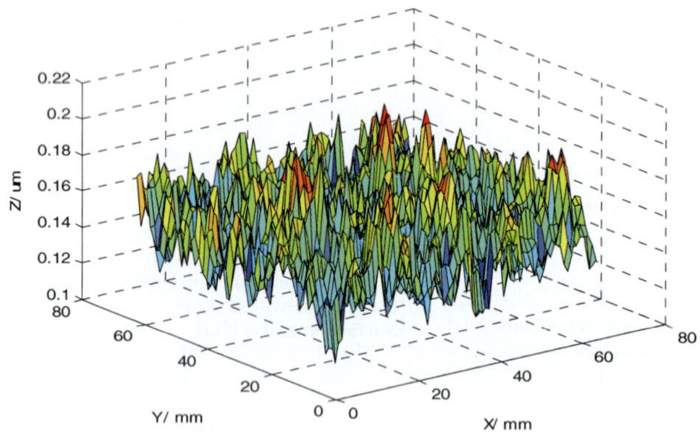

Fig. 4. Non-Gaussian plane.

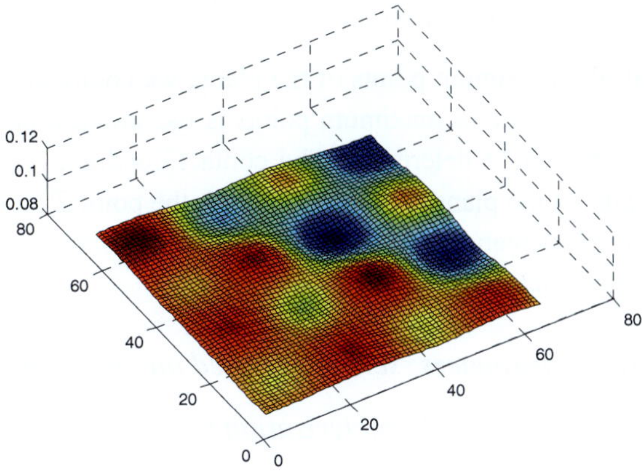

Fig. 5. Form error plane.

2.2. *Assembly simulation*

In this paper, we mainly study the influence of form error on the assembly precision. The form error plane and the ideal plane will be simulated. The parts will be considered as a rigid body. Therefore, the contact plane is considered to contain 3 contact points. The simulation assembly process is shown in Figure 6:

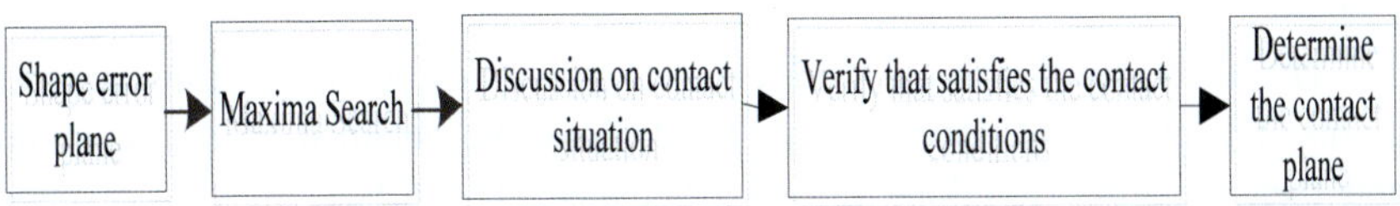

Fig. 6. Assembly simulation flowchart.

2.2.1. *Maxima search*

A violent search method is used to find the maximal value of the plane. The measuring point beginning from the beginning of the first rows and first columns will be searched line by line. For edge points, it is only necessary to compare the measurement points with other points in the surrounding neighborhood.

2.2.2. *Discussion on contact situation*

After getting all the maximum points of the plane, we coordinate the plane with the ideal plane. The plane of maximum points is the smallest possible distance. When the contact point is determined, the contact condition is set, that is, the value of all points on the plane are compared with the point Z on the ideal plane. If all Z values are less than Zi, then the three points of contact are considered as a set of possible contact points.

2.3. *Evaluation parameters' selection based on correlation analysis*

2.3.1. *Selection of surface characterization parameters*

Three-dimensional topography of engineering surface is complex, and each parameter can only describe one aspect of it. According to ISO25187 and surface characterization, the parameters can be grouped into four categories[9].

2.3.2. *Acquisition of assembly precision*

Assembly precision is the actual precision of geometry parameters after product assembly. In this study, assembly precision represents the spatial location of the assembly part's surface relative to the reference plane, that is to say, assembly precision is measured by the size of six degree-of-freedom.

$$\Gamma = [d_x, d_y, d_z, \delta_x, \delta_y, \delta_z] \tag{1}$$

The ideal plane equation Ax+By+Cz+D=0 can be obtained by three contact points. The change of normal vector (A,B,C) can be obtained by the equation, and then the three rotations can be calculated. The offsets in x, y, z three directions can be obtained by comparing the plane equation with the ideal plane.

2.3.3. *Correlation analysis between characterization parameters and assembly precision*

Correlation analysis can establish quantitative indicators to analyze whether the assembly precision is directly related to the characterization parameters. The correlation analysis formula between characterization parameters and assembly precision can be established as follows.

$$\rho_i = \frac{\mathrm{cov}(\mathrm{V},\boldsymbol{\Gamma})}{\sqrt{\mathrm{D}(\mathrm{V})}\cdot\sqrt{\mathrm{D}(\boldsymbol{\Gamma})}}, V=[v_1,v_2,\ldots,v_l], \boldsymbol{\Gamma}=[d_x,d_y,d_z,\delta_x,\delta_y,\delta_z] \quad (2)$$

V represents characterization parameters. And Γ represents assembly precision.

By changing the input parameters, the data of characterization parameters and assembly precision can be obtained. And then correlation analysis can be carried out.

3. Instance Verification

In order to verify the validity of the method for evaluating the form error of the assembly, using MATLAB as a tool, according to the surface parameters of a milling support plane, a computer program is compiled.

Firstly, input parameters, as shown in Table 1; simulate the parts' surface topography and simulate assemble. Secondly, obtain the characterization parameters and assembly precision. Finally, calculate the correlation coefficient between the characterization parameters and assembly accuracy, according to 1000 groups of simulation analysis results. The correlation coefficient table is as shown in Table 2.

Table 1. Input parameters table.

Input parameters	Values
x direction points m	m=64;
y direction points n	n=64
Surface average height mu	$\mathrm{mu} \sim \mathrm{N}(1.6,0.1)$
x direction length Lx	Lx=4
y direction length Ly	Ly=4
Surface standard deviation sd	$\mathrm{sd} \sim \mathrm{N}(1,0.1)$
x directions autocorrelation length bx	$\mathrm{bx} \sim \mathrm{N}(5,1)$
y directions autocorrelation length by	$\mathrm{by} \sim \mathrm{N}(5,1)$
Surface skewness skew	$\mathrm{skew} \sim \mathrm{N}(0.25,0.01)$
Surface kurtosis Kurt	$\mathrm{Kurt} \sim \mathrm{N}(3,0.1)$

Table 2. The correlation coefficient table.

Evaluation parameters	δ_x	δ_y	d_z
root mean square deviation	0.1866	-0.1745	0.9999
Arithmetic mean deviation	0.1866	-0.1745	0.9999
skewness	-0.2091	0.2765	-0.2496
kurtosis	-0.2089	0.2763	-0.2496
ten point height	0.1876	-0.1754	0.9999
Maximum height	0.1876	-0.1754	0.9999
density of summits	0.6496	0.8713	-0.4335
mean values	0.1866	-0.1745	0.9999
variances	0.0061	0.0308	-0.8737
top arithmetic average curvature	0.3550	0.4562	0.4721

As can be seen from the table, the ten parameters show that, density of summits S_{ds} is associated with the rotation of x, y direction; root mean square deviation S_q, arithmetic mean deviation S_a, ten point height S_z, Maximum height S_p, mean values μ, variances σ^2 and Z migration direction are highly correlative.

4. Conclusions

This paper proposes a method to determine the surface form error estimation parameters for assembly precision. The simulation results proved that the method can determine the form error parameters for assembly. Compared with other methods, this method is oriented to the assembly. According to which, we can achieve a quantitative description of the relationship between surface distribution characterization parameters and assembly precision.

Acknowledgment

This research is supported by the National Natural Science Foundation of China (NSFC Grant No. 51275049) and Defense 973 Project (613243). The authors gratefully acknowledge the facilities provided by the Industrial Engineering Laboratory (IEL) at the Beijing Institute of Technology.

References

1. Zhang Zhijing, Jin Xin, etc. Precision micro and small manufacturing theory, technology and application of Mechanical Engineering, 2007, 43 (1): 49-61.

2. Scarr A. Use of the least squares line and plane in the measurement of straightness and flatness. Proceedings of the Institution of Mechanical Engineers, 1967, 182(1): 531-536.

3. Cheraghi S.H., H.S. Lim and S. Motavalli. Straightness and flatness tolerance evaluation: an optimization approach. Precision Engineering, 1996, 18(1): 30-37.

4. Jin X, Zuo F, Zhang T, et al. An entropy-based method to evaluate plane form error for precision assembly[J]. Proceedings of the Institution of Mechanical Engineers, Part B: Journal of Engineering Manufacture, 2013, 227(5): 726-734.

5. Zhang Zhijing, Zuo fuchang, Zhang Tingyu, etc. An Entropy-Based Method to Evaluate Plane Form Error for Precision Assembly, 12th CIRP Conference on Computer Aided Tolerancing, UK.

6. Zhang Zhijing, Zuo fuchang, Jin Xin, etc. Evaluation and Modeling of Geometric Form Error for PrecisionAssembly Analysis, 10th International Symposium of Measurement Technology and IntelligentInstruments, Korea.

7. Tian Ailing, Sun Tao, in Xun, Huang Mengtao non-Gaussian random rough surface of digital simulation [J] Journal of system simulation, 2009, (10): 2840-2842.

8. Jiunn Jong Wu. Simulation of non Gaussian surfaces with FFT [J]. Tribology International (S0301-679X), 2004, 37(4): 339-346.

9. W P Dong, P J Sullivan, K J Stout.Comprehensive study of parameters for characterization 3-D surface topography(v) [J].Wear, 1994, 178 (1): 29-60.

Experimental study of hydraulic actuator for FAST

Yuan Gao

National Astronomical Observatory,
Chinese Academy of Science, Beijing 100012, China
School of Mechanical Engineering,
Guizhou University, Guiyang 550025, China
E-mail: gaoyuan2009@foxmail.com

Qi-Ming Wang and Ming Zhu†

National Astronomical Observatory,
Chinese Academy of Science, Beijing 100012, China
Key Laboratory of Radio Astronomy,
Chinese Academy of Science, Beijing 100012, China
†E-mail: mzhu@bao.ac.cn

Jian-Xing Xue

National Astronomical Observatory,
Chinese Academy of Science, Beijing 100012, China

Qin Huang

School of Mechanical Engineering,
Guizhou University, Guiyang 550025, China

Hydraulic actuator is one of the prime components for Five-hundred-meter Aperture Spherical radio Telescope (FAST). Firstly, the background and key design requirements of hydraulic actuators are detailed discussed. Secondly, working principle of hydraulic actuator is explained according to astronomical observation. Lastly, experimental prototypes and testing platforms are built to verify the motion performance of hydraulic actuators. The results indicate that the actuator scheme is feasible and the maximum position error is 0.2 mm, which can meet the accuracy requirements of hydraulic actuators for FAST.

Keywords: FAST project; design requirement; hydraulic actuator; experimental prototype.

1. Introduction

Five-hundred-meter Aperture Spherical radio Telescope (FAST) will be the largest single-dish radio telescope upon completion in the world[1]. As one of three critical innovations of FAST, the main active reflector system is composed of ring beam, reflector elements, actuators, cables and anchors[2,3]. While, ring beam is the main support structure composed of lattice steel. The cable-net is weaved by 6,670 main cables and these cables are formed by 2,225 cable joints. Each joint connects to a pull-down cable and an actuator. In accordance with the movement of celestial bodies, the shape of reflector must be real-time changed. Figure 1 is the effect graphic of FAST.

Fig. 1. The effect graphic of FAST.

As a scaling mechanism, the actuator has control and position feedback functions, one end of which is fixed with the anchor and another end is hinged at the pull-down cable. Under the condition of controlling signal, actuator can change the length between anchor and pull-down cable by changing its own length to adjust the cable joint's position. Ultimately, the shape of reflector can be changed[4].

In the present study, the hydraulic actuator of main reflector system for FAST is introduced. In section 2, design requirements for actuators are analyzed. In section 3, the operational principle of hydraulic actuator is discussed. In section 4, based on the principle of scheme, the actuator prototypes and experimental platforms are established. In the last section, the scanning tracking experiment is made to verify if the maximum position error of hydraulic actuator could meet the accuracy requirements.

2. Design Requirements

2.1. *Position feedback*

The main function of actuatorsare target positioning, which means the actuator should not only reach the specified location from controlling PC but also provide the actual execution information back to PC. Meanwhile, actuatorsof FASTshould achieve quicker position feedback to adjust the stroke of piston rod.

2.2. *Operational mode*

Hydraulic actuators are pulled by cables and the maximum force is 70 kN. The hydraulic actuators have four operation modes, which are source changing, scanning tracking, running servo and self-locking. The source changing mode means the actuator has to achieve point-to-point motion in 10 minutes. The maximum velocity is 1.6mm/s. The scanning tracking mode refers to hundreds of actuators cooperating to track the target places in closed-loop position control. During the movements, position error of hydraulic actuators must be less than 0.25mm. The running servo mode is used for fault protection, which means the piston rods have to move along cables when some accidents happen. And the self-locking mode can keep the hydraulic actuators in their original positions in case of outages or breaking down.

2.3. *Electro magnetic compatibility (EMC) and condition monitoring*

The frequency of FAST receiver is from 70MHz to 3GHz. Therefore, the EMC should be considered during the design phase of hydraulic actuators. As it is difficult to identify the fault roots manually. When hydraulic actuators are failure, necessary monitoring sensors should be installed.

3. Working Principle of Hydraulic Actuator

Hydraulic actuator of FAST is an integrated equipment including hydraulic cylinder, independent oil source and electronic control system, the scheme is shown in Figure 2[5].

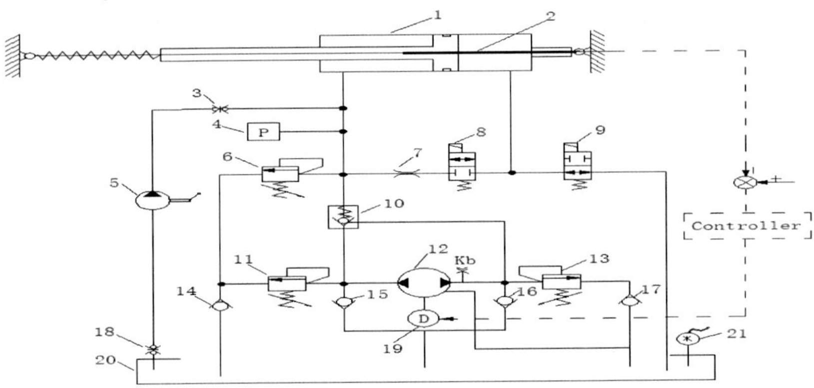

Fig. 2. The hydraulic schematic diagram of hydraulic actuator.

1. Hydraulic cylinder 2. Displacement sensor 3. Pressure tap 4. Pressure sensor 5. Hand pump 6. Overflowvalve 7. Throttling bolt 8,9. Magnetic valve 10. Pilot operated check valve 11. Relief valve 12. Bidirectionalgear pump 13. Counterbalance valve 14,15,16,17. Check valve 18. Quick-change connector 19.Steppeing motor 20. Oil tank 21. Temperature sensor

3.1. *Normal working state*

During loading operation, the cylinder rod of hydraulic actuator is altered along with the changing pulling force. The magnetic valves (8 and 9) are turned off so that the rodless chamber is connected with the oil tank. As piston rod extends, the pump reverses and the outlet pressure of gear pump is gradually increased. When the pressure reaches the set value of counterbalance valve, the pilot operated check valve 10 is opened. Under the effect of load, the hydraulic oil flows from rod chamber to rodless chamber. As piston rod retracts, the pump forwards to transport oil from tank to rod chamber.

During non-load running, two magnetic valves of hydraulic actuator are powered. As piston rod extends, rodless chamber is connected with rod chamber while disconnected with oil tank. The pump reverses to draw oil from oil tank and rod chamber into the rodless chamber. The principle of non-load retraction is same as which of loading.

3.2. *Fault servo and manual manipulation state*

When communication is in a normal state, the magnetic valve 8 is unfolded to connect rod chamber with oil tank. Then, the piston rod is stretch out along the pull-down cable, which is small load servo. On the contrary, when communication is outage, the hydraulic system is overflowed automatically as the system pressure exceeds the set value of overflow valve 6, which is large load servo. Meanwhile, the hydraulic actuator can be controlled manually by linking a hand pump to pressure tap 3, whose operational theory is same as non-load running state.

4. Prototype Testing

4.1. *Experimental prototype*

For easy of disassembling power system from hydraulic cylinder, the experimental prototype of hydraulic actuator is constructed to be integrated, which is shown in Figure 3.

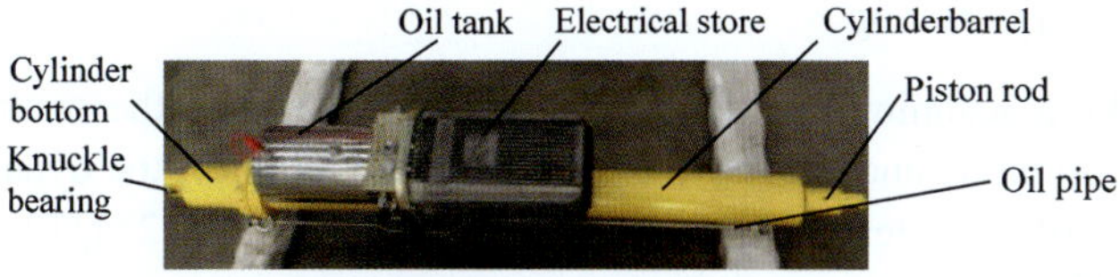

(a) Experimental prototype

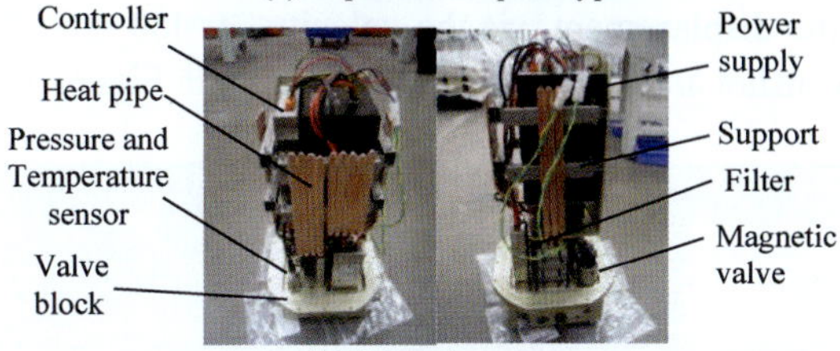

(b) Internal structure

Fig. 3. Experimental prototype.

All of valves are installed on one block for saving space and fitting maintenance. There are only two rigid pipes in the hydraulic actuator prototype to reduce leakage. The total weight of an actuator is less than 120 kg, which is easy to be transported and maintained.

4.2. *Testing Platform*

Testing platforms are divided into constant load platforms and variable load platforms, which are shown as Figure 4. While, Figure 4(a) are three constant load platforms, which consists of three parts, frame, fixed pulley and loads. The weight of the loads is added to hydraulic actuator directly. Figure 4(b) is a variable load platform, which includes frame, two fixed pulleys, one movable pulley and loads. The load of hydraulic actuator is changed by altering the angle composed of fixed pulley, movable pulley and fixed end.

(a) Constant load platforms(b) Variable load platform

Fig. 4. Testing platforms.

4.3. *Prototype testing*

Here, take a scanning tracking test for example. Switch the control system to closed-loop mode and employ the following rule to detect the dynamic response accuracy of the test hydraulic actuator[4], $s = A\sin(2\pi t/T)$, $v = 2\pi A/T \cos(2\pi t/T)$.

Where, s is the displacement, $v$is the velocity. A=480 mm, T=2 400 s. The experimental screenshot and testing results are shown in Figure 5.

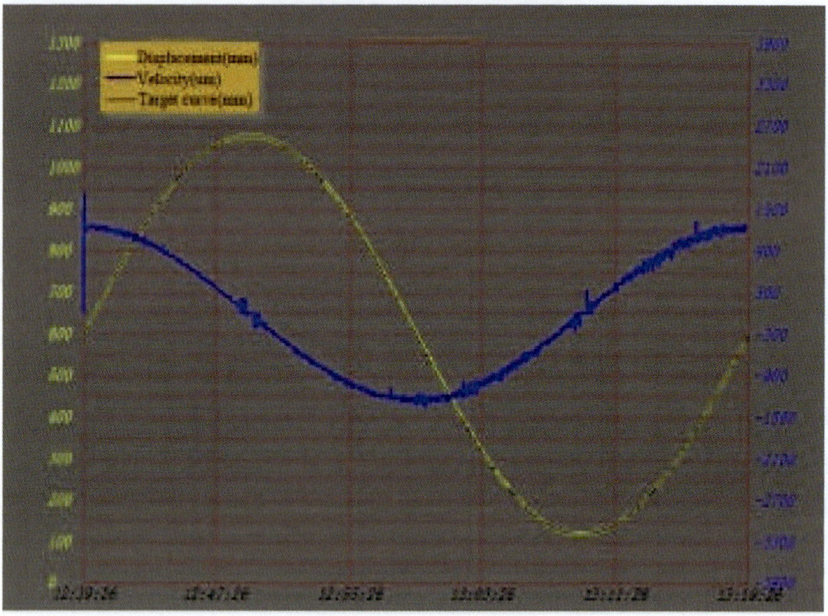

(a) Displacement and velocity

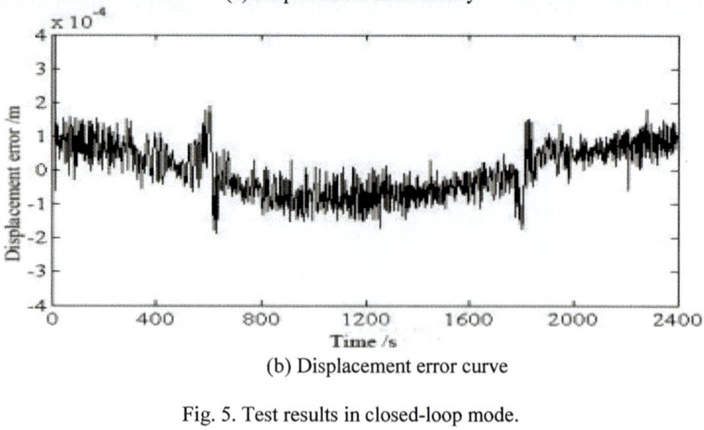

(b) Displacement error curve

Fig. 5. Test results in closed-loop mode.

The results show that the hydraulic actuator can achieve steadily scanning requirement with changing load and the displacement of piston rod is closed totarget curve.The position error presents a cosine curve, while the maximum value is 0.2 mm, which can fulfill the design accuracy requirement of hydraulic actuator for FAST.

5. Conclusions

The article presents some significant aspects of hydraulic actuators of main reflector system for FAST as following,

(1) Designed requirements had been discussed and analyzed in detail.

(2) Working principles of hydraulic actuator were explained thoroughly based on specifically demands of astronomical observation.

(3) Testing prototype and platform were developed. Experiment in closed-loop mode was carried out and the testing results were analyzed. The results indicate that the maximum position error was 0.2 mm, which can meet the accuracy requirements of hydraulic actuators for FAST.

Acknowledgments

Work partially supported by the National Natural Science Foundation of China (11273001, 11503047). Gratefully acknowledges for every researcher of FAST project. Meanwhile, thanks for the cooperating organization, URANUS Hydraulic Machinery Co. Ltd., China.

References

1. NAN Rendong. Five Hundred Meter Aperture Spherical Radio Telescope (FAST)[J]. Science in China: Series G: Physics Mechanics & Astronomy, 2006, 49(2):129-148.
2. WANG Qiming, WU Mingchang, ZHU Ming, Xue Jianxing, The development of the actuator prototypes for the active reflector of FAST, SPIE Astronomical Telescopes + Instrumentation 2012 Symposium in Amsterdam, 2012/7/1-2012/7/6, Amsterdam.
3. ZHU Ming, WANG Qiming, LEI Zheng, WU Mingchang, XUE Jianxing. Research for reliability of the actuator of FAST[A]. Reliability, Maintainability and Safety (ICRMS), 2014 International Conference on[C]. IEEE, 2014:623 - 627.

4. GAO Yuan, WANG Qiming, XUE Jianxing, ZHU Ming. Simulation and Experiment of FAST Hydraulic Actuator[J]. Chinese Hydraulic & Pneumatics, 2015, 12(11): 54-58.
5. Tianjin Uranus Hydraulic Machinery Co., Ltd.. Hydraulic Actuator: China, 103307059A[P]. 2013-09-18.

Modeling of the geometric error stack-up for multistage precision assembly process

Ge Wang, Yan Yan, Fa-Ping Zhang*, Ti-Guang Zhang and Shao-Wei Guo

School of Mechanical Engineering, Beijing Institute of Technology,
Beijing, State ZIP/Zone, China
**E-mail: 974395065@qq.com*

The purpose of this article is to describe the stack up of geometric error. By multiple rigid body kinematics, this article proposes a mathematical model to establish geometric errors accumulation along the multistage assembly process. Firstly, the assembly surface's geometric errors and the caused location change of the second part is given by using the small displacement spin (SDT) to represent each error and the homogeneous matrix to determine the assembly error of single stage. Then, based on the multi-body system theory, assembly errors accumulation among multistage assembly process is established, which provides theoretical basis for precision assembly accuracy analysis and control. Finally, an example is used to validate the established model.

Keywords: Assembly precision; geometric error; the error transformation model.

1. Introduction

In the process of precise assembly, geometric error of parts caused by surface topography is one of the important factors that affect the precision of relative position in the whole machine after assembly. As the development of the assembly process, geometric error of parts continuously propagates and accumulates, which further influences the assembly precision of the following assembly process. Researchers have done lots of research on multistage assembly error transformation model. The authors [1~2] established the state-space model of error propagation by mainly considering the influence of fixture error. For precision assembly, Mantripragada R et al [3] established state transition model. He Boxia et al [4] established the state space model of the whole machine's assembling process. The study above has not involved the effort of geometric errors caused by mating surfaces' topography. Therefore, this paper does some research about this problem.

2. The Representation of Geometric Error

The errors of this article are the key feature error and mating feature error. The key feature error refers to the deviation between the actual surface of the key feature and the ideal surface. Mating features error is decided by the surface of the parts which are mated together.

Each error is represented by the small displacement spin(SDT)which is a tiny displacement vector of a rigid body with six motion components, as shown in Eq. (1).

$$U = [d_x, d_y, d_z, \delta_x, \delta_y, \delta_z]^T \tag{1}$$

In Eq. (1), $d = (d_x, d_y, d_z)^T$ represents three small translation components of error. $\delta = (\delta_x, \delta_y, \delta_z)^T$ represents three small rotational components of error. Because three small rotation angles are small, we can have the following equivalents, $\sin\theta \approx \theta$ $\cos\theta \approx 1$ .Then error can be illustrated by the following homogeneous matrix, as in Eq. (2).

$$T = \begin{bmatrix} 1 & -\delta_z & \delta_y & d_x \\ \delta_z & 1 & -\delta_x & d_y \\ -\delta_y & \delta_x & 1 & d_z \\ 0 & 0 & 0 & 1 \end{bmatrix} \tag{2}$$

3. Modeling of the Geometric Error Transformation in Assembly Process

3.1. *Assembly error of two parts mating*

In Figure 1, as a result of the existence of the form errors of two parts, there will be a matching error after the two parts assembly and then the matching error will cause the posture change of surfaces. The posture change of the second part after the assembly relative to the assembly datum position of the first part can be represented as the posture change of the ideal mating feature surface 2 relative to the ideal mating feature surface 1. The pose change can be represented with the corresponding coordinate system transformation. For convenience, the same mating coordinate system M is established respectively on two mating feature surfaces. In this case, we establish ideal coordinate system O_{M1} , O_{M2} and the actual mating coordinate system $O_{M1'}$, $O_{M2'}$ in the two parts' mating feature surfaces. The three axial of all the coordinate systems are consistent.

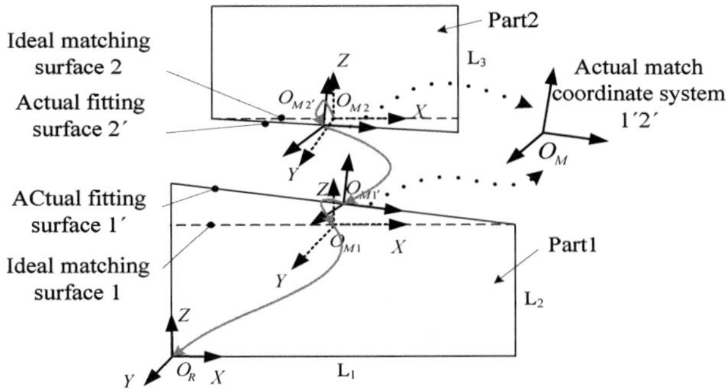

Fig. 1. The two parts assembly and coordinate transformation.

The form errors of the two mating feature surfaces 1' and 2' are respectively U_{M1} and U_{M2} ,The matching error of the two parts is U_{12}. Do the following definition: the posture of coordinate system $\{A\}$ relative to coordinate system $\{B\}$ is T_A^B. According to principle of the homogeneous coordinate transformation and differential transformation of the space object as well as methods of coordinate system mixed transformation and inverse conversion, we can get the following Eq. (3).

$$U_{12} = T_{M2}^{M1} = T_{M1'}^{M1} \cdot T_{M2'}^{M1'} \cdot T_{M2}^{M2'} \tag{3}$$

Ignoring the higher order events, the matrix of mating surface posture change is (4):

$$U_{12} = \begin{bmatrix} 1 & -\delta_{12z} & \delta_{12y} & d_{12x} \\ \delta_{12z} & 1 & -\delta_{12x} & d_{12y} \\ -\delta_{12y} & \delta_{12x} & 1 & d_{12z} \\ 0 & 0 & 0 & 1 \end{bmatrix} \tag{4}$$

$$\begin{cases} \delta_{12x} = \delta_{M1x} - \delta_{M2x} \\ \delta_{12y} = \delta_{M1y} - \delta_{M2y} \\ \delta_{12z} = \delta_{M1z} - \delta_{M2z} \\ d_{12x} = d_{M1x} - d_{M2x} + (\delta_{M1z} - \delta_{M2z}) d_{M2y} - (\delta_{M1y} - \delta_{M2y}) d_{M2z} \\ d_{12y} = d_{M1y} - d_{M2y} - (\delta_{M1z} - \delta_{M2z}) d_{M2x} + (\delta_{M1x} - \delta_{M2x}) d_{M2z} \\ d_{12z} = d_{M1z} - d_{M2z} + (\delta_{M1y} - \delta_{M2y}) d_{M2x} - (\delta_{M1x} - \delta_{M2x}) d_{M2y} \end{cases} \tag{5}$$

In Eq. (5), d_{12x} , d_{12y} , d_{12z} , δ_{12x} , δ_{12y} , δ_{12z} is respectively the three translational components and three rotational components of the matching error U_{12} . d_{M1x} , d_{M1y} , d_{M1z} , δ_{M1x} , δ_{M1y} , δ_{M1z} is the six components of U_{M1} . d_{M2x} , d_{M2y} , d_{M2z} , δ_{M2x} , δ_{M2y} , δ_{M2z} is the six components of U_{M2} .

3.2. *Multistage assembly error transformation modeling*

Form error and matching error transfer diagram is shown in the left of Figure 2, and the right part is the assembly topology structure. According to the assembly sequence, parts involved in the assembly are seen as a low-order body. Founded on the theory of multi-body system, we can establish the error transformation model from the benchmark assembly parts to the final assembly parts.

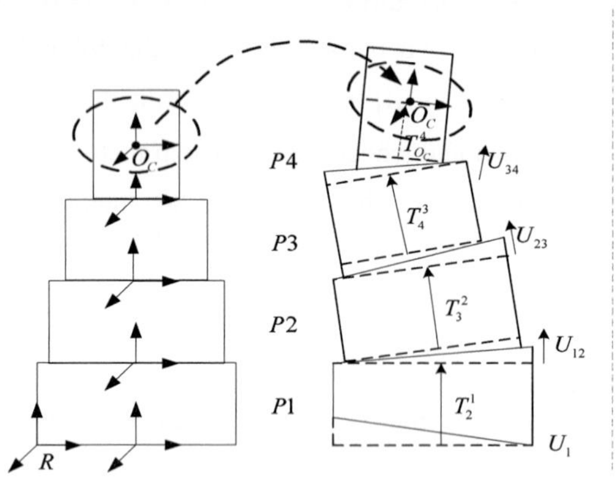

Fig. 2. Multistep assembly process error transfer diagram and assembly topology structure.

Generally, the pose transformation matrix of the system's total error after N parts assembly relatives to the reference coordinate system R is $T_{O_C}^R$:

$$T_{O_C}^R = T_1^R \cdot U_1 \cdot T_2^1 \cdot U_{12} \cdots \cdot U_{(N-1)N} \cdot T_{O_C}^N \tag{6}$$

The total error after the whole assembly precision products is *X*, the corresponding equivalent homogeneous transformation matrix of which is $T_{O_C}^{R}$:

$$T_{O_C}^{R} = \begin{bmatrix} 1 & -\delta_{Xz} & \delta_{Xy} & d_{Xx} \\ \delta_{Xz} & 1 & -\delta_{Xx} & d_{Xy} \\ -\delta_{Xy} & \delta_{Xx} & 1 & d_{Xz} \\ 0 & 0 & 0 & 1 \end{bmatrix} \quad (7)$$

In Eq. (7), $d_{Xx}, d_{Xy}, d_{Xz}, \delta_{Xx}, \delta_{Xy}, \delta_{Xz}$ is the six components of *X*, these six error components are functions of Parts' key features surfaces and mating surfaces, as in Eq. (8).

$$X = F(U_{F1}, \cdots, U_{Fm}, U_{M1}, \cdots, U_{Mn}) \quad (8)$$

$$U_k = \lfloor d_{kx}, d_{ky}, d_{kz}, \delta_{kx}, \delta_{ky}, \delta_{kz} \rfloor \quad (9)$$

In Eq. (8), X is system change in the total error vector, U_{Fi} is feature error vector, $i = 1, 2, \cdots, m$,where m is the number of features; U_{Mj} is mating error vector each assembly surface, $j = 1, 2, \cdots, n$,where n is the number of assembly mating surface. U_k is represented by six error components, as in Eq. (9), $k = 1, 2, \cdots, m + n$.

4. Example Verification

The parts assembly process of precision optical instrument is introduced to validate the model proposed in the paper. The instrument consists of three components: base, bracket and lens cone as shown in Fig. 3.

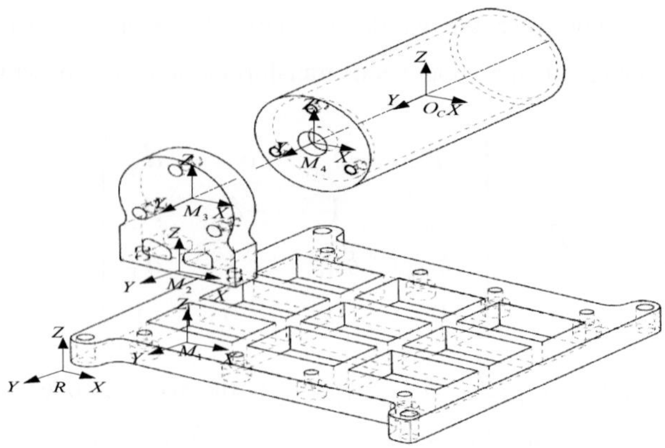

Fig. 3. An optical instrument parts assembly diagram.

Mating feature error can be obtained by measurement data, as shown in Table 1:

Table 1. Each mating feature error.

Mating Feature surface	Error in the form of differential motion vector
Long base surface	$U_{M1}=[0.0015,0.0012,0.0012,7.1174e-05,-8.207e-05,-2.4899e-06]^T$
Support base	$U_{M2}=[0.0862,-0.0165,-0.0184,3.0801e-04,0.0016,-3.066e-06]^T$
Support face	$U_{M3}=[4.5094e-05,-2.5764e-05,0.0029,3.4302e-05,6.0037e-05,8.4884e-19]^T$
The tube face	$U_{M4}=[0.0141,0.078,0.0094,4.6344e-04,-8.3703e-05,5.9085e-08]^T$

The transformation matrix between base, the bracket and lens cone are as follows:

$$T_{M1}^{R}=\begin{bmatrix}1&0&0&62\\0&1&0&-35\\0&0&1&20\\0&0&0&1\end{bmatrix}\quad T_{M3}^{M1}=\begin{bmatrix}1&0&0&0\\0&1&0&-10\\0&0&1&60\\0&0&0&1\end{bmatrix}\quad T_{O_C}^{M_4}=\begin{bmatrix}1&0&0&0\\0&1&0&-90\\0&0&1&0\\0&0&0&1\end{bmatrix} \tag{10}$$

Substituting each mating surface error and the transformation matrix into the error transfer model, we can get the total error X :

$$X=T_{O_C}^{R}=\begin{bmatrix}1&0&0&0\\0&1&0&0.0077\\0&0&1&0\\0&0&0&1\end{bmatrix} \tag{11}$$

5. Conclusion

(1) This article analyses pose change of components due to the mating surface geometric error, which is caused by surface topography when two parts assembly, and then determines the matching error of the single stage;

(2) Based on the theory of multi-body system method, this article established the error propagation model considering the geometric error of all assembly parts;

(3) This article establishes the error transfer model of an entity, and the error of the whole machine (after assembly process) is calculated, which provides an important theoretical basis for precise assembly error analysis.

Acknowledgment

This research is supported by the National Natural Science Foundation of China (NSFC Grant No. 51275049) and Defense 973 Project (613243). The authors gratefully acknowledge the facilities provided by the Industrial Engineering Laboratory (IEL) at the Beijing Institute of Technology.

References

1. Jin J H; Shi J J. State space modeling of sheet metal assembly for dimensional control[J]. ASME Transactions Journal of Manufacturing Science and Engineering, 1999,121(09):756-762.
2. Tian Zhaoqing, Lai Xinmin, Lin Zhongqin. State space model of deviation flow Transfer in multistage sheet assembling[J]. Journal of Mechanical Engineering, 2007, 43(2):202-209.
3. Mantripragada R, Whitney D E. Modeling and controlling variation propagation in mechanical assemblies using state transition models[J]. IEEE Transactions on Robotics and Automation, 1999, 15(1):124-140.
4. He Boxia, Zhang Zhisheng, Dai Min. Mechanical assembly process deviation transfer modeling theory [J]. Journal of Mechanical Engineering, 2008, 44(12):62-68.

Substation coordination technology applied in automatic voltage control system of UHV grid nearby

Wei-Ru Wang*, Xin-Jie Hao, Shu-Yong Song, Hui-Ping Zheng and Ying Qu
Power Grid Technology Center, Shanxi Electric Power Research Institute, Taiyuan, Shanxi, China
*E-mail: hdwangweiru@163.com

Xincong Shi
Dispatch and Control Center of Shanxi Electric Power Company, Taiyuan, Shanxi, China

UHV power flow is heavy in the south of Shanxi power grid, and voltage fluctuation is strenuous. Once the unreasonable reactive flow appeared between substations of near UHV grid, their discrete reactive power equipment can't effectively support the regional voltage, and therewith the depressed reactive power store of the area. In the Automatic Voltage Control (AVC) system of Shanxi power grid, the studied should be focused on the control of near ultra high voltage (UHV) gird. This paper proposed a coordinating technology between 500kV substations in the same voltage area, which use the "active and passive" coordination to realize the refine control of near UHV grid, thus to improve its ability of dynamic reactive reserve and voltage control. 500kV Haihui substation and Jincheng substation of UHV grid nearby are taken for instance to analysis. The results verify that the technology come into effect.

Keywords: Power system; reactive power; voltage; substation coordination technology; ultra high voltage (UHV).

1. Introduction

In the AVC system of Shanxi power grid, we must focus on UHV near region power grid voltage control in the south. Since the power flow is heavy of UHV transmission line, voltage fluctuates remarkably. Only the fast reactive power response from turbine nearby can suppress the voltage fluctuation. The 500kV power plant near zone should have preferable spinning reserve of reactive power to keep sufficient margin. Once the reactive power flow is unreasonable between the substations nearby, discrete reactive power equipment can't effectively support the regional voltage, thus reducing the reactive power spinning reserve near region. Therefore, achieving refine coordination control between UHV grid nearby stations through the study on substation coordination control technology, to make the low capacitance and resistance of their own coordinate in the time sequence control, which is an important means to improve the dynamic reactive power reserve and voltage control capability of UHV grid nearby.

2. Substation Coordination Technology

2.1. *Substation coordination technology concept*

In the three voltage control system based on the soft partition[1-2], the substation coordination is in the level of the secondary control, which is as shown in Figure 1.

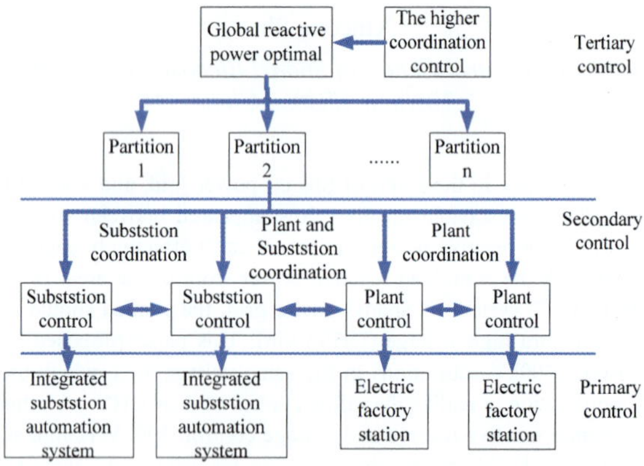

Fig. 1. Station coordination control level.

The overall optimization control objective is given by the optimal power flow from global reactive power optimization in the tertiary control, the control objective is realized by the secondary control. The secondary control is partition control, considering comprehensively all control means and including the power plant and each voltage grade substation. The comprehensive coordination control between plant and plant, plant and substation, substation and substation, make the bus voltage within the partition reach global optimization goal of the tertiary control.

Similar to the station coordinate and plant coordination, because of the different control ability and response characteristics of the substation in the secondary control, it needs coordination in the time sequence control between substations, which is the significance of substation coordinate in the secondary control. Substation coordination consider with the reactive power coordination between different substations in the same voltage partition, belonging to the refine adjustment. Overall, the tertiary control ensures the target voltage optimal, and the coordination control in the secondary ensures that the process to reach the target voltage is optimal.

2.2. *Proposing of Substation coordination technology*

In Shanxi power grid, the major consideration of station coordination technology is the sequential coordination control between the low voltage capacitor and low voltage reactor of 500kV substation, as shown in Figure 2.

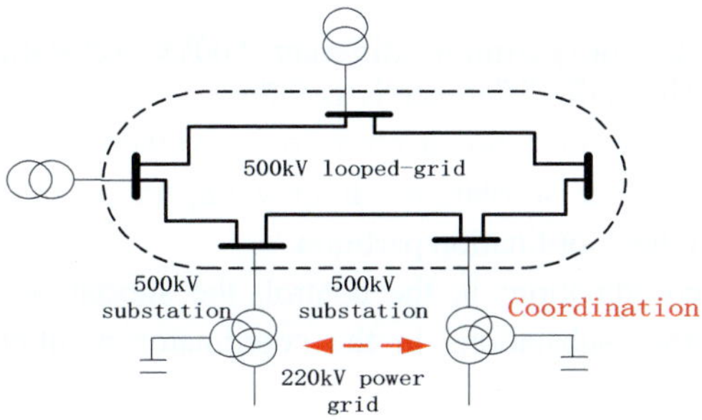

Fig. 2. 500kV looped-grid diagram of the same voltage partition.

In actual operation, the ability to control low voltage reactive power equipment is different from different 500kV substations. The main causes of the different control ability are as follows:

(1) Some substations have put AVC closed-loop control into operation , some not. The substation that has not put AVC closed-loop control into operation depends on manual control, which is difficult to coordinate with the automatic control equipment.

(2) Substations' performance on the response of the control instruction execution is different. Due to reasons such as the communication failure and equipment difference, the control success rate of some substations is low. These need attempts to control successfully.

Due to differences in the control ability, if only to track their own bus voltage in the substation control, it may cause two electric distance closer substations the one putting into capacitor, the other putting into reactor, which will lead to the unreasonable flow of reactive power in a region.

3. Conducting of Substation Coordination Technology

3.1. *Formation of the coordinator partition*

There are two kinds of method of the coordinator partition formation. The one is automatic forming: according to the reactive voltage sensitivity between bus voltage class, the coordinator partition of each voltage grade is formed [3-4]. The other one is artificially specified: according to the requirements of control and regulation, a number of substations form a coordinator partition manually.

According to the selected voltage partition, the 500kV substations within the same voltage partition form the coordinator partition artificially specified in Shanxi power grid.

3.2. *Division of the coordinator roles*

Within the coordinator partition, different 500kV substation has different coordinator role. The main difference is as follow:

(1) Controlled: the substation itself is not controllable, or its control strategy is not affected by other substations. It only impact on the control of other substations within the coordination partition.

(2) Passive coordination: in the control, the substation is only with the voltage of the other substations in the coordinator partition as the control constraints.

(3) Initiative coordination: the substation is not only with the voltage of the other substations in the coordinator partition as the control constraints, bus also it generates its own control strategy to eliminate when the voltage of other substations is out of limit.

(4) Reactive power coordination: for multiple substations specified as reactive power coordination, the system should avoid the reactive power unreasonable flow between the substations in control. That is to say it doesn't allow appear the situation that one substation putting into capacitor, the other putting into reactor.

The role of each of the 500kV substation can be overlapped. It can be "controlled" + "passive coordination", or "controlled" + "passive coordination" + "initiative coordination", etc. "Initiative coordination" will generally consider "passive coordination" at the same time.

3.3. *The overall data flow*

After considering conducting substation coordination, the overall control process of the 500kV substation as shown in the figure below.

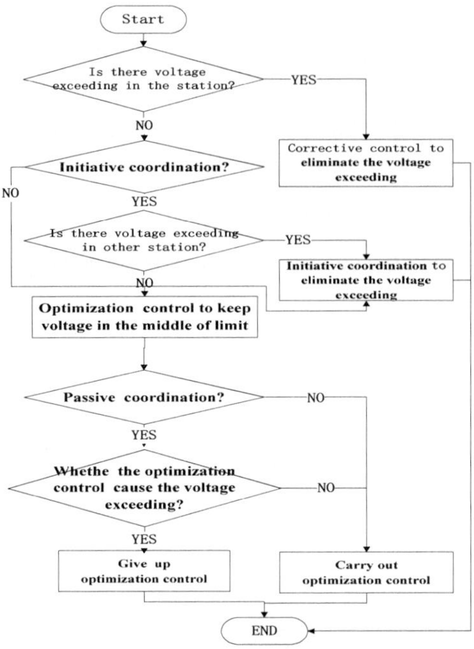

Fig. 3. The overall control process of the substation coordination.

The control process is stated as follows:

The corrective control for eliminating the over-limit voltage within substation is considered firstly. When there is voltage exceeding in the substation, it will generate the control strategy to eliminate the voltage exceeding. Considering that the control target for 500kV substation is mainly 220kV bus voltage, so it is considered the upper and lower voltage operation for 220kV voltage class, the other voltage class main consideration on the upper and lower of the safe operation. When the substation has no voltage exceeding, according to the requirement of substation coordination, considering voltage corrective control is considered within coordination area.

When the substation has been designated as "initiative coordination", checking whether appearing the voltage exceeding in the coordination area, if appear, “active coordination” control process is started. The goal is according to control the reactive power equipment to eliminate the voltage exceeding of adjacent 500kV substation in the coordinate area.

When the coordination bus voltage of this substation and coordination region is qualified, the optimization control strategy is generated.

If the substation is set as "passive coordination", the optimal control strategy generated is checked.

By check, it is verified whether the optimization control strategy of the substation lead to the other bus voltage in the coordination region out of limit. If it may cause the voltage exceeding, the optimization control strategy is given up.

3.4. *Conducting of initiative coordination*

The flow chart of initiative coordination method is shown as Figure 4.

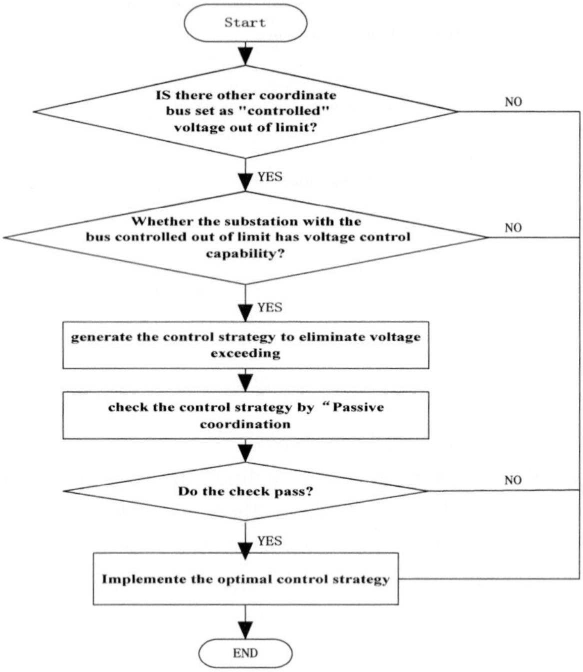

Fig. 4. The flow chart of initiative coordination method.

The control process is stated as follows:

It will be checked that whether there is other coordinate bus set of "controlled" out of limit in the coordination zone. If there is no out of limit, end the process. Otherwise it will continue to execute.

If there is voltage exceeding, it will check whether the substation with the bus out of limit has voltage control capability. If it still has control capability, end the process. Otherwise it will continue to execute.

According to the operation and sensitivity value of reactive power equipment, it will generate the control strategy to eliminate voltage exceeding of coordination bus. The control strategy cannot cause local bus voltage out of limit in the substation.

The control strategy will be checked passively. It means that the control strategy can not cause bus voltage out of limit in reverse in other coordination zone. If check cannot pass, end the process.

After the check passing through, the strategy will be implemented.

3.5. *Conducting of passive coordination*

The flow chart of passive coordination method is shown as Figure 5.

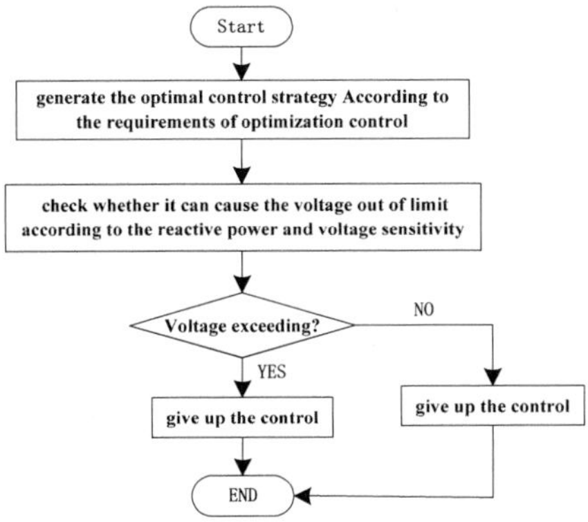

Fig. 5. The flow chart of passive coordination method.

The above process is stated as follows:

According to the requirements of optimization control, the optimal control strategy is generated. The control target is generally the middle value of the largest voltage safety margin[5-6].

According to the reactive power and voltage sensitivity in the coordination area, forecasting the coordination bus voltage value, it is checked that whether it can cause the voltage out of limit. If it may cause the voltage exceeding, give up the control. Otherwise the optimization control will be executed.

4. Control Effect of Substation Coordination Technology

This chapter will take 500kV Haihui and Jincheng substation nearby the UHV grid as an example, to display the control effect of the Substation coordination technology in the coordination region.

500kV Haihui and Jincheng substation both belong to Shanxi southern grid nearby UHV. The two substations is connected directly, electric connection closely.

Figure 6 and Figure 7 respectively give the primary electric wiring diagram of Haihui and Jincheng substation. It can be seen, the two substation are configured the capacitor and reactor on the LV side of main transformer. If it only considers the control objectives of high side bus voltage, and does not consider the sequential coordination control between the two substation, due to the different control responsiveness with different substation, it may cause one substation putting into capacitor, the other side putting into reactor, which will lead to the unreasonable flow of reactive power in a region and reduce the voltage control capability and reactive power reserve rotation of the nearby UHV grid.

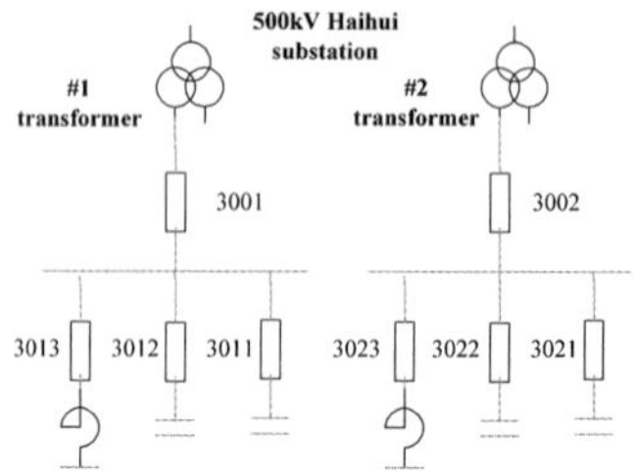

Fig. 6. The primary electric wiring diagram of Haihui substation.

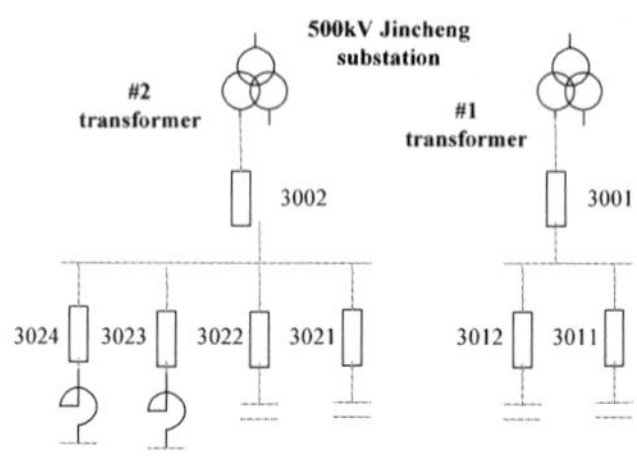

Fig. 7. The primary electric wiring diagram of Jincheng substation.

In the secondary zone control, the Haihui and jincheng substation will be composed of a substation coordination area of “reactive power coordination”. A number of capacitor and reactor switching order will be coordinated to avoid the unreasonable flow of reactive power.

It is given the low voltage reactive power equipment control records of Haihui substation on April 1, 2012 in Table 1. Table 2 gives the low voltage reactive power equipment control records of Jincheng substation on April 1, 2012.

Table 1. The reactive power equipment control records of Haihui substation.

Time	information content	Control type
6:25	#11 reactor of Haihui substation controlled successfully,from operating to hot standby	exit reactor
6:55	#21 reactor of Haihui substation controlled successfully,from operating to hot standby	exit reactor
23:05	#21 reactor of Haihui substation controlled successfully,from hot standby to operating	put in reactor
23:35	#11 reactor of Haihui substation controlled successfully,from hot standby to operating	put in reactor

Table 2. The reactive power equipment control records of Jincheng substation.

Time	information content	Control type
8:25	#6 capacitor of Jincheng substation controlled successfully,from hot standby to operating	put in capacitor
9:25	#1 capacitor of Jincheng substation controlled successfully,from hot standby to operating	put in capacitor
15:35	#1 capacitor of Jincheng substation controlled successfully,from operating to hot standby	exit capacitor
22:35	#6 capacitor of Jincheng substation controlled successfully,from operating to hot standby	exit capacitor

As you can see in Table 1 and Table 2, in the morning rush-hour, after the two reactor of Haihui substation exiting in turn (after 7:00am), the capacitor of Jincheng substation has been input successively according to the requirements of voltage optimization(after 8:25am). In the off-peak hours at night, after the capacitor of Jincheng substation exiting (after 10:40pm), the reactor of Jincheng substation has been input successively.

5. Conclusion

Substation coordination as the secondary coordination control in the process of automatic voltage control, it considers fully the reactive power coordinated control between substations with different control performance. On the basis of achieving voltage control, it control the low voltage reactive power equipment of substation within coordination area on timing sequence, which avoid the unreasonable reactive power flow, and greatly improve the dynamic reactive power reserve of the UHV regional grid.

References

1. SUN Hong-bin, ZHANG Bo-ming, GUO Qing-lai, et al. Design for Global Optimal Voltage Control System Based on Soft Identification of Sec ondary Control Zones. Automation of Electric Power Systems,2003,27 (8) 83-87.
2. SUN Hong-bin, GUO Qing-lai, ZHANG Bo-ming, et al. Automatic Voltage Control Scheme for Large-scale Power Networks. Power System Technology, 2006, 30(S2): 13-18.
3. GUO Qing-lai, SUN Hong-bin, ZHANG Bo-ming, et al. Research on Coordinated Voltage Control of UHV Power Grids. Proceedings of the CSEE, 2009, 29(22): 30-34.
4. GUO Qing-lai, SUN Hong-bin, ZHANG Bo-ming, et al. Study on Coordinated Secondary Voltage Control. Automation of Electric Power Systems, 2005, 29(23): 19-24.
5. CHEN Jianglan, TANG Weidong, XIAO Xiaogang, et al. Coordinated voltage control for Central China Power Grid. Electric Power Automation Equipment , 2011, 31(8): 47-515.
6. DAI Yan. Study and comparison of control strategies for automatic voltage control (AVC) systems. East China Electric Power, 2008, 36(1): 92-95.

Implementation of voltage inverse control scheme for micro grid energy storage inverter based on FPGA

Li-Bin Yang
Key Laboratory of Grid-Connected Photovoltaic Technology,
Electric Power Research Institute of Qinghai Power Grid Corporation,
Xining, 810008, China
E-mail: 18797160393@163.com

Wei-Liang Liu†, Wen-Ying Chen, Yin-Song Wang and Hong-Bo Zhou
Automation Department, North China Electric Power University,
Baoding, 071003, China
†*E-mail: lwlfengzhiying@163.com*

When micro grid is in the state of off grid operation, the voltage stability is maintained by energy storage inverter. Due to the small capacity of micro grid, energy storage level and load changes can easily cause the deterioration of the waveform quality of the system voltage. In order to solve this problem, a new voltage inverse control scheme of energy storage inverter is proposed based on neural network. For the purpose of real-time performance, the design of the neural network special arithmetic unit is implemented using FPGA, and comparison of the implementation schemes for excitation function is carried out. Simulation results show that the proposed scheme is feasible and effective.

Keywords: Energy storage inverter; micro grid; FPGA; inverse control.

1. Introduction

In recent years, the micro grid composed of photovoltaic power generation, wind power and other new energy power generation, has become a research hotspot[1]. Micro grid can work in grid connected mode and off grid operation mode. In off grid operation mode, the system voltage is generally provided by the energy storage inverter. Because of the smaller capacity, the change of the energy storage level and load can affect the waveform quality of the system voltage, and even lead to instability of the system. Conventional control methods like Single-loop control, Dual-loop control[2], Deadbeat control[3] and Repetitive control[4] are not satisfactory for this problem. With the continuous improvement of the quality requirements of the user, this problem need to be solved urgently. In this paper, an inverse control scheme based on neural network is proposed and implemented by FPGA.

2. Structure of Neural Network Inverse Model

According to the working principle of energy storage inverter, the main factors affecting the output voltage of u_o include the port voltage of energy storage unit u_{dc}, load current i_o, filter capacitor voltage u_c and duty ration of the PWM control signal period d, which should be considered when building an inverse model. In the control cycle k, $d(k)$ is the duty ration of the PWM control signal period, sample the u_{dc}, u_c, i_o, u_o when this cycle is coming to an end, and remember them as $u_{dc}(k)$, $u_c(k)$, $i_o(k)$, $u_o(k)$. In order to reflect the dynamic characteristics of the system, seven signals in total are considered to contrast the inverse model, the u_o, i_o in the current cycle and previous cycle, the u_{dc}, u_c in the previous cycle, the d in the previous cycle, namely $u_o(k)$, $i_o(k)$, $u_o(k\text{-}1)$, $i_o(k\text{-}1)$, $u_{dc}(k\text{-}1)$, $u_c(k\text{-}1)$, $d(k\text{-}1)$. Using these seven signals as inputs, a BP network inverse model is constructed. Select the number of hidden layer neurons as 9, take the sigmoid function as the hidden layer neuron transfer function, and take the purline function as the output layer neuron transfer function. The structure of the BP neural network inverse model is shown in Figure 1.

3. Structure of Inverter Voltage Inverse Control System

3.1. *Structure of inverse control system*

The neural network inverse control system of energy storage inverter voltage is mainly composed of three parts, such as the inverter circuit, single chip computer and FPGA operation unit, as shown in Figure 2.

Single chip microcomputer is the main controller of the neural network control system for micro grid energy storage inverter, and it is responsible for data acquisition. Using its 4 A/D acquisition port, the single chip microcomputer can collect the port voltage of the energy storage unit $u_{dc}(k)$, filter capacitor voltage $u_c(k)$, load current $i_o(k)$, inverter output voltage $u_o(k)$, the accuracy is ten bits, and then send them to the FPGA with $u_o{}^*(k+1)$, $\hat{i}_o(k+1)$, $d(k)$. The FPGA controller is a special operation unit for the neural network inverse model, it receives the network input signals sent by single chip microcomputer, calculates the results and send them back to the microcontroller. Microcomputer will execute the PWM control signal to the inverter circuit according to the results calculated by the FPGA.

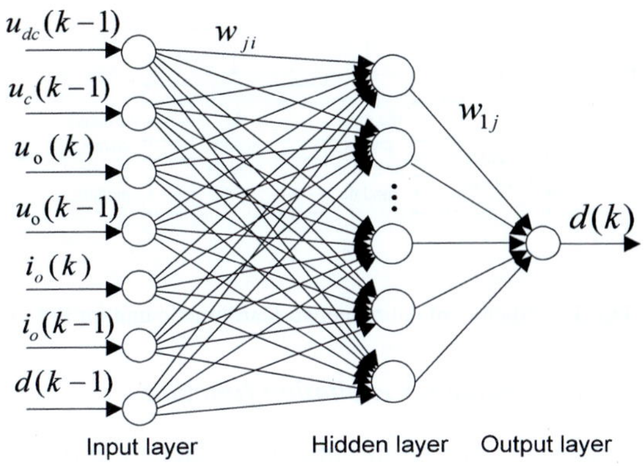

Fig. 1. Topology of BP neural network inverse model.

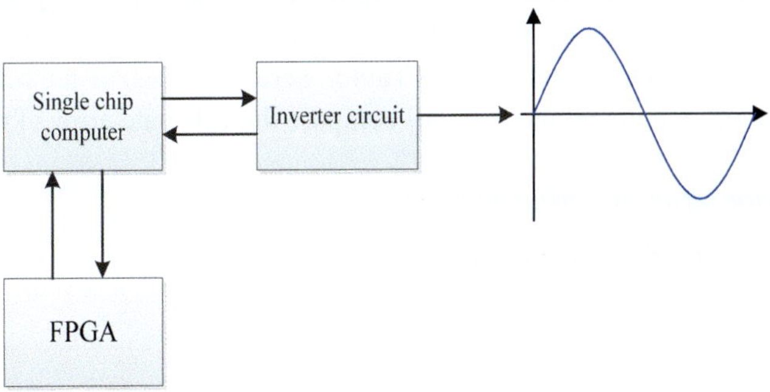

Fig. 2. Novel Control method of voltage.

3.2. *Structure of neural network arithmetic unit for FPGA*

Structure, weights and thresholds of the BP neural network are determined based on a large number of training samples. With the continuous collection of samples, the BP neural network should be trained again to improve the accuracy of the model. Considering the structure, weights and thresholds of the network will be changed after each new training, the design of the neural network control system adopts a modular design ideas. The overall structure is composed of six modules: acquisition module, normalized module, input matrix operation module, neuron module, output matrix operation module and anti normalization module. Its overall structure is shown in Figure 3.

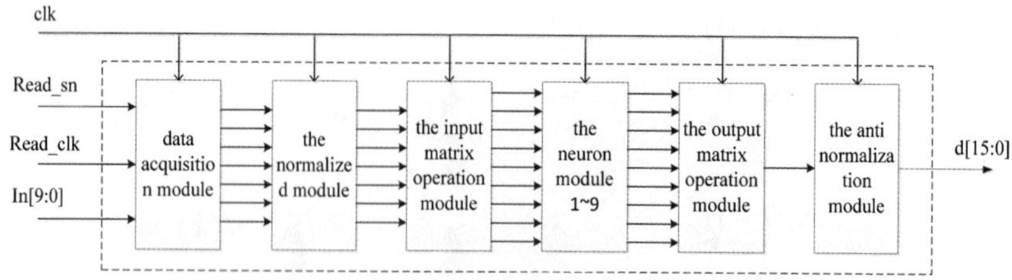

Fig. 3. Structure of neural network arithmetic unit for FPGA.

3.3. *Implementation scheme for excitation function*

The excitation function is one of the important factors to determine the overall performance of artificial neural networks. In this paper, the Sigmoid function is chosen as the hidden layer of the neural network inverse model. How to achieve the high accuracy of the system has a significant impact on the quality of the control system. At present, the realization scheme of the excitation function mainly has two kinds of method, the Taylor Series Expansion method and the Fitting Approximation method, we will compare the calculate error of them.

3.3.1. *Taylor series expansion method*

The expression for the Sigmoid function is

$$y = \frac{1}{1+e^{-x}} = \frac{e^x}{e^x+1} \tag{1}$$

With Taylor series expansion e^x

$$e^x = \lim_{N\to\infty}\sum_{n=0}^{N}\frac{x^n}{n!} = 1+x+\frac{x^2}{2!}+\frac{x^3}{3!}+\ldots\frac{x^N}{N!} \tag{2}$$

Take N=9, namely expanse e^x function to nine order, and the calculation error of the Sigmoid function approximation could be evaluated by MATLAB. The result is shown in Figure 4. It is known that the maximum absolute error between [-10, 10] is about $2.49*10^{-4}$, and the accuracy is satisfied.

3.3.2. *Least Squares fitting approximation method*

The Least Square method often use a basis function to realize curve fitting, such as polynomial, exponential function, trigonometric function and so on. Considering take FPGA to achieve the above basis functions need to consume a lot of logic and a long time, this paper selects linear function as the basis function, and piecewise fit the Sigmoid function. The slope and intercept of each fitted section are stored in the ROM table to call, which can reduce the number of operations per cycle and improve the utilization of the FPGA storage unit. Fit error of the Sigmoid function with piecewise linear approximation is evaluated by MATLAB simulation, as shown in Figure 5.

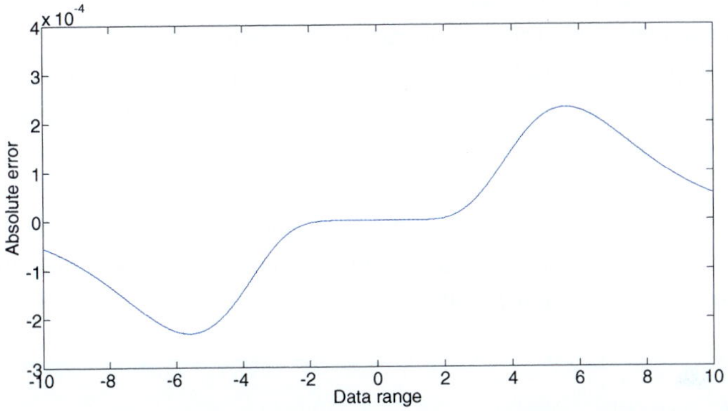

Fig. 4. Calculation Error of nine order expansion.

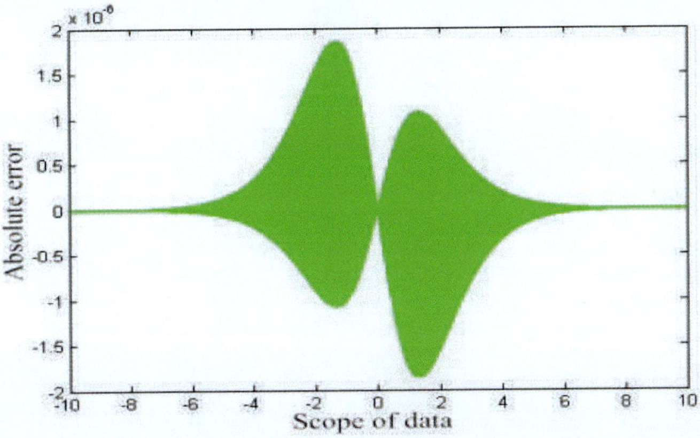

Fig. 5. Fit Error of linear approximation.

It is known the maximum absolute error is about $1.87*10^{-6}$ between [-10,10] for piecewise linear approximation of Sigmoid function, which is greatly reduced compared with the nine order Taylor series expansion.

4. Simulation of Neural Network Inverse Model

A neural network inverse model of storage inverter is built in Quartus using Verilog HDL and its performance is tested in Modelsim-altera simulation software. EP4CE115F type FPGA, made in Altera company, is adopted in the experiment, which has 114,480 logic gates, 3.9M of storage space, 529 I/O ports, and its crystal is 50Mhz.

Time sequence diagram of operation for Neural network inverse model is shown in Figure 6, at each clock rising edge a data is collected, and it need 7 machine cycles for acquisition of the 7 sets of input data. After another machine cycle, the calculation of the input matrix is accomplished; and the calculation result of the duty cycle is obtained at the next machine cycle. The whole computing process needs 10 machine cycles, and each machine cycle is 20ns, so the whole calculation process takes 200ns, which is far lower than the energy storage inverter control period 50us.

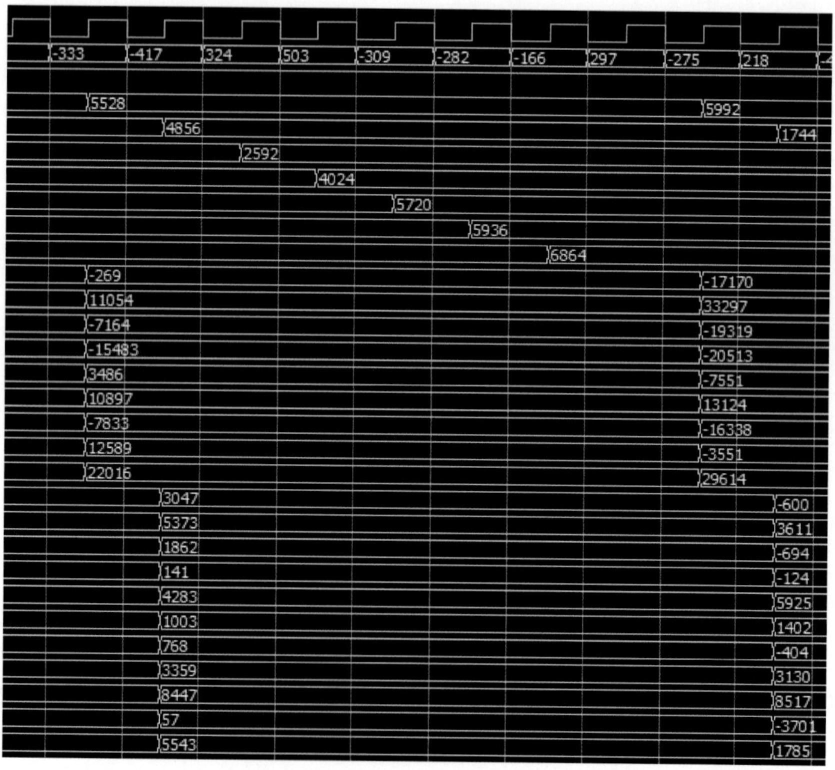

Fig. 6. Time sequence diagram of inverse model of neural network.

In order to verify the accuracy of the inverse model calculated by FPGA, its calculated results are compared with the calculation results of Matlab, which can be considered as completely accurate, and the error curve for the calculated results of FPGA is shown in Figure 7. In most cases, the absolute error is less than 0.005, the maximum absolute error is no more than 0.015, which indicates that the accuracy of the inverse model calculated by FPGA is fully satisfied.

In the Quartus II environment, the inverse model is compiled and the results are shown in Figure 8. It is known that the whole inverse model consumes 37674 logic gates, accounting for 33% of the total logic gate; consumes 28 I/O ports, accounting for 5% of the total I/O ports; consumes memory 518.6Kb, accounting for 13% of the total memory. The resource of EP4CE115F FPGA could easily satisfy the calculation demand of the inverse model.

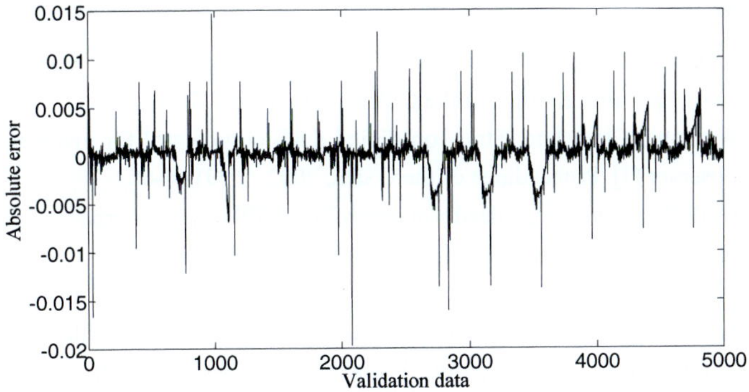

Fig. 7. Absolute error for FPGA.

Flow Summary	
Flow Status	Successful - Mon Jun 15 13:55:39 2015
Quartus II Version	11.0 Build 157 04/27/2011 SJ Full Version
Revision Name	bp_neural_network1
Top-level Entity Name	bp_neural_network1
Family	Cyclone IV E
Device	EP4CE115F29C8
Timing Models	Final
Total logic elements	37,674 / 114,480 (33 %)
Total combinational functions	37,617 / 114,480 (33 %)
Dedicated logic registers	106 / 114,480 (< 1 %)
Total registers	106
Total pins	28 / 529 (5 %)
Total virtual pins	0
Total memory bits	518,646 / 3,981,312 (13 %)
Embedded Multiplier 9-bit elements	0 / 532 (0 %)
Total PLLs	0 / 4 (0 %)

Fig. 8. Compiled results of inverse model.

5. Conclusions

In this paper, a neural network based voltage inverse control scheme for energy storage inverter is proposed. In order to guarantee the real-time performance, the design of the neural network special operation is carried out by using FPGA technology. Simulation experimental results show that the proposed scheme is feasible and effective.

Acknowledgment

This work is supported by Science and Technology Project of State Grid Corporation of China (2014-Z-Y34A). The work of this paper is supported by Science and Technology Project of State Grid Corporation of China (2014-Z-Y34A).

References

1. Yang Xinfa, Su Jian, Lv Zhipeng, Liu Haitao, Li Rui. Overview on Micro-grid Technology [J]. Proceedings of the CSEE, 2014, (1):57-70.
2. Guo Weinong, Chen Jian. Study on digital dual-loop control for inverters based on state-observer [J]. Proceedings of the CSEE, 2002, 22(9):64-68.
3. Mattavelli P. An improved deadbeat control for UPS using disturbance observers[J]. IEEE Transactions on Industrial Electronics, 2005, 52(1):206-212.
4. Liu Xinmin, Zou Xudong, Kang Yong, Liu Bin. An Inverter Based on State-Feedback Integral Control and Repetitive Control Technology With State-Observer [J]. Transactions of China Electrotechnical Society, 2007, 22(1):91-95.

Design of control system for duck shed in paddy field

Ai-Ju Kong[†], Ming-Jin Xin, Cui-Hong Liu, Yu-Qiu Song,
Li-Yan Wu, Hong-Guang Cui and Wen-Tao Ren[‡]
College of Engineering, Shenyang Agricultural University,
Shenyang, 110886, China
[†]*E-mail: kaj9818@126.com*
[‡]*E-mail :rwt@syau.edu.cn*

Automatic control system was developed for duck shed in paddy field. According to the theory of modular design, the control system was divided into automatic door control mould, feeding control mould, watering supply control mould, voice training control mould, human-machine interaction module control, communication module control and power supplying module control. Automatic door, feeding and voice training module were activated in accordance with the time of sunrise and sunset. Information of liquid level of water tank, limit location of automatic door and situation of ducks passing the door were collected and read. The microcontroller sent orders to the pump and the door, so that the control functions were implemented to satisfy the need of ducks in paddy field. Field experiment showed there was no significant difference for the average weight at two kinds of feeding method, however, automatic feeding could save the labor cost of ￥4200 /667m^2, compared with manual feeding. Automatic control system would promote the development of rice-duck eco-farming technology.

Keywords: Automatic control system; rice-duck eco-farming; weight of ducks; economic benefit.

1. Introduction

Rice-duck eco-farming is an ecological farming model which has both good economic and ecological benefits[1-2]. Based on the property that the rice and ducks can grow together, ducks of 10days age were released into paddy field where rice seedlings have been transplanted for about 10 days. Rice-duck eco-farming can control the weeds and pests biologically[3-4]. The intertillage actions of ducks increased oxygen content in paddy field[5-7], and ducks feces enhanced soil fertility[8-9], in addition, ducks actions of touching, pecking and trampling activated the growth of rice [10-12]. Without application of pesticides, fertilizers and herbicides, ecological environment in the field will cycle positively.

The problem which affects application of rice-duck eco-farming is the management of ducks in paddy field. Because of the weather in the north, ducks are released into paddy field to eat weeds and pests during the day, but the ducks are driven to the shed for low temperature and they need to be supplied feed in the evening[13]. At present there are two forms for ducks to spend time in the evening. One method is to drive the ducks from paddy field to home where ducks are supplied feed by human labor and ducks are drove to paddy field in next morning. The other method is to build a simple shed in paddy field and supply feed by human labor. However, the simple shed is open, which can't guarantee the safety of ducks. The weasels, rats, foxes, snakes and other predators will threaten the life of ducks. The traditional rice-duck eco-farming wastes time and human labor, and working condition is poor, so the development of rice-duck eco-farming is limited. Automatic control system for rice-duck eco-farming has been developed, which can realize automatic detection of ducks number in the shed, automatic door opening and closing, automatic feeding, automatic watering and automatic music playing to call ducks back to the shed, so that field management of rice-duck eco-farming can realize automation.

2. Overall Design of Control System

Control system of duck shed in paddy field realized automatic management without manual intervenes. The main functions were as follows:

(1) Setting function: initial values, such as starting time, age of ducks and type of sound, were set.

(2) Timing control: times for door opening and closing, feeding and vocal training were controlled.

(3) Detecting function: information of liquid level of water tank, limit location of automatic door and ducks passing the door were detected.

(4) Real-time control: control commands based on detection result were sent to operating mechanism to accomplish the control action.

(5) Manual control: manual control cooperated with automatic control.

(6) Display function: current operating instruction was displayed.

The system consists of automatic testing system, automatic control system and man-machine interface. Control system schematic was shown as figure 1. To meet the demand of duck shed, the control system was composed of upper computer and lower computer. As upper computer, PC processed the image data of duck shed and analysis whether all ducks had entered the shed. The lower computer was microcontroller, which used for data measurement, data collection and data management. They communicated with each other and microcontroller adjusted the action of door according to the treatment result of image processing. The system detected information of liquid level of watering device, limit location of automatic door and situation of ducks passing the door. Operating mechanism was actuated based on detecting result. All parameters were inputted by keyboard. The system switched between automatic control and manual control to meet the requirement of the measurement and control task.

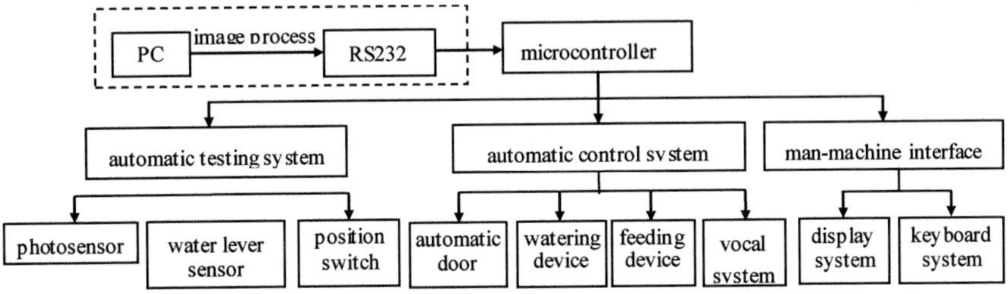

Fig. 1. Duck Shed environment monitoring and control system schematic.

3. Hardware Design of Control System

Block diagram of hardware structure was shown as figure 2. According to the control function, the system was divided into automatic door control mould, feeding control mould, water supplying control mould, vocal control mould, human-machine interaction module control, communication module control and power supplying module control. The auxiliary peripheral components were composed of LED display, keyboard and switch, etc to complete control function. The microcontroller selected the type of STC15F2K40S2. NPN photo

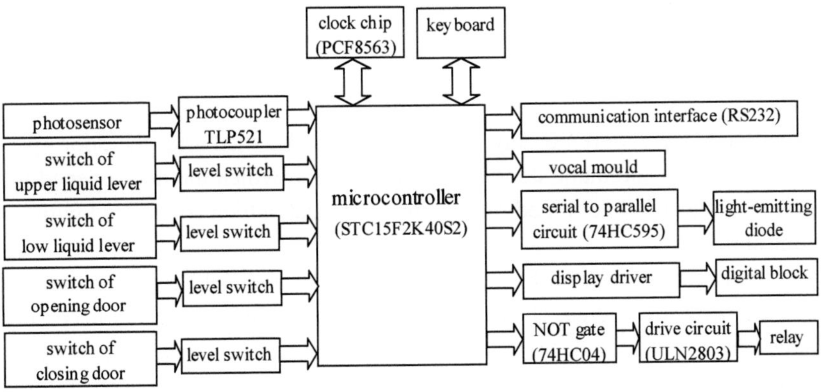

Fig. 2. Block diagram of hardware structure of the system.

sensor, position switch, double float level sensor were selected as sensors of the system. Digital signal processed by level switch was sent to the microcontroller.

4. Software Design of Control System

Software of the duck shed was designed under the method of modularization. The function moulds of vocal, feeding, watering and automatic door were defined as four subprograms and there were program function corresponding with the functions. Main program were operated as follows: as the system was turned on, the microcontroller initialized and monitored every mould. If the program was needed to interrupt, the interrupt order and new program were sent by keyboard and the system continued new task as new program indicated. The parameters of the system would be checked, set and modified. If the parameters needn't to be modified, the system controlled the various components of duck shed according to the detected data and the operation instructions showed through digital display. The main program flow chart was shown as figure 3. In the morning, the automatic door opened to release ducks into paddy field. Vocal control mould initiated before sunset, the automatic door opened again and the feeding system started at the same time. After a certain time interval, the automatic door closed to protect ducks from danger. According to the results of image processing processed by PC, the system judged whether all ducks had return and determined whether the door opened again. All time interval in the program could be adjusted in the range of 0~99min.

4.1. *Automatic door control mould*

According to the seasonal characteristics of Liaoyang region, Liaoning province, the time of sunrise and sunset from mid-June to mid-August were inputted into the program. Based on the inputted time, automatic door adjusted time to open and close in the morning and evening. The time at which the door opened in the morning was determined by the time of sunrise. Judging from previous experience, there were not any ducks staying in the shed after the door opened, at this time the automatic door closed. The time at which the door opened in the evening was also determined by the time of sunset, which cooperated with vocal control mould.

In the evening, affected by the nature of ducks and field environment, ducks could not return to duck shed in a short period of time, so it was necessary to judge the number of ducks back to the shed. Based on image processing, the number of returned ducks was analyzed and sent to microcomputer through communication module. Compared with the number of total ducks which was inputted through keyboard, microcomputer analysis whether all ducks had returned to the shed. If the answer was "yes", the automatic door would maintain close, otherwise, the automatic door would open again to give chance for ducks to return to duck shed.

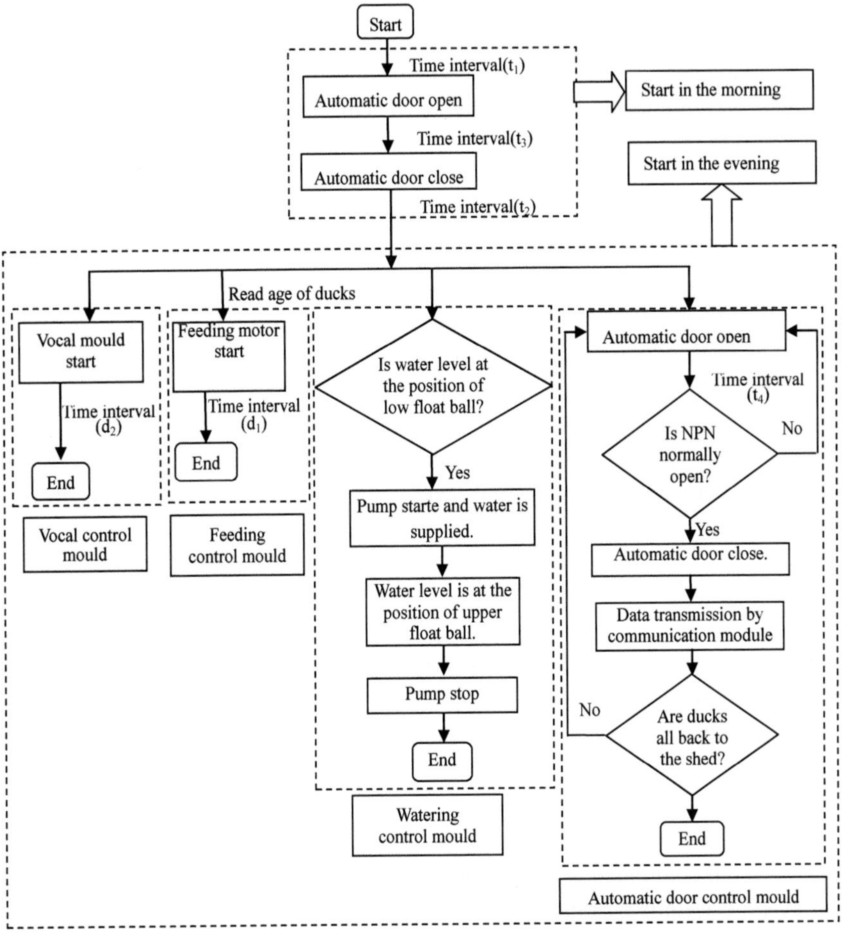

Fig. 3. Main program flow chart of the system.

4.2. *Feeding control mould*

The feeding system worked only one time in every evening during the day a day and the subprogram completed the control action based on the time rule. With the increase of ducks age, the required feed also gradually increased with ducks' age. The relation between the ducks age and required feed was inputted into the subprogram, so that the operating time of feeding motor adjusted automatically. The feed supplied quantitatively, which realized scientific feeding.

4.3. *Vocal control mould*

Vocal control mould started when the automatic door opened in the evening and cooperated with the automatic door. Vocal control mould generated various kind of sound to summon ducks back the shed. Ducks could sense the vocal signal and return to the duck shed gradually.

4.4. *Watering control mould*

The time of ducks drinking water was random, especially at night. Watering control should not be limited by fixed time and could meet the requirement of water supplying. The start of watering control mould didn't influenced by the time, but it was determined by the real water level of the tank. The real water level was detected by double floating ball liquid level control switch which had been selected 304 type to use. The watering control mould prevented the pump from working if the bottom level didn't arrived. The watering control mould realized water supplying at bottom level and prevented water from overflowing the tank.

5. Field Experiment and Economic Benefit Analysis

Experiments were carried out in Haojia village, Yongling town of Xinbin county in Liaoning province. Field experiments had been carried out for 35 days from July1, 2015. The area of paddy field was 667m^2 and water level of paddy field maintained about 5cm. 25 local sheldrakes of four weeks age were selected for experiment. Compared with manual feeding, the experiment studied the effect of automatic feeding on ducks weight. A simple duck shed was built in paddy field for manual feeding and the tasks of feeding ducks and driving duck back to the shed were all completed with human labor. The ducks weight was measured and the average weight was calculated every week. The experiment repeated three times and carried out for 5 weeks. The range of electronic balance was 7.5kg and the accuracy was 0.01g.

The result of experiment was shown as figure 4. Contrast to manual feeding, the ducks weight of increased slowly at the beginning, because ducks needed time to adapt to the control system and set up behavior reflex to vocal signal. The weight would increase more than that of manual feeding ducks until 7 weeks of age. The weight increased 0.05kg at 8 weeks of age and increased 0.08

kg at 9 weeks of age. The variance analysis showed that there was no significant difference for the average weight at two kinds of feeding method (P>0.05), which meant the automatic system could achieve the same result as manual feeding.

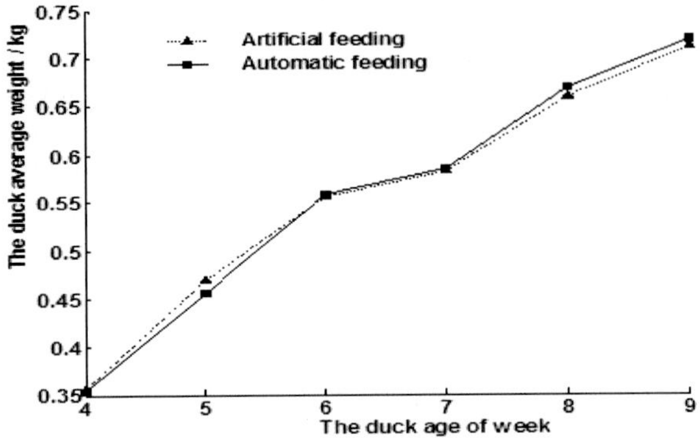

Fig. 4. Comparison for the duck weights.

Compared with manual feeding, the automatic feeding saved labor cost with the same feed consumption. The result of experiment showed labor cost of manual feeding was ￥200 every day and it needed 35 days for ducks living in paddy field, so it cost ￥7000for human labor. For automatic feeding ,the manual maintenance of equipment needed 4 days which would cost ￥800,in addition to the development cost ￥2000, so it cost ￥2880 and saved ￥4200 /667m2. The economic effect was significant compared with manual feeding.

6. Conclusion

(1) The automatic control system applied to rice-duck eco-farming was designed and developed, which realized automatic detection of ducks number in the shed, automatic door opening and closing, automatic feeding, automatic watering and automatic music playing to call ducks back to the shed, so that field management of rice-duck eco-farming can be automated.

(2) Field experiment showed the automatic system could achieve the same result as manual feeding and there was no significant difference for the average weight at two kinds of feeding method. However, compared with manual feeding, automatic feeding could save labor cost of ￥4200 /667m^2, so automatic feeding has economic benefits.

Acknowledgments

This work is supported by Chinese National Natural Science Fund (51081240353). The paper was financially supported by Chinese National Natural Science Fund (51081240353).

References

1. S.S. Islam, M.G. Azam, S.K. Adhikary, Efficiency of integrated rice, fish and duck polyculture as compared to rice and fish culture in a selective area of Khulna district, Pakistan Journal of Biological Sciences (Pakistan, 2004).
2. Qi Chunyan, Zhao Guochen, Hou Ligang, A preliminary study on the ecological benefit of different cropping-breeding coupled mode in rice field of north cold region. Journal of Jilin Agricultural Sciences, (Changchun, China, 2011).
3. Su Sin Teo, Evaluation of different duck varieties for the control of the golden apple snail (Pomacea canaiculata)in transplanted and direct seeded rice, Crop Protection (Kelantan, Malaysia, 2001).
4. Zhong Pingsheng, Liang Guangwen, Zeng Ling, Impacts of releasing duck on the brown planthopper population and its major natural enemies, Plant Protection (Guangdong, China, 2006).
5. Deng Xiao, Liao Xiaolan, Huang Huang, Study on the effect s of rice-duck complex planting and breeding pattern on the amount of methanogens, Environmental Pollution and Control, (Hunan, China, 2004).
6. Gan Dexin, Huang Huang, Jiang Tingjie, The ecological characteristic of no-tillage rice-duck complex system, Journal of Hunan Agricultural University, (Hunan, China, 2008).
7. Shen Jiankai, Huang Huang, Fu Zhiqiang, Nutrition characteristics in root surface and rhizosphere soils of direct-seeding rice in rice-duck eco-farming system, Chinese Journal of Eco-Agriculture, (Hunan, China, 2010).
8. Li Chengfang,Cao Cougui,Wang Jinping, Dynamic variations and losses of N in floodwater of paddy fields in integrated rice-duck and rice-fish ecosystems,Acta Scientiae Circumstantiae Acta Sci Circum, (Hunan, China, 2008).
9. Zhang Fan, Characteristics of nutrient return and uptake in rice-duck mutualism ecosystem of double rice cropping season, Chinese Journal of Eco-Agriculture, (Hunan, China, 2012).
10. Chen Can, Huang Huang, Zheng Huabin, Effects of different mode of ecological planting and raising on rice quality, China Rice, (Hunan, China, 2015).

11. B. D. Goh et al., Effect of Physical Stimulation by Free-ranging Aigamo Ducks on Rice Growth and Yield under a Rice-Duck Farming System, Japanese journal of livestock management, (Japan, 2001).
12. Shaikh Tanveer Hossain, Hideki Sugimoto, Gazi Jashim Uddin Ahmed. Effect of Integrated Rice-Duck Farming on Rice Yield, Farm Productivity, and Rice-Provisioning Ability of Farmers, Asian Journal of Agriculture and Development, (Bangladesh, 2005).
13. Ren Wentao, Wang Yue, Kong Aiju, Development and experiment of automatic duck feeding device with opening way for paddy, Transactions of the Chinese Society of Agricultural Engineering (Shenyang, China, 2016).

A micromechanical silicon DETF resonant accelerometer with single-stage microleverage mechanisms

Shan Cao, Kai-Si Xu, Wei-Guo Su and Wei Zhang*

Institute of Microelectronics, Peking University,
Beijing, 100871, China
**E-mail: weizhang@pku.edu.cn*

This paper presents a silicon-based double-ended tuning fork (DETF) MEMS resonant accelerometer integrated with single-stage microleverage to improve the precision of measurement. Differential output is achieved by two opposing DETFs that are symmetrical distributed. Mechanical and modal analysis have been conducted in ANSYS software. The simulation results demonstrate a considerably high differential sensitivity of 46 Hz/g for the micromechanical resonant accelerometer around a resonant frequency of approximately 43.4 kHz.

Keywords: MEMS; DETF; resonant accelerometer; microleverage.

1. Introduction

The inertial sensor market has been a large and increasing sector of microelectromechanical systems (MEMS) industry in recent decades. In this large sector, MEMS accelerometers have seen significant success in a large spectrum of applications ranging from automotive control and safety systems, consumer electronics and even wearable healthcare hardware for the Internet of Thing (IoT) system. The MEMS accelerometers used in most of these applications penetrate the market by batch manufacturability, small size, low cost and CMOS integration. However, the demands for high sensitivity and accuracy of inertial measurement via MEMS approach is still growing for some promising applications such as pedestrian navigation and seismology [1, 2]. And the micromechanical silicon resonant accelerometers have great potential to realize high-precision inertial measurement due to their high sensitivity, frequency output and large dynamic range [3-6].

Most of resonant accelerometers consist of a proof mass and one or more resonators. The value of acceleration is calculated by the resonant frequency shift resulting from the axial stress produced by the proof mass under external acceleration [7]. Based on previous studies, resonant accelerometers integrated with leverage mechanisms are expected to possess higher sensitivity [8, 9]. Although microleverage have been widely proposed as a mainstream solution by far to amplify the inertial force and improve sensitivity, there is still room for improvement in the amplification factor of microleverage mechanism.

2. Design and Principle

2.1. *Structures*

The structure of the MEMS uniaxial resonant accelerometer with single-stage microleverage is illustrated in Figure 1. The accelerometer consists of a proof mass and a pair of identical sensing structures. Each sensing structure is composed of a pair of single-stage micro leverages and a Double Ended Tuning Fork (DETF) resonator with sensing and exciting electrodes.

In this work, the silicon structure of the resonant accelerometer has an overall area of 3400μm×2400μm and thickness of 55μm. The DETFs were designed with symmetric electrode configuration. The optimized dimensions of resonant beams in DETFs are 6μm×800μm, while the optimized dimensions of single-stage microleverage arm are 40μm×740μm.

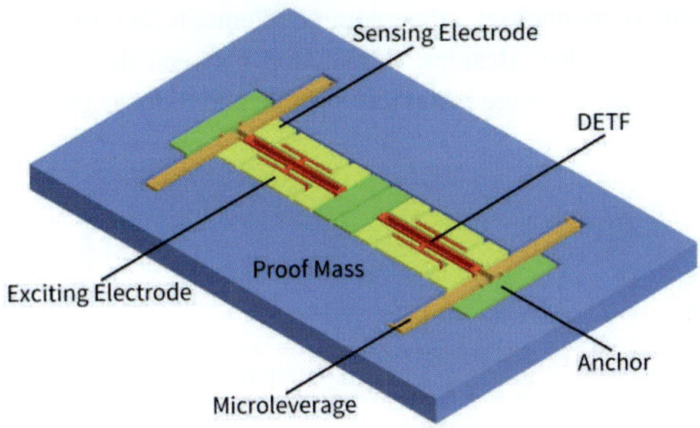

Fig. 1. The scheme of MEMS uniaxial resonant accelerometer with single-stage microleverage.

2.2. *Differential measurement of DETF*

The DETF micromechanical resonant accelerometers is based on two beams resonating in their in-plane fundamental mode shape. A simple design of MEMS resonating accelerometer can be obtained with a pair of opposing DETFs connected to the proof mass at one end and to the ground at the other end. The pair of DETFs are functioned as a force sensor to obtain the proof mass inertial force caused by external acceleration. And the measurement of the acceleration is realized by reading the resonant frequency shift caused under axial load.

As the proof mass is accelerated along the sensing direction, the inertial force will compress one DETF and stretch the opposing DETF. As a result, it leads to a negative and a positive shift in the natural frequency of the beams respectively. With this kind of differential measurement approach, a temperature variation is expected to generate the same amount of frequency shift in both DETFs theoretically. Therefore, the DETFs differential measurement is possible to improve the sensitivity and counteract the influence of temperature variation. And DETF resonator is one of the stress sensitive dynamically balanced structures that reduce the common mode interference.

2.3. *Microleverage mechanisms*

Microleverage mechanism is applied in MEMS devices to transfer and amplify force or displacement from input to output. In the macro-size machines assembled by traditional methods, the leverage rotates freely and is supported rigidly. Owing to bearing, the pivot is rotational around the pin joint. However, the same mechanism may not function well in a microleverage. Limited by current fabrication technology, the required elements are too difficult to be fabricated and assembled. Besides, the stiction between the rotational and the fixed parts makes the leverage behavior unstable and reduces the life time of the leverage greatly.

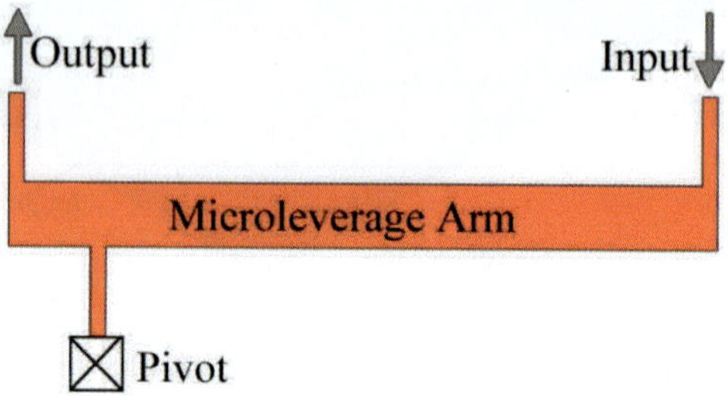

Fig. 2. The schematic of single-stage microleverage.

In microleverage design, the pivot is a planar flexure beam with one end anchored to the substrate, as schematically shown in Figure 2. Compared with scale magnifying traditional macro leverage, the actual amplification factor is usually smaller considering the influence of micro pivot and anchor design.

3. Simulation and Analysis

The performance of MEMS resonant accelerometer with single-stage microleverage is studied by finite element modeling (FEM) in ANSYS software and related physical parameters are listed in Table 1.

Table 1. Simulation related physical parameters.

Parameters	Silion[10]
Density (kg/m^3)	2300
Young Modulus (GPa)	190
Poisson Ratio	0.28

The result of modal simulation is provided by Figure 3, including the first six modes as well as the 11th and 12th modes. From the results, the DETF functions at the 11th and 12th modes driven by the static-electronic force of exciting electrodes.

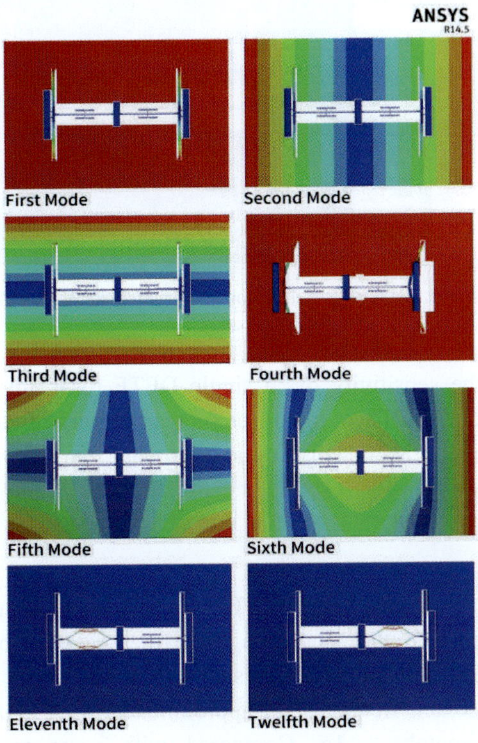

Fig. 3. The modal simulation results of the resonant accelerometer with single-stage microleverage.

The resonant frequency shift under 9.8m/s^2 (1g) static acceleration is simulated then. From Figure 4, the resonant frequency of the resonant beam decreases from 43428Hz to 43405Hz under 1g acceleration.

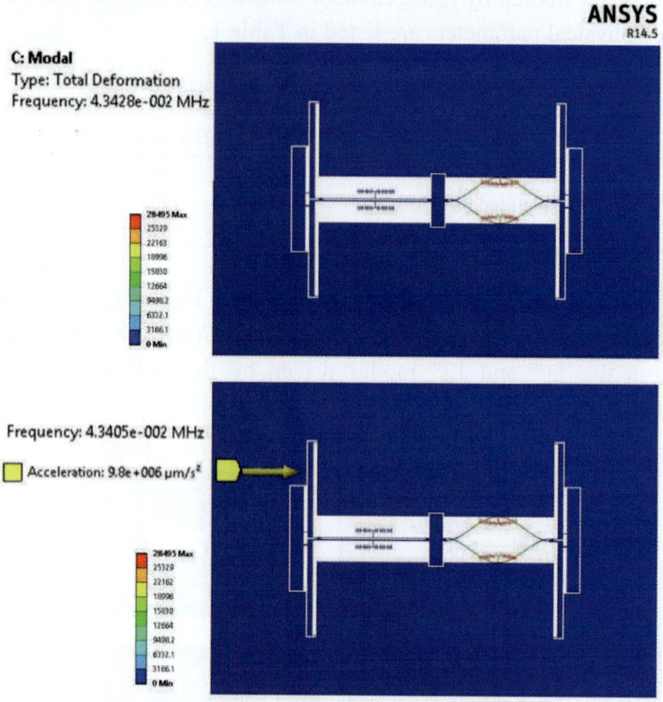

Fig. 4. The modal simulation results of the resonant accelerometer with/without 1g acceleration load.

It indicates the sensitivity of a single DETF resonator is 23Hz/g in the range of 1g acceleration. Then, the differential sensitivity of the MEMS uniaxial resonant accelerometer is 46 Hz/g around a resonant frequency of approximately 43.4 kHz.

4. Conclusions

A highly sensitive silicon-based DETF micromechanical uniaxial resonant accelerometer integrated with single-stage microleverage has been designed and simulated. The FEM simulation results show a high differential sensitivity of 46 Hz/g around a resonant frequency of approximately 43.4 kHz over a dynamic range of approximately +/-1g. Compared with some traditional resonant accelerometers, the present resonant accelerometer has a much higher sensitivity due to the use of single-stage microleverage mechanisms. In addition, this device shall be further developed to be a resonant tilt sensor. Further work will focus on frequency readout circuits for the experimental tests of a complete uniaxial in-plane acceleration measurement system.

Acknowledgments

The authors are grateful to all the colleagues in our team for their precious suggestions and supports.

References

1. S. Y. Cho and C. G. Park, MEMS based pedestrian navigation system, *Journal of Navigation*, vol. 59, pp. 135-153, 2006.
2. J. Laine and D. Mougenot, Benefits of MEMS based seismic accelerometers for oil exploration, in *Proceedings of Transducers'07*, pp. 1473-1477, 2007.
3. T. A. Roessig, R. T. Howe, A. P. Pisano and J. H. Smith, Surface-micromachined resonant accelerometer, in *Proceedings of Transducers'97*, pp. 859-862, 1997.
4. R. Hopkins, J. Borenstein, B. Antkowiak, P. Ward and R Elliott (eds.), The silicon oscillating accelerometer: a MEMS inertial instrument for strategic missile guidance, pp. 45-51, 1998.
5. M. Aikele, K. Bauer, W. Ficker, F. Neubauer, U. Prechtel, J. Schalk and H. Seidel, Resonant accelerometer with self-test, *Sensors and Actuators A: Physical*, vol. 92, pp. 161-167, 2001.
6. A. A. Seshia, M. Palaniapan, T. A. Roessig, R. T. Howe, R. W. Gooch, T. R. Schimert, S. Montague, A vacuum packaged surface micromachined resonant accelerometer, *Journal of Microelectromechanical Systems*, vol. 11, pp. 784-793, 2002.
7. L. He, Y. P. Xu, M. Palaniapan, A CMOS readout circuit for SOI resonant accelerometer with 4-μg bias stability and 20-μg/√Hz resolution, *IEEE Journal of Solid-State Circuits*, vol. 43, pp. 1480-1490, 2008.
8. X. Su, Compliant leverage mechanism design for MEMS applications, University of California, Berkeley, 2001.

9. X. P. S. Su and H. S. Yang, Single-stage microleverage mechanism optimization in a resonant accelerometer, *Structural and Multidisciplinary Optimization*, vol. 21, pp. 246-252, 2001.
10. K. Petersen, Silicon as a mechanical material, *Proceedings of the IEEE*, vol. 70, pp.420-457, 1982.

Optimization study on PCB winding loss of axial-flux reaction flywheel motor

Xiao-wei Wang
Shenzhen Academy of Aerospace Technology, Shenzhen, 518057, China
Guangzhou University, Guangzhou, 510006, China
E-mail: wxwhit@126.com

Yun-tao Wang and Shu-yue Ning
Shenzhen Academy of Aerospace Technology, Shenzhen, 518057, China

Based on the present situation of low space utilization ratio of satellite reaction flywheel, one ultrathin reaction flywheel is put forward on the basis of PCB winding and axial flux, which can greatly improve the flywheel space utilization. In this paper, the loss of different PCB winding structures has been analyzed and studied according to the characteristics of PCB winding: firstly, numerical analysis is used to analyze mechanisms on PCB winding eddy-current loss and circulating-current loss, then according to three-dimensional finite element method, researches on losses have been carried out for different parallel connections of PCB winding conductor and end short-circuit of PCB winding. It can be found by finite element simulation that: in the case of high electricity, PCB winding is suggested to be connected in parallel with multi-conductor. And the effect of inhibiting loss is extremely obvious when it is connected in parallel with multi-conductor in the direction vertical to PCB plane, but the cost of multilayer PCB is relatively high. On condition that PCB be connected in parallel in the same layer, multiple parallel conductors are forbidden to take short circuit at the end of winding. Once single short-circuit structure is used, the electrical angle difference in space among multiple subcircuits with end short circuit is 180°.

Keywords: Axial flux; eddy loss; PCB winding; reaction flywheel.

1. Introduction

Nowadays, with the continuous development of satellite technology, the stronger the satellite functions are, the more demands the payloads of satellite to the automatic control system are put forward. In recent years, reaction flywheel is preferred as an actuator in satellites. In general, reaction flywheel consists of permanent magnet machine and flywheel system. As a power producer of flywheel, radial flux motor with inner rotor or outer rotor is used mostly by the permanent magnet machine. Inertia object of flywheel is connected to motor rotor and extended outward along the radial of motor, then the structure of disk or bow forms. But this structure may result in lower charger ratio and space utilization ratio inside of flywheel and longer axial direction, which is adverse to internal layout of satellite. So some scholars tried to combine the structure of discal flywheel and use permanent magnet synchronous disk motor with axial flux as a power producer [1], [2]. The structure that rotor of discal motor and inertia object of flywheel is unified as a whole makes motor thinner and improves its inner space utilization ratio. In order to achieve the ultrathin design of flywheel, PCB winding technology was adopted in [2], which could simplify motor winding as well as make a further achievement about the ultrathin design of flywheel. Besides, PCB winding has been studied in [3-5] by B Dehez, F Baudart, et al. And it was considered that performance of printed conductor loop after optimization improves 20% comparing with printed conductor loop in the traditional shape of diamond, 35% comparing with traditional compiled winding. The authors also pointed out in the further study that flexible PCB winding technology had higher thermal conductivity than traditional compiled winding and great significance to increase power density of motor. To make hard disk drive efficient and compact, PCB winding was also applied to the micro spindle motor to achieve the ultrathin design of flywheel in [6], [7].

However, the above mentioned documents neglected the loss of PCB in the course of applications and studies of PCB. In fact, for axial flux stator coreless motor, it cannot put winding in stator slot like motor with iron core. Therefore, its stator winding exposes directly in the air-gap magnetic field. When motor rotates, the air-gap magnetic field with high frequency incises stator winding conductor and produces eddy-current loss; on the other hand, PCB winding is limited by manufacturing process, and PCB winding conductor generally has a flat structure, which increases eddy-current loss in PCB winding. The eddy-current loss of stator coreless permanent-magnet motor has been researched in [8-11], but these researches were based on winding compiled by copper wire rather than PCB winding. Thus, this paper makes an optimal analysis for the loss of PCB winding by combing with the structure of PCB winding itself.

2. Flywheel and Structure of PCB Winding

The overall structure of flywheel designed in this paper is shown in Fig. 1. The parts 1, 2, 3, 21 and 31 referred to in Fig. 1 jointly form the drive motor of flywheel. This motor belongs to axial flux motor and stator adopts coreless structure, and stator winding uses PCB winding. What can be seen from the figure is that the PCB belongs to cyclic structure. And part of the graphic is shown in Fig. 2. In this figure, optimal measures were not taken. In the following studies, this model will be made an optimal analysis on loss.

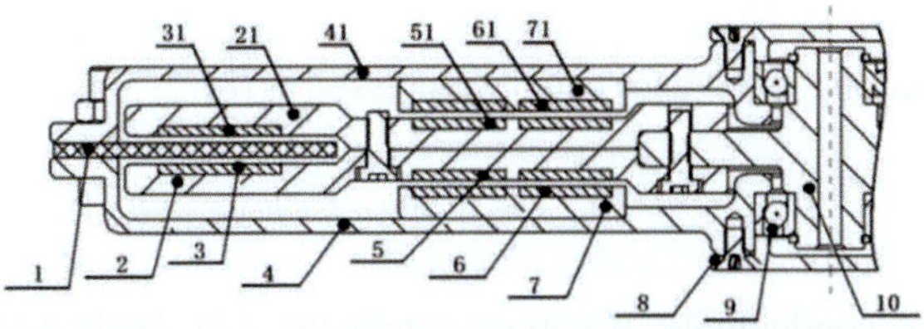

1 PCB winding of flywheel stator, 2 Lower rotor core, 21 Upper rotor core, 3 lower rotor magnetic steel, 31 upper rotor magnetic steel, 4 lower end cover of flywheel disk, 41 upper end cover of flywheel disk, 5 lower magnetic steel with rotation of magnetic bearing, 51 upper magnetic steel with rotation of magnetic bearing, 6 lower magnetic steel with static of magnetic bearing, 61 upper magnetic steel with static of magnetic bearing, 7 Back iron of lower magnetic steel during the static of magnetic bearing, 71 Back iron of upper magnetic steel during the static of magnetic bearing, 8 Bearing end cover, 9 Bearing, 10 Rotor spindle.

Fig. 1. The structure of ultrathin reaction flywheel.

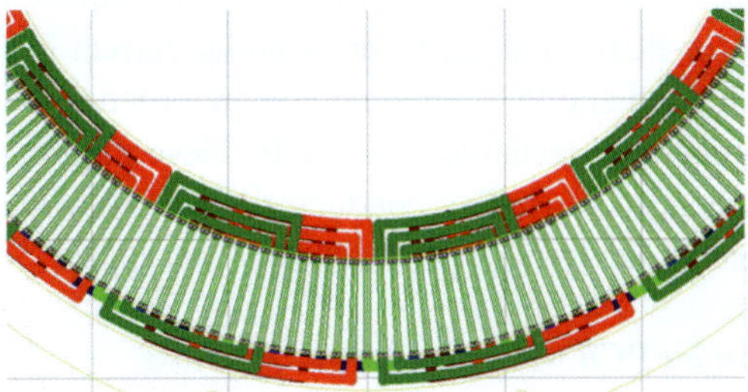

Fig. 2. The diagram of partial PCB winding.

3. Numerical Analysis and Optimization of PCB Winding

3.1. *Eddy-current loss of PCB winding*

The eddy-current loss of PCB winding can be calculated by evaluating eddy-current losses about each conductor and summating all winding conductors. Magnetic fields in different locations in the air-gap have different radial components and tangential components, and each component may include abundant higher-order harmonics. Therefore, the eddy-current loss of winding conductor can be gained by summation of eddy-current losses produced by each harmonic.

For a rectangular PCB conductor with radial length L, tangential width a and axial length b, when it incises magnetic line in the air-gap of permanent-magnet motor with electromagnetic frequency f, the eddy-current loss of this rectangular sectional conductor can be expressed as [8]:

$$\Delta P_e = \frac{\pi^2}{3}\sigma f^2 a^3 bL \sum_{n=1}^{\infty} \lambda^2 [B_{mt\lambda}^2 + B_{ma\lambda}^2] = \frac{\pi^2}{3}\sigma f^2 a^3 bL[B_{mt\lambda}^2 + B_{ma\lambda}^2]\eta_d^2 \tag{1}$$

In (1), σ is conductivity of copper conductor. λ is single wave times. $B^2_{mt\lambda}$ and $B^2_{ma\lambda}$ are tangential peak value and axial peak value of flux density harmonic in λth time respectively. And η_d is distortion coefficient of flux density:

$$\eta_d = \sqrt{1 + \frac{(3B_{mt3})^2 + (3B_{ma3})^2}{B_{mt1}^2 + B_{ma1}^2} + \frac{(5B_{mt5})^2 + (5B_{ma5})^2}{B_{mt1}^2 + B_{ma1}^2} + } \tag{2}$$

What can be seen in (1) is that eddy-current loss is proportional to cube of tangential width a, second power of electromagnetic frequency f, axial length b respectively. But for PCB winding, when winding current is larger, its copper-clad thickness is limited and PCB winding conductor tends to increase its width due to limited processing technology of PCB. Nevertheless, according to the above analysis, when PCB winding conductor is enough wide, it may result in larger eddy-current loss.

3.2. *Optimization on PCB winding eddy-current loss*

In order to optimize PCB winding and decrease its eddy-current loss, parallel connection of multi-conductor is considered at first. As shown in Fig. 3, to simplify the analysis, fan-shaped magnetic steel and winding in the flywheel are simplified as rectangular magnetic steel and winding conductor. In Fig. 3a), there are one phase of winding and a pair of magnetic poles in the partial three-phase electric machine. The winding is a full-pitch winding and its turn number of winding is 2. And the winding conductor is wide. Fig. 3b) shows partial parallel connection of two winding to decrease winding eddy-current loss.

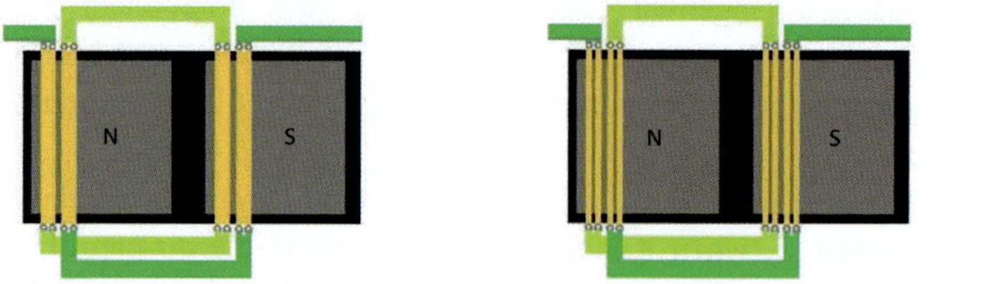

a) PCB winding without parallel connection　　b) PCB winding with partial multiple parallel connections

Fig. 3. The simplified PCB winding diagram.

For the presented structure in Fig. 3b), the eddy-current loss decreases to some extent, but circulating-current is produced between parallel conductors, which causes circulating-current loss. Next the simplified analysis has been done. Based on the numerical analysis graphic in Fig. 3b), for the analysis process as shown in Fig. 4, the following assumptions are made:

1) Neglect brow leakage;

2) Let air-gap magnetic flux density be periodical sinusoidal wave;

3) Make relative winding conductor in the rotor part which is made up of a pair of NS magnetic steels and its magnetic core move to right with the speed v;

4) Let the length, width and depth of single conductor in parallel be l, w and t. Let φ denote phase angle at the positive going zero-crossing point of the first conductor in the left and air-gap magnetic flux density. And let $\Delta\alpha$ denote phase angle between parallel conductors.

If the end of winding conductor is ignored, potentials u_1 to u_8 of 8 conductors in the figure are respectively: $u_1=u_5=B_{max}lv\sin(\varphi)$, $u_2=u_6= B_{max}lv\sin(\varphi+\Delta\alpha)$, $u_3=u_7=B_{max}lv\sin(\varphi+30^\circ)$, $u_4=u_8=B_{max}lv\sin(\varphi+30^\circ+\Delta\alpha)$, in which, the electric potential directions of u_1 to u_4 are lower positive and upper negative, and the electric potential directions of u_5 to u_8 are upper positive and lower negative. It is obvious that $u_2> u_1$, $u_4> u_3$, $u_6> u_5$, $u_8> u_7$. Therefore, it was bound to form circulating-current in these two parallel conductors on the condition that there is short circuit of two parallel conductors by winding ends. And the circulating-current losses produced by u_1 and u_2 in the left two parallel conductors can be expressed as:

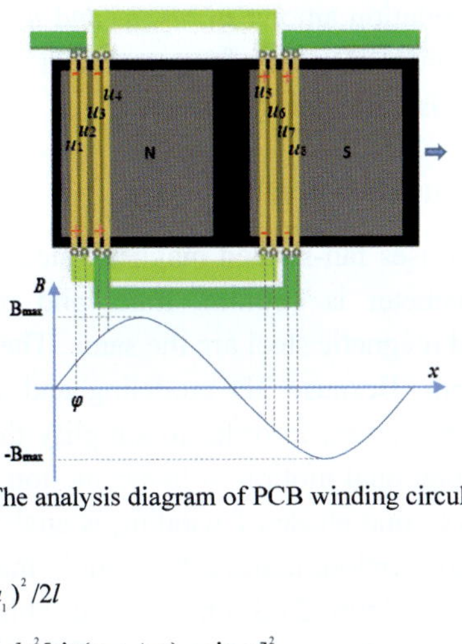

Fig. 4. The analysis diagram of PCB winding circulation.

$$\begin{aligned} P_c &= \sigma wt(u_2 - u_1)^2/2l \\ &= 0.5\sigma wtB_{max}^2 lv^2[\sin(\varphi+\Delta\alpha) - \sin\varphi]^2 \end{aligned} \tag{3}$$

In view of changing forms on this scheme which have six situations as shown in Fig. 5, single end parallel connection is adopted in four schemes a), b), d) and e). And the two schemes c) and f) achieve multi-conductor parallel in a real sense. There is no doubt that only from the winding part as shown in the figure, in these two schemes c) and f), it cannot form circulating-current in parallel conductor.

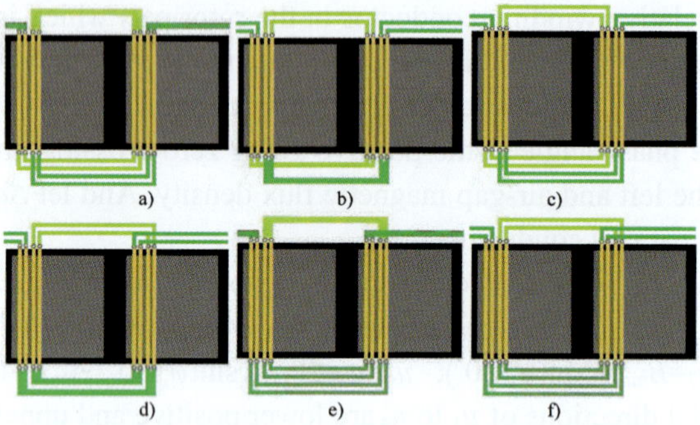

Fig. 5. Six different kinds of PCB winding diagram.

In Fig. 5a), four conductors u_1, u_6, u_5 and u_2 form a return circuit, but for the phase angles differences between u_1 and u_5, u_2 and u_6 are 180°, $u_1+u_2=u_6+u_5$. Hence, there is no circulating-current among four conductors; Fig. 5d) is similar to Fig. 5a), and both figures do not exist circulating-current.

In Fig. 5b), four conductors u_1, u_8, u_7 and u_2 form a return circuit, but for there is no symmetric relation among u_1, u_8, u_7 and u_2, $u_1+u_2 \neq u_6+u_5$. Therefore, there exists circulating-current among four conductors; Fig. 5e) is similar to Fig. 5b), and both figures exist circulating-current.

4. Finite Element Simulation and Modeling

In this paper, flywheel uses fan-shaped magnetic steel whose outer diameter is 196mm and inner diameter is 160mm. Inner and outer diameters of PCB winding conductor and magnetic steel are the same. The highest rotational speed of flywheel is 5000rpm. Because 3D modeling and simulation expend large amounts of hardware resources, in order to simplify the analysis, on one hand this paper simplifies rotational motion to linear motion. On the other hand only one rotational process of one phase of winding is analyzed by finite element in this paper. After the conversion, assume that single magnetic steel is rectangle and its length of a side is 18mm×23.3mm, one of established models is shown in Fig. 6. It is supposed in the figures that the velocity of PCB winding conductor relative to magnetic steel and its iron core at the yoke is 46600.3mm/s (the equal of linear velocity whose flywheel radius is 89mm).

In the model of Ansys Maxwell, NdFe35 is used as magnetic steel, D23-50 as iron core at the yoke, copper as PCB winding. And the bulk conductivity of copper is 5.8E7siemens/m. The graph on air gap magnetic flux density amplitude is shown in Fig. 7 by analysis of magnetostatic field,. What can be seen from the figure is that air gap magnetic flux density is analogous to trapezoidal wave which contains higher-order harmonics. Next different PCB winding are analyzed based on this simulation model.

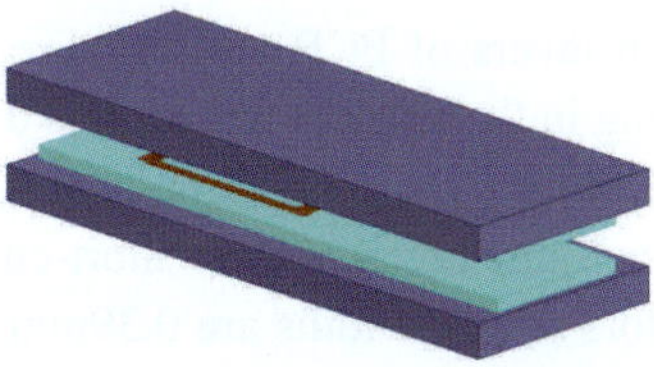

Fig. 6. Simplified three dimensional finite element model.

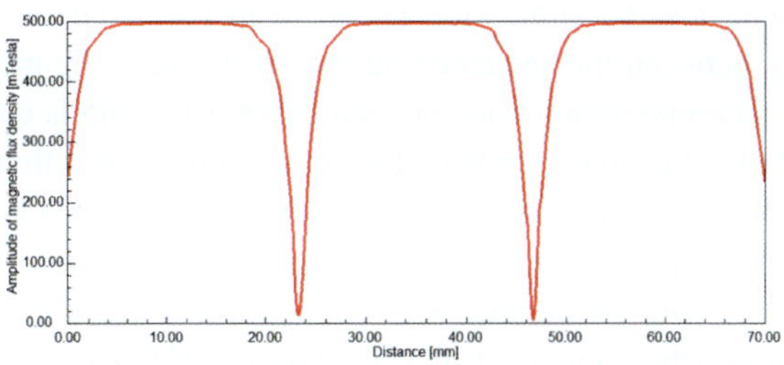

Fig. 7. The graph of air gap magnetic flux density.

5. Calculation on PCB Winding Loss

5.1. *Losses of PCB winding without and with parallel connection*

It can be seen from the above (1) that the wider the PCB conductor is, the larger losses it will produce in motion. So it is required that parallel connection of multi-conductor should be adopted when PCB conductor needs to bear larger current. For special structure of PCB, parallel connection form of multi-conductor is divided into two categories. One is parallel connection of multi-conductor among multilayer boards, and the other is parallel connection of multi-conductor in single layer. To illustrate the problem, transient analysis has been made for single phase within one period of electricity and two turns PCB winding as shown in Fig.3. The spacing between two turns adjacent winding is

0.41mm. During the analysis, make PCB winding do linear motion with the speed of 46600.3mm/s relative to permanent magnet and run for 0.3ms. In order to simplify the analysis, the winding end is ignored in the analysis process, and only PCB conductor in air gap field is analyzed. For PCB conductor, five situations can be analyzed:

1) the width of single PCB conductor is 0.78mm;

2) two PCB conductors whose widths are 0.39mm are horizontally connected in parallel in the same layer, and the spacing between these two parallel conductors is 0.01mm;

3) two PCB conductors whose widths are 0.39mm are connected in parallel between the up and down layers of PCB and the direction which is vertical to PCB plane, and the spacing in the vertical direction is 0.01mm;

4) two PCB conductors whose widths are 0.39mm are horizontally connected in parallel in the same layer, and we short-circuit at the ends;

5) two PCB conductors whose widths are 0.39mm are connected in parallel between the up and down layers of PCB and the direction which is vertical to PCB plane, and we short-circuit at the ends.

PCB winding loss by simulation is shown in Fig.8. There is a one-to-one correspondence between the sequence number of curve in figure and the above five situations. The maximum mark of each curve m1 to m5 is corresponding to curves 1) to 5). What can be seen from the simulation result is that:

1) The maximum values of PCB winding losses occur in 130us. The reason is that two winding of PCB are straddled in the sides of magnetic steel N and S.

2) When PCB winding conductor is not connected in parallel, its maximum loss is 4.109 times (the equivalent of 1.9684W/0.4790W times) larger than that of conductor with 2-branch parallel in the same layer of PCB; if calculation is done according to the above (1), the maximum loss of PCB winding conductor without parallel connection is 4 times (the equivalent of $(2a)^3/a^3$ times) larger than that of conductor with 2-branch parallel. There is only deviation of 2.7% between them, which indicates that these two results are basically coincident. Therefore, it can be inferred by formula (1) that on the condition that the total width of conductor is unchanged, the more the parallel branch number of conductor is, the more obvious the loss control becomes.

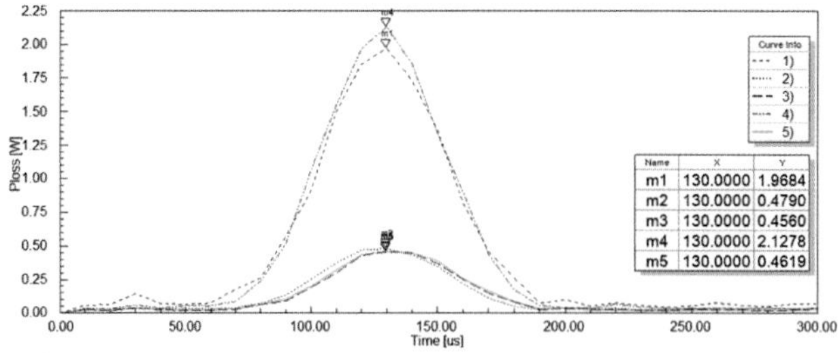

Fig. 8. The loss graph of PCB winding with parallel connection or not.

3) On condition that these two subcircuits are connected in parallel, no matter where these two subcircuits are in the same layer or different layers, the losses would increase when these subcircuits short-circuit in the both ends at the same time. But the difference is that the loss increase of the former is unusually obvious and even surpasses the loss without parallel connection (m4>m1), while the loss of the latter increases a little (m5 is slightly larger than m3). The result shows that after the short circuit at these two ends, these two subcircuits form circulating current. Because the rate of change of tangential component on air-gap magnetic flux density is much larger than that of radial component, which leads to the result that circulating current losses of these two subcircuits in the same layer of PCB are far larger than those of two subcircuits in the different layers of PCB. By combining the manufacturing process and cost of PCB, there is no doubt that vertical parallel connection of multiple subcircuits needs more PCB in more layers and higher cost, while horizontal parallel connection of multiple subcircuits has lower cost of PCB. Besides, given that PCB winding motor does not include stator core, which is beneficial to achieve horizontal parallel connection. But we should be cautious for end short circuit during the horizontal parallel connection.

5.2. *Contrast analysis of situations on horizontal parallel connection*

The situations about multiple subcircuits horizontal parallel connection of PCB conductor is analyzed by combining six structures in Fig. 5. But in view of Fig. 5, they are the same in essence between a) and d), b) and e), c) and f). Hence, three horizontal parallel situations a), b) and c) are modeled and analyzed.

The simulation results are shown in Fig. 9. There is a one-to-one correspondence between the sequence number of curve a), b) and c) in the Fig. 9 and a), b) and c) in Fig. 5). And m1, m2 and m3 are maximum marks of curves a), b) and c). It is found by simulation that:

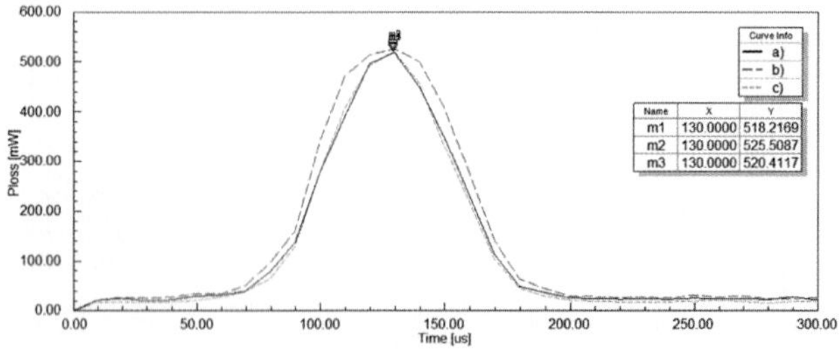

Fig. 9. The loss graph of three kinds of PCB horizontal parallel winding.

1) The maximum values of PCB winding losses still occur in 130us. The maximum losses in these three situations are larger than 479mW in the second situation in section 5.1. The reason is that in the analysis process of this section, the end has not been ignored in winding model. Under the influences of the end magnetic flux leakage and circulating-current, winding end also produces loss.

2) During almost the whole time period of simulation, the loss of PCB winding structure in Fig. 9b) is larger than that of PCB winding structure in Fig. 9a) and c). The loss curves of a) and c) is basically uniform, which also verifies the former numerical analysis. The reason is that in Fig. 9a), the electrical angle difference in space among 4 subcircuits with end short circuit is 180°. The waveform of air-gap magnetic flux density is in the positive and negative half cycle, but it is same about their amplitudes and variation trends as well as generated back EMF amplitudes. Circulating-current and circulating-current loss does not exist among subcircuits; but in Fig. 9b), the electrical angle difference in space among 4 subcircuits with end short circuit is not 180°. The difference of back EMF forms voltage difference inside of these four subcircuits, which results in circulating-current and circulating-current loss. Such situation should be avoid in the practical application if possible.

6. Conclusion

This paper focuses on the research on eddy-current loss and circulating-current loss of PCB winding in ultrathin space reaction flywheel. About these two losses, the structure of PCB winding has been studied. The research shows that: in case of larger current, PCB winding is suggested to be connected in parallel by multi-conductor. And when PCB winding is connected in parallel in the direction which is vertical to PCB plane, the effect of inhibiting loss is the most obvious, but the cost of multilayer PCB is relatively high. In the case of horizontal parallel connection, short circuit of multiple parallel conductors is forbidden at winding end. Once single short-circuit structure is used, the electrical angle difference in space among multiple subcircuits with end short circuit is 180°. This research result has a certain guiding significance not only for flywheel motor which is similar to structure, but also for other motors based on PCB winding structure.

Acknowledgments

This work is supported by China Postdoctoral Science Foundation (2015M582423); the Natural Science Foundation of Guangdong, China (2015A030310486).

References

1. Y. P. Yang, Y. P. Lah, and C. H. Cheung, "Design and control of axial-flux brushless DC wheel motors for electric vehicles-part I: Multiobjective optimal design and analysis," *IEEE Trans. Magn.*, vol. 40, no. 4, pp. 1873-1882, 2004.
2. J. F. Wu, "Design of a miniature axial flux flywheel motor with PCB winding for nanosatellites," in *Proc. IEEE ICOM*, Aug. 23-25, 2012, pp. 544-548.
3. B Dehez, F Baudart, M Markovic, and Y Perriard, "Theoretical and Experimental Investigation of flex-PCB Airgap Windings in Slotless BLDC Machines," *IEEE Trans. Ind. Appl.*, vol. 50, no. 5, pp. 3153-3160, 2014.
4. B Dehez, F Baudart, and Y Perriard, "Analysis of a new topology of flexible PCB winding for slotless BLDC machines," in *Proc. IEEE ICEM*, Sept. 2-5, 2014, pp. 1963-1969.
5. F Baudart, B Dehez, J Denies, M Markovic, and Y Perriard, "Shape optimization of flexible PCB slotless windings in BLDC machines," in *Proc. IEEE ICEMS*, Oct. 26-29, 2013, pp. 943-948.
6. M. Tsai, L. Hsu. "Design of a Miniature Axial Flux Spindle Motor with Rhomboidal PCB Winding," *IEEE Trans. Magn.*, vol. 42, no. 10, pp. 3488-3490, 2006.
7. G. H. Jang, and J. H. Chang, "Development of an axial-gap spindle motor for computer hard disk drives using PCB winding and dual air gaps," *IEEE Trans. Magn.*, vol. 38, no. 5, pp. 3297-3299, 2002.
8. X. L. Wang, "A winding eddy current loss suppression method for power-type axial flux permanent magnet machine with coreless-stator," *Torpedo Technology*, vol. 20, no. 4, pp. 295-300, 2012.
9. P. H. Tang, Y. M. Qi, G. H. Huang, and T. C. Li, "Eddy current loss analysis of ironless flywheel electric machine's winding," *Transactions of China Electrotechnical Society*, vol. 25, no. 3, pp. 27-32, 2010.

10. R. J. Wang, M. J. Kamper, "Calculation of eddy current loss in axial field permanent-magnet machine with coreless stator," *IEEE Trans. Energy Convers.*, vol. 19, no. 3, pp. 532-538, 2004.
11. R Wang, A. J. Kamper, "Evaluation of eddy current losses in axial flux permanent magnet (AFPM) machine with an ironless stator," in *Proc. IEEE Int. Industry Applications Conf.* Industry Applications Conference, vol. 2, Oct. 13-18, 2002, pp. 1289-1294.

Design of intelligent portable instrument for varicose veins

Xiao-Rong Deng

Department of Engineering, Shanghai Second Polytechnic University, Shanghai 201209, P.R. China
Email: xrdeng@sspu.edu.cn

This paper designs an intelligent portable instrument for the treatment of varicose veins based on MSP430 Series Microprocessor .Silicon pressure sensor and micro air pump are applied as the core components. According to user's setting, the instrument for the treatment of varicose veins can adjust the strength of the air pressure and the inflatable mode, automatic implement the inflation of the air bag, the compensation of the air pressure and the conversion of the work mode. Different treatment mode is different from the strength of the air pressure and the duration of the pressure for different patients. The pain of varicose veins can be relieved quickly after the treatment of this kind of instrument, and after long term use patients can achieve better therapeutic effect.

1. Introduction

Varicose veins is a common disease, specially for elderly and long standing people. Normally there are two treatments for varicose veins: conservative treatment and surgical treatment. currently wearing the elastic stockings is common for the conservative treatment, but its disadvantage is to reduce the quality of life. Surgical treatment is implement for severe patients, its disadvantage is pain of operation and high cost of operation. To solve the defects of the above two kinds of treatment, ease the pain of patients with varicose veins, we design a kind of intelligent portable instrument for the treatment of varicose veins, which is proved to be effective. The therapeutic instrument is composed of power module, control module, interface module, air pressure module, charge treatment module and LCD module. The instrument is convenient for the treatment of varicose veins, and has the advantages of easy operation, minimally invasive injury, the remarkable effect of all kinds of varicose disease.

2. Hardware Design

2.1. *Main control unit*

We select micro controllers chip (MSP430F435) as the central micro processing controller because of its character of low power operation. It is very suitable for the portable control system of the fully function. Working voltage 1.8-3.6 V, its current consumption is about 300 uA in its active mode. In addition, it also has a wealth of hardware resources, many i/o ports, including 16 B flash KB+256 memory and 512 BRAM, Integrated 8 channel 12 ADC bit, With internal reference level and automatic scanning, etc. Used to accurately collect pressure signal, also equipped with up to LCD segment of the 160 driver. The block diagram of the system is shown in Fig. 1.

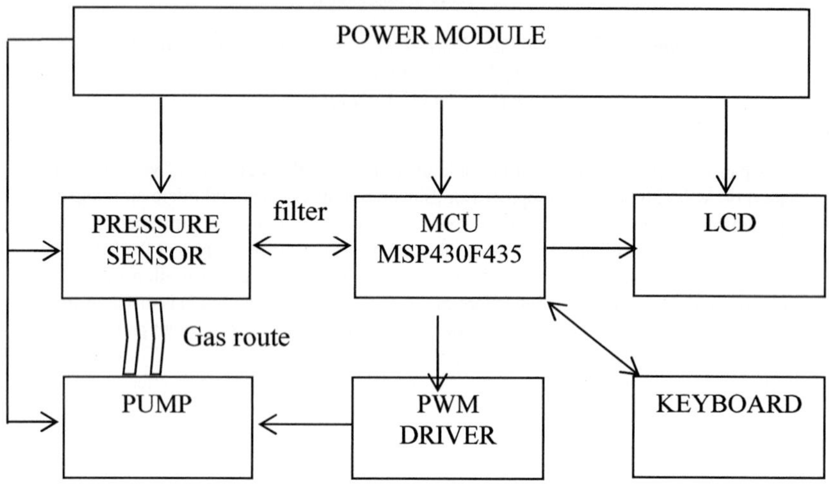

Fig. 1. The block diagram of the system.

2.2. *Pressure sensor*

MPX5050 is a high precision silicon pressure sensor of MPX series produced by MOTOROLA company. Its measuring method is the differential pressure type, and it has excellent linearity, high sensitivity and good repeat-ability because the temperature compensation technique and the laser repair technology controlled by computer are used. The pressure level conversion formula is as follows:
VOUT=Vs*(0018*P+004) ±ERROR

In this formula, Vs=5.0V, the value of P represents the measured air pressure value, VOUT represents the output signal level. In the practical operation, the accurate pressure value can be calculated by adjusting the appropriate ERROR value. The pressure sensor MPX5050 connection diagram is shown in Fig. 2.

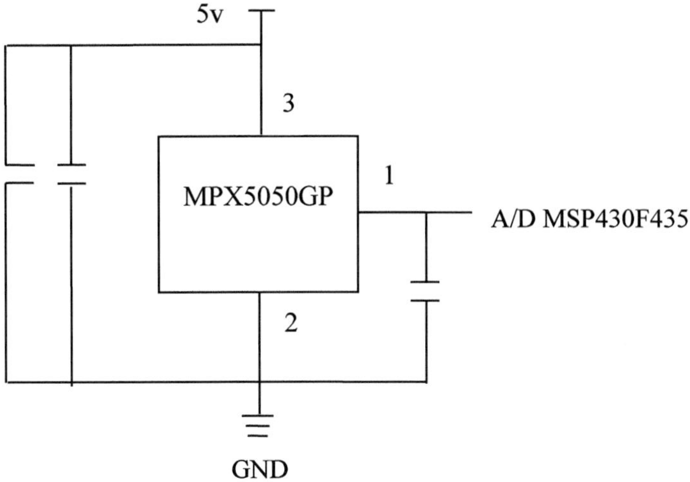

Fig. 2. MPX5050 connection diagram.

2.3. *Signal processing*

The signal of the air pressure that the microprocessor MSP430F435 actual need to monitor is the average pressure signal, but the signals that the pressure sensor measures include all kinds of signals such as pulse wave signal or high frequency noise of circuit itself. So it needs to design a low pass filter before the signal sampling to improve the anti-interference ability of the circuits. The filter designs as two-pole active low pass filter Based on Butter-worth Function.

2.4. *Micro pump and Control*

In the system design, a direct current and low power driving pump is needed. The micro pump produced by American GAST company is used in this design because of its features of low noise, low vibration, small volume, large flow, light weight, and corrosion resistance. Pump is 6 V DC power supplied, the pump body size is 40minx30minx30mm, the maximum flow rate is37L/Min, the largest gas pressure >100 k Pa. All these features are suitable for portable instruments. In practical application, changing the frequency of voltage and power pump will correspondingly change the air flow. The pulse-width modulation signal outputted by the microprocessor MSP430F435 is used to control on-off time, and control the airflow.

3. Software Design

The software design of the system includes the main control program and sub-program of the pressure measuring. The software is written in C language under the MCU development platform of the IAR MSP430 Embedded Work-beach. The flow chart of the whole system software is shown in fig. 3.

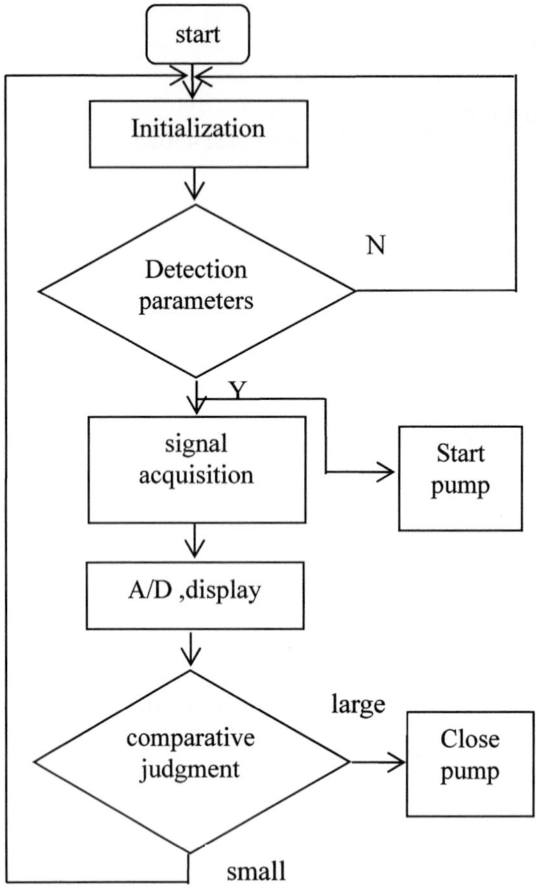

Fig. 3. The flow chart of the whole system software.

4. Conclusion

We design a system of intelligent portable instrument for the treatment of varicose veins based on MSP430 Series Microprocessor. It includes the power supply unit, MCU, operating unit, display unit, air pump, pressure sensor, air treatment unit. Through the internal program design of the MCU, the air pump on-off status can be controlled by the time sequence, and then the working state of gas bag is controlled, therefore it can realize a variety of treatment model by controlling different working states of the gas bag. The beneficial effect of the model is that the mode conversion can be realized by operating unit, which is simple and easy. The device is designed with low power components, and can be charged with lithium battery. It has the characteristics of miniaturization, low power, intelligent, portable and easy to operate.

References

1. Cardner AM, Fox R H. The venous pump of the human for preliminary report[J]. Bristol Medico Med Chir, 1983(98): 109 -112.
2. Laverick M D, M cGivern R C, Grone M D, et a1. A comparison of the effects of electrical calf muscle simulation and the venous foot pump on venous blood flow in the lower leg[J]. Phleb01ogy, 1990 (5): 285-2 9 0.
3. ShouShui Wei, Lei Guo, HongGuang Yang. Design and Realization of the system of plantar arteriovenous pulse for [J]. Electronic Measurement and Instrument, 2007, 21(6): 89-93
4. Xue Feng. Dong, Portable digital blood pressure meter based on MPX5050GP[J]. Electronic World, 2006 (08).
5. http://www.lierda.com/upload/editor/20111130/1322640883.pdf
6. http://wenku.baidu.com/view/92893bb0f111f18583d05ab2.html?re=view

Circuit design and optimization based on EXT, EXTU instruction

Sun-Cheng Xiang*, Min-Xuan Zhang, Zuo-Cheng Xing and Cang Liu
Science and Technology on Parallel and Distributed Processing Laboratory, National University of Defense Technology, Changsha, China
* *xiangsuncheng@163.com, {mxzhang, zcxing, liucang}@nudt.edu.cn*

EXT and EXTU are function instructions based on a chip DSP TMS320C62xx. This paper describes the processes of the circuit hierarchical design method at 130nm logic technology to complete the full-custom circuit design based on EXT, EXTU instructions. The work includes some functional verification in Verilog-level by using NC-Verilog after extracting an equivalent circuit diagram of the functional model, we also make delay analysis and optimization on the circuit diagram based on H-spice and completing the layout in modular design method by checking DRC and LVS successfully. This work also shows the physical design process of circuits by using EDA tools. Experimental result shows that the design circuit can achieve the aimed function effectively, which can be used in floating point unit, conversion unit. At last, some expectations are being put forward in further study.

Keywords: Full-custom circuit design; functional verification; equivalent circuit diagram.

1. Introduction

DSP TMS320C62xx is a kind of chip which is introduced by the United States TI (Texas Instruments), In recent years, with the development of large scale integrated circuit technology, DSP (digital signal processor) replacement rate is accelerating, it's operation speed is further improved and gradually expand the scope of application to communications, computer field, DSP is a unique microprocessor based digital signal to handle large amounts of information [1]. DSP is not only programmable, and in fact, it can deal tens of million complex instructions per second when running the program, much better than general-purpose microprocessor [2]. DSP is an increasingly important computer chip in digital electronic world.

In this paper, the design is a full-custom circuit of the intercepted instruction so we can understand how DSP TMS320C62xx chip works in the computer hardware. The approach of the circuit design affects the circuit's flexibility and adaptability, which has a significant impact on circuit and performance optimization [3].

2. Algorithm of EXT, EXTU Instruction

2.1. *The struction and function of EXT instruction*

Syntax EXT (.unit) src2, csta, cstb, dst

or

EXT (.unit) src2, src1, dst

.unit = .S1 or .S2

Opcode

Constant form is shown in Figure 1.

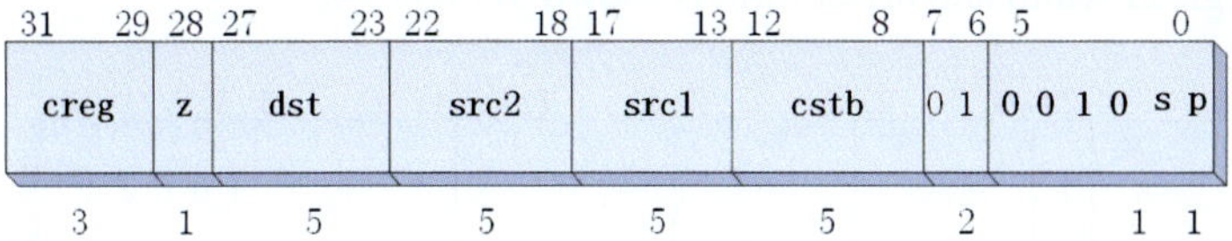

Fig. 1. Constant form.

Register form is shown in Figure 2.

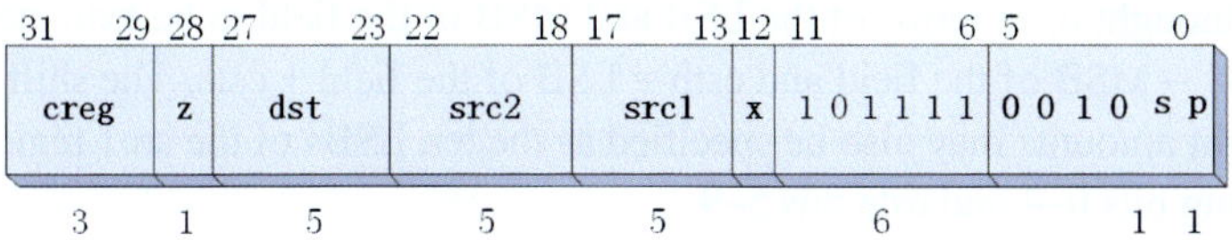

Fig. 2. Register form.

The field in src2, specified by csta and cstb, is extracted and sign-extended to 32 bits. The extract is performed by a shift left followed by a signed shift right. csta and cstb are the shift left amount and shift right amount respectively. This can be thought of in terms of the LSB and MSB of the field to be extracted. Then csta = 31 – MSB of the field and cstb = LSB of the field + csta. The shift left and shift right amounts may also be specified as the ten LSBs of the src1 register with cstb being bits 0–4 and csta bits 5–9.

2.2. *The struction and function of EXTU instruction*

Syntax EXTU (.unit) src2, csta, cstb, dst
or
EXTU (.unit) src2, src1, dst
.unit = .S1 or .S2

Opcode
Constant width and offset form is shown in Figure 3.

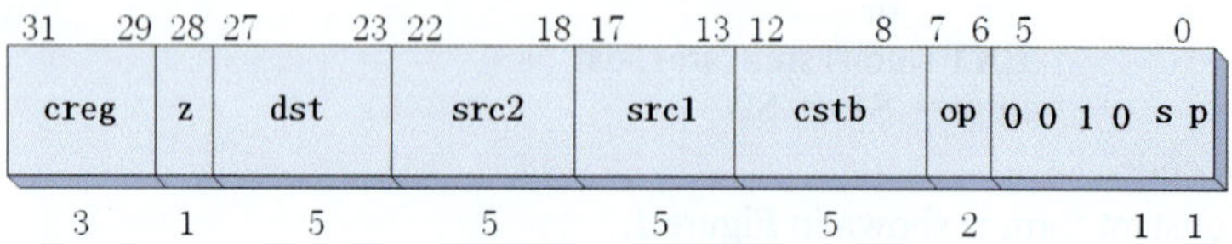

Fig. 3. Constant width and offset form.

Register width and offset form is shown in Figure 4.

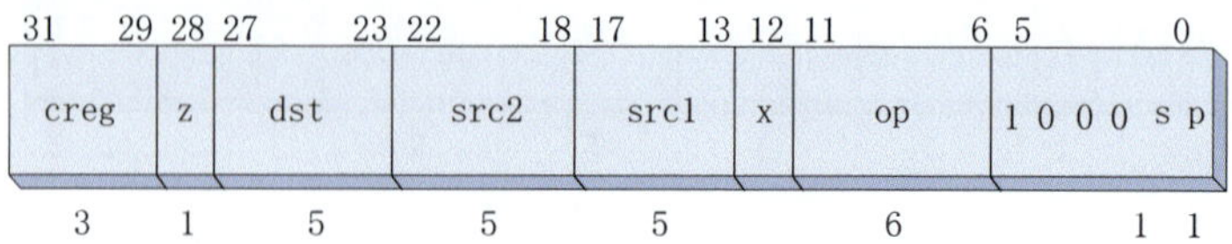

Fig. 4. Register width and offset form.

The field in src2, specified by csta and cstb, is extracted and zero extended to 32 bits. The extract is performed by a shift left followed by an unsigned shift right. csta and cstb are the amounts to shift left and shift right respectively. This can be thought of in terms of the LSB and MSB of the field to be extracted. Then csta = 31 – MSB of the field and cstb = LSB of the field + csta. The shift left and shift right amounts may also be specified as the ten LSBs of the src1 register with cstb being bits 0–4 and csta bits 5–9.

2.3. *The design idea of the implement the circuit*

EXTU instruction is defined to extract unsigned bit field. Instruction shift operation is the key to achieve the function of extracting unsigned bit field. Due to the existence of the left or right shift module in the process of the implementing the function instruction, it is necessary for us to design two 32-bit shift module — the left shift module and the right shift module [4]. The two shift modules are connected to each other after they are designed successfully, which can achieve the basic functions of EXTU instruction. The circuit design use the logarithmic shifter because there is 32 bits input, which is more effective in the area and speed. Logarithmic shifter can be parameterized easily and generated automatically comparing with barrel shifter. At this moment, we can design schematics to implement the 32-bit shift function.

In contrast, EXT instruction is defined to extract signed bit field. It is important to consider how to handle the sign bit in the process of designing the function circuit, which is different from the EXTU instruction [5]. In the experiment of designing the EXT function circuit, it is needed to shift left csta bit firstly, filling with number 0 in the empty position, and then shift right cstb bit, according to the first left to determine the filling number is zero or one, the approach in other places is the same as the way to achieve EXTU instruction. The overall structure of the design idea is shown in Figure 5.

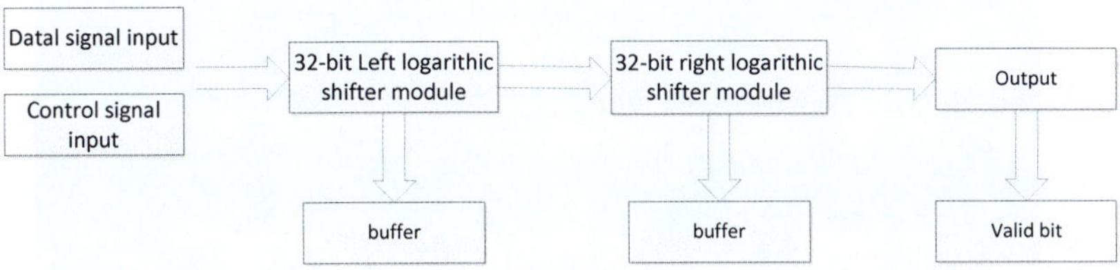

Fig. 5. Overall structure.

Firstly, this paper will define the overall functional and structural description, as in Figure 6, then we introduce the schematic design and circuit simulation result in verification by NC-Verilog, finally, the layout of functional units is designed to meet the requirements of DRC, ERC and LVS inspection. Up to this point, we completed all the steps basically.

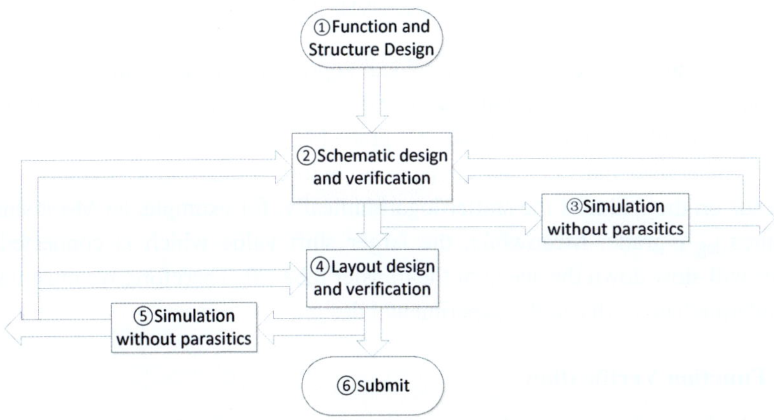

Fig. 6. Full-custom design flow.

3. Circuit-Level Design

As we can see from Figure 7, there is a simple one about left shifter or right shifter which can be controlled by special signal. The input word may be left or right, or remain unchanged, which is depending on the control signal. Many bits shifter can be achieved in a way of connecting in series with a number of such units [6]. When the shift value is more, this approach quickly becomes too complicated, not practical and low speedy. Therefore, we should adopt a more structured approach. This paper uses a logarithmic shifter.

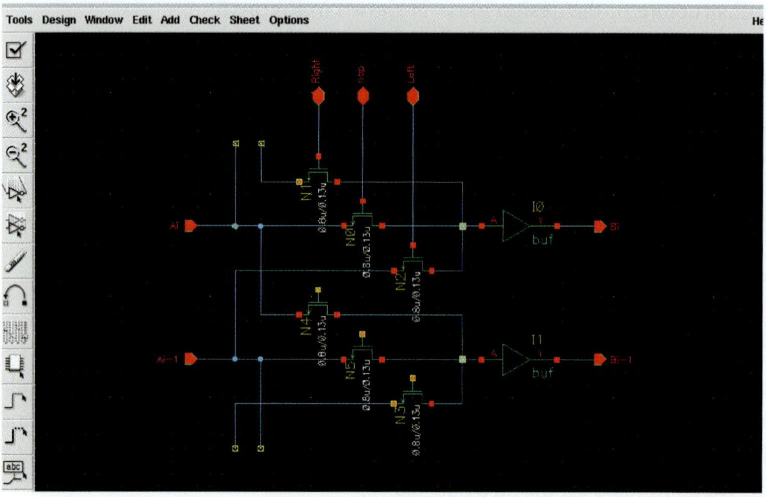

Fig. 7. A controllable shifter.

Logarithmic shifter uses a hierarchical approach. The total value of the shift is broken down into several index values of 2. A logarithmic shifter which has a maximum width of M includes $\log_2 M$ grade [7]. It's i stage can move the data i bits or otherwise transfer data as same as before. Logarithmic shifter's speed depends on the width of the shifter logarithmically, for example, an M-bit shifter requires $\log_2 M$ grade. Meanwhile, the larger shift value which is connected in series will slow down the speed of the transfer tube [8]. Therefore, we must insert the intermediate buffer in the experimental design.

4. Function Verification

With the increasing scale of integrated circuit, new verification requirements are proposed to ensure the correctness and efficiency of the verification. The tool NC-Verilog plays a key role in the process of functional verification, NC-Verilog is Cadence's Verilog digital logic simulator which has many advantages in running speed, high accuracy, debugging powerfully and flexible features, NC-Verilog uses the NCC (Native Compiled Code) technology to improve simulation performance and reduce the usage of memory [9]. In recent years, it has been highly recognized in microelectronics. NC-Verilog operation is divided into two steps: firstly we compile the program, the second is the simulation execution. We use INCA (Interleaved Native Compiled Code) architecture, with a support for multiple languages HDL (Verilog, VHDL), multi-level design (behavior level, RTL level, gate level, switching level) and mixed analog-digital design capability. After extracting an equivalent functional model of circuit diagram, we make functional verification in Verilog-level, including dividing and abstracting functional modules so that completing the most basic functions [10]. We write RTL code and associate verification environment of the function units based on the design rules of the EXT, EXTU instruction. Figure 8 is a waveform of the logic verification examples. The test result shows that the logic function of circuits is verified correctly.

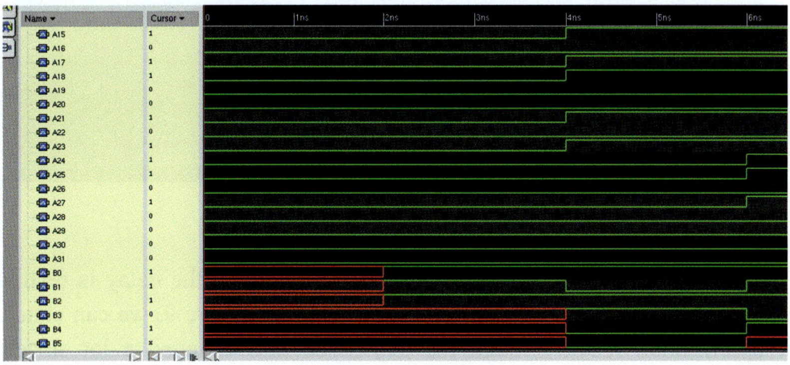

Fig. 8. Functional verification.

5. Timing Optimization

5.1. *Timing analysis*

This work takes a BSIM3v3 model as industry standards to simulate the deep submicron MOSFET transistors model. A complete BSIM3v3 model contains more than 200 parameters, which are relevant to secondary effects model mostly. Berkeley Short-Channel IGFET Model provides a simple analytical model [11]. We use FF, TT, SS, FNSP, SNFP 5 process generations angle in timing analysis, ensuring delay and timing requirements.

5.2. *Timing optimization with buffer*

We have designed a 32-bit left shift module and 32-bit right shift module, the basic structure of the circuit is logarithmic shifter, the speed of the logarithmic shifter depends on the width of the shift in a form of logarithm, A logarithmic shifter which has a width of M requires $\log_2 M$ grade. In the meantime, transfer tubes connected in series will slow down the speed of shifter for larger shift values. Therefore, it is necessary for us to insert the intermediate buffer carefully. In this experiment of optimizing the design, we will divide one shifter module into 5 grades, insert the intermediate buffer between stages, each buffer is constituted by two inverters in series.

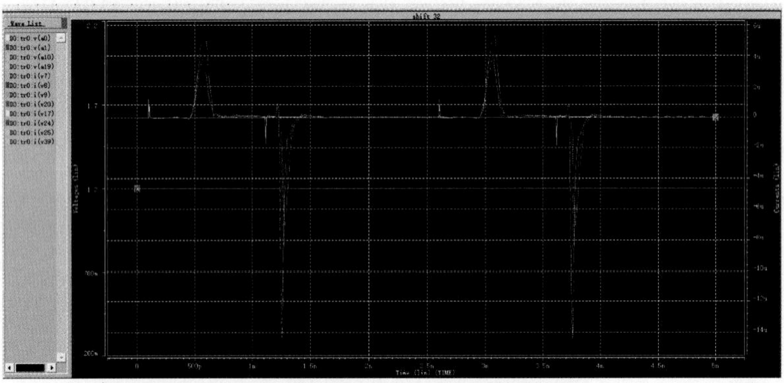

Fig. 9. Timing optimization.

Transfer tube output is not up to standard 1.2V and the delay is relatively large before the circuit optimization. However, from Figure 9, we can conclude that optimization result through the experiment has changed a lot since the completion of the circuit optimization, transfer tube output is up to standard 1.2V and the delay is relatively small. And the transfer tube work faster, the optimization earns a success.

When doing the timing optimization in this circuit inserted with buffer, not only circuit speed becomes faster, but also propagation delay will be reduced.

6. Layout Design and Verification

Layout design is composed of layout drawing and verification, the work of drawing is based on the tool named Virtuoso designed by the company Cadence [12]. We take hierarchical design approach, first complete the work of the various basic drawing unit, and then conduct all necessary verification and inspection based on Calibre tools in company Mentor, including design rule check, LVS, metal density check, antenna effect check, electrical characteristics inspection to ensure correctness of the design. The finished layout is shown in Figure 10.

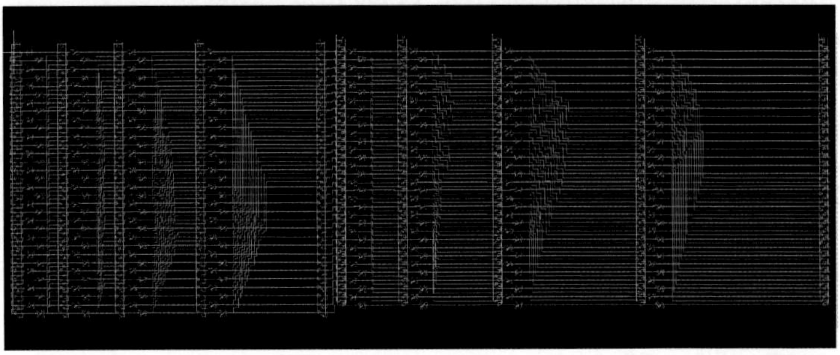

Fig. 10. Layout.

7. Conclusions

In this paper, we presented the circuit design in the approach of full-custom design based on EXT, EXTU instruction. Through the experiment, we can get some information about the layout: the final layout area is 146μm^2, the time required to complete a shift operation in the worst case is 115ps. In contrast, we also did logic synthesis in 130um crafts, the data shows the area in DC (Design Compiler) is 187μm^2, while the time delay reach a point of 168ps, with full-custom design, area and delay have earned respectively a percentage of 22% and 32%.

So we can make some expectations about its application in electronic industrial production: when the performance or density of the design is the major consideration, it seems to be the only choice for us to design the circuit's physical structure manually. High cost of custom-designed essentially means requiring a lot of work and time to deliver the market. Therefore, full-custom design is only to be considered reasonable economically in the following conditions:

- Custom modules can be reused many times.
- Cost can be allocated to the mass production.
- Cost is not the main design restrict in production.

So we have a responsibility to further study that how full-custom circuit achieve optimal performance at low power condition.

References

1. Givojnvic V, Rjiang S, Meyr H, "Compiled simulation of programmable," DSP architectures 1995.

2. Zhao Xiang, “Instruction Set Simulator of High-performance Digital Signal Processor,” (Papers) China IC, March 2015.
3. TMS320C6X, User's manual [Z], Texas Instrument, 1998.
4. Jan M. Rabaey, Anantha Chandrakasan and Borivoje Nikolic, “Digital Integrated Circuits: A Design Perspective, Second Edition,” Electronic Industry Press. October 2014, pp. 438–443.
5. LIN Fang-quan, WU Jian and RUAN Yuan, “Optimization Technology of DSP Instruction Set Simulator,” (Papers) Science Technology and Engineering 2011(29).
6. Sung-Mo Kang, Yusuf Leblebici, “CMOS Digital Intergrated Circuits Analysis and Design (Third Edition),” Beijing, Electronic Industry Press. January 2005， pp. 47-63.
7. C. Piguet *et al.*, Low Power Design of 8-b Embedded Cool Risc Micro-controller Cores,” IEEE Journal of Solid–State Circuits vol. 32, No. 7 July 1997: 1043-1078.
8. Nohl A, Braun G, Schliebusch O, “A univesral technique for fast and flexible instruction set architecture simulation,” 2002.
9. An Jianfeng, Fan Xiaoya and Zhang Shengbing, *et al.*, “Microprocessor Dragon C1 [J],” Computer Engineering and Applications, 0541(24) 83–114.
10. Michael D. Ciletti, “Advanced Digital Design with the verilog HDL,” Beijing, Electronic Industry Press, January 2005， pp. 387-401.
11. University of California Berkeley, BSIM3v3, Manual [M]. USA: “Department of Electrical Engineering and Computer Science,” 1995, pp. 4-18.

Design of a 20Gb/s 4-tap adaptive decision feedback equalizer in 40nm CMOS

Zhong-Liang Han

Radiation Laboratory of Microelectronics,
National University of Defense Technology,
Changsha, China
e-mail: 983414183@qq.com

Yang Guo

Institute of Microelectronics,
National University of Defense Technology,
Changsha, China
e-mail: guoyang@nudt.edu.com

Ze-Hui Guo

Institute of Microelectronics,
National University of Defense Technology,
Changsha, China
e-mail: 1575629695@qq.com

Wen-Liang Xue

Institute of Microelectronics,
National University of Defense Technology,
Changsha, China
e-mail: 335961382@qq.com

In this paper, a 20Gb/s 4-tap adaptive decision feedback equalizer (DFE) is proposed for high speed backplane communication, which includes a high-speed Finite impulse response filter (FIR), two amplifiers and an adaptation unit based on Sign-Sign Least-Mean-Square Error (SSLMS) algorithm. The FIR consists of samplers, an adder and Delay units. In order to realize a high speed, half-rate structure and current mode logic (CML)[6] are employed. Simulation result show that 20Gb/s data rate with horizontal eye opening of outputs is larger than 0.8UI.

Keywords: Adaptive; decision feedback equalizer; finite impulse response filter; half rate structure; current mode logic; sign-sign least-mean-square error.

1. Introduction

With the development of the high-speed serial interface technology, data transmission rate has been as high as dozens of GB per second. The presence of electrolyte loss, reflection, crosstalk and other factors cause the channel to show a low pass characteristic. When the high speed data is transmitted, the low pass characteristic of the channel can lead to inter symbol interference. Inter symbol interference will increase the error rate, reduce the system performance, and even make the system cannot work properly[3]. In order to reduce inter symbol interference, a filter is often connected in series at the receiving end. This filter is often called equalizer.

There are several equalization architectures, such as feed-forward equalizer, DFE and analog equalizer. Among them, feed-forward equalizer is used to eliminate both of the pre-cursor and post-cursor ISI, yet it amplifies the high frequency noises. DFE is capable of removing the post-cursor without amplifying crosstalk and noises[1]. Therefore, feed-forward equalizer and DFE are often used together for better equalization[4]. In this paper, we will present a 20Gb/s 4-tap DFE in 40 nm CMOS technology and under a 1.2V supply.

2. Decision Feedback Equalizer Architecture

The structure block diagram of 4-tap DFE is shown in Fig. 1.

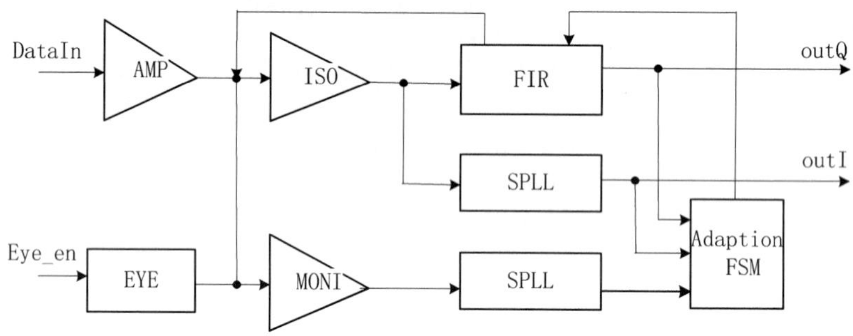

Fig. 1. The proposed 4-tap DFE.

Both FIR and SPLL modules work in half rate mode. To achieve the equalization of the channel, FIR, AMP and ISO together complete the elimination of post Inter symbol interference (ISI). EYE is a digital to analog conversion module, which is used to produce an ideal signal associated with the input signal. The equalized signal and the ideal signal subtract in the MONI module. error between equalized data and the ideal data is defined as[1].

$$e(k) = ID - Y(k) \tag{1}$$

Where ID is the ideal binary, a target digital symbol “0” or “1”, Y (k) is the equalized data[1]. The function of SPLL is to transform the signal from the analog domain to the digital domain. The error signal is converted into a full swing signal through a converter SPLL and fed into the DFE adaptive FSM module as an adaptive input signal. Adaptive FSM according to the input of the error signal, outI,outQ and SSLMS adaptive algorithm to adjust the tap coefficient of the FIR module.

EYE, MONI, data converter SPLL, and the DFE adaptive state machine module under the effect of equalized data together complete the tracking of the channel characteristics.

3. Circuit Design

3.1. *FIR module implementation*

The block diagram of the FIR module is shown in Fig. 2. To realize high speed, a half-rate structure is used and the received signal is split into two data process branches: odd arm and even arm, both of which operate at half-rate clock frequency[1]. The FIR module mainly consists of TAP, LATCH, SPLH three types of modules. TAP is the tap module, and four taps efficiencies can be updated based on the adaptive algorithm for controlling the amount of feedback. In order to optimize the speed, the TAP uses the current mode logic to realize. TAP specific circuit structure is shown in Figure 3. Taps choose the odd path of the output or even depend on the clock CN, CP. The size of the tap coefficient is measured by the size of the tail current. TAPS are programmable current sources. The output junction of four taps achieves a current plus, which is equivalent to the function of an Adder.

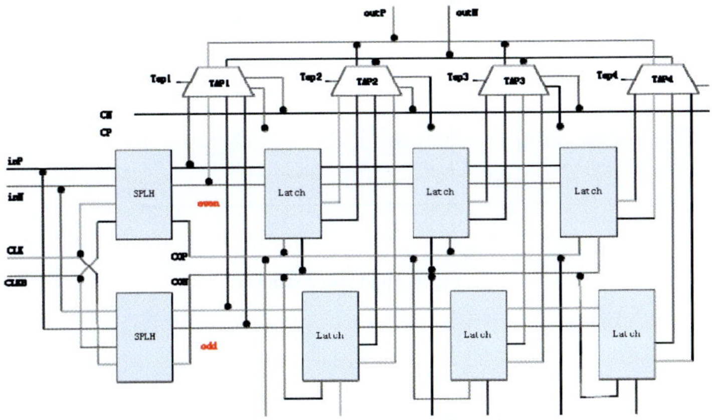

Fig. 2. Half rate working mode FIR.

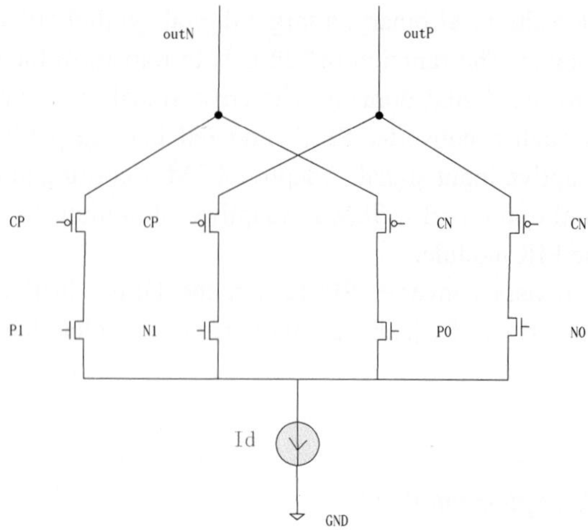

Fig. 3. Current mode logic TAP.

The SPLH module consists of two high speed samplers or comparators and two latchs with block. The specific circuit of the SPLH module is shown in Figure 4 and Figure 5. The function of LATCH acts as a fixed delay unit. The clock COP, CON, CP, CN, CLK, CLKB is the control clock of LATCHs, SPLHs and TAPs, which operates at half the rate of the received signal.

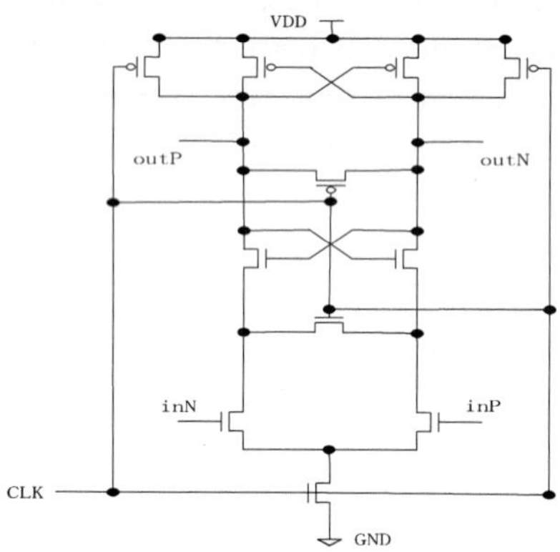

Fig. 4. High speed sampler.

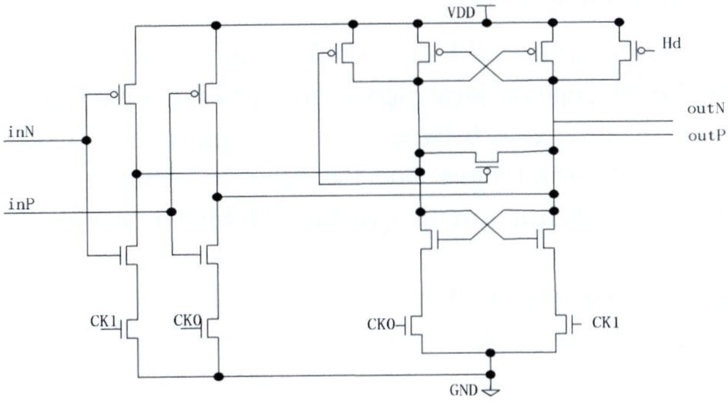

Fig. 5. Latch.

3.2. *AMP and ISO module implementation*

The specific circuit structure of AMP module is shown in Figure 6. The feedback node is connected to the output of the AMP module. There are three main functions of AMP module: Control amplification factor by controlling the tail current, the conversion of current and voltage on the output resistance is realized, the input signal and the feedback signal are used as the difference to get the balanced signal. Under the action of the balanced signal, ISO generates two pairs of differential signals: one pair as the input of the FIR module, and the other pair of signals passing through the high speed converter SPLL produces an output as the output of the whole circuit. The specific circuit structure of ISO module is shown in Figure 7.

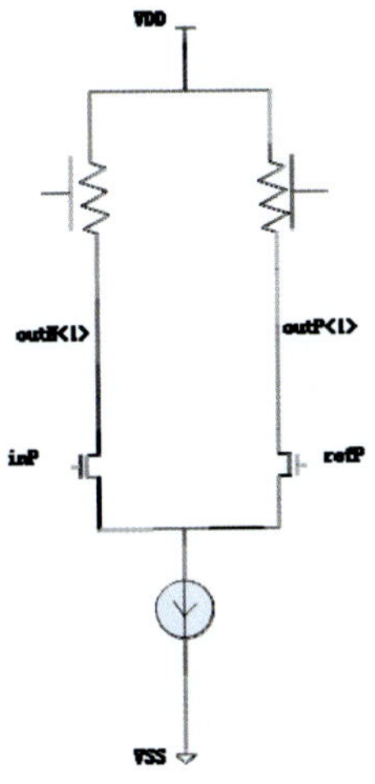

Fig. 6. AMP.

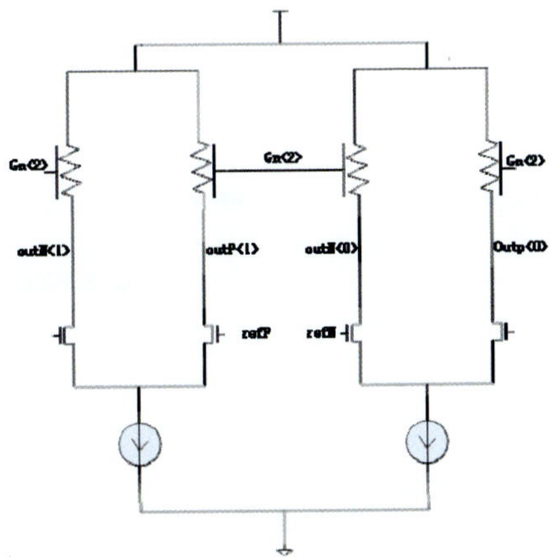

Fig. 7. ISO.

3.3. *EYE module implementation*

The specific circuit structure of EYE module is shown in Figure 8. The EYE module is used to generate an ideal signal related to the input signal. The error signal is equal to the difference between the ideal signal and the balanced signal. The subtraction of the ideal signal and the balanced signal is realized by the MONI module. The specific circuit structure of MONI module is shown in Figure 9.

The adaptive finite state machine module, based on the SSLMS algorithm, is a simple digital block implemented by VERILOG code. The function of adaptive FSM is expressed as following expression.

$$C(k+1)= C(k) + \mu e(k)X(k) \tag{2}$$

Where X(n) is the decided data, at time nT, either one or zero, μ is the update step. C(n) is the tap coefficient at time nT, C(n+1) is the tap coefficient of the next cycle.

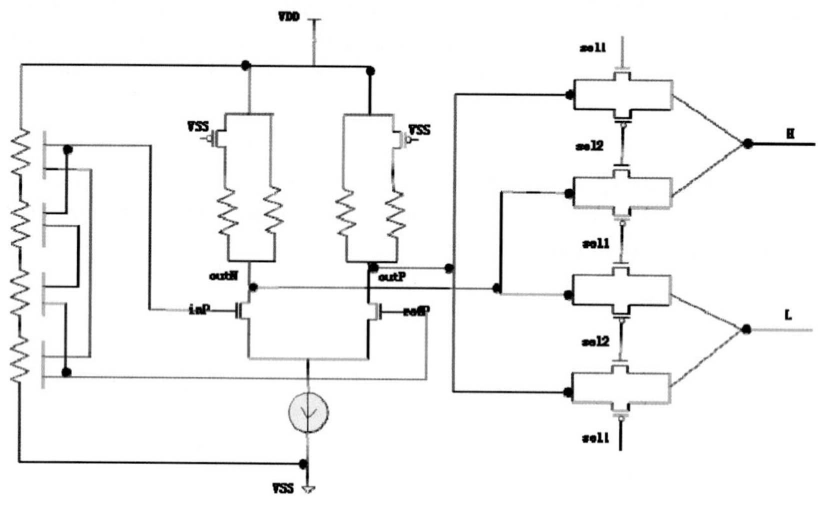

Fig. 8. EYE.

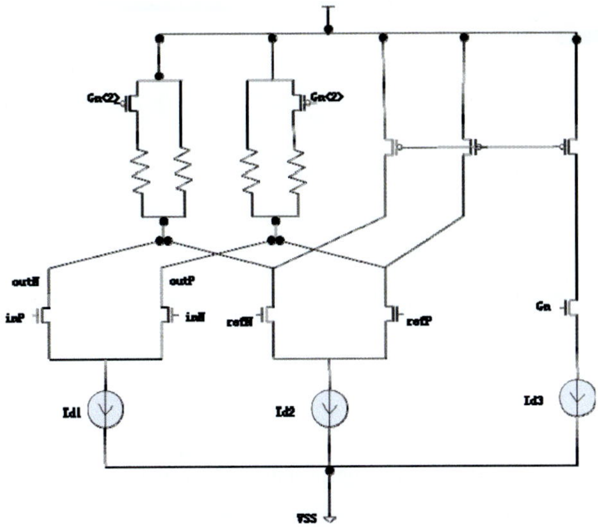

Fig. 9. MONI.

4. Simulation Results

In order to measure the function and performance of DFE, pseudo-random binary sequence (PRBS) is used as the input signal[2]. Output of DFE is a half rate data, because all modules of DFE work at half rate mode. Loss of channel is 42dB in 20GHz[5].The eye diagram before and after DFE are shown in the Fig.10. Schematic simulation result show that 20Gb/s data rate with horizontal eye opening of outputs is larger than 0.8 UI. From the simulation results we can see that both good timing and voltage margins are achieved, which demonstrates the 4-tap DFE could equalize data with serious ISI effectively[2](All the simulation results are obtained under TT corner, 65°C and the power supply voltage 1.2V).

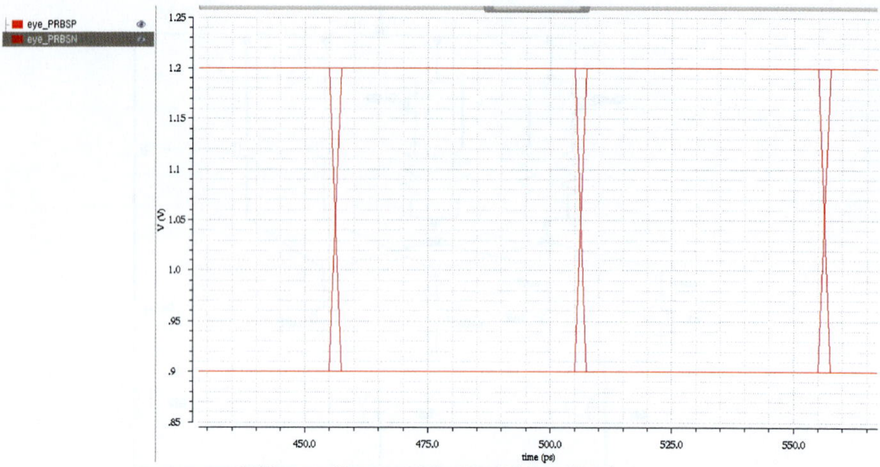

(a)The eye diagram of the PRBS input signal

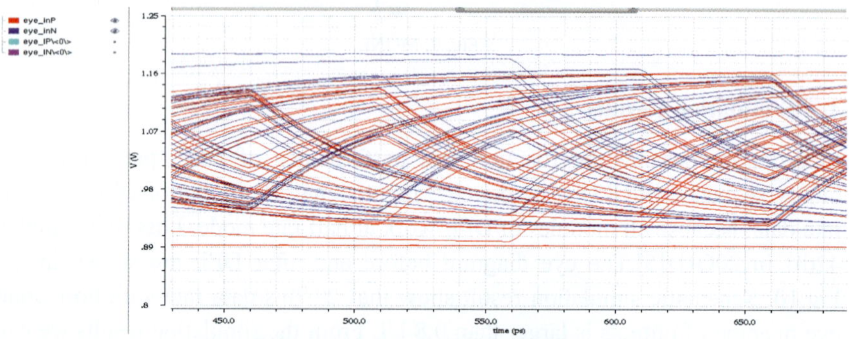

(b)The eye diagram before equalization

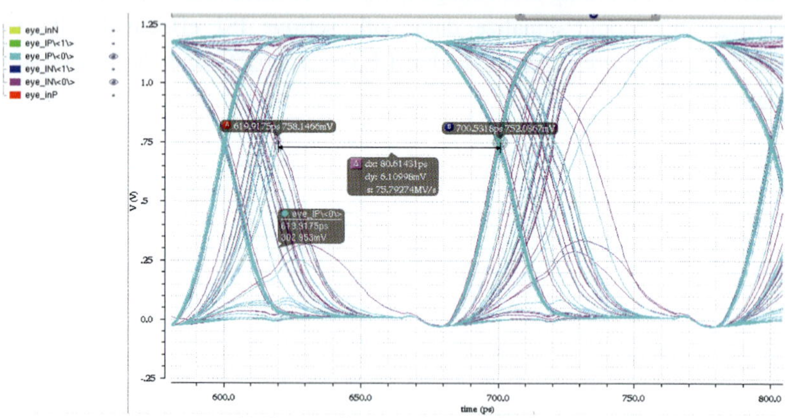

(c) The eye diagram after equalization

Fig. 10. Simulation result.

5. Conclusion

This paper proposed a 20Gb/s 4-tap DFE in 40nmCMOS technology. Half rate FIR modules ease the timing of the circuit, and the current mode logic is used to optimize the speed of the circuit. FIR AMP and ISO together complete the elimination of Inter symbol interference (ISI). EYE, MONI and Adaptive FSM are adopted together to implement the adaptation based on SSLMS algorithm. Thus get a 20Gb/s 4-tap DFE.

References

1 Wu, Xiao, and Qingsheng Hu. "Design of a 6.25 Gb/s adaptive decision feedback equalizer in 0.18μm CMOS technology." Advanced Research and Technology in Industry Applications (WARTIA), 2014 IEEE Workshop on. IEEE, 2014.

2 Cao, Weidong, et al. "A 40Gb/s 39mW 3-tap adaptive closed-loop decision feedback equalizer in 65nm CMOS." Circuits and Systems (MWSCAS), 2015 IEEE 58th International Midwest Symposium on. IEEE, 2015.

3 Cheng, Hung-Yi, et al. "Robust decision feedback equalizer scheme by using sphere-decoding detector." Acoustics, Speech and Signal Processing (ICASSP), 2014 IEEE International Conference on. IEEE, 2014.

4 Choi, Yongsuk, and Yong-Bin Kim. "A 10-Gb/s receiver with a continuous-time linear equalizer and 1-tap decision-feedback equalizer." Circuits and Systems (MWSCAS), 2015 IEEE 58th International Midwest Symposium on. IEEE, 2015.

5 Guo, Jinxing, et al. "A 25Gb/s adaptive decision feedback equalizer for 100G Ethernet in 65NM CMOS technology." Solid-State and Integrated Circuit Technology (ICSICT), 2014 12th IEEE International Conference on. IEEE, 2014.

6 Yuan, Shuai, et al. "A 10Gb/s speculative decision feedback equalizer with a novel implementation of adaption in 65nm CMOS technology." Electron Devices and Solid-State Circuits (EDSSC), 2014 IEEE International Conference on. IEEE, 2014.

Study on Evolution of Machine Tools Based on TRIZ

Ri-Jun Wang†, Ming-Qin Zhang and Wei-Ling Huang
TRIZ Institute, Shandong Jianzhu University,
Jinan, 250101, China
†E-mail: wangrj2011@163.com

Wei-Juan Luan
Laiwu Steel Machinery Manufacturing Co., Ltd.,
Shandong Iron and Steel Group Laiwu, 271104, China
E-mail: firstmark@vip.163.com

TRIZ theory was applied to the innovative research of machine tools. Based on TRIZ directed evolution and in accordance with several evolutional modes such as the increase of controllability and degree of automation, the evolution to microscopic level, high integration then simplification and the enhancement of idealized level, the development of key technologies for machine tools were reviewed. On this basis, the possible future structure of the machine tools was explored and the possible future development direction of the machine tools which included the improvement of controllability and degree of automation and the enhancement of idealized level was proposed in order to provide reference for the innovation and development of machine tools.

Keywords: TRIZ; machine tools; evolution; innovation.

1. Introduction

TRIZ (Theory of Inventive Problem Solving) theory is a systematic theoretical system solving inventive problem which was proposed by the former Soviet Union inventor Genrich S.Altshuller and his team based on the analysis of 2.5 million of the world's high-level patent from 1946. TRIZ theory is the most powerful method solving innovative problems currently, as it allows designers to obtain high quality innovative solutions according to gradual steps clear objectives, just like solving common technical issues[1~3]. Evolution theory is the core of TRIZ, which reveals the basic laws of technical systems' evolution and can guide the designer to develop new products in the right direction, which makes the whole innovation process into a predictable, actionable and effective technological processes[4,5].

Machine tool is one of the core machinery in machinery manufacturing which has a pivotal role for the whole machinery industry. Technological development and social progress put forward higher requirements to machinery manufacturing technology and in order to better achieve processing needs, the performance indicators of machine tool continued developing and its quality and level of automation continuously improved. In this paper, the evolutionary process of machine tool was reviewed and sorted from different perspectives based on TRIZ evolutionary theory and then the possible future development direction of machine tool was explored.

2. Technology Evolution Theory of TRIZ

Technology evolution theory is the theoretical basis of TRIZ which is an important branch of specialized research on technical evolution prediction and and its main point is the evolution of technology systems are not random but follow certain rules. In recent years, the attention to TRIZ technology system evolution theory continued to increase in academia and various research results have been developed, such as: Technical Evolution Theory[6], Technical Evolution Guiding Theory[7], Directed Evolution[8], and Technical Evolution Law[9], etc. Each version is distinctive, but they are all the development of Altshuller's TRIZ traditional evolution theory. Directed Evolution developed by A.Zusman was used to summarize the evolutionary history of machine tool, which included eight kinds of evolutionary patterns[10].

Mode 1 S curve evolution rule of technical systems: generally the growth of technology systems includes four stages: infancy, growth, maturity and recession

Mode 2 Rule of ideation improvement.

Mode 3 Rule of uneven evolution of subsystem.

Mode 4 Rule of evolution towards dynamics and controllability.

Mode 5 Rule of integration increased and then simplified.

Mode 6 Rule of coordinated evolution of subsystem.

Mode 7 Rule of evolution towards micro-level and increasement of the field application.

Mode 8 Rule of evolution towards the reduce of manual intervention.

3. Key Technology Analysis of Machine Tool

Development of machine tool is closely related to the development of mechanical, electronic and computer technology and its evolutionary process is regular which can be analyzed from the following aspects.

3.1. *Increase the degree of automation and reduce manual intervention*

The change of the control system making the machine tool automation continue to improve. Initially machine tool was operated manually, and then Semi-automatic machine which was operated fully automated except the work piece handling and fixed appeared. Now, automatic machine which can achieve fully automated machining movement in accordance with a control command has been widely used. Machine operation requires less human intervention, thus human can be freed from tedious work and to do more mental work.

3.2. *Increase integration and then simplify*

In TRIZ, technical systems tend to be complex structure at first in the development process, that is to improve the characteristics of the system function by increasing the number of system components, and then gradually streamline, with a simple system to achieve the same function or achieve better functionality. In short, to realize the functions as much as possible by the simplest structure.

Fig. 1. Machine center with bamboo hat ATC.

Fig. 2. Multi-axis machine tools.

Traditional machine processing method was single, stand-alone, a knife, a sequence, and then the so-called centralized procedure was used which included following ways: First, many of the machining processes properly focused on a single machine, such as various types of machining centers (Fig. 1) which can simultaneously achieve many kinds of machining processes including milling, boring, drilling, etc. Second, focus on a single machine several stations, such as multi-tool lathes, multi-axis lathes (Fig. 2), multi-broaching and so on. Thus, centralized process processing methods can ensure the expansion of machine tool's capabilities in the premise of the footprint unchanged or little change to achieve the machining of high efficiency, low cost and high quality[11].

3.3. *Improve the Ideal*

Improve the Ideal is the basic rule of the evolution of technology systems which points out the direction of creative problem-solving theory, that is, to improve the Ideal represent the ultimate direction of all the evolution rules. Formula of Ideal is shown in Eq. (1).

$$\text{Ideal} = \frac{\sum \text{Useful function}}{\sum \text{Harmful function+Cost}} \tag{1}$$

Ideal improvement mainly includes the following several evolutionary lines by which the development of the machine was analyzed:

(1) Simplify subsystem

Since the 1950s, the CNC system of machine has experienced tube - transistors - small and medium-scale integrated circuits - LSI - microcomputers - industrial PC, and became smaller gradually. At present, more integrated circuit chip, VLSI components and multi-layer printed circuit board are used, which significantly reduce the footprint of the system.

(2) Simplify operations

CNC machines gradually become a high degree of automation which require less manual intervention and greatly simplified the operation of the workers so that gradually reduce labor intensity. Meanwhile human factors to the precision of parts was minimized which gave more effective guarantee of product quality.

(3) Increase useful function

Since the birth date, the machining accuracy of machine has been improved day by day and the current ultra-precision machining accuracy has reached the nanoscale, so that the quality of machined parts rose to a new level.

(4) Decrease harmful function

From the traditional decentralized process to the current centralized process, the change of processing methods reduced the number of workers as well as the number and footprint of devices, which saved manpower and resources, that is, the machine evolved along the direction of reducing the harmful function.

Thus, the machine developed along with useful functions gradually increased and harmful functions and cost gradually reduced. The evolution of machine is a process of continuous improvement of the Ideal.

4. Predict of Evolution

Currently, with the continuous rise of advanced manufacturing technology, the application of ultra high-speed cutting and ultra-precision processing technology as well as the rapid development of microelectronics and computer technology, modern CNC machine is still in constant development whose development direction can be forecasted by evolution as follows:

4.1. *Enhance controllability and automation level*

Intelligent is the trend of machine development, but current technologies are not mature enough. Intelligent machines need to be further improved, especially in the automatically collect, store, regulate, control and optimization of processing parameters and the reliability, stability and durability after the intelligence, networking and integration[12].

4.2. *Improve the Ideal*

The evolution of any technical system is an ideal ever-increasing process. With the useful effect continuously improved and the harmful effects gradually decreased, it is inevitable that the machine continue to improve along the direction of Ideal increase. Increase of useful effects mainly reflected in the improvement of processing speed and precision, and the decrease of harmful effects mainly included the material savings, the energy consumption reduction and the machine recycling after scrap, etc.

5. Conclusion

(1) The development process of the machine key technologies were studied in accordance with the process of product evolution based on technology evolutionary theory.

(2) The future of the machine were predicted by the application of evolutionary rule which pointed out that improving controllability and the degree of automation and continue to increase the ideal are the possible future development direction of the machine.

Acknowledgments

This work was financially supported by the Natural Science Foundation of Shandong Province (ZR2014EEQ011), University Science and Technology Projects in Shandong Province (J14LB03).

References

1. Niu Zhanwen, Chen Dongliang, He Lihong. Evolutionary analysis of key technologies and innovative design of color sorter based on TRIZ[J]. Cereals and Oils Processing, 2007, (3): 65~67.

2. Petkovic D, Issa M, Pavlovic N D, et al. Application of the TRIZ creativity enhancement approach to design of passively compliant robotic joint[J]. The International Journal of Advanced Manufacturing Technology, 2013, 67 (7): 865~875.
3. Kremer G O, Chiu M C, Lin C Y, et al. Application of axiomatic design, TRIZ, and mixed integer programming to develop innovative designs: a locomotive ballast arrangement case study[J]. The International Journal of Advanced Manufacturing Technology, 2012, 61(7): 827~842.
4. Kim S, Yoon B. Developing a process of concept generation for new product-service systems: a QFD and TRIZ-based approach[J]. Service Business, 2012, 6(9): 323~348.
5. Cho C H, Kim K H. Product development with TRIZ: design evolution of deburring tools for intersecting holes [J]. Journal of Mechanical Science and Technology, 2010, 24(1): 169~173.
6. Savransky S D. Engineering of creativity[M]. New York: CSC Press, 2000.
7. Fry V R, Rivin E I. Guided technology evolution[EB/OL]. http://www.trizgroup.com/article2.html, 1999.
8. Zusman A, Zlotin B, Zainiev G. An application of directed evolution [EB/OL]. http://www.ideationtriz.com/Endoscopic-Case-Study.html
9. Petrov V. The law of system evolution[EB/OL]. http://www.triz-journal.com, 2002.
10. Yang Qingliang. This is the birth of invention: TRIZ theory touching[M]. Beijing: Machinery Industry Press, 2006.
11. Guangzhou Machine Tool Research Institute. History of machine tools development [M]. Guangzhou: Guangdong People's Publishing House, 1997.
12. Shan Yicai, Li Yimin, Liu Shihao, et al. Research present status and development tendency of intelligent machine tools[J]. Tool Technology, 2009, 43(9): 3~6.

A case-based reasoning cost estimating model of hydraulic cylinder remanufacturing

Pin-Pin Qin*, Chun-Jian Liu, Jian-Wei Shu and Ling-Yun Yan
College of Mechanical Engineering, Guangxi University, Nanning 530004, China
**E-mail: qpinpin@gxu.edu.cn*

Hydraulic cylinder remanufacture process has been analyzed and case properties have been extracted, then a case-based reasoning cost estimating model of hydraulic cylinder remanufacturing was built. The three methods, namely, equal weight (EW), gradient descent method (GDM) and analytic hierarchy process (AHP) are used to compute the weights of case attributes. The cost estimating of three different models are compared. The fidelity of CBR model using AHP weights is better than that of the other two models. By solving new projects using experience from historical cases, it is found that CBR was an effective method in cost estimating of hydraulic cylinder remanufacturing.

Keywords: Remanufacturing; hydraulic cylinder; cost estimating; case-based reasoning (CBR); analytic hierarchy process (AHP).

1. Introduction

Remanufacturing is an important measure to saving energy, and it is one of the most effective way to promote recycling economy and green economy for sustainable development. Because remanufacturing of used construction machinery will bring considerable economic benefits to businesses and customers, remanufacturing economic evaluation is now getting more attention. When remanufacturing costs are too high and exceed the value of the new ones, used ones are not worth to be remanufactured. Reasonable remanufacturing cost estimation is important for remanufacturing economy evaluation. The models built in this study are also very helpful to provide effective algorithms for remanufacturing economy evaluation system embedded remanufacturing technology information system, which based on remanufacturing cloud services platform.

With the uncertainty of working conditions and environments, hydraulic cylinder failure conditions are different. Then the remanufacturing technology is

*Corresponding author.

complicated too. In the end, it leads to uncertainty and randomness of manufacturing cost estimating. However, when the type of hydraulic cylinder is the same, failure conditions and remanufacturing technology are similar, there could be similarity and comparability for cost estimating. Considering proximity can resolve the uncertainty and randomness, this study attempts to estimate hydraulic cylinders remanufacturing cost base on case-based reasoning (CBR). By comparing the similarity of the source case and target case among hydraulic cylinder historical data, the target case's cost is the same as the most similar case cost.

2. Methodology

2.1. *CBR*

Case based reasoning is an important reasoning technology which develops from the field of artificial intelligence. CBR is different from rule based reasoning (RBR) and model based reasoning (MBR). In CBR, a reasoned remembers a previous situation similar to the current one and uses that to solve new problems. CBR solving new problems by using the experience of similar solved previous problems to conduct reasoning. CBR has been used successfully in many areas, such as disaster emergency material demand forecasting, crop insect pests forecasting, single disease cost prediction, food demand forecasting, bankruptcy prediction, and finding oil slick.

2.2. *AHP*

Analytic hierarchy process (AHP) was developed by Saaty in the early 1970s to help individuals and groups deal with decision. AHP uses hierarchic structures to represent a decision-making problem and then develops priorities for the alternatives based on the decision maker's judgment throughout the system. As a systematic approach for solving decision-making problems, AHP has been applied in various fields, such as quality decision system [1], cases indexing [2], similarity measurement [3], and cost estimating model [4]. AHP is systemically adopted to determine the relative importance of elements to the goals through pairwise comparison of the elements. In general, AHP modelling consists of three stages: (1) structuring a complex problem in the form of a simple hierarchy; (2)comparing the decision elements using the pairwise method; (3) computing the relative weights of the decision elements.

2.3. *CBR model*

A case-based reasoning cost estimating model of hydraulic cylinder remanufacturing is proposed. Steps to establish CBR model are as followings.
Step 1: According to the hydraulic cylinder remanufacturing process, determine the case attribute and the specific attribute values.
Step 2: According to the CBR technology, represent the case in mathematical equations.

Step 3: Normalize the attribute value. Change qualitative indicators to quantitative indicators. Divided is assembling, cleaning, and inspection indicators into three grades. Take the characteristic factor x_i's maximum and minimum values as x^{imax} and x^{imin}. Processing formula is: when there is a bigger (smaller) factor characteristics and a lower (higher) cost, $x_i = (x^i - x^{imax}) / (x^{imax} - x^{imin})$. when the characteristic factor is smaller (larger) and the cost is lower (higher), $x_i = (x^{imax} - x^i) / (x^{imax} - x^{imin})$.

Step 4: Use AHP to determine the attribute weights of CBR model. First, a hierarchy structure of hydraulic cylinder remanufacturing cost factors was developed. Second, construct the AHP judgmental matrix based on the knowledge of experts. Finally, calculate attribute weightsofcase by AHP.

Step 5: Take nearest neighbor algorithm to compute similarity, find out the greatest similarity as the most similar cases and choose it as the cost of target case.

Step 6: Modify the case to get a new case.

Step 7: Obtain the new case by evaluating results and saved new ones.

To retrieve the most similar case from the case base, the step 5 similarity Eq. (1) is used to calculate the percentage similarity, which indicates the extent of similarity between one or more of the stored cases and the new case.

$$d^{oi} = \left[\sum_{j=1}^{m} w_j (\overline{x_{ij}} - o_j)^2 \right]^{\frac{1}{2}} \tag{1}$$

Where, w_j is the weight of attribute j and affects the size of similarity. $\overline{x_{ij}}$ is the weight of attribute j in the i case history. o_j is the value of target case j 's attribute. d_{oi} is the Euclidean distance between target case o_j and the history case H_i. The d^{oj} is smaller, the similarity between target case and case is bigger.

3. Application

3.1. *Data*

Establishing the historical case base is based on enterprise's remanufacturing condition of key cylinder parts, including 10 attributes, such as recycling old pieces grades, the cost of recycling old pieces, weight (kg), the degree of disassemble, the degree of cleaning, the degree of detection, the complexity of the repair, raw material cost/(RMB), energy consumption cost/(RMB), and the device consumes fee/(RMB). Cost (RMB) is a decision factor. Twenty five history cases and five target cases were calculated. All data are normalized due to involve enterprise's important information.

3.2. *Hierarchy structure based on AHP*

In this study, four layers of hierarchy were structured. Layer 1 was the target, which evaluated the influence of attributes on the cost. Layer 2 contained two main criteria, which are 'quantity criteria' and 'quality criteria'. Layer 3 specified these criteria for specific evaluations.Layer 4, the last layer, contained the decision elements, which are the attributes in the CBR model. All attributes in Layer 4 were related with three criteria (condition of old pieces, old pieces processing,remanufacturing technology);Therefore, all attributes were evaluated with respect to each criterion of layer 3.

Take "expert method" to determine the attribute weights. Expert team members consists of construction machinery, remanufacturing engineering professional and technical personnel and management personnel with an average working experience of 7 years. A questionnaire survey was conducted to collect expert's opinion for evaluating the weights of the attributes. Questionnaires were sent to experienced engineers. They returned a completed questionnaire. Pairwise comparisons using a scale of 1–9 were used to evaluate the relative importance of elements.The relative weights of the decision elements at each layer were computed as the normalized eigenvector associated with the comparison matrix by using the geometric means. Consistency was also checked. When consistency test is qualified, this normalized eigenvector represented the experts' evaluation on attributes weights.

3.3. *Determining the weight of attributes*

To successfully retrieve a case similar to a new case, the keysdepend on the weights. In this study, three kinds of weights are adopted: The first, choose Equal Weight (EW); The second, determined by Gradient Descent Method (GDM) in accordance with the literature;The GDM determines each of the important weights of the attributes as literature; The third, determined by AHP method, as described in 3.2 content. Three kinds of cases' attributeand weights of the elements at layer 4 which influence the cost are computed.

3.4. *Evaluation of performance*

In this study, the relative error rate calculated from Eq. (2) was used. The cost estimating results of different CBR models from the 5 test cases are summarized in Table 1. The results from the CBR model, which determined the weights of attributes using the results of AHP (AHP–CBR), had the minimum average error of 5.54%, and the error of five test samples were less than 8%. The results from the CBR model, which applied the equal weight of attributes (EW–CBR), had the average error of 17.79%. The maximum error was 42.44%, minimum error was 42.44%, and the error fluctuation was the largest too. The results from the CBR model, which determined the weights of attributes by using the GDM (GDM–CBR),had the average error of 31.17%. The maximum error was 31.17%, and minimum error was 7.32%. As a result, error fluctuation was less than EW–CBR model. The model using the GDM was more accurate than the EW–CBR model, and less accurate than AHP–CBR model.

$$Relative\ \ error(\%) = \left| \frac{Actual\ \ Cost - Estimating\ \ Cost}{Actual\ \ Cost} \right| \times 100\% \qquad (2)$$

Table 1. Results of the CBR models.

Cases / Error rate(%) / Models	AHP-CBR	EW-CBR	GDM-CBR
1	3.1	5.2	7.32
2	4.8	10.6	9.8
3	5.5	11.2	12.54
4	6.7	42.44	31.17
5	7.6	17.79	14.36
Average	5.54	17.79	14.36

4. Discussion

The EW–CBR model, the GDM–CBR model and the AHP–CBR model for the test samples gave a relative error of 17.79%, 14.36% and 5.54% respectively. In addition, as shown in Table 1, all relative error of the test samples using AHP–CBR model were within 8%. The EW–CBR model and the GDM–CBR model had over 30% relative error of test samples.

The results of this research showed that the estimating accuracy of the CBR model related to the method for determining the weights of the attributes, and would be advanced by the use of remanufacturing experts' knowledge. The AHP adopted the weights of attributes by using the experience and knowledge of experts was more accurate than the other two CBR models (EW–CBR model and GDM–CBR model). The AHP–CBR model could explain the process of determining the weights of attributes, therefore, users and decision makers can better understand the process of CBR model for estimating remanufacturing cost in the AHP–CBR model. In this way, the AHP–CBR model is better than the EW–CBR, and GDM–CBR model.

5. Conclusion

The CBR hydraulic cylinders remanufacturing cost estimation model is an effective method that reuses the experience of remanufacturing experts from previous cases to obtain the cost of a new case.

The application of hydraulic cylinders remanufacturing cost estimation shows that the AHP–CBR model is more accurate than the EW–CBR model, which applied equal weights for the attributes, and the GDM–CBR model, which determined the weights of attributes using the GDM.

Acknowledgments

The authors gratefully acknowledge the support of the Guangxi Natural Science Found, Project 2015GXNSFAA139280, in supporting all parts of this study.

References

1. Huang W, Jin S, Lai X M, et al. *A decision-making system based on AHP-CBR. Journal of Machinery Design and Manufacture,*Vol. 8 (2008), pp. 205-207.
2. Qu X L, Du J, Sun L F. *A Model for Cases Indexing in CBR System Based on AHP. Journal of Application Research of Computers,* Vol. 4 (2005), pp. 33-34,38.

3. Luo M Zh, Zh M L. *On the method of similarity measurement in retrieval of CBR. Journal of Hefei University of Technology*, Vol. 24 (2001), pp. 1091-1094.
4. Sung Hoon An, Gwang Hee Kim, Kyung In Kang. *A case-based reasoning cost estimating model using experience by analytic hierarchy process. Building and Environment,* Vol. 42 (2007), pp. 2573–2579.
5. Saaty TL. The analytical hierarchy process. (McGraw Hill, 1980).
6. Yau NJ, Yang JB. Applying case-based reasoning technique to retaining wall selection. Automation in Construction, Vol. 7 (1998), pp. 271–83.

The cutting algorithm of spiral bevel gear base on the variable transmission ratio and tilt method

Tao Li and Yuan-Tao Wang*

Materials Engineering, Liaoning Mechatronics College,
Dan Dong, China
E-mail: wangyuantao1985@163.com

In this algorithm, the wheel embryo design process is consistent with the design method of Gleason. The big wheel cutting calculation is the same to Gleason, but the small wheel cutting calculation is not the same to Gleason. Conical shape wheel is used to machining small spiral bevel gear wheels .The transmission ratio between shape wheel and work piece is also a variable. It is mainly applied to NC cutting and grinding machining.

Keywords: Spiral bevel gears; NC.

1. Introduction

When the small wheel of spiral bevel gear is processed, the generating gear and the small wheel are considered as a pair of spiral bevel gears. Assuming the cutting surface of shape wheel was the first surface, the small wheel face was the second surface [1]. If the conical shape wheel has a generating node M, according to the theory of line contact conjugate, at the M point the tooth length and tooth height direction of curvature and geodesic torsion can be obtained in the small wheel. In Fig 1, the cutter forming radius M_r and the cutter tip radius is calculated. When the cutter data is given, the cutting calculation can ensure the tooth length and tooth height direction of curvature and geodesic torsion invariant, but they cannot meet the requirements at the same time. However, this variation can be obtained [2]. In machining process, the variable transmission ratio processing can be modified tooth high curvature. In order to meet the requirements of tooth height direction, the transmission ratio of two order modified coefficient 2C can be solved. Setting tool machining the small wheel can be achieved by using tilt method and variable transmission ratio method at the same time. Gear pitch parameter and parameter adjustment of machine tool are also worked out.

2. The Cutting Algorithm of the Variable Transmission Ratio and Tilt Method

The pinion pitch cone is through the node M and parallel with the root cone plane. The conical surface is called the tooth pitch cone, which node radius is r_{a1} and Pitch cone angle is δ_{f1}. When conical shape wheel is adapted to machining, the tooth pitch cone is not a process of pitch cone [3]. So modified value θ about δ_{f1} is introduced. δ_{f1} can be rewritten as:

$$\delta'_{f1} = \delta_{f1} - \Delta\theta \tag{1}$$

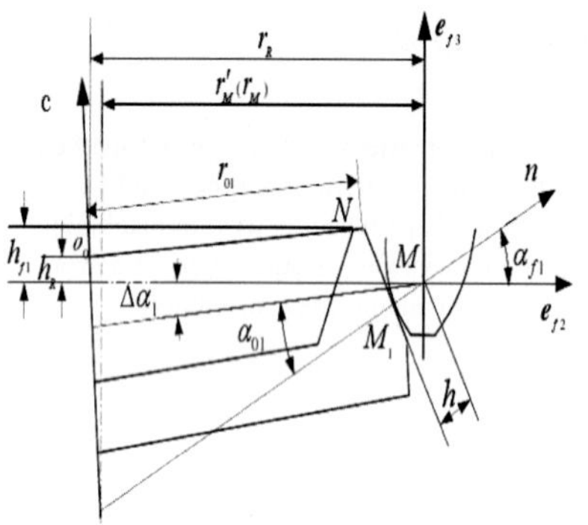

Fig. 1. Cross section of points sub-headings.

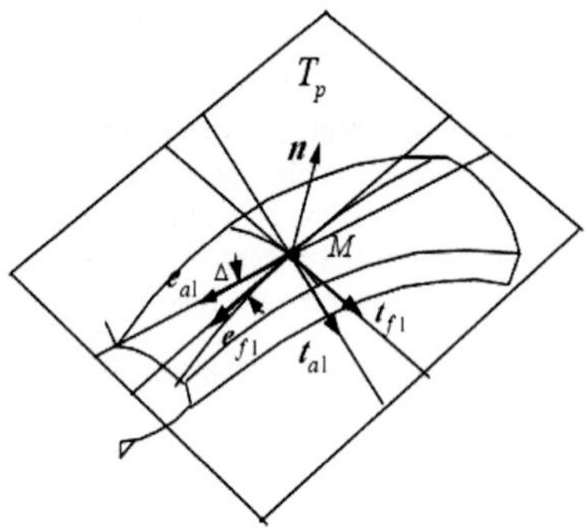

Fig. 2. The midpoint of tooth surface tangent plane.

The initial value of $\Delta\theta$ is zero. The pitch cone of small wheel is:

$$R_{f1} = \frac{r_{\alpha 1}}{\sin \delta'_{f1}} \tag{2}$$

$S_{m1}(M, e_{f1}, e_{f2}, e_{f3})$ is a coordinate system which is based on the M dot. Axis e_{f1} tangents to the tooth line and have the same direction with the relative speed. Axis e_{f3} through M point and has the same direction with n. e_{f1} and e_{f2} are located in the normal plane. Axis e_{f2} is determined by the formula $e_{f2} = e_{f3} \times e_{f1}$.The method of vector n_M can be expressed as:

$$\mathrm{n}_M = \begin{bmatrix} n_{f1} \\ n_{f2} \\ n_{f3} \end{bmatrix} = \begin{bmatrix} \cos\beta_{f1} & \sin\beta_{f1} & 0 \\ -\sin\beta_{f1} & \cos\beta_{f1} & 0 \\ 0 & 0 & 1 \end{bmatrix} \begin{bmatrix} \cos\delta_{f1} & 0 & \sin\delta_{f1} \\ 0 & 1 & 0 \\ -\sin\delta_{f1} & 0 & \cos\delta_{f1} \end{bmatrix} \begin{bmatrix} 0 \\ \cos\alpha_{f1} \\ \sin\alpha_{f1} \end{bmatrix} \tag{3}$$

α_{f1} is the pressure angle, β_{f1} is the helix angle. Equation (3) was the same primary vector in plane S_1. [4]

$$\mathbf{n}_M = \mathbf{n}_{1M} \tag{4}$$

Pressure angle α_{f1} and helix angle β_{f1} can be worked out from equation (4).

$$\sin\alpha_{f1} = \sin\alpha_f \cos(\delta_{\alpha 1} - \delta'_{f1}) - \cos\alpha_f \sin(\delta_{\alpha 1} - \delta'_{f1})\sin\beta_{\alpha 1} \tag{5}$$

$$\cos\beta_{f1} = \frac{\cos\alpha_f \sin\beta_{\alpha 1}}{\cos\alpha_{f1}} \tag{6}$$

Process of pitch cone tooth length direction e_{a1} and tooth pitch cone tooth length direction e_{f1} are located in the M point of tangent plane [5]. In Fig. 2, both e_{a1} and e_{f1} are not coincide. The angle between them is:

$$\sin\Delta = \frac{e_{\alpha 1} \times e_{f1}}{t_{\alpha 1} t_{f1}} = \frac{\cos\beta_{\alpha 1}}{\cos\alpha_{f1}} \sin(\delta_{\alpha 1} - \delta'_{f1}) \tag{7}$$

Along the e_{f1} direction, calculation of points M in the pinion tooth surface curvature is A_{f1}, geodesic torsion is C_{f1}. Along the t_{f1} direction, pinion tooth surface curvature is B_{f1}. By the Euler formula and Bertrand formula, along the e_{f1} direction curvature can be obtained:

$$\begin{aligned}
A_{f1} &= A_{\alpha 1}\cos^2\Delta + B_{\alpha 1}\sin^2\Delta - C_{\alpha 1}\sin 2\Delta \\
B_{f1} &= A_{\alpha 1}\sin^2\Delta + B_{\alpha 1}\cos^2\Delta + C_{\alpha 1}\sin 2\Delta \\
C_{f1} &= (A_{\alpha 1} - B_{\alpha 1})\sin\Delta\cos\Delta + C_{\alpha 1}(\cos^2\Delta - \sin^2\Delta)
\end{aligned} \tag{8}$$

In order to get local conjugate tooth surfaces, we need to modify the radius of gear. Modified value ΔR can be rewritten as:

$$\Delta r = \mp 0.0254 \left(\frac{r_0 \cos\beta_{f2}}{fb} \right)^2 \tag{9}$$

The following equations can be obtained by the induced curvature formula. B'_{f1} can be obtained as follow:

$$B'_{f1} = \frac{C_{f1}^2}{A_{f1} - A_{01}} \tag{10}$$

According to the principle of gear meshing, the increment of $1/B_{f1}$ can be obtained:

$$\Delta\left(\frac{1}{B_{f1}}\right) = \frac{(\varepsilon_2, r_1, n)}{(\partial t)^2} \tag{11}$$

Parameters R_{01}, δ_{01} and β_{01} can be obtained by cutting equation. A rotation angle φ is obtained by determining the two order modified coefficient.

$$\varphi = i_{01}\left(\Delta q - c \cdot \Delta q_1^2\right) \tag{12}$$

3. Conclusion

The tilt method and modified method is used in the traditional Gleason cutting small wheel by different machine tools. Due to the limitations of machinery, the machine tool which has tilt mechanism won't have degeneration mechanism. On the contrary, the machine tool which has degeneration mechanism won't have tilt mechanism. The tilt cutting and variable transmission ratio gear machining can be achieved by the NC machine tool. So it provides the possibility for the establishment of a new algorithm.

Acknowledgments

This work was financially supported by the College Level Scientific Research Fund Project for 2012 (2012009). Design and development of CNC machine tool repair.

References

1. Spiral bevel gear [M]. Beijing: Science Press.
2. Lin Shi. Spiral bevel gear machining key technologies [D]. Master Thesis, Xiamen University, 2008.
3. Gear manual Editorial Board. gear manual (second edition) [M]. Beijing: Mechanical Industry Press, 2006.

4. Yongyun Liu. NC Rough feeds and feeding methods of [J]. 2007.2:62-65 Die Manufacturing Technology
5. Tao Zeng. Spiral bevel gear design and processing [M]. Harbin: Harbin Institute of Technology Press, 1989.

Study on the Kinematic Synchronization of Three Axis Inertial Vibration Exciter

Wei Liu and Ming Su*

School of Mechanical & Electrical Engineering, Guizhou Normal University, Guiyang, 550025, China

E-mail:867500290@qq.com, suming629@163.com

The structure, quality distribution and installation of exciter eccentric affect the vibration trajectory. In the case of considering the phase angle relation of the eccentric device under the static state, the dynamic synchronization process of the three axis inertial vibration exciter is analyzed, establishing a shaker model based on AMESim, making a motion simulation on the change of intermediate rotary shaft eccentric mass and the gear ratio when different initial phase angle conditions. The results show that shaker vibration direction angle size related to the variation of initial installation phase angle and the intermediate shaft eccentric mass, but the trajectory of vibration direction remains unchanged, meanwhile, the vibration trajectory would be composed of different direction ellipses when certain changes in gear ratio.

Keywords: Triaxial exciter; eccentric mass; phase angle; AMESim; motion simulation.

1. Introduction

The function of Inertia vibrator is producing periodic variation of excitation force, which is used to drive to work the body produce continuous vibration, at this stage, the inertia of the exciter mainly used in the screening machinery, according to the number of its drive shaft and the axis of rotation can be divided into uniaxial, biaxial and multiaxial exciter[1]. Single shaft vibration device has simple structure, but the vibration strength is insufficient, only applies to small vibration equipment; Biaxial vibrator work is not stable, normally trajectory is straight line, the ellipse trajectory is difficult to adjust; As for the multiaxial vibrator, when excessive shaft, its complicated structure, the adjustment is inconvenient, therefore, triaxial vibrator frequently used in the large vibration equipment, its advantage lies in working state trajectory is easy to adjust, stability and higher efficiency[2]. Triaxial inertial vibrator can adopt the way of self synchronous and kinematics of synchronization[3-4]. Since the principle of synchronization is driven by three motors, forced synchronous vibration, vibrator three axis synchronous rotation, so as to produce vibration force. Kinematics is a synchronous motor, meshed to transmit energy through the teeth between the gears, this way has higher requirements on design and installation, its advantage is compact structure, reliable operation, strong bearing capacity, etc. Literature [5] theoretical calculation of the system to reach equilibrium requirements under the condition that the initial installation position of the eccentric device is not considered; Literature [6] gives a definition of triaxial shaker force of the heart, it proved the positional relationship between the heart and the center of mass position determines the trajectory of the shaker; Literature [7] analyzed the three motor excitation of self-synchronization shaker's motion effects in different conditions of synchronous phase angle, using Hamilton principle to establish a self-synchronizing condition and stable operating conditions to the shaker; Literature [8] analysis for three axis implementation method of the vibrating screen is not in the same line, and obtain a variety of vibration trajectory straight, oval, linear change, become oval by changing various parameters.

This article in three degrees of freedom dynamic characteristics analysis of triaxial vibrator, using AMESim software modeling and simulation of the kinematics synchronized triaxial inertia type exciter, the results showed that the initial installation of the three groups and the eccentric position of the eccentric mass change has a certain impact on the triaxial vibration exciter direction angle, meanwhile, the vibration change effect obtained when the gear ratio between the three gear change, which provides a reference for the design and installation of three-axis vibration exciter.

2. Triaxial Inertia Type Exciter Structure and Force Analysis

2.1. *Vibrator structure*

Kinematic synchronizing triaxial inertial shakers generally consists of three shaft and the gears, the structure shown in Figure 1, through two pulleys, belt transmits power to the first rotation shafts fixed gear under the action of the motor, after transmission between gear transmission energy, implement three axis of rotation, so as to drive the rotation of the eccentric shaft, centrifugal force, namely exciting force.

Fig. 1. Vibrator structure diagram.

Three axis vibrating screen is based on the basic theory of elliptical and linear shale shaker to design, this paper studies the vibrator in the vibrating screen layout diagram as shown in Figure 2:

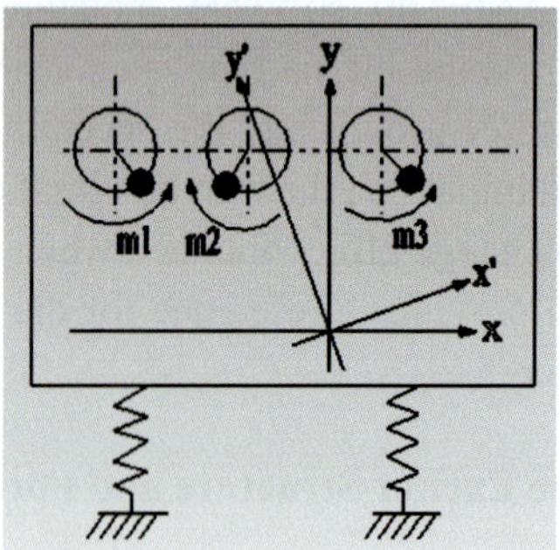

Fig. 2. Vibrator layout diagram.

2.2. *Stress analysis*

From figure 2 can see, through the establishment of coordinate system, vibration device will be broken down into x direction, y direction and ψ direction(torsion movement generated by the eccentric), wherein the x direction is the center of the three axis rotation, y direction is a direction perpendicular to the three axis, ψ direction along the center axis rotation direction.

Shown in Figure 3 is a three-axis vibrator initial installation of the three groups of eccentric block diagram, the intermediate shaft (II axis) center of rotation on its center of mass, left(I axis) and right(III axis) shaft about II axial symmetry distribution, three axis center of mass and the y axis negative direction angle were $\alpha_1, \alpha_2, \alpha_3$, respectively, by changing the point of view, the size can be different running effect[9].

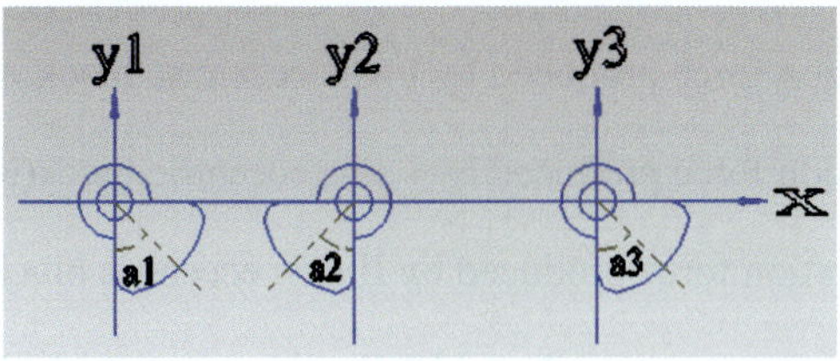

Fig. 3. Eccentric mounting angle schematic.

Three axis eccentric block is connected with each other between the initial installation angle, according to the vibrator in stationary state, can get:

$$m_1 g r_1 \sin\alpha_1 + m_3 g r_3 \sin\alpha_3 = m_2 g r_2 \sin\alpha_2 \tag{1}$$

Type: m_1, m_2, m_3—I axis, II axis, III axis the total mass of the eccentric shaft(kg)

g —Acceleration of gravity(N/kg)

r_1, r_2, r_3 —I axis, II axis, III axis eccentricity(m)

$\alpha_1, \alpha_2, \alpha_3$—I, II, III axis eccentric block the angle between the initial installation direction and negative y direction(rad)

Because in actual design, $\alpha_1 = \alpha_3$, therefore, when an initial phase Angle of the eccentric block is known, the other two eccentric block phase Angle can be calculated.

Three axis vibrator shaft I, II, III axis of horizontal and vertical vibration force can be represented as:

$$\begin{aligned}
&F_{x1} = m_1 r_1 \omega_1^2 \sin\left(\frac{3\pi}{2} + \alpha_1 + \omega_1 t\right), F_{y1} = m_1 r_1 \omega_1^2 \cos\left(\frac{3\pi}{2} + \alpha_1 + \omega_1 t\right) \\
&F_{x2} = m_2 r_2 \omega_2^2 \sin\left(\frac{3\pi}{2} - \alpha_2 - \omega_2 t\right), F_{y2} = m_2 r_2 \omega_2^2 \cos\left(\frac{3\pi}{2} - \alpha_2 - \omega_2 t\right) \\
&F_{x3} = m_3 r_3 \omega_3^2 \sin\left(\frac{3\pi}{2} + \alpha_3 + \omega_3 t\right), F_{y3} = m_3 r_3 \omega_3^2 \cos\left(\frac{3\pi}{2} + \alpha_3 + \omega_3 t\right) \\
&\begin{cases} F_x = F_{x1} + F_{x2} + F_{x3} \\ F_y = F_{y1} + F_{y2} + F_{y3} \end{cases}
\end{aligned} \tag{2}$$

Type: F_x—The x axis vibration force(N)

F_y—The y axis vibration force(N)

F_{x1}—x vibration force produced by I axis eccentric block(N)

F_{y1}—y vibration force produced by I axis eccentric block(N)

F_{x2}—x vibration force produced by II axis eccentric block(N)

F_{y2}—y vibration force produced by II axis eccentric block(N)

F_{x3}—x vibration force produced by III axis eccentric block(N)

F_{y3}—y vibration force produced by III axis eccentric block(N)

$\omega_1, \omega_2, \omega_3$—I, II, III axis eccentric block angular velocity(rad/s)
Around the Z axis of the twist moment (ψ direction) for:

$$\begin{aligned} M_\psi = & m_1 r_1 l_1 \left[\omega_1^2 \sin\left(\omega_1 t + \alpha_1 + \frac{3\pi}{2} + \beta_1\right) - \frac{d\omega_1}{dt}\cos\left(\omega_1 t + \alpha_1 + \frac{3\pi}{2} + \beta_1\right)\right] \\ & - m_2 r_2 l_2 \left[\omega_2^2 \sin\left(\omega_2 t + \alpha_2 - \frac{3\pi}{2} - \beta_2\right) - \frac{d\omega_2}{dt}\cos\left(\omega_2 t + \alpha_2 - \frac{3\pi}{2} - \beta_2\right)\right] \\ & + m_3 r_3 l_3 \left[\omega_3^2 \sin\left(\omega_3 t + \alpha_3 + \frac{3\pi}{2} + \beta_3\right) - \frac{d\omega_3}{dt}\cos\left(\omega_3 t + \alpha_3 + \frac{3\pi}{2} + \beta_3\right)\right] \end{aligned} \tag{3}$$

Type: l_1, l_2, l_3—I, II, III shaft drive shaft rotation center to the system centre of mass of distance(m)

$\frac{d\omega_1}{dt}, \frac{d\omega_2}{dt}, \frac{d\omega_3}{dt}$—I, II, III axis eccentric block angular acceleration(rad/s^2)

$\beta_1, \beta_2, \beta_3$—The angle between the rotary center of I, II, III axis and the positive direction of x axis(rad)

Particularly, when the intermediate shaft through the vibrator centroid, the twist moment $M_\psi = 0$ 。

3. Vibrator Torque Equation

3.1. *Vibration motor mechanical properties*

According to the theory of electric drive, the mechanical properties of the motor excitation practical expression [10]:

$$T = \frac{2T_m}{\frac{s}{s_m} + \frac{s_m}{s}} \tag{4}$$

Type: T —Vibration motor output torque(N.m)

T_m—The maximum torque of the motor vibration(N.m)

S —Vibration motor slip

S_m—Slip ratio of the maximum vibration motor

By querying the motor catalogue, can get on type parameters. Among them:

$$\begin{aligned} T_m &= \lambda_M T_N \\ T_N &= 9550\frac{P_N}{n_N} \\ s &= \frac{n_0 - n}{n_0} \\ s_m &= s_N\left(k_M + \sqrt{k_M^2 - 1}\right) \end{aligned} \tag{5}$$

Type: λ_M —Vibration motor overload factor

T_N—Vibration motor rated torque(N.m)

P_N—Vibration motor rated power(KW)

n_N—Vibration motor rated speed(rpm)

n_0—Vibration motor synchronous speed(rpm)

n—Vibration motor rotor speed(rpm)

s_N—Rated slip

k_M—Vibration motor starting torque ratio

3.2. *Vibrator load torque*

When vibrator is working, the eccentric device on the three axis x direction, y direction of the inertial force, as well as rotation around the center of mass (ψ direction) to produce twist, so as to constitute a system of load torque, reference [11], you can get three axis load torque, respectively is:

$$T_1 = m_1 r_1 \left[\frac{d^2 y_1}{dt^2} \cos\left(\frac{3\pi}{2} + \alpha_1 + \omega_1 t \right) - \frac{d^2 x_1}{dt^2} \sin\left(\frac{3\pi}{2} + \alpha_1 + \omega_1 t \right) + \frac{d^2\psi}{dt^2} l_1 \cos\left(\frac{3\pi}{2} + \alpha_1 + \omega_1 t + \beta_1 \right) \right]$$

$$T_2 = m_2 r_2 \left[\frac{d^2 y_2}{dt^2} \cos\left(\frac{3\pi}{2} - \alpha_2 - \omega_2 t \right) + \frac{d^2 x_2}{dt^2} \sin\left(\frac{3\pi}{2} - \alpha_2 - \omega_2 t \right) + \frac{d^2\psi}{dt^2} l_2 \cos\left(\frac{3\pi}{2} - \alpha_2 - \omega_2 t + \beta_2 \right) \right]$$

$$T_3 = m_3 r_3 \left[\frac{d^2 y_3}{dt^2} \cos\left(\frac{3\pi}{2} + \alpha_3 + \omega_3 t \right) - \frac{d^2 x_3}{dt^2} \sin\left(\frac{3\pi}{2} + \alpha_3 + \omega_3 t \right) + \frac{d^2\psi}{dt^2} l_3 \cos\left(\frac{3\pi}{2} + \alpha_3 + \omega_3 t + \beta_3 \right) \right] \quad (6)$$

Type: T_1, T_2, T_3—I, II, III axle load torque(N.m)

$\frac{d^2 y_1}{dt^2}, \frac{d^2 y_2}{dt^2}, \frac{d^2 y_3}{dt^2}$—I, II, III axis y direction angular acceleration(rad/s^2)

$\frac{d^2 x_1}{dt^2}, \frac{d^2 x_2}{dt^2}, \frac{d^2 x_3}{dt^2}$—I, II, III axis x direction angular acceleration(rad/s^2)

3.3. *Bearing friction torque*

When vibration motor work, the inner bearing can generate friction force when the rotating shaft of the guide motor rotates, then forming a friction torque. Reference [12], I, II, III shaft friction torque is as follows:

$$\begin{aligned} T_{f1} &= T_{v1} + f\omega_1 + m_1 r_1 f \omega_1^2 \frac{d_1}{2} \\ T_{f2} &= T_{v2} + f\omega_2 + m_2 r_2 f \omega_2^2 \frac{d_2}{2} \\ T_{f3} &= T_{v3} + f\omega_3 + m_3 r_3 f \omega_3^2 \frac{d_3}{2} \end{aligned} \quad (7)$$

Type: T_{f1}, T_{f2}, T_{f3}—I axis, II axis, III shaft friction torque(N.m)

T_{v1}, T_{v2}, T_{v3}—I, II, III axis constant drag torque(N.m)

f—The coefficient of friction

d_1, d_2, d_3—I, II, III shaft bearing diameter(m)

3.4. *Three-axis torque equations of vibrator*

Reference [12], the establishment of the I-axis torque equation is as follows:

$$T_0 - T_1 - T_{f1} - F_{21} \bullet \frac{d_{z1}}{2} = J_1 \frac{d\omega_1}{dt} \tag{8}$$

Type: T_0 —I axis input torque(N.m), $T_1 = \frac{1}{n_z} T$

n_z —Motor output to the I shaft transmission ratio

F_{21}—II give I shaft gear on the circumferential force(N)

J_1—The rotational inertia of the I axis(kg.m^2)

d_{z1}—I shaft gear dividing circle diameter(m)

II shaft torque equation is as follows:

$$F_{12} \bullet \frac{d_{z2}}{2} - T_2 - T_{f2} - F_{32} \bullet \frac{d_{z2}}{2} = J_2 \frac{d\omega_2}{dt} \tag{9}$$

Type: F_{12}—I give II shaft gear on the circumferential force(N)

J_2—The rotational inertia of the II axis(kg.m^2)

d_{z2} —II shaft gear dividing circle diameter(m)

F_{32}—III give II shaft gear on the circumferential force(N)

III shaft torque equation is as follows:

$$F_{23} \bullet \frac{d_{z3}}{2} - T_3 - T_{f3} = J_3 \frac{d\omega_3}{dt} \tag{10}$$

Type: F_{23}—II give III shaft gear on the circumferential force(N)

J_3—The rotational inertia of the III axis(kg.m^2)

d_{z3}—III shaft gear dividing circle diameter(m)

F_{21} and F_{12}, F_{32} and F_{23} are two pairs of interaction; Because in the practical work, I, III axis motions must be completely consistent, therefore the design process of a given: $m_3 = m_1$; $\alpha_3 = \alpha_1$; $r_3 = r_1$; $d_{z3} = d_{z1}$; $d_3 = d_1$; $\omega_3 = \omega_1; T_{v3} = T_{v1}; J_3 = J_1$.

Thus, the three-axis joint torque equation can be obtained:

$$T_0-\left(T_1+\frac{d_{z1}}{d_{z2}}T_2+\frac{d_{z1}}{d_{z3}}T_3\right)-\left(T_{f1}+\frac{d_{z1}}{d_{z2}}T_{f2}+\frac{d_{z1}}{d_{z3}}T_{f3}\right)=J_1\frac{d\omega_1}{dt}+J_2\frac{d\omega_2}{dt}+J_3\frac{d\omega_3}{dt} \tag{11}$$

According to the principle of gear transmission, setting I, II, III shaft gear pitch circle diameter ratio for: $d_{z1}:d_{z2}:d_{z3}=1:\zeta:1$, at the same time, on the basis of triaxial linear velocity of the same requirements of gear transmission, can get:

$$v=\omega_1\cdot\frac{d_{z1}}{2}=\omega_2\cdot\frac{d_{z2}}{2}=\omega_3\cdot\frac{d_{z3}}{2},\quad \omega_1:\omega_2:\omega_3=1:\frac{d_{z1}}{d_{z2}}:\frac{d_{z1}}{d_{z3}}=1:\frac{1}{\zeta}:1$$

Based on the above formula for each part of the expression, you can get the system torque equation:

$$\begin{aligned}&\frac{2T_m n_z}{\dfrac{n_0\pi-30\omega n_z}{n_0\pi s_m}+\dfrac{n_0\pi s_m}{n_0\pi-30\omega n_z}}-2m_1r_1\left[\frac{d^2y_1}{dt^2}\cos\left(\frac{3\pi}{2}+\alpha_1+\omega_1t\right)-\frac{d^2x_1}{dt^2}\sin\left(\frac{3\pi}{2}+\alpha_1+\omega_1t\right)\right]\\&-m_1r_1\frac{d^2\psi}{dt^2}l_1\cos\left(\frac{3\pi}{2}+\alpha_1+\omega_1t+\beta_1\right)-m_3r_3\frac{d^2\psi}{dt^2}l_3\cos\left(\frac{3\pi}{2}+\alpha_3+\omega_3t+\beta_3\right)-2T_{v1}-\frac{T_{v2}}{\zeta}\\&-\frac{1}{\zeta}m_2r_2\left[\begin{array}{l}\dfrac{d^2y_2}{dt^2}\cos\left(\dfrac{3\pi}{2}-\alpha_2-\omega_2t\right)+\dfrac{d^2x_2}{dt^2}\sin\left(\dfrac{3\pi}{2}-\alpha_2-\omega_2t\right)\\ \quad+\dfrac{d^2\psi}{dt^2}l_2\cos\left(\dfrac{3\pi}{2}-\alpha_2-\omega_2t+\beta_2\right)\end{array}\right]\\&-\left(2\omega_1+\frac{\omega_2}{\zeta}+m_1r_1\omega_1^2d_1+\frac{m_2r_2\omega_2^2d_2}{2\zeta}\right)f=\left(2J_1+\frac{J_2}{\zeta}\right)\frac{d\omega_1}{dt}\end{aligned} \tag{12}$$

4. Modeling and Simulation of Triaxial Vibration Exciter Based on AMESim

4.1. *Triaxial shaker modeling*

According to kinetic equation, references [13], using AMESim software Simulation library (Simulation) and Signal, Control library (Signal and Control) components in the simulation model to construct the triaxial vibrator, as shown in Figure 4.

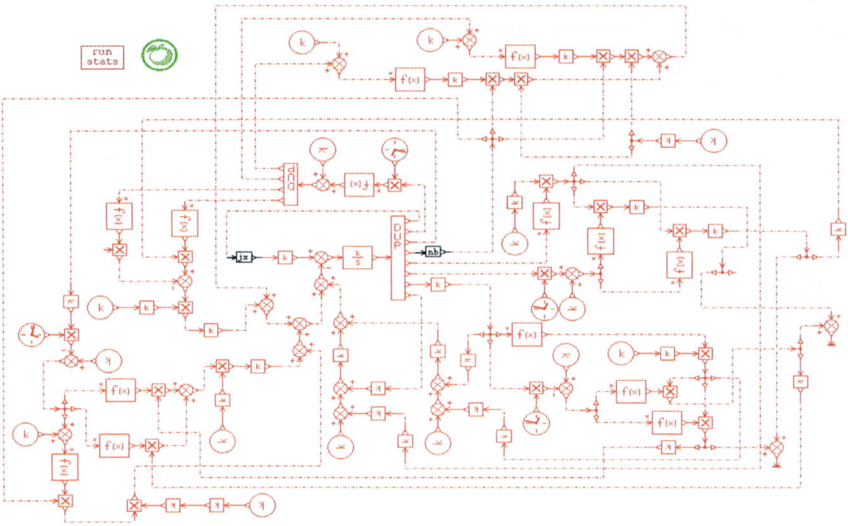

Fig. 4. Triaxial vibrator AMESim model.

Shown in Figure 4 models of small area is related to the box structure of super element, the details of the components as shown in Figure 5, include mechanical properties, the ψ direction of the twist vibration motor torque.

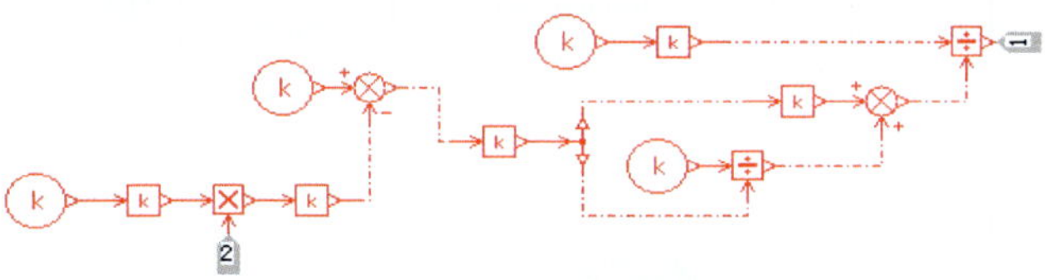

Fig. 5a. Motor mechanical properties.

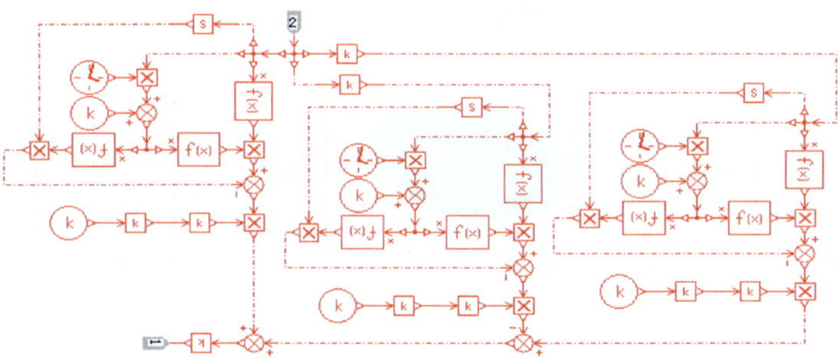

Fig. 5b. Vibrator twist moment.

Fig. 5. The structure of the super element in the model of vibration exciter.

4.2. *Simulation of triaxial shaker*

The model parameters are as follows:

Table 1. Simulation parameter settings.

parameter	value	unit	parameter	value	unit	parameter	value	unit
T_m	494	N.m	m_1	33	kg	f	0.002	/
s_m	0.151	/	m_2	33	kg	β_1	150	°
n_0	1000	rpm	r_1	0.139	m	β_2	120	°
n_z	0.8	/	r_2	0.139	m	β_3	60	°
T_{v1}	0.01	N.m	d_1	0.3	m	ς	1	/
T_{v2}	0.01	N.m	d_2	0.3	m	l_1	3.5	m
J_1	40	$kg.m^2$	d_{z1}	0.56	m	l_2	2.0	m
J_2	40	$kg.m^2$	d_{z2}	0.56	m	l_3	2.0	m

Table 2. Different phase angle of the simulation.

I,III axis α_1 (°)	II axis α_2 (°)	phase difference $\alpha_1-\alpha_2$ (°)	Exciting force image	The angle between the long axis of ellipse and the direction of X axis(°)	The elliptical amplitude(N)	
					The major axis	The short axis
0	0	0	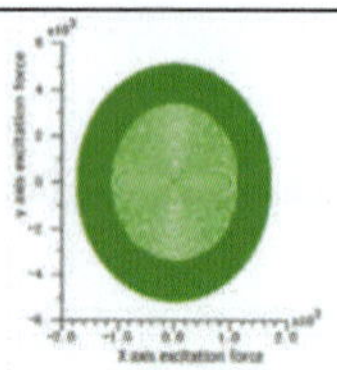	90.0	1.031×10^5	0.344×10^5
5	10.0	-5.0	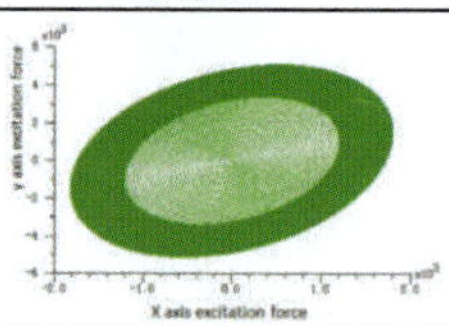	81.3	1.027×10^5	0.346×10^5

10	20.3	-10.3	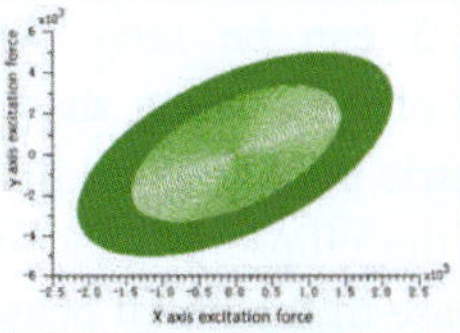	61.4	0.862×10^5	0.596×10^5
15	31.2	-16.2	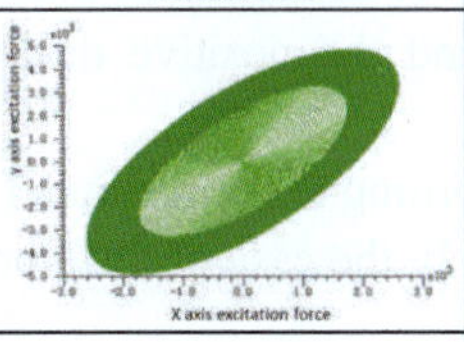	59.2	0.977×10^5	0.483×10^5
20	43.2	-23.2		53.9	1.046×10^5	0.405×10^5
25	57.7	-32.7	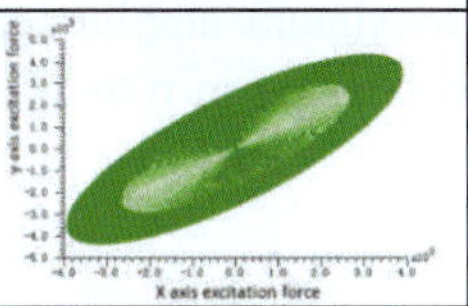	45.7	1.096×10^5	0.397×10^5

The simulation time is set to 200s, the time interval is 0.0002s. According to the formula (1), get: $\sin\alpha_1+\sin\alpha_3=\sin\alpha_2$, $2\sin\alpha_1=\sin\alpha_2$. Simulation for different phase angles, the results shown in Table 2.

Results: in the multiple sets of initial installation phase Angle (0°~30°) experiments, the trajectory is an ellipse and elliptical axle (motion amplitude) between a certain fluctuation. With the increase of the initial phase Angle, long wheelbase is started to shorten, and then increases, the corresponding exists between 5°~15° long axis of the minimum value, three axis eccentric block in the initial phase difference increasing in value, at the same time, the perspective of long axis and the direction of x axis is constantly decreases.

Changing m_2 and set $m_2=50$ kg, according to the formula (1), get: $2m_1\sin\alpha_1=m_2\sin\alpha_2$. The simulation is performed again at a different phase angle conditions, as follows:

Results: when the mass of the eccentric device of the three axis vibration exciter is not quite equal, the exciting force trajectory is still an ellipse, and the amplitude of the vibration is changed with the phase angle. With the increase of the phase angle, the long axis of the ellipse decreases, then increases, and the initial phase difference of the eccentric blocks increases.

Comparison Tables 2 and 3 can be seen: when the eccentric initial installation phase angle (0°~30°) of the rotating shaft is same, the initial phase change width of imbalance eccentric mass is small than eccentric mass balance, in both tables, the long axis of the ellipse are showing a downward opening parabola trajectory with the increase of the initial phase angle, at the same time, the angle between major axis and the positive direction of the x-axis are in constant decrease.

Changing gear ratio ζ, according to the formula (1), get: $2\sin\alpha_1 = \sin\alpha_2$. Setting: $\alpha_1 = 20^\circ$, $\alpha_2 = 43.2^\circ$. In the case of different gear ratio are simulated, as above.

From Table 4 we can see that: when the intermediate shaft gear pitch circle diameter change, the corresponding gear ratio changed from 0.90 to 1.10, the excitation force trajectory is no longer a ellipse, but made up of multiple elliptical cohesion of different vibration direction. If the longest track line through the origin seen as the maximum amplitude, it can be seen that the amplitude of the transmission ratio changes in 0.9×10^5 N or so.

Table 3. Different phase angle simulation situation when changing eccentric block quality.

I,III axis α_1 (°)	II axis α_2 (°)	phase difference $\alpha_1-\alpha_2$ (°)	Exciting force image	The angle between the long axis of ellipse and the direction of X axis(°)	The elliptical amplitude(N)	
					The major axis	The short axis
0	0	0	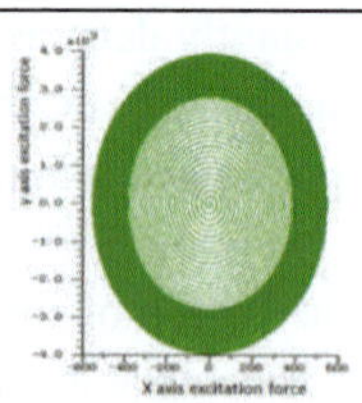	90.0	0.790×10^5	0.110×10^5

5	6.6	-1.6	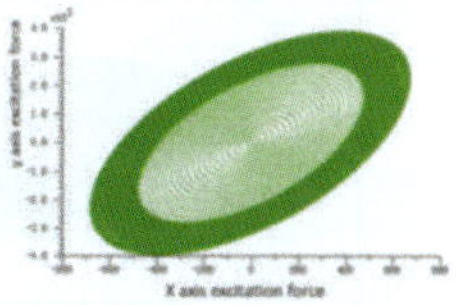	82.0	0.665×10^5	0.420×10^5
10	13.3	-3.3	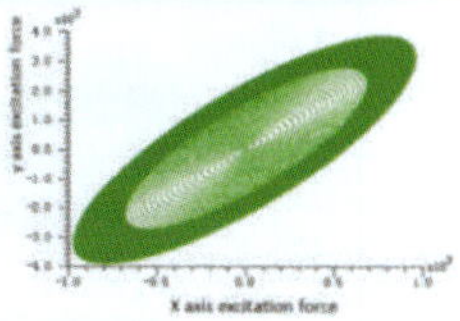	75.1	0.735×10^5	0.277×10^5
15	20.0	-5.0	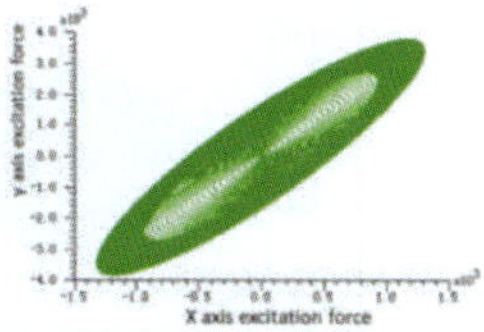	70.3	0.770×10^5	0.219×10^5
20	26.8	-6.8	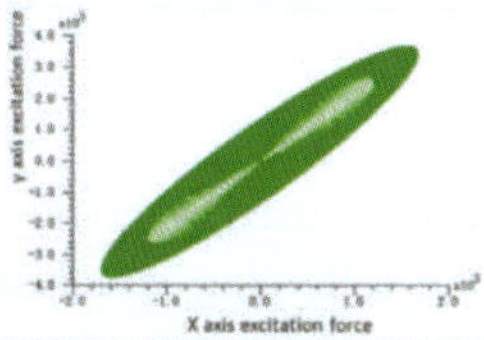	65.0	0.792×10^5	0.177×10^5
25	33.9	-8.9	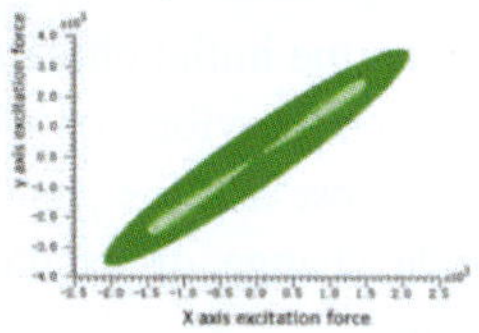	59.4	0.812×10^5	0.148×10^5

Table 4. Simulation of gear transmission ratio.

phase angle			I, II shaft gear transmission ratio ζ	Exciting force image	Maximum amplitude(N)
I,III axis α_1 (°)	II axis α_2 (°)	phase difference $\alpha_1-\alpha_2$(°)			
20	43.2	-23.2	1.10	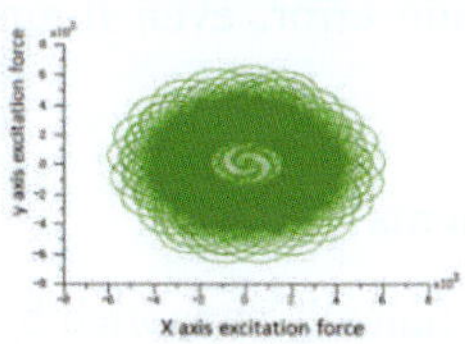	0.893×10^5

			1.05	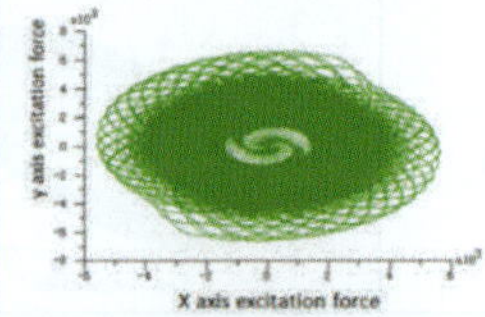	0.886×10^5
			0.95	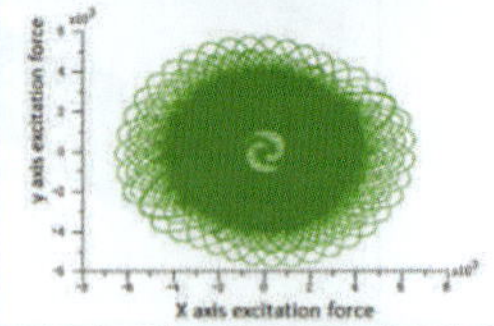	0.912×10^5
			0.90	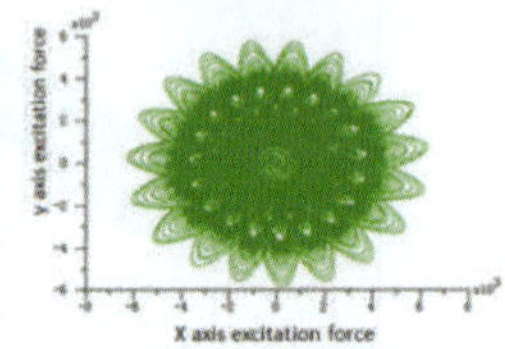	0.910×10^5

5. Conclusion and Discussion

In the experiment, the AMESim software is used to model and simulate the three axis inertial vibration exciter, and the following conclusions can be obtained through the analysis of the results:

(1) For the three axis inertial exciter, the initial phase angle of the eccentric block is the direct factor which influences the vibration direction angle(that is, the angle between the long axis of the ellipse and the positive direction of the x axis), at the same time, the amplitude change of excitation device also has relations with the size of the phase Angle.

(2) Under the eccentric block initial phase angle is determined, exciter shaft of the eccentric mass change determines the amplitude and vibrating direction angle size, but the trajectory of the vibration amplitude and vibration direction remains unchanged.

(3) If the three axis vibration exciter has been given, the gear transmission ratio will affect the vibration effect, and the trajectory is composed of the moving ellipse. It can be seen that, in the actual production, when the gear transmission ratio error, even if small, it will cause a lot of vibration results errors.

Acknowledgments

This work was finically supported by the Foundation of Science and Technology Innovation Talents Team of Guizhou Province ([2014]4013).

References

1. ShiMin Zhang, ChuSheng Liu, HaiPei Zhou, et al. Dynamic Study on Constant Thickness Vibrating Screen of Variable Elliptical Trajectory With Three Axes [J]. Coal Mine Machinery, 2011, 32(06):75-76.
2. BangChun Wen, FengQiao Liu. Theory and Application of mechanical vibration [M]. BeiJing: China Machine Press, 1982:5-16.
3. BangChun Wen. The vibration of the mechanical system synchronization control and synchronization [M]. BeiJing: Science press, 2003.
4. YiQing Chu, FengXiang Mei. Self synchronization effect of single mass double axis parallel inertial vertical vibration conveyor [J]. Coal Mine Machinery, 1980, (03):18-23.
5. JingShi Li, ChunZhi Zhang. The quality of three-axis elliptic vibrating screen vibrating direction angle control [J]. Coal Mine Machinery, 2001, (10):56-57.
6. Jie Ren, YongJun Hou, XianJin Wu. Analysis of Force-center for Three-Axis Inertial Shale Shaker Based on Matlab [J]. Oil field equipment, 2012, 41(03):25-28.
7. YongJun Hou, YongQiang Shen, Ying Tian, et al. Study of Synchronization and Stability of Tri-motor Self-synchronization Balanced Elliptical Motion Shale Shaker on Movement Synthesize [J]. Oil field equipment, 2010, 39(06):1-5.
8. ShiMin Zhang, ChuSheng Liu. Dynamic simulation of three-axised banana vibration screen with variable trajectory based on Simulink [J]. Coal Mine Machinery, 2011, 39(05):85-88.
9. JingShi Li, LiLi Liu. Determination of the vibration direction angle of three axis elliptical vibrating screen and the adjustment of its amplitude [J]. Coal Mine Machinery, 2001, (10):47-48.
10. HaoLie Wu. Motor and electric drive Foundation [M]. Chongqing: Chongqing University press, 2002:46-48.
11. Jian Fan, BangChun Wen. A theoretical study on the control of the vibration of a reverse rotating double vibration exciter [J]. Journal of Vibration Engineering, 1994, 7(04):281-288.
12. GuiZhi Li. Research on the characteristics of single vibration motor system [J]. Coal Mine Machinery, 1996, (08):38-40.
13. YongLing Fu, HaiTao Qi. LMS Imagine. Lab AMESim system modeling and simulation [M]. BeiJing:Beijing University of Aeronautics & Astronautics Press, 2011.

Laser/ultrasonic-vibration combined dressing technique for diamond grinding wheels

Zhi-Bo Yang†, Rui-Yun Yang and Zhen Zhang
Henan Polytechnic University,
Jiaozuo,Henan, China
†*E-mail:yangzhibo2001@163.com*

In order to solve the problems such as low efficiency, bad precision, high cost and high tool wear in the turing of super abrasive grinding wheel, In this paper, a new dressing method of laser heating and ultrasonic vibration assistance is proposed. Under the advantages of ultrasonic vibration dressing methods, aided by laser heating, which can make the grinding wheel surface to soften, plastic, stress reduction of trimming tool, the wear rate is reduced, finishing efficiency. This article contains the design of composite dressing system, building experimental platform and experimental research. Experimental results show that: compared with ordinary mechanical dressing, in choosing the right processing parameters, the truing morphology, single abrasive grains and grinding characteristics of the grinding wheel which in the trimming mode of laser heating and ultrasonic vibration assistance is superior to ordinary trim.

Keywords: Diamond grinding wheel; laser heating and ultrasonic vibration assistance; the right processing parameters.

1. Introduction

With the rapid development of modern science and technology, the difficult-to-process materials with excellent performances, such as optical glasses, high-temperature alloys, titanium alloys, ultra-high-strength steels, ceramics and monocrystalline silicon, are being increasingly applied in many cutting-edge domains mainly including aerospace, aviation, optics, instruments and automobiles. How to efficiently and precisely process these materials has become a research hotspot. Due to the increasingly high requirements on processing precision and the difficult-to-process characteristics of super hard materials, the demands for high-efficiency and precise grinding on these machines, especially, with the use of super hard abrasive grinding wheels, increase significantly [1-3]. The application of super hard abrasive grinding wheels in which the most representative abrasive particles are diamond and cubic boron nitride (CBN) grows at an annual rate of approximately 10% [4]. However, the use of super hard abrasive grinding wheels also exhibits many shortcomings, such as great difficulty, long time and low efficiency in wheel dressing as well as severe wear of dressers. Therefore, the research on high-efficiency and high-quality of wheel dressing technique has become one of the key technologies for achieving the precise/ultra-precise, high-speed and high-efficiency grinding, molding grinding and grinding automation [5].

Currently, there are many high-efficiency dressing methods for super hard abrasive grinding wheels, mainly including point-squeezing dressing, cup wheel dressing, electric spark dressing, electrolytic in-processing dressing (ELID), laser dressing, ultrasonic vibration dressing and etc [6,7]. Additionally, the combination methods of two or more dressing techniques are also applied, such as laser-assisted diamond stylus dressing [8], the combination of mechanical and chemical methods, and the combination of electrolysis and electric spark. Using these methods, the grinding quality can be improved and the wheel wear can be reduced to certain extents; however, there still exist different degrees of limitations in practical applications due to the increasing demands for practicability, economy, convenience and high efficiency as well as the high-hardness characteristics of super-hard abrasive grinding particle wheel [9].

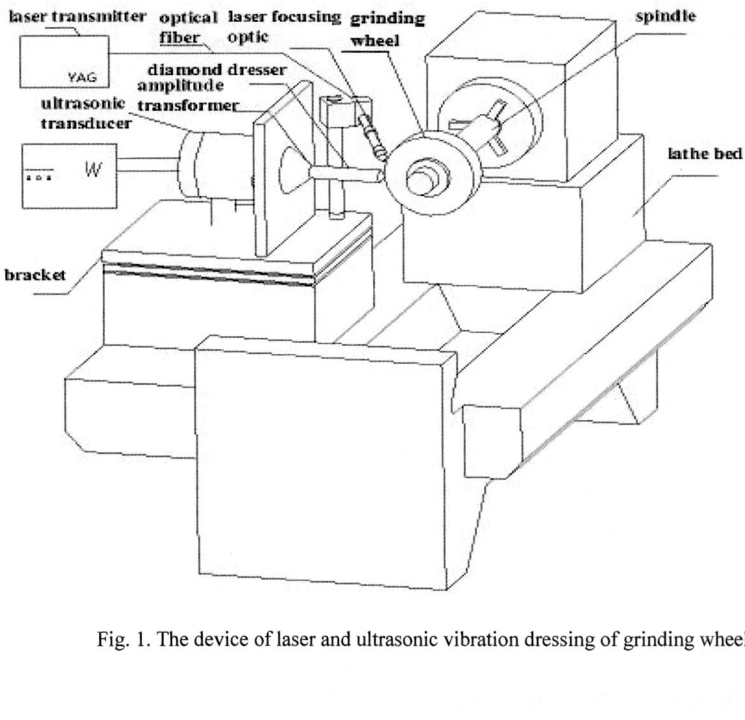

Fig. 1. The device of laser and ultrasonic vibration dressing of grinding wheel.

Table 1. Dressing conditions.

category	project	specifications and parameters
type of grinding wheel	diamond grinding wheel	1A1 150×10×32×5 MBD 120 M 100
the trimming tool	diamond dresser	cone,apex angle is 60 degrees
dressing parameter	spindle speed	25~90rpm
	dressing lead	0.030~0.080mm/rev
	dressing depth	5~20μm/pass
ultrasonic generator	frequency	35KHz
laser parameters	laser transmitter	power 180-240W
	pulse frequency	50-110KHZ
	diameter of laser spot	0.6-2mm

2. Experimental Condition and Scheme

In the present experiments, the two-dimensional ultrasonic vibration dressing system based on laser's thermal effect was used, which consists of ultrasonic generator, transducer, amplitude transformer, laser transmitter, laser head, optical fiber, diamond dressing stylus, dynamometer and wheels. The grinding wheel dressing experiments were conducted on an ultra-precise digital controlled lathe. Fig. 1 displays the ultrasonic-vibration/laser combined dressing device. The super hard abrasive grinding wheel was installed on the machine tool and rotated at a certain speed. The wheel's dressing depth was controlled by the dial indicator. The dynamometer was fixed on the sliding carriage, and the support was installed on the dynamometer platform. The ultrasonic vibration dressing device and laser head were mounted on the support, which were fed axially by the sliding carriage. The vibration was in a two-dimensional elliptical ultrasonic vibration mode.

During the dressing process, the focused laser beam heated the rotating wheel in front of the dressing point with the use of appropriate incident angle, distance, heat-up time and energy density. The diamond dressing stylus cut the wheel surface and fed axially along the wheel. Meanwhile, driven by the ultrasonic vibration equipment, the diamond dressing stylus vibrated regularly, and accordingly, the elliptical ultrasonic vibration was formed at the tip of stylus. Then the dressing stylus shifted upwards along the elliptical orbit and was separated from the wheel. Until the diamond dressing stylus was completely out of touch with the wheel, it could move to the new initial position for the next round of dressing.

Table 1 lists the dressing conditions. To facilitate the observations of wheel using white light interferometer and electron microscope, the bronze-bonded diamond segmental wheel (1A1 150×10×32×5 MBD 120 D 100) was used, and the dressing frequency was set as 35 kHz.

The experimental scheme is described in detail below. Under the same conditions, both the ordinary ultrasonic vibration dressing method and ultrasonic-vibration/laser combined dressing method were conducted, and the dressing performances using different dressing methods were examined by means of microscopic observations and grinding performance tests. Using micros cmorphology and the pattern of single abrasive particle before and after dressing. In dressing performance tests, the wheel's grinding force of ceramic test specimens before and after dressing were compared.

3. Results and Discussion

3.1. *Surface morphology of wheel*

By comparing the SEM photos of the dressed wheel surfaces using ordinary ultrasonic vibration dressing method and ultrasonic-vibration/laser combined dressing method, one can observe that the dressed wheels using these two different dressing methods are entirely different. As shown in Fig. a, for the dressed wheel using ordinary ultrasonic vibration method, the abrasive particles on the surface present severe falling-off, breakage and loosening phenomena, and few surface static abrasive particles exist on the surface. Half of abrasive particles are surrounded by the delta of bond, and the protrusion height of abrasive particles above the base is not significant. Fig. b displays the dressed wheel using the ultrasonic-vibration/laser combined dressing method. One can observe that, for the dressed wheel using the combined dressing method, the surface appearance of wheel is favorable, with little breakage of abrasive particles. Moreover, compared with the condition using ordinary ultrasonic vibration dressing method, more static effective abrasive particles can be observed in the dressed wheel using the combined dressing method, the abrasive particles obviously protrude over the base and exhibit uniform distributions. These abrasive particles, with smooth surface and more edges, have stronger cutting ability, and the bonding ability of bond to abrasive particles are enhanced. The above-described results suggest that, using ordinary ultrasonic vibration dressing method, the material removal on the surface of wheel is based on the breakage of bond and the brittle fracture of abrasive particles; using the ultrasonic-vibration/laser combined dressing method, the material removal on the surface of wheel is based on the ductile removal of bond and the wear of abrasive particles.

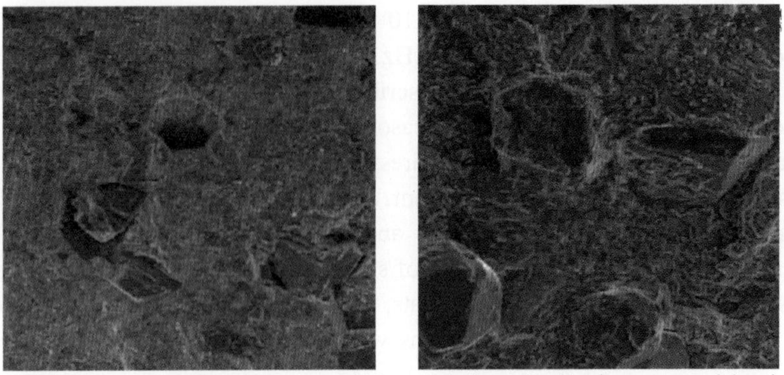

a) general mechanical dressing b) laser and ultrasonic vibration compound dressing

Fig. 2. The topography of grinding wheel.

3.2. *Single abrasive particle pattern on the surface of wheel*

The single abrasive particle pattern on the surface of dressed wheel was observed using the white light interferometer (Talysurf CC I 6000, Taylor Hobson Ltd., and England). Fig. 3 presents the single abrasive particle patterns of the wheel surfaces dressed using different dressing methods, from which we can observe that the grinding particles using these two dressing methods vary greatly in morphology presentation. As shown in Fig. a, using ordinary ultrasonic vibration dressing method, the abrasive particles are half-surrounded by the bond delta, the protrusion height of abrasive particles over base is small and the breakage ratio of edges is high. Additionally, the cutting edge is not sharp enough, leading to a poor grinding ability. As shown in Fig. b, using the ultrasonic-vibration/laser combined dressing method, the abrasive particles, although still have tiny breakage, can basically keep intact. The abrasive particles have smooth surfaces, with significant protrusion height over base and sharp edges, i.e., the abrasive particles have strong grinding ability.

a) general mechanical dressing b) laser and ultrasonic vibration compound dressing

Fig. 3.Single grain morphology of the grinding wheel's surface.

3.3. *Grinding force*

Under the same conditions, the ceramic specimens were ground by the wheels, which were dressed using the ordinary machine dressing method and the ultrasonic-vibration/laser dressing method, respectively. The wheel's grinding force was measured using the grinding-force measuring system as displayed in Fig. 4. Generally, the grinding force can be decomposed into the tangential grinding force, axial grinding force and normal grinding force for further study. These three forces are perpendicular to each other. The axial grinding force is usually small and can be negligible. Since the abrasive particles of wheel have large negative rake angles, the normal grinding force is greatest among these three component forces. The normal grinding force is closely connected with the sharp degree of the abrasive particles. The blunter abrasive particles will lead to the sharp increase of normal friction, and thereby the accelerated wear of wheel and the quality worsening of workpieces; on the contrary, the sharper abrasive particles will reduce the normal friction and normal grinding force and thus can improve the quality of workpieces. Fig. 5 compares the grinding forces of wheels under two different dressing conditions. One can observe that, under the same feed speed and load, the grinding force of wheel dressed using the combined dressing method is smaller than that using the ordinary machine dressing method. Therefore, it can be concluded that, under the premise of the same grinding parameters, using the ultrasonic-vibration/laser combined dressing method, the abrasive particles of wheel can be more protruding and sharp. The dressed wheel using the combined dressing method is superior to the wheel dressed using the ordinary ultrasonic vibration dressing method in terms of the wheel's morphology and grinding performance.

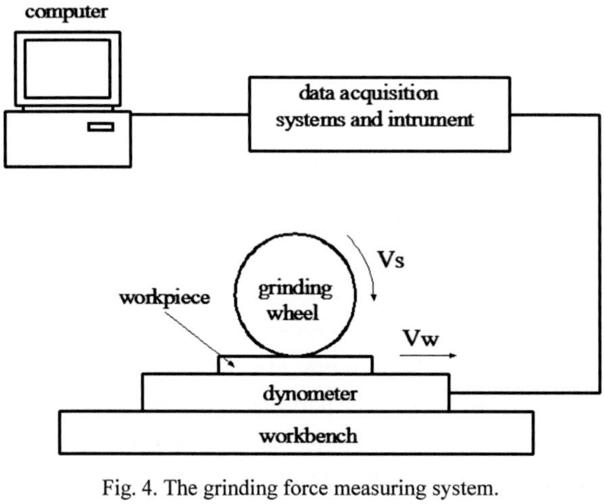

Fig. 4. The grinding force measuring system.

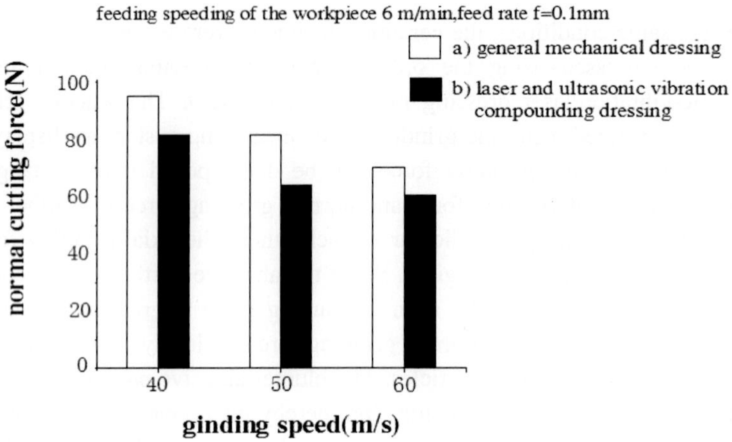

Fig. 5. Grinding force comparison of grinding wheel under different dressing methods.

3.4. *Analysis*

Using the two-dimensional ultrasonic-vibration dressing method, the diamond dressing stylus moves along an ellipse and the dressing depth varies constantly. In combination with laser heating, the wheel surface can be softened and the plasticity is enhanced. The material removal mechanisms on the surface are mainly the tiny breakage of abrasive particles and the ductile removal of bond. The force applied on the dressing tool is reduced, so is the wear. During the dressing process, the diamond stylus is in intermittent contact with the wheel so that the service life can be prolonged. The truing effect can also be achieved during the dressing process. Therefore, the wheel dressed using the combined dressing method is better than the wheel dressed using the ordinary ultrasonic vibration dressing method in terms of the wheel's morphology, single abrasive particle pattern and grinding performance.

4. Conclusion

The dressing experiments indicate that, compared with the condition using ordinary ultrasonic vibration dressing method, the abrasive particles of the wheel dressed using the ultrasonic-vibration/laser combined dressing method maintain favorable patterns under appropriate technological parameters. Using the combined dressing method, more effective abrasive particles and more uniform distributions of abrasive particles can be observed. The abrasive particles have smooth surfaces, large protrusion heights and sharp edges. Additionally, the bonding strength between bond and abrasive particles increases.

Conclusively, the ultrasonic-vibration/laser combined dressing method is preferable than the ordinary ultrasonic vibration dressing method in the pattern of abrasive particles, the morphology of wheel and grinding characteristics.

Acknowledgment

This research was aided by the financial support of The National Natural Science Fund (U1204517), Natural Science Fund of the Education Department of Henan Province, P.R. China (12B460011), and the Opening Project of Key Laboratory of Precision Manufacturing Technology and Engineering, Henan Polytechnic University.

References

1. Chen Genyu, Bu Chun, Deng Hui, Jia Tianyang, Li Shichun. Research on laser and mechanical compound truing and dressing of superabrasive grinding wheels[J]. Chinese Journal of Lasers, 2012, (7):59-65.
2. Liu Shu. Comparative study on dressing methods of diamond grinding wheel [J].Technology Innovation and Application, 2014, (1):118.
3. Zhao Bo, Tang Jun. Development of ultrasonic elliptical vibration machining [J]. Diamond and Abrasives Engineering, 2014, (1):70-78.
4. Li Lianpeng, Liao Denghua, Hu Yonghui, Zhang Feihu. Metal-bonded diamond wheel test analysis of influence of different grinding parameter on grinding quality [J]. Technology Innovation and Application, 2014, (9):29-30.
5. Zhang Bei, Fu Yucan, Su Honghua. Experimental research on dressing of monolayer brazed diamond grinding wheel [J]. China Mechanical Engineering, 2014, (13):1778-1783.
6. Ding Bailu, He Fengbao, Zhang Wei, Li Mingying. Test and study of dressing diamond wheels with super-abrasives wheel dresser [J]. Tool Engineering, 2013, (5):20-22.
7. Deng Jiang. Measurement and control system in the laser dressing of super-abrasive grinding wheel [D]. Hunan University, 2013.
8. Yuan Bo. The research on laser truing of superabrasive grinding wheel [D]. Hunan University, 2013.
9. Zhang Yi, Zhao Jie. Experimental research on the topograph of cBN grinding wheel after mechnaical dressing with ultrasonic vibration assistance [J]. Diamond and Abrasive Engineering, 2012, (2):74-76.

Electrolytic capacitor-less ac-input led driver topology

Meng Zhang
Key Laboratory of Advanced Electrical Engineering and Energy Technology,
Tianjin Polytechnic University,
Tianjin, 300000, China
E-mail: zm2015_happy@163.com

Sheng-Wei Gao
Key Laboratory of Advanced Electrical Engineering and Energy Technology,
Tianjin Polytechnic University,
Tianjin, 300000, China
E-mail: gaoshengwei1978@126.com

Xiao-Fei Liu
Key Laboratory of Advanced Electrical Engineering and Energy Technology,
Tianjin Polytechnic University,
Tianjin, 300000, China
E-mail: 1758182243@qq.com

In order to improve the life and reliability of the drive power supply, the electrolytic capacitor must be removed. This paper proposes a pulsating current LED driver, do not need any electrolytic or complex control circuit to reduce the low frequency output ripple. The circuit is very simple, single switch topology structure simplifies the design of the controller, the circuit can reduce energy storage capacitor to the micro level. We can use thin film capacitor to replace it. A detailed analysis of the operating principle and theoretical of this circuit in this article.

Keyword: Efficiency; electrolytic capacitor; light-emitting diode (LED).

1. Introduction

In order to reduce energy cost in lighting, solid-state lighting such as LED has become a very popular solution to replace the incandescent lamps due to its superior efficacy performance. Unlike fluorescent lamps, LED lamps are also able to ignite instantly and are environmental friendly as they do not contain any mercury content in the structure of the lamp. LED lamps that are currently on the market are comprised of two major parts: the electronic driver circuit and the group of LED semiconductors. In residential or commercial lighting where the ac input line frequency is either 50 or 60 Hz, the harmonics of the ac input current drawn from the lighting equipments such as the LED lamps.

One key issue with the energy storage capacitor is that, in order to meet the size and cost requirements of the LED driver circuits, electrolytic capacitors are commonly used for the energy storage capacitor in current commercial LED lamps due to their high energy density and low cost. A major problem with the use of the electrolytic capacitor in the LED driver circuit is that it can lead to a mismatch between the lifespan of the driver circuit and that of the actual LED devices.

Summarizing the drawbacks and features of the state-of-the art electrolytic capacitor-less LED drivers, most of the works presented require relatively complicated power circuit topologies (i.e., multiple switch or multiple stage) or current control techniques to reduce the size of the energy storage capacitor. The objective of this paper is to develop a simple and reliable electrolytic capacitor-less solution for a high power factor LED driver in residential lighting applications. This paper proposes a high power factor LED driver topology that provides high frequency pulsating current and does not require any electrolytic capacitors to minimize the low-frequency ripple in the LED current. It is a single-stage single-switch topology and does not require any feedback control circuit to regulate the low-frequency ripple in the LED current. The operating stages of the proposed circuit and its characteristics will be explained in details in this paper.

2. Circuit Design

Figure 1 is a traditional principle of the fly back type circuit diagram, the Lp and Ls are primary side and the vice side inductance of transformer respectively. vc as the input voltage, namely the former PFC converter output voltage ip and is are current of transformer primary side and the vice side respectively. ip_pk and is pk are primary side and the vice side current respectively. VLp is for transformer primary side voltage.

According to The recursive rule, we can get the basic unit which in the red dotted line frame from Figure 2. The basic unit is a new fly back circuit.

Secondary winding Ls1 and Lp2 produce magnetic coupling, forming a new fly back converter. This case, when the original winding Lp2 series with a load, the secondary winding Ls1 series with a voltage source, then the load will be driven by a high frequency triangle pulse waveform. The resulting circuit is shown in Figure 4.

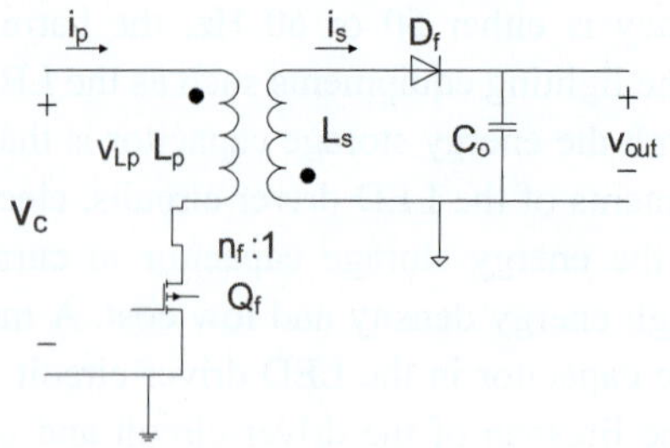

Fig. 1.Typical fly back circuit.

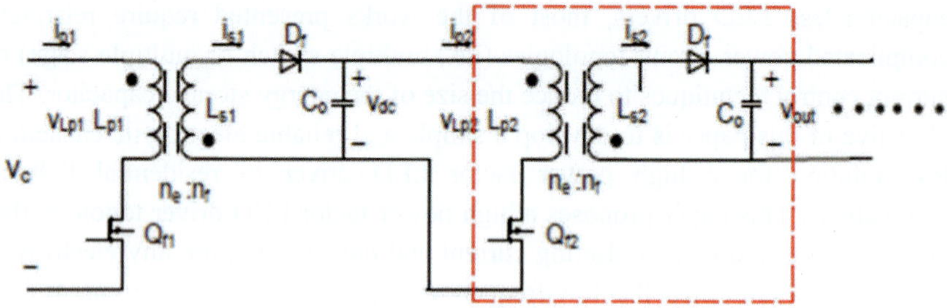

Fig. 2. Cascaded flyback converters.

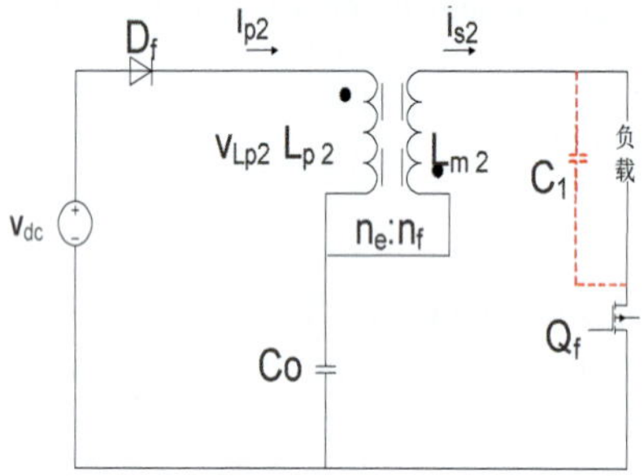

Fig. 3. Simplified pulsating current LED driver.

The Co and C1 is electrolytic capacitors in the original circuit. In the simplified pulsating current LED driver, Co and C1 are Thin film capacitor.

The Figure 5 and Figure 6 shows that using this circuit topology can get a dc wave with a Little fluctuation at a point of “a”. We can get stable flat output waveform, within a small capacitor C1 could filter the ripple. (ω0： switching frequency of MOSFET).

3. Circuit Analysis

There are three stage of the circuit when C1 is omitted and ne equals 1 in a switch cycle. The input voltage source vr represents the rectifier input voltage. The Figure 7 shows the key operations waveform of the circuit.

(Stage 1t0: < t < tl] :When the gate signal (vG) is high, Qf is conducting, iS2 current through the LED increased. The current flows the secondary side of the coupling coil (LS2). If the output voltage of LED is given by (1), The led on-resistance is rD, Vdo is forward voltage or led threshold, m is the number of series-connected of leds. finally this phase occurs at t = dTs and is given by (5).

$$I\left(\left(P-k_i\right)^2, m_i^2, M\right)=\frac{1}{(2\pi)^5}\int\frac{d^3k_i}{2\omega_i}\delta^4\left(P-k_i\right). \tag{1}$$

Punctuation marks are used at the end of equations as if they appeared directly in the text.

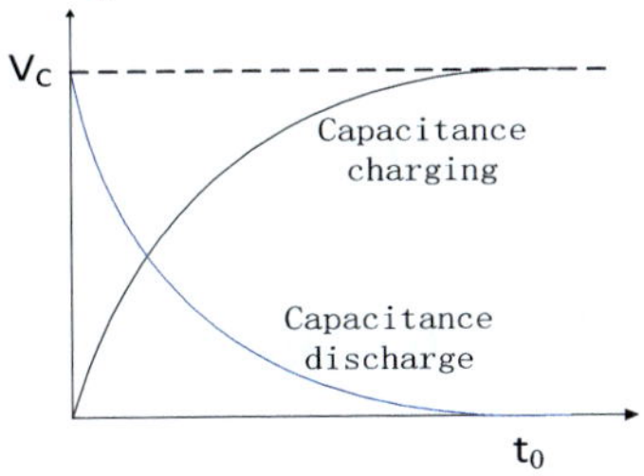

Fig. 4. Capacitance characteristic figure.

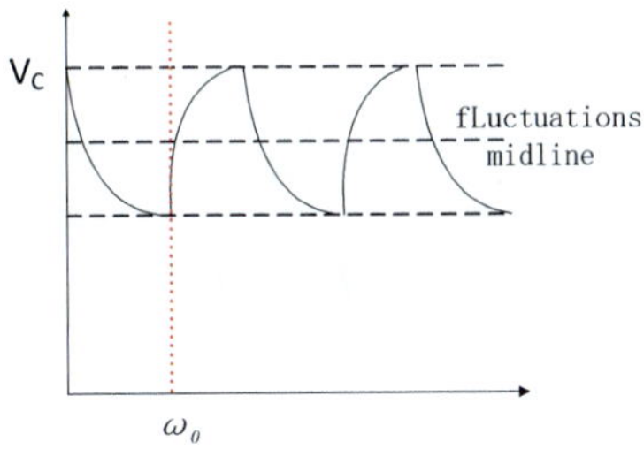

Fig. 5. Capacitance voltage waveform.

3.1. *Maintaining the integrity of the specifications*

The template is used to format your paper and style the text. All margins, column widths, line spaces, and text fonts are prescribed; please do not alter them. You may note peculiarities. For example, the head margin in this template measures proportionately more than is customary. This measurement and others are deliberate, using specifications that anticipate your paper as one part of the entire proceedings, and not as an independent document. Please do not revise any of the current designations.

[Stage 2: t1 < t < t2] : vG is tending to zero and Qf is closed, the voltage of MOSFET (Vds) begins to rise, is 2 drops to zero. Vds is given by (6) . Due to reverse polarity between the two winding L, Ls2 stored energy transfer to the Co and opened by the original edge of Lp2. Since the Vdc is greater than the vr, Df is opening. Therefore, ip2 linear rise as (7), Δ 1 is the release time of inductance's current.

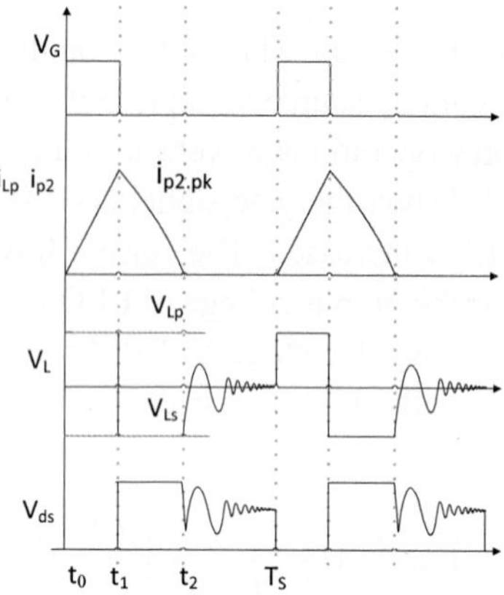

Fig. 6. Key waveforms of the proposed circuit with Co (assuming n=1).

$$v_{out} = m(V_{do} + i_{s2} r_D) \tag{2}$$

$$V_{dc} - v_{out} = (\frac{L_m}{n^2})\frac{di_{s2}(t)}{dt} \qquad 0\langle t \langle dT_s \tag{3a}$$

$$(\frac{L_m}{r_D n^2})\frac{di_{s2}(t)}{dt} + mi_{out} + (\frac{mV_{do} - V_{dc}}{r_D}) = 0 \qquad 0\langle t \langle dT_s \tag{3b}$$

$$i_{s2}(t) = (\frac{V_{dc} - mV_{do}}{mr_D})(1 - e^{-(\frac{mr_D n^2}{L})t}) \qquad 0\langle t \langle dT_s \tag{4}$$

$$i_{s2,pk} = (\frac{V_{dc} - mV_{do}}{mr_D})(1 - e^{-(\frac{mr_D n^2}{L})dT_s}) \tag{5}$$

$$v_{ds} = v_{dc} - v_{L2} - v_{out} \tag{6}$$

$$i_{L1}(t) = \frac{i_{s2,pk}}{n} - (\frac{v_r - V_{dc}}{L_m})(t - T_s)dT_s$$

$$dT_s \langle t \langle (d + \Delta_1)T_s \tag{7}$$

(Stage 3: t2 < t <Ts): Qf and Df are closed, the Vdc remain at the same level. As a result of the existence of C1, L and output capacitance (Coss) provides a current path of the switch.. Due to the L and Coss resonance vLp2

and Vds high-frequency oscillation can be observed. C1 should be very small. In fact, this is enough to reduce the high frequency output voltage ripple. The Vds is obvious oscillation. Vds is shown in (6), at the end of this stage vG is higher again.

Acknowledgments

This work is supported by Science and technology plan projects of Tianjin (14 ZCDGSF00128), Tianjin city outstanding young teachers plan (20130009), the National SME Innovation Program (14c26211200316), and the college students innovation plan (201410058034).

References

1. A Review of the Literature on Light Flicker: Ergonomics, Biological Attributes, Potential Health Effects, and Methods in Which Some LED Lighting May Introduce Flicker, IEEE Standard P1789, Feb. 2010.
2.] A. Wilkins, J. Veitch, and B. Lehman, "LED lighting flicker and potential health concerns: IEEE standard PAR1789 update," in Proc. IEEE Energy Convers. Congr. Expo., 2010, pp. 171–178.
3. S. Li, S.-C. Tan, S. Y. R. Hui, and C. K. Tse, "A review and classific ationof LED Ballasts," in Proc. IEEE Energy Convers. Congr. Expo., 2013, pp. 3102–3109.
4. Report: Lifetime of White LEDs, "Energy eficiency and renewable energy," U.S. Dept. Energy, Washington, DC, Sep. 2009.
5. Manual: Electrolytic Capacitors Application Guide, EvoxRifa, Espoo, Finland, 2001.
6. K. Zhao, P. Ciufo, and S. Perera, "Lifetime analysis of aluminum electrolytic capacitor subject to voltage fluctuations," in Proc. 14th Int. Conf. Harmonics Quality Power, 2010, pp. 1–5.

Design of intelligent greenhouse management machine

Jiang Jiang, Rong-Ling Shi* and Jian Mao
School of Machining and Electrical Engineering,
Xuzhou Institute of Technology,
Jiangsu Xuzhou 221111, China
**Email:849491339@qq.com*

In order to design a new type of electric greenhouse management machine, the paper researched and designed its power transmission system, control system and the machine structure parameters; Using motor as power source, it realizes the green environmental protection society's need for machine's energy conservation and emissions reduction.

Keywords: Greenhouse management machine; intelligent control; green environmental protection; design.

1. Introduction

Because the soil inside the greenhouse cropping index is high, the frequency of the fertilizer, irrigation, farming more than general farmland soil, and high organic matter content, so the soil bulk density is low (about 1.10 ~ 1.139 cm^3) [1], and increased with increasing years of planting extension, viscous soil is bigger. Relevant data show that the thickness of soil magnetism commonly in greenhouses for 15 to 25 cm, 80% ~ 90% of vegetable root distribution among them, the traditional management machine in viscosity larger soil, work ability is low, soil resistance, power consumption increases, and the relative diesel and gasoline engine as power, the greenhouse environment pollution caused by [2].

Therefore, this project adopts the advanced design theory and method, the research of complete drive and control system research and the study of power consumption and motor design, both appearance and other parts of research, make the whole machine is small in size, light weight, compact structure, convenient operation and flexible adjustment, good operation performance, reliability, high durability, no emissions, no noise, low energy consumption.

The whole machine structure of intelligent greenhouse management machine, is mainly composed of motor, drive mechanism, rotary tillage parts, walking mechanism and control system, etc. Motor provides the power, through the coupling connection with reducer, gear reducer choose double output shaft, the side output shaft power by chain wheel drive, pass power to walk round, implementation management machine since go, transition shaft is installed on the clutch, the clutch of meshing and separating, control machines walk and stop; The other side of the output shaft output power directly drive the crank, through claw pass power to handle rotary parts, make it realize rotary tillage operation; Using 89C51 as the core of the control system of thyristor trigger and rotational speed measurement to realize full digital microcomputer control motor double closed loop speed control.

2. Determination of the Main Parameters

2.1. *Motion parameters*

2.1.1. *Unit speed V_m*

Machine speed is one of the main factors influencing the productivity is high and low, according to relevant data, throw soil power increase with the increase of the speed according to the three order functions[3]. Reference to rotary cultivator commonly used line speed (0.55-0.85 m/s) [4], the selected unit speed $V_m = 1.08$ km/h = 0.3 m/s.

2.1.2. *Vehicle speed ratio λ ($\lambda = V_p/V_m$)*

Machine for greenhouse management, λ for rotary tillage blade endpoint peripheral velocity and the unit speed ratio, known as the rotary speed ratio. λ choice should not only guarantee the normal work of the management machine, also need to consider such factors as the efficiency of the whole machine structure, the power consumption. λ value is different, rotary tillage blade in the soil trajectory and the cutting of the shape of the earth it is different[6]. As shown in Figure 1:

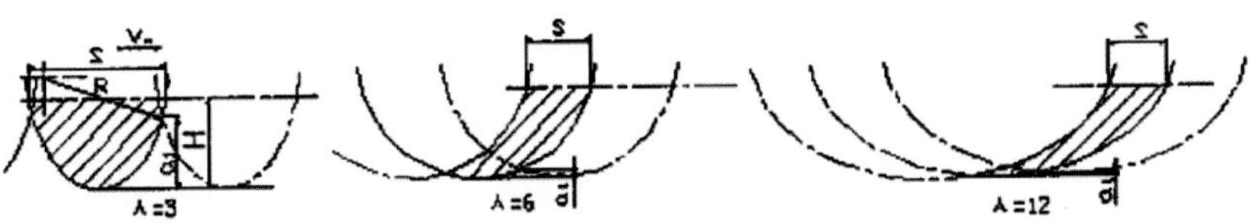

Fig. 1. Different λ rotary cutter trajectory.

S-all soil pitch (cm); R-knife roller radius (mm); Vm-unit speed (m/s); H-a tillage depth (cm); a1-a groove bump height (cm).

Under the same tillage depth, along with the increase of the rotary speed ratio λ pitch cutting soil S decreases, and groove roughness is reduced, and increase power consumption. Faster than λ has important influence on the performance of rotary cultivator, λ choice should not only guarantee the normal work of the rotary cultivator meets the demand of tillage depth, also considering structure of rotary cultivator, power consumption and other factors such as productivity, so the selected $\lambda = 3$. Known $V_m = 1.08$ km/h = 0.3 m/s, $\lambda = 3$, so the driving wheel rim peripheral velocity $V_p = \lambda V_m = 3*0.3 = 0.9$ (m/s). Related research shows that the rotary knife end point peripheral velocity and energy consumption of a quadratic curve relationship [3]. To do this for most types of soil, rotary tillage blade, the peripheral velocity V of the endpoint should be less than 5 m/s, as the selected drive wheel rim peripheral velocity V_p reference standard. Obviously, the selection of the V_p value can satisfy this requirement.

2.1.3. *Driving wheel speed n*

In the case of meet the job requirements, based on the principles of crank rotational speed, the lower the better, the speed of the rotary cultivator usually take 190-280 r/min, so the selected drive wheel speed n = 200r/min.

2.2. *Performance parameters*

2.2.1. *Rotary tillage depth H*

In the greenhouse soil moisture content is higher, generally around 16%, greenhouse soil tillage depth is generally 10-15 cm, determine the rotary tillage depth H=13cm, so we can under the premise of meet the requirements of agronomy, make the digging claws tooth length is not too long.

2.2.2. *Rotary tillage width B*

Easy operation, little space vehicle width B no mission is too large, according to the existing machine width, and the management of machine structure, choose the rotary tillage width for B=500 mm.

2.2.3. *Theory of productivity W*

Management of machine B set to 500 mm in length, so the theory of productivity of the unit

$$W=0.1V_mB=0.1\times1.08\times0.5\approx0.05\text{hm}^2/\text{h}.$$

2.3. *Structural parameters*

2.3.1. *Driving wheel diameter D*

Driving wheel rim peripheral velocity V_p=0.9 m/s, driving wheel speed n=200 r/min, and V_p=ωR (ω for driving wheel angular velocity, R for the driving wheel radius), ω = 2πn / 60, V_p = 2πnR / 60, the

$R=60V_p/2\pi n=(60\times0.9)/(2\times\pi\times200)=0.04297p(m)=4.297cm$

So the driving wheel diameter D = 2 R = 2 * 4.297 = 8.59 cm.

2.3.2. *Walking wheel diameter D '*

Walking wheel diameter position rotary parts of the grave, has assumed that rotary tillage component perpendicular to the ground in front of the grave, handle the rotary claw off the ground height is b, the management machine whole structure shows that walk round radius R '=b -R' (R ' said turn drive cycle of center distance to the bottom edge of the bearing) [7]. Known b=24 cm, r'=6 cm, R '=24-6 =18 cm, so the wheel diameter D '=2R ' =2 * 18 =36 cm.

3. Design of the Power Transmission System

3.1. *Power transmission scheme*

Known to walk round radius R=180 mm, walking round edge at the rate of machine forward velocity V_m, by the formula: V_m=ω 'R '

$\omega'=V_m/R'=0.3/0.18=1.67rad/s$

Walking wheel speed n”:

$n''=60\omega'/2\pi=(60\times1.67)/(2\times3.14)\approx15.95r/min$

The walk to the rotation of the shaft angular velocity ω’.

Motor to drive the rotary tillage parts rotary tillage operations, and to drive wheel rotation, the rotating speed of the motor is 1400 r/min, the speed of the crank for n=140 r/min, walking wheel speed n”=15.95 r/min. So to borrow a reducer will decrease the speed of the motor to the speed of the crank. Choose double output shaft reducer, by the side of the clutch shaft connection work, on the other side through the chain extension transmission power transition through to the walking wheel axis, to avoid the chain size is too large, the triple chain wheel transmission.

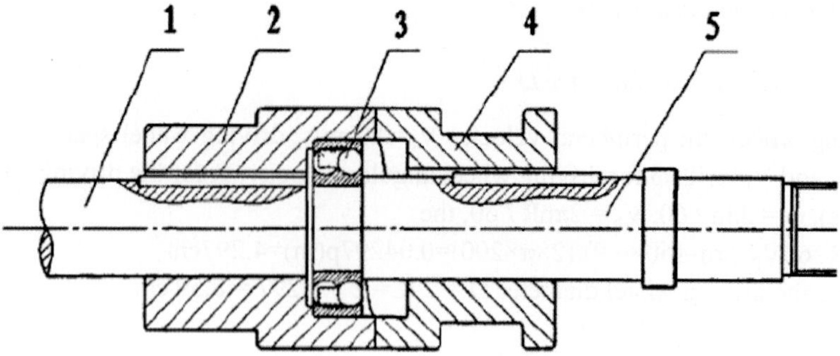

Fig. 2. Work on the shaft power transmission.

1. The reducer output shaft; 2. Take the initiative to clutch; 3. Bearing; 4. The driven clutch; 5. Working shaft

3.2. *The way to work on the shaft of the power of design*

Reducer output shaft was achieved by the clutch shaft connected with work, through the combination of the clutch and the separation control working shaft working condition. Working shaft stability should have at least two a certain distance away from the bearing support to ensure that, in order to machine of the small size as much as possible, avoid using two bearing seat, the left end of the working shaft and set in the clutch of self-aligning ball bearings, clutch is actively and combining with the power transfer effect, also have the effect of bearing seat, near the spline shaft shoulder installation self-aligning ball bearings, through the bearing seat is fixed on the frame, to ensure the operation stability of the working shaft. Compared with using the two bearing seat, and the way of power transmission effectively reduced width of the machine.

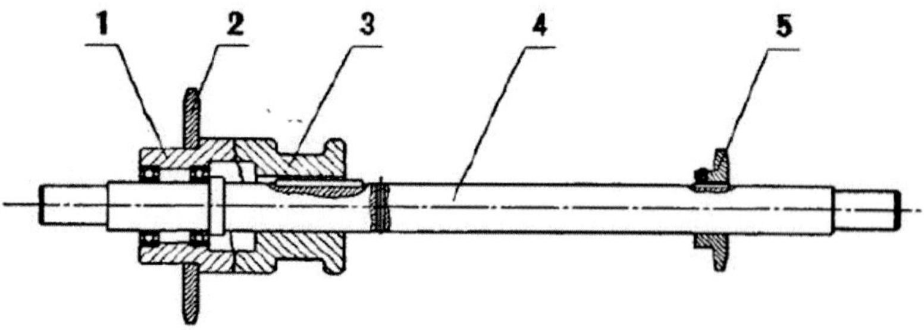

Fig. 3. Walking on the shaft power delivery.

1. Take the initiative to clutch; 2. Chain wheel; 3. Driven clutch; 4. The transition round ; 5. Sprocket.

3.3. *Design of the power transmission way on walking wheel*

As shown in figure 3[5], take the initiative to clutch lining set with two rolling bearings with dust cover, set on the transition shaft l, sprocket and active 2 clutch welded together, and transition axis 1 connection. Reducer output shaft sprocket l, 2, transfer the power of the motor to the chain sprocket 2 turns transition shaft 1, if the clutch and driving shaft meshing, pass through the driven clutch to transition shaft of 1, 3 passed to walk through the sprocket wheel axles, machines and finally realize the walk; If driving clutch and driven clutch, the transition shaft l idling, power cannot pass, stop walking machine.

4. Design of the Control System

4.1 *Principle*

Double closed loop speed regulation system has the advantages of fast dynamic response, strong anti-jamming capability. Its inner loop is current feedback control loop, the outer ring is the motor speed feedback control loop, the regulation of the inner ring and outer ring are made of 89C51 single chip microcomputer software to complete the operation of digital PI control law 2 system has been set up in the speed and current regulator, in order to obtain good static and dynamic performance, a regulator of PI regulator. This kind of dual closed loop controller structure can properly play a role of negative feedback of current cut-off negative feedback and speed.

4.2 *Function*

Double loop system with 89C51 to replace the dc motor speed control device of the current and speed controller and no.6 trigger pulse generating circuit. The whole control program consists of the main program, interrupt service program, PI algorithm and various auxiliary program, program length is 4 KB byte, run over the time of 3.33 ms.

Main program main function is: on the initialization, a stack pointer, and 89C51 and the main chip initialization, the state of the query sensor array and according to the command button to set up corresponding control command symbol, fault detection alarm, and waiting for the sync pulse external interruption, and so on.

5. Power Consumption Calculation

Power consumption management mainly includes the working parts of rotary tillage driving unit power consumption and power consumption.

5.1. *Rotary tillage power consumption calculation*

Rotary tillage power consumption available[6] to calculate the power consumption of the rotary cultivator formula, namely

$N=0.1\ k_\lambda HV_mB$ kW

Type k_λ-one machine is working in soil resistivity (N/cm^2), can be estimated by formula to λ $k_\lambda=k_gk_1k_2k_3k_4$; k_g-the average soil resistivity (N/cm^2); k_1-a plow deep correction coefficient; k_2-a correction coefficient of soil moisture content; k_3-residues vegetation correction coefficient; Correction coefficient k_4 a operation mode; H-a digging depth (cm);V_m- machine speed (m/s); B-a vehicle width (m).

Based on operation conditions within the greenhouses, the look-up table[6] determine: $k_g=16N/cm^2$, $k_1=1.1$, $k_2=0.95$, $k_3=1.2$, $k_4=0.75$

So $k_\lambda=15\times1.1\times0.95\times1.2\times0.75=15.1N/cm^2$

Known rotary tillage depth H=3 cm, unit speed $V_m=0.3$ m/s, rotary tillage width B=0.5 m,

$N=0.1\times15.1\times13\times0.3\times0.5=2.94$kW

5.2. *Driving unit power calculation*

Set the unit in the process of driving by the rolling resistance of soil to F, unit speed, Vm has overcome the consumption of rolling resistance power of [7]:

$N'= FV_m$

And $F=fG_s=fmg$

Type f-in a rolling resistance coefficient; G_s-management machine gravity (N); m-a machine the machine quality management (kg); g-The acceleration of gravity (m/s^2).

A preliminary estimate of greenhouse management machine quality m is 100 kg, f = 0.18 H [7], plug type $F=fG_s=fmg$ as well:

F = 0.18 * 100 * 9.8 = 0.18 N

Known $V_m=0.3$ m/s, plug type $N'= FV_m$, overcome the greenhouse management machine rolling resistance power consumption:

N' = 0.18 * 0.3 = 0.054 kW

Therefore, the total power of the greenhouse management machine consumption N_T is:

$N_T=N+N'=3.26+0.054\approx3.31$kW

5.3. *Determination of motor model*

Considering power loss in the process of the transfer, the power rating of the motor should be greater than the theoretical value of total power consumption management machine, so choose Y112M, 4 motor, the power rating of 4 kw, rated speed of 1400 r/min, the motor starting torque is bigger (1.8 ~ 2.5 times the rated torque), large starting current, overload ability is a bit weak, but the power factor of load is higher, power saving effect is remarkable.

6. Conclusion

In this paper, based on the research of the existing greenhouse management machine, for domestic greenhouse space is narrow, rotary tillage operation such as low efficiency and high energy consumption and environmental pollution problems, combining with the characteristics of greenhouse soil and agronomic and absorb domestic advanced type design concept, put forward the design thought of intelligent greenhouse management machine. Other parts after calculating and checking, as far as possible by using interchangeable parts and standard parts, fully reflects the economic, practical, the design principle of energy saving and environmental protection.

References

1. Chang-ai Zhang, Greenhouse soil physical and chemical and biological conditions. Anhui agricultural science, 2002 (2) : 257-276.
2. Jing Fu etc. China facilities development present situation and the direction of the agricultural machinery [J]. Journal of agricultural machinery, 2006 (6) : 120.
3. Yong-jian Gong,Yong-sen Xiong et al. Facilities in agriculture machinery design research. Journal of jinhua professional technology institute, 2002 (1) : 1-5.
4. Nanjing agricultural university. Agricultural mechanics. China agriculture press, 1996.3.
5. Jiang Jiang, Rong-ling Shi et al. The system design of intelligent greenhouse tillage machine. Agricultural equipment technology magazine, 2016.
6. The academy of agricultural mechanization. Agricultural machinery design manual (I). Machinery industry press, 1988.4:169-186.
7. Lian-xing Gao, Yan Shi et al. Tractors and cars (part II). China agriculture press, 2000.1.

Research on the energy-saving mechanism for the hybrid mechanical loader with parallel power system

Lan Luo, Peng-Yu Wang* and Wen Sun
State Key Laboratory of Automotive Simulation and Control,
Jilin University,
Changchun 130022, China
*E-mail: *wangpy@jlu.edu.cn*

Mao-Lin Ba, Zhi-Heng You and Xiao He
State Key Laboratory of Automotive Simulation and Control
Jilin University,
Changchun 130022, China

To explore the maximum of energy saving potential of hybrid loader and analyze the energy-saving ways of hybrid loader, the researcher took use of AMEsim software to build hybrid loader model. The author simulated on the energy saving ways and made the prototype test at the same time. In the end, with the quantitative approaches, the energy saving potential of energy saving ways were shown in the end.

Keywords: Hybrid mechanical loader; engine; energy-saving way.

1. Introduction

Configuration of hybrid system was divided into three configurations, namely series, parallel, mixed. Taking complex working condition by the loader and high power output into consideration, this article chose the single axis parallel type hybrid to drive system. Parallel power system was the widest form to use at present, which could drive the engine and motor alone or hybrid variety of work mode at the same time. Owing to the truth that the output power of two power source could be stacked, the researcher used smaller power engine. Compared with the cascade system, engine running in parallel system was affected more directly by load condition. The operating range of engine was wider and the output characteristic of the system was closer to the traditional system of engine power. The operating state of the motor depended on how to adjust the engine effectively, thus, the control strategy of parallel system was more complicated. In addition, the parallel system had higher reliability for which electric drive system could still maintain operation when it failed [2].

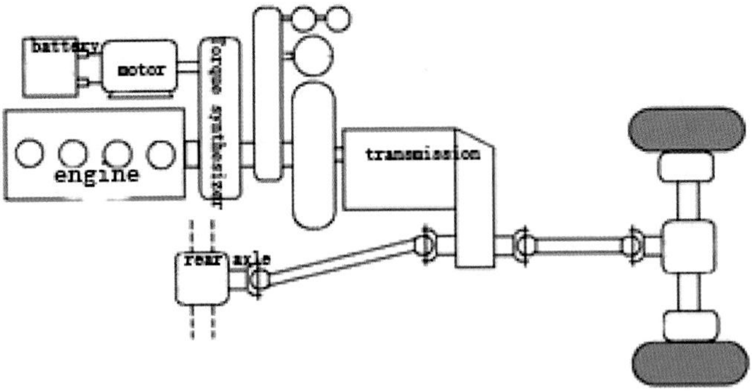

Fig. 1. Solutions of hybrid loader transmission system.

Table 1. The parameters of hybrid power system.

The maximum of engine's output power	144kW@2200rpm	The motor working voltage	360V
The maximum of engine's output torque	950Nm@1400rpm	Motor speed	1450rpm/2800rpm
The motor power	50kW/100kW	The motor torque	340Nm/640Nm
Super capacitance	3000F	The rated voltage	2.7V (DC)

2. Analysis of Hybrid Loader Energy-Saving Way

2.1. *Reduce engine power*

Traditional loaders usually are equipped with high-power engine to adapt to the complex phenomenon. Due to the output of engine power fluctuations, therefore they cannot guarantee the engine to work in high efficient range. According to the working condition, we can choose a smaller engine and drive the engine alone under the most conditions of loader at empty or light load. When there is high power demand , the researcher adopt engine drive and motor power in order to reduce the basis of the engine fuel consumption and improve the whole engine load rate and fuel's utilization rate [3].

2.2. *Help engine speed improve work efficiency*

The process of engine speed from the starting point to the normal working speed can be regarded as a process of which fuel utilization rate is low. When drive power is needed, the truth that the motor helps the engine to reach the speed to work needed quickly, not only can improve the efficiency, also can reduce the instantaneous fuel consumption. After adding the electric drive system, peak power in the hybrid system is greater than the traditional model of the used in the large displacement because the motor's response speeds up quickly. For complex and overloading condition hybrid loader will have a higher response speed, it will improve the efficiency of the entire working cycle and save energy.

2.3. *Adjust the engine working point in effective range*

Under complex conditions, the engine deviates high efficiency range if the load is too small or large, or gear choices are unreasonable. Thus, the engine working point changes widely and fuel efficiency is low. Hybrid loader can regulate engine working in high efficient range by controlling the motor torque, so as to raise the utilization rate of the engine fuel under this kind of circumstance

2.4. *For energy recovery*

Due to the loader's low speed in the work cycle, small action range of loader working device and usage of high unloading situation, recycling this part of the potential energy of hydraulic system was improved and affects the work efficiency, so a hybrid loader does not recover energy.

3. Establishment of Hybrid Loader Model

According to actual structures of the loader, the researcher selected elements in AMESim and set the parameters to establish model of working device and power transmission system and hydraulic system. AMEsim model was controlled through the establishment of controlling strategy in the MATLAB.

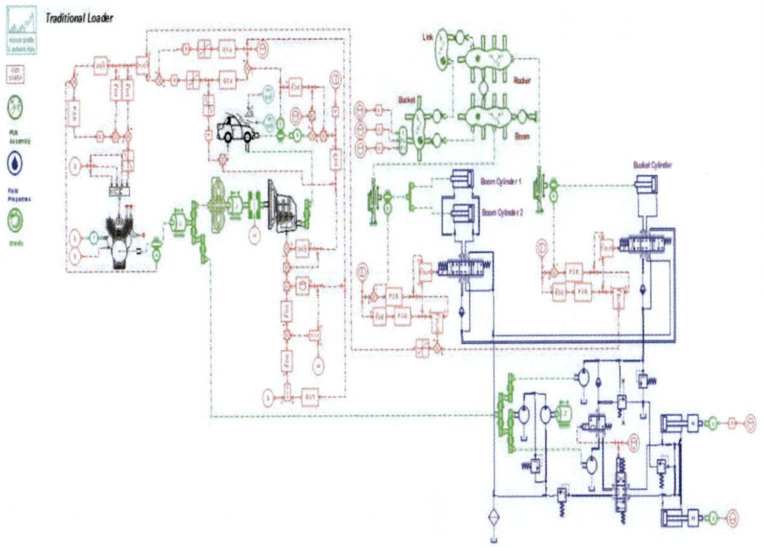

Fig. 2. The system model of hybrid loader power.

4. Analysis for the Results of Simulation

With the simulation analysis of the traditional loader and hybrid loader platform of simulation .In the stage of simulation, we only proved reduction of the engine power sum and adjusted of the engine working point within the range. To reduce the engine power, we only changed the 6.7 L engine into 8.3 L traditional model. The results showed that the overall fuel consumption only with small engine cycle was 283 g. and the highest instantaneous fuel consumption was up to 14 g/s with saving energy about 6.7%.

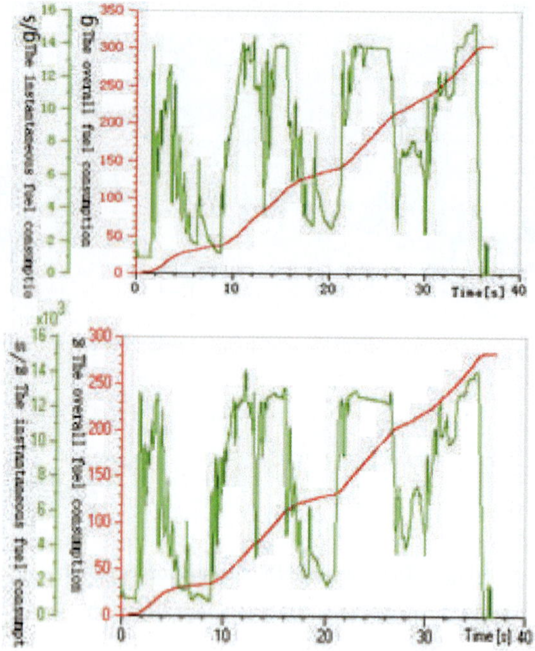

Fig. 3. Consumption curve with 8.3 L and 6.7 L engine for traditional energy.

To adjust the engine working point in high efficiency range in MATLAB, we optimized engine working point through the adjustment of the control strategy. Figure 4 and figure 5 were the distributions of the engine working point before and after of optimization.

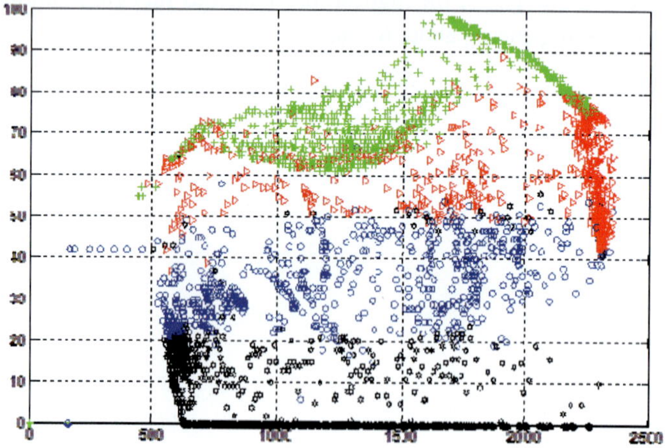

Fig. 4. Before the engine working point distribution optimization.

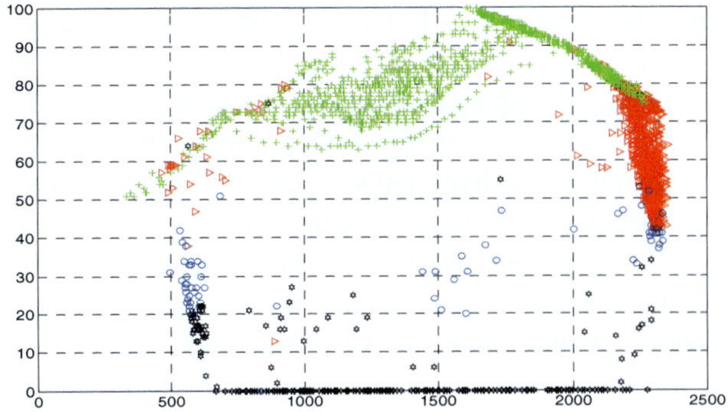

Fig. 5. The engine working point distribution after optimization.

Table 2. Set-point optimization contrast table.

	green	red	blue	black
Traditional models	31%	13%	16%	40%
Hybrid models	59%	4%	3%	34%
The driver's pedal position	θ>85%	50%～85%	20%～50%	θ<20%
levels of the energy consumption	100～106g/kWh	110～130g/kWh	106～110g/kWh	130～140g/kWh

The conclusion could be drawn from the diagram, through adjustment, the original low load point and low efficiency area reduced and most of the working points focused on high load area. About 3% of energy could be saved by estimation.

5. Sample Test

The adjustment process of motor not only affected the efficiency of the loader, but also affected the consumption level indirectly. Therefore we took the sample test to determine its energy saving potential taking the improvement of the work efficiency into considerations, as follows:

Table 3. Working efficiency comparison of the different models.

	Traditional loader	A hybrid loader
material	Loose soil	Loose soil
Driving cycles	"I" shape cycle	"I" shape cycle
workload	30 bucket	50 bucket
gross of fuel	8.77kg	11.524kg
Time consuming	17'52"	28'35"
Fuel consumption	0.2923kg/bucket	0.2305kg/bucket
efficiency	35.72"/bucket	34.3"/bucket

Hybrid models in the same "I" type of energy saving can be up to 21% based on the above data. At the same time, the work efficiency of per bucker improved 1.42. Energy saving by contribution of the efficiency was about 4%.

The following all kinds of energy saving effect were shown in the above Table 4. From the table, the hybrid system engine which speeds up power to improve the efficiency of this item not only improves the working efficiency, but saved up 21.1% of energy, thus it had a good application prospect. Adjustment of the engine working point had the worst effect, because the efficiency of diesel engine range was very large and the process of deliberate adjustment of the engine working point includes the second part of the energy transformation.

However, compared with hybrid cars, secondary energy conversion efficiency would be greatly reduced owing to the truth that hydraulic torque converter or hydraulic system exists between dynamic systems with actuator. In result, the energy saving potential of adjusting the engine working point was not large.

Table 4. Loader statistics about hybrid energy saving effect.

Approaches of energy- saving	Energy saving effect
Reduce the engine power	6.7%
improve work efficiency by Engine speeding up	21.1%
Adjust the engine working point	3%

6. Conclusion

In this paper, the energy saving mechanism was analyzed. The researcher establish parallel hybrid loader model by AMEsim software and had platform joint simulation and a prototype test with control strategy built in MATLAB. In addition, the researcher analyzes hybrid loader three energy-saving ways (to reduce engine power, help improve work efficiency and adjust the energy conserving potential of the engine working point) quantitatively combining with the results of simulation and test. Results showed that the engine speeding up power to improve the energy saving potential had not only the maximum work efficiency, but also could improve the working efficiency, which could be regarded as a hybrid loader researching direction.

References

1. Qi-ming Zhang, Peng-Yu Wang, Cheng Kai et al. Based on the AMESim modeling and simulation of hybrid loader [J]. Journal of mining machinery, 2012, 40 (11):29-33

2. Qi-Ming Zhang. Hybrid loader system design and control strategy research [D]. Jilin University, 2013
3. Peng-Yu Wang, Qing-Nian Wang, Qi-ming Zhang. The Analysis of Energy-saving Mechanism of Hybrid Excavator[C], ICAMME2010, 1657-1662.
4. Shu red, datong, qinjian-junhu. Hybrid control strategy research status and development trend of [J]. Journal of chongqing university (natural science edition), 2011, 24 (6) : 28-31
5. Peak. Energy-saving control technology of hydraulic excavator [D]. Zhejiang University, 2002.

Innovation knowledge acquisition method of product agile industrial design process

Zhi-Xue Sun[†], Chen Chen[‡] and Le Zhang[§]
Department of ME, Shaanxi University of Technology, Hanzhong 723003, China
[†]*Email: sxlgszx@163.com*, [‡]*email:18391609257@163.com*,
[§]*email:18391609368@163.com*

To meet the agile product development needs of enterprise, on the basis of analyzing agile industrial design features, the process of agile industrial design was proposed, which was used to approximate user needs through iteration. Combined with knowledge acquisition process of creative design, the knowledge of creative process was structurally represented by creation knowledge information cell. By constructing template for creative knowledge match, the matching algorithm was designed to control design iteration times. Creative knowledge acquisition in the process of oil tank conceptual design was used as an example to verify the method.

Keywords: Agile industrial design; product design; creative process; knowledge acquisition.

1. Introduction

With the diversification development of consumer demands, the enterprise has to adjust their own pace rapidly and make an agile response[1], thereby, the new challenge is being proposed to industrial design. As the knowledge intensive business industry[2], the changeable marketing environment is determining the success of design which is more based on the multi-disciplinary knowledge to the effective sharing, integration and application. The process of industrial design is exploring the solution of cognitive space to face some problem[3],its essence is exploring the question's optimal solution through creative thinking. However, the creative thinking has great flexibility, and make the process of design more depended on experience, inspiration and intuition of industrial designers. The results of design have a great uncertainty, which are difficult to meet the enterprise's pursuit of design quality. Through acquiring creative process and effective organization and management, it contributes to the designers to reuse the design intellectual quickly and transform it into creative design, and improve the agility of the process of product design.

Through analyzing the features of product agile industrial design, multi-iteration of agile industrial design process is proposed, which makes the design result better approximate user's needs. According to knowledge classification of the different stages of creative process, the creative process knowledge is characterized by creative knowledge information. In order to control the iteration times of agile industrial design effectively, the algorithm of creative knowledge match is proposed which can limit the iteration and acquire dynamic creative knowledge. Ultimately, the method has been verified by taking creative knowledge acquisition in the process of oil tank creative design as an example.

2. Product Agile Industrial Design Features

The industrial design process is influenced by many factors, such as customer needs, technical implementation conditions, social cognition and so on, the direct audience is enterprise, according to the agile environmental feature faced by enterprises[4], which has the feature of product agile industrial design process as follows:

(1) Quick response. The demands for agile development of enterprise determines the agile industrial design process, which should apply to product development planning, and respond to the changeable at any time;

(2) Knowledge-based design. The sharing and reuse of design knowledge are contributed to generate a new plan rapidly and efficiently;

(3) Adaptive. In various stages of product development, the design activity should be adjusted according to design progress, and approach to the design demands.

3. The Process of Agile Industrial Design

Industrial design is a multi-tasking and multi-granularity creative activity, which throughout the entire product development process. The variation of corporate internal and external environment has caused the product development process has a dynamic nature, and made the solution of the design as the basic form of "Program - Iteration - Feedback"[5]. In order to better support agile product development, this paper proposes the creative knowledge acquisition process of product agile industrial design, as shown in Figure1.Based on the industrial design process of "requirement analysis-conceptual design-optimizing design-transition", it makes a multi-granularity segmentation at various stages, the concrete iteration process is as follows:

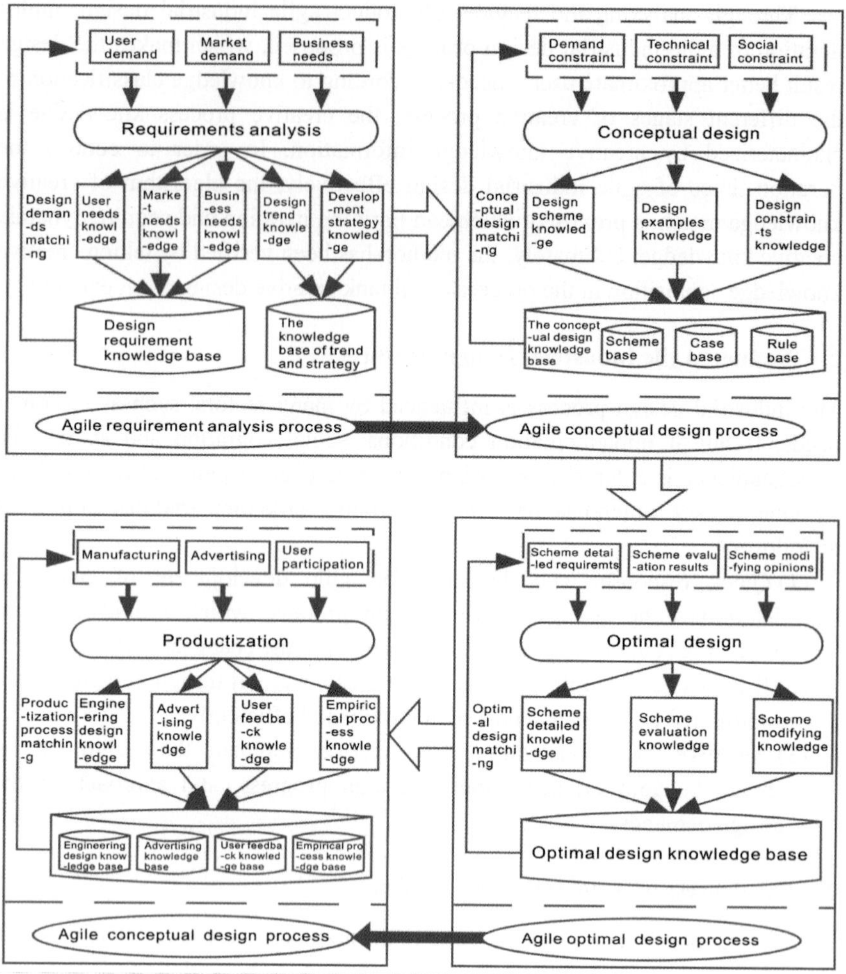

Fig. 1. Product agile industrial design knowledge acquisition process of creative design.

(1) Agile requirement analysis. It divides requirement input into user requirement, market requirement and corporate requirement, forms the design requirement knowledge base, trend and strategy knowledge base through requirement analysis. Match the original input requirements with design requirement knowledge base, if which reaches an agreement then to conceptual design process, or continue with requirement analysis; match the new input requirements with design requirement knowledge base, if which fits there is no need to re-analysis, otherwise, continue to make

requirement analysis. Match trend and strategy with knowledge base, if which goes, we can call the relevant knowledge to conceptual design and regard it as design reference, otherwise, store it in knowledge base;

(2) Agile conceptual design. Based on the requirement constraints, technology constraints and social constraints, we spread the conceptual design, the process and result of which form design scheme base, case base and rule base. Match the original input design constraints with conceptual design knowledge, if which reaches an agreement then to optimizing design process, or continue with conceptual design; match the new input design constraints with conceptual design knowledge, if which fits there is no need to continue with conceptual design, otherwise, continue to make conceptual design process. After which is completed, process knowledge is stored in the conceptual design knowledge base;

(3) Agile design optimization. Adjusting the details of conceptual design scheme with reference to it's detailed requirements, assessment results and revisions. Match the original program requirements with optimized design knowledge, if which goes to transition, or continue to be optimized; then match the new program requirements with optimized design knowledge, if which fits there is no need to re-optimization, otherwise proceed. After the optimized design is completed, process knowledge is stored in the optimized design knowledge base;

(4) Agile transition. Conducting batch trial to the scheme by combining with manufacturing, advertising marketing and user's feedback. Then it produces engineering knowledge, advertising and marketing knowledge, knowledge of user feedback and process knowledge during this process. Match the transition process of the input with the transition knowledge base, if which fits with the above, we can call the relevant knowledge to make it commercialized directly, otherwise, carry out the transition by combining with input knowledge and knowledge of knowledge base. Then process knowledge is stored in the product database after the transition is completed.

4. The Creative Process of Knowledge Acquisition

Creative process knowledge mainly includes the three stages of knowledge about requirements analysis, conceptual design and optimized design of agile industrial design process.

4.1. *Creative process knowledge classification*

According to the agile industrial design process from Fig. 1, the creative process knowledge will be divided into requirement analysis, trends, strategies, conceptual design and optimized design.

(1) Requirement analysis knowledge: including knowledge need, the market demand of knowledge and knowledge of enterprise needs;

(2) Conceptual design knowledge: including design knowledge, case knowledge and knowledge of design constraints;

(3) Trend and strategy of knowledge: including design trends and development strategy;

(4) Optimized design knowledge: including plan detailed, scheme evaluation and plan modification.

4.2. *The Creative process of knowledge acquisition*

According to the process of agile industrial design, the process of knowledge acquisition of different originally stage are as follows:

(1) The agile requirement analysis process of knowledge acquisition. The common methods of demand for knowledge acquisition are questionnaire, interview, verbal protocol, Scenario analysis, Kansei engineering and image scale[6], which are used to explore all kinds of demand information about users, markets and enterprises ,as well as the development trend and strategy of the product. The resulting knowledge exists in photo gallery and text set;

(2) The agile conceptual design process of knowledge acquisition. Integrating the requirement restriction, technology restriction and social constraints which are driven by requirement analysis into the process of conceptual design, at the same time, graphic thinking[4],record of the design process, design analysis, the concept of three-dimensional model building and other methods, which can be used to describe the conceptual scheme. The resulting knowledge are photo gallery, text set, model base, case base, rule base and so on;

(3) The agile optimized design process of knowledge acquisition. That the scheme of conceptual design is submitted to users to review, according to these assessment results, we can put forward solution of amendments and somewhere needed to be adjusted in details, then optimized design is conducted by industrial designers. This process is known as a more detailed conceptual design process, the formative language of which are photo gallery, text set, model base, case base, etc.

Knowledge acquisition process of creative design is shown in Figure 2.

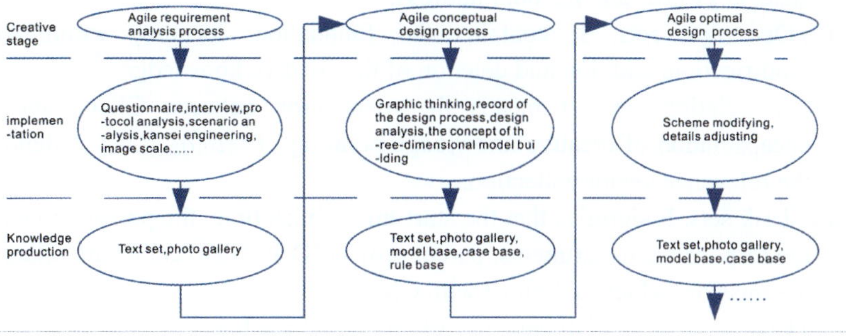

Fig. 2. Knowledge acquisition process of creative design.

5. Agile Industrial Design Supported by Creative Knowledge Matching

5.1. *Creative knowledge representation*

The method of 5W1H (what, when, where, who, why, how)[7] are used to extract the common features of creative process knowledge, and make up creation knowledge information cell. That information cell encapsulates knowledge of the various stages of the creative process, and provides a convenience for knowledge matching, reusing and sharing through the use of effective organization and classification methods.

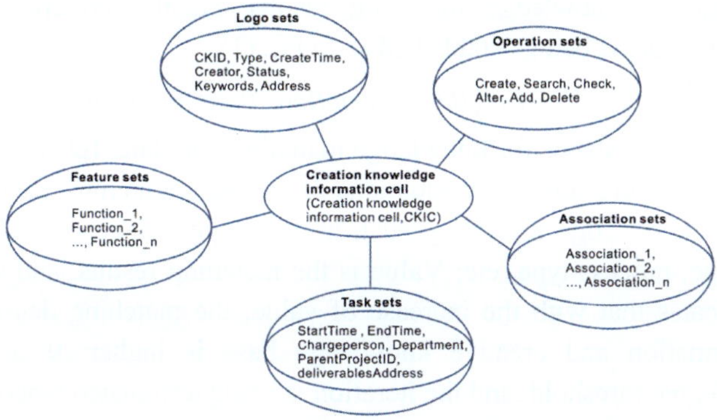

Fig. 3. The composition structure of creative knowledge information cell.

The composition structure of creative knowledge information cell is shown in Figure 3, which consists of logo sets, operation sets, association sets, task sets and characteristic sets. The specific content of the various parts are:

(1) Logo sets. It contains creative knowledge ID, type, creation time, founder, state, keywords and memory address;

(2) Operation sets. It contains the foundation, locating, examination, modification, addition and deletion of the creative knowledge;

(3) Association sets. It means the association between CKIC, and the encapsulation of creative knowledge pointing. $A=\{a_1,a_2,...,a_n\}$ is denoting the relevance vector collection;

(4) Task sets. It signifies that the start time, finish time, the person in charge, the responsible departments, parent project ID and deliverable are related to the mission, which is defined with CKIC;

(5) Characteristic sets. It contains all functional attributes of CKIC.

5.2. *Creative knowledge matching*

It is the process of knowledge retrieval and reusing, and the efficiency, as well as the success rate of the product development can be improved by repeated knowledge matching and creative process of iteration. The number of iterations are controlled by creative knowledge matching results, if the input information matching degree exceeds the preset threshold limit, then there is no need to make further iteration, otherwise, to put input information in the knowledge base and continue to iterate. In order to conduct the knowledge matching process preferably, there has made an encapsulation to the input information of the stages of creative process in accordance with creation knowledge information cell and keywords semantic extension[8] to improve the matching accuracy.

To describe knowledge matching results exactly, creative knowledge matching template is being established as in Eq. (1):

$$T_k=\{ID, Description, Type, Value, \theta\}. \tag{1}$$

In the type, T_k stands for knowledge matching template; ID is the identity of the template; Description is introductions of template matching results; Type is the form of knowledge matching, such as text type, image type, model type, numeric type, process type, etc; Value is the matching results, and $0\leq Value\leq 1$, which indicates that with the increase of value, the matching degree between input information and creative knowledge base is higher; θ is the preset matching degree threshold ,and the iteration is being terminated when value$\geq\theta$.

So the creative knowledge matching algorithm is as follows:

Step1: Information input;

Step 2: Modeling in accordance with creative knowledge information cell, and forming creative knowledge element set$\{CKIC_1, CKIC_2, \ldots, CKIC_m\}$;

Step 3: Making creative knowledge matching to every $CKIC_i$, if Value$\geq\theta$, then to step 5; or else to step 4;

Step 4: Input information is stored in knowledge base, at the same time, to make creative process of iteration, then to step1;

Step 5: The matching is being finished, and the matching results is being exported.

6. Application Examples

Taking wireless mobile Internet terminal U-fi as an example. During the design process, in addition to meet basic wireless Internet hot spot to provide functionality, but also satisfy corporate image positioning, technology embodies, humanistic concern and other requirements, which is a repetitive iteration process of superposition between multiple constraints and a variety of requirements, and produces a large amount of knowledge of the creative process. These knowledge can be organized and applied expediently through the creative process knowledge acquisition system, thereby reducing the repetition and iteration of the design process, and improving the efficiency of the product development. Agile industrial design supported by creative process knowledge acquisition system mainly includes information input, knowledge matching and knowledge searching these three sections.

(1)Information input

As is shown in Figure 4, this part is mainly used for information input of the wireless mobile Internet terminals, including basic information ,text input and material upload these three modules. The basic information module is used for importing project name, starting and ending time, responsible person, project members and other information; the text input is used for the choice of creative process stage, keywords and text; the material upload module is used for uploading all information to the knowledge base, and encoding in terms of CKIC;

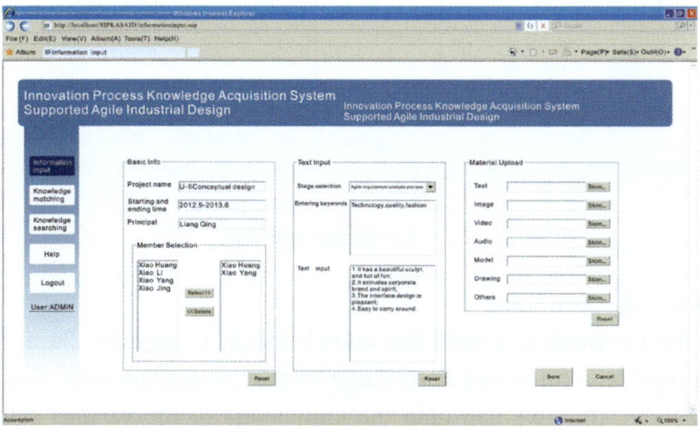

Fig. 4. Information input.

(2) Knowledge matching

As is shown in Figure 5, matching knowledge obtained from "information input" with knowledge in a knowledge base by setting up matching threshold and matching type. The matching results include matching coding, matching rate and matching introduction, at the same time, a detailed comparison is done between input information and matching information according to input project, now arranging the matching degree so as to examine the detailed matching results. If the matching degree is greater than the preset threshold, then there will have a high similarity between input information and information in knowledge base, so the relevant knowledge can be invoked directly to assist creative process; if the matching degree is lesser than the preset threshold, which accounts for input information has more new knowledge ,and need to take iteration into account. Like importing "U-fi conceptual design", keywords and design objects in agile requirement analysis stage, the matching results display the matching degree is 20.4%,which exceeds the threshold of 60%,that is to say having more prior knowledge is enough to support knowledge this agile requirement analysis needed, so there is no need to re-do requirement analysis, so the original demand analysis results can be invoked directly, thus avoiding useless iteration;

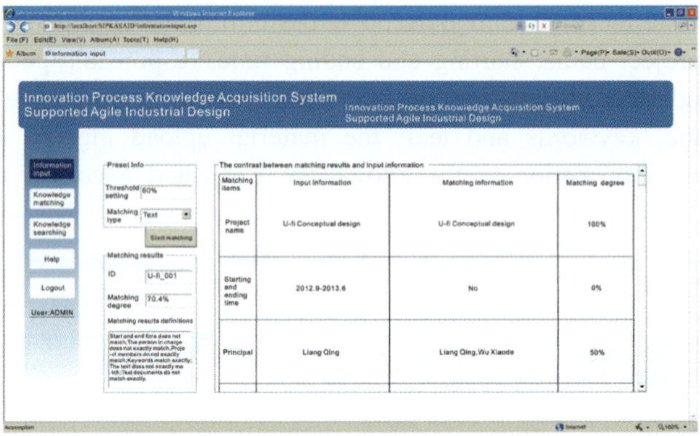

Fig. 5. Knowledge matching.

(3) Knowledge searching

Type the keyword to search the knowledge in knowledge base, the search results are used as a reference knowledge of the creative process. For e.g.: type "U-fi" to search, then we can gain the knowledge of the various stages of the creative process, such as product development trends, product lineup, style positioning, design language, color blending, materials, human factors ergonomics and other related knowledge of agile requirement analysis phase. The results are shown in Figure 6.

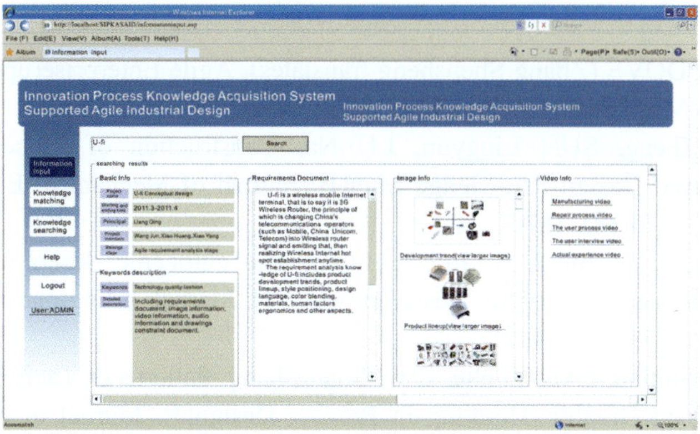

Fig. 6. Knowledge matching.

7. Conclusion

Aiming at product agile development requirements, this paper proposes the creative knowledge acquisition method of product agile industrial design process. By analyzing the characteristics and the process of agile industrial design, and creative knowledge information cell is used to represent creative knowledge, the knowledge matching algorithm is being designed to control the number of iterations of the creative process, and prototype system is being built to access and reuse the creative process knowledge by taking the wireless mobile Internet terminals design as an example. The conclusions are as follows:

(1) Agile industrial design process is dynamic and repeated iteration, hence control the number of iterations reasonably is the key to agile development. The knowledge modeling and matching of creative process contributes to manage the dynamic knowledge of iteration, improve the flexibility and viability of design process;

(2) The creative process knowledge acquisition is being realized through the constant correction and update of information input and knowledge matching, from this making design results approaching user expectation constantly, which is a gradual optimization process based on knowledge acquisition and reuse.

Acknowledgment

This work is supported by County Science and Technology Key Projects Fund.

References

1. RICK DOVE (Zhang Shensheng translates). Agile Enterprise(up, down)[J]. China Mechanical Engineering, 1996 (3,4):22-26;21-27.
2. LIU Zheng, SUN Lingyun, LU Na. Construction of Process-oriented Industrial Design Knowledge Map[J]. Journal of Mechanical Engineering, 2010, 46(8):181-187C.
3. Vajna, S. Clement, S. Jordan, et al. The autogenic design theory: An evolutionary view of the design process [J]. Journal of Engineering Design, 2005, 16(4): 423-440.
4. LUO Shijian, PAN Yunhe, ZHU Shangshang, Patterns of Tacit Knowledge Based on Graphic Thinking in Product Design[J]. Journal of Mechanical Engineering, 2007, 43(6):93-98.
5. A Gunasekaran. Agile manufacturing: a framework for research and development [J]. International Journal of Production Economics, 1999, 62.1/2: 87-105.
6. Yang Gangjun, Chen Guoding, Yu Suihuai, Chu Jianjie. Agile industrial design management based on scrum [C], CAID&CD'2010, Yiwu (China), 2010.
7. Huang Chun-Che, Kuo Chia-Ming. The transformation and search of semi-structured knowledge in organizations [J]. Journal of Knowledge Management, 2003, 17(4):106-123.
8. Ting-Peng Liang, Yung-Fang Yang, Deng-Neng Chen and Yi-Cheng Ku. A semantic-expansion approach to personalized knowledge recommendation [J]. Decision Support Systems, 2008, 45(3):401-402.

Experimental study of optimum parameters on dry ice snow cleaning insulator

Wei Dong

Kunming Power Supply Company of China Southern Power Grid Company, Kunming, China

Chao Gao

School of Electrical Engineering, Wuhan University, Wuhan, China
E-mail: 1515021282@qq.com

Rui-Feng Wei

Kunming Power Supply Company of China Southern Power Grid Company, Kunming, China

Wen-Jun Zhou

School of Electrical Engineering, Wuhan University, Wuhan, China

Yuan Lu

Kunming Power Supply Company of China Southern Power Grid Company, Kunming, China

Yu Xun

Kunming Power Supply Company of China Southern Power Grid Company, Kunming, China

The use of dry ice snow cleaning technique is a new charged cleaning technology. Comparing with normal technologies, it has higher efficiency with no residue and secondary pollution. Analyze the principle of the cleaning technology, determine the evaluation index of the cleaning effect, and then establish a simulation platform. Eventually get the influence of each parameter and the optimal parameters. When the compressed air pressure is 0.6 to 0.7MPa, dry ice snow flow per unit time is 18kg/h, cleaning distance is 15cm to 18cm, cleaning angle is 20° to 25°, cleaning time is 24s to 36s, the experiments show that cleaning effect is good for selected insulator. Due to the different structure and design of different cleaning objects, the corresponding optimal parameters is different. It should be obtained through theoretical analysis and experiments.

Keywords: Dry ice snow cleaning; insulator contamination; optimal parameters; cleaning effect

1. Introduction

The surface of outdoor working device of electric power system will form a certain degree of contamination accumulation for being long exposure to natural environment ,and the pollution flashover accidents occur easily in humid climate conditions [1, 2]. With the transmission voltage and power capacity continuing to improve, the flashover accident of substation electrical equipment external insulation increase seriously [3]. The surface cleaning of electrical device is the main measures taken. The widely used methods to prevent pollution flashover include regular manual cleaning ,reinforcing insulation configuration, using auxiliary umbrella skirt, adjusting the creep age distance of primary equipment, brushing organic coating on surface of insulator, etc [4]. The removal of insulator surface contamination is the main prevention measure, which effectively prevents insulator pollution flashover, guarantees the safe and stable operation of power system [5].

The cleaning of outdoor working device of electric power system is divided into charged sweeping and neutral sweeping. neutral sweeping mostly is Manual sweeping. charged sweeping mostly is charged water sweeping, charged chemicals sweeping, charged mechanical sweeping, charged Dry ice sweeping, etc [2].

Dry ice snow cleaning technique is the liquid carbon dioxide spraying to the surface to be cleaned with compressed air through specific injection device, using liquid carbon dioxide under the high pressure to form high speed flow containing small solid carbon dioxide particles [6], cleaning surface fouling ,oil, residual impurities, and it does not damage the stuff cleaned [7].

Dry ice snow cleaning technique is a new charged cleaning technology. Compared to other live cleaning technology, it has obvious advantages [8], and broad application prospects [9, 10].

2. Dry Ice Snow Cleaning Characteristics and Influence Factors

2.1. *Dry ice snow cleaning characteristics*

The mechanisation of dry ice snow cleaning include the following. Dry ice snow particles in the accelerating high-pressure air have greater kinetic energy, impact the surface and vaporise instantly. The temperature of dry ice snow particles is extremely low (-78°C).The thermal effect reduces the temperature of dirt, and increase brittleness, which reaches the purpose of decontamination. Dry ice snow particles expanse to 800 times while sublimation, resulting in micro explosion at the point of impact, stripping filth with the compressed air flow [11].

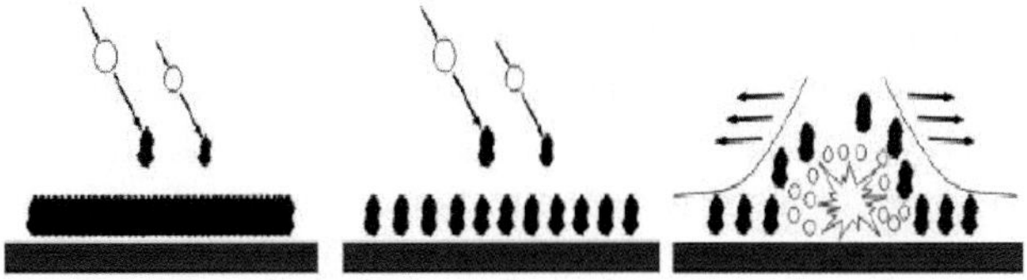

Fig. 1. Schematic diagram of cleaning the contaminant pellets using dry ice snow.
(a) when dry ice snow hitting the contamination layer;
(b) when contamination layer fractures;
(c) when contaminations removed by compressed air stream

The main features of the dry ice cleaning include the following. The cleaning force is strong. The dry ice snow particles have no damage on the surface cleaned. After the cleaning, dry ice snow particles quickly sublime, and has no pollution to the environment around the cleaning medium. When cleaning metal material, the material won't rust and corrode. The dry ice snow cleaning is suitable for a variety of filth.

2.2. *Dry ice snow cleaning influence factors*

The factors that affect dry ice snow cleaning insulator have:

(1) Cleaning distance: Cleaning distance directly affects the kinetic energy of dry ice snow particles and the scattering surface. The shorter is the cleaning distance, the less is the energy loss of dry ice snow particles. The strike force is stronger, but the scattering surface is smaller. The longer is cleaning distance, the more is the energy loss of dry ice snow particles. The strike force is weaker, but the scattering surface is larger.

(2) Cleaning angle: cleaning angle directly affects the vertical force component of the dry ice snow particles. The larger is the vertical force component, the stronger is strike force. The cleaning effect is better. The smaller is the vertical force component, the weaker is strike force. The cleaning effect is worse.

(3) The pressure of compressed air: the process of dry ice snow cleaning is to mix the compressed air and liquid carbon dioxide, then spray it to the surface to be cleaned. Therefore the compressed air pressure will directly determine the strength of the dry ice snow particles strike force, then determine the dry ice cleaning effect.

(4) Dry ice snow flow per unit time: the more is the dry ice snow particles sprayed per unit time, the stronger is the strike force to the surface. The cleaning

effect is better, but the consumption is more, which will lead to the increase of cleaning cost.

(5) Cleaning time: the longer is the cleaning time, the better is the cleaning effect. But the cleaning time is longer, the consumption of liquid carbon dioxide is also more [12].

The pressure of compressed air, liquid carbon dioxide flux, cleaning distance, cleaning angle and cleaning time jointly affect the dry ice snow cleaning. Take the orthogonal test method, decide the influence of each factor according to the variety of cleaning effect.

2.3. *Evaluation method of dry ice snow cleaning*

This experiment adopts the quantitative brushing method [13] to simulate natural contamination. According to the required salt density, dirt density and area of the cleaning surface to calculate the required quality of sodium chloride (purely chemical) and diatomite, and add the appropriate deionized water. In order to simulate natural pollution better, add appropriate dextrin to enhance the cohesive force of the artificial pollution [14]. Stirred the mixture, brush it to the test surface evenly, then dry it indoor.

In order to evaluate the cleaning effect accurately and objectively, the test use salt density and dirt density cleaning rate to comprehensively evaluate dry ice snow cleaning effect. Cleaning rate = (salt density cleaning effect + ash density cleaning effect)/2.

Salt density and dirt density cleaning rate is respectively calculated from the equivalent salt density change rate and the equivalent dirt density change rate before and after cleaning. They are respectively expressed as:

Salt density cleaning rate:

$$\eta_{salt} = \frac{\rho_{s1} - \rho_{s2}}{\rho_{s1}} \times 100\% \tag{1}$$

Dirt density cleaning rate:

$$\eta_{dirt} = \frac{\rho_{d1} - \rho_{d2}}{\rho_{d1}} \times 100\% \tag{2}$$

In the formula, ρ_{s1}, ρ_{s2} is equivalent salt density before and after cleaning. ρ_{d1}, ρ_{d2} is equivalent dirt density before and after cleaning.

3. Dry Ice Snow Cleaning Experiment based on Orthogonal Design

In dry ice snow cleaning experiment, the main factors which affect the cleaning effect are the characteristic of contamination and the cleaning parameters. The equivalent salt density and dirt density of natural formed insulator have large deviation, and it is hard to quantitative analyze. Artificial bedaub is used to simulate the natural contamination. By analyzing the cleaning effect under different cleaning parameters, determine the optimal cleaning parameters.

3.1. *Cleaning test platform*

Test platform of dry ice snow cleaning consists of air compression system, dry ice snow cleaning machine, liquid carbon dioxide cylinder, electric rotary device, test support device and contaminated insulator. Air compression system provide compressed dry air for dry ice snow cleaning machine; Liquid carbon dioxide cylinder provide high pressure liquid carbon dioxide for dry ice snow cleaning machine; Dry ice snow cleaning machine mix the compressed air and liquid carbon dioxide to produce dry ice in a special way, then spray it to the surface of insulator to be cleaned; Electric rotary device is used to place the insulator to be cleaned, and the speed is adjustable; Test brace is used to fix nozzle system, and adjust the distance and height of nozzle system. In the test, the cleaning distance and angle can be changed by adjusting test support device.

3.2. *Test content*

According to the GB/T 26218.1-2010, the pollution is very heavy when salt density is greater than 0.2mg/cm2 and dirt density is greater than 2mg/cm2 [15]. To simulate very heavily polluted areas, equivalent salt density is 0.4mg/cm2 and dust density is 4mg/cm2。 In order to characterize the cohesive force of natural contamination better, 2.3mg/cm2 diatomite and 1.7mg/cm2 dextrin are added. Measure and calculate the surface area of insulator, weigh the corresponding quality of sodium chloride, diatomite and dextrin as required, and brush it to the surface of insulator evenly.

In dry ice snow cleaning test, compressed air pressure (A) is selected 0.3MPa (A1), 0.4MPa (A2), 0.5MPa (A3), 0.6MPa (A4); Dry ice snow flow (B) per unit time is selected 7.7kg/h (B1), 10.3kg/h (B2), 12.9kg/h (B3), 15.4kg/h(B4); Cleaning distance (C) is selected 5cm (C1), 10cm (C2), 15cm (C3), 20cm (C4); Cleaning angle is selected 10° (D1), 15° (D2), 20° (D3), 25° (D4); Cleaning time is selected 6s (E1),12s (E2),18s (E3),24s (E4). The device rotating speed is fixed at 10r/min. The orthogonal test was used to test the influencing factors of cleaning effect.

The orthogonal test is the method taken to replace comprehensive test by part. Select the representative experiments, analyze the result of representative tests, understand the comprehensive test, and optimize the process. It is a scientific method to deal with multi-factor problem effectively [16].

The orthogonal test table as shown in the following.

Table 1. Five factors and four levels orthogonal test table.

	pressure (MPa)	flow (kg/h)	Distance (cm)	angle (°)	time (s)
1	A1(0.3)	B1(7.7)	C1(5)	D1(10)	E1(6)
2	A1(0.3)	B2(10.3)	C2(10)	D2(15)	E2(12)
3	A1(0.3)	B3(12.9)	C3(15)	D3(20)	E3(18)
4	A1(0.3)	B4(15.4)	C4(20)	D4(25)	E4(24)
5	A2(0.4)	B1(7.7)	C2(10)	D3(20)	E4(24)
6	A2(0.4)	B2(10.3)	C1(5)	D4(25)	E3(18)
7	A2(0.4)	B3(12.9)	C4(20)	D1(10)	E2(12)
8	A2(0.4)	B4(15.4)	C3(15)	D2(15)	E1(6)
9	A3(0.5)	B1(7.7)	C3(15)	D4(25)	E2(12)
10	A3(0.5)	B2(10.3)	C4(20)	D3(20)	E1(6)
11	A3(0.5)	B3(12.9)	C1(5)	D2(15)	E4(24)
12	A3(0.5)	B4(15.4)	C2(10)	D1(10)	E3(18)
13	A4(0.6)	B1(7.7)	C4(20)	D2(15)	E3(18)
14	A4(0.6)	B2(10.3)	C3(15)	D1(10)	E4(24)
15	A4(0.6)	B3(12.9)	C2(10)	D4(25)	E1(6)
16	A4(0.6)	B4(15.4)	C1(5)	D3(20)	E2(12)

3.3. *Analysis of test results*

The test results are as follows.

Table 2. The results of orthogonal test.

	Pressure (MPa)	flow (kg/h)	distanc e (cm)	angle (°)	time (s)	Cleaning rate (%)
1	0.3	7.7	5	10	6	9.5
2	0.3	10.3	10	15	12	10.8
3	0.3	12.9	15	20	18	12.6
4	0.3	15.4	20	25	24	13.2
5	0.4	7.7	10	20	24	46.3
6	0.4	10.3	5	25	18	55.7
7	0.4	12.9	20	10	12	26.5
8	0.4	15.4	15	15	6	30.5
9	0.5	7.7	15	25	12	48.6
10	0.5	10.3	20	20	6	25.2
11	0.5	12.9	5	15	24	75.9
12	0.5	15.4	10	10	18	60.4
13	0.6	7.7	20	15	18	38.6
14	0.6	10.3	15	10	24	53.8
15	0.6	12.9	10	25	6	58.4
16	0.6	15.4	5	20	12	81.3

Analyze test results by orthogonal method, and analysis results are as follows.

Table 3. Orthogonal analysis of test results.

	Pressure (A)	Flow (B)	Distance (C)	Angle (D)	Time (E)
k_{i1}/%	11.53	35.75	55.6	37.55	30.90
k_{i2}/%	41.00	36.38	43.98	38.95	41.80
k_{i3}/%	52.53	43.38	36.38	41.35	41.83
k_{i4}/%	58.03	46.35	25.88	43.98	47.30
range	46.5	10.6	29.72	6.43	16.4

In Table 3, kij (i=A,B,C,D,E; j=1,2,3,4) represent the comprehensive average value of the corresponding factors respectively, which reveal the quality in different levels. For instance, kA1 represents the comprehensive value of first level of factor A is 11.53, kA1=(9.5+10.8+12.6+13.2)/4. Range is the difference of the maximum and the minimum, which reveals the influencing degree on cleaning effect of each factor.

Map out the relationship between factors and cleaning rate. Selecting the optimal parameters, and do the validation test.

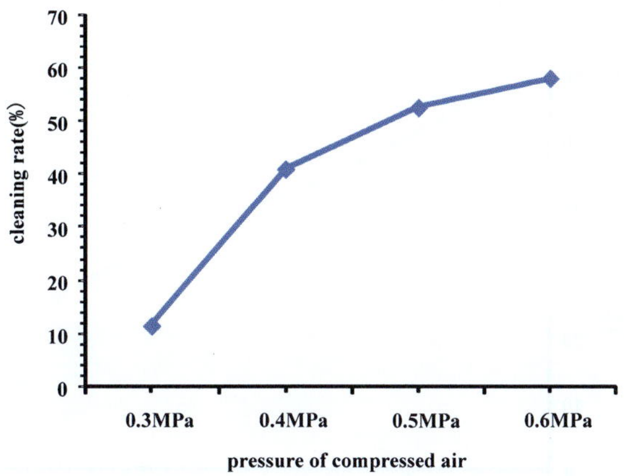

Fig. 2. Cleaning rate curve at different compressed air pressure.

As can be seen from Figure 2, in the test range, the pressure of compressed air and the cleaning rate are positive correlation. With the increase of the pressure of compressed air, the cleaning effect is much better. When the compressed air pressure reached 0.6Mpa, the increasing effect of cleaning is not obvious as the increase of the pressure.

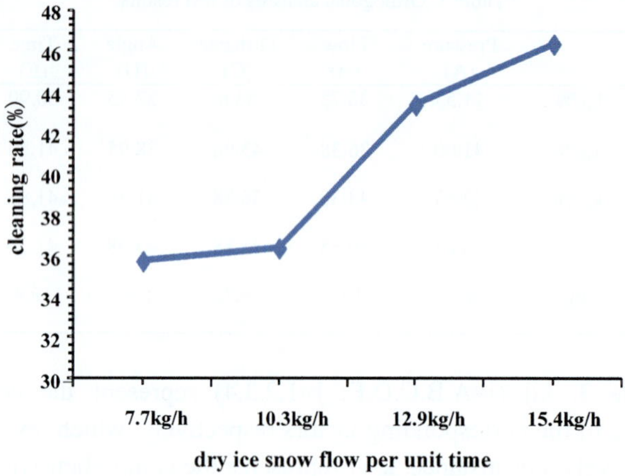

Fig. 3. Cleaning rate curve at different dry ice snow flow per unit time.

As can be seen from Figure 3, in the test range, dry ice snow flow per unit time and cleaning rate are positive correlation. With the increase of dry ice flow per unit time, the cleaning effect is much better. But 7.7kg/h and 10.3kg/h have little difference of cleaning effect.

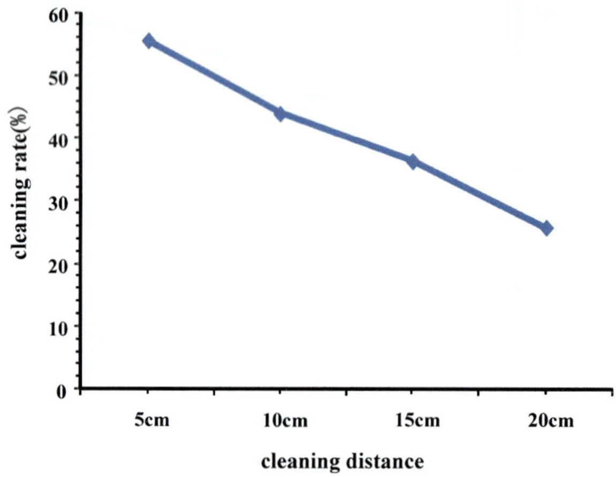

Fig. 4. Cleaning rate curve at different cleaning distance.

As can be seen from Figure 4, in the test range, cleaning distance and cleaning rate was negative correlation. When the cleaning distance becomes larger, the cleaning effect becomes worse.

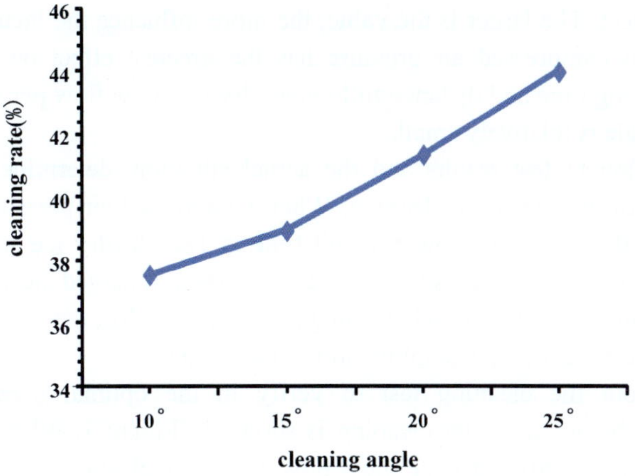

Fig. 5. Cleaning rate curve at different cleaning angle.

As can be seen from Figure 5, in the test range, cleaning angle and cleaning rate are positive correlation. With the increase of cleaning angle, the cleaning effect is better. When cleaning angle is larger, cleaning rate decline as the mutual occlusion between the umbrella skirt does not appear, because the insulator distance is long enough.

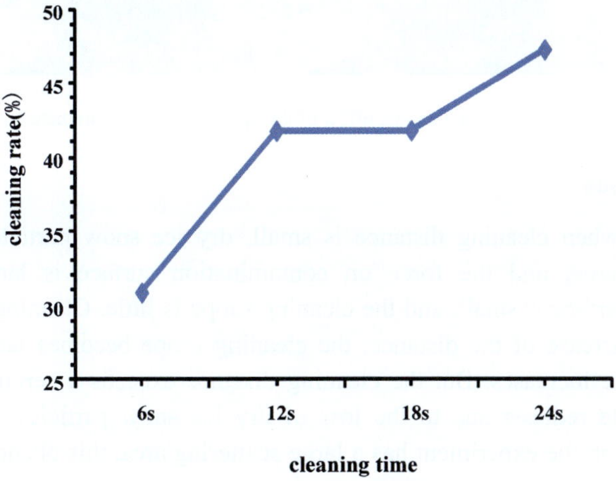

Fig. 6. Cleaning rate curve at different cleaning time.

As can be seen from Figure 6, in the test range, cleaning time and cleaning rate are positive correlation. With the increase of cleaning time, cleaning effect is better. But 12s and 18s have little difference of cleaning effect.

In table 3, the range represents the influence degree of different factors on cleaning effect. The larger is the value, the more influence the factor has. It can be seen that compressed air pressure has the greatest effect on the cleaning effect, cleaning time and distance following, dry ice snow flow per unit time and cleaning angle is relatively small.

According to test results and the actual situation, determine the optimal parameters of dry ice snow cleaning. When compressed air pressure is 0.6MPa to 0.7MPa, dry ice snow flow per unit time is 18kg/h (dry ice snow cleaning machine limited), cleaning distance is 15cm to 18cm (charged cleaning limited), cleaning angle is 20° to 25°and cleaning time is 24s to 36s, the cleaning rate can reach up to 80% under this combination of parameters.

Carry out the cleaning test to verify in the optimal parameters. The comparison before and after cleaning is shown in Figure 7. When compressed air pressure is 0.65MPa, dry ice snow flow is 18kg/h, cleaning distance is 17cm, cleaning angle is 20°and cleaning time is 30s, the cleaning rate can reach up to 85%.

Fig. 7. Insulator cleaning effect of the optimal cleaning parameters.

4. Discussion

In theory, when cleaning distance is small, dry ice snow particles energy is relatively large, and the force on contamination surface is large. But the scattering surface is small, and the cleaning scope is little. Cleaning rate is low. With the increase of the distance, the cleaning scope becomes larger, and the cleaning rate increases. But the cleaning distance exceeds a certain range, the cleaning rate reduces due to the loss of dry ice snow particles. Because the nozzle used in the experiment has a large scattering area, this phenomenon does not appear.

In theory, when the cleaning angle is small, the vertical force of dry ice snow particles on the polluted surface is small, and the cleaning rate is low. As the angle increases, the vertical force gradually increases, and the cleaning rate increases gradually. But the cleaning angle exceeds a certain range, the shielding effect of umbrella skirt leads to the decline of cleaning rate. Because the distance of insulator umbrella skirt is long enough in the experiment, this phenomenon does not appear.

For different types of insulators, the umbrella skirt has different diameters and different adjacent distance, the optimal parameters are different. For different types of cleaning objects, the optimal cleaning parameters are also different due to the different structure and size.

5. Conclusion

Dry ice snow cleaning technique applying in the live cleaning of electrical equipments' external insulation is a new living cleaning technology. The dry ice snow cleaning technique has higher efficiency, and no secondary pollution to the environment. It can ensure the safe operation of the power system by preventing pollution flashover.

The pressure of compressed air, dry ice snow flow per unit time, cleaning distance, cleaning angle and cleaning time jointly influence the dry ice snow cleaning. For the insulator used in this experiment, the optimal parameters is as follows: compressed air pressure is 0.6MPa to 0.7MPa, dry ice snow flow per unit time is 18kg/h (dry ice snow cleaning machine limited), cleaning distance is 15cm to 18cm (charged cleaning limited), cleaning angle is 20° to 25°and cleaning time is 24s to 36s. Then dry ice snow cleaning can achieve good cleaning effect.

Due to the different structure and design of different cleaning objects, the corresponding optimal parameters of dry ice snow cleaning is different. The application should be through the theoretical analysis and the experiment to obtain the corresponding optimal cleaning parameters.

References

1. YAN Zhang, ZHU Deheng. *High voltage and insulation technology*. Beijing, China: China Electric Power Press, 2002:130.
2. YU Huayu, XU Wencheng. *Analysis of filthy flashover technology and prevention*. Electric Safety Technology, 2005, 7(3):28-30.
3. LU Tiecheng. *Hyper-voltage of power systems*.Beijing, China: China Water Power Press, 2009.

4. *Technical Manual on Prevention of filthy flashover*. Nanjing, China: Jiangsu Science and Technology Press, 1993.
5. ZHANG Jin, JI Shengchang, CAO Tao. *Present and Prospect of Power Equipment's Electriferous Cleaning Technique*. Insulating Materials, 2009, 42(2):63-66.
6. Xiong Chucai. *The carbon dioxide snow cleaning technology*. cleaning technology, 2004, 10:64-70.
7. Spur G, Uhlmann E. *Dry-ice blasting for cleaning: process, optimization and application*. International Conference on Erosive and Abrasive Wear (ICEAW). Atlanta, USA: ICEAW, 1999: 402-411.
8. Nakashima M, Yakabe H, Maruyama Y. *Application of semi-automatic robot technology on hot-line maintenance work*. Proceedings of IEEE International Conference on Robotics and Automation. Perth, Australia: IEEE, 1995:843-850.
9. Yano K, Maruyama Y, Morita K. *Development of the semi-automatic hot-line work robot system 'phase-II'*. ESMO-95Proceedings of IEEE Seventh International Conference on Transmission and Distribution Construction, Operation and Live-line Maintenance. Omaha, USA: IEEE, 1995: 212-218.
10. GUAN Genzhi. *High voltage engineering foundation*. Beijing, China: China Electric Power Press, 2005.
11. Liu Ming, Wang Jiali, Ma Xinliang. *Cleaning Experiment Of Substation Insulators by Dry I*ce. High Voltage Technology, 2011(07):1649-1655.
12. Sheng-Chung Yang, Yu-Cheng Lin, Keng-Shiang Huang. Optimization of a pulsed carbon dioxide snoe jet for cleaning CMOS image sensors by using the Taguchi method. Sensors and Actuators, 2007, 139(1):265-271.
13. GB/T 4585—2004 High voltage insulator of AC system artificial contamination test. Beijing, China: China Standards Press, 2004.
14. Salam M A, Ahmad H, Derivation of creepage distance in terms of ESDD and diameter of the contaminated insulator. IEEE Power Engineering Review, 2000, 20(9):44-45.
15. GB/T 26218.1-2010, Selection and dimensioning of high-voltage insulators intended for use in polluted conditions. Part 1: Definitions, information and general principles.
16. Liu Ruijiang, Zhang Yewang, Wen Chongwei, Tang Jian. *Study on design and analysis method of orthogonal test.* experimental technology and management, 2010, 09:52-55.

Partial discharge testing of isolator for high magnetic field facility

Ke Wang*, Nannan Hu*, Hongming Ma*, Cheng Wu†,§,
Wanjiang Pan†, Yi Cao† and Yin-Feng Zhu‡

**Power Research Institute of Yunnan Power Grid Co., Ltd., Kunming 650217, China*

†Institute of Plasma Physics, Chinese Academy of Sciences, Hefei 230031, China

‡School of Mechanical and Electrical Engineering, Anhui Jianzhu University, Hefei 230601, China

§wucheng@ipp.ac.cn

The isolators for high magnetic field facility insulate the helium outlets of the High Temperature Superconducting current leads, which can be at up to 25kV. To evaluate the electrical performance, partial discharge (PD) test was performed in air and helium pressure environment under 16kV alternate current (AC) voltage. To reduce the influence of background noise on PD testing results, narrowband noise was removed by using properly tuned notch filters. The testing results indicate the PD valve of the isolator for high magnetic field facility is lower than allowable valve.

Keywords: Isolator; AC high voltage; partial discharge; background noise.

1. Introduction

The high magnetic field facility is a big scientific device in China, which will be used to test the performance of Tokamak superconducting materials [1], such as critical current test for high-temperature superconducting current leads [2] and critical magnetic field test of low-temperature superconducting material [3]. The goal of high magnetic field facility is to establish a 40T level steady state mixed magnet test device and superconducting magnet test device. The isolators are key component for high magnetic field facility, which play an important role in insulating cooling pipes. The isolator consists of two 316L stainless steel pipes and an insulation pipe [4]. Fig.1 shows the cross section of the isolator, two end

§Corresponding author.

metal pipes act as the high voltage electrode and grounding electrode respectively [5]. The insulation pipe is made from glass fiber reinforced epoxy resin composite material [6]. The 316L stainless steel and glass fiber reinforced epoxy resin composite material have a low permeability, so the materials are chose for the isolator in high magnet field environment due to its magnetism requirement.

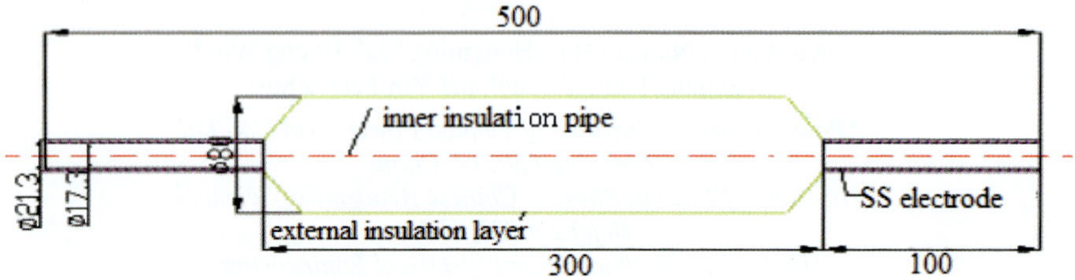

Fig. 1. Cross section of the isolator.

The common assessment methods of insulation performance for the isolators are leakage current, insulation resistance and partial discharge method [7]. Due to void or other defects in glass fiber reinforced epoxy resin composite material, partial discharge will be occurred on insulation pipe, which corresponds to breakdown and extinguish phenomenon under strong electric field intensity. Therefore, partial discharge test has proven to be effective in predictive diagnostics of the insulation on high voltage electrical sample [8-11].

2. Partial Discharge Test for the Isolator

2.1. *Test set-up of partial discharge test in air*

Fig.2 shows the set-up schematic drawing for partial discharge test, transformer supplies the required AC high voltage to the test circuit. The test system is connected to PC (personal computer) by optical cable, and data acquisition unit (MPD 600) is connected to controlling unit by optical cable too. Measuring impedance unit (CPL 542) is jointed with MPD 600 through BNC electric cable. Calibration unit (CAL 542) is jointed with tested object (isolator), which is only used during the standardizing process.

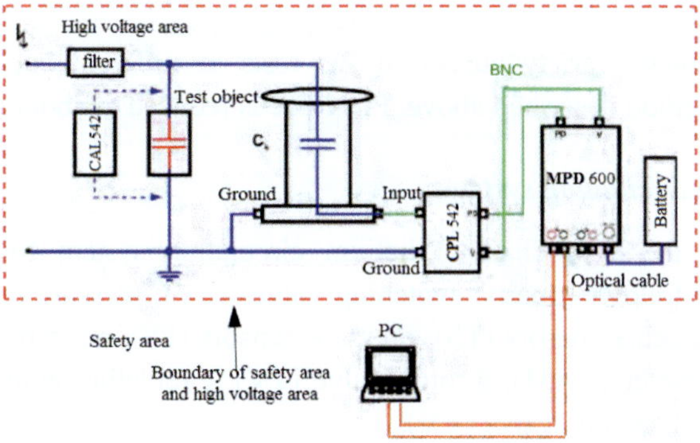

Fig. 2. Set-up schematic drawing for partial discharge test.

The result of partial discharge test has very strict requirement, which is extremely affected and hampered by measurement noise [12]. The background PD value comes from environmental noise. The noise in PD measurements can be classified into wideband noise, narrowband interferences and pulse shaped interferences [13]. In the previous test, the background PD value is high due to some background noise. To obtain satisfying testing results, further research on how to reduce the background noise need to be performed. In the performed PD tests, the narrowband noise is removed by properly tuned notch filters shown in Fig.2. Wideband noise is the measurement background noise, so wideband noise is removed by keeping environment quiet.

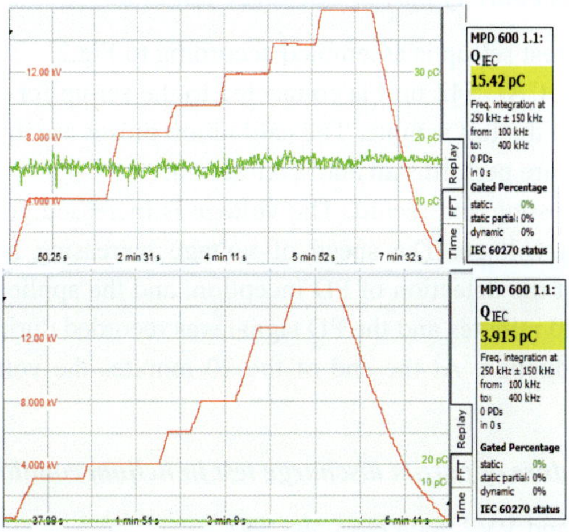

(a) PD curve in former condition (b) PD curve after noise processing

Fig. 3. Comparatively of PD value under different background noise.

Fig. 3 shows variety curves of PD tests in different background, after processing method described above, PD value is reduced to about 4pC.

2.2. *Helium circulation system*

Fig.4 shows the flow chart of helium circulation system for the isolator, including the following parts: switching valves (SV), pressure reducing valves (PRV), temperature sensors (TS), pressure sensors (PS), mass flowmeter (MF), gas analysis meter (GAM), throttle valve (TV), circulating pump (CP), safety valve (SV) and vacuum valve (VV).

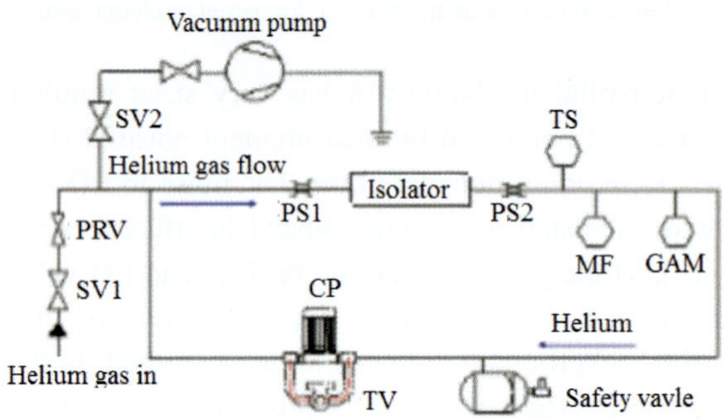

Fig. 4. Flow chart of helium circulation system.

2.3. *Testing procedure of partial discharge test in air*

-Assembly. The test set-up is assembled according to Fig.2.

-Calibration. CAL 542 unit is connected to the set-up for calibration before test, in parallel to the test sample. The calibration pulses are injected and the PD acquisition software performs an auto-calibration.

-Voltage boosting and record. The voltage is increased gradually from 0 to 16 kV AC high voltage. The speed of voltage increasing rate shall be slow enough to permit the detection of PD inception, and the applied voltage shall be maintained for 20 minutes and the PD signal was recorded during this time.

-Voltage reduction. At the end of the 20 minutes the voltage is decreased gradually.

2.4. *Test procedure of partial discharge test in helium condition*

-Open the VV1 and VV2, then close the SV;

-Open the vacuum pump until the pressure inside the helium gas circulating pipes is at 10-2 Pa (vacuum)， then close vacuum pump and VV1;

-Open the SV and PRV until the pressure inside the helium gas circulating pipes is about 1 bar;

-Open the circulating pump, adjust the helium gas flow;

-Remove the plugs of pressure sensors, temperature sensor and vacuum gauge;

-Repeat the process in part 2.3.

3. Partial Discharge Test in Different Condition

Partial discharge tests were done both in air and in helium condition, the isolator was tested in static helium and different helium gas flow. Fig.5 shows partial discharge test for the isolator in helium condition. Fig.6 shows variety curve of partial discharge test in helium condition, in which the red line represents the AC high voltage and the green line represents the value of partial discharge. Table 1 shows the results of partial discharge test in different condition.

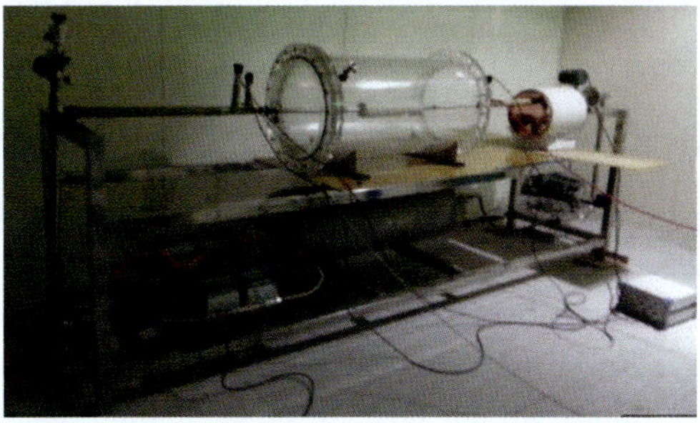

Fig. 5. Partial discharge test in helium condition.

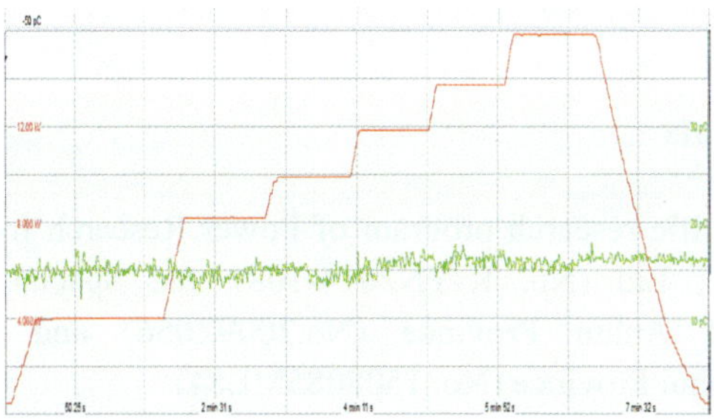

Fig. 6. Variety curve of partial discharge test in helium condition.

Table 1. Result of partial discharge test.

1	AC HV	Static helium	Temperature	Humidity	Background	PD result
IB	16kV	1bar	16.1℃	16%	14.21pc	15.28pc
2	AC HV	Helium flow	Temperature	Humidity	Background	Result
IB	16kV	1bar, 0.7g/s	16.3℃	16%	14.21pc	15.42pc
3	AC HV	Helium flow	Temperature	Humidity	Background	Result
IB	16kV	1bar, 4.5g/s	16.2℃	16%	14.21pc	15.47pc
4	AC HV	air	Temperature	Humidity	Background	Result
IB	16kV	1atm	16.2℃	16%	14.21pc	15.16pc

According to the testing results, the background PD value is about 14pc, the displayed PD value is about 15pc under 16kV AC high voltage and 1bar helium pressure. The difference of two PD values is about 1pc, which is lower than allowable value (5pc), consequently, the electrical performance of the isolator for high magnetic field facility under AC high voltage can satisfy the technical requirements. In addition, the PD values in air and in other helium condition have litter difference, it is proved that the condition of gas has no impact in PD result.

4. Conclusions

In this paper, partial discharge tests were performed to evaluate electric performance of the isolator for high magnetic field facility. The testing results show that the difference between background PD value and testing PD value is about 1pc, which is lower than 5pc and can meet the technical requirement. In addition, the PD values in air and in other helium condition have litter difference, it is proved that the condition of gas has no impact in PD result. Some methods, such as tuned notch filters and keeping quiet, were used to reduce the background noise, after that PD value has a obvious reduction for the isolator.

Acknowledgments

Support by scientific research program of Power Research Institute of Yunnan Power Grid Co., Ltd (No. K-YN2014-108), the Special Fund of Talent Development of Anhui Province (No.2009Z056) and Natural Science Foundation of Anhui Province (No. 1508085ME84).

References

1. Tingzhi Zhou, Kun Lu, Qingxiang Ran, et al. Design and Technology Preparation for the ITER HTSCL [J]. IEEE TRANSACTIONS ON APPLIED SUPERCONDUCTIVITY, 2014, 24(3).
2. Y Bi, et al. R&D towards HTS current leads for ITER. IEEE Tans. Appl. Supercond, 2009, 19(3):1500-1503.
3. Y Song, et al. Status of design and R&D for ITER feeder system procurement in China. presented for MT22 conference 2011. IEEE Trans. Appl. Supercond. ; submitted for publication.
4. Wu Cheng, PAN Wanjiang. Research and Analysis on Tensile and Compressive Fatigue Performance of Cryogenic Axial Insulation breaks. Plasma Science and Technology, 15 (7) (2013), pp.717-720.
5. C Wu, W Pan, Y Zhu, L Fang, Y Cao. High Temperature Baking Test and Study on Axial Insulation Breaks. Plasma Science and Technology, 29 (8) (2015), pp.1-6.
6. Z. X. Wu, H. Zhang, H .H. Yang, X. X. Chu, L. F. Li. Novel radiation-resistant glass fiber/epoxy composite for cryogenic insulation system. Journal of Nuclear Materials, (1-3) (2010), pp.117-120.
7. Foucher de Brandois C, Bauer P, Laurenti A. Room Temperature Insulating breaks (RTIB) technical specifications. 2013, 11/2.0.
8. L. Satish, B. Nazneen. Wavelet-based denoising of partial discharge signals buried in excessive noise and interference. IEEE Trans. Dielectr. Electr. Insul. 10 (4) (2003), pp. 354-367.
9. M. Hasmi, M. Lehtonen, M. Nordman, R. Jabbar, S. Qureshi. Wavelet-based de-noising of on-line PD signals captured by Pearson coil in covered-conductor overhead distribution networks. Electr. Power Energy Syst. 43 (2012), pp. 1185-1192.
10. M. Nafar, T. Nikman, A. Gheisari. Using correlation coefficients for locating partial discharge in power transformer. Electro. Power Energy Syst. 33 (2011), pp. 493-499.
11. R. Sarathi, R. Umamaheswari. Understanding the partial dischargeactivity generated due to particle movement in a composite insulation under AC voltages. Electr. Power Energy Syst. 48 (2013), pp. 1-9.
12. Andre T. Carvalho, Antonio C.S. Lima, Caio F.F.C. Cunha, Mariane Petraglia. Identification of partial discharges immersed in noise in large hydro-generators based on improved wavelet selection methods. Measurement. 75 (2015), pp. 122-133.

13. D. Evagorou, A. Kyprianou, P. Lewin, A. Stavrou, V. Efthymiou, A.Metaxas, G. Georghiou. Feature extraction of partial discharge signals using the wavelet packet transform and classification with a probabilistic neural network, IET Sci. Meas. Technol. 4 (2010), pp. 177-192.

The optimization of thrust rod based on particle swarm optimization

Guo-Sheng Feng*, Peng-Fei Li and Su-Mei Jia
School of Mechanical Engineering,
Shijiazhuang Tiedao University,
Shijiazhuang, China
**E-mail:fgs2005@126.com*

Established the parametric model of a assemble thrust rod using a macro command in CATIA, analysis its strength in the Abaqus, considered the assemble relationship between the various component, used the Isight integrated CATIA and Abaqus based on the particle swarm algorithm to optimized the assemble thrust rod. The optimized result showed that,the thrust rod weight reduced 1.3%, the maximum stress reduced 23.2%.Compared with the traditional optimization, this method uses intelligent algorithms to automatically find the best design, improved product quality and reliability.

Keywords: Thrust rod; particle swarm algorithm; optimized design.

Thrust rod is widely used in multiaxle vehicle balance suspension, for poor working conditions of mine dump truck, thrust rod is damaged easily. So it is very important to carry out the strength check and structural optimization of the thrust rod. At present, the research on the thrust rod is less, and most of the research on the thrust rod is limited to the strength analysis, and the structural optimization of artificial experience. Literature [1]-[2] uses finite element technology to find reasons of a triaxial heavy mining dump truck's thrust rod premature fracture, and puts forward some improving suggestions. In Literature [3], the strength of the type -V thrust rod of a heavy commercial vehicle is checked, and the structure is improved by the experience.

In a western open pit coal mine, the head of a triaxial heavy mining dump truck's thrust rod frequently produce crack or even break, seriously impact on production and operation, this paper uses particle swarm intelligent algorithm to make thrust rod assembly for optimization of the whole structure by using iSIGHT software CATIA and ABAQUS. Compared with the traditional manual optimization method, it uses Intelligent algorithm to find the best scheme, and the simulation process is automatic, the iteration is continue, the data also cover a wide range of design space.

1. Frame Fault Diagnosis

The triaxial heavy mining dump truck's parameters is such as table 1. The frame model is built in CATIA, the 3d model showed in Figure 1. Extracted the middle face of the frame in Hypermesh. The frame modeled by shell element, while the leaf spring and the thrust rod use solid element modeling in this paper. The riveted unit of frame's cross beam and the longitudinal beam and the link plate simulated by cweld unit. The main framework and auxiliary framework are connected with area connectors. Established the beam element along the axis of the rolling ears of leaf spring and the hanging ear, then used the reb3 element to connect the beam with the hanging ear and leaf spring's rolling ears, releasing the axial rotational degrees of freedom. This can achieve the mutual rotation between the hanging ear and the leaf spring's rolling ears, by this to simulate the pin connection of the leaf spring's rolling ears and a hanging ears, the connection of the thrust rod and frame is similar. The mass of the goods, the engine and the cab are simulated by the mass element in the corresponding position, then used the the rbe2 element to connect the frame and the mass element to realize the uniform distribution of the load on the frame. The finite element model of the frame is shown in Figure 2.

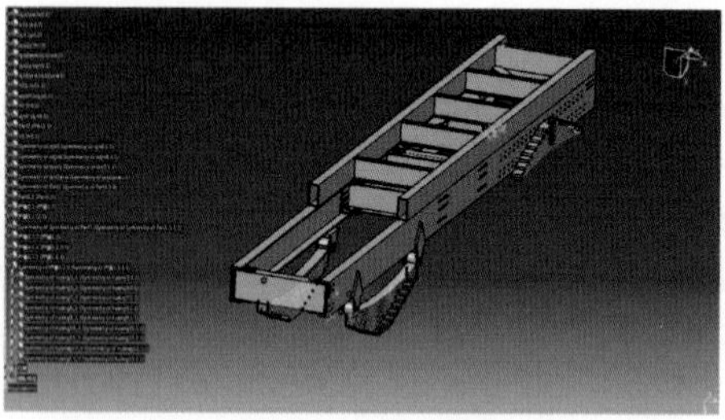

Fig. 1. Frame three-dimensional model.

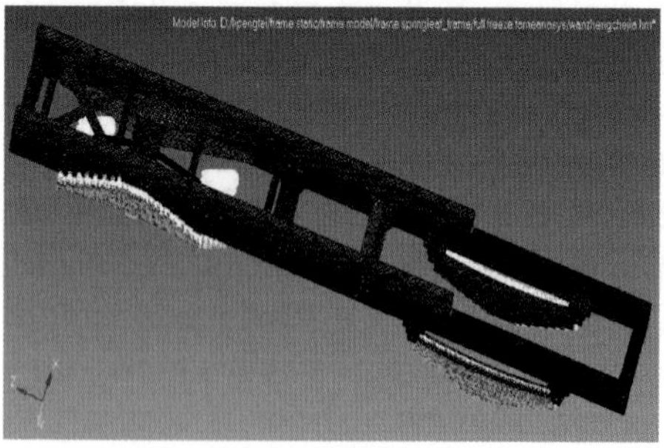

Fig. 2. Frame finite element model.

Table 1. Table of vehicle parameters.

name	*value*
Wheelbase/mm	3500+1400
Overloaded common speed/Km/h	10
No load common speed/Km/h	30
Maximum gradeability	30%
Vehicle equipment mass/t	21
Maximum laden mass/t	55
Front leaf spring's static stiffness/N/cm	6400
Rear leaf spring's static stiffness/N/cm	36159

Working conditions of the frame are mainly such as: bending, torsional, lifting, downhill, brake condition etc. Bending condition is the most common condition, and the mix of bending and torsion conditions is the worst conditions. Therefore, this paper focuses on the analysis of the dynamic strength of the frame under the two conditions. The frame material is made of 500 steel and its yield limit is 500MPa. In order to simulate the triaxial heavy mining dump truck's instantaneous force during working, in bending condition the vertical load should be multiplied by 2.5 times the dynamic load coefficient by experiment and calculation .All degrees of freedom of leaf spring bottom node are constrained. To prevent the leaf spring skidding, constrained the lateral degrees of freedom. With a wheel suspended to simulate the torsional condition, delete the left spring and the constraint remain unchanged. The existing research shows that load coefficient should not exceed 2.5 in the mix of bending and torsional condition of the frame. Under bending condition frame stress cloud picture is shown as in Figure 3. The maximum stress is 360MPa, which is mainly concentrated in the thrust rod, the fourth beam ,the connection place front spring frame and the lifting lug. Under conditions of mix of bending and torsion the frame stress cloud picture is as shown in Figure 4. The maximum stress is 381Mpa, which is mainly concentrated in the connection of the front plate spring frame and the lifting ear. As seen in the Fig. 4, the thrust rod exist a high stress, which is prone to fracture risk.

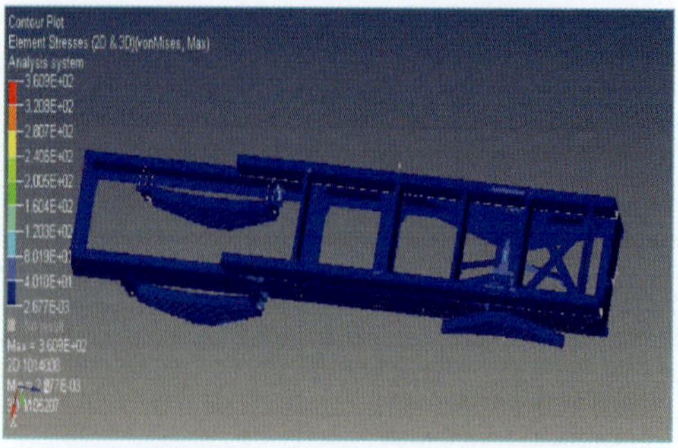

Fig. 3. Bending condition stress cloud.

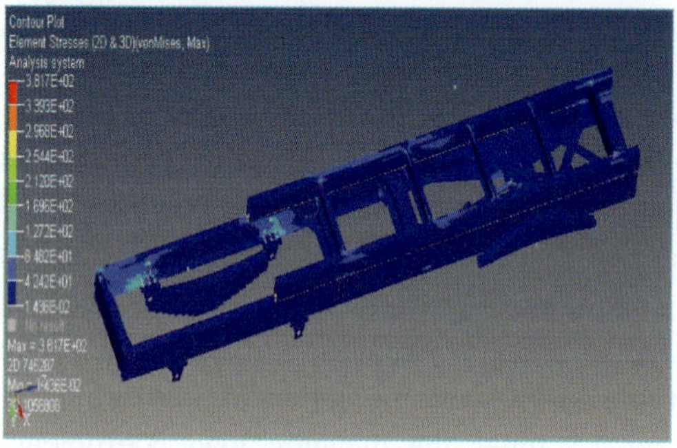

Fig. 4. Torsional condition stress cloud.

2. Particle Swarm Optimization Algorithm

Particle swarm optimization algorithm is a kind of artificial intelligence algorithm which simulates the foraging of birds. It has high efficiency, robustness and generality. It is not sensitive to the problem, and does not require the objective function and the constraint condition. Therefore, it is suitable for the optimization of such complex engineering problems. With the assumption that there are n particles in the particle swarm, in n-dimensional space to search for the optimal solution and the spatial position of the i-the particle indicated by $x_i = (x_{i1}, x_{i2}, ..., x_{in})$, i=1,2,……,N, speed indicated by $v_i = (v_{i1}, v_{i2}, ..., v_{in})$, i=1,2,……,N. The optimal position of the i particle is represented by P_{best}, denoted by $P_i = (P_{i1}, P_{i2}, ..., P_{in})$, i=1,2,……,N. The best individual history in the population is known as the optimal location of the population history, which is denoted by $P_g = (P_{g1}, P_{g2}, ..., P_{gn})$, denoted as g_{best}. Each particle in the search for the best position in the process not only to consider their own history of the optimal position also take into account the optimal position of the group, so the speed and position of the particle at each dimensional (d=1,2, n) is as follows: [4]-[7]:

$$v_{id}(t+1) = w \bullet v_{id}(t) + c_1 \bullet r_1 \bullet (P_{id}(t) - x_{id}(t)) + c_2 \bullet r_2 \bullet (P_{gd}(t) - x_{id}(t)) \quad (1)$$

$$x_{id}(t+1) = x_{id}(t) + v_{id}(t+1) \quad (2)$$

In the equation:

W: the inertia weight, and describes the influence of the last generation speed on the current speed.

C1: a cognitive factor, which indicates that the particle's memory of the best position in the history of its own.

C2: the social factor, which means that the memory of the optimal position of the group is the best.

R1, R2: two independent random number between the [0, 1]

The basic optimization process is shown in Figure 5.

3. The Optimization Model of Thrust Rod

3.1. *The finite element model and stress analysis of thrust rod*

Using the traditional mechanics analysis method for the stress analysis of the thrust rod, firstly calculate the force effect on the driving axle shell, then consider

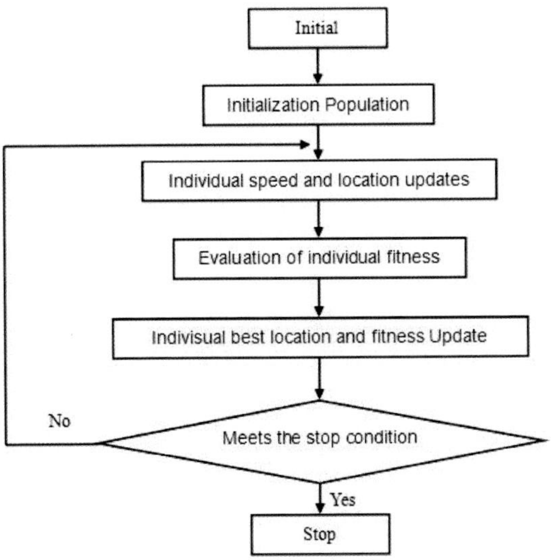

Fig. 5. Particle Swarm algorithm flowchart.

the axle shell and the thrust rods as a system for force analysis. The maximum traction supplied by engine is shown in (3):

$$F_t = \frac{T_t}{r} = \frac{T_e \bullet i_g \bullet i_0 \bullet \eta_T}{r} \tag{3}$$

In the equation:

T_t :Torque on the driving wheel, $N \bullet m$

T_e :The engine torque, $N \bullet m$

i_g :The gear ratio of transmission

i_0 :The gear ratio of main reducer

η_T :The mechanical efficiency of drive train

r :The radius of driving wheel,m

Considering the axle shell and thrust rod as a system for stress analysis, stress analysis diagram is shown in Figure 6,the angle between thrust rods and horizontal plane α,β are shown by drawing, α = 14 °, β = 11 °. According to the torque balance, it will come to [8]-[9]:

$$\sum MA = 0 \tag{4}$$

$$F_2 \times h_1 \times \cos\beta = F_t \times (h_1 + h_2) \tag{5}$$

$$\sum MB = 0 \tag{6}$$

$$F_1 \times \cos\alpha \times h_1 = F_t \times h_2 \tag{7}$$

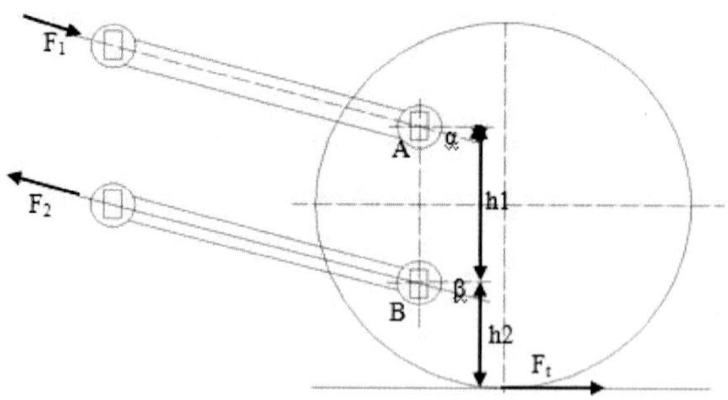

Fig. 6. Thrust rod force analysis diagram.

According to the above formulas and parameters provided by enterprises, it finally comes to: F1=102KN, F2=243KN.

Structure of the thrust rod mainly includes rubber ring, pin shaft, connecting rod, ending cover and other parts. According to the drawings provided by the manufacturer, using macro command in CATIA to establish the model diagram of thrust rod , importing the thrust rod 3d model into Abaqus and using the Python language to generate the finite element model. As is shown in Figure 7. According to the actual working situation, making a fixed constraint on pin.

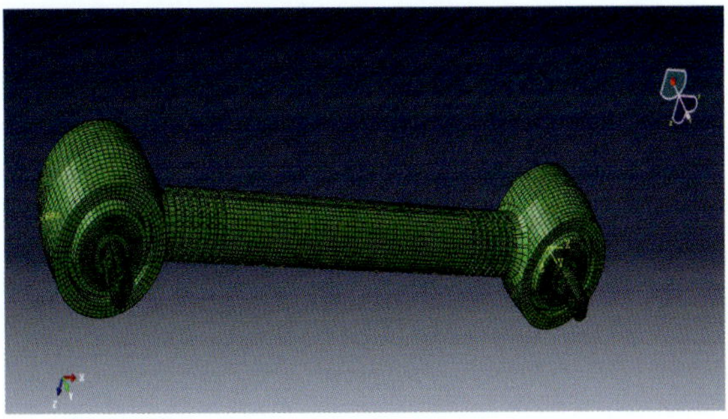

Fig. 7. Thrust rod finite element model.

Shaft of the thrust rod, and applying force to the ending cover of the pin shaft to simulate the pre stressed installation of the thrust rod. Then apply the limit load of work load at the other end of the core-shaft. By the analysis of the results shown in Figure 8, the thrust rod connecting rod head as well as rod tube have larger stress concentration, and the maximum stress value is 369.8 MPa. Currently, the material of the thrust rod is 45steel, and the yield strength is 355Mpa. By the analysis on the maximum stress exceeding the yield strength, the joint of pole head and pole tube is prone to damage which is consistent with actual situation.

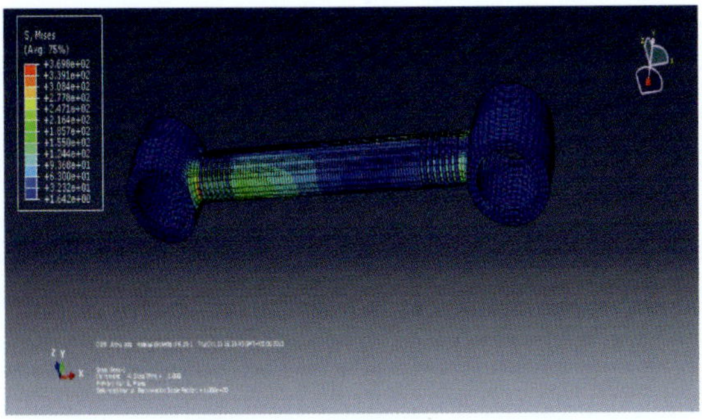

Fig. 8. Thrust rod stress cloud.

3.2. *The thrust rod optimized model*

Due to the crack of thrust rod head in practical engineering, the optimization goal is: the maximum stress and minimum displacement. By literature [9], the larger influence on the thrust rod stress is the size match between pin shaft, rubber ring and rod head thickness. The design variables are pin shaft radius d1, rubber ring inner diameter d2 and rubber ring outer diameter d3, the wall of the head rod radius d4. Thrust rod structure diagram is shown in Figure 9. Each size constraints in the traditional design of parts is within 42≤d1≤50,42≤d2≤50, 50≤d3≤60, 50≤d4≤60.

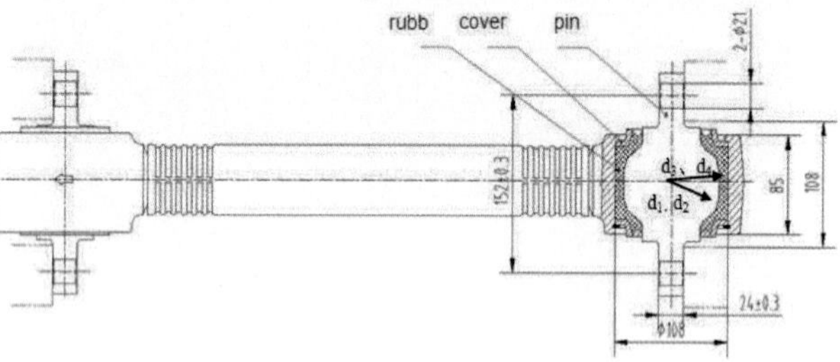

Fig. 9. The structure diagram of thrust rod.

4. Optimization Calculation

In this paper, iSIGHT is used as an integrated platform of CATIA and Abaqus, and the model makes up of seven parts. Use the CATIA's macro command to generate the 3D model of the thrust rod. In order to ensure that the size of the constraint relationship, it need a calculator model to make the variable d3=d4, d1=d2.The 3D model generated by CATIA imported into Abaqus ,and the thrust rod assembly is generated to the finite element model by the Python language for Strength analysis, then generated the .rpt data. The Isight extracted the maximum stress value and the maximum displacement value as the optimization objective. For most optimization problems, The number of particles is enough to take 10.In this paper, the number of particles is 10, the maximum iteration number is 50 times[10].According to this calculation, each particles iteration 50 times, then we need to calculate 500 times. The flow chart of optimize shown in Figure 10, the size before optimization and after optimization are shown in Table 2.

Table 2. The size before and after optimization.

variable	the size before optimization(mm)	the size after optimization(mm)
d1	46	44.932
d2	46	44.932
d3	54	57.461
d4	54	57.461

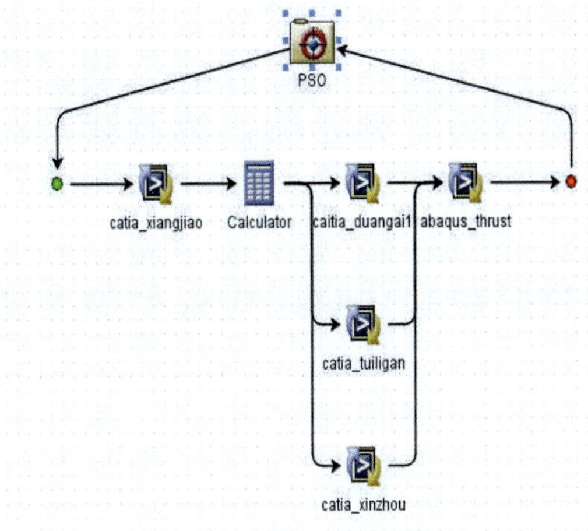

Fig. 10. The flow chart of optimize.

After the optimization, the radius of the pin shaft, and the inner diameter of the rubber ring were reduced by 6.6%, the outer diameter of the rubber ring, and the inner diameter of the thrust rod was increased by 10%.So the thickness of the rubber ring was increased by 4.529mm.Maximum stress from the original 369.8MPa to 284MPa, the maximum stress value decreased by 23.2%.

5. Conclusion

Using particle swarm optimization algorithm to optimize the thrust rod assemble and reduced the mass of the thrust rod as well as the maximum stress of the head of the thrust rod, successfully solving the problem of the head crack of thrust rod due to excessive stress. in this paper, Isight integrated CATIA, Abaqus, calculator realized the multi-component assembly optimization, and it can improve the design efficiency and quality as well as reliability of the product.

Acknowledgment

This project was supported by the National Nature Science Fund of China (Grant No. 11272220).

Reference

1. Feng Guosheng, Jia Sumei, Lin Hao. Crack Analysis of Thrust Bar for 55t Heavy Dump Truck Equalizing Suspension [J]. Journal of Shijiazhuang Tiedao University, 2013, 24 (6).43-46.
2. Feng Guosheng, Jia Sumei, Lin Hao. Analysis and Diagnosis of thrustrod Crack for Mining Heavy Dump Truck [J]. Journal of Vibration, Measurement & Diagnosis, 2014, 34 (3), 553-559.
3. Shi Wenku, Ke Jun, Wang Qian. Limit Load Analysis and Spherical Hinge Structure Optimization for V-Type Thrust Rod[J]. Journal of Xi'an Jiaotong University, 2013, 47(10), 132-136.
4. Pan Feng, Li Weixing, Gao Qi. Particle Swarm Optimizer and Multi-Object Optimization[M]. Beijing: Beijing Institute of Technology Press, 2013.
5. Sun Chaoli. Particle Swarm Optimizer for Optimization Design of Mechanical System[M]. Beijing: China Machine Press, 2012.
6. Bin Yang, Kai-Vwe Bletzinger, Qilin Zhang, et al. Frame Structural Sizing and Topological Optimization via a Parallel Implementation of a Modified Particle Swarm Algorithm[J]. KSCE Journal of Civil Engineering, 2013, 17(6), 1359-1370.
7. S. Kitayama, M. Arakawa, K. Yamazaki, Penalty Function Approach for the Mixed Discrete Nonlinear Problems by Particle Swarm Optimization[J]. Struct Multidise Optim, 2006, 32, 191-202.
8. Zhang Changjiang, Tang Zhigang. Analysis on Optimization Design of Commercial Vehicle Thrust Rod[M]//SAE-China Congress Seclected Papers for 2010[C]. Beijing, 2010,17-23.
9. Wang qian. Structure Optimization and Durability Research of Heavy Commercial Vehicle Thrust Rod [D]. Jilin: Jilin University, 2013.
10. Lai Yuyang. Optimization Theory and Detailed Examples of Isight Parameters[M]. Beijing: Beihang University Press, 2012.

Sparse phase retrieval recovery via iterative hard threshold

Li-Juan Zhu* and Jian-Jun Wang†
School of Mathematics and Statistics,
Southwest University,
Chongqing, China
**zlj1991sdtz@163.com*
†*wjj@swu.edu.cn*

The aim of this paper is to study the recovery performance of the ℓ0 minimization for the sparse problem in the real phase retrieval setting. Though this problem is closely related to the compressive sensing problem, lacking of phase information makes this harder. As cost functions are often at the heart of the problem, we here first study the iterative hard algorithms that are minimizing the ℓ0 penalised cost functions. Herein we propose a new IHT algorithm for the problem of one dimensional (1D) phase retrieval, namely, recovery of 1D sparse signal from the magnitude of its measurement. And we give the theoretical results of error bound and convergence of the algorithm. Simulation results indicate that the proposed algorithm is accurate and more faster than existing techniques.

Keywords: Sparsity; phase retrieval; compressed sensing; iterative thresholding; ℓ0 regularisation.

1. Introduction

The theory of compressed sensing has generated enormous interest in recent years. The goal of compressed sensing is to recover a sparse signal from its linear measurements, where the number of measurements is much smaller than the dimension of the signal. In this paper we consider the phase retrieval [1], [18] for sparse signals, which has important application in X-ray crystallography [7], [8], electron microscopy [10], [11], optics [6], [9], coherence theory and other applications [12]–[14]. In many applications the signals to be reconstructed are sparse.

Herein we consider the phase retrieval problem of reconstructing a vector $x_0 \in {}^n$ from noisy magnitude measurements of the form

$$b_j = \left|\left\langle a_j, x_0 \right\rangle\right| + e_j, \quad j = 1, 2, \ldots, m$$

where $a_j \in \mathbb{R}^n$ is a measurement vector, and $e_j \in \mathbb{R}$ is a noise, phase retrieval consists of recovering x_0 up to a global sign, which has attracted much attention in recent years [2]–[5]. We consider the ℓ0 minimization for the phase retrieval problem where the signal x_0 is approximately k-sparse, which is the ℓ0 minimization problem defined as follows:

$$\min \|x\|_0 \quad st \ \left\| |Ax| - |Ax_0| \right\|_2 \le \varepsilon,$$

where

$$|Ax_0| = \left[\left|\left\langle a_1, x_0 \right\rangle\right|, \left|\left\langle a_2, x_0 \right\rangle\right|, \ldots, \left|\left\langle a_m, x_0 \right\rangle\right|\right]^T$$

and $A = \left[a_1, a_2, \ldots, a_m\right]^T$. This can be written as a standard problem in optimization with a linear penalty and a quadratic and linear inequality constraints. For an appropriate multiplier lambda, the solution to (1) is precisely the solution to the unconstrained optimization problem:

$$\min_x \left\| |Ax| - |Ax_0| \right\|_2^2 + \lambda \|x\|_0,$$

where the Lagrange multiplier λ is a function of A ,b and ε. Then we look at the incarnation of the sparse approximation problem [19]. The ℓ0 regularised optimization problem is defined as follows: for given $|Ax_0|$ and A, find coefficients x minimising the cost function

$$C_{l_0}(x) = \left\| |Ax| - |Ax_0| \right\|_2^2 + \lambda \|x\|_0 \qquad (1)$$

where $\|x\|_0$ is defined as $|\Gamma_1(x)|$ and throughout this paper we use $\Gamma_1(x) = \{x_i : x_i \ne 0\}$ as the set of non-zero coefficients. $|\Gamma_1(x)|$ is the size of this set so that $\|x\|_0$ counts the number of non-zero coefficients. In a previous paper, iterative hard thresholding algorithms [16], [17] were studied. In particular, their convergence to fixed points of ℓ0 regularised (or constrained) cost functions could be proved [15]. In this paper, it is shown similar performance of phase retrieval.

The rest of this paper is organized as follows. In section 2, we proposed an IHT algorithm to solve the ℓ0 regularised optimization problem in the phase retrieval setting and present our main theoretical results, theoretical proofs are given in the rest of section 2. In section 3, we conduct some numerical experiments, which demonstrate that the proposed approach is also efficient and robust in practice. Finally, the conclusion is addressed in section 4.

2. An terative Hard-Thresholding Algorithm for the ℓ0 Regularised Problem

To our knowledge, the first phase retrieval algorithm was proposed by Romberg, Moravec, and Baraniuk in [20] and worked by incorporating an ℓ1 norm constraint into a traditional Fienup-style [21] iterative algorithm. Here we propose a new threshold algorithm in an ℓ0 norm constraint. In this part, we use optimisation transfer by surrogate objective function.

2.1. *Iterative hard thresholding for sparse phase retrieval*

To solve the optimisation problem for the compressive phase retrieval problem

$$\min_x \left\| |Ax| - |Ax_0| \right\|_2^2 + \lambda \|x\|_0 ,$$

we derive the following iterative algorithm

$$x^{[n+1]} = H_{\lambda^{0.5}}(x^{[n]} + (A_I^T A_I - A_{I^C}^T A_{I^C})x_0 - A^T A x^{[n]}),$$

here $A_I = \left[a_j : j \in I \right]$ denotes the sub-matrix of A where only rows with indices in I are kept, for any $x^{[n]}$, x_0 we divide $\{1, \ldots, m\}$ into two groups:

$$I = \left\{ j : sign\left(\langle a_j, x^{[n]} \rangle\right) = sign\left(\langle a_j, x_0 \rangle\right) \right\},$$

$$I^C = \left\{ j : sign\left(\langle a_j, x^{[n]} \rangle\right) = -sign\left(\langle a_j, x_0 \rangle\right) \right\} .$$

$H_{\lambda^{0.5}}$ is the element wise hard thresholding operator

$$H_{\lambda^{0.5}}(x_i)=\begin{cases}0 & if\left|x_i\right|\le\lambda^{0.5},\\ x_i & if\left|x_i\right|>\lambda^{0.5}.\end{cases} \quad (2)$$

We show that the algorithm is guaranteed not to increase (1) and in fact converges to a local minimum of (1) under the condition that $\|A\|_2<1$.

2.2. *Optimisation transfer*

Instead of optimising (1), we introduce a surrogate objective function similar to that in [15].

$$\begin{aligned}C_{l_0}^s(x,y)=&\left\|\left|Ax\right|-\left|Ax_0\right|\right\|_2^2+\lambda\|x\|_0\\&-\|Ax-Ay\|_2^2+\|x-y\|_2^2\end{aligned} \quad (3)$$

It is obvious that $C_{l_0}^s(x,x)=C_{l_0}(x)$, in order for the surrogate cost to majorise the cost function we require that we decrease the original cost function (1) can by optimising (3) with respect to y. The surrogate objective function (3) can be rewritten as

$$\begin{aligned}C_{l_0}^s=&\sum_i[x_i^2-2x_i\left(y_i+\left(A_{I_i}^TA_{I_i}-A_{I_i^c}^TA_{I_i^c}\right)x_0-A^TAy\right)]\\&+\sum_i[\lambda\left|x_i\right|^0]+\left[\|Ax_0\|_2^2+\|y\|_2^2-\|Ay\|_2^2\right]\end{aligned} \quad (4)$$

where $\left|x_i\right|^0$ is one if $x_i\neq 0$ and zero otherwise. The advantage of minimizing (4) in place of (1) is that x_i are decoupled, the minimum of (3) can be calculated by minimising with respect to each x_i individually. Now we ignore the constant terms, if $y_i=0$, the element-wise cost is 0, if $y_i\neq 0$, the cost is $x_i^2-2x_i\left(y_i+\left(A_{I_i}^TA_{I_i}-A_{I_i^c}^TA_{I_i^c}\right)x_0-A^TAy\right)+\lambda$, the minimum of which is thus achieved at $x_i^* = y_i + \left(A_{I_i}^T A_{I_i} - A_{I_i^c}^T A_{I_i^c}\right)x_0 - A^T Ay$. We see that the minimum of (3) is attained at

$$x = H_{\lambda^{0.5}}(y + (A_I^T A_I - A_{I^C}^T A_{I^C})x_0 - A^T Ay),$$

where the hard thresholding operator $H_{\lambda^{0.5}}$ given in (2). Although the minimum need not be unique whenever $y_i + \left(A_{I_i}^T A_{I_i} - A_{I_i^c}^T A_{I_i^c}\right)x_0 - A^T Ay = \lambda^{0.5}$, in the definition of the threshold operator here, a strict inequality is used to ensure a unique update. The iterative hard thresholding algorithm is therefore

$$x^{[n+1]} = H_{\lambda^{0.5}}\left(x^{[n]} + \mu\left((A_I^T A_I - A_{I^C}^T A_{I^C})x_0 - A^T Ax^{[n]}\right)\right).$$

In the next subsection of this paper, we will give the covergence of this algorithm under the condition of $\|A\|_2 < 1$. In this case, the above algorithm converges to a local minimum of the optimisation problem.

We first show that the algorithm reduces the cost function $C_{l_0}(x^n)$ from iteration to iteration. In this subsection an important lemma as follow:

Lemma 2.1: Assume $\|A\|_2 < 1$ and let $x^{[n+1]} = H_\theta(x^{[n]} + (A_I^T A_I$ - $A_{I^C}^T A_{I^C})x_0$ - $A^T Ax^{[n]})$, then the sequences are non-increasing.

Under the condition of $\|A\|_2 < 1$, the first and the third inequality follow, the second inequality is due the fact that $x^{[n+1]}$ is the unique minimiser of $C_{l_o}^s(x, x^{[n]})$. The equations follow because of the definition of surrogate cost function or cost function.

We can see from the lemma, using the algorithm cannot lead to worse results than not using the algorithm, with cost functions chosen appropriately. As the algorithm has multiple fixed points, it is important to analyse these in more detail. Some important lemmas as follow

Lemma 2.2: Define the sets $\mathbf{\Gamma}_0(x) = \{i : x_i^* = 0\}$ and $\mathbf{\Gamma}_1 = \{i : |x_i^*| > \lambda\}$. A necessary and sufficient condition for a point x^* to be a fixed point of algorithm (2) is that for each A_{I_i}, $A_{I_i^c}$,

$$\left|\left(A_{I_i}^T A_{I_i} - A_{I_i^c}^T A_{I_i^c}\right)x_0 - A_i^T A_i x^*\right| \begin{cases} = 0 & if\ i \in \Gamma_1, \\ \leq \lambda^{0.5} & if\ i \in \Gamma_0. \end{cases}$$

Proof: A fixed point is any x^* such that $x^* = T(x^*)$, we have $x_i^* = H_{\lambda^{0.5}}\left(x_i^* + \left(A_{I_i}^T A_{I_i} - A_{I_i^c}^T A_{I_i^c}\right)x_0 - A^T A x^*\right)$. If $i \in \mathbf{\Gamma}_0$, $x_i^* = 0$, this holds if and only if $\left|\left(A_{I_i}^T A_{I_i} - A_{I_i^c}^T A_{I_i^c}\right)x_0 - A^T A x^*\right| \leq \lambda^{0.5}$. If $i \in \Gamma_1$, $x_i^* \neq 0$, we have

$$x_i^* = x_i^* + \left(A_{I_i}^T A_{I_i} - A_{I_i^c}^T A_{I_i^c}\right)x_0 - A^T A x^*.$$

Similarly, this holds if and only if $\left(A_{I_i}^T A_{I_i} - A_{I_i^c}^T A_{I_i^c}\right)x_0 - A^T A x^* = 0$.

Lemma 2.3: Assume $\|A\|_2 < 1$, then a fixed point $x^* = T(x^*)$ is a local minimum of (1).

For an input signal x_0 and a threshold $\lambda^{0.5}$, denote by x^{opt} the global minimum of the optimisation problem (1). Define $\Gamma_0 = \{i : x_i^{opt} = 0\}$ and $\mathbf{\Gamma}_1 = \{i : x_i^{opt} \neq 0\}$. For $\forall i \in \mathbf{\Gamma}_0$, $\left(A_{I_i}^T A_{I_i} - A_{I_i^c}^T A_{I_i^c}\right)x_0 - A^T A x^{opt} \leq \lambda^{0.5}$. For $\forall i \in \mathbf{\Gamma}_1$, $\left|x^{opt}\right| \geq \lambda^{0.5}$ and $\left(A_{I_i}^T A_{I_i} - A_{I_i^c}^T A_{I_i^c}\right)x_0 - A^T A x^{opt} = 0$. Here the threshold $\lambda^{0.5}$ is similar to the threshold in [15].Comparing the conditions to the fixed point conditions of the algorithm in (2.2) we have

Theorem 2.4: Assume $\|A\|_2 < 1$ then the optimal solution to the optimisation problem (1) belongs to the fixed points of the iterative algorithm defined by (2).

2.3. *Convergence of the iterative algorithm*

In this subsection we discuss the convergence property of the sequence $\left(x^{[n]}\right)_n$ defined by (2). The main section is the following theorem:

Theorem 2.5: If $\|A\|_2 < 1$ and if $C_{l_0}^s\left(x^{[0]}\right) < \infty$, then the sequence $\left(x^{[n]}\right)_n$ defined by the iterative procedure in (2) converges to a local minimum of (1).

Convergence is an even more important property of the algorithm. Here under the condition of $C_{l_0}^{s}\left(x^{[0]}\right)<\infty$, it implies that only a finite number of $x_i^{[0]}$ are non-zero at any one time. The proof to this theorem is similar to that of the theorem3 is given in Appendix D of [15].

If the algorithm does not converge to an exact signal representation, then after a finite number of iterations, the algorithm will not change the subset, so from $x^{[n+1]}=x^{[n]}+\left(A_I^T A_I-A_{I^c}^T A_{I^c}\right)x_0-A^T Ax^{[n]}$, we have

$$\begin{aligned} x^{[n+1]}-x &= x^{[n]}+\left(A_I^T A_I-A_{I^c}^T A_{I^c}\right)x_0-A^T Ax^{[n]}-x \\ &=\left(I-A^T A\right)x^{[n]}+A_I^T A_I x+A_{I^c}^T A_{I^c} x-x \\ &=\left(I-A^T A\right)^2\left(x^{[n]}-x\right) \end{aligned}$$

So convergence speed is the linear convergence

$$\left\|x^{[n]}-x^*\right\|_2\le\left\|I-A^T A\right\|_2^{n-m}\left\|x^m-x^*\right\|_2.$$

At the end of this section, a bound on the approximation error is stated in the following lemma.

Lemma 2.6: Assume that $\|A\|_2<1$, that $\{A_i\}$ contains a basis for the signal space, then a tight bound for the approximation error $\left\||Ax_0|-|Ax^*|\right\|_2$ achieved at a fixed point x^* is

$$\left\||Ax_0|-|Ax^*|\right\|_2\le\frac{\sqrt{2}\lambda^{0.5}}{\sqrt{\beta(A)}},$$

Where $(A)>0$, $\beta(A)=\min\left\{\beta^2(A_I),\beta^2\left(A_{I^c}\right)\right\}$, is $\beta(A_I)>0$ such that $\sup_i\left|A_{I_i}^T x\right|\ge\beta(A_I)\|x\|_2$ holds for all x, $\beta\left(A_{I^c}\right)>0$ is such that $\sup_i\left|A_{I_i^c}^T x\right|\ge$ $\beta\left(A_{I^c}\right)\|x\|_2$ holds for all x.

Proof: From (2.2), we have

$$\left|\left(A_{I_i}^T A_{I_i} - A_{I_i^c}^T A_{I_i^c}\right)x_0 - A_i^T A_i x^*\right| \begin{cases} = 0 & if\ i \in \Gamma_1, \\ \le \lambda^{0.5} & if\ i \in \Gamma_0. \end{cases},$$

which means that it suffices to show that $\forall i$,

$$\sup_i \left|\left(A_{I_i}^T A_{I_i} - A_{I_i^c}^T A_{I_i^c}\right)x_0 - \left(A_{I_i}^T A_{I_i} + A_{I_i^c}^T A_{I_i^c}\right)x^*\right| \le \lambda^{0.5},$$

$$\sup_i \left|A_{I_i}^T A_{I_i}\left(x_0 - x^*\right) - A_{I_i^c}^T A_{I_i^c}\left(x_0 + x^*\right)\right| \le \lambda^{0.5},$$

which means that $\exists\, i$, such that

$$\left|A_{I_i}^T A_{I_i}\left(x_0 - x^*\right)\right| + \left|A_{I_i^c}^T A_{I_i^c}\left(x_0 + x^*\right)\right| \le \lambda^{0.5},$$

there is no doubt that the following are ture

$$\left|A_{I_i}^T A_{I_i}\left(x_0 - x^*\right)\right| \le \lambda^{0.5}, \left|A_{I_i^c}^T A_{I_i^c}\left(x_0 + x^*\right)\right| \le \lambda^{0.5}$$

We then have

$$\begin{aligned} & \left\| |Ax_0| - |Ax^*| \right\|_2^2 \\ & = \left\| A_I x_0 - A_I x^* \right\|_2^2 + \left\| A_{I^c} x_0 + A_{I^c} x^* \right\|_2^2 \\ & \le \frac{2\left(\lambda^{0.5}\right)^2}{\beta(A)} \end{aligned}$$

from which the lemma follows.

3. Numerical Simulation

The proposed IHT algorithm for phase retrieval can be seen a natural extension of traditional IHT algorithm for compressive sensing. To evaluate the performance of the IHT algorithm, we provide numerical simulations in this section. This problem is ill-posed since phase information is lost. Therefore, prior information on the signal is needed in order to recover it. In this work we consider the case in which the prior information on the signal is that part of the symbol of Ax is known. In the noiseless setting, a trial is said to succeed if the output x^* satisfies $\|x^* - x_0\|_2/\|x_0\|_2 < 10^{-2}$

3.1. *Experimental settings*

The IHT algorithm which is proposed to solve the induced phase retrieval problem, that includes variants allowing con-stant or variable linesearch choices is summarized in Algorith-m1 as follows:

Algorithm 1 IHT algorithm for Phase retrieval

1: Initialization: Choose initial guess $\mathbf{x}^{[0]}$

2: Input λ, $n = 0$.

3: Repeat

4: General step:

$$x^{[n+1]} = x^{[n]} + A^T diag(sign(A^T x^{[n]}))b - A^T Ax^{[n]}.$$

For every $i = 1, 2, ..., n$, if $\left|x_i^{[n]}\right| \le \sqrt{\lambda}$, then set $x_i^{[n+1]} = 0$. Otherwise, $x_i^{[n+1]} = x_i^{[n]}$.

5: Replace n with $n+1$.

6: Until any of the following stopping criterions are satisfied.

$$\left\|x^{[n+1]} - x^{[n]}\right\|_2^2 \ge 1\times 10^{-16};$$

$$n \le 1000\ ;$$

7: Output $x^{[n+1]}$ as the proximation to x^*.

We compute the initial guess $x^{[0]}$ which is the largest singular vector of the matrix $\frac{1}{m}\sum_j b_j^2 a_j a_j^T$ via a spectral method. Unless otherwise noted, we generated random realizations the true signal vector x_0 as k-sparse length-128 with support chosen uniformly at random and with nonzero coefficients drawn i.i.d zero-mean circular-Gaussian. The A is generated by creating an mn Gaussian matrix with $n = 128$ and $m = 4.5n$, and we set $\|A\|_2 = 0.9 < 1$. With A and x_0, the measurements b are produced by $b = |Ax_0|$. The relative error between the reconstructed signal x^* and the original signal x_0 is denoted as $\|x^* - x_0\|_2 / \|x_0\|_2$. In order to avoid the randomness, we report the average results over 100 independent trails in all experiments. Without special instructions, we set the percentage of prior information is 1%.

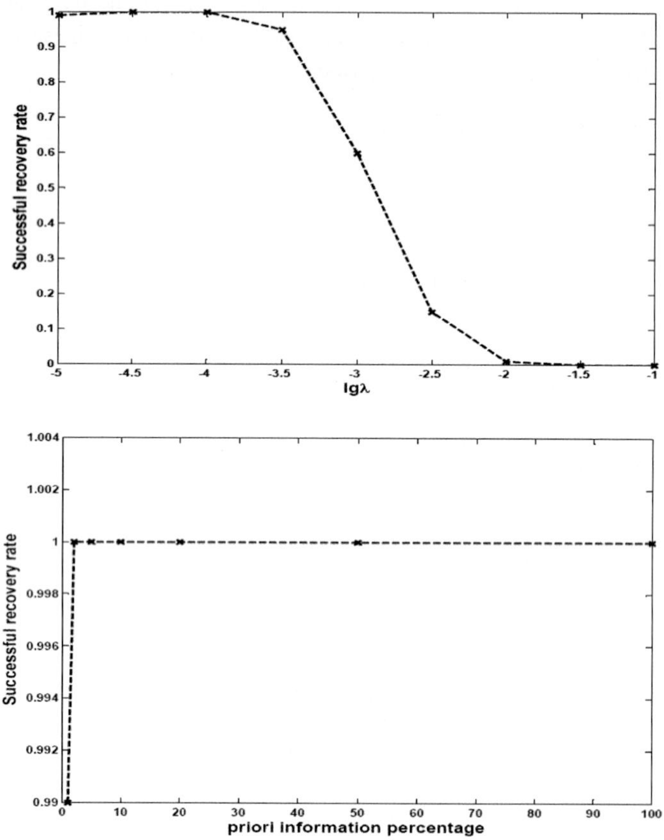

Fig. 1. (Top) Select the best regularization parameter λ among the range from 1 × 10^{-5} to 0.1. (Bottom) Successful rate of recovering real random signals with respect to the priori information percentage.

3.2. *Experimental results*

In order to find the better λ that obtains the minimal relative error, we conducted two sets of trails. Fig. 1 (Top) studies the successful recovery rate with respect to λ. We fix the sparsity level k = 35 and set m and n as before. Without loss of generality, we take $\lambda = 10^{-4}$ as the best regularization parameter. Fig. 1 (Bottom) shows that when the priori information percentage increases to a certain value, the effect of the recovery rate is no longer significant. That is, using our proposed algorithm, a suitable amount of prior information can be achieved with a very good accuracy, which is a very surprising result.

Next, to evaluate the performance of IHT algorithm, the sparse signals are reconstructed from their measurement magnitude by using the proposed algorithm. Then IHT algorithm for phase retrieval are compared to Wirtinger Flow. The mean relative error is analyzed in each case as a function of the number of iterations. These results are shown in the following figures. We set sparsity level k=35.

An instance is presented in Fig. 2 (Top), which carves the recovery of a 35-sparse signal. Obviously, the reconstruction results are good, and the original signal almost coincided with the reconstructed signal. Compared with the Wirtinger Flow, then we qualitatively analyze the convergence performance of the proposed algorithm. The results is depicted in Fig. 2 (Bottom). It is easy to see that these two algorithms are very accurate, however IHT algorithm is faster than Wirtinger Flow.

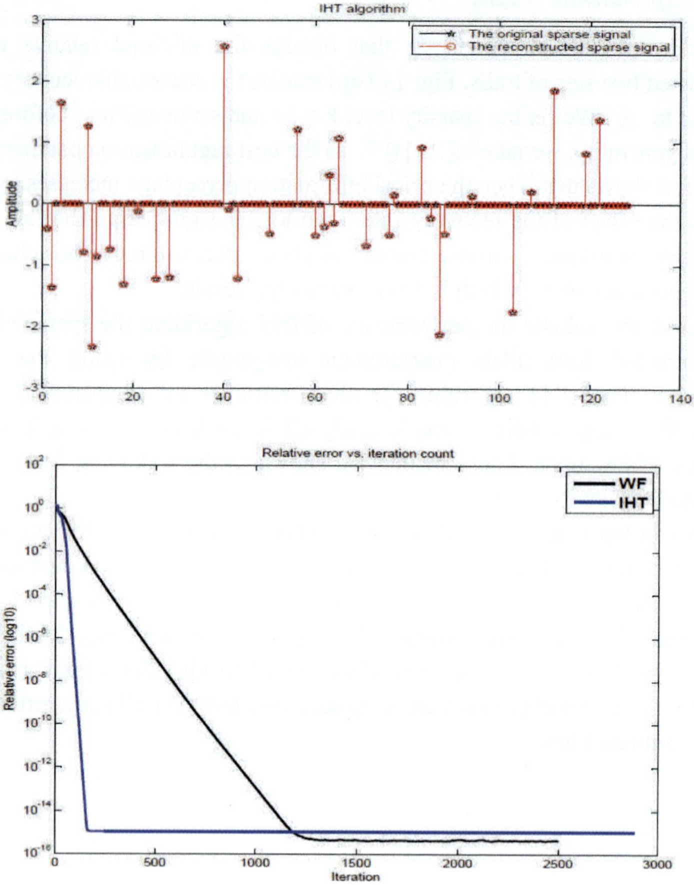

Fig. 2. (Top) A comparison by the relative error versus number of iterations. (Bottom) The performance of the original signal and reconstructed signal.

4. Conclusion

IHT algorithm is proposed and demonstrated to recover a sparse signal from the magnitude of its measurement in this paper. There is no doubt that in classical compressed sensing, it solves the most fundamental problem and the effect is good. However, in the case of unknown phase, it becomes much more complicated. We first give the derivation of the threshold of the algorithm, and give a rigorous theoretical proof. The main theoretical contribution includes the proof of convergence of the algorithm to a local minimum of cost function. And the resultant numerical experiments further support the accuracy of the algorithm as well as its robustness.

Acknowledgment

This work was supported by Natural Science Foundation of China under Grant number 61273020, Fundamental Research Funds for the Central Universities under Grant number XD-JK2015A007.

References

1. M. Moravec, J. Romberg, and R. Baraniuk, Compressive phase retrieval, in Proceedings of SPIE. International Society for Optics and Photonics, 2007.
2. R. Balan, P. Casazza, D. Edidin, On signal reconstruction without phase, Appl. Comput. Harmon. Anal. vol. 20, 2006, pp. 345-356.
3. R. Balan, B. Bodmann, P. Casazza, D. Edidin, Painless reconstruction from Magnitudes of Frame Coefficients, J. Fourier Anal. Appl., vol. 15, 2009, pp. 488-501.
4. E. J. Candès, T. Strohmer, V. Voroninski, PhaseLift: Exact and Stable Signal Recovery from Magnitude Measurements via Convex Programming, Communications on Pure and Applied Mathematics. vol. 66, 2013, pp. 1241-1274.
5. E. J. Candès, Y. Eldar, T. Strohmer, V. Voroninski, Phase Retrieval via Matrix Completion Problem, SIAM Journal on Imaging Sciences, vol. 6, 2013, pp. 199-225.
6. A. Walther, The question of phase retrieval in optics. Opt. Acta. 1963, pp. 41-49.
7. R.W. Harrison, Phase problem in crystallography. J. Opt. Soc. Am. A.vol. 10, 1993, pp. 1045-1055.
8. J. Miao, P. Charalambous, J. Kirz, and D. Sayre. Extending the methodlogy of x-ray crystallography to allow imaging of micrometer sized noncrystalline specimens. Nature, 1999, 400(6742): 342-344.
9. R. P. Millane, Phase retrieval in crystallography and optics, J. Opt. Soc. Amer. A, 1990, 7(3): 394-411.

10. D. Misell. A method for the solution of the phase problem in electron microscopy. Journal of Physics D: Applied Physics, 1973, 6(1): L6.
11. A. Szoke, Holographic microscopy with a complicated reference, J. Imag. Sci. Technol. 1997, 41: 332-341.
12. H.M. Quiney, Coherent diffractive imaging using short wavelength light sources: A tutorial review, J. Modern Opt. 2010 57: 1109-1149.
13. J. V. Corbett, The Pauli problem, state reconstruction and quantum-real numbers, Rep. Math. Phys.2006, 57(1): 53-68.
14. T. Heinosaari, L. Mazzarella, and M. M. Wolf, Quantum tomography under prior information, Commun. Math. Phys. 2013, 318(2): 355-374.
15. Thomas Blumensath, Mike E. Davies, Iterative Thresholding for Sparse Approximations, J. Fourier Anal. Appl.2008, 14(5): 629-654.
16. T. Blumensath and M. E. Davies, Iterative hard thresholding for compressed sensing, Appl. Comput. Harmonic Anal., 2009, 27(3): 265-274.
17. T. Blumensath and M. E. Davies, Normalized iterative hard threshold-ing: Guaranteed stability and performance, IEEE J. Sel. Topics Signal Process. 2010, 4(2): 298-309.
18. Yang Wang, Zhiqiang Xu, Phase retrieval for sparse signals. Appl. Comput. Harmon. Anal. 2014, 37(3): 531-544.
19. Tropp, J.A.: Greed is good: algorithmic results for sparse approxima-tion. IEEE Trans. Inf. Theory. 2004, 50(10): 2231-2242
20. M. L. Moravec, J. K. Romberg, and R. Baraniuk, Compressive phase retrieval, in Proc. SPIE Conf. Ser., San Diego, CA, Aug. 2007, vol. 6701.
21. J. R. Fienup, Phase retrieval algorithms: A comparison, Appl. Opt., 1982, 21: 2758-2769.

Design of a novel radial adjustable pipeline robot

Lei Zhang and Xiao Wang*
Engineering Institute,
Ocean University of China,
Qingdao, China
**e-mail:980165373@qq.com*

Ji-Zhong Zhang
Electromechanical Engineering Institute,
Qingdao University,
Qingdao, China

This paper develops a cable-less pipeline robot which can adapt to the change of pipe diameter. It introduces the composition and function of robot mechanism as well as control system in detail. In this study, the robot employs the crawler traveling mechanism with symmetrical distribution in three groups. Through the rotating of ball screw and the change of the quadrilateral structure both driven by the stepper motor, the radial dimension of robot will be changed. The control system of pipeline robot adopts the modularized design method, takes stm32f103 as its embedded processor, provides a variety of hardware functions as well as the expansion of the circuit interfaces, has a high versatility and applicability. Through verification, the designed pipeline robot can work in various environments of pipeline, realizing in-pipe detection, data acquisition as well as other functions.

Keywords: Pipeline robot; crawler type; radial adjustable; control system; sensor.

1. Introduction

In the use of engineering pipeline, corrosion, rupture and damage are inevitable due to various factors. At the same time, it is likely to cause accidents and even some unnecessary losses without inspection and maintenance. What's more, the environment of pipelines is difficult to arrive directly or enter into, so it is necessary to study on the mobile platform which can be stably working in the pipeline[1].

Pipeline robot belongs to the field of the specialized robot[2]. Starting from 1990s, the research of pipeline robot has become a hot direction, a number of universities and research institutions had made great contributions to the study of the pipe-line robot[3]. At abroad, the series of THES pipeline robots were developed by Tokyo Institute of Technology [4]. Meanwhile, Toshiba reported a micro pipeline robot that can walk in a 1 inch pipeline[5]. In China, Harbin Institute of Technology had developed a variety of wheel type robots based on the active adaptive of mechanism[6]. The Shenyang Institute of Automation developed a spiral pipeline robot from the form of natural environment adaptability[7]. Tsinghua University had also developed a new type of wheel pipeline robot[8]. The developed robots are mostly in the experimental stage, because the mechanism design is relatively complex, the costs are high and the control system is simple, together with poor functional design and being lacking of instantaneity and expansibility, all that are not beneficial to the further development. Therefore, this paper proposes a new kind of pipeline robot with simple mechanism, climbing ability and large load capacity, can realize the adjustment of the radial size. According to the functional requirements of robot, the control system was designed, which can be used to communicate with the host computer in time, obtain the information of the robot's gesture, radial pressure and the pipeline environment by a variety of interface circuits. Meanwhile, expansion interfaces are also reserved for audio, video, sensors and other functional modules. The new designed pipeline robot has strong adaptability, operation and versatility, which can provide useful information and basis for the research of pipeline robot.

2. The Design Mechanism of Pipeline Robot

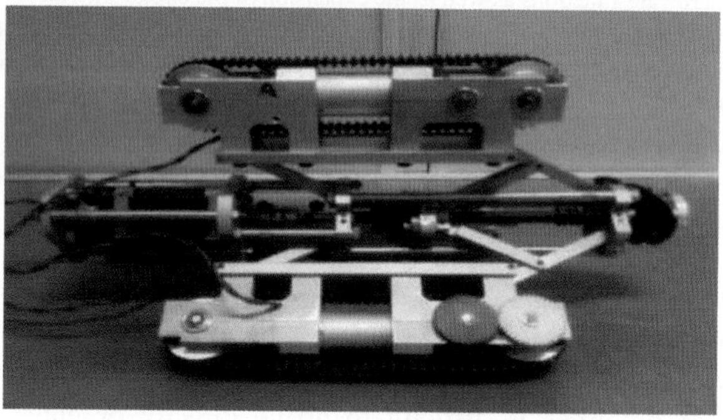

Fig. 1. Physical crawler pipeline robot.

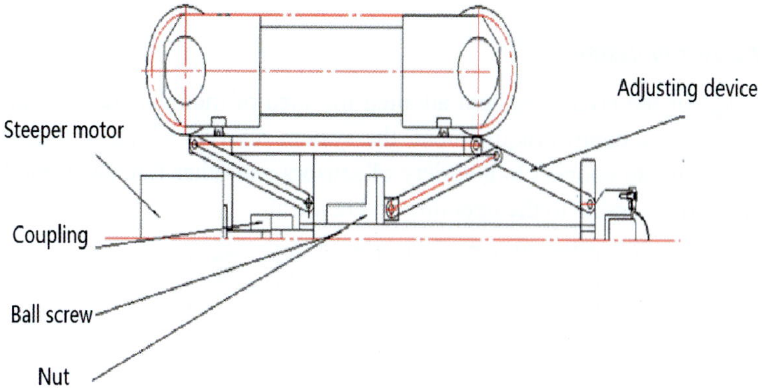

Fig. 2. Diagram of the mechanism.

According to different motion mechanisms, the pipeline robot can be divided into 6 types, such as wheel type, rack type, leg type, crawling type, joint type and pig type[9]. Nowadays, the wheel type and track type are commonly used. Specifically, the wheel type has a better flexibility and speed, but its requirements on the motion environment are relatively high; the track type can provide greater traction and better adhesion. Therefore, when running in the pipeline, the overall performance of track type is better than the wheel type considering the stability and obstacle-crossing ability. Thus, the track type is adopted in this study, and the physical pipeline robot is shown in Figure 1.

As shown in Figure 2, the support of the robot in the pipeline is realized by three-shaft mechanisms which are 120° apart in space. Each group of the track shaft is driven independently, so the robot can work in the pipeline through three-shaft differential driving[10]. In order to meet the requirement of active pipe diameter adaptation, the mechanism of changing cylinder diameter achieve the active adjustment by the step motor with the rotation of the ball screw, and the adjustment range is 200mm~300mm. The mechanism of parallel four-bar linkage can realize the synchronous expansion or contraction of the three shafts. The pressure sensor is positioned between the screw nut and connecting piece, which is used to get the radial pressure value of the robot in real time.

The heavy cable will have a serious interference to the robot's operation when the robot works in a long distance or turns with more curves. In some cases, the robot's ability of overcoming the cable's interference even exceeds the driving capability, causing the robot to get stuck and other adverse consequences[11]. For solving the interference from the cable, the load module was designed, which is mainly used to carry power and some sensors.

3. The Design and Implement of Control System

3.1. *The overall design*

The design of the control system adopted the idea of modular function design. According to the control requirement, the system reserves expansion interfaces and can be able to connect to a variety of sensors and other functional modules, which is also favorable to the updating and research.

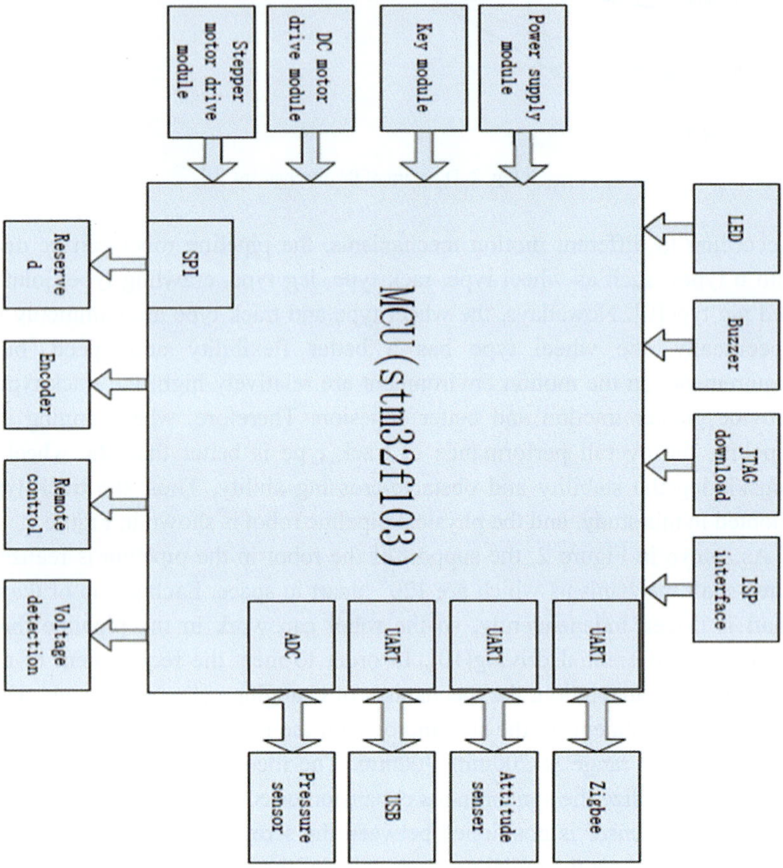

Fig. 3. Overall design scheme.

Fig. 4. Physical controller.

The overall design of the robot's control system is shown in Figure 3 and the physical controller is shown in Figure 4. The core processor usesSTM32F103, which is cost-effective with lower power consumption, excellent performance and other advantages[12]. In the meantime, it is equipped with underlying database without operating the assembly code, which has greatly optimized the development of the system and owns strong portability. The system adopts the mode of centralized control, providing effective decision-making for the whole system, which includes main controller module, power supply module, motor drive module, communication module, remote control module and external module.

Table 1. The functional design of the controller.

Module	Sub-module	Module	Sub-module
Power supply module	A1 Power supply for controller	Main controller module	D1 STM32 processor
	A2 Voltage detection		D2 LED module
	A3 Stabilizing module		D3 Key module
	A4 DC motor power		D4 Buzzer module
	A5 Stepper motor power		D5Download switch module
	A6 Power output	Communication module	E1 Zigbee module
Motor driving module	B1 DC motor		E2 Remote control module
	B2 Stepper motor		E3 USB to serial module
Sensor module	C1 Pressure sensor module		E4 SD card module
	C2 Attitude sensor module		E5 JTAG download
	C3 Encoder module		E6 ISP download

3.2. *Main modules of control system*

Parts of modules are as follows:

(1) Power supply and stabilized voltage module:

Power module mainly provides various required voltage for the control system and the external devices. No external cable is needed, which greatly improves the working efficiency of the robot. The voltage stabilizing module is mainly used to improve the stability of the power supply in the system. Meanwhile, because the robot uses its own power supply, it is necessary to keep the power of robot in a safe value in order to ensure the accuracy of the robot control system and the safety of robot's return. Thus, the system has a voltage detection module, which can transmit the feedback information of power voltage in real time. When the system voltage is lower than the safety value, the system would automatically alert so that the robot can work in a safety state.

(2) Motor drive module:

The system includes DC motor drive module, which can be used to drive various DC motors to control the motion state of robot. The external port of stepper motor is led out from main controller, which can be used to connect to the stepper motor driver to adjust the radial diameter of robot.

(3) Sensor module

The attitude sensor module is used to obtain the information of three-axis attitude and the information of three-axis acceleration in real time. And the pressure sensor module is used to obtain the information of radial pressure.

(4) Communication module

Build the communication between the robot and the master computer.

5) External modules

The external module includes the wireless remote control module, camera module and the reserved SPI interface. The wireless remote control can be used to control the motors, and we can change the motion state of the robot. Because the in-pipe environment is complex and uncertain, the robot needs to be equipped with camera, which can be used to observe the in-pipe environment in real time.

4. The Functional Implementation of the Robot

The software design of the control system was based on the Keil development platform. This tool provides friendly user's interface and powerful debugging system, which is convenient to work with. The software design of control system adopts module design method, enhancing the maintainability of the code and improving the expansibility. The bottom level driver includes voltage detection, controller initialization, sensor initialization, date communication, wireless remote control initialization, motor motion control, robot alarm and so on. The following provides a summary of these modules.

Voltage detection: detect the value of the system voltage periodically and give an alarm when the voltage is lower than the default value

Controller initialization: initialize the controller's control parameters of the peripherals

Communication: initialize the communication module and get information

Motor control: the motor is controlled by the wireless remote control according to the communication protocol

Processing of sensor information: check the current state of the sensor and process the information

Alarm indication: set the level of possible errors during the process of robot's working and take the corresponding warning and treatment

Based on the control of each bottom level driver, many of the higher level control tasks can be accomplished, such as diameter adjustment, forward, backward, turning, obstacle crossing, information acquisition as well as path planning. This study is taking the forward motion control as an example.

The pipeline robot has two working modus: 1) manual mode: people can obtain the current environment information by some built-in sensors, such as camera, gesture sensor and pressure sensor. 2) automatic mode: in a simple environment, people can get the feedback parameters from the motion, then adjust the speed of three shafts and switch the running mode through the built kinematics model.

The planning process of two modes is shown in Figure 5.

The traction of pipeline robot is created by the extrusion from the track shaft to the pipe surface. So the pressure should be in a suitable range. When the pressure is not within this range, people need to adjust the diameter of the robot. Meanwhile the pipeline robot adopts the differential driving type when turn the corner, and the ratio of the speed of the three shafts is equal to the ratio of the curvature radius of the three shafts. In addition, in order to avoid overturning and other uncontrolled phenomena, the information of the robot's attitude need to be acquired in real time.

5. Performance Test of Robot

We made an experiment in a PE pipe whose diameter is 260mm.The results prove that the robot can achieve a variety of motor functions, such as forward, backward, turning, diameter adjustment and so on. Through the attitude sensor, we can see that the robot's attitude almost has no change in the straight pipeline. Moreover, when the slope of pipeline is increased, the robot can still work stably

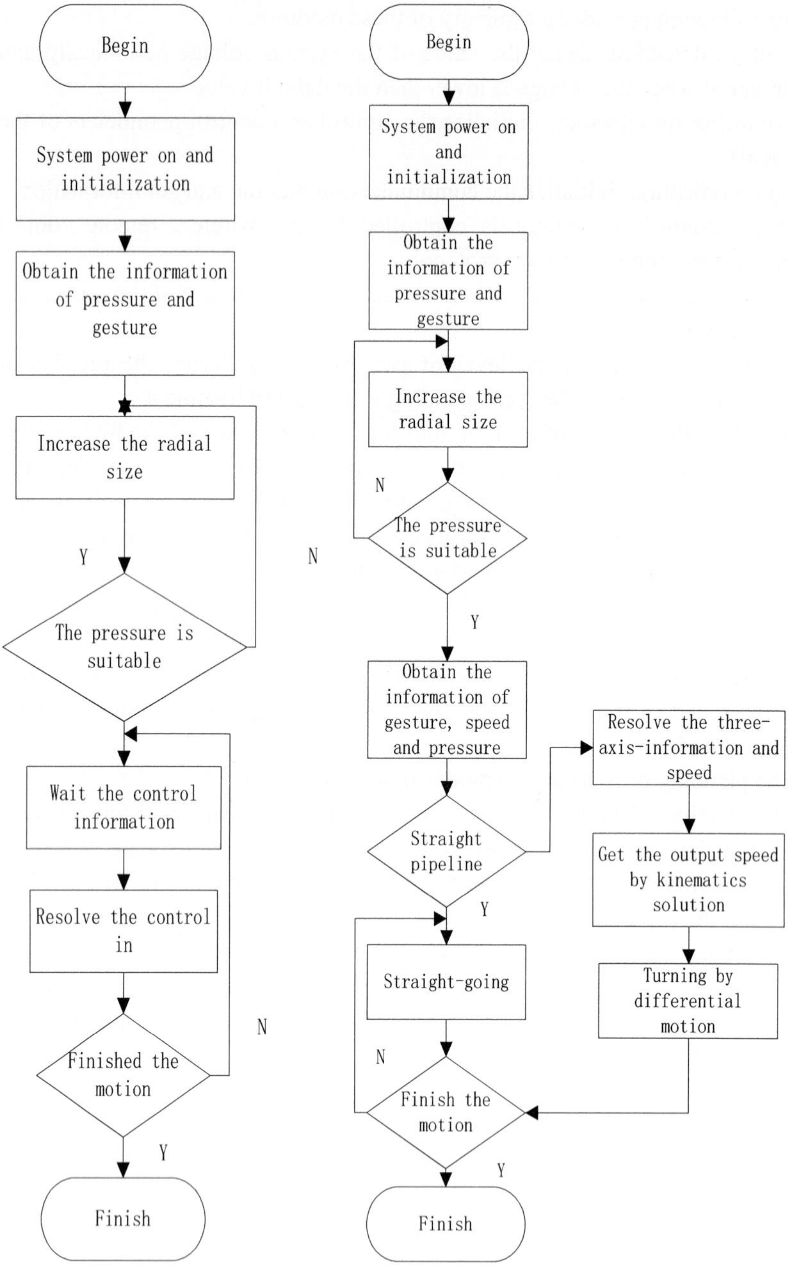

Fig. 5. The flowchart of the manual mode and the automatic mode.

until the inclination is up to 60°. When reaching its maximum climbing angle, the robot can't move because the output torque of the DC motor is not enough. However, if it is provided with a motor with large torque, the problem can be probably solved. The results also present that the designed robot can work in a variety of pipeline environments and obtain the parameters of the robot during its working process.

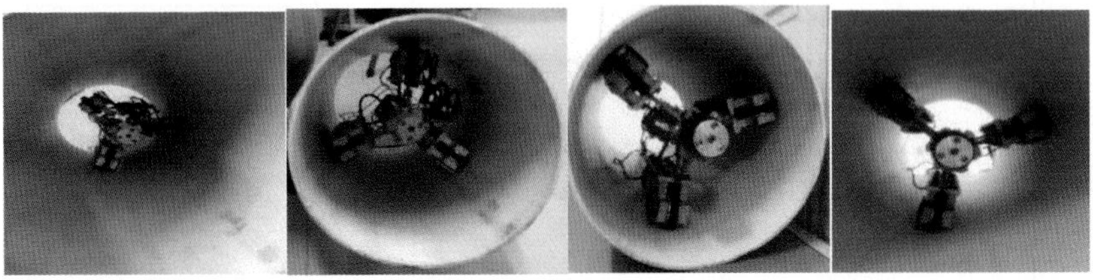

Fig. 6. The process of work in the pipeline.

5.1. *Test of motion*

The corresponding relationship between the velocity of motors and motion states is shown in the Table 2, and the accurate value of speed is related to the curvature radius.

Table 2. The velocity of motor and motion states.

No.1 motor	NO.2 motor	No.3 motor	State
Unchanged	Unchanged	Unchanged	Forward
Increased	Decreased	Increased	Turn left
Increased	Increased	Decreased	Turn right
Decreased	Increased	Increased	Upward
Increased	Decreased	Decreased	Downward

Fig. 7. Backview.

5.2. *Acquisition of attitude information*

According to the obtained attitude information, the robot's state can be analyzed so as to complete corresponding operational optimization. Setting the robot moving in constant velocity, the attitude information during different motion states are shown in Figure 8 and Figure 9.

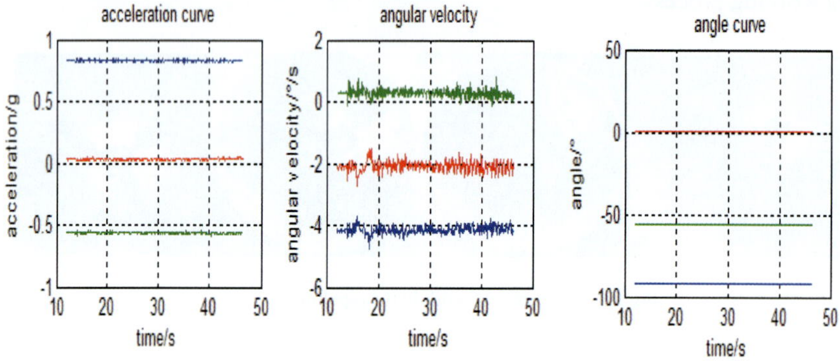

Fig. 8. Horizontal movement.

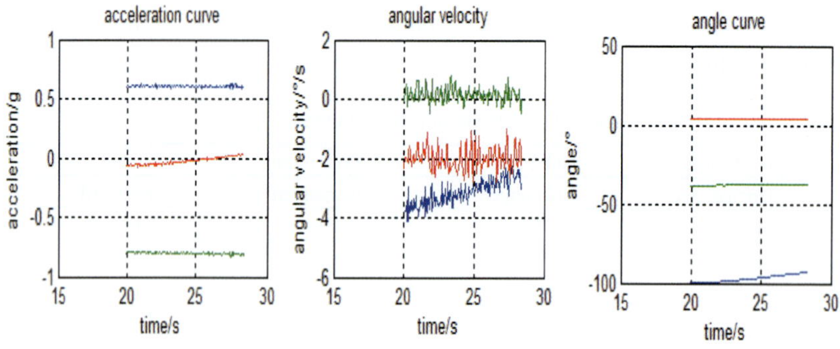

Fig. 9. Climbing.

6. Conclusion

After testing and improvement, we successfully controlled the robot with the wireless remote control and obtained the information of attitude and radial pressure, and the designed pipeline robot can work well in the pipe type of straight, elbow, 60 degrees inclined pipeline. Further more, it can get rid of the motion interference caused by the cable with the help of the built-in power. The experiment in this study verifies the designed pipeline robot can work stably in various environments, which achieves the anticipant goal and provides some basic information for further research.

Acknowledgment

This work was supported by the Major Program of Natural Science Foundation of China (No. 51279185).

References

1. Wang Dianju, Li Runping, Huang Guangming. The research progress of pipeline robot[J]. Machine Tool & Hydraulic, 2008, 36(4): 185-187.
2. Gan Xiaoming, Xu Binshi, Dong Shiyun, et al. The development of pipeline robot[J]. Robot Technique and Application, 2003, (6):5-10.
3. Zhang Yunwei, Yan Guozheng. In-pipe inspection robot with active pipe-diameter adaptability and automatic tractive force adjusting [J].Mechanism and Machine Theory, 2007, 42(12):1618-1631.
4. S. Hirose, H. Ohno, T. Mitsui, K. Suyama. Design of In-Pipe Inspection Vehicles for Φ25, Φ50, Φ150 Pipes, in Proceedings IEEE International Conference on Robotics and Automation. 1999. Detroit, USA. IEEE. 1999: 2309-2314
5. K. Suzumori, T. Miyagawa, M. Kimura, Y. Hasegawa. Micro Inspection Robot for 1-in Pipes. IEEE/ASME Transactions on Mechatronics, 1999, 4(3):286-292
6. Deng Zongquan, Chen Jun, Jiang Shengyuan, Sun Dongchang. Traction robot driven by six independent wheels for inspection inside pipeline. Chinese Journal of Mechanical Engineering, 2005, 41(9):67-72
7. li Peng, Ma Shugen, Li Bin, et al. Design and Motion Analysis of an in-pipe Robot with Adaptability to Pipe Diameters[J]. Chinese Journal of Mechanical Engineering, 2009, 45(1):154-161
8. Zhang Xiuli, ZhengHaojun. A small pipe-inspection robot. Robot, 2001, 23(7):626-629
9. Wang Songtao, Zhu Hua. Kinematic analysis on obstacle-crossing behavior of tracked rescue robot in colliery[J]. Mining & Processing Equipment, 2012, 40(8):97-100
10. Wu Yan, Wang Caidong, Wang Xinjie, NiuZhijun. Design and research on variable-diametered pipeline robots[J]Machine China. 2013, 41(4):124-126.

11. Roh S G, Choi H R. Differential-Drive In-Pipe Robot for Moving Inside Urban Gas Pipelines[J]. IEEE Transactions on Robotics, 2005, 21(1):1-17.
12. Sun Shuying, Sun Zhijia, Kou Chao. The Developing and Application of the New Generation of Embedded MPU-STM32F103[J]. Microcomputer Applications, 2010, 31(12):59-64.

Design of a new type of welding robot

Yi-Lun Han, Pei Chen*, Zhi-Heng Li, Yan-Gao Liang, Ming-Bo Li

College of Transportation Shandong University of Science and Technology, Qingdao 266510, China

E-mail: chenpei9528@126.com

The emergence of the automatic welding robot, improved the working conditions of manual welding, and improved the welding quality and welding efficiency. But for pipe welding robot technology is not very perfect, to a certain extent, increased labor costs on the contrary. In order to solve the lack of manual welding in the process of welded pipe, and make up for some defects of the existing welding robot, this paper designed a kind of high efficiency, high welding quality, can achieve automatic welding of various kinds of pipe fittings of new type automatic welding robot.

Keywords: Welding robot; symmetric structure; drive motor.

1. Introduction

Welding is a material connection technology, is a basic production methods of modern manufacturing industry. With the rapid development of modern manufacturing, welding has become indispensable key processing means of mechanical, nuclear industry, aerospace, energy, transportation, oil chemical industry, construction and electronics and other industries [1]. As a comprehensive application technology, welding technology has the characteristics of multidisciplinary cross fusion. Welding technology of automation, flexibility and intelligence is inevitable trend for the future development of welding technology.

Robot is a kind of can according to the internal procedures or control algorithm through the recognition ability to control their own behavior and the feel of the machine of artificial intelligence. Robot has a lot of types, it can be divided into industrial robots, exploring robot, service robot and military robots according to the application [2], however, and most mature application is industrial robots. Industrial robots more commonly used for welding in actual production, handling, spraying, assembly, glue, etc. Welding robot as an important branch of industrial robots, robot technology is mature, application range is wide.

With the development of welding technology, the continuous application of intelligent robot, and the modern production of the welding technology is put forward to further improve efficiency, optimize the quality, improve labor conditions and other requirements, automatic welding robot arises at the historic moment. Although have automatic welding robot on the market the technology, but also has its limitations, especially large-scale pipe which on the one hand, and large pipe has a very wide range of applications. Therefore, the existing welding robot remains to be further improvement and development.

2. The Structure and Working Principle of the New Type of Welding Robot

In order to solve welding defects of manual welding pipe fittings in the process, and make up for the existing defects of the part of the robot welding. This paper designed a kind of high efficiency, high welding quality, can achieve automatic welding of various kinds of pipe fittings of new type automatic welding robot. The structure of the robot is shown in Figure 1.

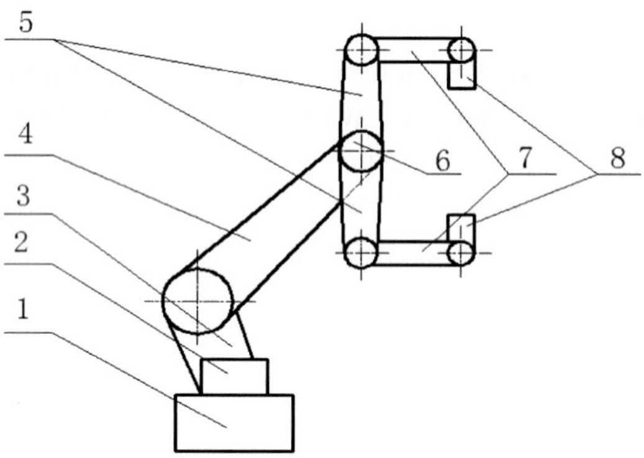

1 base 2 waist 3 buffer device 4 arms 5 brackets 6 bar 7 forearm 8 wrist

Fig. 1. The structure of the new type of welding robot.

Known from the figure, the welding robot has a single body double wrist mechanism, and the support arm, forearm and wrist is symmetrical structure.

The working principle of the work process is as follows: the welding robot is in shutdown state, as shown in Figure 1 of the state. On the upper and lower two wrist installed after welding torch, started up power source, in the structure of motor and speed reducer drive, the bracket on the rotating to the right, on the wrist and forearm, welding torch on the whole to the right, move to the top of the work piece after the specified location, the small Angle on the wrist and on gun rotate to the left, makes the welding gun aimed at the right point on the level of work piece diameter; Instead, under the support arm to rotate to the left, the left forearm and wrist and welding gun under the overall movement, and then under the wrist and welding torch small Angle of rotation to the right, make the welding torch on the most left point on the level of work piece diameter. Then by PLC and computer control, welding torch along the semicircle arc welding to the left, on the welding torch along the semicircle arc welding to the right, in the process of gradually welding, two welding torch and two wrist to rotate at the same time, the two forearm to move at the same time, two bracket to simultaneously rotate in opposite directions. When the torch arrived at the most left point on the horizontal diameter of work piece, the torch arrived at the right point on the horizontal diameter of work piece, the welding process is over.

3. New Welding Robot Configuration Designs

Welding robot configuration design is an important part of its body structure design. For the purpose of the robot configuration design [3], is to determine the number of degrees of freedom and its distribution, the movement of each joint forms and their combination, and determine the structure size of components. These designs are greatly affects the quality of the robot, stiffness and the scope of work, and also have great influence to the welding precision of the robot end. So the research of welding robot the robot configuration has the vital significance and practical value.

3.1. *Arm configuration design*

The end of the robot welding torch is through the various position of the arm to complete the work; it is a direct artifact for welding work. In theory, should think of space anywhere, arm will need at least three degrees of freedom. There are four common forms of arm configuration [4], as shown in Table 1.

Table 1. Arm configuration form.

	rectangular coordinate type	cylindrical coordinates type	spherical coordinates type	joint type
principle	the spatial position of welding torch is determined by the displacement of three vertical direction	the use of joint rotation and movement combination to determine the position of the wrist	consisting of two revolute joints and one prismatic joint	posture adjustment by three to achieve the end of the rotating joint
advantages	high structure rigidity and high precision work	Work faster	compact structure	compact structure, flexible operation
disadvantages	the overall size and slow working speed	the operation is not flexible	poor flexibility	control is more complicated

The principles and the advantages and disadvantages of the various forms of configuration from the table above, this paper designs a new type of welding robot arm had better choose articulated arm. This arm takes up less space, work space, flexible action, and process operation of high precision [5].

3.2. *Wrist configuration design*

Wrist is the connecting part of the arm and hand, plays a role of supporting hands and change hand gestures, in order to make the hand in any direction in space, require that the wrist can realize the rotation of space three coordinate axes X, Y, Z, that is a deflection, bluffing and slewing three degrees of freedom. Now commonly used wrist structure has the following several kinds: BBR, RBR, RRR and BRR, B for the bending pitch in the form of joint structure, and R for rotating the rotational structure form of joints [6]. As shown in Figure 2.

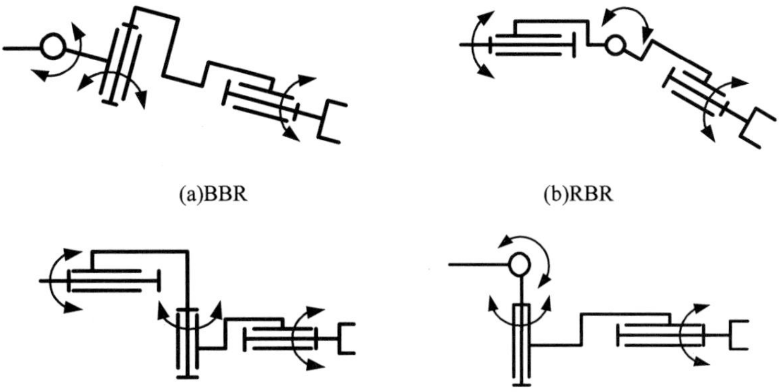

(a)BBR (b)RBR

(c)RRR (d)BRR

Fig. 2. Wrist configuration diagram.

Usually the flip is called the Roll of the wrist, using R; The Pitch is called the Pitch of the wrist, with P; the deflection is called Yaw of the wrist, with Y said. This paper designs a new type of welding robot wrist using remote transmission type RBR wrist, so RBR type wrist output a twisted axis three degrees of freedom in the sport [7], and the advantages of this kind of long distance transmission can be larger driving source in all size, weight, away from the wrist, sometimes on the arm of the back-end for balance weight, this not only reduces the overall weight of the wrist, and improved the balance of the overall structure of the robot.

4 The Calculation of Drive Motor

4.1. *A small arm calculation of driving motor*

First assume that the wrist load is m_1 = 6 kg, wrist quality is m_2 = 2 kg, lower quality is m_3 = 1 kg.

So can calculate the motor to drive for resistance

$$F=(m_1+m_2+m_3)g=(6+2+1)\times 9.8N=882N \tag{1}$$

Assuming the rotary speed of the small arm is $\omega = 1.5r/s$, the efficiency of the transmission system is $\eta = 0.98$

The motor power is calculated as follows:

$$\Pr = \frac{Fv}{1000\eta} = \frac{88.2 \times 2\pi \times 1.5}{1000 \times 0.98} KW = 0.85KW \quad (2)$$

Considering the safety coefficient and takes $S = 1.2$
Then the final power of motor

$$\mathrm{P_r}' = S \cdot \mathrm{P_r} = 1.2 \times 0.85KW = 1.02KW \quad (3)$$

Check the mechanical designer's handbook; select the drive motor model for YY90S4 forearm need, various parameters such as Table 2.

Table 2. The parameters of small arm motor.

model	power	current	voltage	frequency	speed
YY90S4	1.1KW	7.14A	220V	50	1400r/s
power factor	locked-rotor torque/ rated torque	locked-rotor current	maximum torque/ rated torque	quality	efficiency
0.95	0.32	30A	1.7	17.5kg	77%

4.2. *A big arm calculation of driving motor*

According to the calculation of the drive motor with a small arm, a big arm quality m_4=4kg, can get the parameters of the drive motor of a big arm as shown in Table 3.

Table 3. The parameters of big arm motor.

model	power	current	voltage	frequency	speed
YY90L2	2.2KW	13.67A	220V	50	2800r/s
power factor	locked-rotor torque/ rated torque	locked-rotor current	maximum torque/ rated torque	quality	efficiency
0.95	0.3	65A	1.7	21kg	71%

5. Conclusions

This paper designed a new type of welding robot have the following advantages:

(1) The new welding robot structure stability, stable welding process, to ensure the welding quality, improves the production efficiency;

(2) It can not move the work piece under the situation of all position welding, for now use a wide range of pipe fittings, the welding robot of this kind of welding is a very effective way of welding;

(3) In the welding process without the need for manual operation, save the artificial production cost, improve the working environment;

(4) The upper and lower welding at the same time, save working time, speed up the welding speed, indirectly increased the production efficiency.

Acknowledgment

This work is supported by KJZD-13-25-NSH of the research on the automobile assistant safety driving system based on Information Fusion Technology and by 2014-1-26 an experimental study on the vehicle safety driving system.

References

1. Lin Shang-yang, Chen Ben-shan, et al. Welding robot and its application. Beijing: Mechanical Industry Press, 2000.
2. Cai Zi-xing. Beijing: Tsinghua University Press, 2000, 2~28.
3. Bi Ying-li. The research of international advanced technology and development of automatic welding equipment and process. Journal of Jiamusi Institute of Education, 2013, 10:433 ~ 434.
4. Li Zhi-hui. Analysis and design of body structure of 6R type spraying robot kinematics[D]. Nanning: Mechanical manufacturing and automation of Guang-xi University master thesis, 2010:6-12.
5. William E. Ford. What is an Open Architecture Robot Controller, International Symposium on Intelligent Control, 1994,16-18:27~31.
6. Guo Hai. Six DOF welding robot motion control accuracy analysis and research [D]. Hangzhou: Master's thesis of Zhejiang University of Technology of mechanical and electronic engineering, 2011:5-6.
7. Tan Yijiong, Zhou Fangming, Wang Jiangchao, Huang Zhijie. Current situation and development trend of welding robot technology. Welding, 2006, 36 (3):6 ~ 10.

Analysis and research on propulsion speed of rigid robotic fish fin

Rong-Qiang Guan
Automotive Engineering,
Jilin Teachers Institute of Engineering and Technology,
Changchun, Jilin, China
E-mail: dongnana860901@163.com

On the basis of previous studies, this article uses three joint robotic fish rigidity fin as the study object, researches and analysises the robotic fish propeller thrust and speed, experimental studies the caudal fin swinging underwater robot speed and motion parameters, and uses MATLAB to simulate, as a function of thrust, gets the parameters of its efficiency and motion, provides a new way of thinking for the study of high efficiency, high maneuverability and low noise underwater vehicle.

Keywords: Rigid fin; the speed of advancement; hydrodynamics; craft.

1. Introduction

In all fish propulsion mode, swinging tail fin propulsion mode is to promote the fastest, most efficient propulsion mode [1, 2]. These fish are characterized by having a streamlined body, very rigid, crescent-shaped tail fin. In the course of swimming the caudal fin does the bigger amplitude of the oscillation, and the body is just a small swing or fluctuation, or even remain rigid [3]. With this progression, there are sharks, tuna and so on. Shape of the tail fin and motion parameters is important performance parameters of the new tail fin swinging underwater propulsion.

2. Navigation Structure

The main structure of vehicle[4] is shown in figure 1, the size of 600mm×200mm ×100mm; the main driving force is provided by the caudal fin; maximum frequency of 2Hz; swing range of ± 200o; the portion of flexible body has three joints[5], each joint is consists of a servo control. The shape of flexible body is constituted by the steel ring set on a black rubber skin, at a certain distance it is supported by one bone plate which is attached to the fish body so that the fish skin is presented in the shape of a bellows, skin thickness is approximately 0.8mm, bellows structure in the swinging expands naturally, and additional resistance at the time suffered a small swing. Caudal fin dimensions is shown in figure 2, and it chooses high aspect ratio caudal fin, the chord length C = 37.22mm, span length B = 130, sweep angle $\alpha = 47°$.

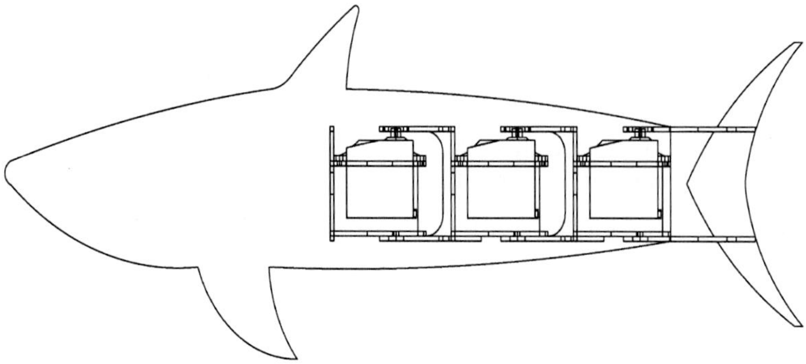

Fig. 1.   Aircraft structure.

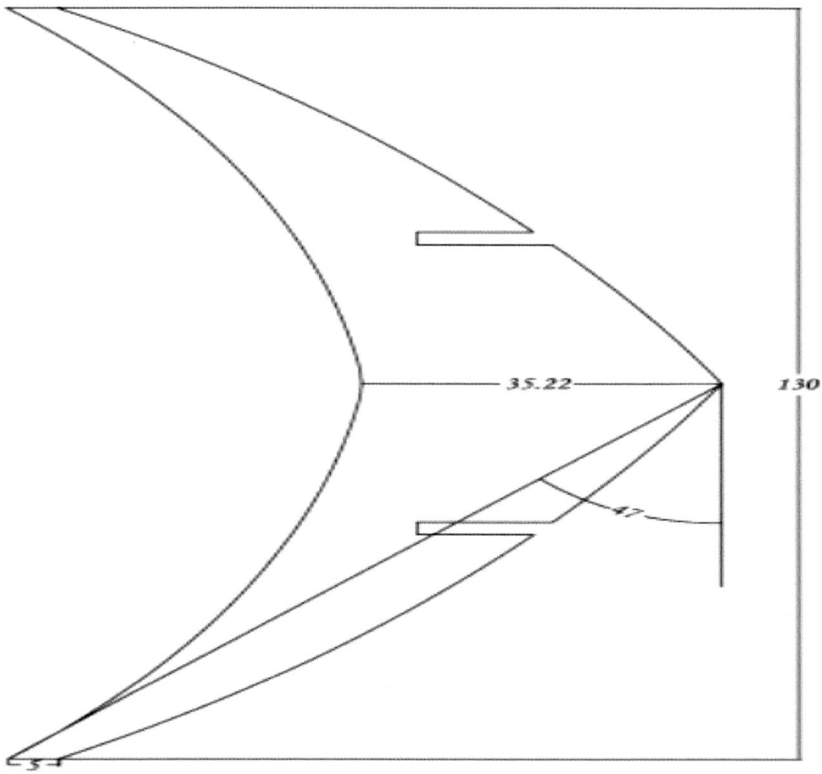

Fig. 2. The size of tail fin.

3. Caudal Fin Thruster Mechanical Model

The mechanical model of caudal fin thruster is shown in figure 3, firstly we use upward swimming fish as the reference system, calculate the amount of fluid on the tail fin and vortex acceleration, in Figure 3 the tail fin OA around O points does rotation producing the angle of attack , OA relative to the flow direction is the swing angle α , swing angle amplitude is α_0 ; and caudal fin along the direction of perpendicular to the axis of the center for OB do flapping around, flapping displacement is h, amplitude is h_0 . Water flow speed is v, lateral flapping speed is v_a. At a time Δt, the area caudal fin swept is the parallelogram $OAA'O'$ enveloping area , according to the parallelogram theorem, it can be obtained that $S_{OAA'O'} = S_{OBCD}$, so that the volume of fluid V' and the fluid mass swept tail fin m can be represented by formula (1) and (2), in it S is the area of the tail fin, i.e., the area is represented by the line OA.

$$V' = S\cos\alpha(v_a - \tan\alpha v)\Delta t \tag{1}$$

$$m = \rho V' = \rho S \cos\alpha(v_a - \tan\alpha v)\Delta t \tag{2}$$

The acceleration of the portion of the fluid which is swept by caudal acts at the lee side of the leading edge of the tail fin in (the left side of line OA shown in figure 3), can be understood as a transportation of a low pressure area formed to the role of fluid . Here you can approximate that: Incremental leading edge vortex is caused by the fluid backward speed, that reasoning leads to induced velocity is equal to the lateral (direction HH') of the tail fin flapping speed v_a (without considering coupling the first half period off).

The formula (2) is substituted into the momentum theorem, then we can get formula (3):

$$F\Delta t = mv_a = \rho S \cos\alpha(v_a - \tan\alpha v)\Delta t v_a \tag{3}$$

The thrust of Caudal producing is as follows:

$$F = \rho S \cos\alpha(v_a - \tan\alpha v)v_a \tag{4}$$

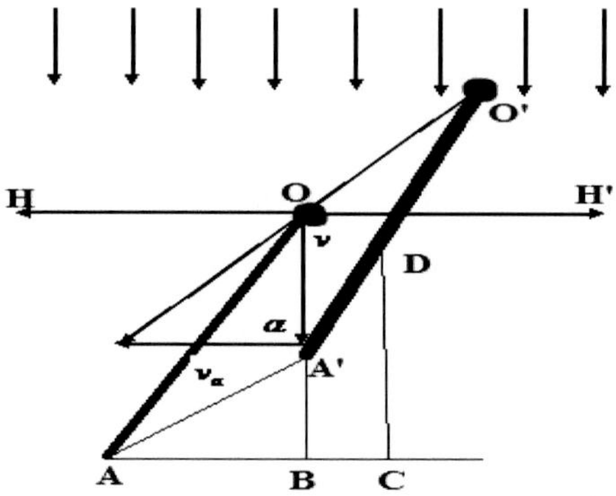

Fig. 3. Mechanical Model caudal fin propeller.

Next we discuss the relationship between flow rate and lateral flapping speeds. The basic law of the tail fin motion as known is:

$$\begin{cases} h = h_0 \cos \omega t \\ \alpha = \alpha_0 \cos(\omega t - \phi) \end{cases} \tag{5}$$

In formula (5) $\omega = 2\pi f$, f is the frequency of flapping, the rocking change of the attack angle than the main swing slow φ , then:

$$v_a = h = 2\pi f h_0 \sin(2\pi f t) \tag{6}$$

We also set up $St' = fh_0 / v$, so that $v = fh_0 / St'$, put the formula (6) into equation (4) , it can be obtained:

$$F = \rho S \cos \alpha \left[2\pi f h_0 \sin(2\pi f t) - \tan \frac{fh_0}{St'} \right] \times 2\pi f h_0 \sin(2\pi f t) \tag{7}$$

The formula (5) is substituted into formula (7), then we can get:

$$\begin{aligned} F = {} & \rho S \cos[\alpha_0 \cos(2\pi f t - \varphi] \times \\ & \left\{ h_0 2\pi f \sin(2\pi f t) - \tan[\alpha_0 \cos(2\pi f t - \varphi)] \frac{fh_0}{St'} \right\} \times h_0 2\pi f \sin(2\pi f t) \end{aligned} \tag{8}$$

4. The Rigid Fins forward Speed

The deformation of fins in the longitudinal is negligible, so that it can be seen as similar to a straight line, that is rigid fins. Then we can derive the speed formula of advancing fluctuations of the rigid fins, establish the coordinate system. The curve of top fin projected on a plane OXY is similar to a sine curve, we assume curve equation is formula (9), where A is the amplitude; ω is the fluctuation frequency; wave number $k = 2\pi / \lambda$; λ is the wavelength.

$$y(x,t) = A\sin(kx + \omega t) \tag{9}$$

When fins start instantly, that is T≈0, the formula (9) can be simplified as:

$$y(x,0) = A\sin(kx) \tag{10}$$

If we ignored the longitudinal bending deformation portion of fins, i.e. assumed that the amplitude of fluctuation fins along the Z-axis is linear. In a wavelength fluctuation of fins, the quality of the enveloping fluid is formula (11), where L is the length of the longitudinal fins, ρ is the density of water.

$$M_{water} = \frac{1}{2}\rho\int_0^{\lambda}|y|\sqrt{L^2 - y^2}\,dx \tag{11}$$

Put the formula (10) into equation (11), we can get the mass of fluid within a wavelength is:

$$\begin{aligned} M_{water} &= \frac{1}{2}\rho\int_0^{\lambda}|y|\sqrt{L^2 - y^2}\,dx \\ &= \frac{1}{2}\rho\int_0^{\lambda}|A\sin(kx)|\sqrt{L^2 - A^2\sin^2(kx)}\,dx \\ &= \rho\int_0^{\lambda/2}A\sin(kx)\sqrt{L^2 - A^2\sin^2(kx)}\,dx \end{aligned} \tag{12}$$

According to a variant relation, we can get formula (13), θ is the angle between the surface of the fins and Z axes:

$$\sqrt{L^2 - A^2\sin^2(kx)} = L\cos\theta \tag{13}$$

Put formula (13) into formula (12), we can get:

$$M_{water} = \frac{2\rho AL\cos\theta}{k}\sin^2(\frac{k\lambda}{4}) \tag{14}$$

Enveloping fluid velocity relative to the fins motion curve is wavenumber V. In an instant start pressing fins pushed relative fins reach wavenumber V, ground-reference system, provided the speed of fins relative to the ground is V_{fin}, accelerated speed relative to the ground as part of the fluid is V_{water}, there are:

$$V = V_{water} + V_{fin} \tag{15}$$

According to the law of conservation of momentum, there are:

$$M_{water} \times V_{water} = M_{fin} \times V_{fin} \tag{16}$$

Put the formula (16) into formula (15), we can get:

$$V_{fin} = V - V_{water} = \frac{M_{water}}{M_{water} + M_{fin}} V \tag{17}$$

Put $V = \lambda f = \frac{\omega\lambda}{2\pi}$ to the formula (17), we can get:

$$V_{fin} = \frac{\frac{2\rho AL\cos\theta}{k}\sin^2(\frac{\lambda k}{4})}{\frac{2\rho AL\cos\theta}{k}\sin^2(\frac{\lambda k}{4}) + M_{fin}} \cdot \frac{\omega\lambda}{2\pi} \tag{18}$$

We order $V_{fin} = \Re \cdot \frac{\omega\lambda}{2\pi}$ is the driven feature coefficient, then:

$$V_{fin} = \Re \cdot \frac{\omega\lambda}{2\pi} \tag{19}$$

Then the formula (18) can be written as:

$$V_{fin} = \Re \cdot \frac{\omega\lambda}{2\pi} \quad (0 < \Re < 1) \tag{20}$$

By the formula (20) it can be shown that: V_{fin} and ω is proportional, by increasing the angular frequency ω, V_{fin} can be improved.

5. Experimental Results and Conclusions

The parameters of rigid fins are supposed to the following values, and then substituted respectively into the above formula, simulated on MATLAB, then we can obtain the curve shown in figure 4.

The longitudinal length of the fins: L = 0.2m

Wavelength: λ = 0.314m

Angular frequency fluctuation: $\omega = 2\pi$

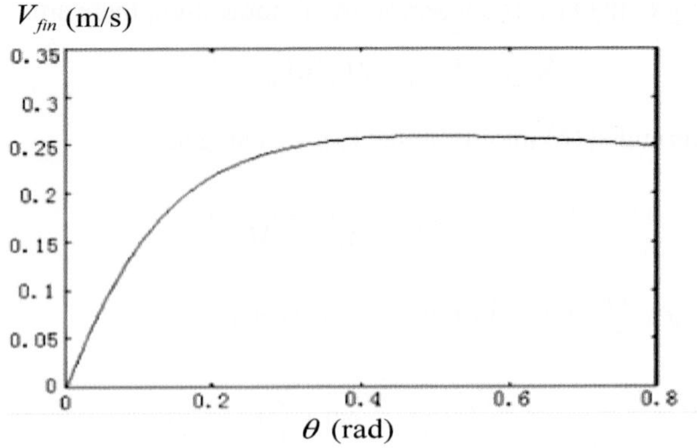

Fig. 4. Simulation curve.

Thereby we can draw conclusions of this paper as follows:

(1) Rigid crescent-shaped caudal fin propulsion mode is a novel for propulsion of micro underwater robots and diving underwater device.

(2) Using the momentum theorem and over propeller theory, we can build the relationship between the caudal fin motion parameters and propulsion, obtain estimation method of caudal fin propulsion.

References

1. Wang Tianmiao, Liang Jianhong. *Estimates based on caudal fin thrust and efficiency over the pusher Theory*[J]. Chinese Journal of Mechanical Engineering, 2005, 41(8):1～6.
2. Wang Shuo, Tan Min. Robofish[M]. Beijing University of Posts and Telecommunications Press, 2006.
3. Liang Jianhong, Wang Tianmiao, Wei Hongxing. *Underwater ROBOFISH Progress I-fish propulsion mechanism*[J]. Robot, 2002, 24(2):107~111.
4. Cho Kyu-Jin, HAsada. *Multi-axis sma acetuator array for driving anthropomorphic robod hand* [C]. International Conference on Robotics and Automation, Bareelona, Spain, 2005.
5. Wang Sai. *Three joint robotic fish swimming free kinetic analysis of latent liter.* Xi'an University of Technology, 2009.

Study on the influence of safety of air spring failure for straddle-type monorail vehicle

Zi-Xue Du, Dan-Ting Cao†, Xiao-Xia Wen and Qin-Hong He

Research Institute of Rail Transit, Chongqing Jiaotong University, Chongqing, China

†*E-mail:1107671181@qq.com*

The air spring rubber capsule rupture of straddle-type monorail vehicle model was established. Analysis on the safety and vehicle system with air spring failure, focusing on the overturning stability coefficient and horizontal wheel's radial force and rolling angle. Analysis result shows that vehicle driving safety become worse than the former with failure of air spring, and guide wheels appear derailment phenomenon, affecting the safety of vehicle. In addition, when the vehicle speed reaches 43km/h, the stability of the vehicle with different air spring failure condition become poor, and it is necessary to reduce the speed of the vehicle.

Keywords: Straddle-type monorail vehicle; air spring; capsizal coefficient; derailment stability; lateral force.

1. Introduction

Straddle-type monorail vehicle air spring is an important part of the central suspension system. It can connect the body with the bogie in vertical, longitudinal and radial respectively to transfer load and attenuate the vibration caused by the road, in order to ensure the safety and comfort of monorail vehicle. When the air spring trouble happened, the suspension system stiffness will change which will affect the running quality of the monorail vehicle. This paper research on the influence of failure air spring for monorail vehicle safety performance by analyzing the air spring airless failure process and establishing failure state dynamics model.

2. Analysis of Air Springs Leakage Process of Monorail Vehicle

Monorail vehicle air springs consist of a daqu bag and a laminated spring, as shown in Figure 1. Generally, wear plates above the lower bearing enhance the bogies' performance of passing through the curve, which the bogies can normally operate under the condition of airless of air springs[1]. If air spring on the side of the bogie is airless because of balloon bursting, a air spring airbag leakage will lead to the process that the other one appears airless as the left and right air springs interconnect, the stiffness of the secondary suspension will be converted from normal linear stiffness to other characteristic stiffness, vertical stiffness becomes laminated spring vertical stiffness, radial stiffness of the air spring is replaced by the stiffness of friction unit and laminated spring in series, the upper cover plate contacts the wear plate plane and slides relatively when the left and right air springs are completely airless, the phenomena of vertical instantaneous detachment may occur. The monorail vehicle is in the unstable state this moment, the dramatic change of the stiffness of the secondary suspension will cause traffic safety problems. Based on the analysis above, this paper established a monorail vehicle dynamics model that air springs are airless. It studies the influence trends of the stability of the monorail of a different number of air spring failure condition vehicle with vehicle dynamics simulation model.

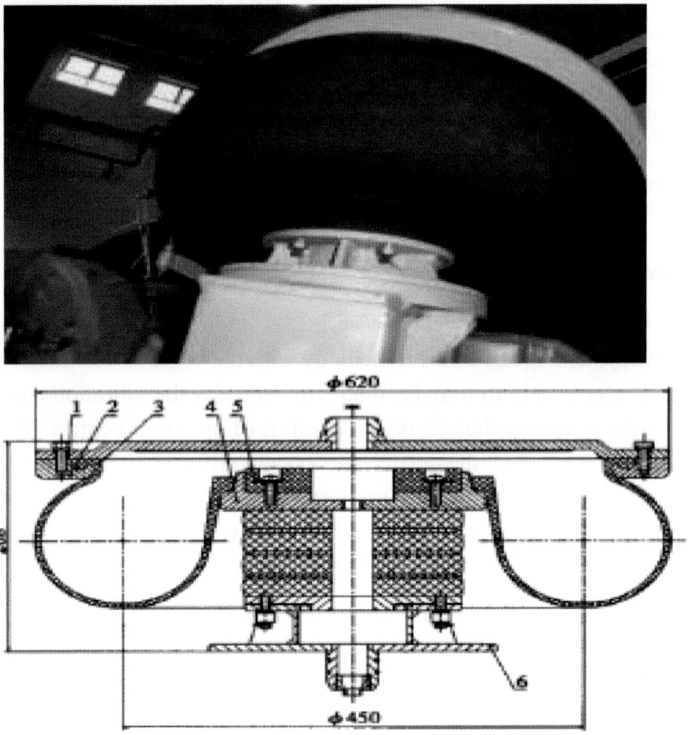

1: upper cover plate; 2: ring; 3: rubber air bag; 4: laminated spring;5: wear plate;6-base

Fig. 1.Structures of air spring.

3. Dynamics Model of Air Spring Failure for Straddle-Type Monorail Vehicle

A coupling dynamics simulation model of vehicle tire track beam was established.When the air spring airless failure, building the friction force and the radial deformation force of the laminated spring between the upper cover plate and the inner supporting surface. The radial elastic retaining force installed between the center pin and the bogie frame was simulated by single side impact force to prevent excessive radial movement in the operation of the body. Friction force is simulated by contact force to reflect frictional action of plane direction.And three direction rigidity of laminated spring was simulated by bushing. As the air spring balloon burst failure, the radial deformation displacement is mainly composed of the friction force and the laminated spring stiffness, providing the radial stability of the vehicle. As shown in Figure 2, it is a model of air spring failure for straddle-type monorail vehicle. SYS540H4[2] type air spring main mechanical simulation parameters are shown in Table 1.

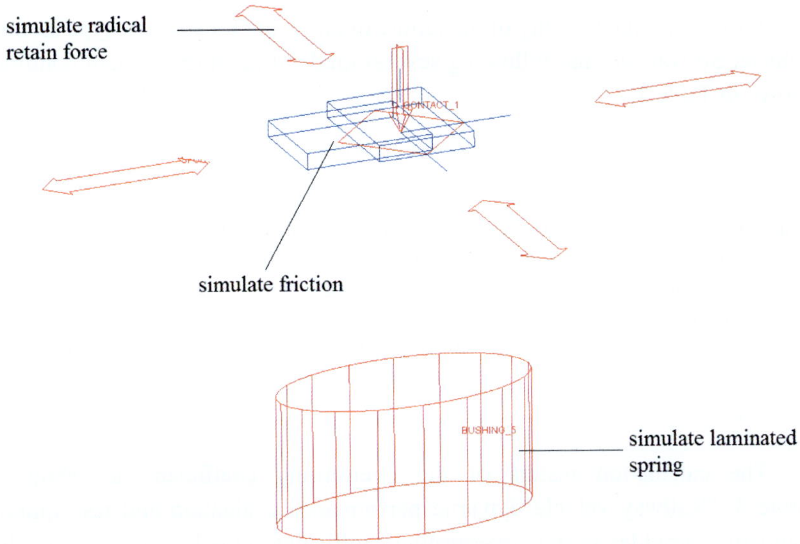

Fig. 2. Failure model of air spring.

Table 1. Simulation characteristic parameters of air spring mechanical components.

Item	Data
Normal air spring longitudinal stiffness	138700(N/m)
Normal air spring lateral stiffness	138700(N/m)
Normal air spring vertical stiffness	307000(N/m)
Laminated spring longitudinal stiffness	1070000(N/m)
Laminated spring lateral stiffness	1070000(N/m)
Laminated spring vertical stiffness	10400000(N/m)
Static and kinetic friction coefficients	0.081/0.143

4. Dynamic Simulation Analysis of Air Spring after Failure for Straddle-Type Monorail Vehicle

4.1. *Overturning stability analysis of monorail vehicle when the air spring failure*

According to study on the anti-overturning stability of the locomotive, make some simple changes in the parameters of the overturning coefficient—the walking wheel's vertical load instead of the locomotive wheel's vertical load—to evaluate the anti-overturning stability of the failure of air spring for straddle-type monorail vehicle. So consider the following several kinds of air spring failure conditions shown in Table 2.

Table 2. Failure mode of air spring.

Condition	Failure position
the first condition	front bogie on the right side of air spring failure
the second condition	rear bogie on the right side of air spring failure
the third condition	front bogie bilateral air spring failure
The fourth condition	rear bogie bilateral air spring failure
The fifth condition	front bogie on the right side and rear bogie bilateral air spring failure
The sixth condition	front bogie bilateral and rear bogie on the right side of air spring failure

The calculation results of the overturning coefficient are shown in Table 3. "Railway vehicle dynamic performance evaluation and test appraisal standards" provides vehicle overturning coefficient –D is less than 0.8 in China [3].

$$D=\frac{P_2-P_1}{P_2+P_1}$$

P1-The vertical force of the load reducing side of wheel

P2-The vertical force of the load increasing side of wheel

Table 3. Overturning coefficient under different conditions.

	Front axle of front bogie D_1	Rear axle of front bogie D_2	Front axle of rear bogie D_3	Rear axle of rear bogie D_4
Normal condition	0.31	0.31	0.31	0.31
The first condition	0.475	0.397	0.387	0.364
The second condition	0.387	0.385	0.325	0.424
The third condition	0.417	0.455	0.397	0.373
The fourth condition	0.365	0.364	0.386	0.386
The fifth condition	0.387	0.352	0.362	0.343
The sixth condition	0.420	0.433	0.360	0.443

By calculating different air spring failure conditions, which meet the requirements of allowable overturning coefficient of the locomotive of GB5599-85. But compared with the overturning coefficient of the normal working condition of the air spring, the numerical value is increased. The main reason is that the radial and vertical stiffness of laminated spring is relatively larger than that of the normal air spring, and between the wheel and rail has a large radial and vertical impact when turning, which leads to the increase of the overturning coefficient.

4.2. *Derailment stability analysis of monorail vehicle when air spring failure*

Due to the influence of the road conditions, such as the curve radius, the ultra-high and pavement of incentives.[4] On the curve, the guide wheelsor stable wheels will appear instantaneous transverse force less than 0.2 times of preloading, but this phenomenon is not enough to determine the safety of vehicle operation. In extreme cases, the lateral force of the guide wheels can be reduced to zero. If there is only stable tires on track, then no matter how stable wheel lateral stiffness, the movement of the vehicle is in an unstable state; If only guide post on track, whether the vehicle has the stable orientation ability, which is related to the lateral stiffness of the guide wheel and the radial stiffness of the walking wheel. To sum up, this paper believes that when the vehicle through the curve the change of lateral force is bigger. When the front and rear bogie side of the two horizontal wheel at the same time there is a lateral pressure changed to zero, and another horizontal wheel lateral force is less than 0.2 times the preloading, which as a derailment.[5] Figure 3 for the failure of the air spring. When the vehicle running at the speed of 36km/h and the pre pressure of the guide wheel and the stable wheel are 5750N, the results of its lateral force simulation.

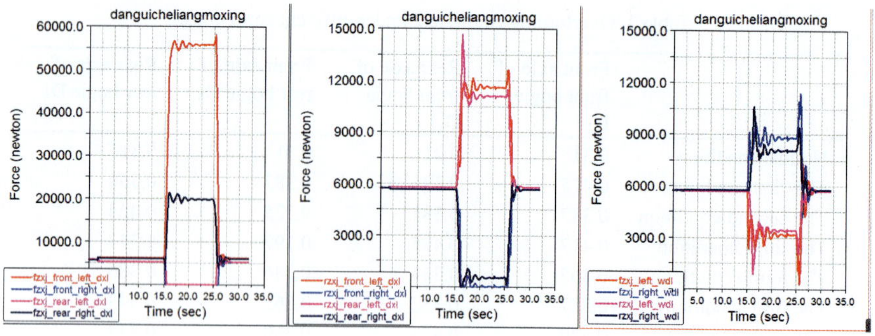

(a) Lateral force of guide wheel and stable wheel (the third condition)

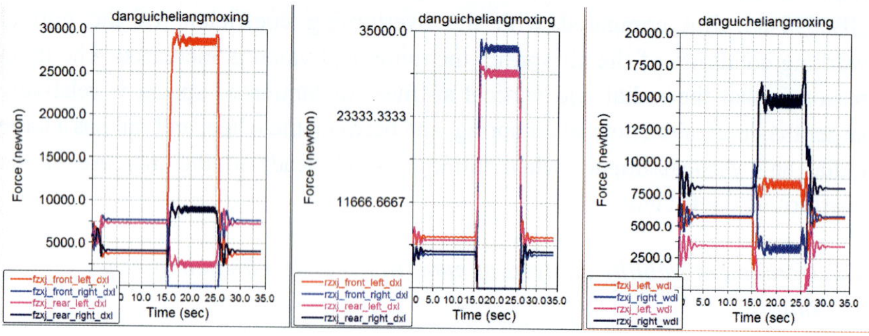

(b) Lateral force of guide wheel and stable wheel (the fifth condition)

Fig. 3. Horizontal wheel lateral force.

In the Figure, fzxj_front_left_dxl said left front guide wheel lateral force of the front bogie, and the rest followed by analogy.

As can be seen from Figure 3 when two air spring fail on the front bogie and vehicle runs to curve, two guide wheels on the right side of the rear bogie of lateral force to reduce to zero. Under the condition of five, the lateral force of the front right guide wheel and the left stable wheel of the rear bogie is changed to 0, the derailment happened. When the vehicle is passing through the curve at the speed of 36km/h, there are two kinds of working condition of air spring failure derailment phenomenon, so vehicle safety is poor. And other working condition of air spring failure despite guide wheel lateral force becomes zero which is not consistent with the derailment evaluation criterion, but does not exclude the presence of derailment trend.

4.3. *The air spring failure of monorail vehicle roll angle*

Monorail vehicle has the characteristics of distance between center of mass and rail surface and small tread, which make it has a "neutral and instability mode". In the steady state, the roll angle of the air spring is much larger than that of the metro vehicle and the stability of the air spring failure cannot be ignored.

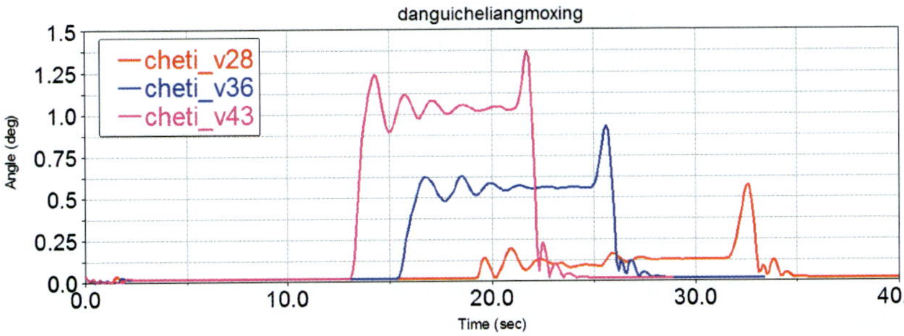

Fig. 4. Under three kinds of speed of the body roll angle.

In the straight line, the body roll angle is very small when air spring fails. While the vehicle travels to the curve, the body begins to roll out. When v = 28 km/h, body roll angle is 0.62 ° and v = 43 km/h body roll angle has been increased to 1.4°. This shows that with the increase of vehicle speed, the body with the increase of speed, making the centrifugal force increases and the body happens outward tilt, body roll angle gradually to the outside dumping.When the vehicle speed reaches 43km/h, the body roll angle is the biggest and the maximum roll angle is 1.4 degrees.

5. Conclusion

Through monorail vehicle dynamics simulation, the monorail vehicle safety performance deteriorates after the failure of air springs. To reduce the speed make operation security, the body is carried by the rubber heaps that have large lateral and vertical stiffness after the airbag rupture of air springs, resulting in irregularity shock on the condition of irregularity road. So air springs of low transverse stiffness with a laminated rubber stack can compensate the lack of displacement of capsule itself when the vehicle body and the bogie generate a large displacement to some extent, act as the vertical damping and have a certain degree of flexibility under the condition of air spring capsule failure, which maintains the dynamic performance of the minimum and ensures vehicle safety.

Acknowledgment

This work is supported by The National Natural Science Fund Project (item number: 51475062).

References

1. Hao Gao, Ren Luo, Mao-Ru Chi, Huan-Yun Dai. Safety analysis of vehicle system which air springs are airless[J]. Journal of Traffic and Transportation Engineering, 2012.
2. Yan-Feng Pang, Lei Yu, Hong-Guang Song. Study of SYS400C type air springof large horizontal displacement of [J]. Railway Vehicles, 2012, 50(10):4-7.
3. Dynamic performance evaluation and experimental qualification specifications of Railway vehicle-GB 5599-85 [S]. Beijing: China Standards Press, 1986.
4. Wei-Hua Ma, Zi-Li Wang. Evaluation Method and examples of resisting derailment stability[J]. Railway Vehicles, 2004, 42(8):1-3.
5. Zi-Xue Du, Ning Li. Simulative analysis of steer wheels force of monorail vehicle[J]. Railway locomotive vehicle, 2011, 31(4).

An authentication control system of payment terminal based on identity management and control card

Wen-Jie Li
East Inner Mongolia Electric Power Co., Ltd. Power Research Institute, Hohhot, Inner Mongolia, China
E-mail: wenjieli163@163.com

Hong-Wei Xie
East Inner Mongolia Electric Power Co., Ltd., Hohhot, Inner Mongolia, China
Email:xiehongwei2@sina.cn

Xin Fan
East Inner Mongolia Electric Power Co., Ltd. Power Research Institute, Hohhot, Inner Mongolia, China
E-mail:fanxin@163.com

Bing-Shu Li and Chao Zhou
East Inner Mongolia Electric Power Co., Ltd., Hohhot, Inner Mongolia, China
E-mail: md_zhouchao@sina.cn

Hong-Bo Ma
Beijing KeDong Electric Power Control System Co., Ltd. Beijing, China
E-mail:mahongbo2@sgepri.sgcc.com.cn

Hua-Min Wang
East Inner Mongolia Electric Power Co., Ltd., Hohhot, Inner Mongolia, China
E-mail: wanghuaming@sgeime.sgcc.com.cn

Lei Xing
East Inner Mongolia Electric Power Co., Ltd., Hohhot, Inner Mongolia, China
E-mail:xinglei@sgcc.com.cn

Establish a diversified platform and intelligent mobile payment terminal has become the trend of electric power applications. In this paper, an authentication control system of payment terminal based on identity management and control card is proposed. By introducing identity cards we design a security escort procedures, greatly improved the security of the existing solutions in the section of the identity authentication when pays the bill, and reducing human errors and potential safety hazards may arise in the process of payment.

Keywords: Identity card; authentication and control system; electric power.

1. Introduction

With the rapid development of smart grid technology, the establishment of a diversified payment platform and smart mobile payment terminal has become the trend of electric power applications. Due to the power company belongs to the non-financial sector, currently in charge of the payment does not have the safety management mechanism in the intelligent payment terminal deployment. The national power grid intelligent payment terminal lack of a variety of means of payment support, then the user is mostly in cash payment. So the process of the remittance inevitably exist all kinds of security problems and human error, besides the lack of technical control means easy to operate business management loopholes, causing damage to power.

The research is on the method of the electronic payment based on the identity authentication and control mechanism of the payment terminal. On the one hand, it can provide technical and management basis for the application of power payment solution, on the other hand, it can provide reliable technical means for the large-scale promotion of remote intelligent payment. Through the introduction of identity authentication and control card equipment and build a payment terminal identity authentication security management system, greatly improved the security of the existing solutions in the section of the identity authentication when he is paying the bill, reducing the process of the solution may occur in human error and potential security risks.

2. Related Technology

2.1. *IC card*

IC card (Integrated Circuit Card) also known as smart card, intelligent card, microcircuit card or micro chip card, etc. It is a micro chip embedded ISO 7816 compliant card, a card made form. IC card due to its inherent information security, portability, perfect the standardization, etc., in the field of identity authentication, banking.

2.2. *Electronic remittance*

In the public business often encounter cash paid, and you're not going to the company to pay cash, then you need to go to the bank to fill out the cash remittance, the money through the bank to the company account. Then with a single receipt, to the company for the remaining procedures and avoid the risk of carrying large amounts of cash. It is the Electronic remittance that informatization of the payment with the help of electronic information platform, achieving online payment.

3. Overall Scheme of the System

3.1. *Topology of the scheme*

Payment terminal identity authentication security management and control system can be used to centralized deployment, the company in the configuration of the production database server, application server, interface server. The physical deployment of the system is shown in the following Figure 1:

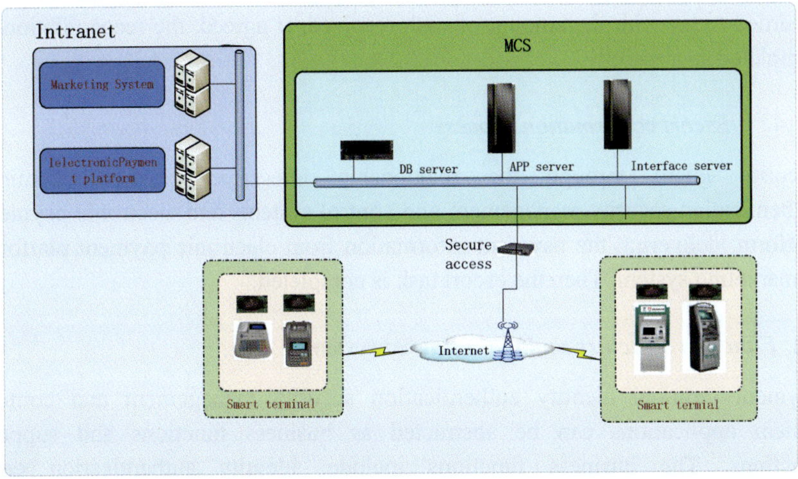

Fig. 1. Physical structure.

MCS: identity authentication security management and control system.

The specific steps during the escort task in the payment terminal security guard, as follows:

3.1.1. *User login process*

When the security guard began to escort mission of the day, using the authentication control card bond to security guard on the need for escort operations payment terminal to identify, and enter the password, waiting for verification.

3.1.2. *Authentication Process*

Payment terminal password and other relevant information will be passed to the payment terminal authentication security management and control systems. Management and control system would be check cardholder information, the binding relationship between control card and payment terminals, then comparison of transmission over the encrypted password and the system has been stored in database. If both exactly the same the identity authentication is succeed, security guard can successfully log on to escort operation page.

3.1.3. *Verify the amounts escort process*

Security guard comparative the amount of payment in the payment terminal recorded today with the amount of cash received, If agreed, the reconciliation is completed successfully.

3.1.4. *Escort confirmation process*

Security guard through the relationship integrated payment terminal authentication security management and control systems and electronic payment platform, delivering the payment information from electronic payment platform to marketing system. Then the escort task is completed.

3.2. *Function structure and main flow of system*

Payment terminal identity authentication security management and control system applications can be abstracted as business functions and support functions. The business functions include: identity authentication card information management, identity card and security guard bind, identity card and terminal bind, security guard authentication. Function structure shown in Figure 2.

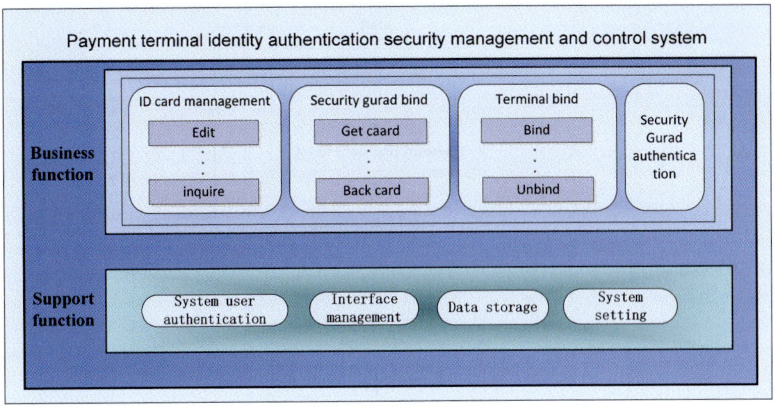

Fig. 2. Function structure.

3.2.1. *Identity card information management module*

The desired information is written into the authentication control card through a card reader, and manages the control card information, including card number information, password information, which the card state and so on. The system can be written information in the control card flexible, including security guard personnel identity information.

3.2.2. *Identity card and security guard binding module*

This module is used to manage Identity card and cardholder information. Besides, making rules to restrict the access of the Identity card, including the same operator can obtain the same card every time. Mainly functions as the following:

a. Record the time to take card and back card. Recording time to take cards in the moment of binding, and recording the time back card when unbind the relationship between Identity card and cardholder.

b. Identity card is available or not. When an exception occurs after the Identity card and the cardholder has bound, it can modify the state of the Identity card is unusable. These make the card to stop using, in order to ensure the follow-up to handle exceptions.

c. Monitor status between the Identity card and the cardholder. Monitoring the state of cardholders and Identity card, in the event of abnormal using, you can unbind associations between them to ensure the safety of the system.

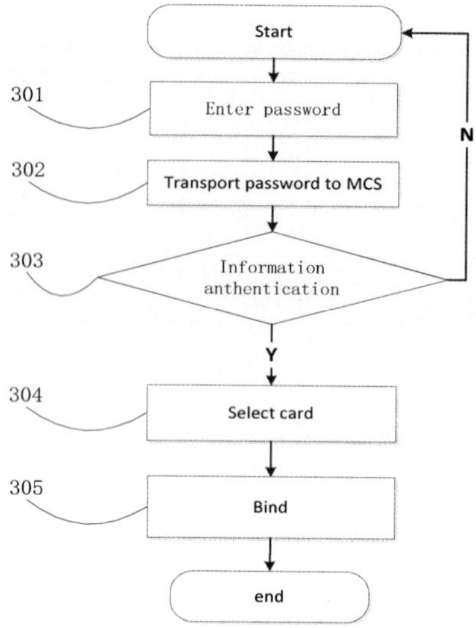

Fig. 3. The flow of bind security guard.

Step 301: This step is a security guard to enter a password, for security guard log preliminary authentication. Ensure legitimate users are using the system.

Step 302: Transfer password to control systems. For password encryption to ensure information security during transmission, prevents eavesdropping and tampering.

Step 303: Compare the security guard encrypted passwords and encrypted information in the database. Verification the process continues, If not through backing to the start.

Step 304: Select the identity card. Security guard can only choose an identity card, and only select the available identity card.

Step 305: Binding. Write security guards information to identity card, and record binding information to the database.

3.2.3. *Identity card and terminal bind module*

This module is used to manage the relationship of terminal and Identity card binding, unbinding, inquiries. While making certain restrictions of binding rules, such as, a card can bind multiple terminals, the same card and terminal can be bound once, and only the Identity card has been received can be selected. And the main flow as follows:

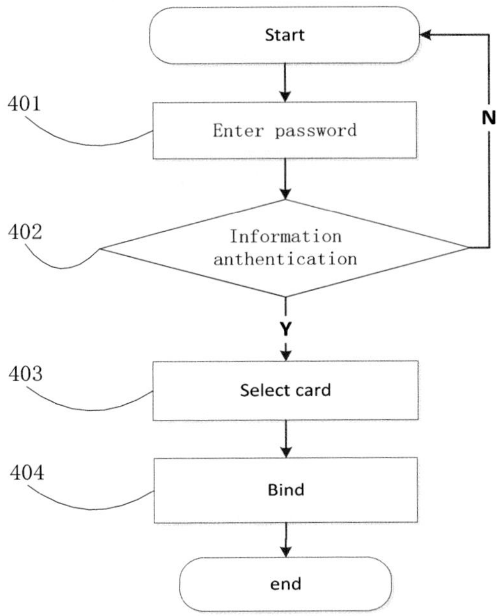

Fig. 4. The flow of bind terminal.

Step 301: This step is to enter a password in terminal, ensure that the terminal is a legitimate use.

Step 302: Confirmation of the information of terminal, including the terminal number, the terminal location, and the terminal model. Then Compare the terminal encrypted passwords and encrypted information in the database. Verification the process continues, If not through backing to the start.

Step 303: Select the identity card. The terminal can only choose an identity card, and only select the available identity card.

Step 304: Binding. Write terminal information to identity card, and record binding information to the database.

Through the above steps will be able to bind security guard, identity card, payment terminal relationships together.

3.2.4. *Security guard authentication*

This model is used to verity the match of security guard, Identity card and terminal. After the terminal is inserted Identity card and input card key, the terminal will call Identity card authentication interface and sends a request message, payment terminal authentication security management and control system at the request of the relationship between information and check identity card number and the card key of the terminal, and return the verification results. The flow chart is as follows:

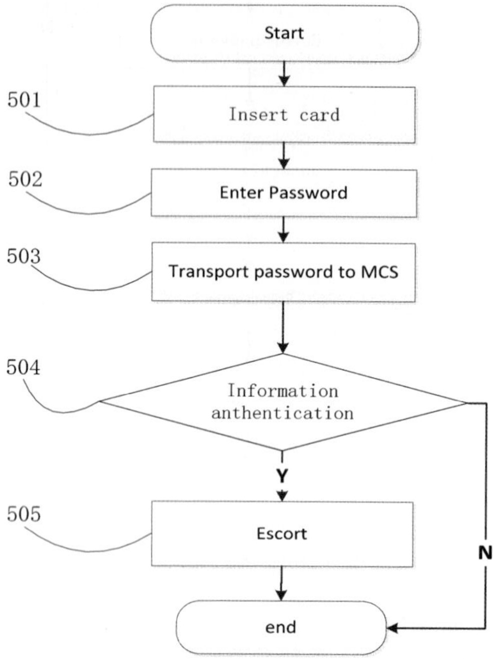

Fig. 5. The flow of authentication.

Step 501: This step is a security guard inserted the identification card into the terminal.

Step 502: Security guard to enter the user password and the terminal password.

Step 503: Transfer password to control systems.

Step 504: Compare the security guard encrypted passwords and encrypted information in the database, in particular, the relationship between Identify card, terminal and security guard. Verification the process continues, If not through go to the end.

Step 505: Start the escort process.

3.3. *Security design*

Security payment terminal authentication security management and control systems should pay attention to the system running safe, to avoid security risk of proliferation. Concrete should from the network security, application security, data security three aspects for deployment.

Network security. Network traffic security refers to the business data through the network transmission need to take security measures to prevent it may be intercepted, tampered, or deleted. And the system uses the following methods to ensure network security:

3.3.1. *Using IPSEC VPN*

Adopt asymmetric encryption algorithm to prevented sensitive information from the illegal interception when transmitted over the network; with a digital signature means to ensure the integrity of information and it is not being modified.

3.3.2. *Deployment of IDS (Intrusion Detection System)*

Intrusion detection system deployed in all major servers resides subnets for network monitoring, and recording invasion source IP address, destination IP address, the type of attacks, the time of attacks and other information.

3.3.3. *Data security*

Its protection measures including: data classification protection, data backup and data encryption storage, etc. Specific protective measures are as follows:

3.3.4. *Data classify protection*

Adopting strict authorization management, make the data and programs from illegal users wantonly change. For every step of the key business data record the operation log. In the process of transmission use the encryption and safe transportation. System provides the function that record operation log of the system, the audit of modify the abnormal flow of critical data processing, and record the operation greatly impact on system performance.

3.3.5. *Data backup*

Regular backup data, when the data of the formal system is damaged or abnormal, the formal system can quickly recover data with backup data.

3.3.6. *Stored encrypted data*

In the terminal you need to encrypt data stored, including: key information, terminal Identity information, APN dialing and other sensitive information.

3.3.7. *Application security*

System application security include: user authentication, access control, application security audits, key information confidentiality and integrity protection, prevention and other malicious code.

4. Summary

By introducing the authentication control cards, the process of security guard authentication provides a new security measure. And encrypt the user password with mixed MD5 + 3DES, effectively improved the security of payment information during transmission. At the same time, by constructing payment terminal authentication security management and control systems to achieve purposes that the data transfer of escort task automation. And by means of the payment terminal authentication security management and control system can be flexibly configured to the information of authentication control cards, control cards and identity security guard binding information, authentication control card and payment terminals binding information. Make electronic payment has multiple security mechanism, but also facilitate the dynamic allocation between the security guard and escort missions. It is important for security guard to improve the security of the payment process.

References

1. Manz, Edgar, Fink, "A hybrid authentication and authorization process for control system networks," 2010 Sixth International Conference on Information Assurance and Security (IAS 2010), p 36-9, 2010.
2. Rzasa, Rzonca, Stec, Trybus, "Analysis of challenge-response authentication in a networked control system," Communications in Computer and Information Science, v 291 CCIS, p 271-279, 2012, Computer Networks - 19th International Conference, CN 2012, Proceedings.
3. Zou Zhe-qiang, "Research of authentication method of intrinsically safe interconnection of coal mine monitoring and control system," Industry and Mine Automation, v 36, n 7, p 26-9, July 2010 Language: Chinese.
4. Wu Junjun，Zhu Wenxuan, "Research on application-independent identity authentication and access control system," Control & Management, n 2, p 41-3, 2007.

The application of a kind of controllable drogue in the control of shaking motion of aerial refueling hose-drogue

Jin Zhang[†], Suo-Zhong Yuan and Quan-Quan Gong
College of Automation Engineering, Nanjing University, of Aeronautics and Astronautics, Nanjing, China
[†]*E-mail: 869384791@qq.com*

The bow wave effect on hose-drogue could increase the difficulty of coupling with a refueling drogue for a receiver plane safely and accurately. A mathematical model of shaking motion of aerial refueling hose drogue is presented by adding the concept of controllable drogue to the model of rigid multibody dynamics. Then the bow wave model of the receiver aircraft is built by Rankine half body model, and the effect on hose-drogue in different flight conditions are analyzed. To reduce the movement amplitude of the hose-drogue in the air, the LQR (linear-quadratic-regulator) controller to the motion of the hose-drogue is designed. The simulation results show the effectiveness of the control to the shaking motion of the hose-drogue, and then it can reduce the difficulty of aerial refueling docking enhancing the security of aerial refueling.

Keywords: Aerial refueling; bow wave; controllable Drogue; LQR controller.

1. Introduction

When the receiver aircraft approaches the hose-drogue, the drogue will deviate from the nose of the receiver aircraft by the bow wave [1] of the receiver aircraft which increases the difficulty of coupling with a refueling drogue for a receiver aircraft safely and accurately or even results in the probe damage and severe accident. Thus, there are urgent needs to improve the Probe-Drogue refueling process to ensure the overall safety of the operation for both manned and unmanned missions. In the aspect of the modeling, Kamman's [2] modeling method is a lumped-mass, finite-segment approach in which a hose is modeled by a series of rigid links connected with ball-and-socket joints [3].The approach of modeling of Kamman is simple, but it's too simple for the drogue modeling. Walton R. Williamson [4] get the experimental data from wind tunnel test to model the drogue's dynamic, but his modeling for the hose is simple. It's lack of the study for the control of the motion of the hose-drogue when it come across the bow wave of the receiver aircraft.

In this paper, a mathematical model of shaking motion of aerial refueling hose-drogue is presented by adding the concept of controllable drogue to the model of rigid multibody dynamics. Then the bow wave model is built to simulate and analyse the hose-drogue dynamics with the approaching receiver aircraft in the process of contact. The no-linear model is linearized and reduced to design LQR controller to control the shaking motion of the hose-drogue under the atmospheric turbulence and the bow wave of the receiver aircraft.

2. Modeling of the Shaking Motion of Aerial Refueling Hose-Drogue

2.1. *Assumptions of modeling*

The hose-drogue assembly is modeled as a link-connected system, where the hose consists of a finite number of cylindrically shaped rigid links connected with frictionless spherical joints. The masses and loads associated with each link are concentrated at the joints. The joint between the drogue and the end of the hose is treated as un-rotatable. The other end of the assembly is connected to a refueling pod (towing point) .

2.2. *Controllable drogue*

The model of the hose could be found in [5]. At the end of the hose where the drogue is connected, the resultant external force is

$$\boldsymbol{Q}_N = (m_N + m_{dro})\boldsymbol{g} + \boldsymbol{D}_{L,N}/2 + \boldsymbol{D}_{dro} \tag{1}$$

In Eq(1), m_N is one-half of the mass of the last hose section, m_{dro} is the mass of the refueling drogue. $\boldsymbol{D}_{l,N}$ is aerodynamic force acting on link N. $\boldsymbol{D}_{dro}$ is the aerodynamic force acting on controllable drogue. For controlling the drogue, the drogue is modified as shown in Fig.1.

In Fig. 1(a), (b) are the side view and rear view of the controllable drogue respectively. The drogue canopy is attached to the coupling using an outer ring of leading-edge strut arms that connect to the outer edge of the canopy and trailing-edge strut arms that connect to the inner ring of the single-lobe canopy. Eight leading-edge and eight trailing-edge strut arms were modified to permit manipulation of the canopy by expanding and contacting the angle between the leading and trailing strut arms. Rigid strut braces were used to hold the leading-edge and trailing-edge strut arms at a fixed angle of separation. Two strut braces were used per quadrant of the drogue for a total of eight. The strut locations on the drogue are referred to as the starboard brace (actuator1), the port-side braces (actuator3), the upper brace (actuactor2) aligned with the vertical, and the lower brace (actuator4) aligned with the vertical, as shown in Fig. 1. Manipulating the angle between the leading-edge and eight trailing-edge strut arms can change the aerodynamic force the canopy [4, 6] subjected in the air flow. The aerodynamic force may be expressed as

$$\boldsymbol{D}_{dro} = \overline{q}f(u_{act}, \alpha, \beta) \approx \overline{q}\left(C_0 + C_u(u_{act} - u_0) + \begin{bmatrix} C_\alpha & C_\beta \end{bmatrix}\begin{bmatrix} \alpha - \alpha_0 \\ \beta \end{bmatrix}\right) \tag{2}$$

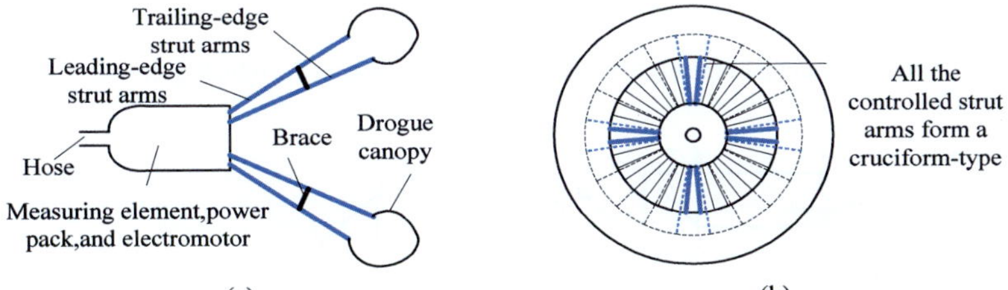

Fig. 1. Diagram of the controllable drogue.

In this case, $\overline{q}$ is the dynamic pressure. C_0 is the baseline 3-element vector containing the bias values for the drogue, C_u is a 3 by 4 element matrix containing the forces parameters for each of the four actuators with C_{ui} for actuator i. The vector u_{act} is the 4 by 1 vector of the actuator angles, u_0 is the angle about which the system is linearized, C_α is the 3 by 1 vector of force sensitivities to angle of attack α and C_β is the 3 by 1 vector of force sensitivities to angle of sideslip β , The trim angle of attack α_0 was based on running steady-state simulations.

3. Modeling of the Bow Wave of the Receiver Aircraft

Due to the high speed motion of the aircraft, strong airflow known as bow wave is produced around the nose of the receiver aircraft. In fact, this turbulence is the major factor of the failure for the contact as shown in Fig. 2.

In the process of contact, the motion region of the refueling drogue is confined to the nearby region of the nose, the application of Rankine half body model [7,8] of the hydromechanics in 3-D could simulate the airflow around the nose of the receiver aircraft roughly. As shown in Fig. 3(a), which is the diagram for any point $C(x, y, z)$ of the space around the nose of the receiver aircraft relative to the origin point O. Assume that the nose of the receiver aircraft is

revolving solid. To solve the velocity of the bow wave, the Rankine half body model is applied on the plane *OBC*.

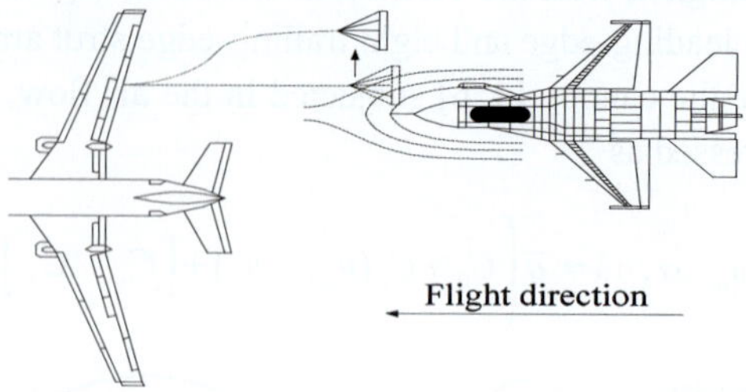

Fig. 2. Diagram of the F-16 receiver aircraft approaching the refueling drogue.

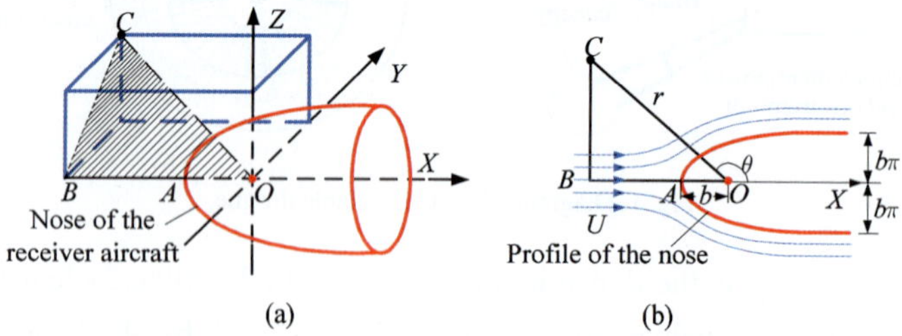

Fig. 3. Diagram of the Rankine half body model.

As shown in Fig. 3(b). The flow field may be treated as a point source which intensity is $Q\,(Q>0)$ seated at the origin of coordinates superimposed on the uniform flow which flow velocity is U and direction is positive of axis X.

the stream function of the flow around a Rankine half body may be expressed as

$$\psi = Ur\sin\theta + \frac{Q}{2\pi}\theta \tag{3}$$

By this stream function, the flow velocity component in polar coordinates may be expressed as

$$v_r = \frac{1}{r}\frac{\partial\psi}{\partial\theta} = U\cos\theta + \frac{Q}{2\pi r},\; v_\theta = -\frac{\partial\psi}{\partial r} = -U\sin\theta \tag{4}$$

That is the flow velocity component v_r, v_θ of bow wave in polar coordinates at any point on plane *OBC* could be solved by the nose radius and the speed of the receiver aircraft [7].

The radial velocity vector and circumferential velocity vector are as follows

$$\boldsymbol{v_r} = v_r cos\theta\, \boldsymbol{i} + v_r sin\theta\, \boldsymbol{j}\,, \boldsymbol{v_\theta} = -v_\theta\, sin\theta\, \boldsymbol{i} + v_\theta\, cos\theta\, \boldsymbol{j} \tag{5}$$

In Eq. (5) $\boldsymbol{i}$ and $\boldsymbol{j}$ are the orthonormal vector in the rectangular coordinate system. The flow velocity component of bow wave in rectangular coordinates on plane OBC are

$$v_x = v_r cos\theta - v_\theta\, sin\theta \ , \ v_y = v_r sin\theta + v_\theta\, cos\theta \tag{6}$$

The three velocity component of the bow wave of C point in Fig. 3(a) are

$$vbow_x = v_x\,, vbow_y = v_y \frac{y}{\sqrt{y^2+z^2}}\,, vbow_z = v_y \frac{z}{\sqrt{y^2+z^2}} \tag{7}$$

The bow wave of the F-16 receiver aircraft is simulated. The velocity of F-16 is assumed to be $100\,m/s$ and the maximum radius of the nose of F-16 is assumed to be $0.4\,m$, as shown in Fig.4 which is the section plan along the symmetry axis of the nose of the receiver aircraft and shows that the Rankine half body model could simulate the bow wave of the receiver aircraft approximatively.

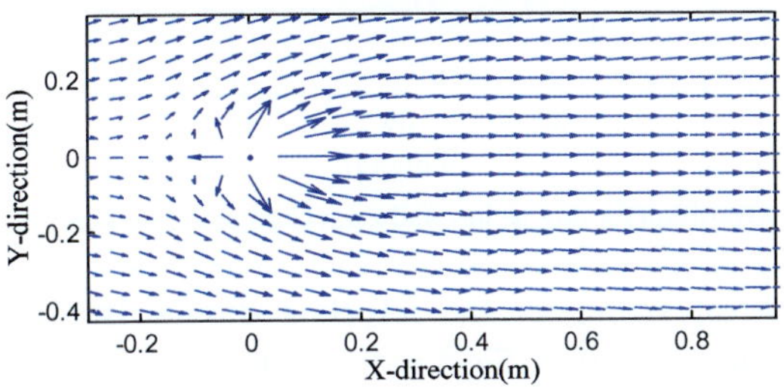

Fig. 4. The diagram of the result of the bow wave.

4. Simulation of the Shaking Motion of the Hose-Drogue

The example of a kind of tanker is taken to simulate the shaking motion of the hose-drogue. The weight is 60000 *kg*, the wing span is 30 *m*, the length of the refueling hose is 14.4 m, the external diameter is 0.066 *m*, the whole weight of the hose is 80 *kg*, the diameter of the drogue is 0.61 *m*, and the weight of the drogue is 29 *kg*. The value of N is 20 and the towing point is selected beneath the half of the right wing.

Assume that the probe of F-16 is aim at the center of the refueling drogue horizontally. Here the interference source is just the bow wave(without atmospheric turbulence). The whole process of simulation last 60 *s*. And the

F-16 receiver aircraft waits at the refueling drogue 25 m behind in the first 5 s, then the F-16 receiver aircraft approaches the refueling drogue with the close speeds of 0.6 m/s, 1.5 m/s, 2.1 m/s, 3 m/s relative to the tanker horizontally in the last 55 s. When the probe of F-16 receiver aircraft arrives at the initial position of the refueling drogue, the receiver aircraft and the tanker will hold the relative position, as shown in Fig. 5.

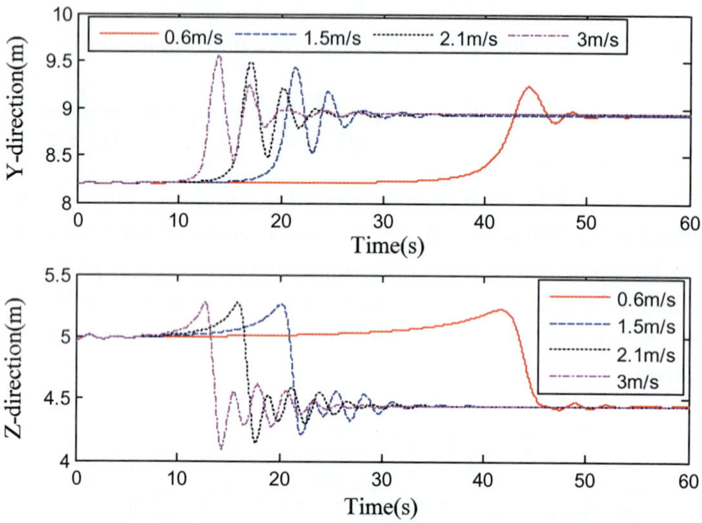

Fig. 5. The relative position for drogue in y and z direction with four kinds of close speeds.

The amplitude of swing for the refueling drogue is about 1.5 m in y direction and 1 m in z direction. Fig.5 indicates that when the refueling drogue approaches the nose of the receiver aircraft, the refueling drogue will be pushed out of the nose by the bow wave of receiver aircraft. But due to the dynamic effects of the hose-drogue, the drogue will swing back quickly. This bow wave effect increases with the close speed increase. If the receiver aircraft comes in too slowly, the probe will miss the drogue due to the swing of refueling drogue. On the other hand, if the receiver aircraft comes in too fast, the probe will lock into the refueling drogue with an impulsive impact, and as a consequence, a violent hose whip [9,10] may occur, resulting in a catastrophic rupturing of the hose or a broken probe. In either case, events such as these are undesirable.

5. Control of the Shaking Motion of the Hose-Drogue

5.1. *Structure of the control system*

As has been noted in [5], the deviation is the error between the drogue position relative to the tanker and the reference position. Then the error signal input the LQR controller to control the servomotor which could drive the strut braces to expand or contact the angle between the leading and trailing strut arms for manipulating the canopy, changing the aerodynamic force on the drogue for realizing the control of the drogue position relative to the tanker. Considering the flight condition that the flight speed of the tanker is $100\,m/s$, flight altitude is $3000\,m$, the angle of attack and pitch angle are 1.5°, the tanker flies at the level of the cruise condition, the technology of actively stabilized drogue refueling system (ASDRS) is adopted as shown in Fig. 6.

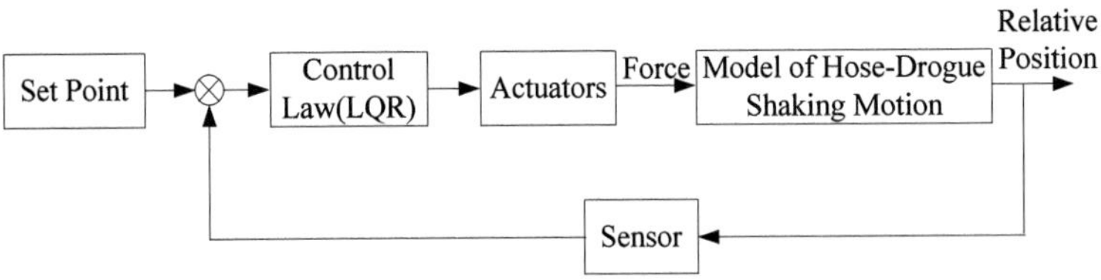

Fig. 6. Block diagram of Hose-Drogue shaking motion control system.

5.2. *Design for the LQR controller*

In the model of the controllable drogue for the hose –drogue in section 2, all four actuators are treated as inputs u_{act} to manipulate the canopy. As mentioned above, there are 20 lumped masses on the hose-drogue, and θ_{K1}, θ_{K2}, $\dot{\theta}_{K1}$, $\dot{\theta}_{K2}$ ($K = 1 \sim 20$, totality is 80) are chosen to be treated as the states of the system. The outputs are the position of the drogue in y-axis, the position of the drogue in z-axis, the speed of the drogue in y-axis, and the speed of the drogue in z-axis (totality is 4) under the tanker path coordinate frame. The no-linear model is linearized and reduced to a 4-order incremental system which may be expressed as

$$\dot{\boldsymbol{x}} = \boldsymbol{A}_r \boldsymbol{x} + \boldsymbol{B}_r \boldsymbol{u} \quad y = \boldsymbol{C}_r \boldsymbol{x} \tag{8}$$

In Eq. (8) $\boldsymbol{A}_r$ is system matrix of the reduced order system. $\boldsymbol{B}_r$ is input matrix of the reduced order system. In order to make the states measurable to design LQR controller conveniently, let $\boldsymbol{x} = \boldsymbol{C}_r^{-1}\boldsymbol{y}$ and substitute in Eq. (8). Then

$$\dot{y} = \boldsymbol{C}_r \boldsymbol{A}_r \boldsymbol{C}_r^{-1} y + \boldsymbol{C}_r \boldsymbol{B}_r u \quad y = \boldsymbol{I}y \tag{9}$$

That is

$$\dot{\tilde{x}} = \tilde{A}\tilde{x} + \tilde{B}u \quad y = I\tilde{x} \tag{10}$$

In Eq. (10) $\boldsymbol{I}$ is 4-order identity matrix. The state $\tilde{x}$ consists of the error Δy which is between the drogue position relative to the tanker in y direction and the reference position in y direction under the tanker path coordinate frame and the error Δz ,and the speed of the drogue v_y in y direction under the tanker path coordinate frame, and the speed v_z . To eliminate the deviation with LQR better, let $\dot{z} = \tilde{x}$ and the system is constructed as follows

$$\begin{bmatrix} \dot{\tilde{x}} \\ \dot{z} \end{bmatrix} = \begin{bmatrix} \tilde{A} & O \\ I & O \end{bmatrix} \begin{bmatrix} \tilde{x} \\ z \end{bmatrix} + \begin{bmatrix} \tilde{B} \\ O \end{bmatrix} u \quad \tilde{y} = \begin{bmatrix} I & O \\ O & I \end{bmatrix} \begin{bmatrix} \tilde{x} \\ z \end{bmatrix} \tag{11}$$

In Eq. (11) $\boldsymbol{O}$ is 4-order zero matrix.

Using the linearized dynamics, an LQR is designed to track relative position. The LQR chooses the control that minimizes the cost criterion

$$J = \frac{1}{2}\int_0^{\infty} \left\{ \begin{bmatrix} \tilde{x}^T & z^T \end{bmatrix} Q \begin{bmatrix} \tilde{x} \\ z \end{bmatrix} + u^T R u \right\} dt \tag{12}$$

Subject to the linear dynamic constraints Eq.(11).By the optimal control theory [11], in linear time invariant system, to make the J minimum, $\boldsymbol{Q}$ is weight matrix, and $\boldsymbol{R}$ is weight matrix. The optimal input should be

$$u^* = -R^{-1}B^T P \begin{bmatrix} \tilde{x} \\ z \end{bmatrix} \tag{13}$$

In which $P = P^T > 0$ and satisfies the Riccati Equation.

5.3. *Simulation of the control for the hose-drogue shaking motion*

The LQR controller designed is applied into the no-linear model of the hose-drogue shaking motion. A Dryden atmospheric turbulence [5] model for 60 s is added into the model of the hose-drogue shaking motion, at meantime, 1.5 m/s of close speed for the contact is considered by adding the bow wave model of the receiver aircraft into the model of the hose-drogue shaking motion. The F-16 receiver aircraft waits at the refueling drogue 25 m behind in the first 5 s , then the F-16 receiver aircraft approaches the refueling drogue with the close speed of 1.5 m/s relative to the tanker horizontally in the last 55 s. And the probe of F-16 is aim at the center of the refueling drogue horizontally. When the probe of F-16 receiver aircraft arrives at the initial position of the refueling drogue, the receiver aircraft and the tanker will hold the relative position. The whole process is to simulate the condition that the receiver aircraft approaches the refueling drogue influenced by the bow wave of receiver aircraft, the atmospheric turbulence in the air. The effectiveness of the control to the shaking motion of the hose-drogue is verified, as shown in Fig. 7. From Fig. 7 the trajectory of the refueling drogue is in confusion under the effect of bow wave and atmospheric turbulence without ASDRS, and the amplitude of swing is about 1.5 m which increases the difficulty of coupling with a refueling drogue for a receiver aircraft safely and accurately. With the ASDRS, the amplitude of swing is greatly reduced and the amplitude of swing is about 10 cm as shown in Fig. 7 which meets the requirements [12] of the aerial refueling technology. From Fig. 8, with the ASDRS, the trajectory of the refueling drogue is more focused which decreases the difficulty of coupling with a refueling drogue for a receiver aircraft safely and accurately.

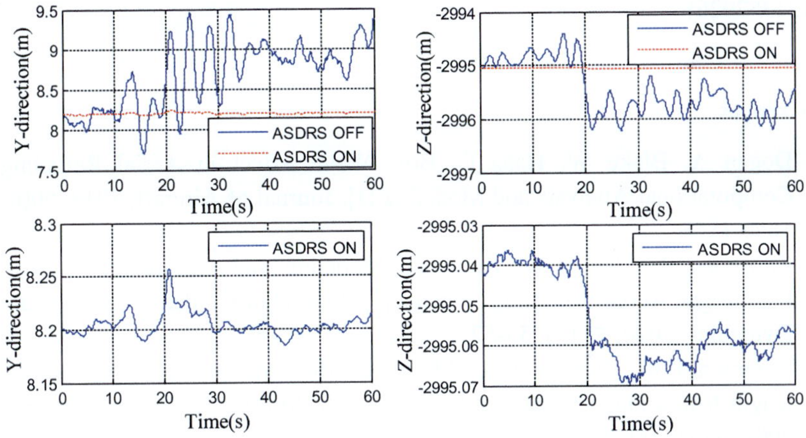

Fig. 7. The effectiveness of the controllable drogue under the bow wave and atmospheric turbulence.

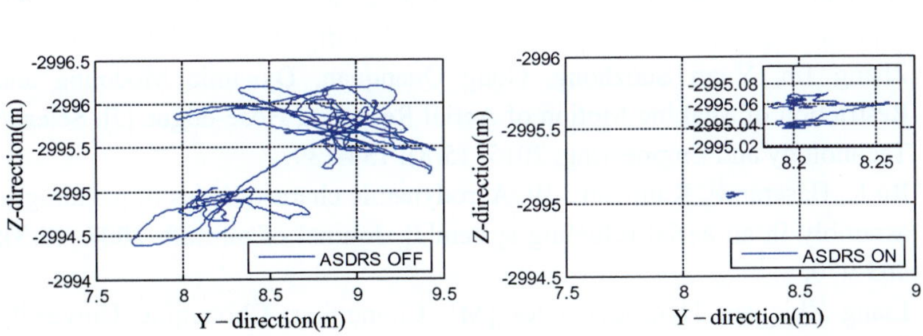

(a) (b)

Fig. 8. The trajectory of the drogue in rear view.

6. Conclusion

A mathematical model of shaking motion of aerial refueling hose-drogue is presented by adding the concept of controllable drogue to the model of rigid multibody dynamics. And the bow wave effect is analyzed by the model of Rankine half body model. The LQR controller designed shows the effectiveness of the control to the shaking motion of the hose-drogue due to the bow wave effect and the atmospheric turbulence.

Acknowledgments

This work is supported by Jin Zhang, Suo-Zhong Yuan, Quan-Quan Gong. Work partially supported by the National Natural Science Foundation of China (No. 61273050) and the Aeronautical Science Foundation of China (No. 20121352026).

References

1. Dogan A, Blake W, Haag C. Bow Wave Effect in Aerial Refueling: Computational Analysis and Modeling [J]. Journal of Aircraft, 2013, 50(6): 1856-1868.
2. Kapseong Ro, James W. Kamman. Modeling and Simulation of Hose-Paradrogue Aerial Refueling Systems [J]. Journal of Guidance, Control, and Dynamics, 2010, 33 (1): 53-57.
3. Kamman J W, Hustopn R L. Multibody dynamics modeling of variable length towed and tethered cable systems [J]. Multibody System Dynamic, 2001, 5 (3): 211-221.
4. Williamson W R, Reed E, Glenn G J, et al. Controllable drogue for automated aerial refueling [J]. Journal of Aircraft, 2010, 47 (2): 515-527.
5. Zhang Jin, Yuan Suozhong, Gong Quanquan. Dynamic Modeling and Restriction of Shaking Motion of Aerial Refueling Hose-dogue [J]. Science Technology and Engineering, 2015, 15(8): 134-139.
6. Ro K, Basaran E, Kamman J W. Aerodynamic characteristics of paradrogue assembly in an aerial refueling system[J]. Journal of aircraft, 2007, 44(3): 963-970.
7. Liang Zhiquan. Hydrodynamics [M]. Chongqing: Chongqing University Press, 2002: 112-121.

8. Bhandari U, Thomas P R, Bullock S, et al. Bow Wave Effect in Probe and Drogue Aerial Refueling [C]//AIAA Guidance, Navigation, and Control Conference. University of Bath, 2013.
9. Vassberg J C, Yeh D T, Blair A J, et al. Numerical simulations of KC-10 in-flight refueling hose-drogue dynamics with an approaching F/A-18D receiver aircraft[C]//23rd Applied Aerodynamics Conference and Exhibit, Toronto, Ontario, Canada, AIAA. 2005, 4605.
10. Vassberg J C, Yeh D T, Blair A J, et al. Dynamic characteristics of a KC-10 wing-pod refueling hose by numerical simulation[C]//20th AIAA Applied Aerodynamics Conference. 2002.
11. Hu Shousong, Wang Zhiquan, Hu Weili. Optimal control theory and system [M]. Beijing: Science Press, 2005: 162-180.
12. Kapseong Ro, Taeseung Kuk, James W. Kamman. Active Control of Aerial Refueling Hose-Drogue Systems[C]//AIAA Guidance, Navigation, and Control Conference. Toronto, Ontario Canada, 2-5August, 2010: 9-10.

Design of explicit model-following flight control system based on LQR for unmanned helicopter

Liang Xue*, Sai-Fei Wu, Sen Jia and Xin-Hua Wang
College of Automation Engineering,
Nanjing University of Aeronautics and Astronautics,
**E-mail: lightxue11@163.com*

Liang XUE
Flight Control Laboratory, 29 Jiangjun Road Nanjing,
Nanjing, Province of JiangSu, China
E-mail: lightxue11@163.com

The small scale unmanned helicopters with the characteristics of unstable and strong coupling are difficult to control. To solve the problem, a control method of explicit model-following based on LQR was proposed in this paper. Then a flight control system for small-scale helicopters was designed on the base of the method. The explicit model, control matrix and LQR feedback control matrix of the control system were described in detail. The simulation results showed that the proposed flight control system had good dynamic tracking performance and decoupling capability.

Keywords: Small unmanned helicopter; flight control system; model-following; LQR.

1. Introduction

Among variety of Unmanned Air Vehicles (UAVs), unmanned helicopters is one of the most versatile and agile platforms. A helicopter can operate in different flight modes, such as vertical take-off/landing, hovering, longitudinal/lateral flight. Owing to these natural advantages, they are widely used in military and civil domains[1,2].

However, the helicopter systems are characterized by the highly nonlinear coupling effect between the rotational moments and the translational accelerations. Moreover, a small-scale helicopter is more difficult to control than its full scale counterparts because it is more sensitive to control inputs and it has faster dynamic response and stronger coupling[3]. How to effectively reduce the coupling and achieve the best control performance are important issues in the control of the autonomous helicopters. In literature [4], it uses $H\infty$ theory to reduce coupling. But the effect is not satisfying and the tracking performance is bad. In literature [5], the explicit model tracking control is used to decouple. And the system has decoupling and tracking ability. However, the effect is not as best as expected.

In this paper, the explicit model tracking control method based on LQR was presented, which combined the characteristics of LQR optimal control and decoupling of explicit model following control. A flight control system for the small scale unmanned helicopter was designed, which made each channels have better dynamic tracking and decoupling performance.

2. Helicopter Dynamic Model

The crucial issue of small scale unmanned helicopter flight control system is dynamic model. Among other UAV, unmanned helicopter has specific characteristic. Due to the complex nature and physical structure of the helicopter system, the precise mathematic model is quite difficult to obtain[6,7]. The unmanned helicopter studied in this paper is a small scale unmanned helicopter, the state equation of the linearized dynamic model is

$$\begin{cases} \Delta\dot{X} = A\Delta X + B\Delta U \\ Y = C\Delta X \end{cases} \tag{1}$$

where, $\Delta X = \left[\Delta u\ \Delta v\ \Delta w\ \Delta\theta\ \Delta\varphi\ \Delta\psi\ \Delta q\ \Delta p\ \Delta r\right]^T$ is the state vector of the helicopter. Respectively, $[\Delta u\ \Delta v\ \Delta w]$ is the increment of the translational velocities of the helicopter in the trim points, $[\Delta\theta\ \Delta\varphi\ \Delta\psi]$ is the increment of the Euler angles describing helicopter attitudes in the trim points, and $[\Delta q\ \Delta p\ \Delta r]$ is the increment of the angular velocity vector in the trim points. $\Delta U = [\Delta u_e\ \Delta u_a\ \Delta u_r\ \Delta u_c]^T$ is the increment of the input of the helicopter in trim points. It represents longitudinal cyclic pitch, lateral cyclic pitch, tail rotor pitch and collective respectively. A is the system dynamic state matrix of 9*9; B is the control matrix of 9*4. C is the output matrix of 9*4.

3. Flight Control System Design

In this section, the explicit model-following controller based on LQR was proposed for the full dynamic model of a small-scale helicopter. And the implementation details of the flight control system were described. The flight control system designed in this paper can be shown as follows.

3.1. *Flight control system*

The control method of flight control system is explicit model-following control based on LQR, which controls the attitude angle directly. The purpose of the system is to make each channel has a better dynamic tracking and excellent decoupling performance. Its structure is shown in Figure 1.

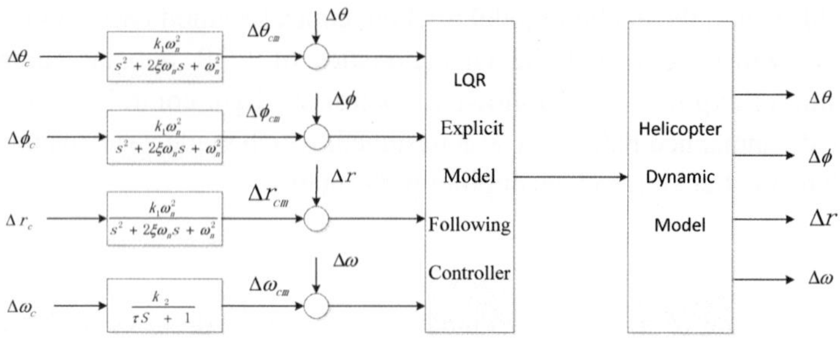

Fig. 1. The structure of flight control system.

3.2. *LQR explicit model-following controller design*

The concrete structure of LQR explicit model-following controller is shown as Figure 2.

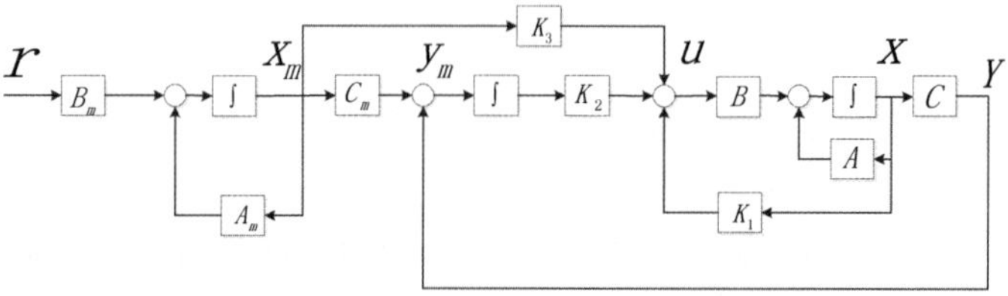

Fig. 2. LQR explicit model-following controller.

Assume that the state equation of the explicit model to track is

$$\begin{aligned} \dot{x}_m &= A_m x_m + B_m r \\ y_m &= C_m x_m \end{aligned} \tag{2}$$

where, the value of A_m , B_m depend on the requirements of the response quality for attitude angle. R represents the instruction signal of pitch angle, roll angle, yaw rate and the collective pitch of helicopter, $r = [\Delta\theta_c\ \Delta\varphi_c\ \Delta r_c\ \Delta\omega_c]^T$.

Then, LQR controller is designed. The requirement of LQR controller is to make the system stable, which can control output $y_{(t)}$ of the plant dynamically to track the explicit model $y_{m(t)}$ respectively and make steady-state error of tracking must be zero.

First of all, the output error is defined by

$$e(t) \quad y_{m.}(t) \quad y(t) \tag{3}$$

In order to improve the tracking accuracy of the system, the integral of steady-state error is introduced.

Let,

$$\dot{z}(t) = e(t) = -Cx + C_m x_m \tag{4}$$

From (2), (3) and (4), we can obtain equation (5).

$$\begin{bmatrix} \dot{x} \\ \dot{z} \\ \dot{x}_m \end{bmatrix} = \begin{bmatrix} A & 0 & 0 \\ -C & 0 & C_m \\ 0 & 0 & A_m \end{bmatrix} \begin{bmatrix} x \\ z \\ x_m \end{bmatrix} + \begin{bmatrix} B \\ 0 \\ 0 \end{bmatrix} u \tag{5}$$

The equation (5) can be simplified as equation (6).

$$\dot{X} = A_L X + B_L u \tag{6}$$

For equation (6), if (A, B) matrix was controllable, we could design control matrix by using feedback of LQR state. The function is defined as equation (7).

$$J = \frac{1}{2} \int (X^T Q X + u^T R u) dt \tag{7}$$

Therefore, the feedback control will be defined as

$$u = KX \tag{8}$$

where, the gain matrix K of feedback is

$$K = -R^{-1} B^T P \tag{9}$$

In the equation (9), P was the solution of the algebraic Riccati equation (10).

$$A^T P + PA + Q - PBR^{-1} B^T P = 0 \tag{10}$$

where, $Q \geq 0, P \geq 0$.

From equation (5) and (8), u can be written in another form as below.

$$u = KX = K_1 x + K_2 z + K_3 x_m = K_1 x + K_2 \int e(t) dt + K_3 x_m \tag{11}$$

According to equations(8)~(11), the control matrix can be obtain. And the design of LQR explicit model-following controller is completed.

The explicit model of pitch, roll, and yaw rate are two order linear dynamic models. The form is $k_1\omega_n^2/(s^2+2\xi\omega_n s+\omega_n^2)$. The model of collective pitch is $k_2/(\tau s+1)$.

3.3. *Performance evaluation*

To analyze the performance of explicit model-following control system based on LQR quantitatively, we propose performance criteria of system. The performance of dynamic tracking and decoupling of system are considered mainly.

First, the tracking performance criterion is defined. When the step signal can be added as any input of the four channels to the system, the error between state variables of the model and the value of actual helicopter state is obtained by calculation. During N time units, we can get: $\sum_{i=1}^{N}|e_i|$.

e_{ii} is denoted as the sensitivity coefficient of model, which means the steady-state output of the explicit model when the input of the corresponding control channel is set as step signal. The average of the error vector during the working time is as below.

$$|\bar{e}|=\frac{1}{e_{ii}N}\sum_{i=1}^{N}|e_i| \tag{12}$$

where, generally $N \le t_S/T$ and t_S is the regulating time, which could be computed approximatively by $t_S = 4.4/\omega_n$.

The performance of model tracking is defined as:

$$Q=1-|\bar{e}|$$

Then, the decoupling performance criterion is defined. In a similar way, the performance of decoupling of each channel is evaluated. For example, when roll channel with an input of step signal, the decoupling performance of pitch channel for roll input is defined as:

$$Q_{\theta_\varphi}=1-\frac{1}{e_{11}(20)}\sum_{i=1}^{20}|e_{\theta_i}| \tag{13}$$

In the formula (12), the cumulative error is calculated during roll input with step signal. This error is the output error of pitch channel due to coupling.

4. Simulation Results and Discussion

In this section, numerical simulation results were presented to investigate the performance of the proposed LQR explicit model-following controller of small scale unmanned helicopter.

The mass of helicopter studied in this paper is 10kg. The helicopter has a single electric rotor and a electric tail rotor. The A, B, C of equation (1) for the helicopter are shown as below.

In the simulation, we respectively suppose the step instruction signal as desired input signal of one of channels, and the input of other channels will be zero, which is in order to observe the decoupling and tracking performance of the system. The simulation curves are shown as below.

The transfer function of model for each channels are shown in table1, which are obtained from nature characteristics of the helicopter.

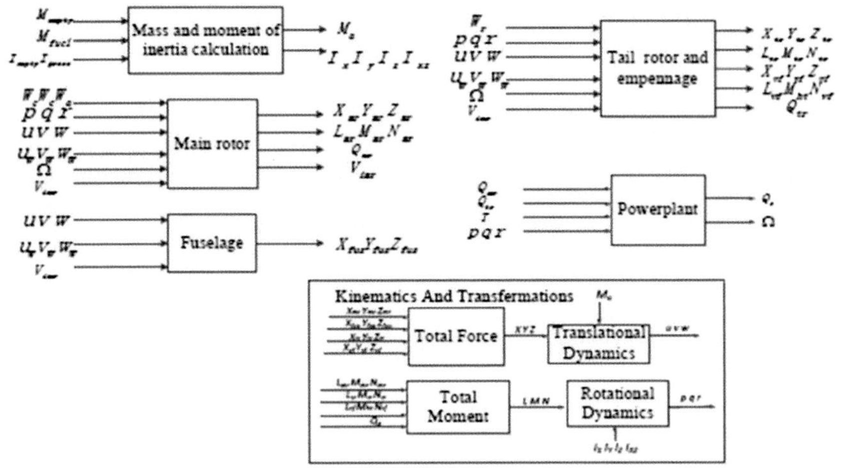

Fig. 3. Structure diagram of comprehensive nonlinear mathematical model for helicopter.

$$A = \begin{bmatrix} -0.0160 & -0.0000 & -0.0086 & -9.7759 & 0 & 0 & -0.0007 & 0 & 0.0038 \\ -0.0011 & -0.0643 & 0.0062 & 0.0009 & 9.7573 & 0 & 0.0010 & -0.0017 & -0.5840 \\ -0.3370 & -0.0013 & -0.9234 & 0.0149 & -0.6023 & 0 & 1.0041 & -0.0038 & 0 \\ 0 & 0 & 0 & 0.0000 & 0 & 0 & 0.9981 & 0 & -0.0616 \\ 0 & 0 & 0 & 0.0000 & -0.0000 & 0 & -0.0001 & 1.0000 & -0.0015 \\ 0 & 0 & 0 & -0.0000 & 0.0000 & 0 & 0.0616 & 0 & 0.9981 \\ -0.0260 & -0.0000 & 0.0073 & 0 & 0 & 0 & 0.0063 & -0.0000 & 0.0000 \\ 0.0113 & -0.1259 & 0.0642 & 0 & 0 & 0 & 0.0038 & -0.0094 & 1.6137 \\ -0.0900 & -0.9236 & -0.0242 & 0 & 0 & 0 & -0.0276 & 0.0688 & -11.8282 \end{bmatrix}$$

$$B=\begin{bmatrix} 0 & 0 & 0 & -0.0180 \\ 0 & 0 & -1.5600 & 0.1547 \\ 0 & 0 & 0 & -24.8950 \\ 0 & 0 & 0 & 0 \\ 0 & 0 & 0 & 0 \\ 0 & 0 & 0 & 0 \\ 5.48 & 0 & 0 & 0.0096 \\ 0 & 0.83 & -5.9961 & 1.7485 \\ 0 & 0 & 43.9508 & -0.0026 \end{bmatrix}, \quad C=\begin{bmatrix} 1&0&0&0&0&0&0&0&0 \\ 0&1&0&0&0&0&0&0&0 \\ 0&0&1&0&0&0&0&0&0 \\ 0&0&0&1&0&0&0&0&0 \\ 0&0&0&0&1&0&0&0&0 \\ 0&0&0&0&0&1&0&0&0 \\ 0&0&0&0&0&0&1&0&0 \\ 0&0&0&0&0&0&0&1&0 \\ 0&0&0&0&0&0&0&0&1 \end{bmatrix}$$

Table 1. Transfer function of model for each channel.

Channel	Transfer function
Pitch	$\frac{54}{s^2+4.2s+9}$
Roll	$\frac{108}{s^2+4.2s+9}$
Yaw Rate	$\frac{250}{s^2+7s+25}$
Collective Pitch	$\frac{-2}{0.25s+1}$

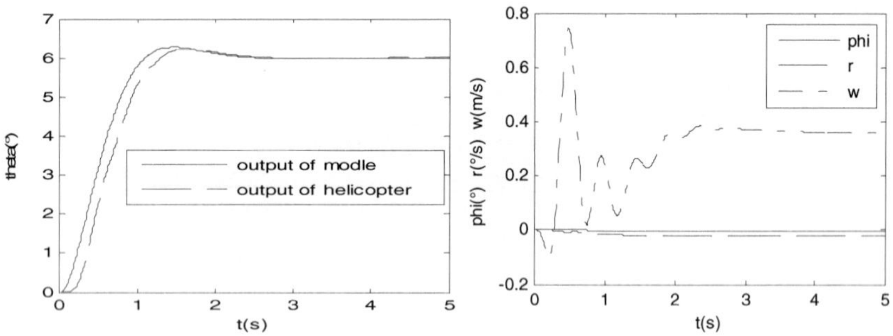

Fig. 4. The response of pitch channel with input of step signal ($\Delta\theta_c=1, \Delta\varphi_c=0, \Delta r_c=0, \Delta\omega_c=0$).

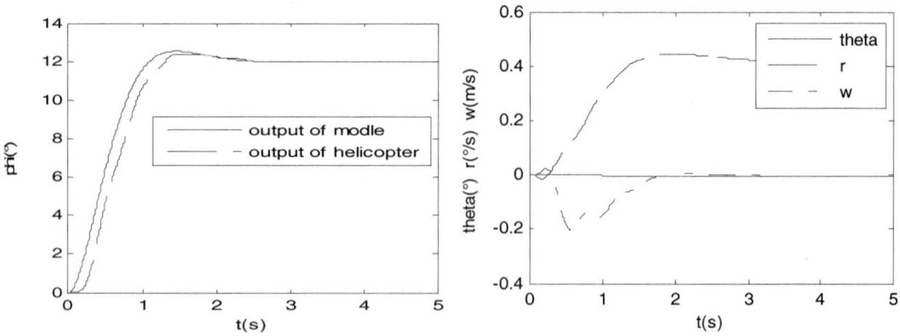

Fig. 5. The response of roll channel with input of step signal ($\Delta\theta_c=0, \Delta\varphi_c=1, \Delta r_c=0, \Delta\omega_c=0$).

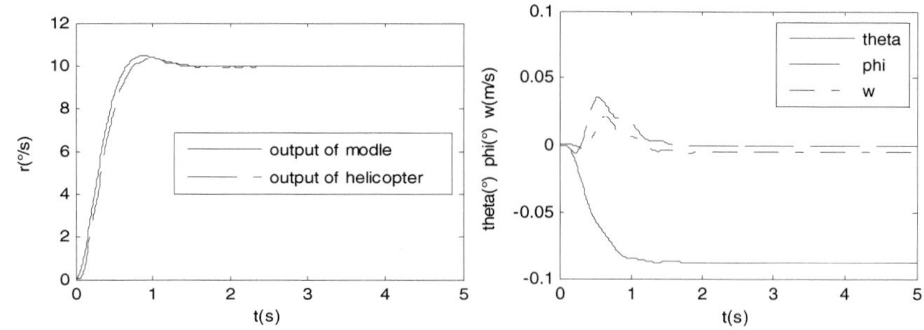

Fig. 6. The response of yaw rate with input of step signal ($\Delta\theta_c=0, \Delta\varphi_c=0, \Delta r_c=1, \Delta\omega_c=0$).

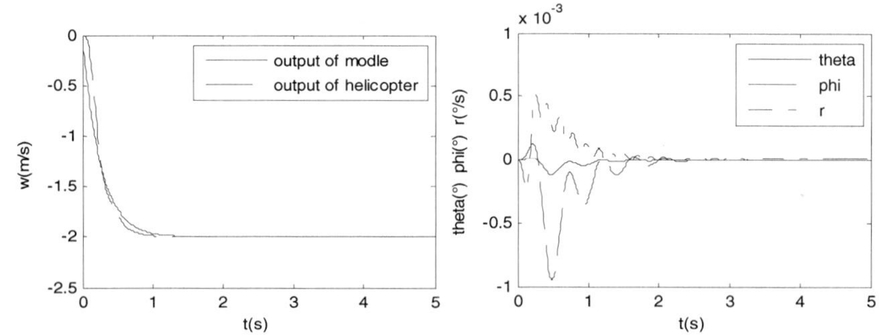

Fig. 7. The response of collective pitch channel with input of step signal ($\Delta\theta_c=0, \Delta\varphi_c=0, \Delta r_c=0, \Delta\omega_c=1$).

Table 2. The evaluation of tracking and decoupling performance of each channel.

	Q_θ	Q_ϕ	Q_r	Q_w
$\Delta\theta_c$	0.9120	0.9995	0.9975	0.9642
$\Delta\varphi_c$	0.9989	0.9021	0.9736	0.9885
Δr_c	0.9844	0.9936	0.9664	0.9996
$\Delta\omega_c$	0.9987	0.9934	0.9951	0.9867

From the simulation curves, we can see that the outputs of the pitch, roll, yaw rate and collective pitch four channels track the output of explicit model well. When one of channels input is set as step signal while the others are set to zero, the output of other channels changed slightly and wouldn't be affected.

In Table 2, for example, if the input of system is $\Delta\theta_c$, the cross term of Q_θ and $\Delta\theta_c$ represents tracking performance and other terms show the decoupling degree. The larger the value in Table 2 is, the better tracking and decoupling performance the system can have.

Therefore, from simulation curves and evaluations of tracking and decoupling performance of each channel, we could know that the system is better in tracking and the decoupling.

5. Conclusions

An explicit model-following control method based on LQR was proposed to solve the strong coupling and difficult controlling of small scale helicopter in the paper. The results shown this flight control system of helicopter designed had good dynamic tracking performance. The most important thing was the decoupling performance of system was improved. As a future work we will implement this control method in a real helicopter. A new vehicle is being built, which will include appropriate control hardware to compute control signals.

References

1. G. Cai, B. Chen, K. Peng, M. Dong, and T. H. Lee, Modeling and control of the yaw channel of a UAV helicopter[J]. IEEE.
2. H. Liu, Y. Yu, G. Lu, and Y. Zhong, Robust LQR attitude control of a 3DOF lab helicopter[C]. in Proc. 29th Chin. Control Conf., Beijing, China, 2010, pp. 529–534.
3. Wang Hong-qiang, Wang Da-bo, Duan Hai-bin. Robust Tracking Decoupled Flight Control of an Unmanned Helicopter[J]. Mechanical Science and Technology for Aerospace Engineering. April Vol. 282009 No. 4. C. Jarlskog, *CP Violation* (World Scientific, Singapore, 1988).
4. WANG Yong, GAO Jin-yuan. Decoupling control of helicopter basing on H∞ theory. FLIGHT DYNAMICS[J]. Vol. 18 No. 3.Sep. 2000.
5. ZHENG Feng-ying, YANG Yi-dong. Flight control system design of explicit model-following for helicopter based on control distribution matrix decouple[J]. JOURNAL OF NAVAL AERONAUTICAL ENGINEERING INSTITUTE. Vol. 22 No.1. Jan. 2007.
6. ZENG Li-lan, WANG Dao-bo, FAN Chun-xia. An Unmanned Helicopter Decoupling Control Based on the Fuzzy PID Control Method[J]. JOURNAL OF APPLIED SCIENCES. Vol. 22, No. 4 Dec. 2004.
7. Yang Yi-dong. Helicopter Flight Control[M]. National Defense Industry Press, 2011.

Static analysis on the serial-parallel leg of a novel quadruped robot

Jian-She Gao*, Ming-Xiang LI† and Bao-Tang Wang‡
School of Mechanical Engineering, Zhengzhou University, Zhengzhou, Henan, China, 450001
*E-mail: *gao_jianshe@163.com, †li_mxiang@163.com, ‡wangbaotangzzumech@163.com*

A novel serial-parallel hybrid quadruped robot is presented to solve the problem that the robot based on the directly serial leg mechanism has great weight/load ratio and weak static performance, and the static analysis on the serial-parallel leg of the robot has been carried out. Considering the mapping of constraint and actuation forces, the statics equilibrium equation was established based on the vector theory and virtual work principle. Global stiffness index was defined, which can measure quantitatively the deformation within the workspace of the leg. Giving out the parameters of the hybrid leg, according to the static model, force and torque curves of hybrid leg's four drive motors were drawn by MATLAB when its foot moved along a given trajectory with 100 N force exerted upon it along three different axes. Result of the simulation shows that the hybrid leg has best carrying capacity of force in the x-axis direction, while in z-axis direction is the worst. This research is of great significance to a series of further studies upon dynamic properties analysis, mechanisms optimal design and system control of the novel serial-parallel quadruped walking robots.

Keywords: Serial-parallel; static analysis; global stiffness index.

1. Introduction

Walking robots can adjust walking posture flexibly and choose supporting points in the reachable domain of toe points properly due to its footholds are discrete, and thus they have higher abilities of obstacle avoidance and obstacle crossing [1]. Among all the walking robots, quadruped walking robots are much more stable than two-legged robots and simpler structure than hexapod robots, which makes it become the focus on the field of robots researches at present [2, 3].

With the in-depth research into the application of quadruped robot on special topography, people put forward higher requirements for the speed, maneuverability and adaptability, as a consequence, a robot characterized by high speed, strong bearing, low energy consumption and self-bring power source has become the research emphasis in quadruped robot field [4]. While leg of quadruped robot, as the most important fundamental motion component, play a decisive role in the kinematic, dynamic and static characteristics of quadruped robot [5], therefore, leg design has become an important influential factor of quadruped robot. At present, quadruped robot's leg adopts either serial structure [6, 7] or parallel one [8], for instance, the famous BigDog [6] designed by Boston Dynamics. However, these two kinds of structure have their limitations due to their structure features. Serial structure leg has simple structure, easy control, but it has high weight-load ratio and low bearing capacity [9]. On the contrary, parallel structure leg has low weight-load ratio, high bearing capacity and high stability, but it has complex structure and small workspace. In order to break through their limitations, researchers began to turn their attention to serial-parallel hybrid leg structure and great advancements have been achieved. Zhou Yulin [10] et al. brought hybrid mechanism into humanoid robot firstly and its mechanism configuration was constructed. Tang Yanhua [11] designed a new high-burden robot with hybrid structure, and analyzed its statics and kinematics characters. Tian Xinghua and Gao Feng [12] summarized the advantages and disadvantages of existing robot's leg structure, and three different hybrid mechanism configurations are proposed and analyzed.

To improve static stiffness and static performance of quadruped robot, a novel serial-parallel hybrid leg mechanism was presented. Statics analysis was performed first and statics equilibrium equation was established. Then global stiffness index of hybrid robot was defined based on the result of statics analysis. Finally, according to the static model, force and torque curves of hybrid leg's four drive motors were drawn by MATLAB when its foot moved along a given trajectory with 100 N forces exerted upon it along three different axes through concrete example. These study lay a foundation for further study on the serial-parallel hybrid leg.

2. Mechanism Description

The novel quadruped robot presented in this paper consists of four identical serial-parallel legs and one carrying-platform, as shown in Fig. 1. The hybrid leg shown in Fig. 2 is connected by a hip joint, a thigh and a calf sequentially, and the hip joint is a 3-RRR spherical parallel mechanism. To decrease movement inertia and improve reaction rate and motion stability, one driving motor is mounted beside the fixed platform of the hip. Knee joint is driven by a linear motor with the purpose of decreasing rotational inertia imposed upon knee joint by leg mechanism. Due to the serial-parallel hybrid leg mechanism, the quadruped robot has both big workspace of serial robot and high bearing capacities of parallel robot at the same time.

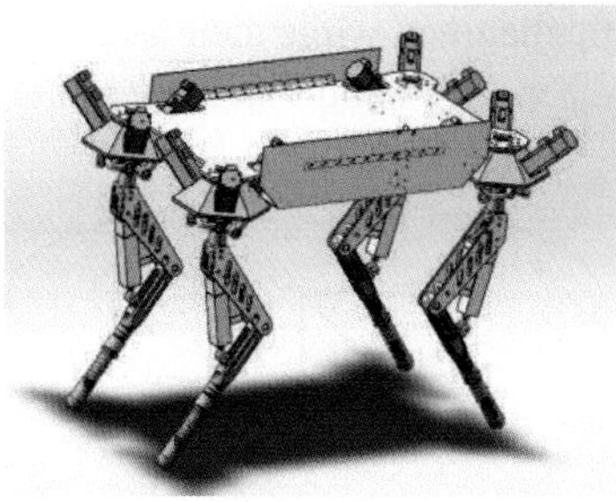

Fig. 1. A novel quadruped walking robot.

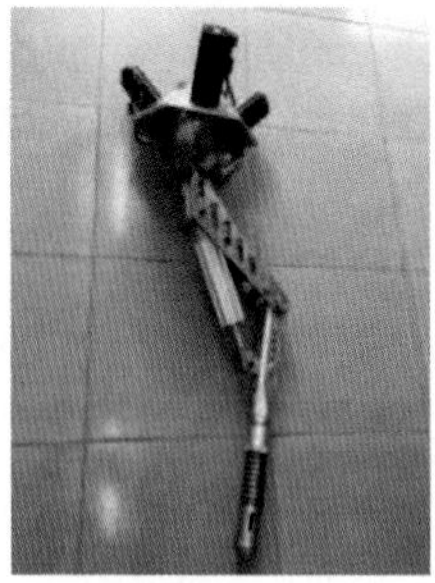

Fig. 2. Leg prototype.

3. Static Analysis

Viewing the hip joint as a spherical joint, we should analyze the mapping relationship between generalized force and torque imposed on moving platform of the hip joint and linear motor first. Then study the mapping relationship between torque vectors exerted on moving platform and driving motors. Finally the statics character of the leg is gained combining the above analysis.

3.1. *Statics analysis on the whole leg*

Viewing the hip joint as a 3-DOF spherical parallel mechanism, the hybrid leg can be seen as a *SR* serial mechanism, as shown in Fig. 3. The length of the thigh and calf are L_1 and L_2 respectively, and the distance from the origin of $\{\boldsymbol{C}\}$ to the hinged joints $\boldsymbol{D}_1$ and $\boldsymbol{D}_2$ are l_1 and l_2 separately. Fixed and moving coordinate systems $\boldsymbol{O}x_0y_0z_0$ and $\boldsymbol{O}xyz$ are established at the rotation center $\boldsymbol{O}$ of the hip joint separately, systems $\boldsymbol{C}x_cy_cz_c$ and $\boldsymbol{P}x_py_pz_p$ are set at rotation axis of knee and foot point $\boldsymbol{P}$ respectively. In system $\boldsymbol{O}x_0y_0z_0$, x_0-axis is perpendicular to the paper and point outside, and z_0-axis point downward. In the initial position, systems $\boldsymbol{O}xyz$ and $\boldsymbol{O}x_0y_0z_0$ share the same posture, and z-axis point from $\boldsymbol{O}$ to $\boldsymbol{C}$. In addition, x_p-axis, x_c-axis and x-axis share the same direction perpendicular to $\boldsymbol{OCP}$ plane, z_p-axis and y_p-axis is in the same direction of vector $\boldsymbol{CP}$ as well. In Fig.3, vector $\boldsymbol{F}$ and $\boldsymbol{M}$ represent the force and torque applied on the hip joint, vector $\boldsymbol{F_p}$ represents the amount external force impose on foot $\boldsymbol{P}$, and vector $\boldsymbol{F_L}$ represents driving force provided by linear motor.

The hip joint is a 3-RRR spherical parallel mechanism, and its coordinate systems are built as shown in Fig.4. Driving motors are installed at the first rotation joint of each branch, $\boldsymbol{O}x_{ij}y_{ij}z_{ij}$ represents the coordinate system of rotation joint j of branch i, $(i, j$=1, 2, 3), and vectors $\boldsymbol{u}_i$, $\boldsymbol{v}_i$ and $\boldsymbol{w}_i$ donate the unit vector of first, second and third revolute joint of branch i respectively. Owing to the entire rotation axis intersecting at point O, imagine each systems is shifted into their centers to simplify the schematic diagram, while their origins are at the center O in fact. Systems $\boldsymbol{Oxyz}$ and $\boldsymbol{Ox_0y_0z_0}$ in Fig. 4 and Fig. 3 are same.

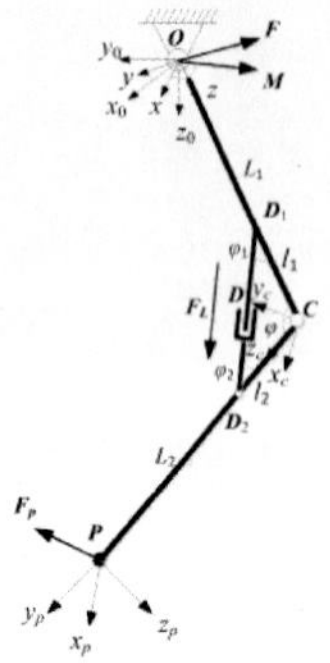

Fig. 3. Static analysis diagram of the leg.

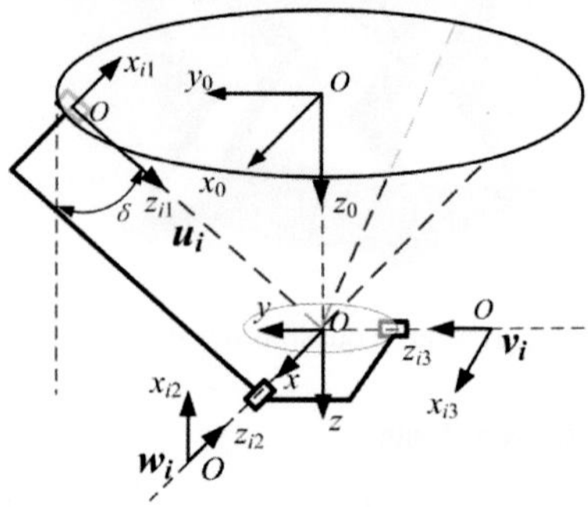

Fig. 4. Coordinated systems of the hip joint.

Through static analysis of the hybrid leg, equation (1) is gained,

$$\begin{cases} \boldsymbol{F} + \boldsymbol{F}_L + \boldsymbol{F}_p = 0 \\ \boldsymbol{M} + (\boldsymbol{L}_1 - \boldsymbol{l}_1) \times \boldsymbol{F}_L + \boldsymbol{OP} \times \boldsymbol{F}_p = 0 \end{cases}. \tag{1}$$

The hip joint has three rotational freedoms, so vector $\boldsymbol{F}$ must be balanced by constraining force of the leg, thus what we need consider is the mapping relationship of vector $\boldsymbol{M}$ and $\boldsymbol{F}_P$, rewritten equation (1) as

$$-\boldsymbol{OP} \times \boldsymbol{F}_p = \boldsymbol{M} + (\boldsymbol{L}_1 - \boldsymbol{l}_1) \times \boldsymbol{F}_L . \tag{2}$$

Let $\boldsymbol{e}_x$, $\boldsymbol{e}_y$, $\boldsymbol{e}_z$, and $\boldsymbol{e}_L$ represent unit vector in the direction of x-axis, y-axis, z-axis and vector $\boldsymbol{F}_L$ respectively, and coordinate value of P is set as (p_x, p_y, p_z). Suppose f_L is the scalar value of vector $\boldsymbol{F}_L$, and vector $\boldsymbol{M}$ is set as $(m_x, m_y, m_z)^{\mathrm{T}}$, then equation (2) can be rewritten as

$$\boldsymbol{F}_p = \boldsymbol{G}_p [\boldsymbol{M} \quad f_L]^{\mathrm{T}}, \tag{3}$$

where, matrix $\boldsymbol{P}$ is the antisymmetric matrix of vector $\boldsymbol{OP}$, and

$$\boldsymbol{G}_p = \boldsymbol{P}^{-1} \left[\boldsymbol{e}_x \quad \boldsymbol{e}_y \quad \boldsymbol{e}_z \quad (\boldsymbol{L}_1 - \boldsymbol{l}_1) \times \boldsymbol{e}_L \right].$$

In equation (3), there are four unknown variables, while it only can be rewritten as three equations, so one constraint equation is needed to establish to get the four unknown variables.

Knee joint $\boldsymbol{C}$ is a driven joint, so torque posed on it in the direction of its rotational axis is zero, so the constraint equation is established as

$$\boldsymbol{F}_p z_p L_2 + f_L \boldsymbol{e}_L \cdot \boldsymbol{l}_2 = |\boldsymbol{M}_c| = 0. \tag{4}$$

Combining equations (3) and (4), the statics equilibrium equation of the SR serial mechanism is established as

$$\begin{bmatrix} \boldsymbol{F}_P & \boldsymbol{F}_P \cdot z_p \end{bmatrix}^{\mathrm{T}} = \boldsymbol{G}_h \begin{bmatrix} \boldsymbol{M} & f_L \end{bmatrix}^{\mathrm{T}}, \tag{5}$$

where,

$$\boldsymbol{G}_h = \begin{bmatrix} \boldsymbol{G}_p & \boldsymbol{G}_F \end{bmatrix}^{\mathrm{T}},$$

$$\boldsymbol{G}_F = \begin{bmatrix} 0 & 0 & 0 & -\dfrac{\boldsymbol{e}_L \cdot \boldsymbol{l}_2}{L_2} \end{bmatrix}.$$

If matrix $\boldsymbol{G}_h$ is invertible, equation (5) can be rewritten as

$$\begin{bmatrix} \boldsymbol{M} & f_L \end{bmatrix}^{\mathrm{T}} = \boldsymbol{G}_h^{-1} \begin{bmatrix} \boldsymbol{F}_P & \boldsymbol{F}_P \cdot z_p \end{bmatrix}^{\mathrm{T}}. \tag{6}$$

3.2. *Statics analysis on the hip joint*

From the structural features of the hip joint, constraint equation of the hip joint is built as

$$\boldsymbol{w}_i \cdot \boldsymbol{v}_i = 0 \quad (i=1, 2, 3). \tag{7}$$

Take a derivative with respect to time of equation (7), and then equation (8) is gained

$$\dot{\boldsymbol{w}}_i \cdot \boldsymbol{v}_i + \boldsymbol{w}_i \cdot \dot{\boldsymbol{v}}_i = 0 \cdot \tag{8}$$

Rotation of $\boldsymbol{v}_1$, $\boldsymbol{v}_2$ and $\boldsymbol{v}_3$ can be seen as a rotation around a same axis with angular velocity $\boldsymbol{\omega}$ because their rotational axes intersect at point O. Differential of vector $\boldsymbol{v}_i$ can be expressed as

$$\dot{\boldsymbol{v}}_i = \boldsymbol{\omega} \times \boldsymbol{v}_i \cdot \tag{9}$$

Differential of vector $\boldsymbol{w}_i$ can be gained as shown in equation (10)

$$\dot{\boldsymbol{w}}_i = \dot{\theta}_i \cdot \boldsymbol{u}_i \times \boldsymbol{w}_i, \tag{10}$$

where, θ_i represents the angular vector of driving motor i.

Combining equations (8), (9) and (10), equation (11) is obtained

$$\dot{\theta}_i \cdot (\boldsymbol{u}_i \times \boldsymbol{w}_i) \cdot \boldsymbol{v}_i = (\boldsymbol{w}_i \times \boldsymbol{v}_i) \cdot \boldsymbol{\omega} \cdot \tag{11}$$

Simplifying equation (11), inverse kinematic solution is gained in equation (12)

$$\dot{\boldsymbol{\theta}} = \left[\frac{(\boldsymbol{w}_i \times \boldsymbol{v}_i)}{(\boldsymbol{u}_i \times \boldsymbol{w}_i) \cdot \boldsymbol{v}_i} \right]^{\mathrm{T}} \cdot \boldsymbol{\omega}\,, \tag{12}$$

Because the hip joint has three rotational freedoms, so vector $\boldsymbol{F}$ applied on the hip joint must be balanced by its constraining force. According to principle of virtual work, all the virtual work of driving forces and generalized forces is equal, and then equation (13) is formatted

$$\boldsymbol{M}_q^{\mathrm{T}} \dot{\boldsymbol{\theta}} = \boldsymbol{M}^{\mathrm{T}} \boldsymbol{\omega} \cdot \tag{13}$$

Bringing equation (12) into equation (13), the statics equilibrium equation is formatted as

$$\boldsymbol{M} = \boldsymbol{G}_k \boldsymbol{M}_q\,, \tag{14}$$

where, $\boldsymbol{G}_k$ represents the Jacobean matrix of the hip joint, and $\boldsymbol{G}_k \in R^{3\times3}$.

If matrix $\boldsymbol{G}_k$ is invertible, equation (15) can be gained

$$\boldsymbol{M}_q = \boldsymbol{G}_k^{-1} \boldsymbol{M}\,. \tag{15}$$

With equation (6) and (15), the mapping relationship of force $\boldsymbol{F}_p$ applied on foot of the hybrid leg and driving torque m_1, m_2, m_3 and driving force f_L is gained.

4. Global Stiffness Index

The hybrid leg will have a deformation when force is exerted on it. In order to measure the deformation degree, global stiffness index k is defined.

We define an operating flexibility matrix C to demonstrate the relationship between the deformation $\boldsymbol{s}$ and the external force $\boldsymbol{F}$:

$$\boldsymbol{s} = \boldsymbol{C}\boldsymbol{F}_p\,. \tag{16}$$

The square of deformation $\boldsymbol{s}$ is

$$\boldsymbol{s}^{\mathrm{T}} \boldsymbol{s} = \boldsymbol{F}_p^{\mathrm{T}} \boldsymbol{C}^{\mathrm{T}} \boldsymbol{C} \boldsymbol{F}_p\,. \tag{17}$$

Here we define matrix $\boldsymbol{K}=\boldsymbol{C}^{\mathrm{T}}\boldsymbol{C}$, then equation (17) can be rewritten as

$$\boldsymbol{s}^{\mathrm{T}}\boldsymbol{s}=\boldsymbol{F}_p^{\ \mathrm{T}}\boldsymbol{K}\boldsymbol{F}_p\,. \tag{18}$$

If the external force is normalized to 1, that is, $\boldsymbol{F}_p^{\mathrm{T}}\boldsymbol{F}_p=1$, the maximum deformation of the hybrid leg could be obtained as:

$$\|s_{\max}\|=\sqrt{\max(\sigma)}\,, \tag{19}$$

where, σ denotes the eigenvalues of the matrix $\boldsymbol{K}$ [13].

Then the global stiffness index can be defined as

$$k=\int_W\|s_{\max}\|dW\Big/\int_W dW\,. \tag{20}$$

This index indicates the maximum deformation within the workspace when the foot is imposed an external unit force, and the performance of the hybrid leg is better while the index is smaller. In addition, the index provides optimizing criterion for subsequent structural improvement of the leg.

5. Specific Example Analysis

The parameters of the leg is set as follow: $\delta=45^{o}$, $L_1=426$ mm, $L_2=448$ mm, $l_1=300$ mm and $l_2=100$ mm. Owing to the hybrid leg can be equivalent to a *SR* serial robot, when the position of $\boldsymbol{P}$ is given, the leg has many postures. While different posture is corresponding to different matrix $\boldsymbol{G}_h$ and $\boldsymbol{G}_k$, so the matrix $\boldsymbol{G}_h$ and $\boldsymbol{G}_k$ is uncertain if the posture of the leg is not given. Taking the least energy-consuming posture as the optimal posture, that is, the axis of knee joint is horizontal and the angular relating the moving platform to the fixed platform is zero all the time.

Now a movement curve of the foot point $\boldsymbol{P}$ for10 seconds is given as

$$\begin{cases} X=0 \\ Y=245-10t\,. \\ Z=671-10t \end{cases} \tag{21}$$

Based on the static model established above, graphs of driving torques m_1, m_2 and m_3 and the driving force f_L when external force of 100 N is exerted on the foot point $\boldsymbol{P}$ in three coordinate axis directions can be figured out by MATLAB software, as shown in Fig. 5.

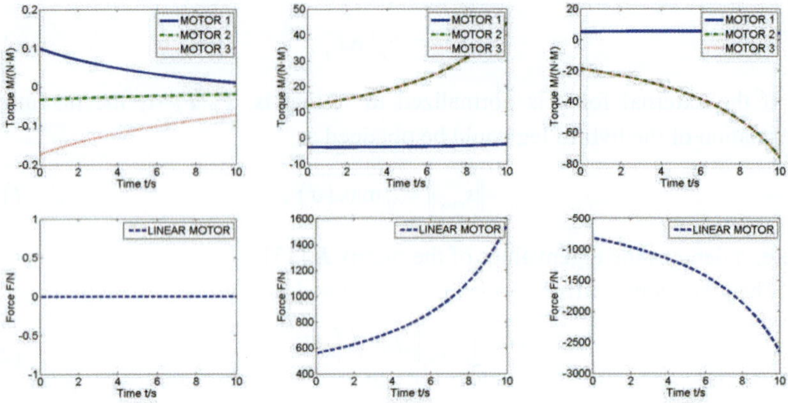

(a) F=[100 0 0]N (b) F=[0 100 0]N (c) F=[0 0 100]N

Fig. 5. Diagram of statics analysis.

Comparing graphs (a), (b) and (c) of Fig. 5, at the same time driving torques m_1, m_2 and m_3 and the driving force f_L are smallest when external force is applied in x-axis direction and biggest in z-axis direction. The simulation results have great significance for the later research of footstep planning.

6. Conclusions

(1) A novel quadruped robot with four identical serial-parallel hybrid legs is presented. The hybrid leg has four degrees of freedom: three rotational degrees around the rotation center of the hip and one rotational degree around the axis of knee joint.

(2) Based on the vector method and differential transform method, statics equilibrium equation of the hybrid leg is established and its static performance is analyzed.

(3) Global stiffness index is defined to measure the deformation quantitatively within the workspace when external unit force is exerted on the foot point, and the index provides optimizing criterion for subsequent structural improvement of the hybrid leg.

(4) Simulation has been performed on the static model of the hybrid leg, and results indicate that the leg has best capacity in x-axis, and it has smallest capacity in z-axis. The simulation results have great significance for the later research of footstep planning.

Acknowledgments

This work is supported by the National Natural Science Foundation (No. U1304510) and Outstanding Young Teachers Development Foundation of Zhengzhou University (No. 1421321076).

References

1. Ishikawa T, Makino K, Imani J, et al. *Journal of Advanced Computational Intelligence and Intelligent Informatics*, 2014, 18(2): 135-139.
2. Mantian Li, Zhenyu Jiang, Wei Guo, et al. *Robot*, 2014, 36(1): 21-28 (in Chinese).
3. Geonea I, Ungureanu A., Dumitru N., et al. *Mechanisms and Machine Science*, 2015, 24: 147-155.
4. Yibin Li, Bin Li, Jiuhong Ruan, et al. Research of mammal boinic quadruped robots: A review [C]//*IEEE Conference on Robotics, Automation and Mechatronics. Piscataway, USA: IEEE*, 2011:166-171.
5. Ananthanarayanan A, Azadi M, Kim S. *Bioinspiration & Biomimetics*, 2012, 7(4):1-12.
6. Player R, Buehler M, Raibert M. BigDog[C]//*Unmanned Systems. Technology VIII*. Bellingham, USA: SPIE, 2006: No. 62302O.
7. Hui Chai, Jian Meng, Xuewen Rong, et al. *Robot*, 2014, 36(4): 385-391.
8. Yu Rong, Zhenlin Jin, Mengke Qu. *Optics and Precision Engineering*, 2012, 20(7):1532-1541 (in Chinese).
9. Guanjiao Ren, Weihai Chen, Bin Chen, et al. *Journal of Mechanical Engineering*, 2011, 47(11): 14-22.
10. Yulin Zhou, Feng Gao. *Journal of* Mechanical *Engineering*, 2006, 42(11): 66-74 (in Chinese).
11. Yanhua Tang. *Applied Mechanics and Materials*, 2014, 635-637: 145-149.
12. Xinghua Tian, Feng Gao, Xianbao Chen, et al.*Journal of Mechanical Engineering*, 2013, 49(6): 81-88 (in Chinese).
13. Jianshe Gao, Mingxiang Li, Allison, G, et al. *Chinese Journal of Mechanical Engineering*. 2015, 28(4): 684-690.

Design and analysis on the live working robot system

Yin-Bo Du*, Da-Hong Wang†, Li Tan* and Cheng-Jiang Wang†,‡

*State Grid Chongqing Electric Power Co. Electric Power Research Institute, Chongqing, 401123, China

†Shenzhen Institute of Advanced Smart Technology, Shenzhen, 518000, China

‡Email: 375020040@qq.com

Based on the requirements of 10 kV live line work process, and the characteristics of the current operating environment, we have developed a high-voltage electric field operations robot. The high-voltage live working robot is composed of two six hydraulic manipulator arm free who share identical configuration, which aim to achieve remote operation of the master-slave arms. The mentioned robot mainly consists of chassis, work bucket, working arm, the hydraulic control system and safety protection devices and other components. This robot can safely and reliably carry out live working, reducing operating time and improving work efficiency. Moreover, it can liberate the operator from dangerous, heavy, mental stress work environment. The successful development of live working robots will be better for people's production and life, and provide better service for electricity users.

Keywords: Live working robot; hydraulic control system; master-slave manipulator.

1. Introduction

At present, the high voltage live working robot has become an important equipment to ensure the safety and reliability of power supply equipment, and to improve the economic benefit and service quality of power network. But the way of live work in China is still artificial live operation, operation personnel in high pressure, high altitude environment, and it is easy to cause personal casualty accident.[1] So it is very important to develop a kind of live working robot which is used in high voltage line.

The high voltage live working robot is a high-tech product which uses the technology of high voltage insulation, mechanical, hydraulic, control and computer science.[2,3] It can improve the live working level of automation and safety, reduce the operator's labor intensity, and reduce the personal harm of the strong electromagnetic field to the operator. Therefore, the popularization and application of the high voltage live working robot has great social significance and economic benefits. From the 80's of last century, many countries have carried out research on the high voltage live working robot, such as Japan, Spain, the United States, Canada, France and so on.[4,5,6]

Early in January 2002, Shandong Electric Power Research Institute (EPRI) developed the domestic first set of high voltage live working robot, and the robot have greatly improved in the aspects of safety, reliability, practicality and flexibility.[7] The robot has completed the design of the master slave manipulator, special lift system for robot design and processing, tool system design and processing, the insulation protection system design and processing; and the robot has achieved the task of live-wire breaking, live-T wiring, live installing the shield, live-nut breaking and so on.

According to the requirement of 10 kV line live working process and the characteristics of the working environment, a high voltage live working robot which can satisfy the requirements of the field work environment is developed. The high voltage live working robot meets the process requirements of the high-voltage line live working, and is adapted to the working environment, and complete the assigned tasks. The robot can shorten the time of live working hours, and improve the efficiency of live working, reduce the labor intensity of the workers; and the robot can make the operation personnel and the high-voltage electric field completely isolated, to maximize the safety of the operator.

2. The Structure Characteristics of High Voltage Live Working Robot

Up to now, there are two main ways of working on high voltage live working robot. The first way is that the operator and the operating manipulator work together in a high altitude operation platform, the operator through the operation of the main hand to carry out the operation of the slave arms. The first generation of live working robot named I Phase developed by Japan is the use of this approach.[8] The other is that the operator is located on the ground or the operating room installed in the mobile car, through the robot's vision system to provide the data for the operation of the robot arm to complete the operation. The second generation and the third generation of live working robot developed by Japan take this way.[5] The operating efficiency of the former is relatively high, but there are security risks; the latter can ensure the operator to complete the operation in a more secure way while maintaining a certain working efficiency. And with the development of the technology of the vision system, the second methods will be the trend of the future live working robot. At present, most of the live working robot in most countries adopt the six degrees of freedom of the dual robot manipulator.[9,10] Hydraulic drive and motor drive is the common two driving methods, Hydraulic driven mechanical arm can be easier to achieve insulation, and because of the compact structure, high load capacity ratio, strong load capacity and other characteristics are more suitable for live working.

The live working robot adopts the design scheme that two six free hydraulic robot arm which has the same configuration is used as the working robot arm to carry out live work from the hand and to realize the main hand to the remote operation of the slave hand. In combination with the existing design and manufacturing standards of aerial vehicle, a high voltage live working robot is designed. The high voltage live working robot is mainly composed of the chassis, the working bucket, the working arm, the hydraulic control system and the safety protection device, and the robot uses the current advanced technology in the chassis design, upper apart design, working bucket design, control system design and other aspects. The chassis enables the mechanical arm to reach the specified operating area and height; by controlling the master hand, the operator can complete the corresponding action in real time and accurately from the slave hand; the hydraulic control system can control the end position and attitude of the robot arm, and feedback the current information of the robot arm to the main hand control device in the same time; special tools can help the robot to finish the task better; safety protection device ensure the safety of operation.

3. Overall Structure Design

The live working robot is designed according to the crank arm type self walking robot for high altitude operation, and introduces the technical specification of live working technology and the high voltage insulation technology into the design, which is safe, flexible and reliable. Its overall design schematic diagram shown in Figure 1.

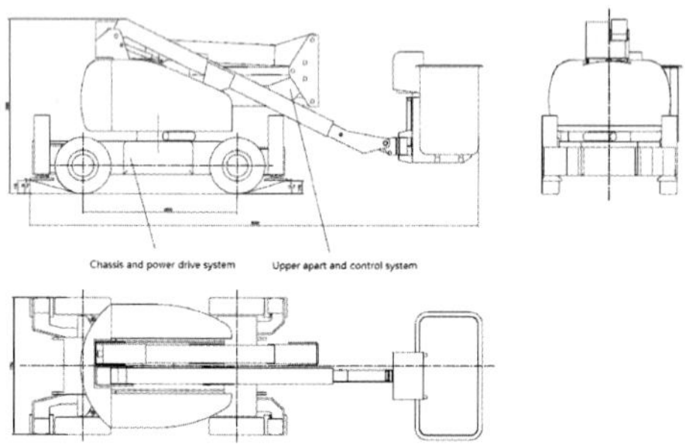

Fig. 1. Schematic diagram of the robot overall design, followed by the side view, the rear view and top view.

As shown in Figure 1, the robot is composed of a chassis, a working bucket, a working arm, a hydraulic control system and a safety protection device. The upper apart telescopic arm and the end working bucket are made of insulating material, which can ensure the insulation level of 10 thousand volts. The working bucket part can move 50 cm in the vertical direction, and can satisfy the layout of various high voltage lines.

3.1. *Chassis*

The chassis of the vehicle is a JY14 aerial work vehicle chassis, which is added with a leveling supporting foot and a hydraulic steering power assisting device.

3.2. *Working bucket*

The high-voltage live line work robot configurations double single bucket type insulation bucket; insulated bucket has the function of high withstand voltage, power frequency withstand voltage is 10 kV/1min no breakdown; leakage current is 300 μA (10 kV); the rated load of double single bucket twin type 200 kg; working bucket level rotation angle is 120 degrees, vertical direction can move up to 50 cm.

The working bucket has the automatic level adjustment function. The leveling mechanism ensures that the working position is in a horizontal state, the angle between the bottom surface and the horizontal plane is not more than 5 degrees. The leveling process is stable and reliable, and does not appear vibration, impact, slipping and so on. When the working bucket is raised at the rated load, the brake can be reliably at any position, and the subsidence quantity of the working bucket after the brake is not more than 0.5% of the height of the working bucket under that working condition.

The working bucket and the chassis are respectively equipped with a quick hydraulic interface with pressure limitation (the caliber of the interface and the pressure of the interface according to the standard configuration), which can be connected with the hydraulic tool.

3.3. *Working arm*

The working arm of the high-voltage live working robot is divided into two types, namely, a folding arm and a telescopic type. The effective insulation length of the upper arm insulation part if 3.2 m, the power frequency withstand voltage is 10 kV/1min; the lower arm is the folding arm, the fixed insulation section effective insulation length is 0.2 m, the power frequency withstand voltage is 10 kV/1min; The leakage current of whole vehicle is less than or equal to 500μA. The insulation part of the upper arm meets the requirements of the B class live working car in the United States National A92.2 ANSI standard.

Insulating arm outer surface coated with a moisture-proof, anti UV coating; insulating arm inner surface takes the hydrophobic measures and its surface is smooth, bright and clean, no dent, pits phenomenon.

3.4. *Safety protection device*

Work basket, a turntable respectively is equipped with a set of interlocking control device, the installation position of the working bucket control device is suitable for operator control, and the turntable control device can prior to operation hopper control, and a single full range operating handle is used in the working bucket. When the control handle is loosened, it should be automatically reset, and the operation direction is consistent with the direction of motion control. Each operation should not interfere with each other and cause false operation, the operation should be light and flexible, accurate and reliable. The robot has a device that prevents the isolation of the arm and the working bucket from falling freely when the system is in trouble.

The four supporting leg and control system of the high-voltage live working robot has a locking device, which comprises a working arm and supporting leg interlocking device and a supporting leg moving alarm device. Limit device is arranged at the end position of each action.

At the end of the work, the working arm is restored to its original position, and there should be a fastening device. When the bucket arm vehicle in running condition, the outrigger retractable agencies should ensure the supporting legs securely fixed in a predetermined position, rotating part should not have the relative motion.

Operation bucket, turntable are equipped with emergency stop device, emergency return device. When abnormal motor, battery power supply interruption or hydraulic system is abnormal, the working bucket and insulating arm emergency system can automatically recover, to ensure that when the bucket arm car is abnormal, the insulation bucket staff can safely on the ground. The emergency system is equipped with a vehicle DC power emergency hydraulic pump, and the electric system is also equipped with a switch off the power supply.

4. Brief Introduction of Structure and Function

4.1. *Chassis and power drive system*

As the basis for the installation of the vehicle, the chassis needs stable structure, and its strength and deformation need to meet the requirements of bucket arm vehicle. The power drive system adopts the battery power supply, vehicle driving depends on motor drive, the rise of crank arm and the vehicle adjustment need hydraulic drive, and through the hydraulic system to achieve the various functional parts of the movement. The basic structure of this system is shown in Figure 2.

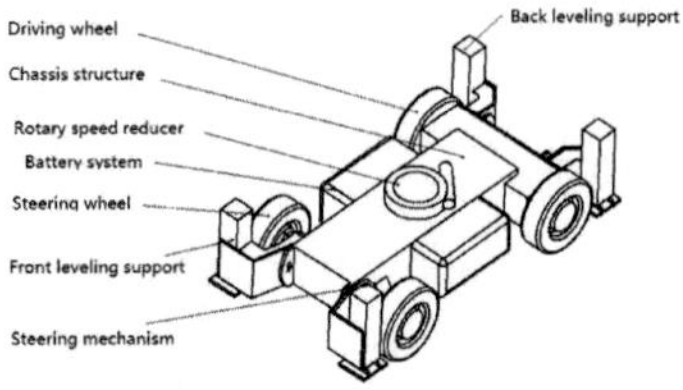

Fig. 2. Chassis and power drive system diagram.

4.2. *Upper apart assembly*

The upper apart assembly comprises a rotary table, a lower folding arm, an upper folding arm, a basic arm, a telescopic insulating arm, an insulating working platform and a hydraulic motion system. The upper apart assembly is the main function component of the bucket arm vehicle, it is required to work stably, safely and stably without jitter. All parts are required to carry out safety checks, and according to the provisions of the safety margin to retain sufficient. As shown in Figure 3.

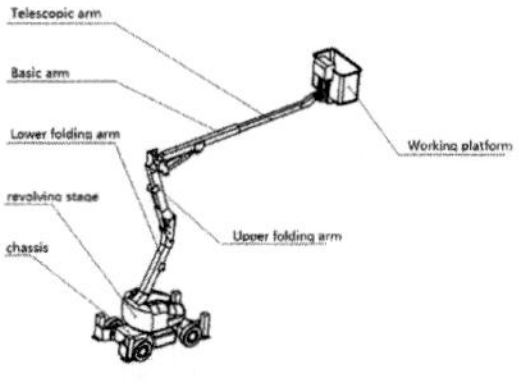

Fig. 3. Schematic diagram of the upper apart assembly.

4.3. *Electric control system*

Electrical control is the use of PLC control and battery powered solution, and including the electrical system of turntable and chassis, and the electrical system of platform. The two systems take the electrical system of platform as priority. According to GB/T 9465-2008 ("Aerial work truck"), when people work on the platform, the function of control platform on the chassis is disabled. At the same time, it is equipped with tow line control box, which is convenient to move and adjust the position of the vehicle.

The platform is provided with a foot pedal switch, when the switch is operated, the operation vehicle can only be operated on the platform, the vehicle can not be controlled under the control console, and the safety of personnel on the platform can be ensured.

4.4. *Hydraulic control system*

The high voltage live working robot is powered by batteries, and the hydraulic system performs the function of movement. In addition to walking, the operating vehicle is driven by hydraulic pressure in the direction of the steering, the rotation, the folding lifting, and the telescopic movement.

The hydraulic control system of the high-voltage live working robot design in accordance with the provisions of GB3766-88 the hydraulic system general technical conditions", and is provided with a safety device to prevent overload and hydraulic shock. The hydraulic control system is also equipped with servo hydraulic step speed control system and hydraulic control insurance trinity single handle control, and is equipped with remote working bucket the upper and the lower part of the engine start and stop system. The hydraulic control system is capable of continuous rise and fall, turning and running mechanism of balance, reliable, no shaking shock, and even at the time of the maximum load, there is no leakage in hydraulic system, the oil temperature is normal. Operating vehicle hydraulic control system schematic diagram shown in Figure 4.

5. Concluding Remarks

In this paper, a high voltage live working robot is designed, which consists of a chassis, a working bucket, a working arm, a hydraulic control system and a safety protection device. This robot can safely and reliably carry out live work, save the working time, improve the work efficiency, and the operator is liberated from the dangerous, heavy, mental tension in the work environment. The successful development of live working robot will better serve the people's production and life, and will better serve the power users.

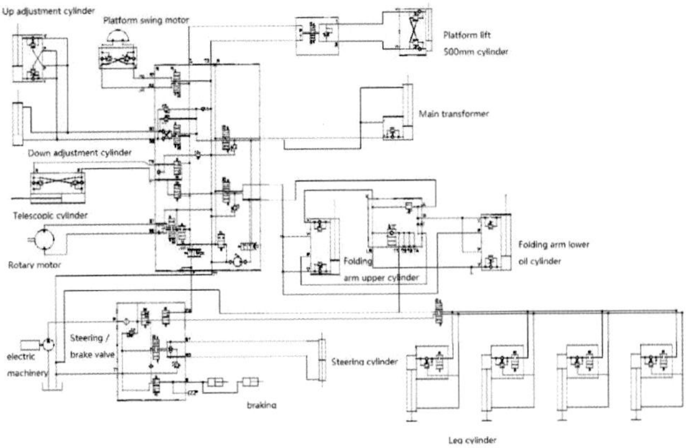

Fig. 4. Principle diagram of hydraulic control system.

References

1. Lu Shouyin, et al. *Research on the High Voltage Electric Power Live Line Working Robot* [D]. Shanghai: Shanghai Jiao Tong University, 2003.
2. Sawada J, Kusumoto K, Maikawaa Y, et al. *A Mobile Robot for Inspection of Power Transmition Lines.* IEEEtrans on Power Delilery, 1991, 6(1):309-315.
3. Yano K, Maruyama Y, Morita K, et al. *Development of the Semi-automatic Hot-line Work Robot Systems "Phase 2".* In: *Proc of the 7th Inter Conf on Transmission and Distribution Construction and Live Maintenance.* 1995, 212-218.
4. Santamaria A, Aracil R, Tuduri A, et al. *Teleoperated Robots for Live Power Lines Maintenance(ROBTET).* In: *Proc of 14th International Conference and Exhibition on Electricity Distribution, Part 1 Contributions, Vol 3,* 1997. 31/1-5.
5. Moriyuki NAKASHIMA, Hideo YAKABE, Yoshinaga MARUYAMA, et al. *Application of Semi-Automatic Robot Technology on Hot-Line Maintenance Work[C].* IEEE International Conference on Robotics and Automation, 1995, 843-850.
6. R. Aracil, M. Ferre, M. Hernando, E. Pinto, J.M. Sebastian. *Telerobotic system for live-power line mainatenance: ROBTET [J].* Control Engineering Practice, 2002, 10: 1271-1281.
7. LU Shouyin, et al. *DWR-I Telerobot for High Voltage Electric Power Line Live Working Maintenance [J].* Journal of Shanghai Jiao Tong University. 2005, 39(6): 910-915.
8. Moriyuki NAKASHIMA, Hideo YAKABE, Yoshinaga MARUYAMA, et al. *Application of Semi-Automatic Robot Technology on Hot-Line*

Maintenance Work[C]. IEEE International Conference on Robotics and Automation, 1995，843-850.

9. R. Aracil, M. Ferre, M. Hernando, E. Pinto, J.M. Sebastian. *Telerobotic system for live-power line mainatenance: ROBTET [J].* Control Engineering Practice，2002,10: 1271-1281.
10. Li Xiaotang. *A Study on Some Issues for Inverse Kinematics of 6 DOF Serial Manipulator [D].* Beijing: Beijing University of Posts and Telecommunications, 2006.

Trajectory planning of 3DOF power tower climb robot

Wei Song*, Yong-Bing Li†,‡, Qing Zhou*, Hai-Bing Zhang*,
Ling-Yun Wan* and Wen-Fu Xu†,§

*State Grid Chongqing Electric Power Co. Electric Power Research Institute,
Chongqing, 401123, China
†Shenzhen Graduate School of Harbin Institute of Technology,
Shenzhen, 518055, China
‡Shenzhen Institute of Advanced Smart Technology,
Shenzhen, 518000, China
§Email:15889502286@163.com

In this paper, we design a 3DOF climbing robot in the software SolidWorks, and build the positive kinematics equation by the DH method; we use the csape function planted in the MATLAB to compute the curve of the robot's gait which is appropriate. Then, we import the simplified model to the ADAMS software. After exerting constraints and drives, we get the simulation of the robot's motion. We also define some markers to measure the angle, which can help us get the inverse kinematic solutions. Through fitting the curve in the MATLAB and the simulation in the ADAMS, we get the comfortable trajectory of the robot and the inverse kinematic solutions of the robot.

Keywords: Climbing robot; kinematics; trajectory planning; ADAMS simulation.

1. Introduction

In this paper, we design one 3DOFs serial robot, which can climb the power tower, by the software of SolidWorks [1]. In order to climb to the deserved place on the power tower, and overpass the barriers of the power tower, it is necessary to build the kinematics equation and design the motion trajectory. Therefore, we use the analysis software Adams, which is a powerful dynamic analysis software, widely used in the fields of machinery, vehicles and aerospace [2]. Besides, designer can build three-dimensional model directly, can also import the model built in other software to Adams to analyze the kinematics of the model. In the paper, we build the model in the software of SolidWorks firstly, then import the model to Adams to build the kinematics simulation and design the motion trajectory. In this way, we can save a lot of time and decrease the high cost of research. In the way we realize the high efficiency and low cost design.

2. Robot Introduction

Monitoring and maintenance of the high voltage power tower is a complex and difficult task to the power inspection department. So it is very necessary to design an auto equipment to replace human beings for inspection of power tower [3]. The structure of power tower is special, the main body is made up of angle steel truss. There are some short steel trusses between the main bodies, they are used for increasing the stability of the power tower; and there are some screw bolts on the main body, the function of them is to connect the angle steel trusses. So the robot should not only have the function of climbing, but also can overpass the barriers on the power tower. Based on the analysis, we design a three-dimensional climbing robot [4].

The structure design of the robot mainly includes two parts, the design of the motion system and the design of the adsorption device. Robot adopts electromagnetic adsorption device, so the robot can reliably adsorb on the truss; by using the linear drive of worm gear and worm drive and screw motor, the robot can climb up to the top of the truss.

According to the overall structure of the robot, the robot consists of two parts. The two parts are composed of an adsorption device, a worm wheel and worm mechanism driven by a stepping motor, and a connecting rod. A screw motor is connected to the left arm of the robot; to fit the screw nut on the connecting arm of the right end of the robot. The three-dimensional structure of the robot is shown in Figure 1.

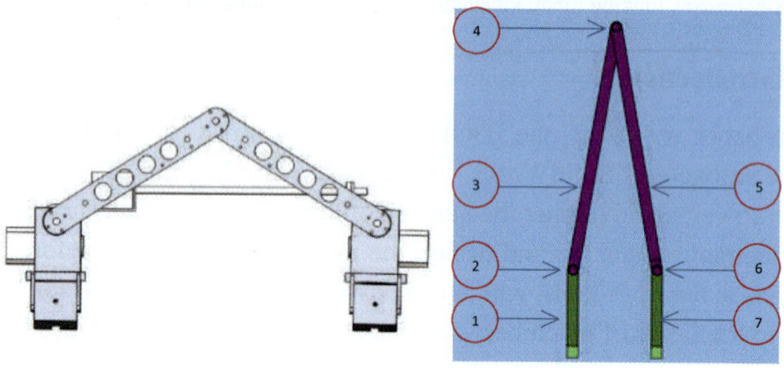

Fig. 1. The structure of 3DOF climbing robot and the simplified model.

For the convenience of analysis, the simplified structure is designed in SolidWorks, and the length of the connecting rod and the height of the pedestal on the left are the same as that in the right.

In the simplified model, each part of the model is named as: 1 corresponding to the left pedestal, 3 corresponding to the left connecting rod, 5 corresponding to the right connecting rod, and 7 corresponding to the right pedestal; The connection point of each connecting piece is the joint of the robot, which is 2 corresponding to joint 1, 4 corresponding to joint 2 and 6 corresponding to joint 3. The analysis of this article is based on setting here.

3. Robot Forward Kinematics

A robot's motion cycle can be broken down into two steps: First, the left pedestal is fixed, the joint 1, the joint 2 and the joint 3 turn a certain angle to make the right pedestal to reach a certain position; Second, the right pedestal is fixed, the joint 1, the joint 2 and the joint 3 turn a certain angle to make the left pedestal to reach a certain position, so that the robot can achieve a stepping motion. The two steps of the research method is exactly the same, limited to the length of this article only the first step of the movement is analyzed.

According to the simplified model to solve the forward kinematics of the robot, the relationship between the robot trajectory and the joint angles can be obtained. The forward kinematics of the robot is analyzed by D-H method [5]. The D-H parameter table of the robot is established as shown in Table 1.

Table 1. The D-H parameter table.

Joint i	θ_i	α_i	a_i	d_i
1	θ_1	0	l_1	0
2	θ_2	0	l_2	0
3	θ_3	0	l_3	0

The range of each joint angle is shown in Table 2:

Table 2. The rotation range of each joint angle.

Joint angle	θ_1	θ_2	θ_3
Range of rotation	10°~80°	-160°~ -20°	-80°~ -10°

The transformation matrix between adjacent coordinate systems can be obtained by formula (1.1).

$$^{i-1}A_i = \begin{bmatrix} c\theta_i & -s\theta_i * c\alpha_i & s\theta_i * s\alpha_i & a_i * c\theta_i \\ s\theta_i & c\theta_i * c\alpha_i & -c\theta_i * s\alpha_i & a_i * s\theta_i \\ 0 & s\alpha_i & c\alpha_i & d_i \\ 0 & 0 & 0 & 1 \end{bmatrix}. \tag{3.1}$$

The parameters of the D-H parameter table are brought into the formula (3.1), which can obtain the coordinate transformation matrix of the adjacent joint.

$$^{0}A_2 = {}^{0}A_1 * {}^{1}A_2, {}^{0}A_3 = {}^{0}A_1 * {}^{1}A_2 * {}^{2}A_3. \tag{3.2}$$

By using the coordinate system of the base joint as the base coordinate system, the transformation matrix $^{0}A_2$ of the robot joint 3 and the end of the robot can be obtained by multiplying the adjacent joint transform matrix.

According to the coordinate transformation matrix of the formula, the relationship between the robot end position and the angle of the joint is obtained:

$$\begin{bmatrix} p_x \\ p_y \\ p_z \end{bmatrix} = \begin{bmatrix} l_1 c\theta_1 + l_2 c\theta_{12} + l_3 c\theta_{123} \\ l_1 s\theta_1 + l_2 s\theta_{12} + l_3 s\theta_{123} \\ 0 \end{bmatrix}. \tag{3.3}$$

In these formulas, $\theta_{12}=\theta_1+\theta_2$, $\theta_{123}=\theta_1+\theta_2+\theta_3$.

4. Robot Trajectory Analysis

The trajectory of the robot is analyzed by using MATLAB, SolidWorks and ADAMS. The whole analysis process is shown in Figure 2. In this paper, the axis position of the robot's joint 3 is used as the research object, and the trajectory curve of the joint 3 is obtained by fitting the function.

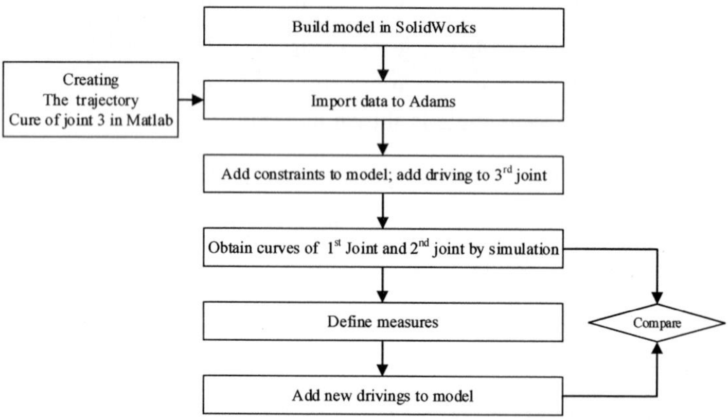

Fig. 2. Motion simulation flow chart.

In robot control process, in order to avoid rigid impact, polynomial interpolation method is adopted to avoid the acceleration mutation at the end of the trajectory. This method can meet first-order and second-order derivative and continuous of the trajectory, but due to excessive sampling will produce shock. The trajectory of the robot end obtained by using the method of cubic spline interpolation can satisfy first-order and second-order derivative and continuous of the trajectory curve, and the trajectory is smooth, and the order is lower [6].

According to the demand of this paper, we first fit the position curve of the joint 3 by spline function, and the related data is shown in Table 3.

Table 3. Curve fitting parameter table.

i	0	1	2	3	4
T(i)	0	0.2	0.4	0.8	1
Xi(t)	0	147.814	282.856	337.304	356.905
Yi(t)	0	57.965	144.475	109.534	0
V(t)	0				0
V(y)	0				0

By using the csape spline function in the MATLAB toolbox, the position of the robot joint 3 is analyzed by cubic spline interpolation, and the trajectory curve is shown in Figure 3, Figure 4 [7]. The curve x (t) and y (t) is discretized using MATLAB and saved as *.txt file, then import into the ADAMS to generate a spline curve can be called as the joint 3 drive.

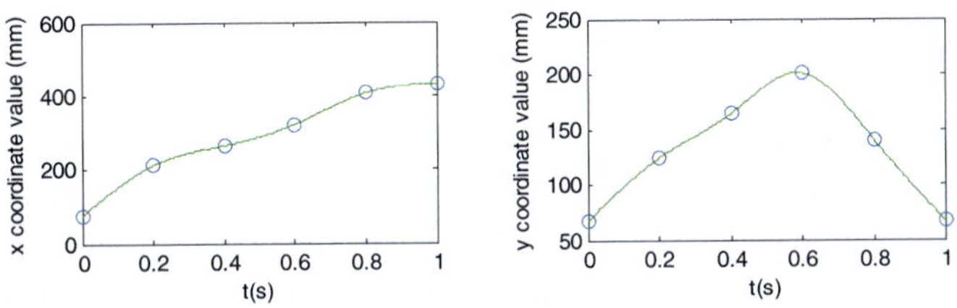

Fig. 3. Joint 3 axis X, axis Y coordinate value curve.

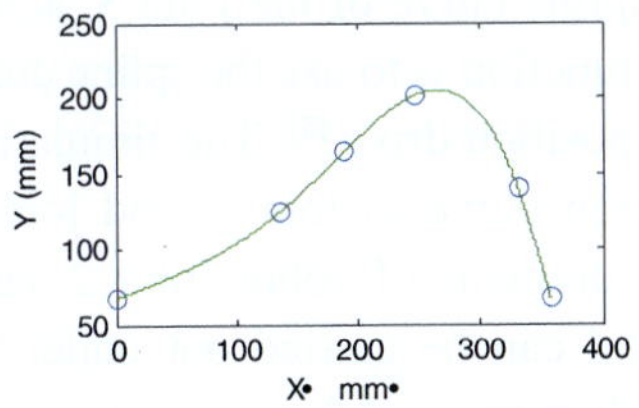

Fig. 4. Trajectory curve of joint 3.

5. Robot Kinematics Simulation

In the assembly environment of SolidWorks, the various parts of the simplified model are saved as *.x_t file, and then import them to ADAMS. After adding the constraint to the various components the model is shown in Figure 5. Based on the end control of the robot, the left pedestal is fixed as the fixed constraint, and the joint 3 is used as the locus of the trajectory, and the control of the end position can be achieved by adding the appropriate drive to the joint 3.

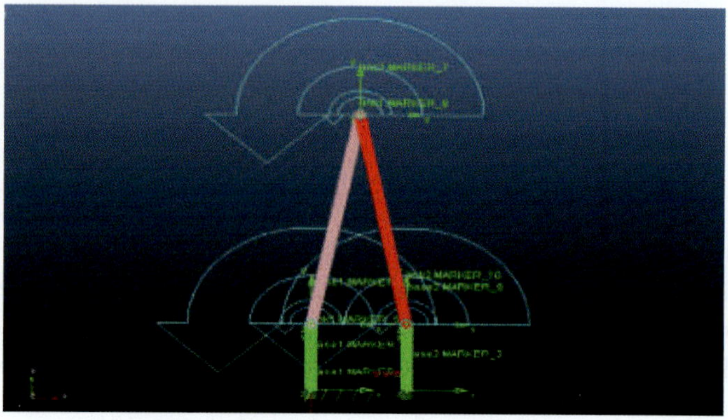

Fig. 5. The simplified model in ADAMS.

In the definition of the robot's joint 3 drive, the *.txt file generated by MATLAB discretization is imported to the ADAMS, to generate the spline trajectory. The specific methods are as follows: in the ADAMS, select File menu, point to import..., point to File Type, point to Test Data (*. *), point to Create Spline, point to file to read, select files (*.txt) that have been created, point to OK. So the system will automatically generate a spline called spline_1.

After adding the spline curve of the joint 3, add the point drive function in the joint 3, the driving function is to use the spline curve before as the X direction and Y direction of the position drive [8]. The simulation time is set to 1s, the step size is 1000. The motion curve of joint 1 and joint 2 can be obtained by the definition of the measurement of robot joint 2 and joint 1. The kinematics simulation of the model can be carried out under the condition of the model without constraint or other errors. After the end of the simulation, open the post-processing module to get the corner of the curve of the joint 1, the joint 2 as shown in Figure 6, Figure 7.

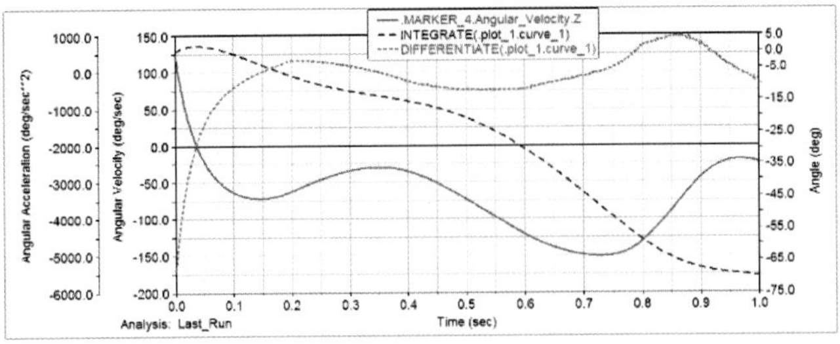

Fig. 6. Rotation angle, angular velocity and angular acceleration curve of joint 1.

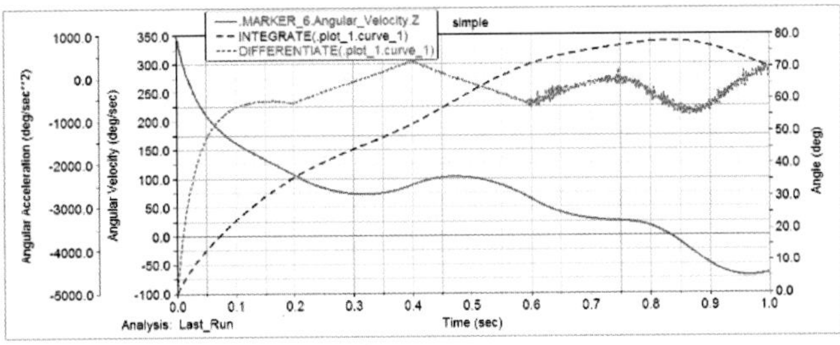

Fig. 7. Rotation angle, angular velocity and angular acceleration curve of joint 2.

ADAMS computing processing power is very powerful, we can turn the joint angle curve output *.txt format for other analysis of the text. Using the data processing tools of post-processing module we can directly save the rotating angle curve as joint1_spline and joint2_spline numerical data as the drive of the joints. Because the unit of the corner in the drive function is radian, so the joint 1, joint 2 should add the drive functions: AKISPL (time, 0, joint1_spline, 0)*pi/180, AKISPL (time, 0, joint2_spline, 0)*pi/180. The joint 3 should add the new drive function: 70d*sin (pi/2*time). Thus, the desired end trajectory can be obtained by the second simulation, as shown in Figure 8.

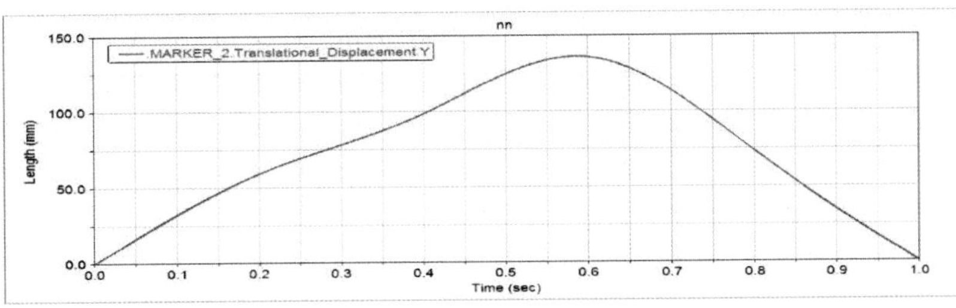

Fig. 8. The end trajectory curve.

By comparing the motion trajectory of the two joint 3 in the Y axis direction, the two curves exactly match, which verifies the correctness of the inverse kinematics solution of the robot by using ADAMS. So we can get the needed data of the robot's end trajectory and the each joint angle through the ADAMS motion simulation, which has brought great convenience for the robot's kinematics research.

6. Conclusion

In this paper, we design a three-dimensional robot model of the power tower in the software of SolidWorks. In order to carry out the kinematics analysis of the robot in ADAMS, the model is simplified. The forward kinematics of the robot is obtained by the DH method, in a reasonable working range, the trajectory points of the joint 3 are set and the trajectory of the joint 3 is fitted by the three spline curve. The obtained data are introduced into ADAMS and the kinematics simulation is carried out, the rotation angle data of 1 and 2 joints were obtained in the way. Finally, the motion trajectory of the robot's end is obtained by adding proper drive to each joint.

References

1. Kwaki, J.S. and T. W. Kim. A Review of Adhesion and Friction Models for Gecko Feet [J]. International Journal of Precision Engineering and Manufacturing, 2010. 11(1), pp. 171-186.
2. Song Shaoyun, Yin Fang. The application of ADAMS in mechanical design [M]. National Defence Industry Press, 2005.
3. Cao Zhihua, Lu Xiaolong, Zhao Shiping. Gait Analysis of Power Tower Climbing Robot [J]. Journal of Xi'an Jiaotong University .2011, 45 (8), pp. 67-72.
4. Giuk Lee, Hwang Kim, Kunchan Seo. MultiTrack: A multi-linked track robot with suction adhesion for climbing and transition [J]. Robotics and Autonomous Systems, 2015 (72), pp. 207–216.
5. Cai Zixing. Robotics [M]. Beijing: Tsinghua University Press, 2000.
6. Yan Sangbin, Han Baoling, Luo Qingsheng. Research and Simulation of the gait trajectory of the six legged walking robot [J]. computer simulation,2007,10(24), pp. 156-160.

7. Zhu Xiaofeng. Research on single leg structure design and compliant control of six legged bionic robot [D]. Master Thesis of Harbin Institute of Technology, 2010.
8. Liu Hongwei, Xue Dangqin. Simulation of kinematics and inverse solution of five degree of freedom series robot [J].New Technology & New Arocess, 2009. (2), pp. 64-67.

Decoupling control of flatness and gauge based on chaos PSO neutral network

Jian-Xin Zhou*, Jian-Kun Li
School of Electrical Engineering,
North China University of Science and Technology,
Tangshan,063000 , China
**E-mail: zhoujianxin1977@126.com*

Usually, the parameters of flatness and gauge coupling control system varied in a certain range in the hot rolling process. For this situation, an improved chaotic particle swarm algorithm was proposed and applied in this study. The neural network weights and thresholds were optimized using this improved approach, and a PID neural network controller was also designed on basis of the particle swarm optimization. The proposed algorithm was improved based on the combination between neural network controller and RBF model identification. The simulation results show that effects of parameter variation on sheet shape and gauge in the coupling control system was eliminated, and thus a satisfactory effect of decoupling control was realized.

Keywords: Chaos algorithm; particle swarm algorithm; decoupling control; neural networks.

1. Introduction

Usually, the flatness and gauge are considered as two important indexes to measure the steel strip quality. However, attributing to the coupling, nonlinear and time-varying characteristics from automatic gauge control (AGC) and automatic flatness control (AFC) systems, the mechanism actions from one of pressure and roll bending may be influenced by the parameters from another, which hinders the integrated control of AFC-AGC system. Therefore, it is necessary for the rolling process automation to investigate the decoupling control of flatness and gauge.

In the traditional control system of flatness and gauge, AGC and AFC systems were independently controlled as a local, single and static system, ignoring the coupling effects among the control functions. Moreover, the expected control effects usually cannot be obtained from the single input-single output (SISO) system design which even leads to an unstable system. Hence, the modeling and control of nonlinear multivariable objects has been recognized as one of the difficulties in control fields[1].

PID neural network (PIDNN), as a combination of PID control regulations and neural network, exhibits the proportion, integral and differential functions which were respectively from different hidden layer units of the neural network. The establishment of PIDNN shows that a new multilayer feed forward neural network is produced. According to the available literatures, PIDNN was modified by the error back-propagation algorithm and a gradient learning method was used to the PID neural network, in which weights fall easily into the local optimum in the learning process [2]. In addition, the parameters of PIDNN were also optimized using the particle swarm optimization algorithm (PSO) with a satisfactory effect. However, some defects such as loss of diversity were reflected the PSO algorithm when the particle swarmed into the local optimum [3]. The main objectives in this study are to improve the PSO algorithm, and then its effectiveness will be verified through the simulation of hot rolling AFC-AGC system.

2. Mathematical Model of AGC-AFC System

The transfer function of AGC-ASC synthetic system can be shown .

$$\begin{bmatrix} \Delta\sigma_1 \\ \Delta h \end{bmatrix} = \begin{bmatrix} G_{11}(s) & G_{12}(s) \\ G_{21}(s) & G_{22}(s) \end{bmatrix} \begin{bmatrix} \Delta F \\ \Delta S \end{bmatrix} + \begin{bmatrix} \eta_1 \\ \eta_2 \end{bmatrix} \tag{1}$$

$$\begin{cases} G_{11}(s) = \dfrac{E}{h} \cdot \left(\dfrac{1}{K_F} - \dfrac{1}{K_P} \cdot \dfrac{C_P Q}{C_P + Q} \cdot \dfrac{1}{C_F} \right) \cdot \dfrac{K_1}{1+T_1 s} e^{-\tau s} \\ G_{12}(s) = -\dfrac{E}{h} \cdot \dfrac{C_P Q}{K_P(C_P + Q)} \cdot \dfrac{K_2}{1+T_2 s} e^{-\tau s} \\ G_{21}(s) = \dfrac{C_P}{(C_P + Q) \cdot C_F} \cdot \dfrac{K_1}{1+T_1 s} e^{-\tau s} \\ G_{22}(s) = \dfrac{C_P}{C_P + Q} \cdot \dfrac{K_2}{1+T_2 s} e^{-\tau s} \end{cases} \tag{2}$$

Where, S , C_P, h ,F , C_F K_F , K_P ,σ_1, E ,η_1 and η_2 were the initial roll gap, longitudinal stiffness coefficient of rolling mill, actual rolling thickness of plate, roll bending force, longitudinal stiffness coefficient of bending roller, the transverse stiffness coefficient of roller, horizontal stiffness coefficient of rolling mill, the transverse tensile force difference, the elastic modulus of exit plates, the interference coefficients resulted from disturbance from shape and the interference coefficients resulted from thickness uneven of strip, respectively.

3. Chaotic PSO Neural Network PID Type Single Neuron Multivariable Decoupling Control

3.1. *Standard PSO*

The basic idea of PSO algorithm is based on the followed basic idea. A group of particles without mass and volume are randomly initialized and each particle represents a potential solution. Advantages and disadvantages of the particles were evaluated by the fitness function that it was set in advance. Each particle will move in feasible solution space. The moving direction and distance of each particle are determined by its velocity variable. Usually, the current optimal particles are followed by other particles, and the optimal solution is obtained via the search with generation by generation finally. In each generation, two extremes are tracked by the particles. They are the searched optimal solutions of particle itself and the whole particle swarm, respectively. Provided that a group including m particles flights at a constant speed in a D dimension search space, the i particle properties in the state at time t are set as follows. Location:

$$x_i^t = (x_{i1}^t, x_{i2}^t, \cdots x_{id}^t)^T \qquad x_{id}^t \in [L_d, U_d]$$

L_d and U_d represent the lower and upper limits of search space, respectively.

Speed:

$$v_{id}^t = (v_{i1}, v_{i2}, \cdots, v_{id})^T \qquad v_{id}^t \in [v_{\min,d}, v_{\max,d}]$$

$v_{\min}$ and $v_{\max}$ represent the minimum and maximum speeds, respectively.

Individual optimal position:

$$p_i^t = (p_{i1}^t, p_{i2}^t, \quad , p_{iD}^t)^T$$

Global optimal position:

$$p_g^t = (p_{g1}^t, p_{g2}^t, \cdots, p_{gD}^t)^T \qquad 1 \le d \le D;\ 1 \le i \le M$$

The position of particle at time $t+1$ is obtained by the updates of formulas (3) and (4).

$$v_{id}^{t+1} = wv_{id}^t + c_1 r_1 (p_{id}^t - x_{id}^t) + c_2 r_2 (p_{gd}^t - x_{id}^t) \tag{3}$$

$$x_{id}^{t+1} = x_{id}^t + v_{id}^{t+1} \tag{4}$$

Where, r_1 and r_2 are random numbers uniformly distributed the interval of (0, 1) to maintain the diversity of group. c_1 and c_2 were the learning factors, also known as acceleration factors which endow the particle swarm with the ability of self-summing up and learning from the excellent individuals in the group. w was known as the inertia weight which indicated the contribution to the current speed of particles.

3.2. *Improved particle swarm optimization (IPSO)*

Based on the searched global optimal location in current group and the ergodicity of chaotic variables, a chaotic sequence was generated to prevent some particles from stagnation in the iteration. The current position of a particle was substituted by the optimal position of these particles in the current particle swarm, and thus the algorithm premature problem attributing to particle stagnation was solved. The chaotic mutation operator was introduced in the algorithm in this study. Particle swarm was considered to be trapped into local optimum and its diversity was lost when more than 50% of the particles have an approximate or identical solution. In this case, the variability of optimal particle needs to be performed with the aim of prompting other particles to jump out of the local optimum as soon as possible.

A variation Logistic chaotic mutation was adopted in this paper and the variation formula was as follows.

$$x'_{gbest} = \mu * x'_{gbest} * (1 - x'_{gbest}) \tag{5}$$

Where, the control parameter $\mu = 4$. The control system was completely in chaotic state with $\mu = 4$, and it would be conducive to escape from the local optima. g'_{best} is the position of optimal particle after variation. After the recalculation of fitness value F', x_{gbest} and g_{best} can be updated if a better solution is obtained.

In the basic PSO algorithm, the modification of individual behavior for all particles is only determined by its individual and global extremum. However, the information of other particles is ignored.

The lack of interactivity is presented in searching solution space of particle swarm. In order to increase the interaction among the particles, the current flight velocity of particles was amended by the parameters of other particles with high fitness according to the method which was firstly introduced by Wang[5]. formula. The velocity of particles is updated in accordance with the following formula:

$$v_{id}^{(k+1)} = wv_{id}^{(k)} + c_1 r_1 (p_{id}^{(k)} - x_{id}^{(k)}) + c_2 \frac{1}{n} \sum_{j=1}^{n} r_{2j} (q_{jd}^{(k)} - x_{id}^{(k)}) \tag{6}$$

Where, r_{2j} is a random number. being uniformly distributed in the interval range of [0,1]. q_{jd} is the position of the first N particles of all particles. The corresponding fitness values of their individual extremes are sorted in ascending order.

In formula (6), the velocity correction for each particle is not dependent on the optimal location of a single particle in the particle population, but depends on more positions of the optimal particles. That is to say, the effect of more particles on a particle is taken into account. This method can expand the search direction of the particle in the solution space so that the search process becomes smoother and the global optimal solution can be obtained by faster convergence. At the same time, the first-increase-then-decrease inertia weight strategy is adopted to improve the defects in the decreasing strategy. The formula is as follows.

$$w(t) = \begin{cases} 1\times\dfrac{t}{t_{\max}}+1.4 & 0\le\dfrac{t}{t_{\max}}\le 0.5 \\ -1\times\dfrac{t}{t_{\max}}+1.4 & 0.5\le\dfrac{t}{t_{\max}}\le 1 \end{cases} \tag{7}$$

4. Single Neuron PID Multivariable Decoupling Control Based on IPSO-NN

The structure of the decoupling control system is shown in Fig. 1. In neural network decoupling control, decoupling control system is composed of two PID neural network controller, two RBF neural network controllers and a controlled coupling object for coupling the flatness and thickness. As a controller, the single neuron PID can learn of inverse dynamics of the controlled object. As an identification model of controlled object, the radial basis function (RBF) network can learn the positive dynamic characteristics.

The control network is composed of two identical sub-networks. The network input layer is composed of the given amounts and the actual output of the system in together. The three PID parameters including proportional, integral and differential elements are composed of the network hidden layer which is embodied by the network weighted coefficient w .The sub-networks are cross-coupled from the hidden layer to the output layer and then formed a control input of the controlled object. Finally, multivariable decoupling control of the system can be completed.

The network forward calculation formulas at any sampling time k are as follows.

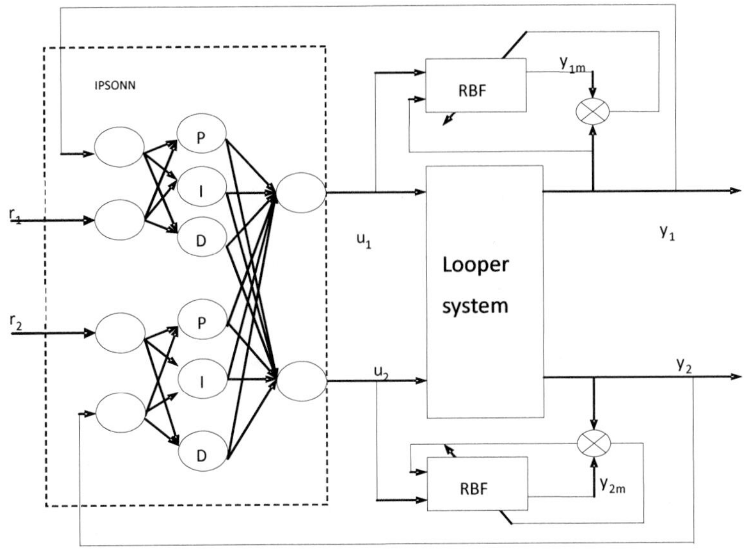

Fig. 1. System of looper multi-variable decoupling.

The input layer: the input is as follows. $u_{s1}(k)=r_s(k)$, $u_{s2}(k)=y_s(k)$, the output is as follows. $x_{si}(k)=u_{si}(k)$.

The hidden layer: the input is as follows. $u_{si}^{'}(k)=\sum_{i=1}^{2}w_{sij}x_{si}(k)$, the outputs of proportional, integral and differential elements are as follows, respectively. $x_{s1}^{'}(k)=u_{s1}^{'}(k)$, $x_{s2}^{'}(k)=x_{s2}^{'}(k-1)+u_{s2}^{'}(k)$ and $x_{s3}^{'}(k)=u_{s3}^{'}(k-1)$

The output layer: the input and output are as follows, respectively.

$$u_h^{''}(k)=\sum_{s=1}^{2}\sum_{j=1}^{3}\omega_{sjh}^{'}x_{sj}(k) \text{ and } x_h^{''}(k)=u_h^{''}(k)$$

Where, s is the subnet number, $s=1,2$; i is the serial number of input layer, $i=1,2$; j is the serial number of hidden layer, $j=1,2,3$; h is the serial number of output layer, $h=1,2$.

BP algorithm has been proved to be an effective neural network learning algorithm. However, there are some problems of slow convergence rate and being trapped into a local minimum and so on about this algorithm. The emergence of particle swarm optimization (PSO) provides a new approach to solve the above problem. In IPSO algorithm, connection weights and thresholds of the neural networks are expressed by the particle position vectors. The total numbers of weights and thresholds are represented the dimensions of particles.

In order to obtain a satisfactory system dynamics, the controller and the object are considered as a generalized network-integration for a minimal objective function which suggests the perfect characteristics of dynamic. The objective function is as follows.

$$J=\sum_{s=1}^{n}E_h=\frac{1}{l}\sum_{s=1}^{n}\sum_{k=1}^{l}[r_s(k)-y_s(k)]^2 \tag{8}$$

Where, l and n are the numbers of sampling points and controlled objects, respectively.

The specific steps of this algorithm are shown below.

(1) According to the control system requirements, the numbers of neural network layers and the numbers of neurons for every layer are determined. The weights of $w_{sij}(0)$ and $w_{sij}^{'}(0)$ are initialized. Meanwhile, the population size, learning factor and iterated times of the particle swarm are also initialized. The initial velocity of particles is set to zero.

(2) The fitness values for each particle are calculated. The individual and global optimal values are updated and compared with p_{best} and g_{best} at the same time. If the results are better than them, it is necessary to update p_{best} and g_{best}.

(3) For all particles, the corresponding fitness values of their individual extremes P are sorted in ascending order. The N particles with the exception of the researched particle are considered as the learning objects of it. Simultaneously, the current positions of the N particles are recorded.

(4) To analyze the diversity of particle swarm, the following step (5) was performed if some particles are trapped into a local optimum, otherwise, velocity and position of the particles are re-updated according to formula (6) and (4).

(5) According to formula (5) on the optimal locations of particles are optimized using logistic chaotic optimization. The global locations are carried out based on chaotic search. The extreme values X_g^* are If the updated if the search solution is better than them.

(6) Maximal iterative times or convergence accuracy is achieved, the optimal solution would be output. Otherwise, the step (2) will be re-performed.

5. Simulation Test

To verify effectiveness of the improved algorithm, the simulation of AGC-AFC system decoupling control was carried out in this study. The structure of PID neural network was 2-3-1, and $\{r(k), y(k)\}$ was set as its input value. The structure of RBF neural network was 3-5-1, and $\{u(k), y(k), y(k-1)\}$ was set as its input value. The momentum factor $\alpha = 0.05$ and learning rate $\eta = 0.5$ were set, respectively. The learning factors of particle swarm were $c_1 = 2.0$, $c_2 = 2.0$. The particle swarm population and maximum Iterations were 50 and 1500, respectively.

When thickness deviation of 1 mm was exerted as a step-function signal, the output curves of AGC and AFC simulations were showed in Figure 2 and Figure 3, respectively. The results showed the adjustment time of the roll gap was short with the adjustment of pressure. The output curves of plate thickness variation has almost no overshoot. The effect of plate thickness adjustment on its shape was very small with a magnitude of 10^{-4}. The flatness was restored to its original state after the short term fluctuations and adjusting time was about 0.3 seconds. The quality requirements of flatness was satisfied.

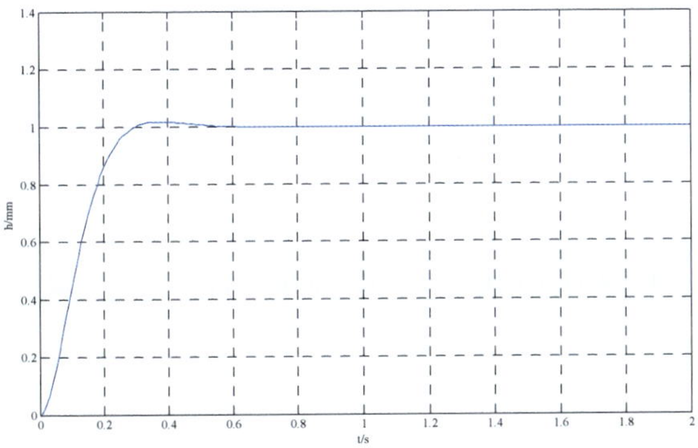

Fig. 2. Output curve of AGC simulation when $h = 1mm$.

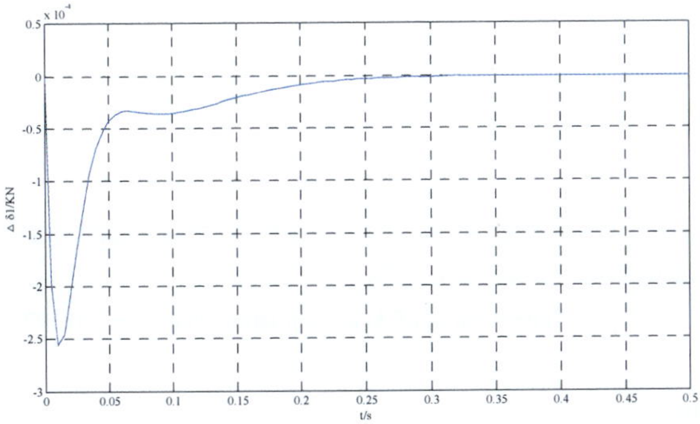

Fig. 3. Output curve of AFC simulation $h = 1mm$.

When $\sigma_1 = 0.1KN$ was exerted as a step-function signal, the output curves of AGC and AFC simulations were showed in Figure 4 and Figure 5, respectively. The overshoot was small of bending force change. The adjustment time of the thickness was about 0.2 seconds. The effect of bending force adjustment on its thickness was very small with a magnitude of 10^{-8}.The thickness was restored to its original state after the short term fluctuations and decoupling control was achieved.

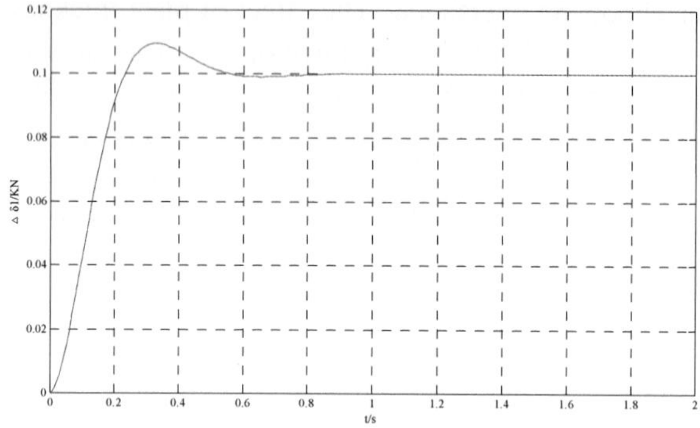

Fig. 4. Output curve of AFC simulation when $\Delta\sigma_1 = 0.1KN$.

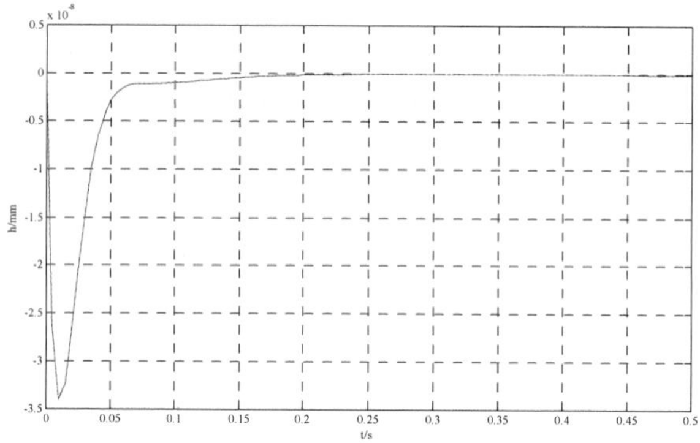

Fig. 5. Output curve of AGC simulation when $\Delta\sigma_1 = 0.1KN$.

6. Conclusion

In the case of unknown AGC-AFC system model and coupling characteristic, the decoupling control strategy based on chaotic PSO-optimized neural network was proposed. The control system was identified by RBF neural network. The tracking time slow and low convergence precision were overcome by improved control scheme. Seen from the simulation results, a better control effect for AGC-AFC system can be achieved after decoupling, which showed the proposed control strategy was effective.

Acknowledgment

This work is supported by key project in Hebei Province Department of Education, (No.ZD2015059).

References

1. LINGHU Kezhi, HE Anrui, YANG Quan, et al, Feedback-decoupling for a combined shape and gauge control system in hot strip rolling, *Journal of University of Science and Technology Beijing,* 2007, Vol. 29(3), pp. 338-341.
2. JIAO Zhu-qing, QU Bai-da and XU Bao-guo, Decoupling Design for Multivariate PID Controller Based on RBF Neural Network, *Journal of System Simulation*, 2008, Vol. 20(3), pp. 627-630.
3. ZHOU Xi-feng, LIN Ying-ying and GUO Qian-gang, PID Neural Network Decoupling Control Based on Particle Swarm Optimization, *Computer Technology and Development*, 2013, Vol. 23(9), pp. 158-161.
4. HUANG Min, ZHU Qi-bing and CUI Bao-tong, Strip flatness and gauge decoupling predictive control based on wavelet neural networks, *Control Theory &Applications*, 2008, Vol. 25(6), pp. 1027-1031.
5. WANG Xiao-xia, WANG Tao and GU Gen-dai, Neural network based on modified particle swarm optimization and its application, *Journal of North China Electric Power University*, 2009, Vol. 36(5), pp. 99-102.
6. LI Bo-qun, FU Jian and ZHANG Rui-cheng, Decoupling control of hot rolling looper height and tension system. *Journal of University of Science and Technology Beijing*, 2005, Vol. 27(5), pp. 596-599.

Research on energy shaping control strategy of current source inverter

Chao-Chao LI[*]. Jing Bai[*,‡]. Yang LI[*], Shuo Zhan[*] and Hua-De Li[†]

[*]*College of Electric Information Engineering,*
Beihua University, Jilin, Jilin, China
[†]*College of Mechanical Engineering*
University of Science & Technology Beijing, Beijing, China.
[‡]*E-mail: jlbyj@163.com*

In order to make a better output waveform of Current Source Inverter (CSI) and to reduce the influence of system performance when the parameters change. A energy shaping control strategy for the three-phase CSI is proposed in this paper. The port Controlled Hamiltonian (PCH) model of three-phase CSI is established in dq coordinate. The controller is designed in order to control the output of system. The solution for the Interconnection and Damping Assignment (IDA) is given through the analysis of the Lyapunov stability conditions. The rationality and validity of the proposal is verified through the simulation results.

Keywords: Current source inverter; PCH system; energy shaping; IDA-PBC.

1. Introduction

Current source converter is widely used in high power applications mainly due to it has the advantages of high power, four quadrant operation, good load voltage waveform, reliable short-circuit protection and low cost. With the application and development of low temperature superconducting technology in the power converter, and with the continuous improvement of energy storage inductor efficiency, the high power density, high efficiency and high performance of power electronic converter application in various fields has been get more research [1-2].

The inverter can be divided into two kinds of voltage type and current type according to the different types of DC power supply. Compared to it's counterpart, the current source inverter has boost characteristics and has no middle part of DC-DC chopper. But there are still some problems in the current source inverter. A single-phase current mode PWM inverter of one-cycle control strategy with limited storage inductance current is proposed in [3] in order to realize active control of inductor current. The result is the addition of a free wheeling loop consists by the bypass switch which makes the system become more complex and difficult to control. The upper and lower bridge arm of current type PWM inverter must be one and only one switch tube conduction at any time. And the DC side is not allowed to open because the current source characteristics. These make the modulation become more complex [4]. The interconnection and damping assignment passivity controller for a three-phase front end converter is proposed in [5] to control the output voltage of the inverter. The controller can improve the dynamics performance of the system and reduce the weakened impact of the change of system parameters.

In recent years, with the development of nolinear control theory, the nolinear robust control method of energy shaping based on passivity theory has been successfully applied in the field of power electronic system. In order to get a better output waveform and the better static and dynamic characteristics, this paper presents a passivity control strategy for a three-phase CSI based on the study of the previous work [6-7]. The designed controller outputs sinusoidal waveform and has a good robustness for the whole system. Simulation experiments are done in this paper to verify the correctness of the proposed control strategy, and the experimental results show that the designed controller is simple and effective.

2. PCH Modeling

2.1. *PCH system*

PCH system model for typical switching system can be written as,

$$\begin{cases} \dot{x} = [J(x) - R(x)]\dfrac{\partial H}{\partial x}(x) + g(x)u + \xi \\ y = g^T(x)\dfrac{\partial H}{\partial x}(x), \end{cases} \tag{1}$$

where $\boldsymbol{x} \in \boldsymbol{R}^n$ is the state vector, continuous function $H(x)$: $\boldsymbol{R}^n \rightarrow \boldsymbol{R}$ represents the total energy storage, $\boldsymbol{u}, \boldsymbol{y} \in \boldsymbol{R}^n$ is the power variable of port, ξ is an external disturbance. Port variable of $\boldsymbol{u}$ and $\boldsymbol{y}$ are conjugate variables, in a sense, their ranks of the product reflects the exchange between system and environment. For example, the current and voltage in the circuit, the force and speed in mechanical system. Interconnection structure is reflected by $n \times n$ skew symmetric matrix $\boldsymbol{J(x)} = - \boldsymbol{J^T(x)}$ and $n \times m$ matrix $\boldsymbol{g(x)}$, and it represents the dissipative matrix when $\boldsymbol{R(x)} = \boldsymbol{R}^{\mathrm{T}}\boldsymbol{(x)} \geq 0$. All of these matrix depends on the value of $\boldsymbol{x}$.

2.2. *Current topology of three-phase CSI*

As it can be seen in Fig. 1, the three-phase CSI is constituted by a large inductance L_{dc}, six IGBTs, the three-phase filtering capacitor C_f, the three-phase filtering inducter L_f and load RL. The constant DC current i_{dc} can be obtained by variable DC current which is flatted by a large inducter L_{dc}. Then i_{dc} is converted to three-phase current with the cooperation of switching tube. Finally, the three-phase current supplies to load via filter.

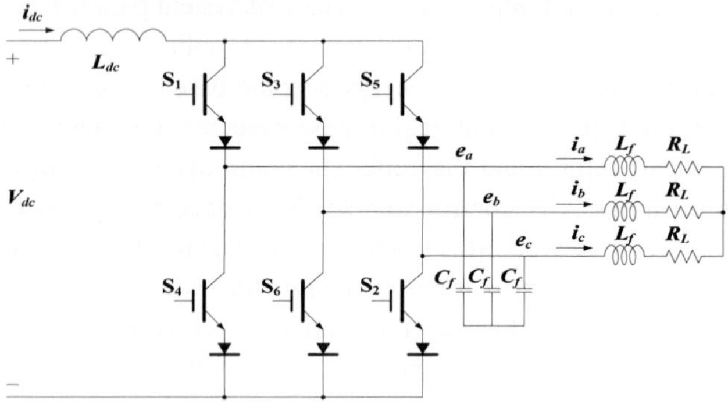

Fig. 1. three-phase CSI circuit.

According to the circuit theorem, we can get the state equation of system in *dq* coordinates by using the state average method for the lumped parameter network system with a storage unit.

$$\begin{cases} C_f \dfrac{de_d}{dt} = s_d i_s - i_d + \omega C_f e_q \\ C_f \dfrac{de_q}{dt} = s_q i_s - i_q - \omega C_f e_d \\ L\dfrac{di_d}{dt} = e_d - Ri_d + \omega L i_q \\ L\dfrac{di_q}{dt} = e_q - Ri_q - \omega L i_d \\ L_d \dfrac{di_s}{dt} = v_{dc} - s_d i_d - s_q i_q \end{cases} \Rightarrow \begin{bmatrix} C_f \dot{e}_d \\ C_f \dot{e}_q \\ L\dot{i}_d \\ L\dot{i}_q \\ L_d \dot{i}_s \end{bmatrix} = \begin{bmatrix} 0 & \omega C_f & -1 & 0 & s_d \\ -\omega C_f & 0 & 0 & -1 & s_q \\ 1 & 0 & -R & \omega L & 0 \\ 0 & 1 & -\omega L & -R & 0 \\ -s_d & -s_q & 0 & 0 & 0 \end{bmatrix} \begin{bmatrix} e_d \\ e_q \\ i_d \\ i_q \\ i_s \end{bmatrix} + \begin{bmatrix} 0 \\ 0 \\ 0 \\ 0 \\ v_{dc} \end{bmatrix}. \tag{2}$$

where i_s is the current of DC inductance, e_d and e_q are *dq* component of voltage *e*, i_d and i_q are *dq* component of load current *i*, s_d and s_q are duty ratio function in *dq* coordinate and ω is angular velocity of *dq* coordinate.

2.3. *PCH model of three-phase CSI*

The current source inverter can be regarded as a port controlled power converter system. The total energy of system is the sum of storing energy of the energy storage unit, and it can be expressed as,

$$H(x)=\frac{1}{2}(C_f e_d{}^2+C_f e_q{}^2+Li_d{}^2+Li_q{}^2+L_d i_s{}^2). \tag{3}$$

The system's state variables are chosen as the voltage and and current variables to describe the state of system storage unit. The state variables are multiplied their corresponding parameters in order to calculate conveniently, so the system's state vector is written as follows:

$$x=\begin{bmatrix}C_f e_d & C_f e_q & Li_d & Li_q & L_d i_s\end{bmatrix}^T. \tag{4}$$

Eq. (3) and Eq. (4) can be written as the form of vector, that is,

$$H(x)=x^T D^{-1} x, \tag{5}$$

where $\boldsymbol{D}$=diag(C_f, C_f, L, L, L_d).

Compared with Eq. (1) and Eq. (2), the input vector and output matrix are,

$$u=\begin{bmatrix}s_d & s_q\end{bmatrix}^T;\ g(x=\begin{bmatrix}i_s & 0 & 0 & 0 & -e_d\\ 0 & i_s & 0 & 0 & -e_q\end{bmatrix}^T, \tag{6}$$

the interconnection and damping matrix of system is,

$$J=\begin{bmatrix}0 & \omega C_f & -1 & 0 & 0\\ -\omega C_f & 0 & 0 & -1 & 0\\ 1 & 0 & 0 & \omega L & 0\\ 0 & 1 & -\omega L & 0 & 0\\ 0 & 0 & 0 & 0 & 0\end{bmatrix};\mathrm{R}=\begin{bmatrix}0 & 0 & 0 & 0 & 0\\ 0 & 0 & 0 & 0 & 0\\ 0 & 0 & R & 0 & 0\\ 0 & 0 & 0 & R & 0\\ 0 & 0 & 0 & 0 & 0\end{bmatrix}. \tag{7}$$

3. Design of Energy Shaping Controller

There are many methods to achieve energy shaping control. The method of interconnection and damping assignment, which is proposed by Mexico Ortega and other scholars, is simple and effective to solve the equation. The method is proposed to find a static state feedback control input $\boldsymbol{u=\beta(x)}$, so that the closed-loop characteristic is dissipative PCH system, that is,

$$\dot{x} = [J_d(x) - R_d(x)]\frac{\partial H_d}{\partial x}(x). \tag{8}$$

The new energy function $H_d(x)$ takes strict local minimum at the desired equilibrium point.

The design steps of passive controller for three-phase CSI are as follows:

Step 1: Determine the parameters of the closed-loop system

Eq. (8) is the form of closed system, its energy function is,

$$H_d(x) = \frac{(x - x^*)^T D^{-1}(x - x^*)}{2}, \tag{9}$$

where $\boldsymbol{J_d(x)}$ and $\boldsymbol{R_d(x)}$ are interconnection and damping matrix of system, they can be written as,

$$J_d(x) = J(x) + J_a(x), \tag{10}$$

$$R_d(x) = R(x) + R_a(x), \tag{11}$$

where $\boldsymbol{J_a(x)}$ and $\boldsymbol{R_a(x)}$ are matrix to synthesize the desired control storage.

Due to the IDA-PBC strategy draws a parameterized partial differential equation through the selection of matrix, and the solution of partial differential equations show that the energy function can be assigned [5]. Therefore, we can select matrix $\boldsymbol{J_a(x)}$ to eliminate the coupling between state variables. $\boldsymbol{J_a(x)}$ and $\boldsymbol{R_a(x)}$ can be written as,

$$J_a = \begin{bmatrix} 0 & -\omega C_f & 1 & 0 & 0 \\ \omega C_f & 0 & 0 & 1 & 0 \\ -1 & 0 & 0 & -\omega L & 0 \\ 0 & -1 & \omega L & 0 & 0 \\ 0 & 0 & 0 & 0 & 0 \end{bmatrix}, R_a = \begin{bmatrix} R_1 & 0 & 0 & 0 & 0 \\ 0 & R_2 & 0 & 0 & 0 \\ 0 & 0 & R_3 & 0 & 0 \\ 0 & 0 & 0 & R_4 & 0 \\ 0 & 0 & 0 & 0 & R_5 \end{bmatrix}. \tag{12}$$

Step 2: Determine the desired equilibrium point

The control target of the inverter is state variables can be stable, that is,

$$x'\Big|_{x=x^*} = 0 \tag{13}$$

therefore,

$$i_s^* = I_s, e_d^* = E_d, e_q^* = E_q. \tag{14}$$

Using (2), (4), (13) and (14), i^*_d and i^*_q can be obtained,

$$\begin{cases} i^*_d = \dfrac{RE_d + \omega LE_q}{R^2 + \omega^2 L^2} \\ i^*_q = \dfrac{RE_q - \omega LE_d}{R^2 + \omega^2 L^2} \end{cases}. \tag{15}$$

Then, the equilibrium point of the system is,

$$x^* = \begin{bmatrix} C_f e^*_d & C_f e^*_q & Li^*_d & Li^*_q & L_d i^*_s \end{bmatrix}^T. \tag{16}$$

Step 3: Determine the control law

Using (1) and (8), the structural equation can be obtained,

$$[J(x) - R(x)]\frac{\partial H}{\partial x}(x) + g(x)u + \xi = [J_d(x) - R_d(x)]\frac{\partial H_d}{\partial x}(x). \tag{17}$$

Putting (2)-(5), (7), (9)-(16) into (17), the solution of (17) in the control laws for s_d and s_q is,

$$\begin{bmatrix} s_d & s_q \end{bmatrix}^T = \begin{bmatrix} \dfrac{i_d - \omega C_f e_d - R_1(e_d - e_d^*)}{i_s} \\ \dfrac{i_q + \omega C_f e_q - R_2(e_q - e_q^*)}{i_s} \end{bmatrix}. \tag{18}$$

4. Simulation Analysis

In order to verify the effectiveness of the proposed control strategy, the three-phase CSI is simulated in Simulink according to the principle diagram as shown in Fig. 2.

The simulation system can be divided into four parts for the DC control circuit, main converter circuit, main controller and PWM generator. The controller outputs switching function according to the control law, the the switching function is converted into PWM pulse to trigger the switch tube of main circuit.

Fig. 3 is the current waveform of DC side, it shows the DC current can quickly track the reference current in the control of current feedback control loop, and the DC current can reach a stable value in a relatively short period of time.

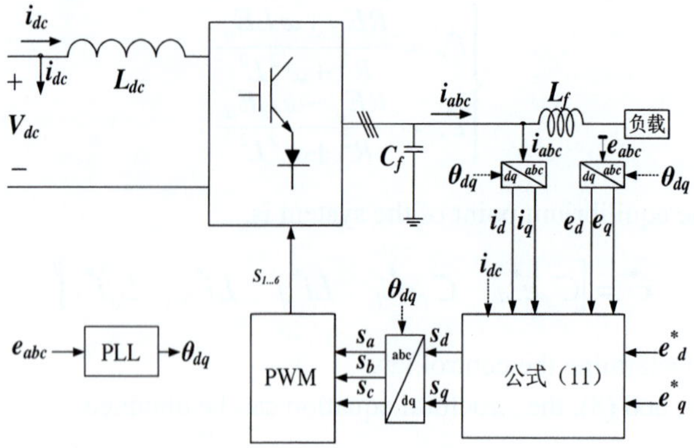

Fig. 2. The principle diagram of IDA-PBC for three-phase CSI.

The waveform of switching function (s_d and s_q) is shown as Fig. 4, the active and reactive power can be respectively controlled by adjusting the interconnection and damping parameters (R_1 and R_2), the power decoupling control is achieved.

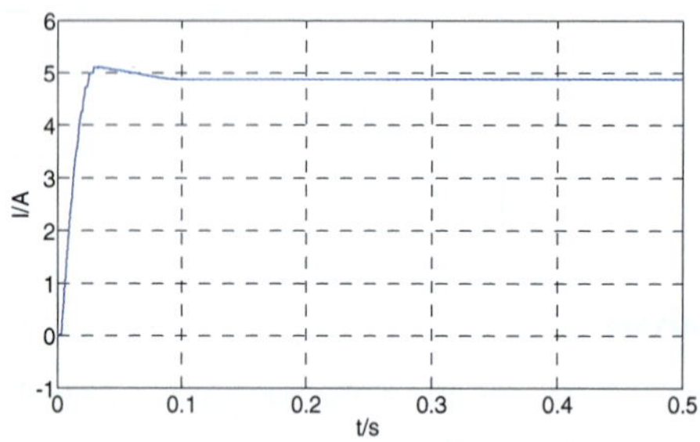

Fig. 3. The DC side current waveform.

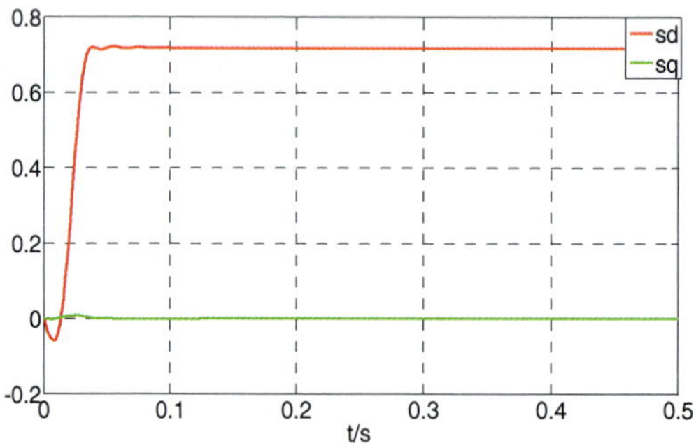

Fig. 4. Switching function waveform.

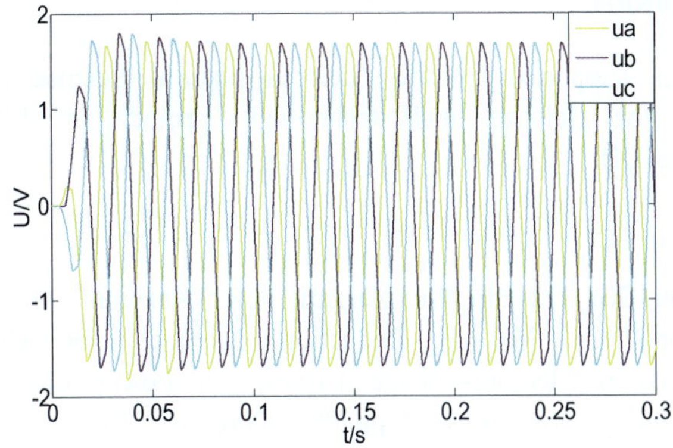

Fig. 5. The three-phase load voltage waveform.

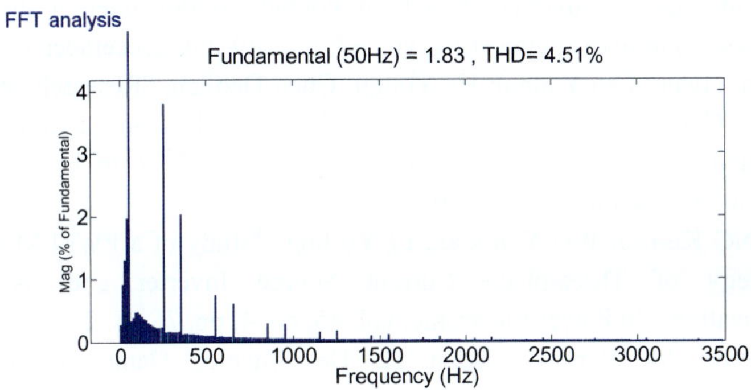

Fig. 6. The distortion spectrum of voltage.

The three-phase load voltage waveform is shown an Fig. 5 and the total distortion rate of load voltage is shown in Fig. 6. It can be seen from these figure, voltage wave form has a good sine degree and the total distortion rate is only 4.51%.

5. Conclusion

In the light of the complex system of three-phase CSI, this paper designed the energy shaping controller based on the theory of the PCH system. The decoupling of the state variables is realized through the configuration of the interconnect structure parameters, and the system output sinusoidal waveform. The structure of the controller is simple and easy to realize. Simulation results show that the designed controller is feasible and effective.

Acknowledgments

This work is supported by the National Natural Science Fund Project (51477011), the science and technology project in Jilin Province (20140204008GX).

References

1. Ning Zhu; Xu, D.; Bin Wu; Zargari, N.R.; Kazerani, M.; Fangrui Liu, "Common-Mode Voltage Reduction Methods for Current-Source Converters in Medium-Voltage Drives," in Power Electronics, IEEE Transactions on , vol. 28, no. 2, pp. 995-1006, Feb. 2013.
2. Bai Jing; Liu Jian; Lu Shi-Qi; Li Chao-Chao, "Harmonic characteristic of current source converters based on double Fourier integral theory," in Chinese Journal of Engineering, vol. 37, pp. 249-254, December 1, 2015.
3. Chen Yiwen; Qiu Yanhui; He Yongji; Chen Daolian, "Research on Single-phase Current Mode PWM Inverter of One-Cycle Control Strategy with Limited Storage Inductance Current," in Journal of Electrician Technique, vol. 30, no. 14, pp. 311-319, 2015.
4. WANG Ke-rou; WU Yan-xian; LI Yu-ling, "Study of a PWM Modulation Strategy of Three-phase Current Source Inverter and its Digital Realization." in Power Electronic, vol. 45, no. 4, pp. 76-78, 2011.
5. Federico M. Serra, Cristian H. De Angelo, Daniel G. Forchetti. Interconnection and damping assignment control of a three-phase front end converter. Electrical Power and Energy Systems, 60 (2014) 317–324.
6. Ortega, R., & Spong, M. (1989). Adaptive motion control of rigid robots: A tutorial. Automatica, 25(6), 877-888.
7. Ortega, R., Arjan van der Schaft, Bernhard Maschke, Gerardo Escobar (2002). Interconnection and damping assignment passivity-based control of port-controlled Hamiltonian systems. Automatica, 25(6), 877-888.

Design of a ship glass defogging system based on STM32 and an improved digital PID algorithm

Peng Lv*, Rui Zhang and Zhi-Jun Li
School of Automation, Wuhan University of Technology,
Wuhan 430070, P.R. China
**E-mail: lvpeng@whut.edu.cn*

In order to avoid the bad influence on ship sailing from fogging on the ship glass, a ship glass defogging system is introduced in the paper. The system was composed of a PC machine and a STM32 microcontroller system, they constituted a master-slave structure by RS485 serial communication. It was based on the root cause of fogging, real-time monitoring the dew point in the cockpit and the temperature of the ship glass, controlling the work status of resistance wire on the glass by an improved digital PID algorithm to defog the ship glass. The system is characterized with fast response speed, high control accuracy, energy conservation, consumption reduction and other advantages.

Keywords: Ship glass defogging; STM32; dew point ; an improved digital PID algorithm.

1. Introduction

As we all know, the marine environment is unpredictable. Sail view is obstructed and visibility is reduced caused by fogging arising from the change of environment, which makes the voyage in extremely serious grave danger. So to keep a good line of sight during the voyage and minimize the environmental impact on ships has become a very important research topic [1].

Dew point is a special value of temperature. It's the minimum temperature that air should reach to keep a certain humidity. When the temperature of the ship glass below the dew point, air can't hold too much moisture, moisture will turn into fog on the ship glass [2]. The system is based on the root causes of fogging, real-time monitoring the dew point in the cockpit and the temperature of the ship's glass, using the improved digital PID algorithm to process sample data, controlling the temperature of glass quickly and precisely to make it in or above the critical state of fogging, thereby avoiding fogging on the ship glass.

2. General Design of the System

The system is a master-slave structure, consisting of a PC and the STM32 microcontroller by serial communication with RS485, in order to monitor the dew point in the cockpit and the temperature of the ship glass in real time. This system used humidity and temperature sensor SHT15 and temperature sensor PT100 to collect the dew point and the temperature respectively. The signal will be transformed, amplified and filtered before sent to STM32. The chip processes them by the improved digital PID algorithm and driver module to change the work status of resistance wire on the glass through the heating device by judging the relationship of the dew point in the cockpit and the temperature of ship glass. The diagram of the overall structure of the system is shown in Figure 1.

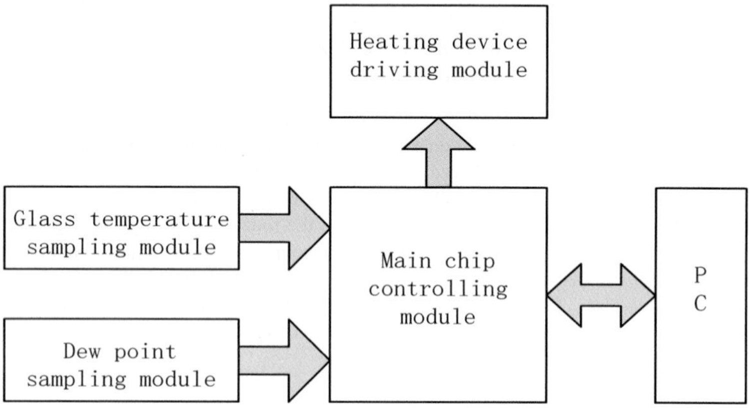

Fig. 1. Diagram of overall structure of system.

3. Design of System Hardware Circuit

3.1. *Main chip control module*

The module is comprised of single-chip microcontroller minimum system and serial communication. We chose STM32F103CBT6 which is produced by STMicroelectronics NV as the master chip. In addition to resetting, clock and power circuit, the single-chip microcontroller minimum system has a four-wire SWD debugging circuit and internal AD acquisition circuit [3]. It aims to receive and process collected data, unload them to PC, perform the algorithm and output control signals. The circuit of the microcontroller minimum system is illustrated in Figure 2.

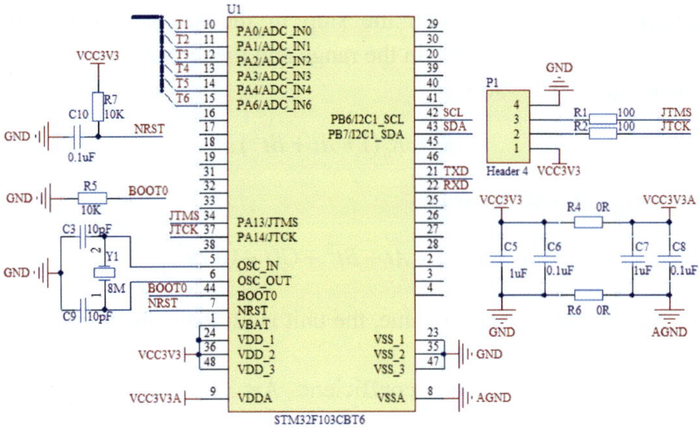

Fig. 2. Circuit of the microcontroller minimum system.

The function of serial communication module is to provide the data access interface to the PC monitoring software, which is used to monitor the dew point in the cockpit and the temperature of the ship glass. MAX3485 was selected to be the level conversion chip [1]. The circuit of serial communication module is shown in Figure 3.

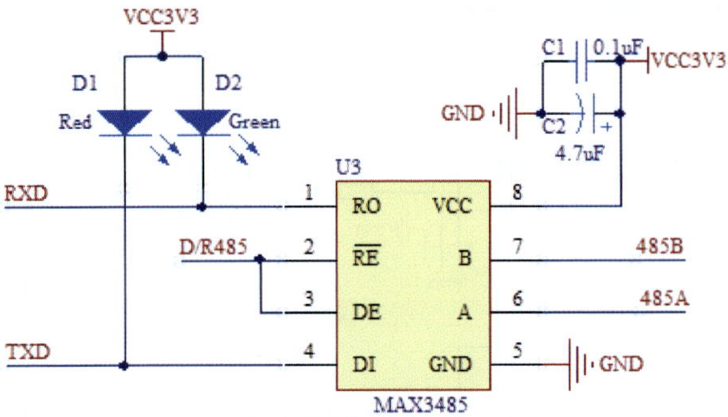

Fig. 3. Circuit of serial communication module.

3.2. *Glass temperature sampling module*

Thermocouple PT100 was selected as the temperature sensor. It gets the advantages of anti-vibration, strong stability, good linearity, high accuracy and so on. When the temperature changes from 0 to 100°C, the maximum nonlinear deviation is less than 0.5°C. In the range of –50°C -600°C, the performance of any other temperature sensor cannot exceed it. Taking into account the environmental temperature when the ship in the sea is –40°C -50°C. The equations of the resistance value in the range and the temperature are as follows:

In the range of 0°C -50°C:

$$R_t = R_0(1 + At + Bt^2). \tag{1}$$

In the range of –40 °C -0°C:

$$R_t = R_0\left[1 + At + Bt^2 + C(t-100)t^3\right]. \tag{2}$$

R_t is the PT100 resistance value, the unit is Ω; Ro is the resistance value of PT100 at 0°C, the value is 100Ω.

A: First-order temperature coefficient, A= 3.90802×10^{-3} °C;B: Second-order temperature coefficient, B=-5.802×10^{-7} °C; C: Third-order temperature coefficient , C=-4.27350×10^{-12} °C.

The output signal of the sensor is relatively weak, while the external interference can cause the error of the output signal. So a specific amplifying and filtering circuit is needed to modify and improve the signal. The circuit of the temperature sensor is shown in Figure 4 [4].

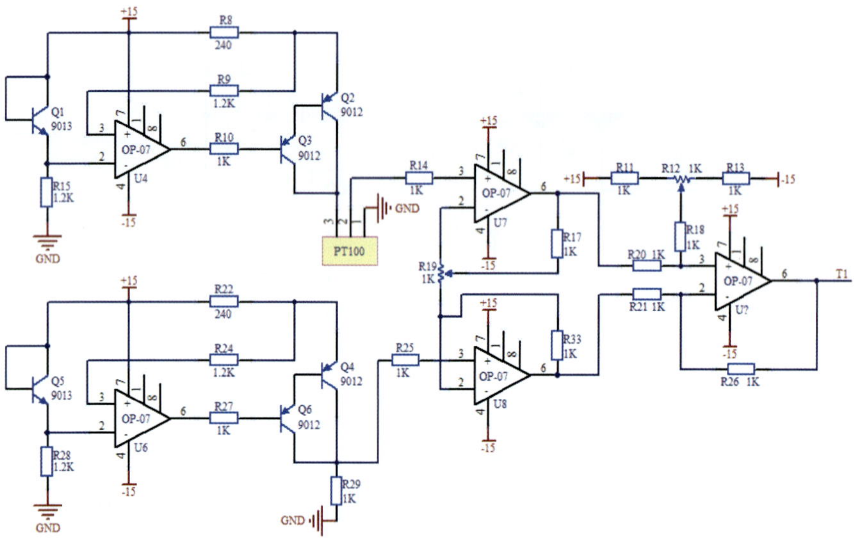

Fig. 4. Circuit of the temperature sensor.

3.3. *Dew point sampling module*

SHT15 is a digital temperature and humidity sensor chip produced by Sensirion Company in Switzerland. The internal temperature sensor is developed on the basis of band gap material PTAT (proportional to absolute temperature) which has excellent linearity. The digital output of temperature SO_t was converted to temperature value by Eq. (3). To obtain accurate measured data, humidity value was transformed from the digital output of humidity SO_{RH} through Eq. (4). Because of the significant difference between the actual temperature and the reference testing temperature 25°C, Eq. (5) was used for temperature compensation in order to get the accurate humidity. There are many calculation methods of getting dew point (T_d), most of them are very complex. When it came to the temperature range between –40 °C and 50°C in the system, we used Eq. (6) to get better measuring accuracy.

$$T = d_1 + d_2 \cdot SO_T \,. \tag{3}$$

$$RH_{linear} = c_1 + c_2 \cdot SO_{RH} + c_3 \cdot SO_{RH}{}^2 \,. \tag{4}$$

$$RH_{true} = (T-25)\cdot(t_1 + t_2 \cdot SO_{RH}) + RH_{linear} \,. \tag{5}$$

$$T_d(RH,T) = T_n \cdot \frac{\ln(\frac{RH}{100\%}) + \frac{m \cdot T}{T_n + T}}{m - \mathrm{n}(\frac{RH}{100\%}) - \frac{m\cdot T}{T_n + T}} \,. \tag{6}$$

In these equations, d_1, d_2 are the coefficients of the temperature's conversion and d_1=–40.1 °C, d_2=0.04°C; c_1, c_2, c_3 are the coefficients of the humidity's conversion, and $c_1 = -2.0468$, c_2=0.0367, $c_3 = -1.5955\times10^{-6}$. t_1, t_2 are the compensation-coefficients of temperature, t_1=0.01, t_2 =0.0008. T_n and m are calculation parameters of dew point, when the temperature is between 0 and 50°C, T_n=243.12°C, m=17.62, when the temperature is from –40°C to 0°C, T_n=272.62°C, m=22.46 [5].

SHTI5 communicates with the main chip through I^2C bus by serial data (SDA) line and serial clock (SCL) line [6]. The circuit of SHT15 module is illustrated in Figure 5.

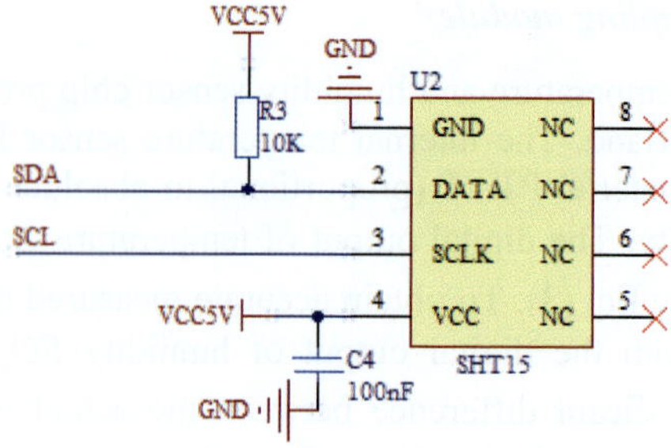

Fig. 5. Circuit of SHT15 module.

3.4. *Circuit of heating means driver*

In order to guarantee the device's reliability, there must be an optical coupler to insulate the core chip from the strong-current sides, which can prevent strong-current from flowing back. In this paper, TLP-627 was used as optical coupler. When the control system works, a signal will be produced by one of the STM32's I/O pins, and then it will be conditioned by resistances and diodes, after that it will be filtered by capacitances. Finally, it will be received by TLP-627 optical coupler, and then controlling the relay's on-off by the Darlington transistor inside to make the resistance wire on the glass heat or not. The circuit of heating device driver is shown in Figure 6.

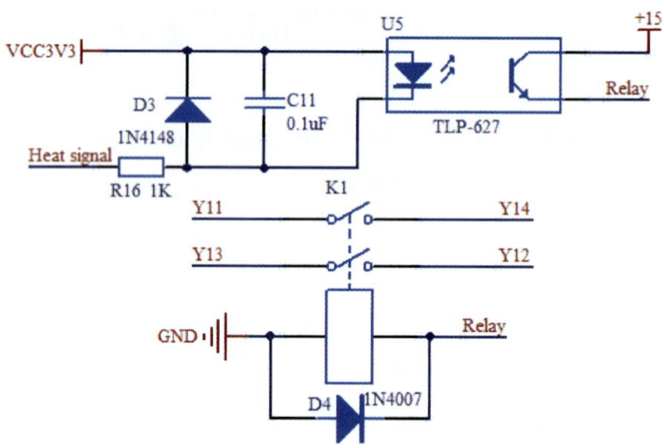

Fig. 6. Circuit of heating device driver.

4. Design of Software

4.1. *Analysis of control method*

In theory, when the temperature of the glass is less than the dew point of the cockpit, the glass will fog. But in the laboratory, a lot of experiments had been done by using air conditioners, humidifiers, hair dryer and other auxiliary equipment, and also a certain amount of literature had been consulted , in fact, when the difference value between the temperature of the glass and the dew point is less than 1.3°C, the glass will fog. So the function of the control algorithm is to make the glass temperature always in a state which above the dew point to more than 1.3°C, then it reach defogging effect, and also saves energy. When the system is turned on or just there is a sudden great change in environment, that may occur that the glass temperature is much smaller than the dew point in the cockpit, in order to ensure safety of navigation, we need rapid response and heat the ship glass rapidly, when it's time to steady-state that their difference value reached 1.3°C, the system needs more precise control methods to ensure system's accuracy.

Although the Bang-Bang control algorithm has high speed, its output often fluctuates up and down near the expected value, the steady-state performance is poor, so it is often combined with other algorithms which have high-stability. PID control algorithm has a simple structure, high robustness and high-stability, but it cannot get rapid control response [7].

In consideration of the actual needs, the system should retain the advantages of both PID control and rapid control response, so we took use of an improved digital PID algorithm what is Bang-Bang control works before PID control to eliminate the large error. Taking the difference value between glass temperature and the dew point in the cockpit as the final sample input value, the set value is 1.3, the difference value between the current value and the set value at the k time is defined as $e(k)$, the overall control system structure is shown in Eq.(7).

$$\begin{cases} Bang-Bang \quad control \quad stage, \quad e(k)<0 \\ \quad PID \qquad control \quad stage, \quad e(k)\geq 0 \end{cases}. \tag{7}$$

4.2. *Bang-Bang control algorithm*

Bang-Bang control is actually a switch control, also called time-optimal control, refers to the state space is divided into two regions, a region corresponding to the control variable takes the positive maximum value, another region corresponding to the control variable takes the negative maximum value, within the allowable control range, the system switches from one state to another state in the shortest time [8]. Combined with this paper, when $e(k)<0$, the outputs of Bang-Bang control reaches the maximum, and the resistance heats the glass vessels in maximal power, when $e(k)\geq 0$, the output of Bang-Bang control is zero. Expression is shown in Eq. (8).

$$u(k)=\begin{cases} u_{\max}, & e(k)<0 \\ 0, & e(k)\geq 0 \end{cases}. \tag{8}$$

4.3. *Digital PID control algorithm*

Digital PID control system is a discrete system of time, the control system is intermittently controlled by the controller, that is, within each sampling period, the measured sensor data is translated into a unified standard signal and input to the controller, getting the deviation by comparing with the set point in controller, then using the PID control operation to get the controlled amount and output the controlled amount to complete the corresponding adjustment task. We chose adding algorithm which has smaller deviation [9]. It is shown in Eq. (9).

$$\begin{aligned} &\Delta u(k)=u(k)-u(k-1)=K_p\left[e(k)-e(k-1)\right]+ \\ &\frac{K_p}{T_i}e(k)+\frac{K_pT_d}{T}\left[e(k)-2e(k-1)+e(k-2)\right] \end{aligned}. \tag{9}$$

Adding algorithm outputs increment $\Delta u(k)$ of control variables, without accumulated sum of calculation deviation, it only uses sampling values in the last three times, so the calculation amount is reduced, and previous conditions do not affect this output. K_p, T, T_i and T_d are comprehensively determined by parameters, the system can adopt theoretical calculation to determine these parameters, without unnecessary details here.

4.4. *Design of system software*

Programming of STM32 lower computer is divided into three parts. They are the primary program, serial communication program and control algorithm program. Main program achieves the coordination of each program and makes the system run continuously. The flowchart of serial communication main program is shown in Figure 7. The main function of control algorithm program: when the final sample input value which just is the difference value between the ship glass temperature and dew point in the cockpit is input to the STM32, calculating the difference value $e(k)$ between the current value and the set value, when $e(k)<0$,Bang-Bang control works; when $e(k)\geq 0$, Digital PID control works. The flowchart of the control algorithm main program is shown in Figure 8.

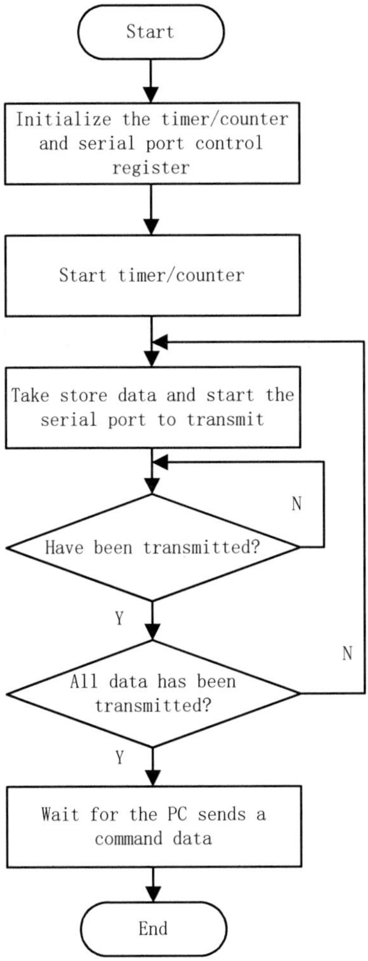

Fig. 7. Flowchart of serial communication main program.

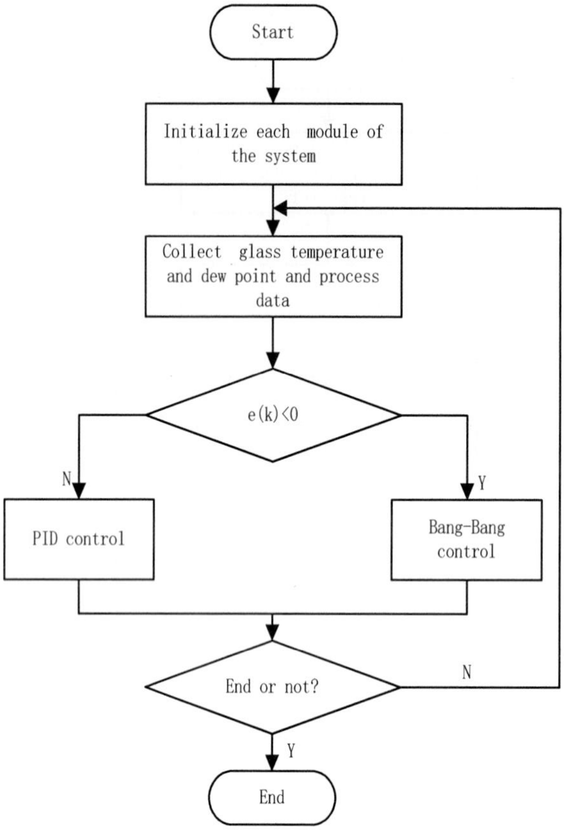

Fig. 8. Flowchart of control algorithm main program.

5. Conclusions

Traditional glass defogging system is to set the threshold of temperature in the FLASH controller, which is compared with the temperature of the glass. Then, the system heats glass to defog. Although this method can meet the requirements, it is blind and consumes too much energy. Starting with the fundamental reason of fogging, the system can detect the temperature of ship glass and the dew point in the cockpit in real time. Based on Bang-Bang+ digital PID control algorithm, the system is characterized with a fast response speed, high control accuracy, energy conservation and consumption reduction and other advantages. Based on Hardware circuit with strong anti-interference ability, this system performs well without extra hardware overhead. Compared with traditional intelligent temperature control system, the system is applicable to the place where the environment is complex, and requiring high precision and strong ability of anti-interference.

Acknowledgments

In this paper, the research was supported by Nature Science and Technology Major Project of China (2011ZX4002-21) and The Fundamental Research Funds for the Central Universities (WUT:2013-IV-056).

References

1. Zhu Penghao. Research and Design of glass defogging system based on large ships[D]. Wuhan: Wuhan University of Technology, 2013.
2. Wen Xiangang, Ma Shuqing, Du Bo, et al. Indicative Analysis of Grass Temperature and Dew-Point Temperature to the Occurrence of Dew and Frost [J].Meteorological, 2012, 38(6): 745-750.
3. Zhou Lina, Zhou Jianping, Xu Yan, et al. Oil well pressure monitoring system based on STM32F103CBT6 and Zigbee [J]. Internet of Things Technologies, 2014, 6: 38-41.
4. Chen Jie, Huang Hong. Sensors and detection technology [M]. Beijing: CHINA MACHINE PRESS, 2014.
5. The SENSIRION Company's global website. SHT1x Technical Manual (English). http://www.sensirion.com.cn/fileadmin/user_upload/customers/sensirion/Documents/Humidity/Sensirion_Humidity_SHT1x_Datasheet_V5.pdf.
6. Xu Huifang. Design of intelligent temperature and humidity data acquisition system based on SHT15 [J]. Journal of Chinese Agricultural Mechanization, 2015, 36(5):237-239.
7. Suo Yangyang, Zhao Xin. Implementation of Bang-Bang + PID Two-mode Control Algorithm in Heavy Duty Transportation Vehicle Steering Control System [J]. Beijing: Journal Of Beijing Polytechnic College, 2015, 14(4): 39-42.
8. Qiang Minghui, Ma Wenxue, Ma Yongwei. Improved Bang-Bang Control Algorithm in Temperature Control Based on S7-200[J].Automation and instrumentation, 2013, 6:97-99.
9. Zhang Xianlei, Yuan Zhuogang, Zhang Qiang. Fuzzy PID control to cement vertical roller mill based on bang-bang [J].Jinan: Journal of University of Jinan (Sic. &Tech.), 2014, 29(2):100-106.

Improvement and implementation of adaptive line-spectrum enhancement

Li-Wei Li

Weapon Engineering Department of Naval University of Engineering, Wuhan 430033, China

E-mail: 1320053911@qq.com

Enhancing the line spetrum of ship-radiated noise by Adaptive Line-spectrum Enhancement is good to detect and extract line spectra. An improved adaptive line enhancement method of line-spectrum detection is used .To solve the problem that adaptive line enhancer implemented by FPGA through bttom layer HDL coding has a poor development efficiency, a model of improved adaptive line enhancer is established by DSP Builder and files which can be used by FPGA are generated automatically. Then the improved adaptive line enhancer is simulated by the simulation model, and the feasibility of the model is validated . It is easy and simple to design adaptive line enhancer by using this method.

Keywords: Ship-radiated noise; line spectrum; simulink; DSP builder; adaptive line-spectrum enhancement; FPGA.

1. Introduction

Ships' radiation signal contains abundant line-spectrum and the detection and extraction of radiated noise of line-spectrum has important significance for ship target detection and recognition. Enhancing narrow-band line-spectrum by adaptive line enhancer is good to extract line-spectrum features. In this paper an improved adaptive line enhancement method of line-spectrum detection method is used. Based on the analysis of minimum mean square error (LMS) adaptive algorithm, use DSP Builder and a variety of EDA tools to design the improved adaptive line enhancement, which greatly improves the design efficiency and device performance.

2. The Principle of Adaptive Line Enhancement

Adaptive line enhancement (ALE) consists of a programmable filter and adaptive algorithm and the composition is shown in figure 1. The input signal $x(n)$ include signal $S_N(n)$ and noise $S_B(n)$. After a certain delay, the input signal input programmable filter. After a delay, because of the signal correlation, the signal remains but the noise attenuates, in this way, the signal and noise separated. The weights of filter are adjusted by the the difference between the input signal and output signal.

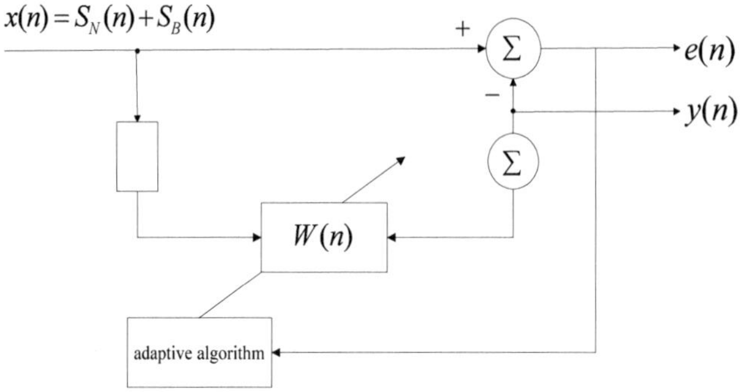

Fig. 1. The principle of adaptive line enhancement.

Suppose at the moment of n ,the input signal is $x(n)$, the output signal is $y(n)$, so $y(n)=w^T(n)x(n-\Delta)$,then define the error signal is the difference between the input signal and the actual output signal, that is to say, $e(n)=x(n)-y(n)$ .So the weight vector iterative formula of the LMS algorithm is $W(n+1)=W(n)+2\mu e(n)X(n-\Delta)$.

Among, μ is the step size factor, controls the stability and convergence speed of the adaptive control system through the adjustment coefficient, to make the mean square error minimum.

3. The Improved Adaptive Line Enhancement

Sound field has not only sound pressure but also vibrational velocity. In the far field, sound pressure and vibration velocity of ship radiated noise can be considered to be relevant, and environmental sound field can be regarded as isotropic noise field.Take the advantage of vector hydrophone which synchronously measure the sound pressure signal $p(n)$ and vibrational velocity signal $v_x(n)$, to improve the adaptive line enhancement as double input structure, as shown in Figure 2:

4. The Establishment and Simulation of the Model

4.1. *The design process of DSP Builder*

Figure 3 is the process diagram of DSP system design by DSP Builder. Design process is established based on MATLAB and then Simulink model file (MDL)

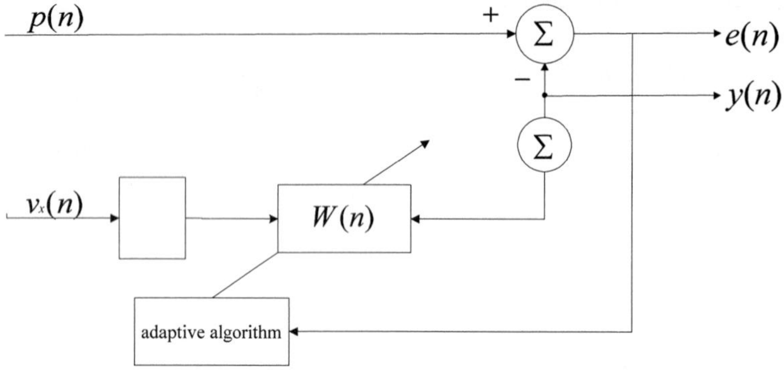

Fig. 2. The principle of improved adaptive line enhancement.

is converted into hardware description language (VHDL) design documents (VHD) by SignalCompiler. Compile and synthesize the RTL code and simulation of VHDL file. At last, use the programming files produced by Quartus ii to complete the FPGA configuration.

4.2. *Modeling and simulation of the system*

The Figure 1 shows that the LMS adaptive line enhancement is mainly consist of delay module, addition module, multiplication module and coefficient of update module. Figure 4 is the model of two kinds of fourth-order ALE. The upper is the improved ALE and the other is the usual ALE.The input signal is got by a sine signal superimposed on the random noise signal.The delay module, addition module and multiplication module are directly called from the module in the DSP Builder. Through the Scope module of Simulink, the output signal waveform can be observed.The coefficient of update module is to complete the renewal of the weight parameter adjustment of the filter, and the filter module makes the implementation of the fourth-order FIR filter.This two modules are both into subsystems whicn can be easily called, as shown in Figure 5, Figure 6.

Figure 7 shows the simulink simulation results, μ =0. 004. From the result we can see that the output of the improved ALE achieved an more ideal result that the noise signal attenuates much more and the useful signal remains as much as possible.

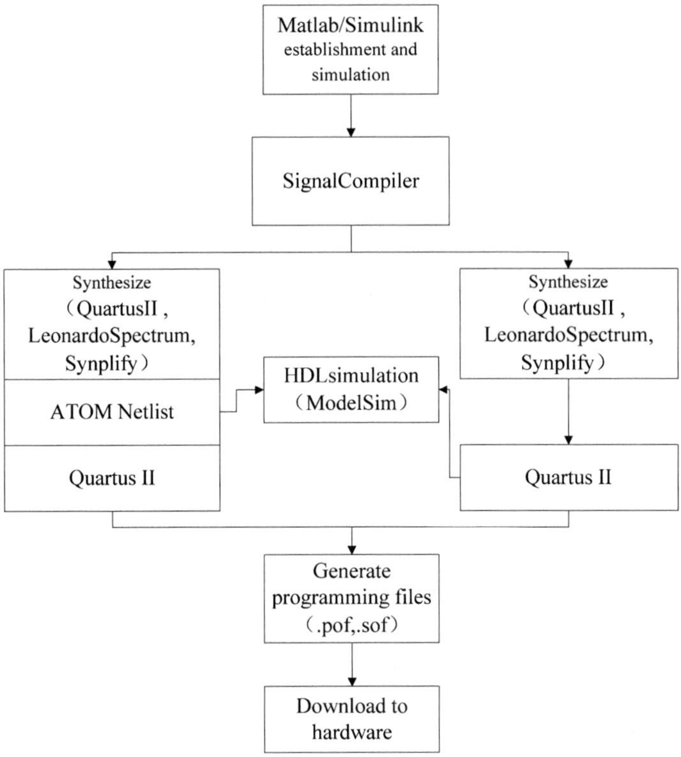

Fig. 3. The process diagram of DSP system design by DSP Builder.

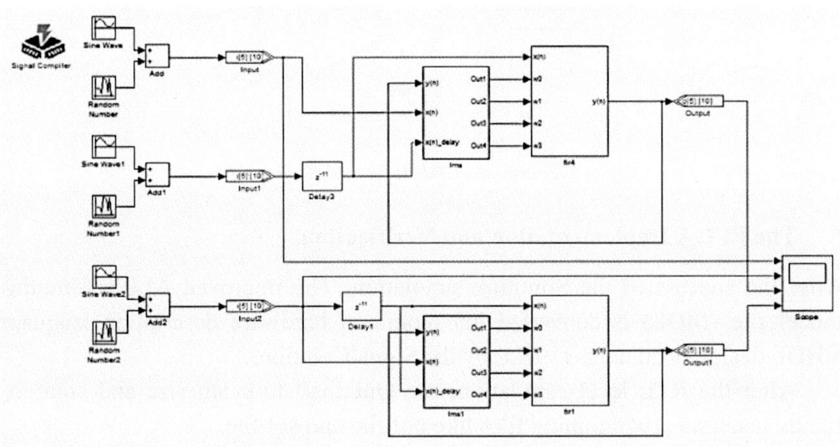

Fig. 4. The model of two kinds of fourth-order ALE.

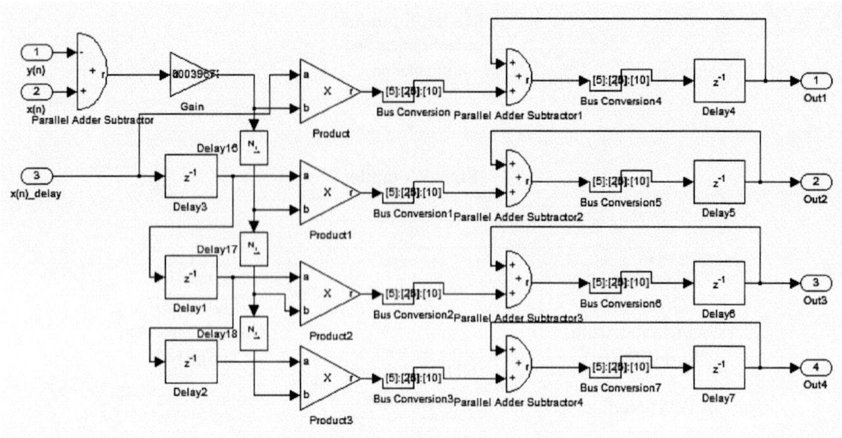

Fig. 5. Coefficient update module diagram.

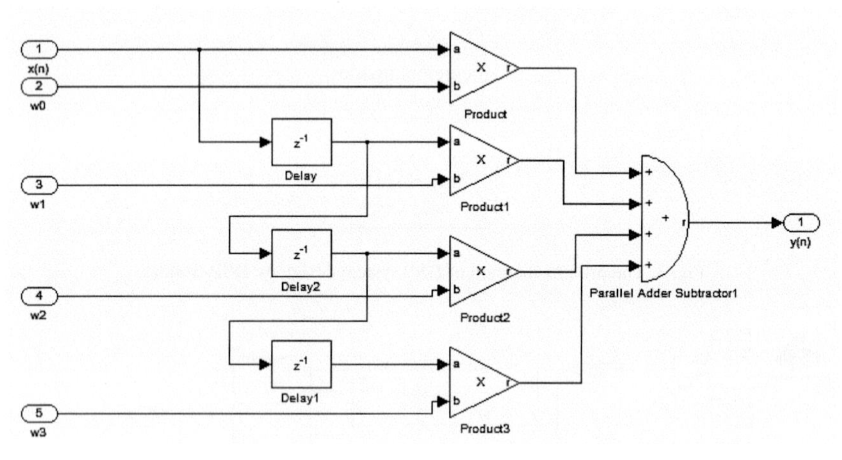

Fig. 6. Fourth-order FIR filter module diagram.

5. The FPGA Implementation and Verification

After the success of the Simulink simulation, The improved ALE's Simulink model file (MDL) is converted into common hardware description language VHDL design documents (VHD) by the SignalCompiler.

After the RTL level simulation, use QuartusII to synthesize and compile, finally generate programming files like pof file and sof file.

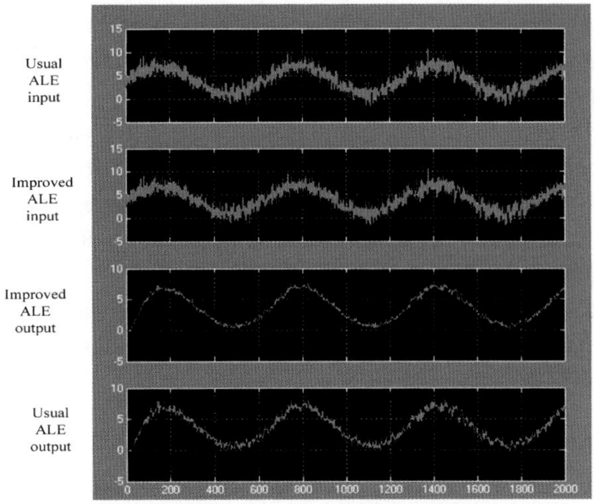

Fig. 7. The comparation of Simulink simulation diagram.

6. Conclusion

In this paper, based on the analysis of the minimum mean square error algorithm, complete the design of the improved ALE's model by the FPGA DSP development tools DSP Builder. This method makes the system level tools and the RTL design tools connected to take the largest advantage of these tools and makes the design flexible, efficient and shortens the development cycle, has the very high application value.

References

1. SHAN Guangchao, ZHAO Hanbo. *Detection and Analysis of Ship Radiated-noise Line Spectrum.* Ship Electronic Engineering, 2014 (10):119-122
2. SUN Xi, LI Bing, FANG Yi. *Modified Adaptive Line-spectrum Enhancement Algorithm to the Detection of Line—spectrum Signal*[J]. China Science and Technology Information, 2012 (8):62-62.
3. LIU Xiong fei, GAO Jin ding, QI Hai bing.*A New Way on FPGA Implementation of LMS Adaptive Filter*[J] Piezoelectric and acoustooptic, 2007, 29(1):87-89.
4. LEI Neng-fang. *FPGA design of digital signal processor based on DSP Builder*.Electronic Design Engineering.

5. MENG Sha-sha, DONG Hui-ning, DONG Jian. *DSP Builder-based Model of Adaptive Digital Filter. Communications Technology*, 2011, 44(8): 146-148.
6. XU Zhongxian. *Simple Analysis on Adaptive Line-spectrum Enhancement Algorithm. Science and Technology Information*, 2010 (15):96-97.
7. ZHANG Xiaoyong. *Modified Adaptive Line-spectrum Enhancement Algorithm to extract Line—spectrum Signal* [J]. Research on Telecommunication Technology, 2014 (4):16-20.
8. LIU Kaijian, WU Guangmin, ZHANG Haibo. *Design and Implementation of FPGA Self-adapting Filter Based on LMS*. Instrumentation. Analysis. Monitoring, 2008 (4):10-12.

Speed sensorless control of PMSG based on virtual instantaneous power phase locked loop

Tao Wu
College of Electrical Engineering,
Northwest University for Nationalities, Lanzhou, China
E-mail:wutao826@126.com

Hong-Tao Liu
College of Electrical Engineering,
Northwest University for Nationalities, Lanzhou, China;
E-mail:lht8778@126.com

A speed detection technology based on virtual instantaneous power phase locked loop is used on PMSG' control strategy in this paper, and the speed sensor-less control technology based on the stator voltage vector orientation is realized in generator-side converter. With the space vector pulse width modulation (SVPWM) method and grid voltage synthetic vector orientation, the power decoupling control and the maximum power capture are realized in grid-side converter. The MATLAB simulation results show the system is of a fast dynamic response, capable of accurately measuring the speed and the rotor position of the generator together with the function of MPPT and power decoupling control. The feasibility and correctness of the control strategy are verified by MATLAB simulation.

Keywords: PMSG; virtual instantaneous power PLL; power decoupling control; speed sensorless control method.

1. Introduction

Direct-driven permanent magnet synchronous wind power generation (D-PMSG) system is one of the research focuses of wind power technology [1]. D-PMSG system is characterized by variable-speed operation, variable-pitch regulation, low speed, high efficiency and high power factor [2]. Therefore, the research on this system is significant to improve the efficiency and reliability of wind power generation.

In the traditional wind power generation control system, the speed and location of the generator rotor must be checked so as to complete speed control and magnetic linkage orientation. The traditional control method based on speed test will give rise to several problems [3], such as increasing the system maintenance cost and reducing the system reliability. Therefore, more and more attention is put on the speed sensor-less control method, which, on the whole, can be divided into open-loop estimation method and closed-loop estimation method. In the open-loop estimation method, the system is of quick dynamic response and lower accuracy. In the location estimation method controlled by closed loop, the system accuracy is improved, but the algorithm is too complex, unfavorable to engineering application. Therefore, simple and reliable algorithm research is still one of the important research aspects of the speed sensor-less control.

The main system of D-PMSG is composed of wind turbine, direct-drive permanent magnet generator, generator-side converter, DC-side energy-storage capacitor, grid-side converter and AC-side filter reactor.

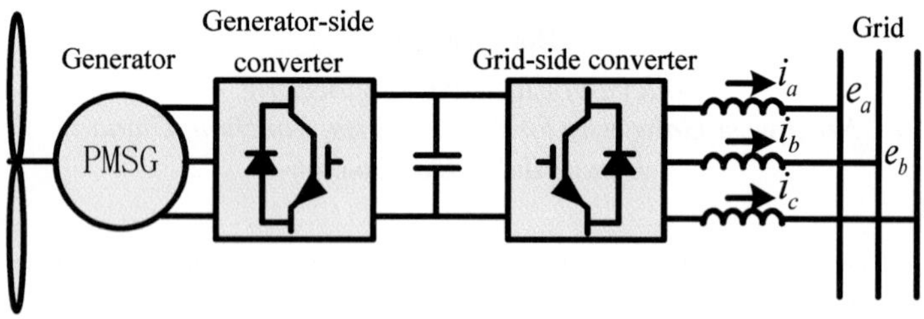

Fig. 1. Structure of D-PMSG system.

The converter in the paper is of "back-to-back" structure [4]· Since the generator-side converter is controlled by PWM, the rotor current harmonics of the generator can be effectively reduced, thus effectively reducing the torque ripple of the generator. Simultaneously, MPPT control is completed by the speed sensor-less control method based on virtual instantaneous power phase locked principle. The classical PI regulator is used to control the pitch angle at high wind speed, realizing the variable-pitch constant-power control. With the space vector pulse width modulation (SVPWM) mode and the control strategy of grid voltage synthetic vector orientation, the decoupling control of active power and reactive power is realized in the grid-side converter. The feasibility and correctness of the scheme has been proved by MATLAB simulation.

2. Speed Sensor-Less Control Strategy

When the permanent magnet synchronous generator in the megawatt direct driven system is operated for power generation at the lowest cut-in wind speed point, its stator voltage is of some frequency and amplitude value, to provide guarantee for the speed sensor-less control technology [5-6]. A speed sensor-less control method based on virtual instantaneous power phase locked principle is adopted in the paper, with its basic steps as follows:

Virtual instantaneous power (p) is defined by three-phase circuit power. Its phasor expression is:

$$p = u_a i_a + u_b i_b + u_c i_c \tag{1}$$

$i_a + i_b + i_c = 0$ when the system is of three-phase three-wire system, from which:

$$p = u_{ab} i_a + u_{cb} i_c \tag{2}$$

In consideration of the phase locked loop applied in the generator side, three-phase voltage and three-phase current are symmetrical. From Eq. (1), the instantaneous power is zero under the condition that u_a is orthogonal to i_a, that is to say, the phase difference of u_a and i_a is $\pi/2$. The virtual power phase-locked loop is as follows:

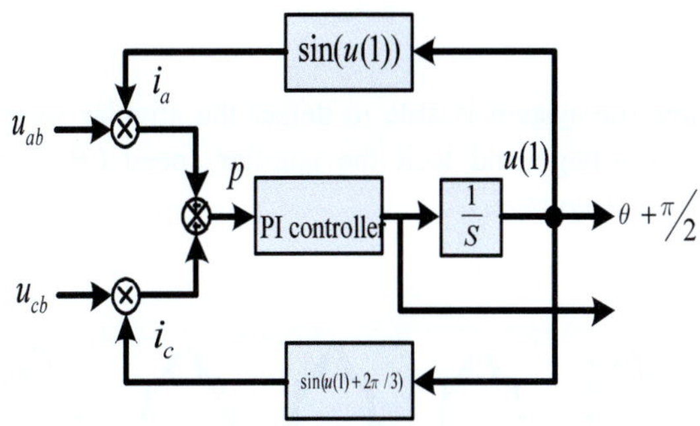

Fig. 2. Virtual instantaneous power phase locked loop.

Only when the input p of the variable parameter PI controller is stable, namely, $p = 0$, PLL circuit is at its stable working point. The constraint condition of the stable working point is that ω equals to the grid frequency and i_a is orthogonal to the measured three-phase voltage. There is just only one stable working point. That is to say, the reference current output is $\pi/2$ ahead of phase A fundamental positive sequence voltage. When the reference current output is $\pi/2$ lagged behind phase A fundamental positive sequence voltage, the working point is not stable. At the working point, if an accidental disturbance leads to a tiny reduction in system frequency, namely, the reduction in the frequency of three-phase voltage, the following conditions will occur: the fundamental positive-sequence component is rotated slower than the reference current output i_a and its angle with i_a will be less than $\pi/2$; the virtual instantaneous active power $p \geq 0$; the output of PI controller (ω) is increased. Thus, the fundamental positive-sequence component is rotated slower than the reference current output i_a. Such positive feedback characteristic is the unstable characteristic of the working point. When the reference current output is $\pi/2$ ahead of phase A fundamental positive sequence voltage, for the same disturbance, i.e., the reduction in the frequency of three-phase voltage, the following conditions will occur: the fundamental positive-sequence component is rotated slower than the reference current output i_a and its angle with i_a will be more than $\pi/2$; the virtual instantaneous active power $p \leq 0$; the output of PI controller (ω) is decreased. Thus, i_a is kept to be orthogonal to three-phase voltage, achieving to the purpose of tracking the system frequency and phase. Therefore, after the phase locked output is stable, the following formula constraint is met:

$$\begin{cases} \hat{\omega} = \omega_{u_{a1}} \\ \hat{\theta} = \theta_{u_{a1}} + \pi/2 \end{cases} \tag{3}$$

That is to say, the system is able to detect the angular speed ($\omega_{u_{a1}}$) of the generator terminal voltage and lock the angular speed ($\theta_{u_{a1}} + \pi/2$) which is $\pi/2$ ahead of phase A voltage.

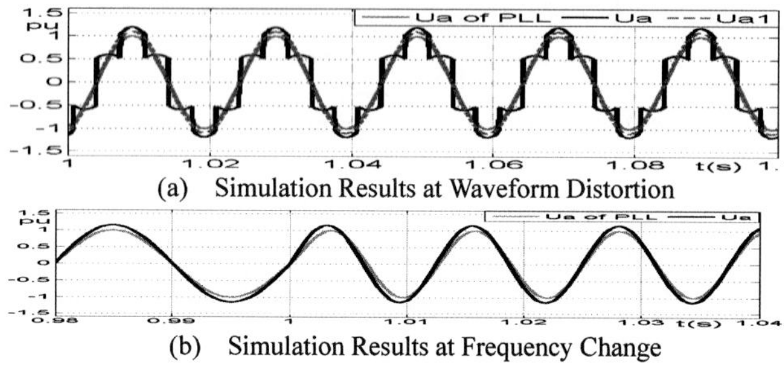

Fig. 3. Simulation results of virtual instantaneous power phase-locked loop.

Seen from Fig. 3, when the voltage is subjected from severe distortion, phase A fundamental voltage can be reliably tracked by this phase locked loop. Seen from Fig. 4, when the frequency is increased to 80Hz from 50Hz at 1s, the tracking process can be completed by this phase locked loop within one and half cycles, i.e., about 0.01875s, fully reflecting the stability and dynamic performance of this phase locked loop algorithm. Simultaneously, this phase locked loop is characterized by simple algorithm and convenient realization.

3. Control Strategy of Generator-side Converter

In the paper, the generator-side converter is controlled by the unit power factor orientated based on generator terminal voltage. Its advantages are to complete the maximum power capture of the wind generator, increase the power factor of the generator and lower the distortion rate of the voltage harmonics.

Known from the mathematical model of PMSG, the equation of generator power in the synchronous rotating coordinate system is [7]:

$$\begin{cases} P = u_d i_d + u_q i_q \\ Q = u_d i_q - u_q i_d \end{cases} \tag{4}$$

When the orientation is made in generator terminal voltage, $u_q = 0$, Eq. (4) is simplified as:

$$\begin{cases} P = u_d i_d \\ Q = u_d i_q \end{cases} \tag{5}$$

When the unit power factor is adopted for control, $Q^* = 0$, the control system drawing of the grid-side controller is shown below:

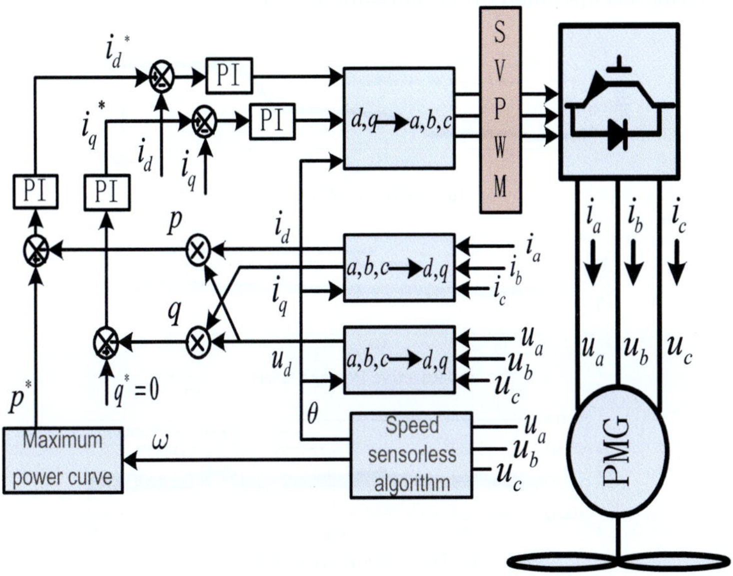

Fig. 4. Generator-side converter control.

Known from Fig. 3, through the detection on the terminal voltage and current of the motor, the output power of the motor is calculated from Eq. (5). The active power is given by the main controller based on the maximum power curve while the given reactive power is 0, thus completing the unit power factor control. The given value is compared with the calculated value to form outer power loop, so as to control the active output of the generator. The given AC/DC axis current (i_d^*) is obtained through PI regulator. i_q^* is compared with the measured values i_d and i_q to complete the inner current loop control. The converter is controlled by space vector pulse width modulation (SVPWM).

4. Simulation based on MATLAB

The simulation system is designed in MATLAB/SIMULINK as follows: the wind field is of PMSG capacity of 10MW, terminal voltage of 600V, frequency (f) of 60Hz, rated DC-side voltage of 1,100V, DC bus capacitor of 0.08F and rated wind speed of 13m/s.

Speed senor control and speed sensor-less control processes are compared in the simulation. In simulation, the grid-side converter is required to be operated at the unit power factor, so $Q^* = 0$ is assumed. The initial wind speed is 6m/s and the wind speed is increased to 8m/s at t=4s, to 11m/s at t=8s and to 15m/s at t=14s. The simulation duration is 20s. Fig. 5 is the change of various curves during the operation of the simulation system.

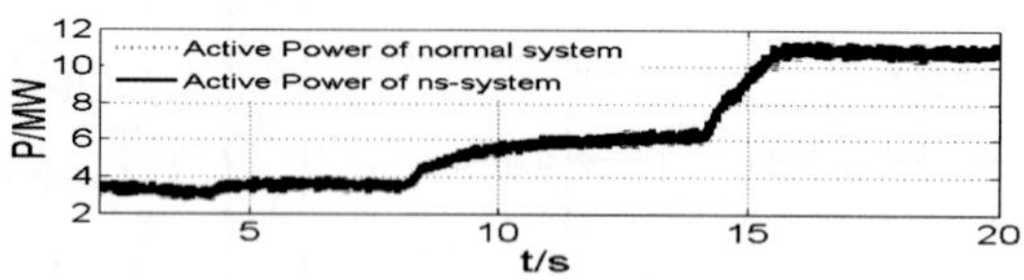

(a) Active Power Curve

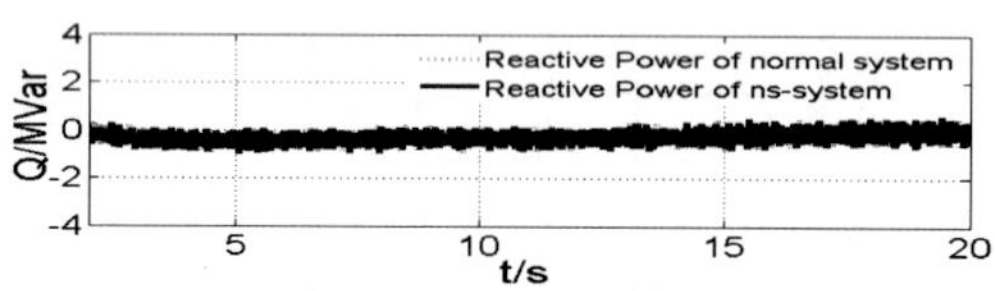

(b) Reactive Power Curve

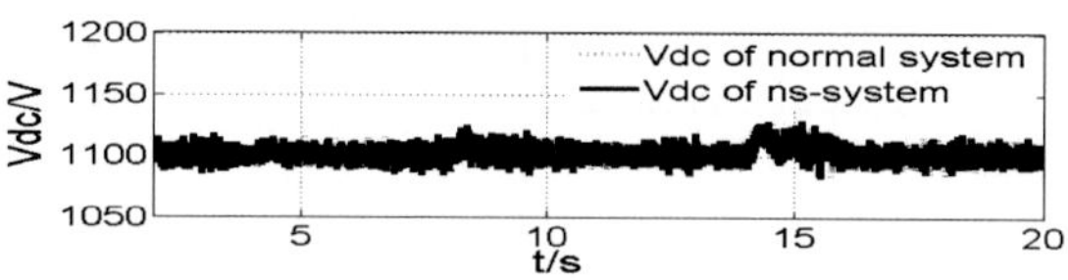

(c) DC-side Voltage Curve

Fig. 5. Generator operation status curve.

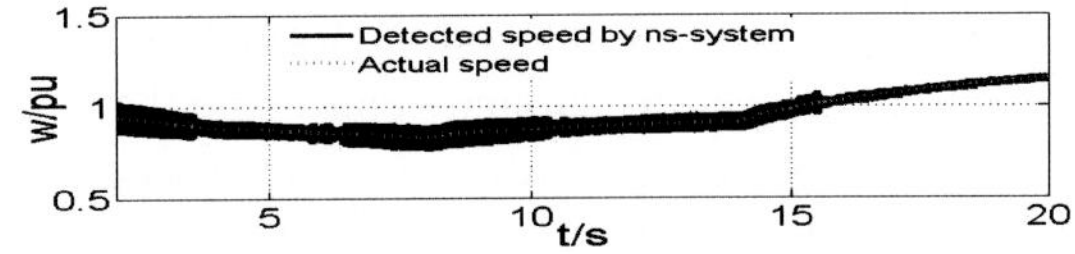

(d) Detected Speed and Actual Speed Curves

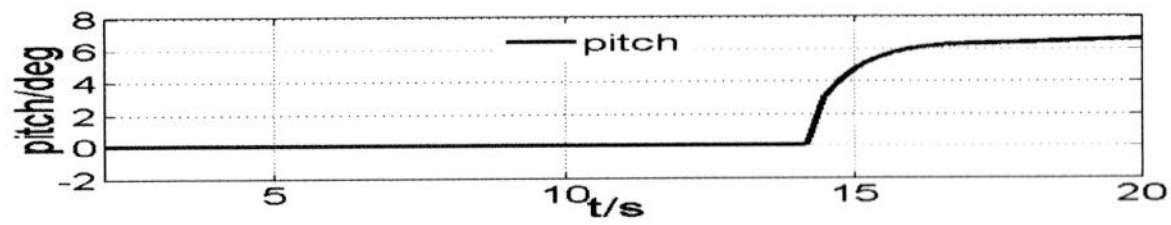

(e) Pitch Angle Curve

Fig. 5. (*Continued*)

Seen from Simulation Figure 5 (a, b and c), the control effects in the speed senor control and speed sensor-less control processes are almost the same (the curves of power, DC-side voltage and generator speed are almost superpositioned completely). In Fig. 5(d), it is reflected that the measured speed based on the virtual instantaneous power phase locked principle is fully consistent with the actual speed of the generator and fluctuated around the actual speed value. But the interference will not affect the normal operation of the system.

Seen from Simulation Figure 5 (a and b), the generator speed at low wind speed section will be adjusted in time with the change of the wind speed for variable-speed operation. The active power will be quickly adjusted with the change of the motor speed, to capture the maximum wind energy. Seen from Simulation Figure 5 (e), when the wind speed exceeds the rated wind speed, the generator will be of variable-pitch adjustment; the pitch angle will be adjusted in time with the change of the wind speed; the active power will be stabilized around the rated power of the generator, to realize constant-power control. Seen from Simulation Figure 5 (b), the output reactive power of the generator is always fluctuated around 0 since the unit power factor control is adopted. Seen from Simulation Figure 5, the DC-side voltage can be stably controlled to be 1,100V when the operation status of the generator is changed.

5. Conclusions

Through the research on the control method of the D-PMSG in the paper, the speed sensor-less control based on virtual instantaneous power phase locked principle is adopted for the generator-side converter under the condition the disadvantages of the sensor are overcome. It is concluded from the simulation test that the principle based on the virtual instantaneous power phase locked loop is of good properties, is capable of measuring the actual speed of the generator accurately and also of integrating the traditional maximum power capture control with the power decoupling control and the variable-pitch control properly, to provide theoretical basis for the speed sensor-less control technology of PMSG system.

Acknowledgments

This paper is supported by the Fundamental Research Funds for the Central Uni versities of Northwest University for Nationalities (grant 31920130057;grant 31920150064).

References

1. Haque M E, Negnevitsky M, Muttaqi K M. A novel control strategy for a variable speed wind turbine with a permanent-magnet synchronous generator[J]. IEEE Trans on Industrial Applications, 2010, 46(1): 331-339.
2. Mukhtiar S, Ambrish C. Application of Adaptive Network-Based Fuzzy Inference System for Sensorless Control of PMSG Based Wind Turbine With Nonlinear Load Compensation Capabilities[J]. IEEE Trans on Power Electronics, 2011, 26(1): 165-175.
3. Yao Jun,Liao Yong,Zhuang Kai. Coordinated control strategy of back-to-back PWM converter for permanent magnet direct-driven wind turbine[J]. Automation of Electric Power System, 2008, 32(20):88-92.
4. Xia Yihui,Liu Jianbo,Zhang Xiaofeng,et al. Design and Analysis of Digital Phase Looked Loop Based on Virtual Instantaneous Power[J]. Proceedings of the CSU-EPSA, 2012, 24(4): 43-47.
5. Katsumi Nishida, Tarek Ahmed, Mutsuo Nakaoka.A Novel Current Control System Using PLL Circuit For Interior Permanent Magnet Synchronous Generator[A].Power Electronics Specialists Conference.Rhodes: 15-19 June 2008: 4717-4722.
6. Li S, Timothy A, Richard P. Optimal and Direct-Current Vector Control of Direct Driven PMSG Wind Turbines[J]. IEEE Trans on Power Electronics, 2012, 27(5): 2325-2337.
7. YANG Yong, RUAN Yi. Grid-connected Inverter for Wind Power Generation System[J]. Journal of Shanghai University, 2009, 13(1): 51-56.

Chapter 2

Modeling and Simulation

Surface region recurrence pick-up algorithm research based on STL model

Bo Qian* and Ming-Yi Wang†

School of Mechanical and Power Engineering,
East China University of Science and Technology,
Shanghai, 200237, China
**E-mail: qianbo@ecust.edu.cn*
†*E-mail: wangmingyi1892@126.com*

The picking up surface region algorithm is slowly and inefficient based on 3-Dimension model of STL (Structural Text Language) file in traditional, and it's not easy to deal with the crack of STL model. To solve this problem, this article presents a rapid recurrence search algorithm of picking up surface region. The algorithm improved the building rate highly because of using triangle adjacent and border overlapping to recurrence and only once searching for every eligible triangle. The algorithm efficiency is $O(n)$. It's able to easily identify the STL file cracks, and other non-normal conditions, it also auto build 3-Dimension boundary rings of region contour. This algorithm have applied to support building of multi-rapid prototyping (about stereolithography, selected laser sintering), the result proved the algorithm not only enhanced calculation speed but also improved process quality, and achieved good effect.

Keywords: STL model; region pick-up; crack identify; recurrence algorithm; rapid prototyping.

1. Information

Rapid prototyping is a kind of manufacture technology which is driven by CAD (Computer aided design) model and can generate rapidly three-dimensional solid in arbitral complex shape. Based on manufacture principle of discrete layer, three-dimensional CAD model will be divided into a series of sequential two-dimensional slices along a certain direction(usually along Z direction) which are manufactured by prototyping machine in turn and are connected automatically, thus three-dimensional physical solid is generated 1.

STL (Structural Text Language) file is an input file format of three-dimensional model in rapid prototyping field, which has stored three vertex coordinates and outer normal vectors of discrete triangles. These triangles are formed by the triangularization of CAD model surfaces, whose storage order is arbitrary [6][7][8][9]. Although STL file is described by some discrete triangular grids, its validity depends on the internal implied topological relationship. The correct data model should meet the following rules: (1) there is only one public edge between two adjacent triangles, namely, adjacent triangles must share two vertices; (2) every edge of the triangle is connected by only two triangles 1.

In the paper, STL model refers to product model in STL file format which has been widely applied into rapid prototyping technology 2. Especially in the early PR technics process, the model should be adding support, dimensioning and other operations, however, because of thousands of slices in STL model, pick-up needs a long time to traverse all triangles in two layered model by traditional pick-up method, which brings many adverse effects on the technics process. As shown in figure 1, a dragon crafts model is composed of more than ninety thousands of triangles, which needs half an hour to add support, thus it can not reflect the meaning "speediness" of rapid prototyping. As for this problem, the author put forwards a kind of recursive searching pick-up algorithm based on the research of surface region rapid pick-up algorithm on STL model. And the result shows that the algorithm improve pick-up speed greatly.

2. Pick-up Conception

Region pick-up of the model surface 3 is to extract all the triangles on STL model which meet the conditions as follows: they are the area, shape of triangles, angle between triangles and horizontal plane and so on. As long as triangles meet the specified conditions, they can be identified from STL model. And then these triangles are merged, namely, the public edges of these adjacent triangles are deleted, and other edges are preserved and sorted by certain clockwise, consequently, one or more three-dimensional contours are formed in which the merged triangles are used to fill the surface region. The formed three-dimensional region is pick-up region. Moreover, the merged triangles are considered to be contains attributes of the region.

The traditional pick-up region method 4is as follows: on STL model, the information of all triangles is discrete, separate, and has no topological relationship between the two. On the above merged process, every triangle can search adjacent triangles by traversing all the information, therefore, as for the number of triangles is n on the STL model, in theory the efficiency of the algorithm is $O(n^2)$. However, in the actual test the efficiency of the algorithm is low (as shown in Table 1) which seriously affects the overall performance of the software operation, therefore, it is necessary to increase pick-up speed especially improve the merged algorithm of triangles.

3. Rapid Recurrence Pick-up

Recurrence search algorithm is put forward to solve such kind of problems. In order to achieve the aim, the author considers that firstly should construct topological relationships 5 of all the triangles on reading STL model file. The specific situations can be shown by the following definitions:

$T(n)$——definition (1) , is the triangle whose sequence number is n on STL model, and $n \in (0, N-1)$, moreover, N stands for the sum of triangles on STL model; $L(n,m)$——definition (2), is one edge whose sequence number is m in the definition $T(n)$, and $m \in (0,2)$;

$b = I(n)$——function(1), returning to STL model and checking whether the triangle whose sequence number is n has been searched in the recurrence. If b is $True$ which shows being searched, otherwise, b is $False$;

$n_2 = F_1(n_1, m_1)$——function(2), returning to the sequence number of triangle which is adjacent to $T(n_1)$ and has the common edge with $L(n_1, m_1)$, and $n_2 \in (0, N-1)$; if $n_2 < 0$, it shows that there is no triangle which has the common edge with $L(n_1, m_1)$ in $T(n_1)$. This kind of situation usually appears when there is irregular geometry next to $T(n_1)$ with cracks or leaks, etc.

$m_2 = F_2(n_1, m_1, n_2)$——function (3), returning to the sequence number of one edge in $T(n_2)$, and $L(n_1, m_1) = L(n_2, m_2)$, namely, the edge has the common edge with $L(n_1, m_1)$.

It is easy to construct topological relationships of all triangles in mathematical expression through the above functions. The specific steps are as follows: firstly, the seed triangle meeting the above conditions should be found which is considered to be source triangle; secondly, its three edges are indexed to three adjacent triangles, and the triangles are considered to be destination triangles. Thirdly, it judges whether the destination triangles are searched and meet the pick-up conditions or not. If meeting the conditions, these destination triangles will be conversed into source triangles; conversely, the remained edges should be indexed after returning the source triangles. When all the triangles are searched through the above steps, the recurrence will conclude. The core idea is to find a qualified region immediately through a triangle meeting the conditions, and its simulation is like a drop of water spreading on the smooth plane until forming a closed region with boundary. The above ideas can be shown by the following procedures:

a) Traverse all the triangles on STL model whose sign 0.
b) Traverse all remained triangles, whose sign is 0, and find a qualified triangle, meanwhile, traverse the triangles whose sign is 1. If having not found a qualified triangle, conclude the recurrence.
c) Traverse three edges of the triangle, and check whether three edges are traversed or not; if not, choose edge which doesn't traverse and

enter the next step; if being traversed, return to the second step and continue it.

d) Find the adjacent triangle by edge topological relationship, and check whether adjacent triangle meets the pick-up conditions or not; if meeting the conditions, check whether it is searched or not (the sign is 0), and then enter the next step if being searched, otherwise, add the aforementioned edges into region contours boundary array, and then return to the third step.

e) Set the adjacent triangle as sign 1, which is added into triangles of pick-up region, and then enter the third step to traverse the triangles.

On the above mentioned processes, the triangles are judged whether being searched or not by the affiliated sign. If the sign is 0, it means being not searched; if the sign is 1, it means being searching. In addition, when a triangle is traversed or index, the searching numbers (Count) will be added 1, and then Count will used to judge whether all the triangles have been traversed. The algorithm will be concluded after finishing the traversal.

Fig. 1. Dragon STL model type.

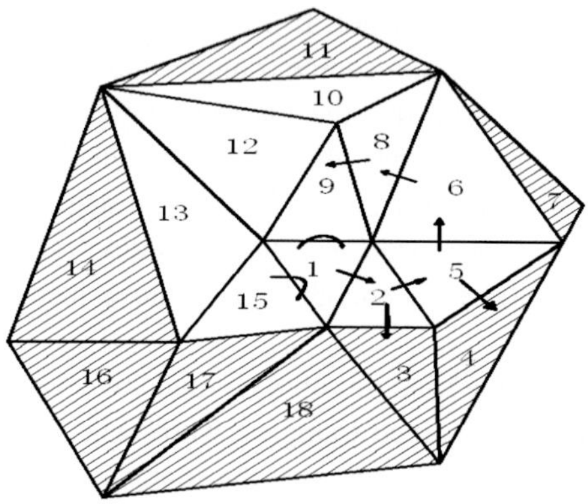

Fig. 2. Recursive search diagram.

The specific pick-up procedures are described by the Figures 1 and 2 as follows:

a Firstly, search out the first condition $T(1)$ which is the angle less than <30°between triangle vector and Z axis, and do the recurrence by $T(1)$;

b Secondly, find $T(2)$ through the function $F_1(1,0)$, if $T(2)$ meets the above conditions and r= $False$, and then do the recurrence by $T(2)$;

c Get the serial number 0 of the edge in $T(2)$ through the function $F_2(1,0,2)$;

d Index two remained edges $L(2,1)$ and $L(2,2)$ of $T(2)$;

e Get the value 3 through the function $F_1(2,1)$, however, if $T(3)$ cannot meet the conditions, it doesn't enter the recurrence;

f Get the value 5 through the function $F_1(2,2)$, at this time $T(5)$ meets the conditions and $I(5) = False$, so do the recurrence by $T(5)$;

....

When source triangle is $T(15)$, the value is 1 through the function $F_1(15,2)$; because $I(1) = True$, then return to $I(1) = True$; similarly, it gets $T(9)$ through the function $F_1(1,2)$; because $I(1) = True$, then return to $T(1)$. All the contours boundaries are found out by the above principle, and then end-to-end boundary is formed by concatenating the coordinates of head-end points on the boundary.

On the region pick-up procedure, cracks can be identified automatically by using recursive search algorithm through topological relationships. The specific idea is as follows: on constructing topological relationships of STL model, if it doesn't meet the consistency rule and gets the value $n_2 = F_1(n_1, m_1)$ ($n_2 < 0$), it judges that $L(n_1, m_1)$ is the adjacent edge of one crack; consequently, values n_1, m_1 are stored into the attributes of pick-up region in order to subsequent application.

4. Analysis of Algorithm Complexity

From the above specific implementation, it shows that this algorithm adopts recursive generation thoughts, every triangle is searched only once and its algorithm efficiency is $O(n)$. In the actual operation of the algorithm, it shows the algorithm has obvious advantages compared with traditional search algorithm (its efficiency is $O(n^2)$).

The instance analysis of the algorithm is illustrated through stereolithography rapid prototyping technics support generation as follows: in the stereolithography technics support generation, it needs to select on the basis of triangle vectors of STL model. Usually it judges by the angle between vectors and coordinate system Z axis. If the angle is less than a certain degree (usually 30 degrees), the triangle should be picked up; otherwise, it doesn't need to be picked up.

In view of various kinds of STL models, the author simply made a classification of all STL models on the basis of its surface smooth degree: Fold type, namely, its surface is scraggy, ravines are interlaced, curvature changes rapidly, and region numbers after picking up are great, while two-dimensional projection area of every region is small (as shown in Figures 1 and 3); smooth type namely, its surface is smooth, curvature changes slowly, and region numbers after picking up are few, while two-dimensional projection area of every region is large (as shown in Figure 4); transitional type are the model in the middle of fold type and smooth type (as shown in Figure 5).

On the process of manufacturing technics support, various parameters (including numbers of triangles, traditional algorithm time-consuming and rapid recursive algorithm time-consuming) are shown in Table.

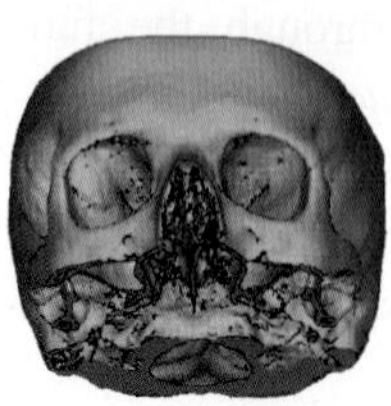

Fig. 3. Skull-fold type.

Fig. 4. Telephone-smooth.

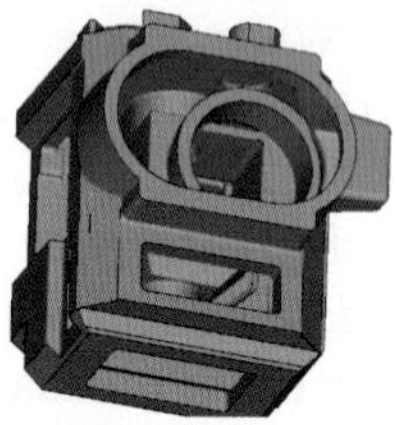

Fig. 5. Engine cylinder.

Table 1. STL model region pick-up data on adding technics support.

STL model	Numbers of triangles	Region numbers after pick-up	Traditional algorithm time-consuming(s)	Recursive algorithm time-consuming(s)
Dragon	89424	200	10.3	0.06
Skull	353444	2506	55.6	0.53
Telephone	20094	21	2.2	0.03
Testware	892	1	0.1	<0.01
Engine cylinder	47700	97	4.9	0.05
Engine exhaust	19282	116	2.2	0.02

According to Table 1, it obtains the curve of triangles numbers and computing time. The result shows that traditional algorithm time-consuming grows curvedly, while recursive pick-up algorithm time-consuming grows linearly. The consuming time is within one second, which is acceptable on rapid prototyping technics support operations.

5. Conclusion

Based on recurrence search pick-up algorithm on triangles, it makes full use of topological information and cracks information of STL model surfaces, effectively realizes surface region pick-up operation on STL model, easily identifies abnormal characteristics such as cracks and so on, and ensures the quickness and correctness of the PR. The algorithm has applied into the laboratory developed HRP series prototyping systems such as stereolithography, powder sintering and so on. It is stable and reliable. Almost all STL file can pick up surface region within a very short period according to the conditions. The long-term application result shows that through using the algorithm it not only increases computing speed, but also greatly improve technics quality, which has achieved good effects.

Acknowledgments

This work is supported by Beijing Natural Science Foundation, No. 3132007 and National Natural Science Foundation of China (Grant No. 21206152).

References

1. L.C Zhang, M Han, S.H Huang. Fault tolerant slicing of STL file based on crack tracking technology. Forging machinery[J], 2002, 37(2):52-54.
2. Q.S Wei, Y.S Shi, W.C Dong, S.H Huang. Three dimensional mesh based on horizontal scan line algorithm for STL model [J].Journal of Huazhong University of Science and Technology (NATURAL SCIENCE EDITION), 2006, 34(4):86-87.
3. X.S Xu. Fast picking technique of surface point of STL model [J]. Journal of Engineering Graphics, 2005, 26(3):18~22.
4. B Qian, J.H Mo, Lichao Zhang. Study on Stereolithography net shaped support generation algorithm [J]. Forging equipment and manufacturing technology, 2005, 40(4):122~124.
5. K.B Guo, L.C Zhang, C.J Wang, S.H Huang. Realization of Boolean operation of STL model [J].Journal of Huazhong University of Science and Technology (NATURAL SCIENCE EDITION), 2006, 34(7):96-99.
6. L.C Zhang, M Han, S.H Huang. An effective error-tolerance slicing algorithm for STL file[J]. International Journal of Advanced Manufacturing Technology, 2002, 20(10): 363-373.

7. V Chandru, S Manober. Voxel-based modeling for layered manufacturing[J], IEEE CG & A, 1995, 15(6):42-47.
8. J.M Zhang, J Zhu, G.J Sun. Voxel based 3D modeling for rapid prototyping manufacturing[J]. Journal of Xi'An Jiaotong University, 2000, 34(10):62-66.
9. C.C Kai, G.G Jacob, T Mei. Interface between CAD and rapid prototyping systems, Part 1: a study of existing interface[J]. Advanced Manufacturing Technology, 1997, 13(8): 566-570.

Calculation of takeoff track of aircraft

Zhong-Hu Jia, Xu Zhang* and Wei-Gang Guo

Department of Aircraft Engineering,
Naval Aeronautical and Astronautical University,
Yantai Shandong 264001, China
**E-mail: zhang101xu101@163.com*

According to the CG motion equation of the vertical of plane, the calculation model of ground running and climbing is built depending on the mission requirements, takeoff running distance and track of climbing are calculated in two different status The calculation result shows that the great differences of tracks of takeoff can be found in two different conditions. The method of calculation and the results of takeoff tracks provide theoretical foundation for takeoff safety. It indicates that the model is reasonable, the method is proved to be correct and the data is precise.

Key words: Takeoff track; calculation model; flight.

0. Introduction

This accelerating process is takeoff process, which is from the aircraft takeoff running to a safe altitude over the airport. Since the thrust weight ratio of modern aircraft is relatively large, generally when the plane is off the ground after the ground running into accelerated, thus takeoff process can be divided into the ground accelerated running process and accelerated climbing process (in Figure 1 [1-8]. The takeoff running distance (L_1) is defined which distance is from the starting of takeoff running to leaving ground. The horizontal distance from leaving ground to the safe height is called rising going forward distance (L_2). The takeoff distance is the total of takeoff running distance and rising going forward distance. The takeoff distance is an important index of the takeoff performance and it is directly related to the required length of the airport runway and the area of airport.

The takeoff track is a trajectory which is corresponding the height of aircraft takeoff and the takeoff distance, and the takeoff track computing is an important part of takeoff performance calculation. In the calculation process, we should set up a calculation model according to characteristic in different development stages, in order to ensure the accuracy of the calculation results.

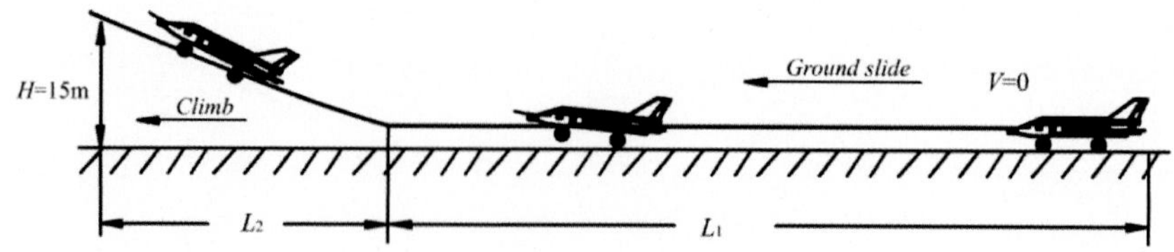

Fig. 1. Takeoff process.

1. Calculation Model

1.1. *Ground running calculation model*

In the ground running process, when the front wheel lifts up the gesture of aircraft is different, so we should consider the difference of angle of attack, lift coefficients and drag coefficients. According to the height ($H = 0$) from the ground to the aircraft, trajectory angle ($\theta = 0$) and its force, the centroid motion equation in vertical plane can be rewritten as differential equations [9]

$$\begin{cases} \dfrac{dV}{dt} = \dfrac{1}{m}\left\{[P\cos(\alpha + \varphi_P) - fmg] - (C_D - fC_L)\dfrac{\rho SV^2}{2} - mg\sin\phi\right\} \\ \dfrac{dL}{dt} = V - U \end{cases} \tag{1}$$

From differential equations (1), when the aircraft is running on the ground, state variables are air speed V, horizontal distance L and fuel consumption m_c, the control variable is the angle of attack α.

1.2. *Ascent stage calculation model*

The takeoff process is actually a climb question, thus we need establish a climb computing model

$$\begin{cases} m\dfrac{dV}{dt} = P\cos(\alpha + \varphi_P) - C_D\dfrac{\rho SV^2}{2} - mg\sin\theta \\ mV\dfrac{d\theta}{dt} = P\sin(\alpha + \varphi_P) + C_L\dfrac{\rho SV^2}{2} - mg\cos\theta \end{cases} \tag{2}$$

2. Calculation Processing Method

2.1. *Ground takeoff running stage*

According to equation (1), takeoff running distance can be calculated, firstly it must determine the control rule and the aerodynamic characteristics of aircraft, such as the lift co-efficient C_{L}, drag coefficient C_{D}, as well as the engine characteristic data, such as thrust P, fuel consumption C_e.

Determining the control rule of angle of attack and lift coefficient C_{L}, drag coefficient C_{D}

Takeoff running process can be divided into three stages: the first one is from the aircraft at a standstill to the lifted front wheels, the second one is from the lifted front wheels to the leaving ground angle of attack, the third one is from the leaving ground angle of attack to the takeoff leaving ground speed. The downtime angle of attack α_{ij} is the three-point running angle of attack of first stage, the aircraft is two-point running in the second and third stages, it is assumed that the change of angle of attack in proportion to the velocity in the second stage, the angle of attack is defined the leaving ground angle of attack α_{ld} in the third stage. When we know the angle of attack, we can use the lift coefficient curve and polar curves (considering ground effect) to check out the lift coefficient C_{L} and drag co-efficient C_{D}.

Determination of thrust P and fuel consumption per hour C_e

We can obtain the thrust and fuel consumption under the standard atmospheric condition at sea level by the characteristics data of the engine such as the characteristic of height and speed and rotate speed characteristic. Basic conditions of calculation are a given engine condition, aircraft speed and altitude. Then the thrust and fuel consumption can be revised according to the actual situation.

Once we have the control rule of angle of attack, the aerodynamic characteristics of the aircraft and the data of engine characteristics, differential equations group (1) can be solved by the Runge - Kutta method, you can get speed, distance, fuel consumption per hour data changes with the course of time.

2.2. *Ascent stage*

After the aircraft off the ground, both speed and altitude decrease. In order to ensure safety and to achieve speed and altitude as quickly as possible, generally we take the approach of accelerating, and during the ascent, cleaning up the aircraft. The angle of attack changes continuously during the ascent, it's difficult to use the direct application of the angle of attack as a control variable to calculate the rising trajectory. According to the method described in the literature [10], it introduced a constant $\frac{dV}{dH}$ in sub-control mode. To determine the angle of attack control mode under the control using constant $\frac{dV}{dH}$, so turn the (2) to (3), namely

$$
\begin{cases}
V_y = \dfrac{\left[P\cos(\alpha + \varphi_P) - C_D \dfrac{\rho V^2 S}{2} \right]}{mg} \cdot \dfrac{V}{1 + \dfrac{V}{g} \cdot \dfrac{dV}{dH}} \\
C_L = \dfrac{mV^2 \sin\theta \dfrac{d\theta}{dH} + mg\cos\theta - P\sin(\alpha + \varphi_P)}{\dfrac{\rho V^2 S}{2}}
\end{cases} \tag{3}
$$

After determining the α and θ by the iterative method, we can find out the speed, height, distance, and fuel consumption per hour in the Runge - Kutta method according to the equation (2).

3. Calculation Results

According to the calculation model and calculation processing method, calculation program can be compiled, the ground effect, landing gear position and the flap position in the data files of program will affect the lift coefficient C_L and drag co-efficient C_D, and the air density always changes with altitude in the calculation process. According to mission requirements, we calculate a fighter takeoff path in two different conditions respectively.

The results are plotted. Figure 2 shows that track curve is relatively steep under standard conditions, while it is relatively flat track under standard conditions, because higher altitude of airports, takeoff weight, the uphill runway, following wind etc., which have a negatively effect on the takeoff path under nonstandard conditions, the takeoff paths under the two conditions are quite different.

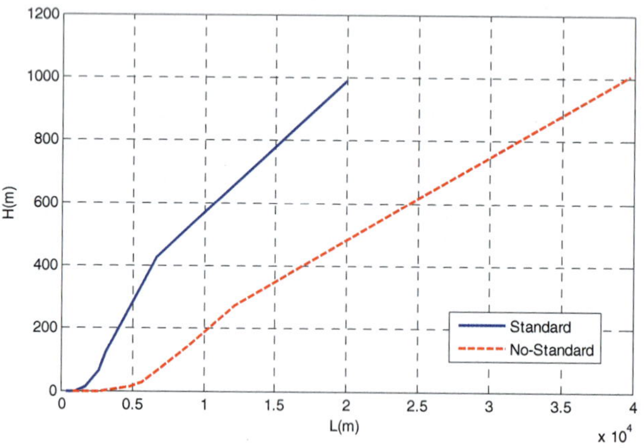

Fig. 2. Takeoff path comparison curve.

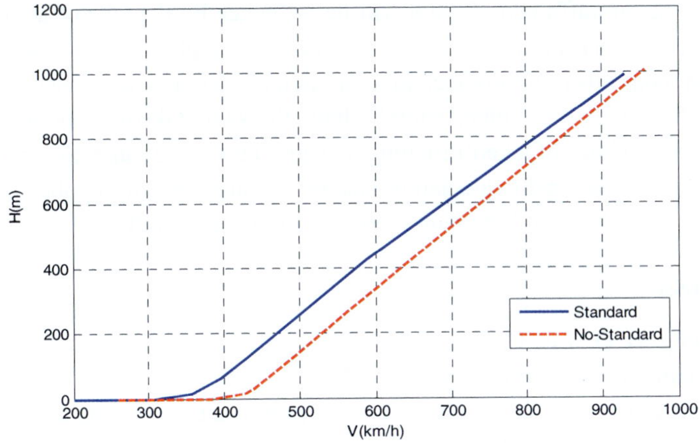

Fig. 3. Flight altitude-speed comparison curve.

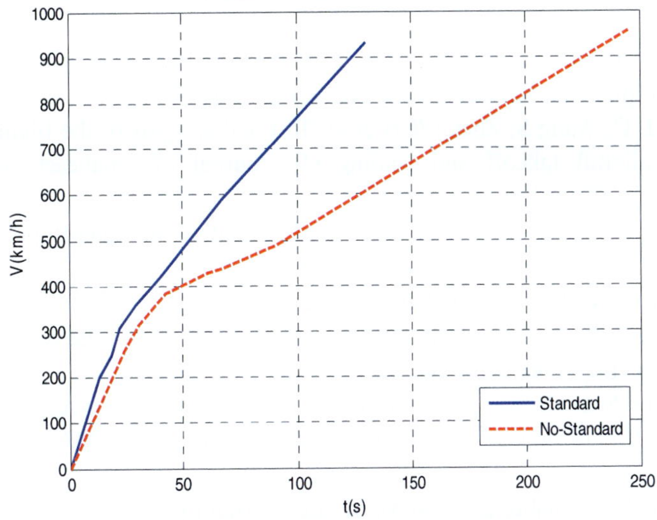

Fig. 4. Flight speed-time comparison curve.

4. Conclusion

The aircraft takeoff process is actually a process of increasing speed, which is affected by the engine thrust, aerodynamics, gravity and ground surface (roll segment) frictional force. Therefore, all the external forces will affect the aircraft's takeoff performance. In the calculation process of aircraft takeoff trajectory, calculation models should be separately established to calculate based on the characteristics of different stages. From the calculation results, we can find that the aircraft's takeoff path will be influence by the takeoff atmospheric conditions, higher altitude of airports, takeoff weight, runway slope, wind direction and other environmental conditions. The active service bombers, military transports and fighter aircrafts have the same takeoff principle and the calculation methods of takeoff running and aerial track are same. So, the method of aircraft's takeoff path calculation and results are not only applicable to the fighter, but also have an important reference for other aircraft.

References

1. Xiong H Q, Liu C, Zheng B W. Airplane flight dynamics[M].Beijing: Aviation Industry Press, 1990:81-84.
2. Fang Z P, Chen W C, Zhang S G. Aviation aircraft flight dynamics[M]. Beijing University of Aeronautics and Astronautics Press, 2005:70-72.
3. Song H Y, Cai L C. Calculation method of engine thrust in aircraft takeoff track calculation [J]. College of Transportation Engineering, 2010, 10 (2): 59-63.
4. Zhang F, Yan G H. Study on the calculation method of aircraft takeoff track [J]. Equipment Manufacturing Technology, 2011,11: 15-17.
5. Cai L C, Wang S, Zheng R H et al. Test and analysis of the flight track of the aircraft takeoff and landing [J]. Journal of Southeast University, 2001,32 (2):264-267.
6. Li X Y, Xing X. Design and implementation of takeoff performance calculation program for civil aircraft [J]. Journal of Civil Aviation University of China, 2008:26 (1):8-12.
7. Michael V. Cook. Flight Dynamics Principles [M], Elsevier aerospace engineering, 2007:320-330.
8. Pinder S D, Crave T G, Nikiforuk P N. A practical investigation of a takeoff performance monitor for turboprop aircraft[R]. AIAA-2001-4170, 2001: 1-8.
9. Chen T N. Quality and control of aircraft flight performance [M]. Beijing: National Defense Industry Press, 2010:42-46.
10. Airplane flight performance computing manual. Xi'an: Journal of flight mechanics, 1987.

Analysis on magnetic field characteristics of an electromagnetic actuator for laser beam control

Ling Tong*, Jia-Qi Jin† and Feng Sun‡

School of Mechanical Engineering, Shenyang University of Technology, No.111, Shenliao West Road, Economic & Technological Development Zone, Shenyang, 110870, P.R.CHINA

**Email: tonglingsy@163.com*

†Email: jinjq612@126.com

‡Email: sunfeng2009@gmail.com

This paper presents an electromagnetic actuator for controlling the laser beam of laser cutting machine. In this actuator, the permanent magnet supporting force is provided by 8 pairs of permanent magnets, and 3 pairs of electromagnets drive the suspended part to move in three degrees of freedom. This paper introduces the structure of the electromagnetic and generates the FEM mesh. Then, the magnetic field characteristic of the suspended part and the supporting frame is analyzed. Finally, the magnetic field characteristic of the actuator without control current is analyzed. The analysis results indicate that the magnetic field produced by permanent magnets can be ignored in the controlling process.

Keywords: Finite element method (FEM); magnetic induction; electromagnet; permalloy; permanent magnet.

1. Introduction

Laser cutting is one of typical special processing technologies, which cuts the work-piece by a laser beam. However, the cutting efficiency of laser cutting is limited by the position between the focus of the laser beam and the axis of nozzle. It is proved that the off-axis nozzle can improve the cutting efficiency [1]. Considering the aerodynamic features of the gas flow, in the process of machining, the cutting efficiency can also be improved that the axis of the gas jet is always in the back of the cutting direction of the laser focus. Specifically, the electromagnetic actuator can be used to control and change the position of the laser focus.

Many researchers have used electromagnetic suspension technology in many applications, such as magnetic suspension train, magnetic suspension bearing, magnetic suspension actuator, etc [2-6]. Furthermore, in order to improve the efficiency of the laser cutting machine, Professor Zhang, Tokyo Institute of Technology, has developed a magnetic drive actuator to control the position of the laser focus [7].

An electromagnetic actuator is proposed to control the position of the laser focus in this paper. The repulsive force of permanent magnets is used for the contact-less supporting force in vertical direction, and the attractive force of electromagnets is used for controlling the actuator, which can realize high-speed positioning and high-precision in 3-DOF in horizontal direction.

2. Structure of the Actuator

The basic structure of the electromagnetic levitation actuator is shown in the Fig.1. It consists of the driving electromagnets, a suspended part and a supporting frame part. The suspended part in the gravity direction adopts the way of passive suspended that like poles repel. The permanent magnet supporting force is provided by 8 pairs of permanent magnets on the top and bottom surfaces of the suspended part, which averagely arranged on the moving part. The one fixed on the moving part is rectangular permanent magnet, and the other one fixed on the supporting frame part is ring permanent magnet. Due to the repulsion between the rectangular permanent magnets and the ring permanent magnets, the disturbing forces are produced by each pair of permanent magnets when the supporting force is produced in the vertical direction. In horizontal direction, 3 pairs of electromagnets drive the suspended part to move in three degrees of freedom that include two moving and one rotation. These electromagnets arranged on every side of the suspended part are used to overcome the disturbing force, meanwhile controlling the suspended part to move freely and stably in the horizontal plane by changing the air gap and current. The magnetic forces produced by electromagnet are equal to the disturbing forces so that the suspended part can move stably.

3. The FEM Mesh Generation

In order to analyze the magnetic induction of the electromagnet levitation actuator in all modes, the ANSYS is used to analyze the magnetic field characteristic.

The finite element meshes are generated by the meth tool which is shown in Fig. 2, and the minimum length of all modes is 2mm. The numbers of the node and element are 322534 and 181722 respectively when the finite element model of the suspended part is generated separately; the numbers of the node and element are 393603 and 226032 respectively when both of the suspended part and the supporting frame are analyzed; the numbers of the node and element are 417782 and 229872 respectively when the whole actuator is analyzed.

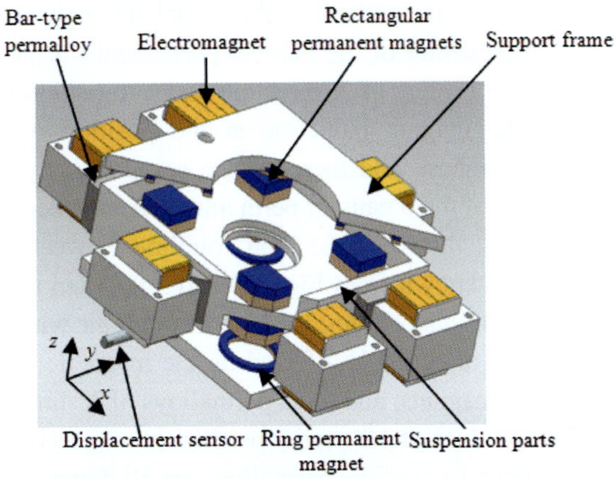

Fig. 1. Basic structure of the electromagnetic actuator.

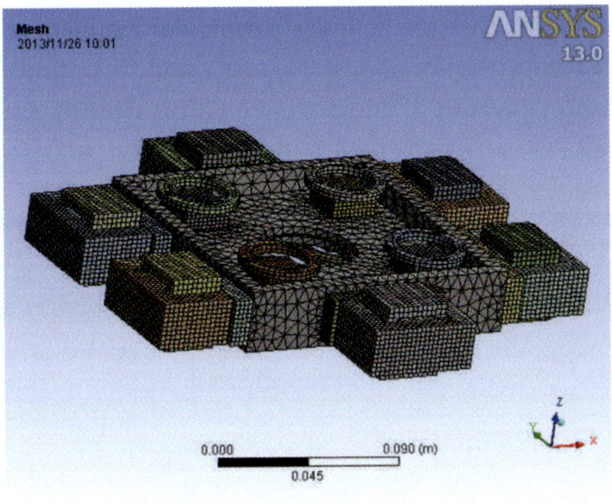

Fig. 2. The finite element mesh model.

4. The Analysis on the Magnetic Field Characteristic

4.1. *Magnetic field analysis on the suspended part*

In order to understand how magnetic field that produced by permanent magnet affects the electromagnet, it is necessary to analyze the magnetic induction of the suspended part on different position. To be easy to observe the change of the magnetic field of the suspended part on different position and status, the analytical plate is loaded on the model to indicate the direction of the magnetic lines of force and the magnetic strength. There into, the direction of the magnetic lines of force and the magnetic strength of the model are analyzed. Fig. 3a) is the result of the plate of the center of the vertical permanent magnets. Fig. 3b) and 3c) are the results of the centers of X direction and Y direction respectively. As shown in Fig. 3a), the magnet induction at the position of the 4 pairs of permanent magnets is highest, and there is also comparatively lower magnetic induction at the position of the 6 pairs of perm alloys. And except for the effect of each pair of perm alloys, the magnetic field coupling between each permanent magnet is low. It can be known from Fig. 3b) and 3c), when the magnetic field produced by 8 permanent magnets form a loop through air with the permalloys, the each permalloy of Y direction is affected by the magnetic field produced by the 4 nearby permanent magnets, and the 2 permalloys of X direction are affected by the magnetic field produced by 2 nearby permanent magnets. Furthermore, the directions of the magnetic lines of the permalloys are all downward. From Fig. 3a) and 3b) it can be known that the magnet induction of the center of the permalloy in Y direction as well as the side near the permanent magnets is highest, and it is comparatively lower in the locations. Fig. 3a) and 3c) show that the magnet induction of the permalloys in X direction is well-distributed, but it is higher than that in Y direction.

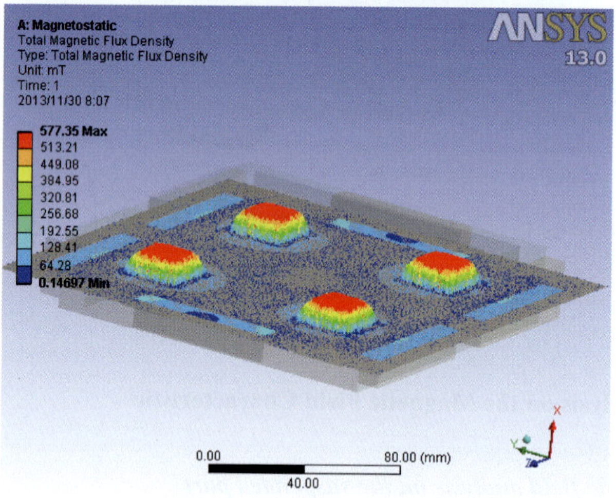

a)The horizontal center section

Fig. 3. The magnetic induction of suspended part in the drive.

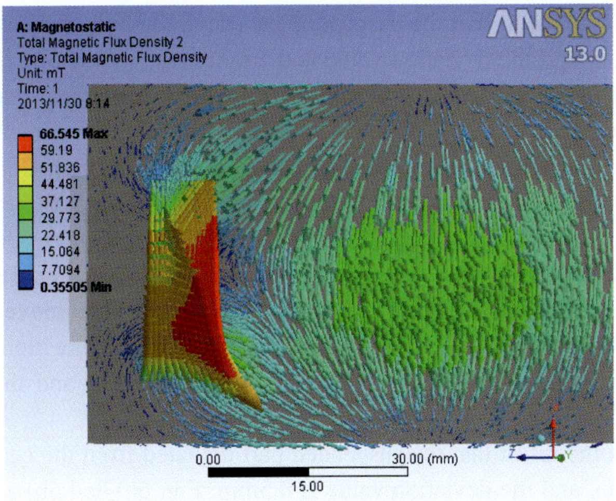

b)The center in X direction

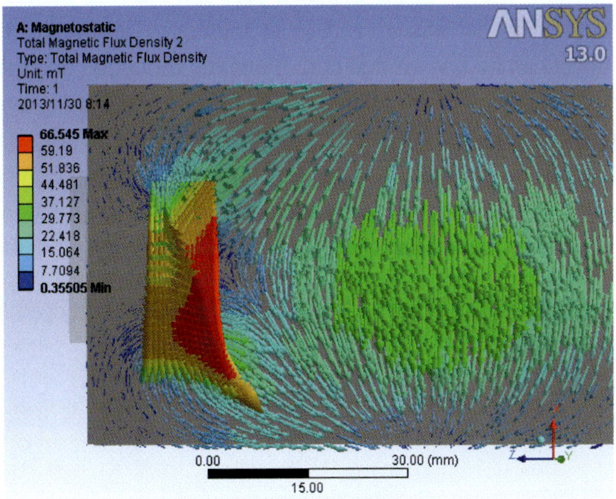

c) The center in Y direction

Fig. 3. (*Continued*)

4.2. *Magnetic field analysis on the suspended part and the supporting frame*

Fig. 4 is the result of magnet induction that the square permanent magnets center of the suspended part coincides with the center of the ring permanent magnets. Comparing with Fig. 3, it is can be known that, the magnet induction heightens because 8 ring permanent magnets are added, and the direction of the magnet line unchanged.

Fig. 5 is the result that the suspended part deviated from the central position in X direction, and the deviation value is 0.5mm. Fig. 5a) and 5b) are the results of the center of the suspended part and that of square permanent magnets on the side of X direction. In Fig. 5a) the distance between the ring permanent magnets and the permalloy is decreased, and it is increased in Fig. 5b). Comparing Fig. 5a) and 5b) it can be known that both magnetic induction and the direction of magnetic line are basically the same in the ring permanent magnets and the square permanent magnets. The direction of magnet line in Fig. 5a) and 5b) is basically the same with that in Fig. 4b). Above all, the movement of the suspended part in X direction doesn't apparently affects the direction of magnetic line of the permalloys when there are only suspended part and the supporting frame.

Fig. 6 is the result that the suspended part deviated from the central position in Y direction, and the deviation value is 0.5mm. Fig. 6a) and 6b) are the results of the center of the suspended part and that of square permanent magnets on the side of Y direction. In Fig.6b) the distance between the ring permanent magnets and the permalloy is decreased, and it is increased in Fig. 6b). Comparing Fig. 6a) and 6b) it can be known that both magnetic induction and the direction of magnetic line are basically the same in the ring permanent magnets and the square permanent magnets. The direction of magnet line in Fig. 6a) and 6b) is basically the same with that in Fig. 4b). Above all, the movement of the suspended part in X direction doesn't apparently affects the direction of magnetic line of the permalloy when there are only suspended part and the supporting frame.

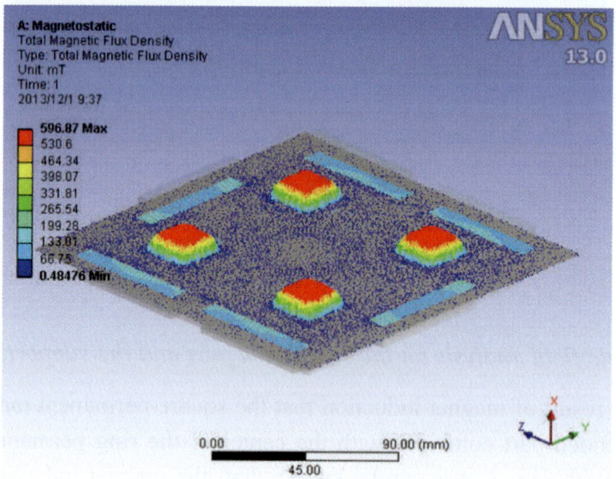

a) The horizontal center section

Fig. 4. The magnetic induction that suspended part and supporting part are in the central location.

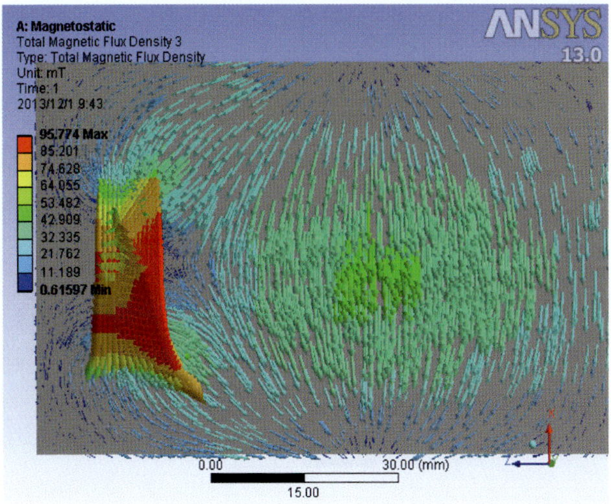

b) The center in Y direction

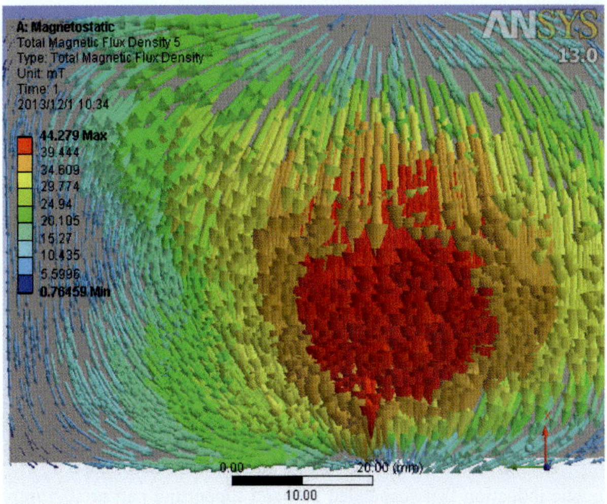

c) The center in Y direction

Fig. 4. (*Continued*)

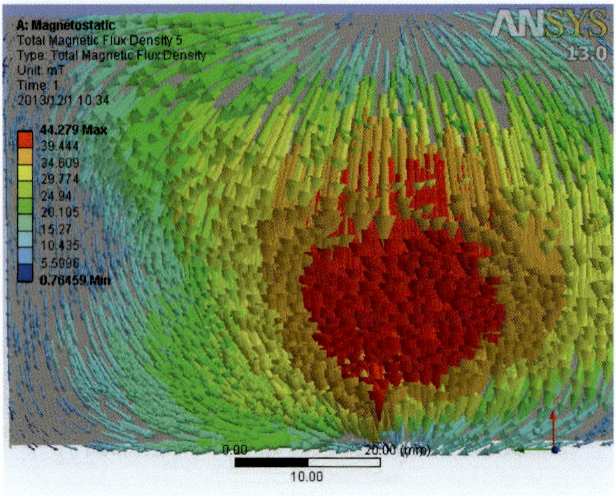

a) The air gap decreased side of center section

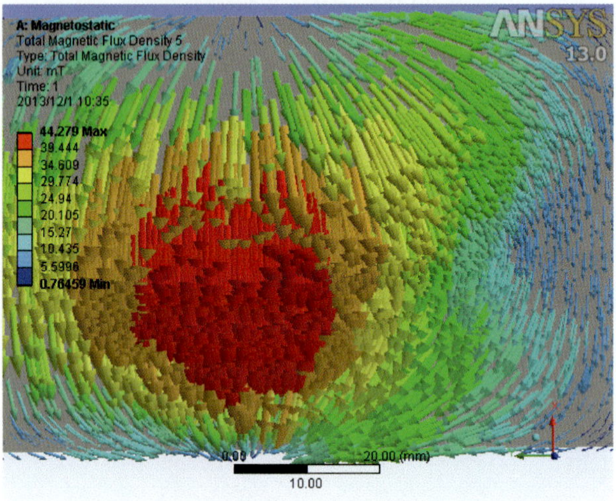

b) The air gap increased side of center section

Fig. 5. The magnetic induction that suspended part and supporting part are 0.5 mm off center in X direction.

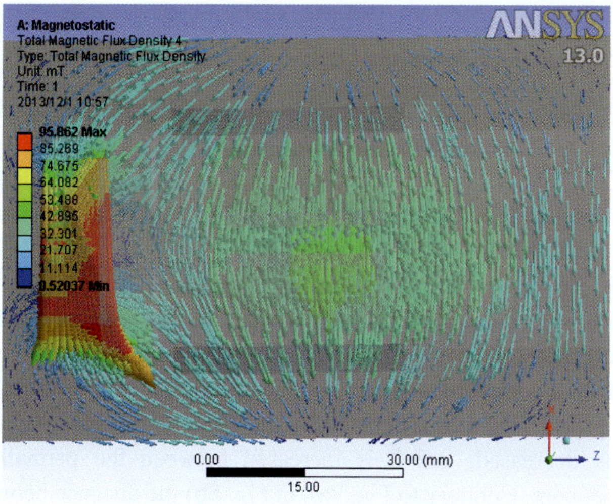

a) The air gap decreased side of center section

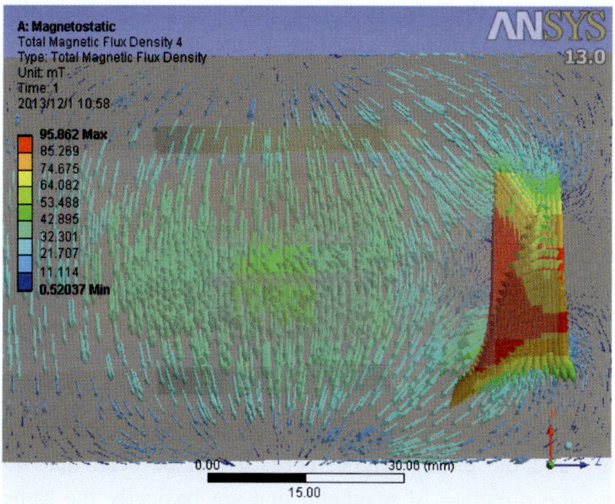

b) The air gap increased side of center section

Fig. 6. The magnetic induction that suspended part and supporting part are 0.5 mm off center in Y direction.

4.3 *Magnetic field analysis on the actuator without control current*

Fig. 7 is the analysis on the magnetic induction of the actuator without control current at the central position. Comparing with Fig. 4, there are traces of the magnetic field in the electromagnet core because of the adding of the 6 electromagnets, and the magnetic induction and the direction of the magnetic line in the other parts changed little.

Fig. 8 is the result that the suspended part deviated from the central position in X direction when there is no control current, and the deviation value is 0.5mm. Because the distance between the ring permanent magnet is different from the distance between permalloy and electromagnets core, a contrastive analysis is be done. In Fig. 8a) the distance between the ring permanent magnets and the permalloy is decreased, and it is increased between the permalloy and the electromagnet core. Contrary to Fig. 8a), in Fig. 8b) the distance between the ring permanent magnets and the permalloy is increased, and it is decreased between the permalloy and the electromagnet core. It can be seen that both magnetic induction and the direction of magnetic line are basically the same in the ring permanent magnets, the square permanent magnets, permalloy and the electromagnet core. Comparing with Fig. 7, the direction of the magnetic line basically unchanged. Above all, the movement of the suspended part in X direction doesn't apparently affects the direction of magnetic line of the permalloy when there is no control current in the actuator.

Fig. 9 is the result that the suspended part deviated from the central position in Y direction when there is no control current, and the deviation value is 0.5mm. Because the distance between the ring permanent magnet is different from the distance between permalloy and electromagnets core, a contrastive analysis is be done. In Fig. 9a) the distance between the ring permanent magnets and the permalloy is decreased, and it is increased between the permalloy and the electromagnet core. Contrary to Fig. 9a), in Fig. 9b) the distance between the ring permanent magnets and the permalloy is increased, and it is decreased between the permalloy and the electromagnet core. It can be seen that both magnetic induction and the direction of magnetic line are basically the same in the ring permanent magnets, the square permanent magnets, permalloy and the electromagnet core. Comparing with Fig. 7, the direction of the magnetic line basically unchanged.

Above all, the movement of the suspended part in Y direction doesn't apparently affects the direction of magnetic line of the permalloy when there is no control current in the actuator.

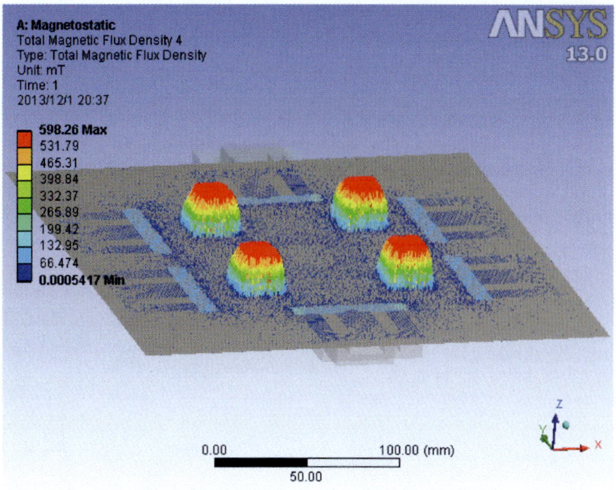

a) The horizontal center section

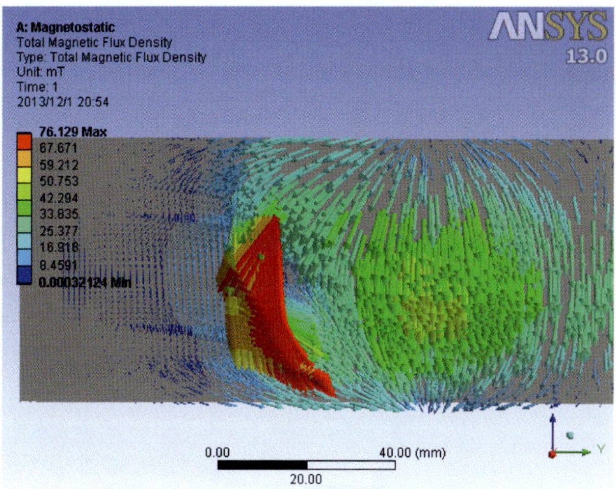

b) The center in Y direction

Fig.7. The magnetic induction that actuator is in the central location without control current.

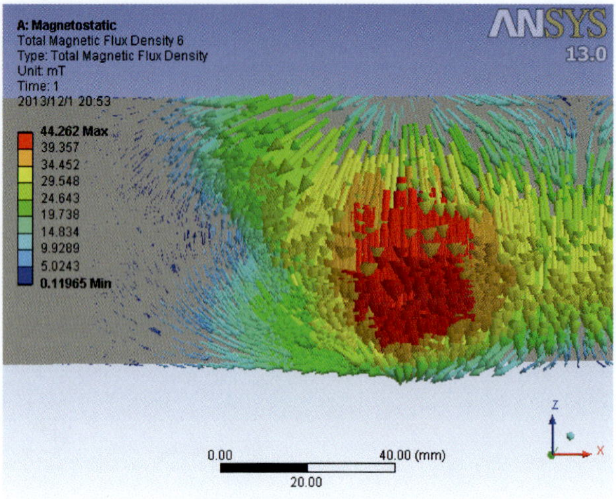

c) The center in Y direction

Fig. 7. (*Continued*)

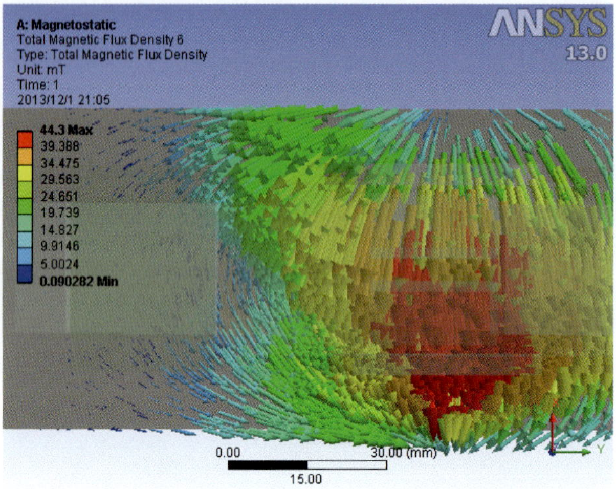

a) The air gap decreased side of center section

Fig. 8. The magnetic induction that actuator without control current is 0.5mm off center position in the X direction.

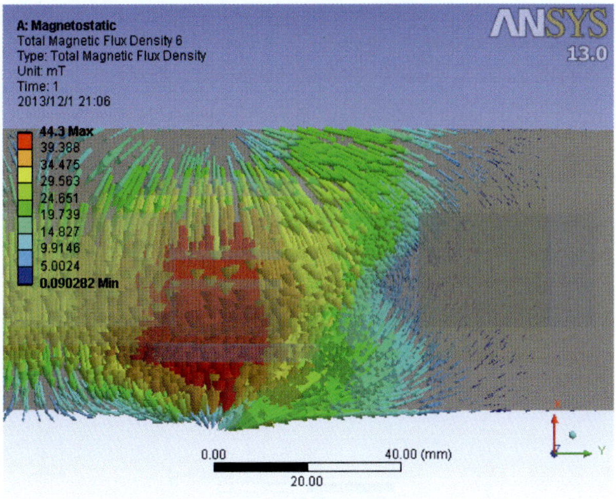

b) The air gap increased side of center section

Fig. 8. (*Continued*)

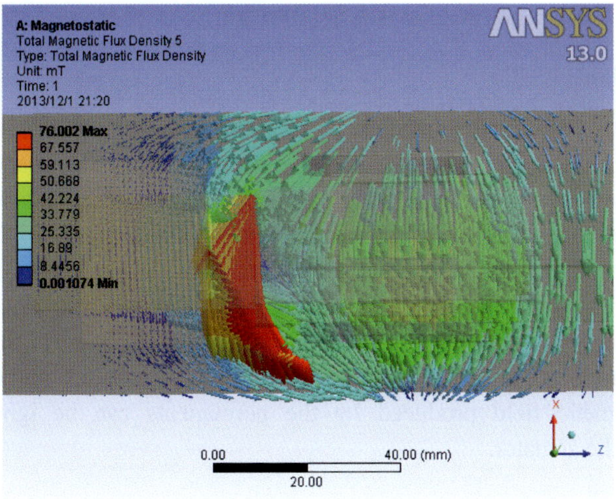

a) The air gap decreased side of center section

Fig. 9. The magnetic induction that actuator without control current is 0.5mm off center position in the Y direction.

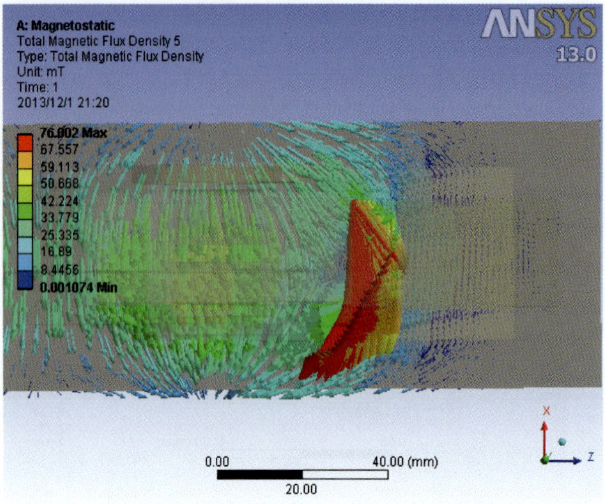

b) The air gap increased side of center section

Fig. 9. (*Continued*)

5. Conclusion

The magnetic field characteristics of the electromagnet actuator in different condition including different distances between the square permanent magnets and the permalloys in the suspended part as well as different distances between the ring permanent magnets and the electromagnet core in driving part are analyzed in this paper. The results show that when the permanent magnets are added, the magnetic induction heightens, and the direction of the magnetic lines keeps unchanged. When the electromagnets are added, there is tiny magnetic induction produced, and it affects the driving process infinitesimally. The effect of the magnetic field produced by the permanents can be ignored during controlling the actuator.

Acknowledgments

Research supported by National Natural Science Fund of China (Grant No. 51105257, No. 51310105025), and the colleges and universities of Liaoning Province outstanding young scholars growth plan (Grant No. LQJ2014012).

References

1. F. Quintero, J. Pou and J.L Fernandez, Optimization of an off-axis nozzle for assist gas injection in laser fusion cutting[J].Optics and Lasers in Engineering 44 (2006) 1158–1171.
2. Wang Weiming, Yu Chi, Ma Shuyuan, Xie Hu, Electromagnetic force FEM analysis of magnetic levitation motion stage[J]. Micrometers, 2011, Vol. 44, No. 10, pp. 22-24.
3. Song Wenrong, Yu Guofei, Sun Baoyu, et al, Research on the structure design of a precision stage based on magnetic levitation technology and used in micro- electron manufacturing field[J]. Optics and Precision Engineering, 2002, Vol. 10, No. 3, pp. 271-275.
4. Chen Benyong, Lei Yong, Li Yanbao, et al, Theoretical analysis and modeling of a magnetic levitation micro motion stage, Chinese Journal of Scientific Instrument, 2006, Vol. 27, No. 9, pp. 1125-1128.
5. Shang Lingyan, Zhou Jianggang , Research on position control of vertical steel strip by magnetic force[J], Computer Simulation, 2007, Vol. 24, No. 6, pp. 307-310.
6. Zhang Yunpeng, Liu Shuqin, Li Hongwei, et al, Calculation of Radial Electromagnetic Force of Axial Hybrid Magnetic Bearing Based on Magnetic Circuit Analysis[J]. Transactions of China Electro technical Society, 2012, Vol. 27, No. 5, pp.137-142.
7. Xiaoyou Zhang, Deqiang Chu, Tadahiko Shinshi, Precision Control of Magnetic Drive Actuator Used for Laser Cutting with Compensation of Zero Point of Displacement Sensor[J]. Applied Mechanics and Materials, 2012, Vol. 220-223, pp. 978-98.

Optimization study about time adjusting between gap protection used in transformer and circuit reclosing

Xin-Jie Hao*, Wei-Ru Wang, Shu-Yong Song, Hui-Ping Zheng, Xue-Ting Cheng and Ying Qu

Shanxi Electrical Power Research Institute, Taiyuan, Shanxi, China

**E-mail: haoxinjie@126.com*

Operation modes of two transformers in 220kV power substations are usually different. One is directly neutrally grounded and another is grounded with a gap. When an instantaneous failure occurs on a transmission line, which connected to the substation, The gap grounded transformer will be removed by the mal-operation of zero-sequence protection at the high voltage side, and the power supply of the substation will be disturbed. A field case of fault is studied in this paper, and the fault procedure is reproduced by software simulation. The optimal time cooperation between gap protection used in transformer and circuit reclosing is obtained by time sequence adjustment.

Keywords: Neutral of main transformer; gap grounded; instantaneous fault; time sequence cooperation.

1. Introduction

The power transformer is one of the most important equipment in substations, and its normal operation is vital for the reliability and stability of the power system. Hence, multiple protections are deployed for the main transformer, and the gap protection is used as grounded fault backup for the half-insolated transformer, which is indirectly grounded[1,2]. The gap protection results in the overvoltage at the neutral point of the transformer and zero-sequence current after insulation breakdown when external fault occurs. There are some gap protection mal-operation events in power systems in recent years. For example, a B-phase fault results in main transformer breakdown by gap protection at a substation in Fujian province at November 16, 2010. An external fault lead to No. 2 main transformer shutdown by gap protection mal-operation at a substation in Shanxi province at October 21, 2013. This kind of events make the main transformer stop and the reliability and stability of the power system are also severely disturbed[3,4].

The extension of gap zero-sequence protection and its influence are qualitatively analyzed in [5,6]. The mismatch between the gap protection and the zero sequence protection are confronted by using zero sequence and banning gap protection in [7]. A protection scheme, which uses different time delay for zero sequence voltage and gap zero current protection is proposed in [8]. However, the simulation validations are rarely studied in relevant literatures.

A mal-operation of the main transformer's gap protection, which is caused by the noncooperation between the gap current protection and the circuit reclosing at a 220kV power substation in Shanxi grid, is analyzed in this paper. By means of simulation analysis, the reason of the fault is reproduced and an optimal time cooperation scheme of gap protection and circuit reclosing is proposed.

2. The Configuration of the Gap Protection of Main Transformer and Circuit Reclosing

The neutral winding of the multiple insulation transformer has weaker isolation than the other parts. To prevent the insulation damage of the neutral point of the main transformer, the principle diagram of the configuration of the gap protection [9] is shown in Fig. 1. A discharging gap is settled at the neutral point of the transformer to connect to the ground. There will be zero sequence current at neutral point of the main transformer when the gap is punctured by the neutral point overvoltage.

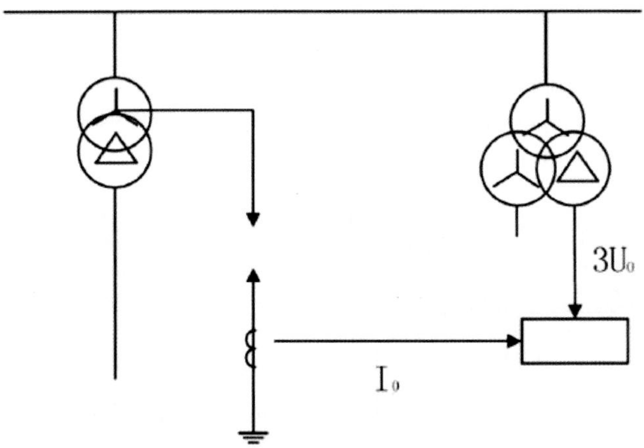

Fig. 1. Principle diagram of gap protection.

The logic diagram of gap protection is shown in Fig. 2. The gap protection will trip and open the breaker of the main transformer with a certain time delay when the zero sequence voltage ($3U_0$) or the zero sequence current ($3I_0$) exceed the threshold (I_{op} or U_{op}).

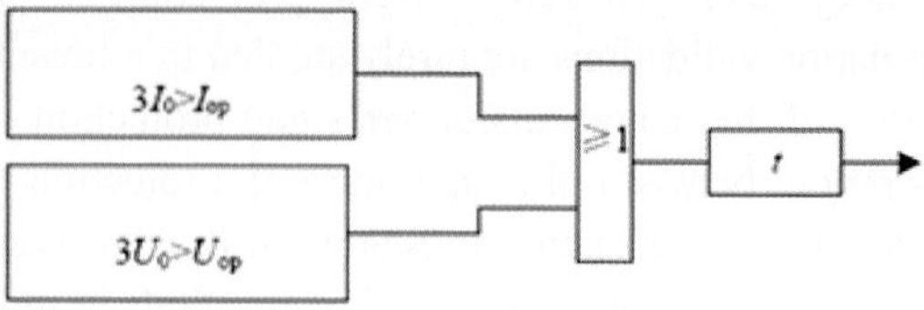

Fig. 2. Logic diagram of gap protection.

The most common fault on 220kV and above transmission lines is single phase instantaneous fault. Hence, circuit reclosing is usually deployed to confront instantaneous fault. After circuit reclosing, the transmission line will continue working as usual if the fault is instantaneous. The breaker will be accelerated to open if the fault is permanent.

3. Fault Analysis and Simulation Study

3.1. *Fault analysis*

A phase A to ground fault occurred on Puxin transmission line at 11:52:49 on October 21, Two pilots both operated fast to open phase A breaker, remove the fault and then successfully reclose the circuit. At the same time, the high-voltage-side neutral gap of No. 2 main transformer was punctured and the gap zero sequence current protection operated. The No. 202 breaker at the high-voltage-side was opened and No. 2 main transformer was shutdown. The Equivalent diagram is shown in Fig. 3.

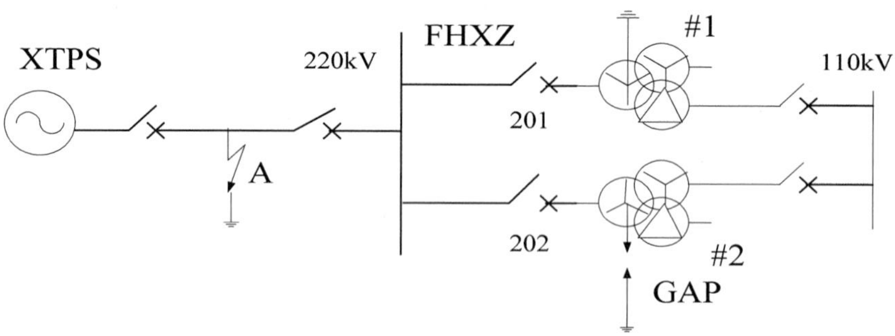

Fig. 3. Equivalent diagram.

The fault recorder is shown in Fig. 4. The waveforms are successively phase A, B, C and zero sequence bus voltages from up to down.

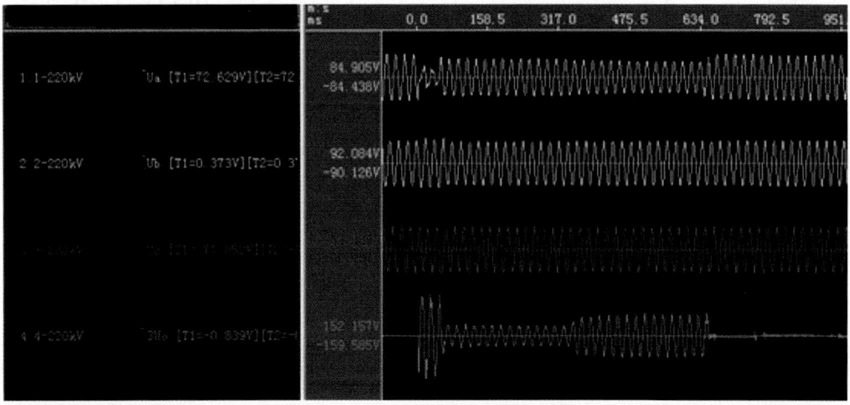

Fig. 4. Fault-recorded waveform.

Table 1. Parameters of main equipments.

Equipments	Voltage level(kV)	Parameters	Values
No. 1 transformer	220/121/10.5 (YN, yn0,d11)	Impedance voltage (%) Uk12, Uk13, Uk23	10.25/35.78/22.78
		Capacity(MVA)	150/150/75
		Current of no-load (%)	0.37
		Loss of no-load(kW)	95.32
		Zero sequence impedance (p.u.)	0.713/-0.11/1.609
No. 2 transformer	220/121/10.5 (YN, yn0,d11)	Impedance voltage (%) Uk12, Uk13, Uk23	10.07/21.28/9
		Capacity(MVA)	150/150/75
		Current of no-load (%)	0.26
		Loss of no-load(kW)	104.9
		Zero sequence impedance (p.u.)	0.713/-0.11/1.609
Puxin transmission line	220	Length (km)	24.917
		Fault location from Xinzhuang substation (km)	17.55
		Positive sequence impedance (p.u.)	R1:1.295 X1:7.587
		Zero sequence impedance (p.u.)	R0:6.56 X0:23.95

3.2. *Simulation study*

No. 2 main transformer is located at Xinzhuang power substation. A detailed simulation model is constructed and analyzed thoroughly[10,11]. As Pushang substation is multiply connected to the main grid, it can be equivalent by an ideal voltage source. As Puxin transmission line is short, it can be equivalent by a PI model transmission line. The configuration of Xinzhuang substation is set as real and the transformers use the practical parameters. The detailed parameters of the model are shown in Table 1.

The main procedure of the fault is shown in Table 2 based on fault recorder. The following simulation is based on this table and the fault time is set as 1.0s.

Table 2. Time list of the main procedure of the fault.

time	0 ms	52ms	356 ms	651ms
operation	Phase A to ground	Open phase A	No. 2 transformer zero sequence protection operate	Reclosing successfully

Zero sequence bus voltage calculation during fault is shown in Fig. 5. The simulation waveform is similar with the fault recorder.

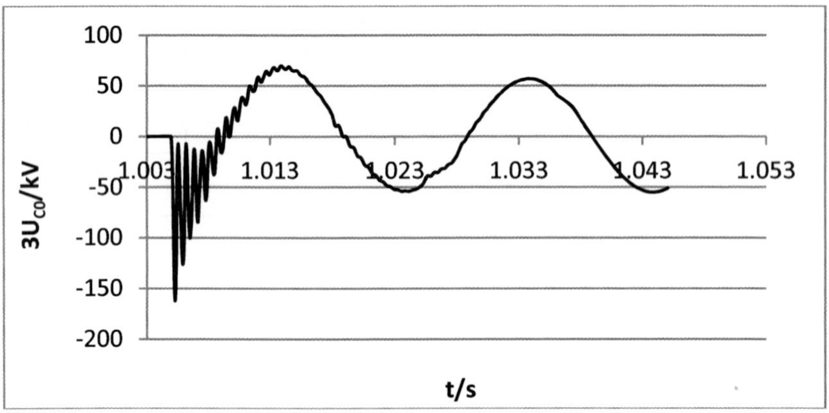

Fig. 5. Zero sequence bus voltage .

There is a high voltage peak (156kV) at the neutral point of No. 2 transformer after phase A to ground fault. The length of the gap is 285mm, and the mean power frequency puncture voltage peak is 159kV. As the gap discharge is uncertain, the puncture will take place at the voltage peak range of

144kV~172kV. The voltage peak in simulation is in this range, so the gap puncture of No. 2 transformer is caused by instantaneous fault with zero voltage peak after the fault.

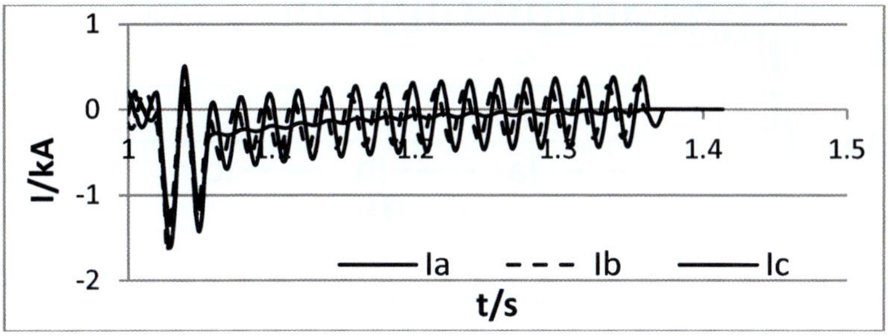

Fig. 6. High-voltage-side current of No. 2 main transformer.

As can be seen in Fig. 6, After gap puncture, the current pass through the neutral point of the transformer exceed the threshold , and then No. 202 breaker at the high-voltage-side opened with a time delay. Thus, No. 2 main transformer is shut down and the reliability of power systems cannot be guaranteed.

3.3. *Simulation study*

To avoid the situationmentioned above, A novel protection scheme, in which the action time of the zero sequence protection is set longer than circuit reclosing time, is proposed in this paper. A similar scheme has been introduced in [12], however, the consequence of the scheme is not simulated and discussed in detail. The action time of zero sequence protection should be longer than circuit reclosing and should make the time of non-complete phase operation of transformer as shorter as possible. According to the analysis above, adjust the procedure after fault as Table 3, and the simulation result is shown in Fig. 7.

Table 3. Time list of the main procedure of the fault.

time	0 ms	52ms	651ms	751ms
operation	Phase A to ground	Open phase A	Reclosing successfully	No. 2 transformer zero sequence protection operate

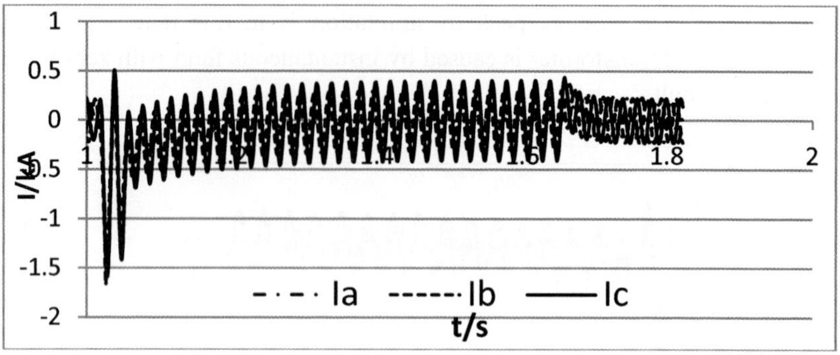

Fig. 7. Current waveform after adjustment (main transformer is not shut down).

The current of No. 2 main transformer at high-voltage-side during fault and after circuit reclosing is shown in Fig. 7. It can be seen that after the proposed adjustment scheme, although the transformer will be non-complete phase operated for a longer time, the transformer can be operated normally after removing the fault, and then the reliability of the system can be improved.

4. Conclusions

The external fault in power systems may lead to the mal-operation of gap zero sequence protection and shut down the transformer improperly. This phenomenon is reproduced successfully by simulation and the time cooperation between gap zero sequence protection and circuit reclosing is discussed. The following conclusions can be drawn.

(1) The gap puncture of No. 2 transformer is caused by instantaneous single phase to ground fault with too high zero sequence voltage peak.

(2) If the action time of gap zero sequence current protection can be set longer than the circuit reclosing time, the main transformer will not be shut down improperly and the reliability of the system can be improved. If the fault is permanent, the zero sequence protection should operate and shut down the transformer to avoid the second time voltage shock to the transformer.

References

1. DONG Kun. The principal of intermittent earth fault protection[J]. Relay, 2007, 35(Supplement 1):138-140.
2. ZHOU Hongyun, CHEN Hongyou. Analysis of neutral point over-voltage protection for 110kV transformer[J]. Transformer, 2008, 45(5):44-47.

3. JI Xijun, HUANG Shengchun, FAN Ping, et al. Analysis and disposal of primary transformer protection misoperation[J].Electric Power Automation Equipment, 2007, 27(1):64-67.
4. ZHANG Peng, ZHANG Hong, Liu Weilin, et al. Analysis of null sequence relay operation to neutral gap of transformer[J]. Inner Mongolia Electric Power, 2010, 28(3):11-13.
5. WU Wancai, SONG Shijun, ZHANG Xiaolei. Mal-operation Analysis and Suggestion on Zero-sequence Relay Caused by Transforms Neutral Point Gap Breakdown[J]. Automation of Electrical Power System, 2012, 36(14): 90-94.
6. LIN Haiyuan, WUXiaomei, LU Tingqin, Cao Kehui. Analysis of main transformer gap protection misoperation and countermeasure[J]. Electric Power Automation Equipment, 2012, 36(6):150-152.
7. Zhang Youliang, Yan Zhigang, Zhao Yanmei. Solving Measures for Mismatch in Main transformer Gap Protection and Zero Sequence Protection[J]. INNER MONGOLIA ELECTRIC POWER, 2012, 30(4): 23-26.
8. SHEN Yanhua, YUAN Wenjia, LI Fang.Thinking of main transformer gap protection[J]. Power System Protection and Control, 2010, 38(10):145-146.
9. National Power and Communication Center. Relay Protection Practical Technique Question and Answer of the Power System[M]. Beijing: China Electric Power Press, 1997.
10. MA Hui, GUO Xiaolong. Neutral over-voltage protection for 220kV main transformer[J].High Voltage Engineering, 2001, 27(Supplement 1):60-61.
11. XIE Guangrun. The Over-voltage of Electric Power System[M]. Beijing: WaterConservancy and ElectricityPress,1985.
12. PENG Xiangyuang, ZHONG Dingzhu, LI Qian, et al. Analysis and advice on gap protection problem of main transformers neutral point[J]. China southern Power Grid Technology Research, 2005, 1(2):39-43.

Fuzzy control methods applied for micro-pump system

Feng-Ling Li* and Fang-Yi Li
College of Automobile and Mechanical Engineering,
Changsha University of Science & Technology,
Changsha,410076, China,
**E-mail: flli4789@126.com*

A new fuzzy method was proposed used as micro-pump controller of water-position driver with the uncertainty math model. Takagi–Sugeno fuzzy if-then ruler numbers are adaptive with the control requirements and data space. The ruler is identified by the input position difference and difference rate, and the output of controller is the driving voltage of micro-pump. A recursive kalman filtering algorithm is presented in order to avoid the difficulty of the inverse matrix calculation. The simulation of step plus test demonstrated the stated method is effective and can meet the dynamic control requirement.

Keywords: Micro-pump; fuzzy control; if-then rule; water position control.

1. Introduction

Micro- pump system is widely used in many field such as micro fluid, micro fabricate and bio-mechanical engineering [1-3]. Although PID controlleris one of the most widely controller for many industrial process control applications [4]. Comparing with PID control, fuzzy control has many advantages. It does not need accurate mathematics model. In recent years, Takagi–Sugeno (T–S) fuzzy models have been proved to be effective in uncertain or non- linear process.

Fuzzy control theory was first introduced by Zadeh [5]. Generation of fuzzy rules from numerical data is important for fuzzy system. In most systems fuzzy if-then rules were typically derived from operator's experience and/or expert's knowledge of the system. For better performance, generating fuzzy rules such as gradient descent method [6], particle swarm approaches [7] and the extended fuzzy entropy [8] have been proposed.

2. System Description and Micro-Pump Mathematical Model

The whole control system majorly includes a micro-pump, pipe, camera and drive circuit of camera and micro-pump, shown as Fig. 1. The camera is used as water position sensing part, which is worked as semi-physical simulation system. The position data can be sent to the PC using USB interface. The control data also can be sent to the physical system by USB interface. The fuzzy controller is programmed using Matlab s function.

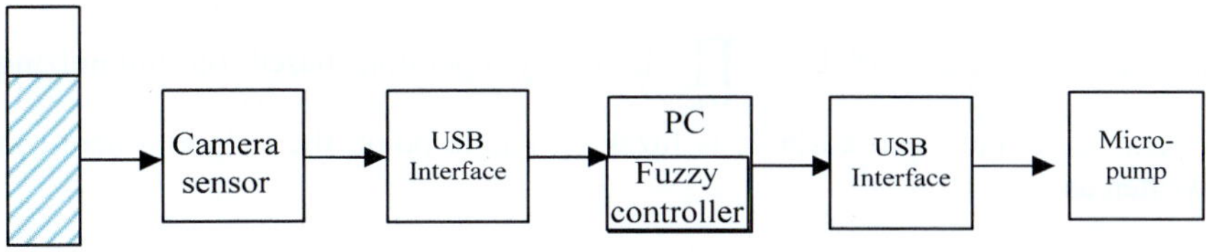

Fig. 1. The whole physical control system structure.

The micro-pump is direct voltage drive pump. The approximating polynomial model is described in many data sheet. The resulting approximating polynomial takes the form:

$$P = \rho(K(C_0 - C_1 q) - C_2 q^2 - C_3(q_1 - q)^2) \tag{1}$$

Where, P Presents pressure, K is Correction factor. ρ is fluid density, q Pump volumetric , q_1 is Pump nominal delivery, C0, C1, C2, and C3 are coefficients ,determined for a specific fluid and a specific angular velocity of the pump's driving shaft. If we adopted the PID controller, we must explore massive experiment for test the above multi-coefficients. And the T-S fuzzy controller is easier with the whole system input-output experiment.

3. Fuzzy Controller Design

T-S model is the development of linguistic variable until 1973. T–S fuzzy model is used to describe nonlinear systems [10], suitable to the nonlinear systems [11]. The typical T-S model is shown as

$$\begin{aligned} &R_j : if \quad x_1\ is\ A_1^j \quad x_2\ is\ A_2^j \ldots x_n \quad isA_n^j \quad then \\ &y = p_{j0} + \sum_{i=1}^{n} p_{ji} x_i, \quad j = 1, 2, \cdots, M \end{aligned} \tag{2}$$

Where x_i is the input variables set, A_i^j is the subset of fuzzy set, M is the number of fuzzy rules.

By the method of weighted mean defuzzification, to convert the vector into a single value, the output of MISO can written as

$$y = \sum_{i=1}^{c} (\omega_i y_i \Big/ \sum_{i=1}^{c} \omega_i) \tag{3}$$

$$\omega_i = \prod_{k\in I} \mu A_{kj}(x_k) \tag{4}$$

Where, $\bar{\omega}_i = \omega_i \Big/ \sum_{i=1}^{c} \omega_i$, $i = 1,2,\cdots,c$, $\mu A_{kj}(x_k)$ is calculated by the fuzzy partition of input variables, $\prod_{k\in I}$ is Fuzzy operator, based on minimizing operation. If the input variable is fuzzy by single point, the output system can written as

$$\begin{aligned} y &= \sum_{i=1}^{c} (\bar{\omega}_i y_i) = \sum_{i=1}^{c} \bar{\omega}_i (h_0^i + h_1^i x_1 + h_2^i x_2 + \cdots + h_r^i x_r) \\ &= \sum_{i=1}^{c} [\bar{\omega}_i \ \bar{\omega}_i x_1 \ \bar{\omega}_i x_2 \cdots \bar{\omega}_i x_r] [h_0^i \ h_1^i \ h_2^i \cdots h_r^i]^T \end{aligned} \tag{5}$$

The matrix description of the fuzzy system

$$Y = XP \tag{6}$$

Where, $L = crI$, Y is $N \times I$, X is $N \times L$; P is $L \times I$.

Therefore, P is estimated by standard least square method

$$P^* = (X^T X)^{-1} X^T Y \tag{7}$$

The estimation of P^* must be calculated through the inverse matrix of input variable x. A recursive kalman filtering algorithm is constructed to resolve the difficulty of inverse matrix computation [12]. The algorithm is written as

$$P_{i+1} = P_i + \frac{s_{i+1} \cdot x_{i+1}^T \cdot (y_{i+1} - x_{i+1}^T \cdot P_i)}{1 + x_{i+1} \cdot s_i \cdot x_{i+1}^T} \tag{8}$$

$$s_{i+1} = s_i - \frac{s_i \cdot x_{i+1}^T \cdot x_{i+1} \cdot s_i)}{1 + x_{i+1} \cdot s_i \cdot x_{i+1}^T} \tag{9}$$

Where, x_i^T is the ith row vector of X, y_i is the ith row vector of Y , $i = 0,1,2,\cdots,N-1$.

The adaptive algorithm for extracting fuzzy rules is following:

(1) Select the input variables and fuzzy partition the input space with the initial value ($c = 2$)

(2) According to the formula(4),calculate the parameter μA_{kj} (x)

(3) Combined the formula (7), the input variables X will be obtained

(4) According to the formula(9) (10),calculate the data matrix P

(5) if the system control error meet the design, turn to the step(6),otherwise $c = c + 1$,and return to the step(2)
(6) if the algorithm is satisfied with engineer standard , the program is finished.

4. Simulation and Test Results of Fuzzy Controller

The experiment includes three parts. Firstly, collect the water position and micro-pump data using the system shown as Fig. 1. Given the micro-pump a voltage signal ranged from 3.76 to 3.9V, then measure the position variable of water. The test curve is shown as Fig. 2.

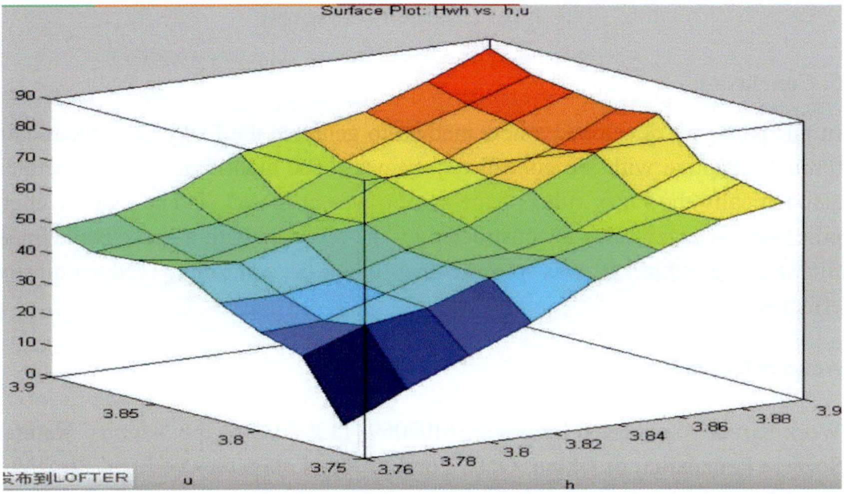

Fig. 2. The collected water position and micro-pump voltage on open loop test system.

The second part is program the T-S controlling model based the input-output data. The input variables of T-S controlling are error e(k) and its increment △e(k) .The error of water position is ranged to [-1 1], and the range of rate of water position is [-0.1 0.1], and the control variable is normalized to [-1 1]. According the stated algorithm, program the T-S model, and transfer to the s-function embedded to the Simulink model. At last, given the system a step input signal, the water position curve is shown as Fig. 3.

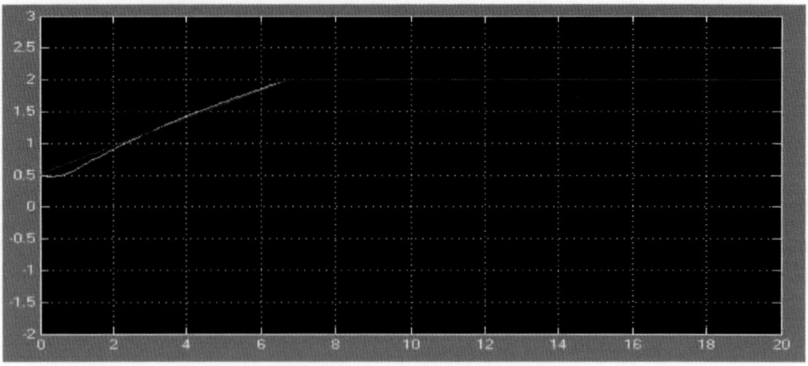

Fig. 3. The test result of T-S controller.

5. Conclusion

In this paper we developed a new method to generate the fuzzy rules from data through adaptive with the control requirement. The algorithm which generates automatically the fuzzy rules of the system was proposed. T The identification procedure of fuzzy rules is constructed by the test data with a recursive kalman filtering algorithm. It was proved that the T-S controller is dynamically effective.

Acknowledgments

Work partially supported by grant 51408069, 11302033 of the National Natural Science Foundation of China, Open Fund of Key Laboratory of Road Structure and Material of Ministry of Transport (Changsha University of Science & Technology) (kfj130204). The project was supported by Open Fund of Key Laboratory of Road Structure and Material of Ministry of Transport (Changsha University of Science & Technology) (kfj130204).

References

1. Homsy A, Koster S, Eijkel J C, et al. A high current density DC magneto hydrodynamic (MHD) micropump.[J]. Lab on a Chip, 2005, 5(4):466-471.
2. Sabashnikov A, Popov A F, Bowles C T, et al. First Experience with the Synergy Micro-Pump in Patients in INTERMACS Class 1–2 as a Bridge to Transplantation: Pushing the Limits [J]. Artificial Organs, 2015, 39(2):118–125.

3. Salari A, Navi M, Dalton C. A novel alternating current multiple array electrothermal micropump for lab-on-a-chip applications [J]. Biomicrofluidics, 2015, 9(1):053003-1406.
4. Astrom K J, Hagglund T. PID controller, theory design and tuning. NC: Instrument Society of America Research Triangle Park, 1995.
5. Zadeh L A. Zadeh, L.A.: The concept of a linguistic variable and its application to approximate reasoning. Part 1 Inform. Sci. 8, 199-249; Part 2. Inform. Sci. 8, 301-353; Part 3. Inform. Sci. 9, 43-80 [J]. Information Sciences, 1975, 8(3):199-249.
6. Brtrenji, H.R., Khedkar, P. Learning and tuning fuzzy logic controllers through reinforcements. IEEE Trans. Neural Networks 3(5), 724–740 (1992).
7. Ernesto Araujo, Leandro dos S. Coelho. Particle swarm approaches using Lozi map chaotic sequences to fuzzy modeling of an experimental thermal-vacuum system [J]. Applied Soft Computing, 2008 (8): 1354–1364.
8. Park I K, Choi G S, Park J J. Generating Optimal Fuzzy If-Then Rules Using the Partition of Fuzzy Input Space [M] Multimedia, Computer Graphics and Broadcasting. Springer Berlin Heidelberg, 2011:45-53.
9. T.G. Hicks, T.W. Edwards, Pump Application Engineering, McGraw-Hill, NY, 1971.
10. Takagi T, Sugeno M. Fuzzy identification of systems and its applications to modeling and control [J].
11. IEEE Trans Syst Man Cybern, 1985, 15:116–132.
12. Tanaka K, Wang H O. Fuzzy control systems design and analysis: a linear matrix inequality approach [M]. 2001, Wiley, New York.
13. Zhang X, Zhang L M, Li S X. Rapid federal Kalman filtering based on the inverse matrix recursive[J]. Electronic Measurement Technology, 2011.

Identification of driving intention for shift schedule optimization

Long Zhang*, Xiao-Feng Yin, Xiao-Hua Wu and Xu Chen

Sichuan Provincial Key Lab of Automotive Engineering, Xihua University, Jinniu District , Chengdu, China

*E-mail: zhangl_0129@163.com

With respect to the common used recognition methods for driving intention, there is a lack of effective means to apply the recognition results to calculate shift schedule which can reflect the driving style. This paper demonstrates a recognition method of driving intention which can be used to optimize the shift schedule and can also reflect the driving style of individual driver. Combined with the operation experience from excellent drivers, the relationship model between the engine speed, the throttle opening, the change rate of throttle opening and the driver's shifting performance expectations has been established and quantified using fuzzy logic. The simulation results show that the recognition results of the model are consistent with the driving habits of experienced drivers. The method overcomes the limitation of the current shift schedule of the vehicles with automatic transmission which mainly considers a single best performance, and meanwhile makes the gear schedule be more personalized.

Keywords: Fuzzy logic; driving intention; recognition method; shift schedule; model.

1. Introduction

With the development of the automobile industry and the improvement of the intelligent control technologies, the shift decision-making for the vehicle equipped with automatic transmission has entered into the stage of intelligent research and application [1][2]. The traditional shift schedule cannot fully meet the changes of the driving conditions, so it is difficult to make a balance between power performance, fuel economy and emission when vehicle operates in acceleration or deceleration, climbing hill, turning corner and other complicated road conditions. In addition, it is not only difficult to correctly reflect the driving intention but also easy to cause frequent and unexpected shift. Therefore, the intelligent shift decision-making will be an important direction for automatic transmission development, and the method for making intelligent shift schedule is the key to intelligent shift, moreover, the critical features of the intelligent shift decision-making system are reflected by its adaptability to drivers, automobiles and environment, therefore one of the core technology is the recognition method for driving intention.

However, the existing formulation methods for shift schedule of vehicles with automatic transmission rarely consider the driving intention in the optimization of shift schedule.

In this research, we focus on the identification of driving intention for shift schedule optimization in which the power performance, fuel economy and emission are considered together. Though analyzing the relationship between the driver's operation behaviors (i.e., the throttle opening, the change rate of throttle opening, etc.), the running status of vehicle (i.e., engine speed, etc.) and the corresponding performance, combining with control experience of excellent driver, making fuzzy rules of driving intention identification. it is possible to use the MIMO fuzzy inference system to quantify driver performance expectations that can be used directly in the shift schedule optimization.

2. Expression of Driving Intention

By making a comprehensive analysis of the acquired running status of vehicle and the traveling environment, the driver makes judgments, based on his driving experience, to make judgments to operate the accelerator pedal, the brake pedal and the steering wheel corresponding operations when the vehicles are in the running process, this awareness is called the driving intention [3]. It reflects the driver's driving experience and style in essence. In this paper, identification of driving intention for shift schedule optimization which power performance, fuel economy performance and emission are all taken into account by analyzing the driving control parameters and the vehicle running status, etc. For example, if the driver presses the accelerator pedal to its lowest position as fast as possible, he or she expects the vehicle could get the optimal power performance with the fuel economy in acceptable scope. Reasonable and accurate identification of driving intention has significant meaning for realizing personalized intelligent shift.

2.1. *Fuzzy inference system*

Due to the ambiguity of the driving intention, it is difficult to describe it by constructing a precise mathematical model, but the fuzzy inference can depict the approximate and imprecise features in the real world. Therefore, in this paper, the fuzzy logic is used to describe the driving intention.

We built two fuzzy logic inference systems with three inputs and one output, in which the throttle opening, the change rate of throttle opening and the engine speed are used as input, while for each system the driver's expectation on power performance or fuel economy is used as output.

2.1.1. *Power performance expectation inference system*

Taking the engine speed, the throttle opening and its rate of change as inputs of the fuzzy logic reasoning and the driver's expectations of the power performance as output, a fuzzy inference system of power performance with three inputs and one output is established, as shown in Fig. 1.

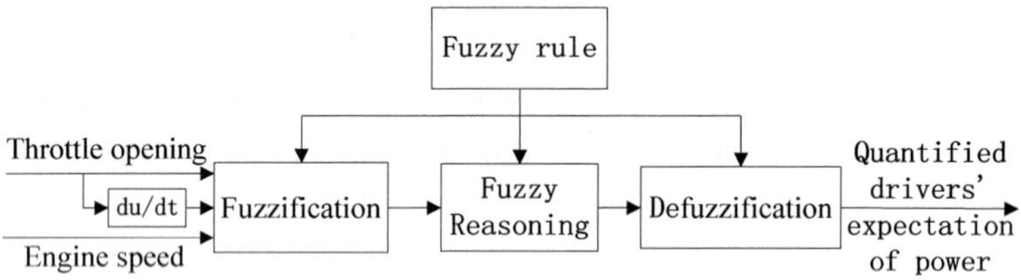

Fig. 1. Fuzzy inference system of the power performance.

(1) *Membership function of input variables*

According to the operation experience of excellent drivers , the domain of the throttle opening is divided into 5 fuzzy subsets, {Very Small (*VS*), Small (*S*), Medium (*M*), Big (*B*), Very Big (*VB*)}, and the change rate of throttle opening in the field of [-10,10] is divided into 5 fuzzy subsets as well, {Negative Big (*NB*), Negative Small (*NS*), Hold (*O*), Positive Small (*PS*), Positive Big (*PB*)}, here *O* means keeping the pedal position unchanged, negative number indicates the speed of the accelerator pedal, that is to say the greater the absolute value is, the faster the pedal release speed is. The engine speed is also divided into 5 fuzzy subsets, {Very Small (*VS*), Small (*S*), Medium (*M*), Big (*B*), Very Big (*VB*)}.

(2) *Membership function of output variables*

As shown in Table 2, driver's expectation of power performance is divided into 7 fuzzy subsets, {Very Low (*VL*), Low (*L*), Little Low (*LL*), Medium (*M*), Little High (*LH*), High (*H*), Very High (*VH*)}.

Depending on the operation experience of excellent drivers, expert knowledge and characteristics of the engine, 125 fuzzy inference rules for power performance expectation are developed, as shown in Table 1.

Taking the first rule as an example, it means that when the throttle opening is very small, the change rate of throttle opening is negatively big, and the engine speed is very small, the driver's expectation on power performance is little low.

Table 1. Fuzzy inference rules for power performance expectation.

d_a/d_t	a \ n_e	VS	S	M	B	VB
NB	VS	LL	LL	LL	LL	L
	S	LL	LL	LL	L	L
	M	LL	LL	L	L	L
	B	LL	LL	L	L	VL
	VB	LL	LL	L	L	VL
NS	VS	M	LL	LL	LL	L
	S	M	LL	LL	LL	L
	M	M	LL	LL	L	L
	B	M	LL	L	L	L
	VB	M	LL	L	L	L
O	VS	VL	LL	LH	H	VH
	S	VL	LL	LH	H	VH
	M	VL	L	LH	LH	H
	B	VL	L	M	LH	H
	VB	VL	L	M	LH	H
PS	VS	L	LL	LH	H	VH
	S	L	LL	LH	H	VH
	M	L	LL	M	H	VH
	B	VL	LL	M	LH	H
	VB	VL	L	LL	LH	H
PB	VS	L	M	H	H	VH
	S	L	LL	LH	H	VH
	M	L	LL	LH	H	VH
	B	L	LL	M	H	VH
	VB	VL	LL	M	H	H

Table 2. Fuzzy inference rules for fuel economy expectation.

d_a/d_t	a \ n_e	VS	S	M	B	VB
NB	VS	M	M	LH	LH	LH
	S	M	M	LH	LH	LH
	M	M	LH	LH	LH	H
	B	M	LH	LH	H	VH
	VB	M	LH	LH	H	VH
NS	VS	M	M	LH	LH	LH
	S	M	M	LH	LH	LH
	M	M	M	LH	LH	H
	B	M	LH	LH	H	H
	VB	M	LH	LH	H	H
O	VS	LH	M	L	VL	VL
	S	LH	LH	L	VL	VL
	M	LH	LH	LL	L	VL
	B	H	LH	LL	L	L
	VB	H	LH	M	LL	L
PS	VS	LH	M	L	VL	VL
	S	LH	M	LL	VL	VL
	M	LH	M	M	VL	VL
	B	H	LH	M	L	VL
	VB	H	LH	M	L	L
PB	VS	LH	LL	VL	VL	VL
	S	LH	M	L	VL	VL
	M	LH	LH	LL	VL	VL
	B	LH	LH	LL	VL	VL
	VB	H	LH	M	L	VL

2.1.2. *Fuel economy performance expectation inference system*

Similarly, 125 fuzzy inference rules for fuel economy expectation are developed, as shown in Table 2.

2.2. *Computation model for driver's performance expectation*

Three dimensional effect graphs of the expectation of power performance and fuel economy are shown in Fig. 2 and Fig. 3.

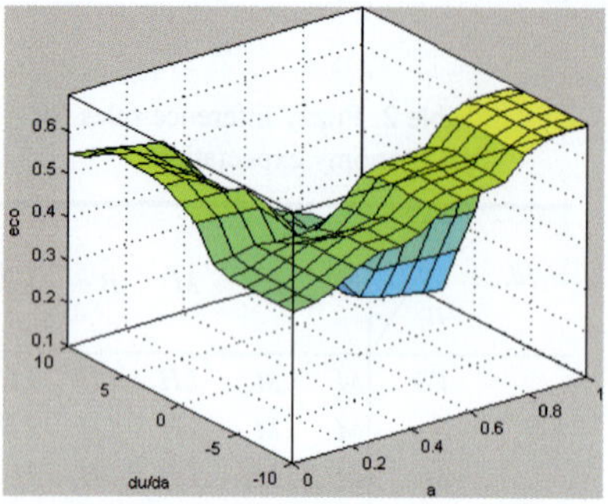

Fig. 2. Power performance expectation.

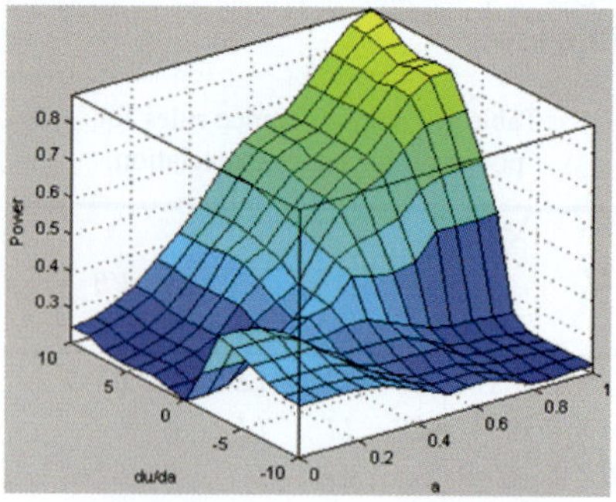

Fig. 3. Fuel economy expectation.

Since the shifting time extremely short, a normal hardware is difficult to collect the real-time driver's operation behaviors and the information of vehicle's running condition for calculating the shift schedule online. Therefore, we take the throttle opening (20%, 40%, 60%, 80% and 100%), the change rate of the throttle opening(-8 -4, 0, 4 and 8) and the engine speed (1500, 2500, 3500, 4500 and 5500) combined as input of the fuzzy identification model of driving intention in turn, then calculate 125 groups of shift schedule and store in TCU. And then by collecting the real-time information, call the appropriate shift schedule in the TCU based on the principle of proximity, if the parameter just in the middle of two numbers, take big one. For example, if the throttle opening at a certain time is 25%, the change rate of the throttle opening is 5 and the engine speed is 3200r/min, the parameter set [25%, 5, 3200] will be used to find the pre-stored shift schedule which is further used to determine the optimal gear position. The computation model of driver's performance expectation is shown in Fig. 4.

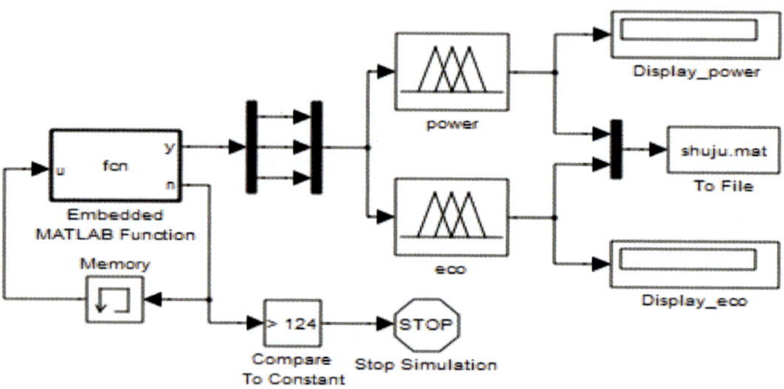

Fig. 4. Computation model for driver's performance expectations.

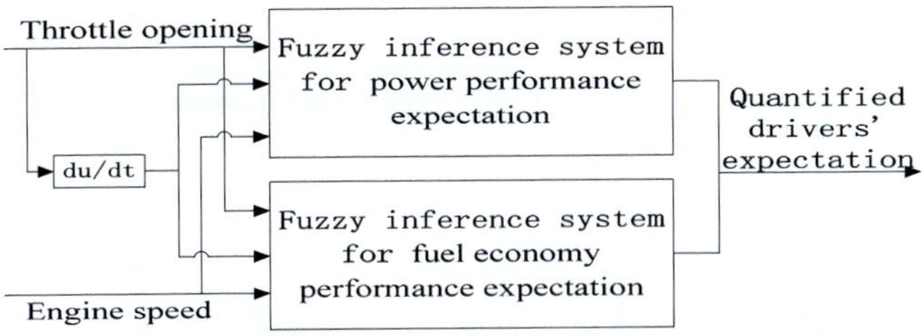

Fig. 5. Fuzzy inference system for identification of driving intention.

According to the driver's expectation on power performance and fuel economy, the driver's expectation on emission can be calculated, as in Eq. (1).

$$\omega_{co} = \omega_{hc} = \omega_{nox} = (1 - \omega_d - \omega_{fc})/3 . \tag{1}$$

Here, ω_d, ω_{fc}, ω_{co}, ω_{hc} and ω_{nox} are the weights corresponding to power performance, fuel economy performance, CO emission, HC emission, and NOX emission respectively, which could reflect the preference of the driver.

3. Simulation Result

To verify the recognition method of driving intention, we have established a simulation model using MATLAB/Simulink software, in which the engine sp-eed, the throttle opening and its change rate are provided by the ECE cycle, and driver's expectation on various performance are the output, as shown in Fig. 5.

Simulation was carried out to trace the speed profile specified by the ECE urban driving cycle including idle, acceleration, constant speed, deceleration and other 15 condition with different speed and load, as shown in Fig. 6. Recognition results of driving intention are shown in Fig. 7.

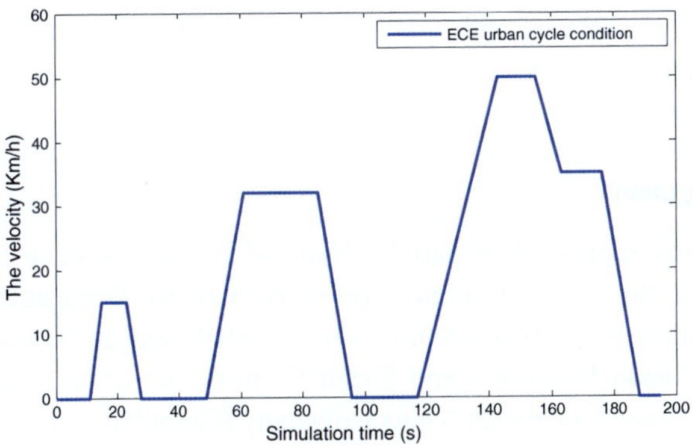

Fig. 6. ECE urban driving cycle.

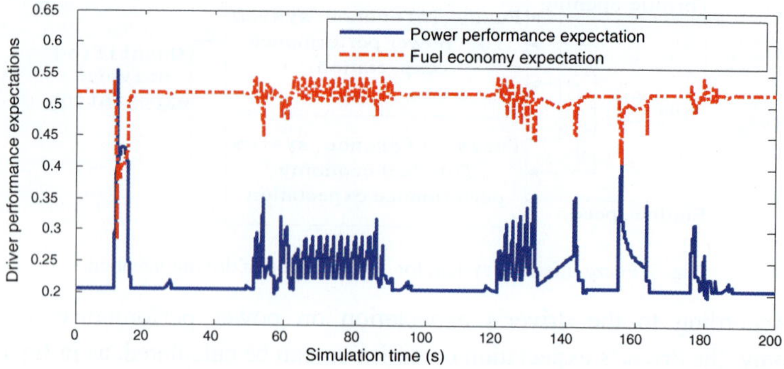

Fig. 7. Recognition results.

The simulation results show fuel economy expectation is in general higher than the power expectation under ECE urban driving cycle, that is to say, the driver expect to reduce fuel consumption. This is in accordance with driver's driving habits under the urban cycle in which the pursuit of fuel economy is the main goal.

4. Conclusion

In this paper, through taking into account the power performance, the fuel economy and emissions together, we proposed a recognition method for driving intention, which can be used to optimize the shift schedule and can also reflect individual driving style by using the fuzzy logic inference system. The recognition system for driving intention is simulated in the ECE urban driving cycle in which the maximum speed does not exceed 60km/h. The simulation results are in accordance with driver's driving habits under the urban cycle. It also shows that the fuzzy inference method of driving intention proposed in this paper is reasonable.

Acknowledgments

The work was supported in part by National Natural Science Foundation of China (Grant No. 51375402), Science & Technology Department of Sichuan Province (Grant Nos. 2015TD0021, 2016HH0010, and 2013JY0088), Sichuan Provincial Human Resources and Social Security Department, and Innovation Fund for Graduate Students at Xihua University (Grant No. ycjj2015149).

References

1. W. Xing, B. Yang, Research on Automatic Transmission Shifting Rules of Domestic Situation, *Equipment Manufacturing Technology Magazine*. 283-286, 3 (2014).
2. H. Jin, A. Ge, H. Chen, Development Trend of Automatic Transmission Technology of Motor Vehicles, *Automobile Technology*. 2007(2).
3. S. Zhu, J. Hu, Driving Intention Based Optimal Shift Control for Automatic Transmission, in *Proc. 4th International Conference on Mechanical Science and Engineering*, (Sanya, China, 2014).

The comparison of neural network and case-based reasoning in cost estimating modeling of remanufacturing hydraulic cylinder

Pin-Pin Qin, Yun Sun, Chun-Lan Yang* and Yu Xie
College of Mechanical Engineering, Guangxi University,
Nanning 530004, China
**E-mail: qpinpin@gxu.edu.cn*

Yun Sun
Beijing Normal University,
Macao University of Science and Technology Group,
Zhuhai 519085, China
E-mail: 409502537@qq.com

The current research sets the BP neural network and CBR as factors of engineering machinery hydraulic cylinder remanufacturing cost prediction. Take the differences between the new product manufacturing and remanufacturing and the process of hydraulic cylinder remanufacturing into consideration, we extract 10 feature factors and case attributes and establish BP prediction model and CBR model respectively. The comparison of two models indicates that the BP model and CBR model are both feasible ways to estimate the manufacturing cost of engineering machinery hydraulic cylinder and the BP model is more accurate. Study provides a feasible method for engineering machinery hydraulic cylinder remanufacturing production cost prediction, as well as effective measures for remanufacturing economic efficiency.

Keywords: Remanufacturing; hydraulic cylinder; cost estimating; BP neural network; case-based reasoning (CBR).

1. Introduction

Circular economy mode, which has beingpromoted currently by China's government, is an advanced economy mode that aims to increase economic benefits, reduce resource consumption, reduce environmental pollution, and promote employment[1]. Remanufacturing is highly corresponding to the Nation's strategic needs of structuring circular economy and is an effective way

---

*Corresponding author.

to realize the sustainable development of circular economy, energy conservation, and emissions reduction[2]. If remanufacturing production cost becomes too high, or even to exceed the price of new product, the remanufacturing will lose its value. Conducting scientific assessment for re-manufacturing economic efficiency, predicting components for remanufacturing production cost, reducing the inaccuracy and randomness for remanufacturing benefit evaluation, and improving economic efficiency of remanufacturing businesses has significance meaning to the remanufacturing cost management [3-5].

2. The BP Prediction Model and the CBR Prediction Model

2.1. *BP prediction model for hydraulic cylinder remanufacturing production cost*

We establish the process of Hydraulic cylinder remanufacturing production.According to the relationship between the characteristics and the cost of production, we take the Old parts recycle grades, Old parts recycle prices, Weight, Difficulty of disassembly, Difficulty of cleaning, Difficulty of inspection, Complicity of repairing technology, The complicity of repairing technology, Material costs, Utility expenses, and Equipment costs these 10 factors as network input factorsand remanufacturing cost as the output of the network to set up BP model . BP model network diagram is a single hidden layer BP network, the number of hidden layer nodes are initially set at 31 and only one output for every layer nodes. The best valueof hidden layer the can be determined through calculation [6].

2.2. *CBR prediction model for hydraulic cylinder remanufacturing production cost*

We establish the process of hydraulic cylinder remanufacturing production, analysis and extract the case attributes, use AHP to assignattribute weights, use CBR to search for the most similar case, and finallyset up the CBR forecasting model of hydraulic cylinder remanufacturing production cost. The procedure for setting up CBR forecasting model is as follow.

(1) Determine the case attributes based on the hydraulic cylinder remanufacturing production processes.

(2) Mathematical expression based on CBR cases.

(3) Normalized the attribute values. Change the qualitative indicators into quantitative indicators, such as to set the disassembly, cleaning, testing indexes into three levels. Set the maximum and minimum factors as $x_{i\max}$ and $x_{i\min}$, respectively, with the processing formula: When features become larger (smaller)and the cost becomes lower (higher), the normalization formula isas in Eq. (1)[7]:

$$\bar{x}_i = (x_i - x_{i\min}) / (x_{i\max} - x_{i\min}) \quad (1)$$

When features become smaller (larger) and the cost becomes lower (higher), the normalization formula isas in Eq. (2)[8]:

$$\bar{x}_i = (x_{i\max} - x_i) / (x_{i\max} - x_{i\min}) \quad (2)$$

(4) Base on AHP to apply the CBR case attributes weights. Firstly, we use the AHP model to signifythe hydraulic cylinder remanufacturing production cost factors. Secondly, we build up AHP judgment matrix by using the expert's experience knowledge. At last, we use APH to calculate the case attribute weights002E

(5) Using the nearest neighbor algorithm.We take the most similar case from the case library and use the cost of it as the predictive results of the target case production cost.

(6) Revise the history case and form a new case.

(7) Form the conclusion of the new cases by fine tune the results.

The calculation formula in step (5) isas in Eq. (3) [9]:

$$d_{oi} = \left[\sum_{j=1}^{m} w_j \left(\bar{X}_{ij} - O_j \right)^2 \right]^{\frac{1}{2}} \quad (3)$$

w_j is the weight for factor j, it affects the magnitude of similarity。 $\bar{X}_{ij}$ represents the i th historical case j th factor value; O_j represents target case's j th factor value; d_{oi} represents the Euclidean distance between target case O_j and i th history case; the lower the value of d_{oi}, the more similar it is between the history case and target case.

3. The Application

3.1. *The data*

We use one engineering machinery remanufacturing enterprise's historical data to establish case history base and find out the case attributes include 10 characteristics. There are 25 historical cases and five target sample. Since the data are sensitive enterprise production data, so we normalize them before undergoing further tasting. By normalizing, we can get the predictive production cost (Yuan).

3.2. *The BP network testing*

We test the BP network repeatedly formulating through MATLA. When the testing process either reaches 10000 times or error precision reaches 0.0001, the test ends. And thus the network structure parameters of the weights and thresholds were completely set.

After 10000 times of testing, the testing error sum of squares of the BP network E is equal to 0.0002. BP network testing actual value curve and the predicted curve are calculated. Results show that the production cost of the actual value and predicted value are well fitted, which indicates that BP prediction model can be used in the cost prediction of the target sample of hydraulic cylinder.

3.3. *Determine the CBR case attribute weights*

Based on the history case attributes of remanufacturing production, we set up four levels hierarchical structure of remanufacturing production cost factors. At the top layer is the target layer, the factors of hydraulic cylinder remanufacturing production cost. The second layer is the qualitative indexes and quantitative indexes. The third layer contains three assessment criteria: the condition of old parts, the handling of old parts, and the repairing technology. The fourth layer is the specific impact index, namely the CBR case attributes of the model.

Attribute weights are determined by the "experts method"[10]. The expert team includes engineering and remanufacturing engineering management personnel, and professional and technical personnel, who all have over seven years of relevant experience. We use questionnaire to collect experts' views of case attributes. Proportion of 1-9 scaling method was applied to construct comparative judgment matrix and determine the relative significant between each element and calculate the eigenvector and eigenvalue for the consistency check [11]. Through the consistency check, the characteristics of the normalized vector-valued thus can represent experts' assessment of case attributes significant. Each of 10 case attributes are respectively 0.033,0.033,0.033,0.017, 0.017,0.017,0.171,0.313,0.181and0.186.

3.4. *The evaluation of BP and CBR forecasting model*

The results of BP and CBR forecasting model are shown in table 1. As indicated, the average error of BP model is 3.04%, which is less than the CBR model. And the errors of BP model from all five test samples are less than 5%, with the maximum of 4.65%. Through the normalization, the actual and forecast costs are differ by 129 Yuan. CBR model average error is 5.54%, which is greater than the BP model. And errors from five test sample are all less than 8% with maximum of 7.23%. After the normalization, the actual cost and forecast cost are differ by 169 Yuan.

Table 1. The predicted results of BP and CBR Model.

Numbers	Actual	BPModel		CBRModel	
		Predicted	Error/%	Predicted	Error/%
01	1.334	1.288	3.55	1.244	7.23
02	1.099	1.064	3.33	1.183	7.10
03	1.423	1.360	4.65	1.377	3.34
04	1.372	1.326	3.46	1.448	5.25
05	1.263	1.257	0.46	1.326	4.75
Average Error/%		3.09		5.54	

4. Conclusions

We introduce the artificial neural network for remanufacturing engineering machinery hydraulic cylinder production cost prediction. Then we combine with the characteristics of the hydraulic cylinder remanufacturing production process to extract 10 remanufacturing production cost characteristic factors and establish the relationship between characteristic factors and the cost of production. With characteristic factors as input and the hydraulic cylinder remanufacturing production cost as the output, we then establish the BP neural network prediction model. We then test the model until the prediction error is less than the specified value and apply the model for the new type of hydraulic cylinder remanufacturing production cost prediction. Results indicate that it is a feasible method.

We also introduce CBR for engineering machinery hydraulic cylinder remanufacturing production cost prediction. Combine with the characteristics of the hydraulic cylinder remanufacturing production process, we extract remanufacturing production cost case characteristic factors and establish the CBR model. The CBR model forecast new type hydraulic cylinder of the

remanufacturing production cost by repeatedly using expert knowledge. The results prove that it is a feasible method too.
The comparison results between BP and CBR forecasting model shows that the results of prediction of BP model are better than those of CBR model.

Acknowledgments

The authors gratefully acknowledge the support of Guangxi Natural Science Found, Project 2015GXNSFAA139280, in supporting all parts of this study.

References

1. Xu B. *Recent progress of remanufacture industry and technology in China. Journal of thermal spray technology,* Vol. 2 (2010), pp. 1-6.(in Chinese).
2. Cheng-Han Wu, *"OEM product design in a price competition with remanufactured product," Omega,* Vol. 41 (2013), pp. 287-298.
3. Geyer, R., Van Wassenhove, and L. N., & Atasu, A. *The economics of remanufacturing under limited component durability and finite life cycles on remanufacturing profit, Management science,* Vol. 53 (2001), pp. 88-100.
4. Guide Jr, and V. D. R. Jr. *Production planning and control for remanufacturing: Industry practice and research needs, Journal of Operations management,* Vol.18 (2000), pp. 467-468.
5. Serra Caner Bulmus, Stuart X. Zhu, and Ruud Teunter. *Competition for cores in remanufacturing, European journal of operational research,* Vol. 233, 2014, pp. 105-113.
6. N. Karayiannis. *Theory of backpropagation neural network. International joint conference on neural networks* (Baltimore, MD, 1992).
7. Qu X L, Du J, Sun L F. *A model for cases indexing in CBR system based on AHP.Journal of application research of computers ,* Vol. 4 (2005), pp. 33-34,38. (in Chinese).
8. Luo M Zh, Zh M L. *On the method of similarity measurement in retrieval of CBR. Journal of Hefei university of technology,* Vol. 24 (2001), pp. 1091-1094. (in Chinese).
9. Fu ZH Y, Cheng J. *Research on emergency material demand forecast model in disaster. Journal of logistics technology,* Vol. 10 (2009), pp.11-13.(in Chinese).
10. Sung Hoon An, Gwang Hee Kim,Kyung In Kang. *A case-based reasoning cost estimating model using experience by analytic hierarchy process. Building and Environment,* Vol. 42 (2007), pp. 2573–2579.
11. Saaty TL. *The Analytical Hierarchy Process.* (McGraw Hill, 1980).

Multi-software co-simulation of electromagnetic radiation of DC Link in electric vehicle drive system

Ya-Hui Su* and Guo-Qiang Chen

School of Mechanical and Power Engineering,
Henan Polytechnic University,
Jiaozuo, 454000, China
**E-mail:hpusyh@163.com*

Aiming at the problem of general software in the limitation of large and complex structure's modeling in general electromagnetic simulation software package, a multi-software co-simulation method is proposed. The vehicle body is meshed in the Hypermesh package that is a pre-processing one that can import and mesh vehicle body with high quality. With only the interference of the direct current link current being considered, the electromagnetic radiation distribution characteristic is simulated and analyzed by using combination of the method of moments and the transmission line method. The methods and conclusions provide excellent reference for automobile EMC simulation modeling and assessment.

Keywords: Vehicle modeling; software co-simulation; electromagnetic radiation.

1. Introduction

In recent years, environmental issues attracted wide attention of the whole society, people's requirement for environmental protection become higher and higher. Because of the electric vehicle's unique advantages of energy saving and environment protection, it has broad development prospect [1-2]. In order to meet the requirements of safety, comfort, the high power electronic devices are integrated in the electric vehicle [3-4]. A variety of electronic devices in the electric vehicle are highly sensitive to receiving weak signals, and the frequency band of electronic devices are very wide [5-7]. When these electronic devices are working, they will transmit electromagnetic waves to the background. The radiation of the power cables is the most serious. The electromagnetic radiation is produced by these electromagnetic waves. It not only affects the normal operation of the electronic devices, but also affects human health. Therefore, the modeling and optimization of electromagnetic radiation in the electric vehicle drive system is of important theoretical significance.

At present, the research on the simulation of automobile EMC (Electro Magnetic Compatibility) is mainly focused on components and subsystems. There is little research in the whole vehicle system level EMC. The main reason is that the structure of the whole vehicle is very complicated. It is difficult in building a vehicle EMC model, which can guarantee an acceptable accuracy and computational efficiency. Combination of the CATIA and Hypermesh software packages is proposed to rapidly build the model of the vehicle body in the paper. The balance between the fine model of EMC of the vehicle, the computer memory requirement and the computation time is solved. An effective method is provided to quickly and efficiently obtain the vehicle EMC model.

2. Radiated EMI Model of the Whole Vehicle

The seat and interior of the vehicle are dielectric material which has little effect on the electromagnetic radiation. These dielectric materials in the 3D model can be ignored. Small voids, seams of the vehicle body are fully filled. A simplified model of the vehicle body is built in CATIA. The size of the simplified vehicle body is 3570×1300×1440(mm). It is shown in Fig. 2(a)

The automobile geometry model contains a large number of the local details. A lot of details have little effects on the electromagnetic radiation. The model is impossible to be built completely, because it may result in a large number of grids that cannot be calculated quickly. Therefore, it is necessary to simplify and optimize the model of the complex structure of the vehicle [8].

The CAD model of the vehicle contains a lot of small structures, fillets and small geometric features. The small geometric features are shown in Fig. 1. These small features will produce a lot of small triangular meshes during the meshing process in the Hypermesh software package. These small features not only increase the computation time and memory during the simulation, but also produce abnormal grids and affect the calculation accuracy.

(a) The small area (b) The chamfer

Fig. 1. The small geometric features.

These small geometric features should be cleaned up. The small structure is independent, so it won't affect the results when it is cleaned up. These chamfers are usually connected to a few surfaces. It can't be cleaned up directly. These chamfers should be treated as the small area. Some edges of the small area don't affect the basic structure. These edges can be deleted to form a larger area.

Most of the small geometric features are cleaned up after this step. But some lines forming into abnormal shape will be left. So it needs to be further cleaned. These remaining lines are removed by the commands of toggle edge to form a larger area structure. The after cleaning geometric features model of the vehicle is shown in Fig. 2(a). Then triangular mesh should be meshed by the command of automesh tool. The mesh size is 200mm. The meshed model is shown in Fig. 2(b).

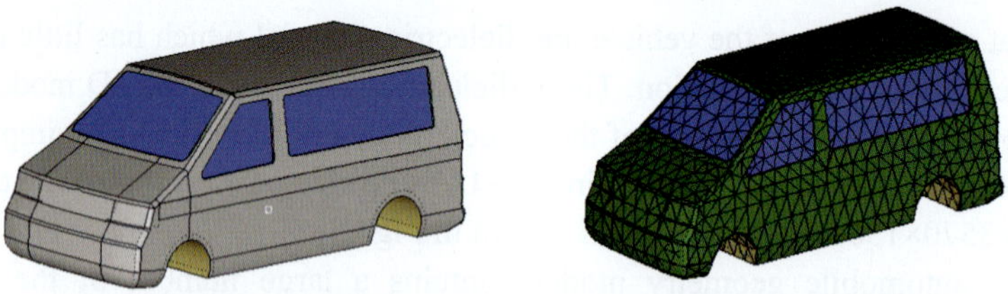

(a) After cleaning geometric features (b) Mesh model of the vehicle

Fig. 2. The cleaning geometric features and meshing model of the vehicle.

The mesh model is imported into the FEKO software and the DC link cable is modeled using the transmission line method. The radiated EMI model of the vehicle is shown in Fig. 3.

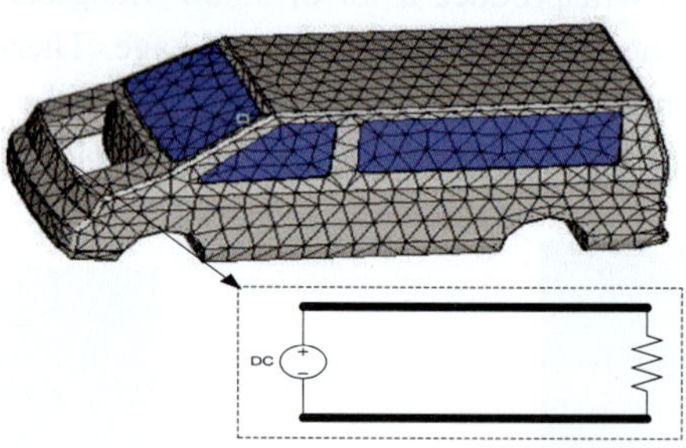

Fig. 3. The radiated EMI model of the vehicle.

3. Parameter Setting of the Simulation Model

The DC link is actually composed of many fine copper wires. The fine copper wires are the same in diameter and material. The fine copper wires are the same in electromagnetic radiation characteristics. So the single copper wire can be considered as the research object when the electromagnetic radiation characteristic of the DC link is studied.

The transmission line method is used in the circuit model of the DC link. The cable structure is processed approximately. The inner conductor of a plurality of soft copper wire is processed into a solid. The solid radius and the plurality of soft copper wire radius are equal [9]. The DC link cable length is 100mm, and the cross sectional area of the single stranded cable wire is 10mm^2. The DC link is not shielded. The material of the inner insulation layer is polyvinyl fluoride. The inner insulation layer radius is 1.1 mm. The DC link current is used as the excitation source of electromagnetic radiation. The amplitude frequency characteristic is shown in Fig. 4. The motor resistance is 15 ohm [10].

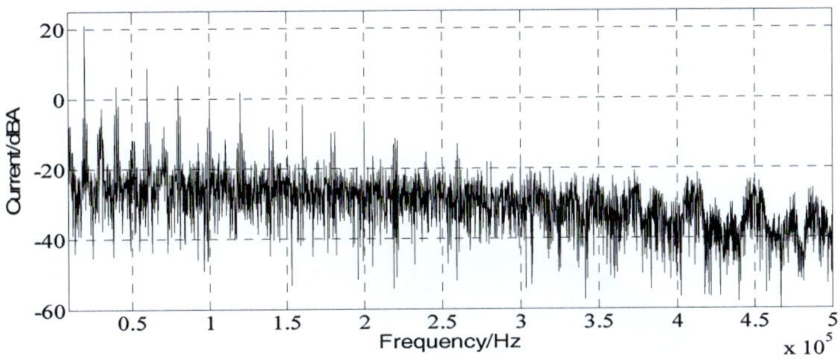

Fig. 4. The amplitude frequency characteristic of DC link.

The material of the vehicle body material is stainless steel. The relative magnetic permeability μ_r is 1, and the conductivity σ is 1.1×10^6. The DC link material is copper and the copper relative dielectric constant ε_r is 1, the conductivity σ is 5.8×10^7 S/m. The DC link line distances from the rear and the front of the vehicle are 3120mm and 350mm [11]. The frequency range is 9kHz~500kHz.

4. Analysis of Simulation Result

The probes are set at the position A and B. A is the position of the driver's head. B is the position of the passenger's head. B is behind A. The distance between A and B is 1.5m. The frequency domain distribution of the magnetic field is shown in Fig. 5 given that the specified reference speed is 1200rad/s. The switching frequency is 10000Hz. From Fig. 5, the magnetic field decreases with the increasing of frequency. The magnetic field of A is the strongest. The value of A is compared with the value of GB/T18387—2008. The simulation results are within the standard values.

The frequency domain distribution of the space electric field is shown in Fig. 6 given that the specified reference speed is 1200rad/s. The switching frequency is 10000Hz. From Fig. 6, the space electric field decreases with the increasing of the distance from the interference source.

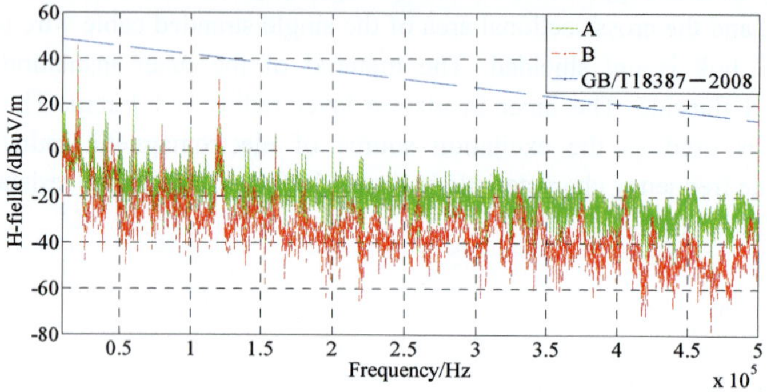

Fig. 5. The frequency domain distribution of the magnetic field.

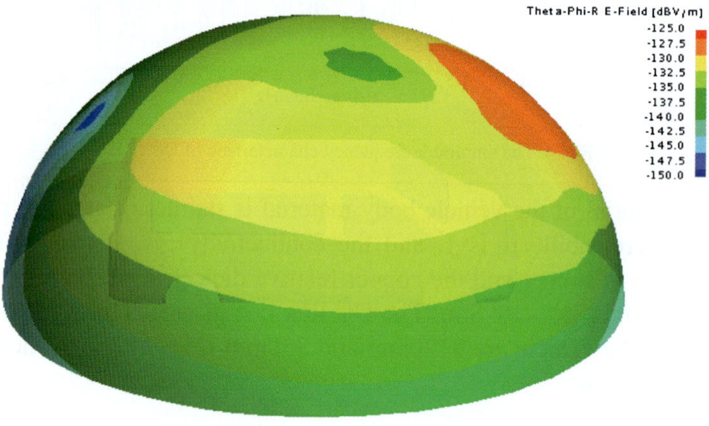

Fig. 6. The distribution of the space electric field.

The simulation results of the far field radiation are shown in Fig.7 given that the specified reference speed is 1200rad/s. The switching frequency is 10000Hz. From Fig.7, the gain of the Theta direction is -226.53dBi. The gain of the Phi direction is -227.95dBi. The DC link is perpendicular to the Theta direction but parallel to the Phi direction. So, the gain of the Theta direction is bigger than the Phi direction.

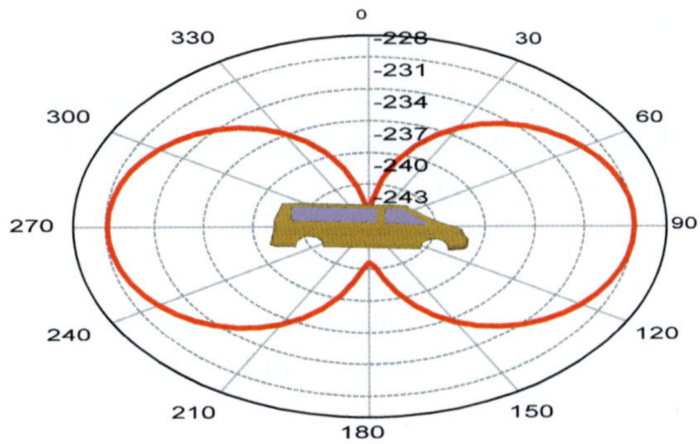

(a) The gain of the Phi direction

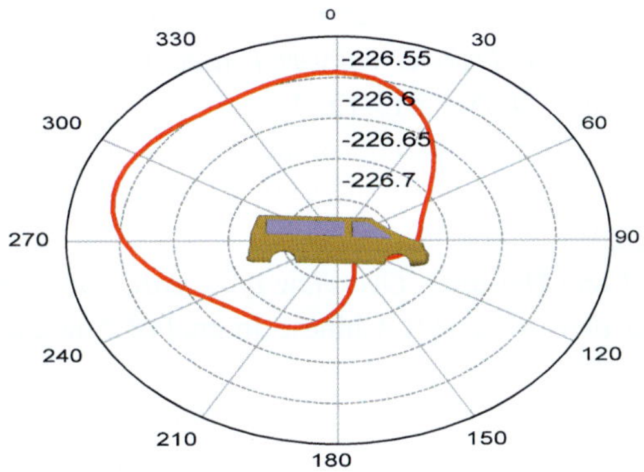

(b) The gain of the Theta direction

Fig. 7. The gain of the Phi direction and Theta direction.

5. Conclusions

A multi-software co-simulation method is proposed. According to the characteristics of CATIA, Hypermesh, and FEKO software packages, the vehicle CAD model is built. The method has greatly improved the efficiency in vehicle EMC simulation and analysis. The simulation result shows the amplitude of the magnetic field decreases with the increasing of frequency. The space electric field decreases with the increasing of the distance from the interference source. According to the electromagnetic radiation distribution characteristic, the electromagnetic compatibility in the electric vehicle drive system can be optimized.

Acknowledgment

This work is supported by National Science Foundation of China (No. U1304525). Work partially supported by Scholarship Fund of China Scholarship Council (File No. 201408410031).

References

1. Kai-Wei Hu and Chang-Ming Liaw, *Incorporated Operation Control of DC Microgrid and Electric Vehicle*, IEEE Transactions on Industrial Electronics 63, 202 (2016).
2. Mahran Quraan, Taejung Yeo and Pietro Tricoli, *Design and Control of Modular Multilevel Converters for Battery Electric Vehicles*, IEEE Transactions on Power Electronics 31, 507 (2016).
3. Wei Tian-yi, Shen Yu-zhuo, Zhang Yi-cheng, Yao Yong-tao. *EMI analysis and EMC design of DC/DC converter used in electric vehicle*[J]. Low Voltage Apparatus, 2005, 05:13-15+2.
4. Yan Jun-ping, Zhao You-qun. *Simulation of Conducted Emission of DC/DC converter in Electric Vehicle*[J]. Agricultural Equipment & Vehicle Engineering, 2011, (12):8-11.
5. Zheng Ling, Long Hai-qing. *Overview of research on EMC of the electric vehicle motor drive system*[J]. Chinese Journal of Automotive Engineering, 2014, 05:319-327.
6. Wang Hong-wu, Zhang Ji, Yang Teng-fei. *Status quo of the EMC research on drive system in electric vehicle*[J]. Electronic Measurement Technology, 2011, 34(6):18-22+59.
7. Kremer Frank, Frei Stephan. *Simulation of Emissions of Power Electronic Devices in Electrical and Hybrid Electrical Vehicles* [C]. 2010 Asia-Pacific Symposium on Electromagnetic Compatibility, APEMC 2010: 691–694.
8. Yu Ji-hui, Zheng Ya-li, Zou Zhi-xing. *Simulation study on crosstalk and radiation of wire inside automobile*[J]. Journal of System Simulation, 2006, 20(17):4737-4739.
9. Long Hai-qing. *Study on the EMC of PWM drive motor system of electric vehicle*[D]. Chongqing University, 2014.
10. Wang Quan-di, Liang Ling-hong, An Zong-yu, Peng He-meng, Liu Qing-song. *Finite element modeling and multi-software co-simulation in automobile EMC prediction*[J]. Journal of Chongqing University, 2013, 02:7- 7-11+36.
11. Du Ming-lei, Xu Zhong-ming, Ding Liang-xu, *Simulation and experimental study on the electromagnetic radiation interference of power cables in hybrid electric cars*[J]. Automotive Engineering, 2014, 36(6):734 -739.

Improved volumetric error modeling of machine tools based on differential operators

Zhen-Ya He[a] and Xian-Min Zhang[b]

School of Mechanical and Automotive Engineering, South China University of Technology, Guangzhou 510640, China

E-mail: [a]mezhyhe@scut.edu.cn, [b]zhangxm@scut.edu.cn

Jian-Zhong Fu

State Key Laboratory of Fluid Power and Mechatronic Systems, Zhejiang University, Hangzhou 310027, China

E-mail: fjz@zju.edu.cn

This paper proposes an improved volumetric error modeling based on differential operators for machine tools. Firstly, the traditional modeling method based on the rigid body kinematics was introduced. Secondly, a further analysis of the volumetric errors was carried out, and an improved modeling was established based on differential operators. Finally, the simulation experiment on a 5-axis machine tool was conducted to verify the feasibility of the improved modeling. The results shows that the volumetric errors calculated based on the improved method is consistent with that calculated based on the traditional modeling, and the maximum difference is 0.0014μm. Hence, the presented modeling of machine tools in this study is feasible. Compared to the traditional modeling, it is more concise. Moreover, it is a constructive method and could be used to analyze the error sensitivity of each movement axis to machining precision.

Keywords: Machine tool; volumetric error modeling; differential operators.

1. Introduction

Machining precision is an important index to evaluate the performance of CNC machine tools[1]. There are many factors that affect the machining precision of CNC machine tools, such as the original manufacture errors, geometric errors, thermal deformation errors, the errors caused by cutting load and machine tool's vibration, test errors of detection system, etc[2]. More than 60% of the total errors are the geometric error and thermal deformation error [3]. Therefore, the volumetric errors in this paper are defined as the total errors of the geometric errors and thermal deformation errors.

Error prevention and error compensation are the basic methods that could improve the machining precision of machine tools [4]. At present, the error compensation has become one of the most important ways to improve the machine performance.

However, analyzing and establishing volumetric error modeling of machine tools is the first step to carry out the error compensation. Therefore, extensive research has been performed to establish error modeling. There are the vector method, rigid body kinematics, multi-body system modeling, D-H homogeneous transformation method (HTM), etc [5-9]. So far, however, a standardized, universal, and integrated modeling theory has not yet been formed. Hence, this paper will focus on how to establish a universal volumetric error modeling of machine tools, in order to provide the basis for error compensation.

2. Volumetric Error Modeling

Assume that the movement components of machine tools are rigid bodies, the HTM ${}^{i-1}_{i}\boldsymbol{T}$ could be used to describe the transformation relations between the adjacent movement component i and $(i-1)$, namely ${}^{i-1}\boldsymbol{P} = {}^{i-1}_{i}\boldsymbol{T} \cdot {}^{i}\boldsymbol{P}$, where ${}^{i}\boldsymbol{P}$ represents the coordinate values of reference point P in the coordinate frame $\{i\}$, attached to the movement component i. Assume that there is only one kinematic chain--spindle kinematic chain, the ideal transformation relation of the reference point P, attached to last movement component, can be expressed as follows,

$$ {}^{0}\boldsymbol{P} = {}^{0}_{n}\boldsymbol{T} \cdot {}^{n}\boldsymbol{P}; \quad {}^{0}_{n}\boldsymbol{T} = \prod_{i=1}^{n} {}^{i-1}_{i}\boldsymbol{T} \tag{1} $$

where, ${}^{0}\boldsymbol{P}$ represents position of reference point P in the inertial coordinated frame $\{0\}$.

However, there are six errors induced by the each movement axis, including 1 linear displacement errors, 2 straightness errors and 3 angle errors. Therefore, the error transformations should be considered. The actual transformation relation can be expressed as ${}^{i-1}_{i}\boldsymbol{T}' = {}^{i-1}_{i}\boldsymbol{T} \cdot {}^{i-1}_{i}\Delta\boldsymbol{T}$. See Ref. [4], the error transformation matrixcan be expressed as

$$ {}^{i-1}_{i}\Delta\boldsymbol{T} = \begin{bmatrix} 1 & -\varepsilon_{zi} & \varepsilon_{yi} & \delta_{xi} \\ \varepsilon_{zi} & 1 & -\varepsilon_{xi} & \delta_{yi} \\ -\varepsilon_{yi} & \varepsilon_{xi} & 1 & \delta_{zi} \\ 0 & 0 & 0 & 1 \end{bmatrix} \tag{2} $$

where, δ is the linear error, ε is the angle error, the first subscript represents the error direction, and the second represents the movement axis.

The actual kinematical equation of the machine tools can be calculated as

$$ {}^{0}\boldsymbol{P}' = {}^{0}_{n}\boldsymbol{T}' \cdot {}^{n}\boldsymbol{P};\; {}^{0}_{n}\boldsymbol{T}' = \prod_{i=1}^{n} ({}^{i-1}_{i}\boldsymbol{T} \cdot {}^{i-1}_{i}\Delta\boldsymbol{T}) \tag{3} $$

where, ${}^{0}_{n}\boldsymbol{T}'$ is the total transformation matric with volumetric errors, expressed by matrix multiplication format. According to Eqs. (1) and (2), the error modeling of machine tools can be expressed as follows,

$$ {}^{0}\boldsymbol{E} = {}^{0}\boldsymbol{P}' - {}^{0}\boldsymbol{P} = \left[\prod_{i=1}^{n} ({}^{i-1}_{i}\boldsymbol{T} \cdot {}^{i-1}_{i}\Delta\boldsymbol{T}) - \prod_{i=1}^{n} {}^{i-1}_{i}\boldsymbol{T} \right] \cdot {}^{n}\boldsymbol{P} \tag{4} $$

3. Improved Error Modeling

Due to that the sixerrors of the each movement axis are very small, the actual movement of movement axis can be decomposed two movements, namely on the basis of the ideal movement, a micro-motion is added. Therefore, the actual transformation matric can be expressed as two parts, one is ideal transformation matrix, and the other is error component transformation matrix.

$$ {}^{i-1}_{i}\boldsymbol{T}' = {}^{i-1}_{i}\boldsymbol{T}(\boldsymbol{I}_{4\times4} + {}^{i-1}_{i}\boldsymbol{\Delta}) = {}^{i-1}_{i}\boldsymbol{T} + {}^{i-1}_{i}\boldsymbol{T} \cdot {}^{i-1}_{i}\boldsymbol{\Delta} \tag{5} $$

where,

$$ {}^{i-1}_{i}\boldsymbol{\Delta} = {}^{i-1}_{i}\Delta\boldsymbol{T} - \boldsymbol{I}_{4\times4} = \begin{bmatrix} 0 & -\varepsilon_{zi} & \varepsilon_{yi} & \delta_{xi} \\ \varepsilon_{zi} & 0 & -\varepsilon_{xi} & \delta_{yi} \\ -\varepsilon_{yi} & \varepsilon_{xi} & 0 & \delta_{zi} \\ 0 & 0 & 0 & 0 \end{bmatrix} \tag{6} $$

${}^{i-1}_{i}\boldsymbol{\Delta}$ is named as the differential operator of the error transformation matrix.

According to Eq. (5), the actual total transformation matrix of the reference point attached to the last movement axis can be expressed as follows,

$$ {}^{0}_{n}\boldsymbol{T}' = \prod_{i=1}^{n} {}^{i-1}_{i}\boldsymbol{T}' = \prod_{i=1}^{n} ({}^{i-1}_{i}\boldsymbol{T} + {}^{i-1}_{i}\boldsymbol{T} \cdot {}^{i-1}_{i}\boldsymbol{\Delta}) \tag{7} $$

The errors are far less than the values of the axis's movement, so the second order term and high-order term can be ignored.

$$ {}^{0}_{n}\boldsymbol{T}' \approx \prod_{i=1}^{n} {}^{i-1}_{i}\boldsymbol{T} + \sum_{i=1}^{n} d\,{}^{0}_{n}\boldsymbol{T}_{i} = {}^{0}_{n}\boldsymbol{T} + d\,{}^{0}_{n}\boldsymbol{T} \tag{8} $$

where,

$$d{}_{n}^{0}\boldsymbol{T} = \sum_{i=1}^{n} d{}_{n}^{0}\boldsymbol{T}_i = \sum_{i=1}^{n} ({}_{1}^{0}\boldsymbol{T} \cdot \cdots \cdot {}_{i}^{i-1}\boldsymbol{T} \cdot {}_{i}^{i-1}\boldsymbol{\Delta} \cdot {}_{i+1}^{i}\boldsymbol{T} \cdot \cdots \cdot {}_{n}^{n-1}\boldsymbol{T}) \tag{9}$$

$d{}_{n}^{0}\boldsymbol{T}_i$ represents the error component transformation matrix of i axis. According to Eqs. (8) and (9), it can be seen that the actual total transformation matrix ${}_{n}^{0}\boldsymbol{T}'$ can be decomposed into the summation of the ideal transformation matrix and error component transformation matrix of each movement axis. Therefore, the Error modeling of machine tools (see Eq. (4)) can be improved as follows,

$${}^{0}\boldsymbol{E} = {}^{0}\boldsymbol{P}' - {}^{0}\boldsymbol{P} = d{}_{n}^{0}\boldsymbol{T} \cdot {}^{n}\boldsymbol{P} \tag{10}$$

Compared with Eq. (4), the improved error modeling has the following advantages: it is more concise and intuitional to express the influence of the each movement axis to the volumetric errors. Furthermore, it has good universality and belongs to a constructive method. According to Eq. (9) the error modeling can be easily established by only analyzing the transfer relationship of the machine tool kinematic chain.

4. Simulation and Results

To verify the feasibility of the improved volumetric error modeling, let us take a 5-axis machine tool as an example.

4.1. *Structure analysis*

Fig. 1 illustrates the schematic structural diagram of a 5-axis machine tool. The topological structure of the 5-axis machine tool is shown in Fig. 2. There are two open-loop kinematic chains: (a)Machine bed--Y-axis--Z-axis--Spindle--Tool; (b)Machine bed--X-axis--A-axis--C-axis--Workpiece.

4.2. *Modeling of volumetric error*

Let ${}^{t}\boldsymbol{P}$ be the tool tip position of reference point P_1 described in the tool coordinate frame {T}. When the movement value $\boldsymbol{u} = [x \ y \ z \ \gamma \ \alpha]^T$, P_1 gets to the another reference point P_2, the position in the workpiece coordinate frame {W} can be described as follows

$${}^{w}\boldsymbol{P} = \begin{bmatrix} p_{wx} & p_{wy} & p_{wz} & 1 \end{bmatrix}^T = {}_{t}^{w}\boldsymbol{T} \cdot {}^{t}\boldsymbol{P} \tag{11}$$

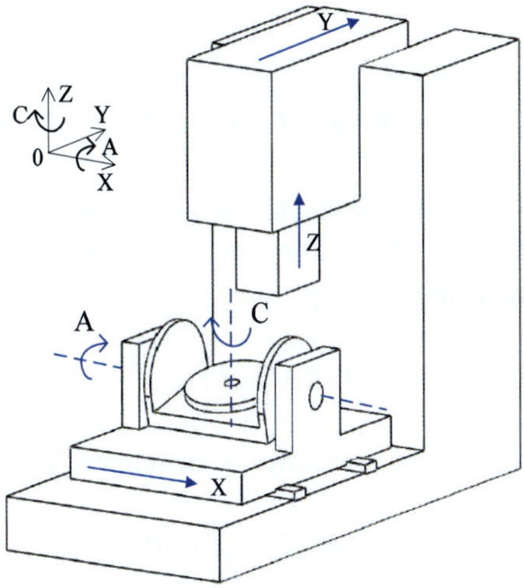

Fig. 1. Structural diagram of a 5-axis machine tool.

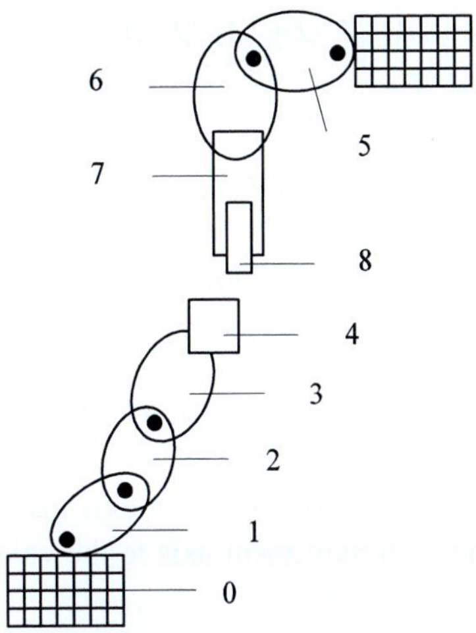

0. machine bed, 1. x-axis, 2. A-axis, 3.C-axis, 4. workpiece, 5. y-axis, 6. z-axis, 7. spindle, 8. tool

Fig. 2. Topological structure of the 5-axismachine tool.

$$ {}^{w}_{t}\boldsymbol{T} = {}^{4}_{8}\boldsymbol{T} = ({}^{0}_{1}\boldsymbol{T}\cdot{}^{1}_{2}\boldsymbol{T}\cdot{}^{2}_{3}\boldsymbol{T}\cdot{}^{3}_{4}\boldsymbol{T})^{-1}({}^{0}_{5}\boldsymbol{T}\cdot{}^{5}_{6}\boldsymbol{T}\cdot{}^{6}_{7}\boldsymbol{T}\cdot{}^{7}_{8}\boldsymbol{T}) \quad (12) $$

The details of the HTMs could be found in Ref. [10]. The actual transformation matrixfrom the cutting point of the tool to the workpiece coordinate frame can be described by

$$ {}^{w}_{t}\boldsymbol{T}' = ({}^{0}_{1}\boldsymbol{T}\cdot{}^{0}_{1}\Delta\boldsymbol{T}\cdot{}^{1}_{2}\boldsymbol{T}\cdot{}^{1}_{2}\Delta\boldsymbol{T}\cdot{}^{2}_{3}\boldsymbol{T}\cdot{}^{2}_{3}\Delta\boldsymbol{T}\cdot{}^{3}_{4}\boldsymbol{T})^{-1}({}^{0}_{5}\boldsymbol{T}\cdot{}^{0}_{5}\Delta\boldsymbol{T}\cdot{}^{5}_{6}\boldsymbol{T}\cdot{}^{5}_{6}\Delta\boldsymbol{T}\cdot{}^{6}_{7}\boldsymbol{T}\cdot{}^{7}_{8}\boldsymbol{T}) \quad (13) $$

Therefore, the traditional error modeling can be reduced as

$$ \boldsymbol{E} = ({}^{w}_{t}\boldsymbol{T}' - {}^{w}_{t}\boldsymbol{T})\cdot{}^{t}\boldsymbol{P} \quad (14) $$

Then ${}^{w}_{t}\boldsymbol{T}'$ is split into many polynomialsbased on differential operators,

$$ \begin{aligned} {}^{w}_{t}\boldsymbol{T}' &= [{}^{0}_{1}\boldsymbol{T}\cdot(\boldsymbol{I}+{}^{0}_{1}\boldsymbol{\Delta})\cdot{}^{1}_{2}\boldsymbol{T}\cdot(\boldsymbol{I}+{}^{1}_{2}\boldsymbol{\Delta})\cdot{}^{2}_{3}\boldsymbol{T}\cdot(\boldsymbol{I}+{}^{2}_{3}\boldsymbol{\Delta})\cdot{}^{3}_{4}\boldsymbol{T}]^{-1} \\ &\quad [{}^{0}_{5}\boldsymbol{T}\cdot(\boldsymbol{I}+{}^{0}_{5}\boldsymbol{\Delta})\cdot{}^{5}_{6}\boldsymbol{T}\cdot(\boldsymbol{I}+{}^{5}_{6}\boldsymbol{\Delta})\cdot{}^{6}_{7}\boldsymbol{T}\cdot{}^{7}_{8}\boldsymbol{T}] \\ &\approx {}^{w}_{t}\boldsymbol{T} + d\,{}^{w}_{t}\boldsymbol{T} \end{aligned} \quad (15) $$

$$ d\,{}^{w}_{t}\boldsymbol{T} = d\,{}^{w}_{t}\boldsymbol{T}_1 + d\,{}^{w}_{t}\boldsymbol{T}_2 + d\,{}^{w}_{t}\boldsymbol{T}_3 + d\,{}^{w}_{t}\boldsymbol{T}_5 + d\,{}^{w}_{t}\boldsymbol{T}_6 \quad (16) $$

where, $d\,{}^{w}_{t}\boldsymbol{T}_i$ represents the error transformation matrix from the single movement axis's errors to the tool cutting point, respectively.

$$ \boldsymbol{E} = ({}^{w}_{t}\boldsymbol{T}' - {}^{w}_{t}\boldsymbol{T})\cdot{}^{t}\boldsymbol{P} = d\,{}^{w}_{t}\boldsymbol{T}\cdot{}^{t}\boldsymbol{P} \quad (17) $$

Eq. (17) is the improved error modeling of the 5-axis machine tool, which can reflect the influences of the each movement axis to the volumetric errors of the machine tool, respectively.

4.3. *Simulation*

To verify the feasibility of the improved modeling, a simulation experiment was conducted.20 sampling points in the machine workspace are chosen, as shown in Fig. 3. All error parameters of machine tool axis generated randomly by computer, as shown in Fig. 4.

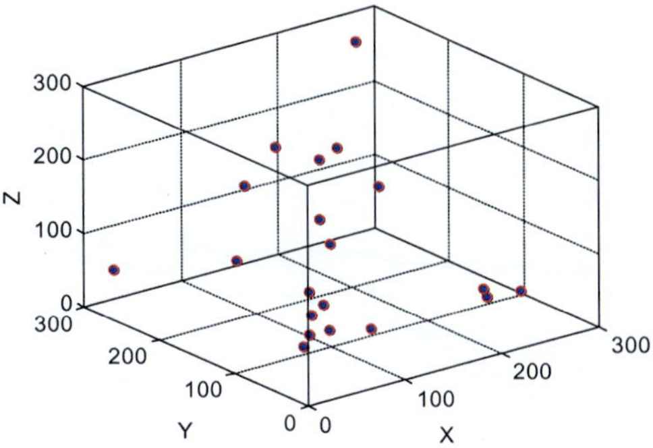

Fig. 3. Distribution of samples in the workspace of machine tool.

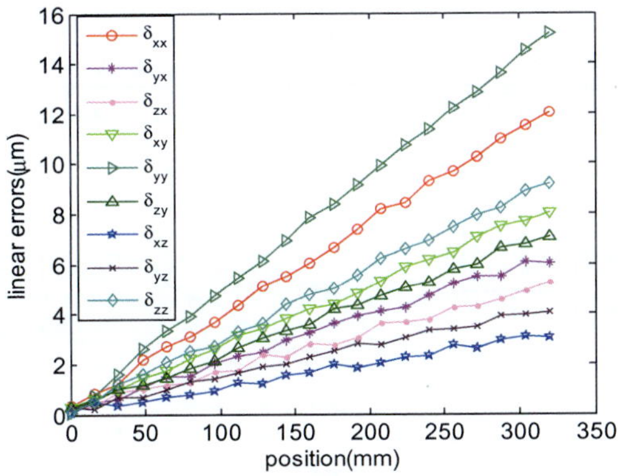

(a)

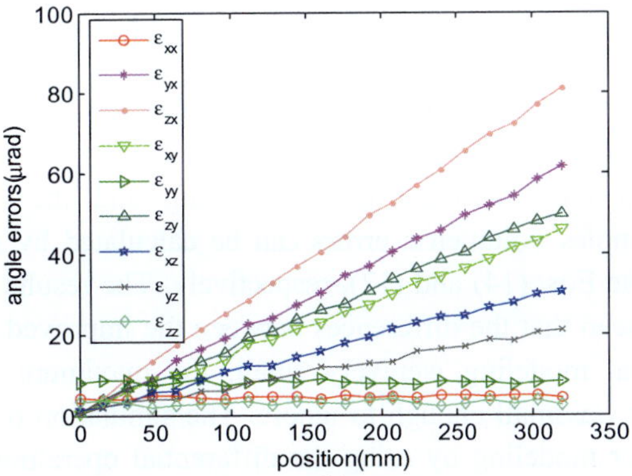

(b)

Fig. 4. Error parameters of machine tool axis: (a)linear errors of X-axis, Y-axis and Z-axis, (b)angle errors of X-axis, Y-axis and Z-axis, (c)linear errors and angle errors of A-axis, and (d)linear errors and angle errors of C-axis.

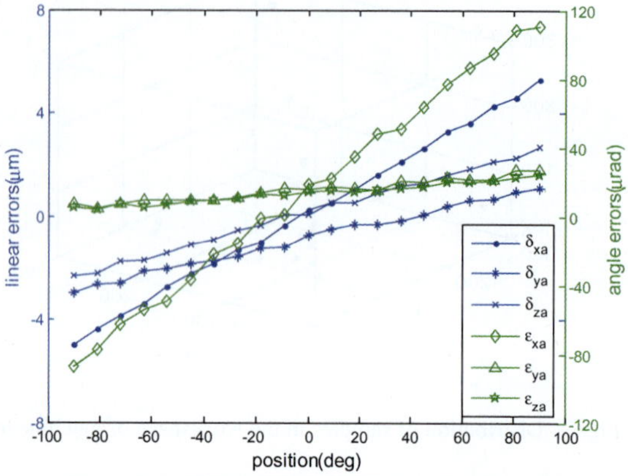

(c)

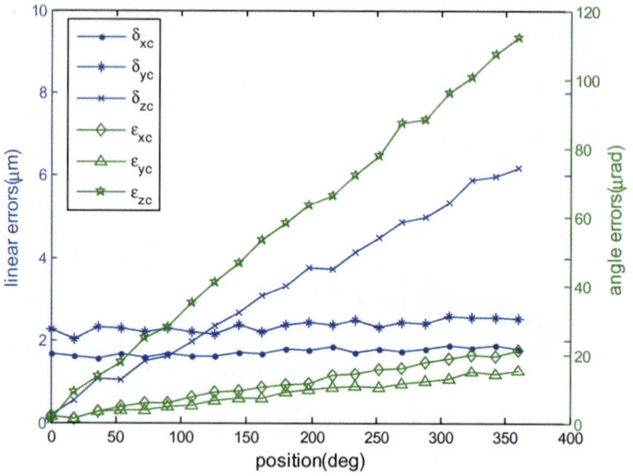

(d)

Fig. 4. (*Continued*)

According to the traditional error modeling and the improved error modeling, the samples volumetric errors can be calculated by substituting the parameters into the Eqs. (14) and (17), respectively. The results were shown in Fig. 5. It can be seen that the differences between the improved error modeling and the traditional modeling werevery small, the maximum difference was 0.0014μm, which wassmall enough to ignore. The simulation results show that the improved error modeling by using the differential operators to decompose the transformation matrix are feasibility, and it has good universality and belongs to a constructive method.

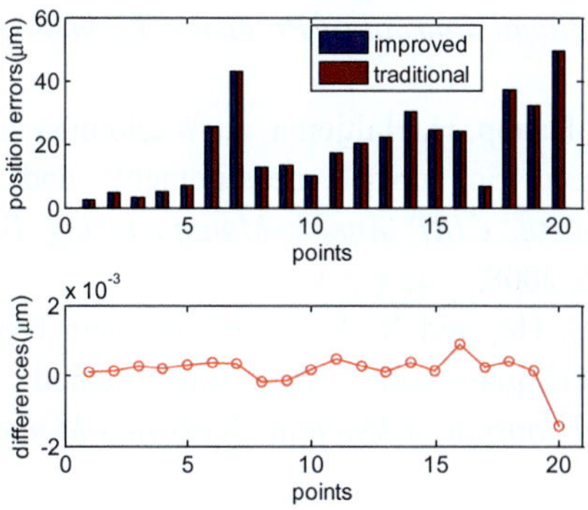

Fig. 5. Position errors of tool cutting points and differences between the improved modeling and the traditional modeling.

5. Conclusions

This paper presented an improved volumetric error modeling based on differential operators for machine tools. Firstly, the traditional modeling method of machine tools based on rigid body kinematics was introduced, and then an improved error modeling was proposed on the basis of traditional modeling using the differential operators. To verify the feasibility of the presented method, a simulation experiment on a 5-axis machine tool was carried out. The results show that the volumetric errors calculated based on the improved method is consistent with that calculated based on the traditional modeling, and the maximum difference is 0.0014μm. Hence, the presented error modeling of machine tools is feasible. And compared to the traditional modeling, it is more concise. Moreover, it is a constructive method and could be used to analyze the error sensitivity of each movement axis to machining precision.

Acknowledgments

This work was financially supported by the China Postdoctoral Science Foundation (No. 2015M580718), the Open Foundation of the State Key Laboratory of Fluid Power and Mechatronic Systems (No. GZKF-201506), the Fundamental Research Funds for the Central Universities (No. 2015ZM043), and the Natural Science Foundation of Guangdong Province (No. S2013030013355).

References

1. S. Mekid, and T. Ogedengbe. A review of machine tool accuracy enhancement through error compensation in serial and parallel kinematic

machines. *International Journal of Precision Technology,* vol. 1, no. 3, pp. 251-286, 2010.

2. H. Schwenke, W. Knapp, H. Haitjema, A. Weckenmann, R. Schmitt, and F. Delbressine. Geometric error measurement and compensation of machines—an update. *CIRP Annals-Manufacturing Technology,* vol. 57, no. 2, pp. 660-675, 2008.
3. H. Shen, J. Fu, Y. He, and X. Yao. On-line asynchronous compensation methods for static/quasi-static error implemented on CNC machine tools.*International Journal of Machine Tools and Manufacture,* vol. 60, pp. 14-26, 2012.
4. Z. Y. He, J. Fu, and X. H. Yao. Volumetric Error Modeling and Analysis for CNC Machine Tool Based on Multi-Body System.*Key Engineering Materials,* vol. 426, pp. 441-446, 2010.
5. Y. Hsu, and S. Wang. A new compensation method for geometry errors of five-axis machine tools.*International Journal of Machine Tools and Manufacture,* vol. 47, no. 2, pp. 352-360, 2007.
6. S. Zhu, G. Ding, S. Qin, J. Lei, L. Zhuang, and K. Yan. Integrated geometric error modeling, identification and compensation of CNC machine tools.*International Journal of Machine Tools and Manufacture,* vol. 52, no. 1, pp. 24-29, 2012.
7. A. W. Khan, and W. Chen. A methodology for systematic geometric error compensation in five-axis machine tools.*The International Journal of Advanced Manufacturing Technology,* vol. 53, no. 5-8, pp. 615-628, 2011.
8. Z. He, X. Yao, J. Fu, and Z. Chen. Volumetric Error Prediction and Compensation of NC Machine Tool Based on Least Square Support Vector Machine.*Advanced Science Letters,* vol. 4, no. 6-7, pp. 1-5, 2011.
9. M. S. Uddin, S. Ibaraki, A. Matsubara, and T. Matsushita. Prediction and compensation of machining geometric errors of five-axis machining centers with kinematic errors.*Precision engineering,* vol. 33, no. 2, pp. 194-201, 2009.
10. A. W. Khan, and C. Wuyi. Systematic geometric error modeling for workspace volumetric calibration of a 5-axis turbine blade grinding machine.*Chinese Journal of Aeronautics,* vol. 23, no. 5, pp. 604-615, 2010.

Gasoline engine combustion analysis system based on labview

Guan-Jie Xu and Ai-Ping Shi*
School of Energy and Power Engineering, Jiangsu University, Zhenjiang, Jiangsu province, China
E-mail: 1329389523@qq.com

In order to have a depth research for the combustion process and the status of the engine emissions ,it is necessary to develop a excellent combustion analysis system. The data collection, data processing, data computing and other display functions were realized by the self-developed combustion analysis system software , and some relative analyses of cylinder dynamic combustion were done. At the same time, it is so important for designers have an optimization of engine if they use the P-Φ indicator diagrams. Pmax indicator diagrams, air-fuel ratio(ξ)and fuel consumption variation data and so on. The test system for an internal engine combustion analysis through continuous testing and use, is feasible. No doubt, Better user interface and lower costs of this combustion system is pretty good for tester analysis the related data. Designers could facilitate the continuous development and utilization, so it has a certain value in engineering.

Keywords: Combustion analysis; pressure changes; engine; test.

1. Introduction

As is known to all, emissions of major pollutants source in the engine, if we can better control the combustion process of the engine such as the fuel injection quantity, mixed gas formation time, ignition timing, etc[1-6]. we can makes the combustion more fully, fewer emissions, to achieve energy saving and environmental protection requirement, so for the engine combustion analysis is particularly important. A lot of test and analysis instrument used to study the change of engine combustion has been developed abroad. For example, the United States PEI company with advanced computer technology, developed by DAB-500 and DAB-504, Japan's Onokazu company launched CB-466, 467, 566, and Austria AVL company's related products, etc.. But its price is often too expensive, and the test interface is not in line with the operating habits of domestic experimental researchers, the maintenance cost is high. In this independent development of combustion analysis software based on the

LABVIEW, and in the experimental engine test and application, get more accurate experimental data is particularly important.

2. The Design of the System

The test system includes 2613B crank angle sensor,6117B pressure sensor,5011B charge amplifier and so on. In the whole system, the first data analysis system carry out a number of relatively simple analysis, the next analysis of the results of the processing carry out the second data analysis. At the same time, the system also set up a main part to deal with the accidence. The operation, with the data acquisition and measurement analysis system to get the actual running conditions, according to the set of processing standard to check the working conditions. When an exception occurs or when the judgment is about to appear abnormal, the information is fed back to the optimized matching center. On the one hand, it could ensure the optimization of the automatic, on the other hand, it could ensure the safety of the engine itself. Setting the system optimization matching database, to ensure that in different working conditions are able to quickly find the contrast analysis of the source, and finally gets the constraints of the data of analysis, pressure with crank angle changes, in particular the relationship as shown in Figure 1.

3. System Software Design

With the help of Labview, engine combustion test analysis software is designed, it mainly using the graphical programming language and it has the data acquisition instrument control characteristics of the function of library functions and related development tools. The preparation of the software is divided into five steps: file read - file save - combustion analysis - format management - data validation. And software of eight modules (parameter setting, data acquisition, data processing, image generation, image processing, data searching, file storage, help module) is originally written to keep independent of each other, of which a module parameters or variables do not influence other related module, the natural increase of the whole test system reliability and capability of processing. Throughout the analysis process, due to the large number of data collected, a huge number, the need for a comprehensive processing and classification. The combustion analysis software can be simple point 0 after correction data files, digital filtering processing data files and cycle mean of data files, file parameters and environmental documents for classification, a good solution to this problem. According to different analysis needs, can set up and choose. Finally, the end of the combustion analysis, will be in accordance with

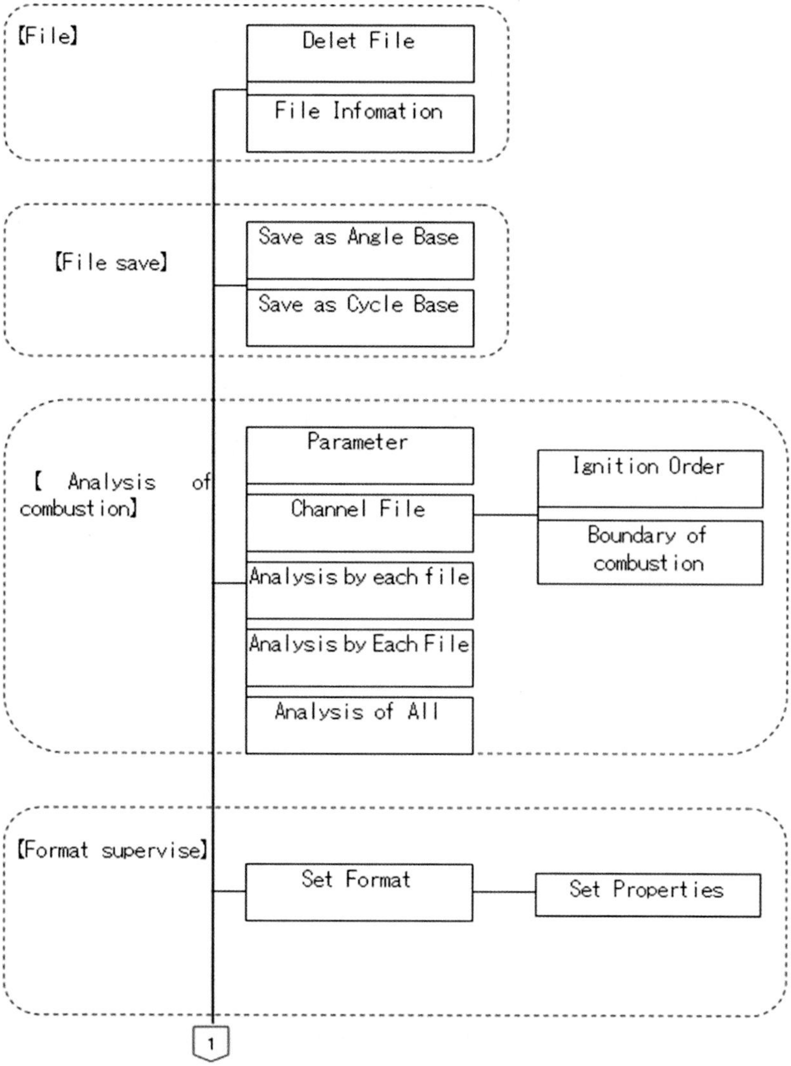

(a)

Fig. 1. Graph relationship analysis.

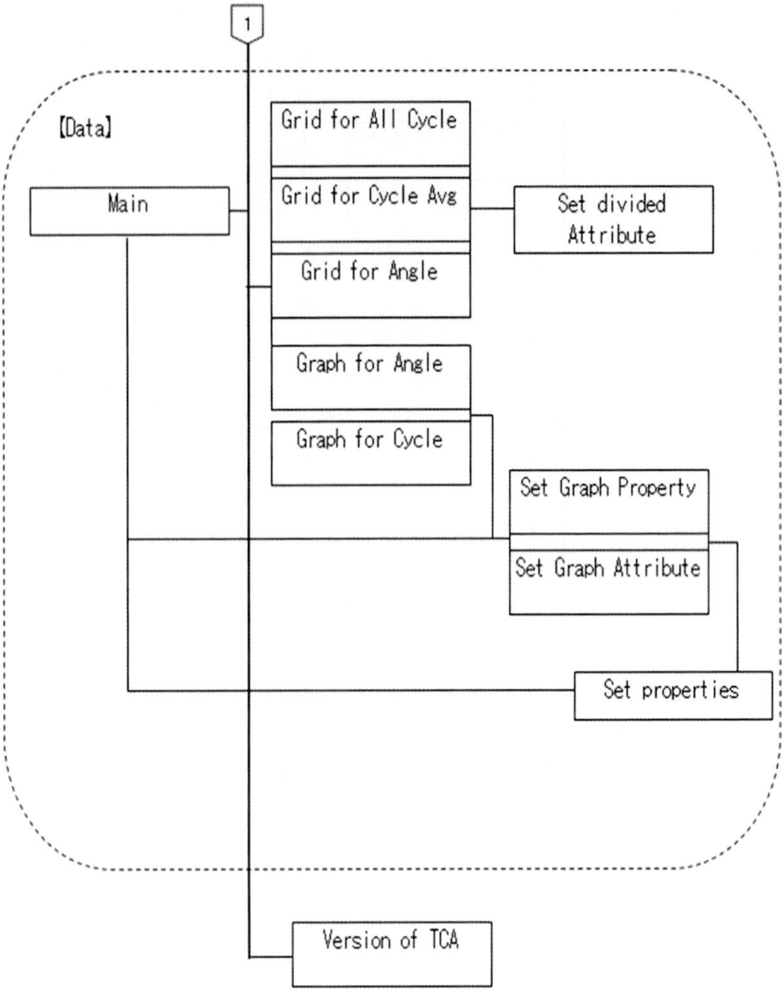

(b)

Fig. 1. (*Continued*)

the requirements of the ability to store 1-8 basic documents, as well as the 1-10 cycle file for reference.

3.1. *Combustion residue analysis module*

In the actual operation of the engine, the fuel is burned, there will be a part of the residual gas in the internal combustion process before the next combustion process, this part of the internal combustion engine will have an impact on the next cycle. Usually, the internal combustion engine measurement and analysis system, generally do not take into account the influence of the residual gas in the cylinder on the next combustion, naturally lead to the internal combustion engine combustion parameters calculation errors, which seriously affect the accuracy of the calculation. The experimental results show that the influence of this error on the heat transfer of the air intake is about -30%, the heat transfer in the cylinder is about 1%, and the low load is about 5%[7-8]. The advantages of the present combustion analysis software: Comprehensive analysis obtained actual residues condition, combined with the next acquisition data, the air / fuel ratio calculating the dynamic change process, using the dynamic high precision air fuel ratio analysis calculation and the analysis accuracy than the general analysis software. The idea is to make use of least square method and five point three times smoothing formula (assuming data collected by the 2n+1 equal interval points)[9-10]. The specific processing process is shown in Figure 2.

3.2. *Operational test interface*

According to collection experiments are finally after filtering process get a large number of data files, the combustion analysis software can file data and effective classification management, for different file for artificial set and revised. With a part of the experimental data as an example, in the main interface of the software, click the right mouse button appears as shown in Figure 3. The diagram shows that this software can set the type and number of files, you can also set a number of important files, encryption, which is a lot of software do not have. From which we can easily find many important information such as the number of cylinders, the total number of cycles, the number of data and so on. Table 1 is detailed in Figure 4, the number of the circle in the circle corresponds to the icon code.

3.3. *Experimental analysis*

The engine parameters are shown in Table 2.

As shown in Figure 4, the right side of the graph is file name, as well as the name of the project to be analyzed, it can be selected. The horizontal coordinate is the crank angle, the vertical coordinate is the pressure value. Left side analysis software to draw the pattern and the right data into the graphics generated by origin almost identical. Although the overall changes in the same, but there are differences between the parts of the cylinder. The reason is that there is relationship between the sensor and the loss degree of each cylinder injector.

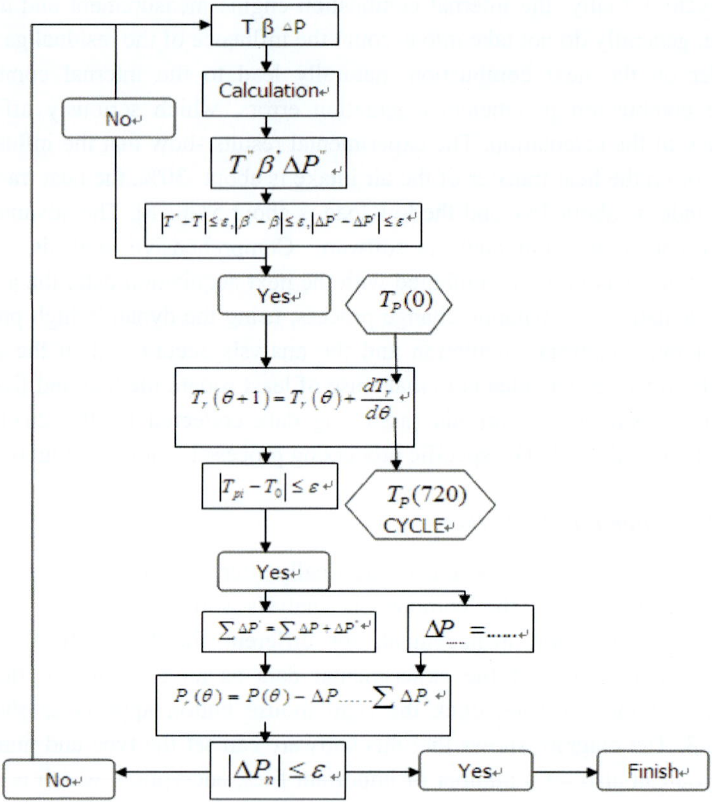

Fig. 2. Combustion analysis flow chart.

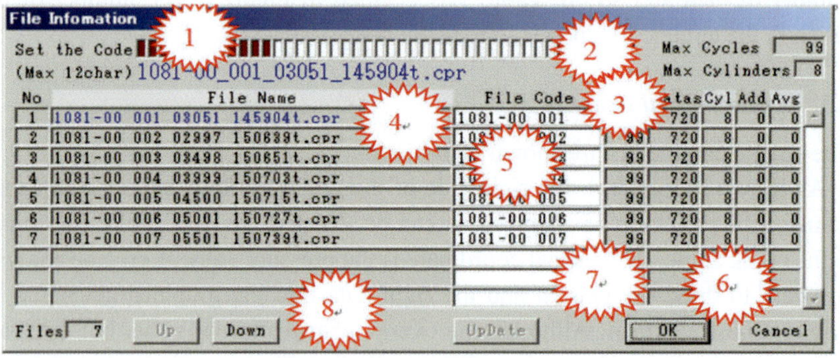

Fig. 3. File information operation interface.

Table 1. File information detail table.

Icon code	content	Icon code	content
1	Set the Code	5	File Code
2	Maximum Cycle number in all files	6	Total number of cycles
3	Maximum number of cylinders in all files	7	Number of loop test data
4	File Name	8	Number of Autobak files

Table 2. Engine parameters.

content	technical parameter	content	Technical parameter
Engine style	4 cylinder, electronically controlled gasoline engine	Compression ratio(ξ)	11.4
Cylinder diameter（D）	85mm	Electronic controlled injection system	Single point, packet injection
Piston stroke (S)	55.3mm	Firing order	1-3-4-2
Rated speed (n)	4995r/min	Length of connecting rod	137.95mm
displacement (ml)	1998mL	model	3433-06

This is consistent with the initial expected, the test system is reliable and stable. At the same time, the data can be processed and analyzed according to the different analysis angles, and it can be specified for a particular cylinder or an analytical project (eight cylinder). As in Figures 5, 6, 7, overall pressure ranges are relatively concentrated, and the change of cylinder is not uniform, there may is due to each cylinder of the internal combustion engine spark plug and the nozzle of the product carbon degree different, atomization and ignition condition is different, natural late each cylinder residual gas volume inconsistent, affected the engine with the actual inflation rate, changes arising from the difference in the performance of the engine. Similarly, in addition to the above curve, the test software can also be a statistical summary of the important data in each cylinder. As shown in Figure8 , the first column can be cylinder wall temperature (Tw), by comparing each cylinder temperature to the average temperature, to determine whether there is one cylinder abnormal, can repair and optimization of the different proportion; the fuel consumption rate of the situation (10%, 50%, 90%), can improve the injection strategy, weaken the differences between each other, can make the engine more stable operation, improve fuel economy; the air-fuel ratio (E) a situation can monitor the engine running at the same time, reflect the adequacy and stability of combustion; the calorific value (Q) to analyze and judge the engine fuel combustion efficiency. There are other different experimental parameters to consider the whole combustion process, which provides a more accurate reference for the later optimization.

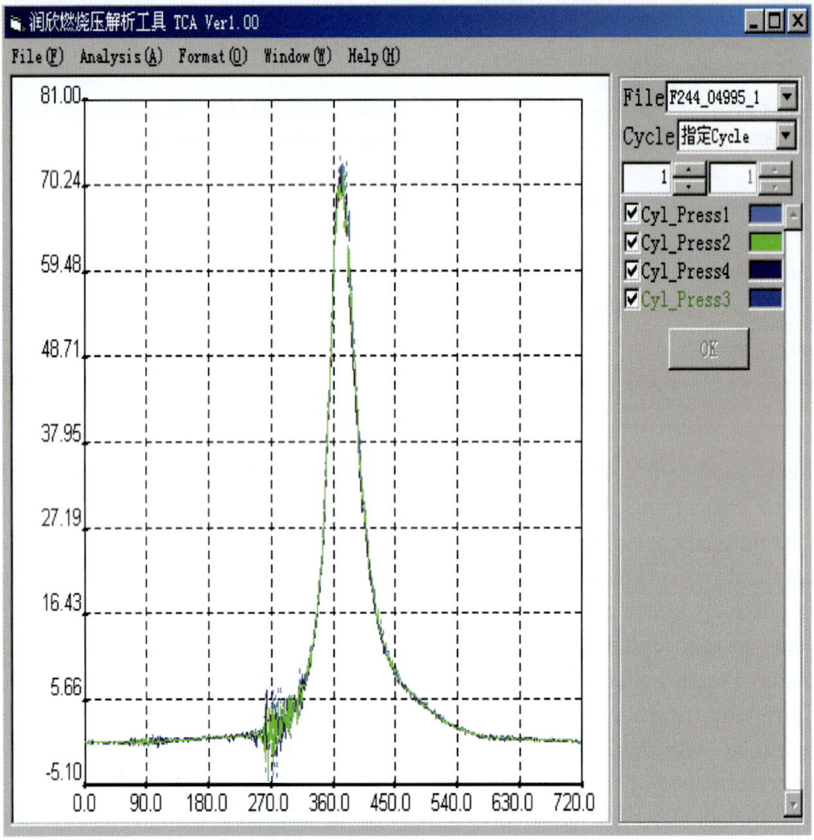

Fig. 4. Indicator diagram.

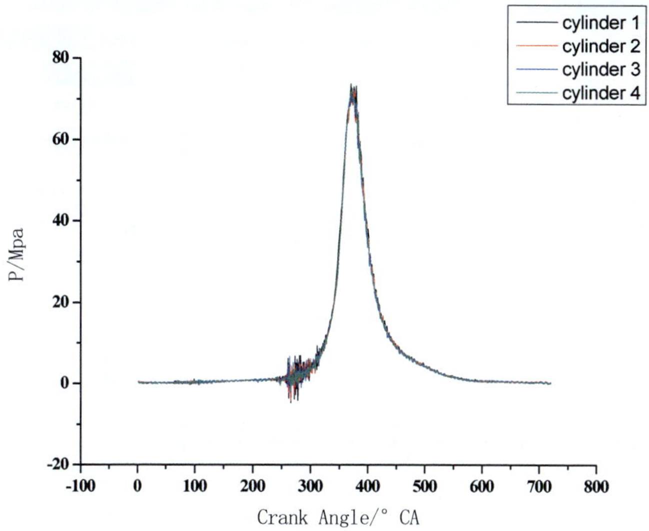

Fig. 5. P-Φ graph.

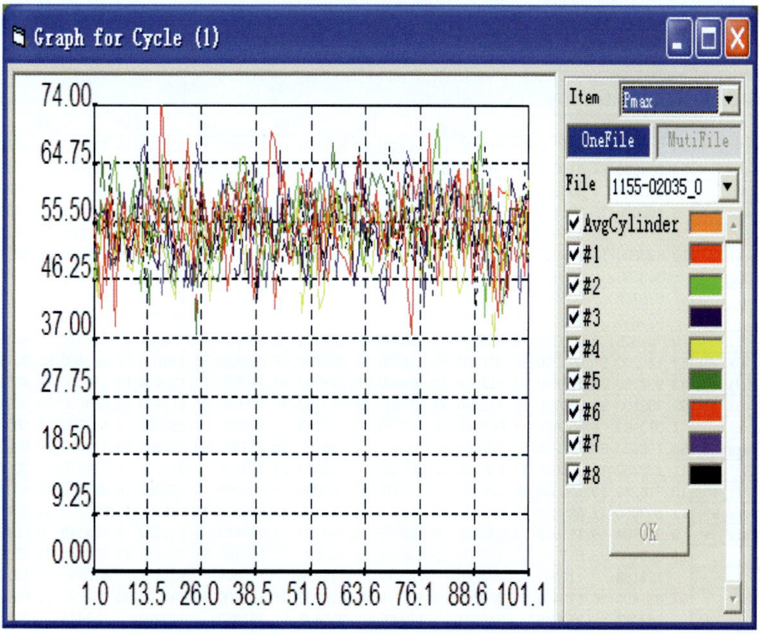

Fig. 6. Change of Maximum pressure in cylinder.

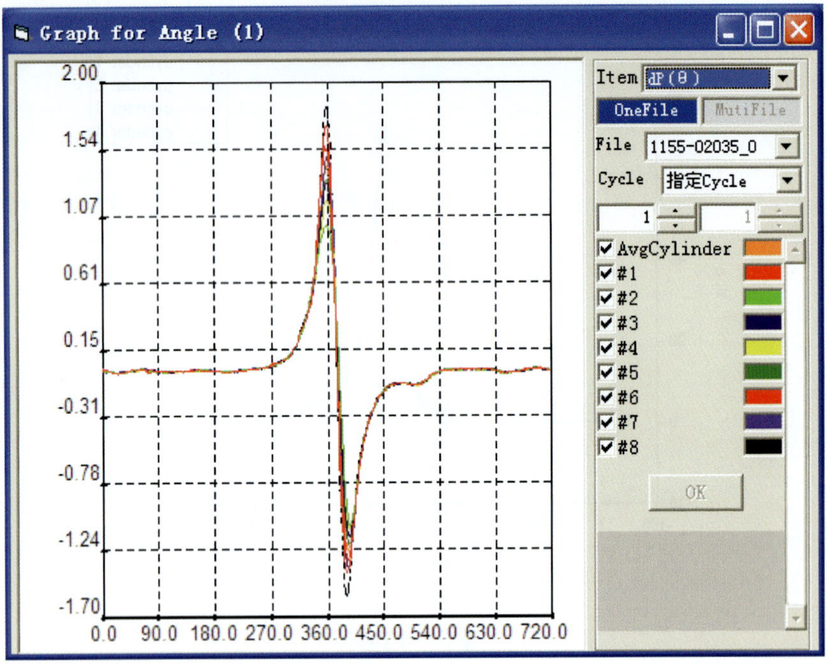

Fig. 7. Pressure change rate curve.

Grid for Cycle (1)

	Avg Cylin	#1	#2	#3	#4	#5	#6	#7	#8
IMEP	10.365900	10.475770	10.119720	10.343680	10.507100	10.175280	10.422070	10.576780	10.306840
Pmax	54.804890	54.813840	55.380470	53.437460	52.354470	55.901290	55.914470	55.421880	55.215260
dPMax	1.573489	1.550629	1.627261	1.476778	1.416586	1.697968	1.666060	1.572386	1.580243
Apmax	75.498800	75.891100	74.584200	75.881200	76.455400	74.802000	75.267300	75.762400	75.346500
θ cs	36.670800	36.336600	37.475300	36.485100	34.445600	37.118800	37.950500	36.613900	36.940600
θ ce	18.690600	18.613900	18.435600	18.108900	18.366300	19.732700	19.118800	18.722800	18.425800
PLOSS	-.003388	.009866	-.000780	-.009674	-.002390	.006470	-.033053	-.014153	.016609
Q	.876760	.886036	.855444	.878134	.893183	.858158	.878216	.894471	.870436
E	.899256	.895487	.903130	.893676	.881310	.916453	.912259	.895759	.895975
U	9.781325	9.938378	9.555579	9.853363	10.127510	9.330654	9.826757	10.009370	9.608994
AI10%	57.149200	57.594500	56.071700	57.914200	58.980200	55.804800	56.436700	57.423600	56.967900
AI50%	73.938700	74.321600	73.119000	74.768400	75.718500	72.540500	73.138400	74.197100	73.705800
AI90%	92.753100	92.830700	92.442200	93.637100	94.414400	91.889000	92.118400	92.663000	92.030400
Rmax	1.573489	1.550629	1.627261	1.476778	1.416586	1.697968	1.666060	1.572386	1.580243
ARmax	58.738900	58.574200	58.207900	58.782200	59.188100	58.663400	58.920800	58.772300	58.802000
P330	11.132750	11.265670	11.106130	11.218380	11.240620	10.708110	10.879150	11.339970	11.304010
ε eff_t	10.516240	10.426390	10.602880	10.357370	10.095080	10.902150	10.802960	10.459160	10.483900
ε eff_b	11.803550	11.829930	11.860140	11.750870	11.833290	11.678930	11.783120	11.837350	11.854780
ε eff	9.209704	9.151818	9.332792	9.030725	8.864230	9.447675	9.444603	9.184548	9.221245
Tw	23.556400	23.817300	23.923500	23.446500	23.268800	22.818700	23.326100	23.897200	23.953200
β	.123188	.118541	.122210	.125943	.117720	.125804	.127015	.123240	.125032
Tin	97.821400	86.179400	99.080800	03.467700	92.694000	96.660300	96.598500	95.232200	12.658800
G1	.349357	.353046	.340881	.349905	.355837	.341953	.349970	.356418	.346844

Fig. 8. Cylinder project detail drawing.

4. Conclusions

(1) With the help of virtual instrument architecture developed for the engine combustion analysis system, it has a friendly man-machine interface and also improves the efficiency of the program, the whole engine performance test provides effective help. In the whole cycle, the variation of the pressure with the crank angle is different, especially after the arrival of the top dead center. The minimum pressure process was improved, so as to achieve the maximum pressure process, the uneven is weakened effectively. If the operation of the engine was improved, it would have a positive significance on gasoline engine performance and fuel economy. No doubt, the gasoline engine working performance will get one step closer to enhance.

(2) The test system with the pros and cons of twice filtering, smoothing and convergent analysis and consider combustion residues. It has been a greater assurance of the results and the accuracy of the measurement. The data is more consistent with the actual operating condition of engine, for improving its performance.

(3) From the test and analysis, we get a lot combustion experiment data and the image. For example, relationship between the crank angle and pressure, mean effective pressure, rate of pressure change, the maximum combustion pressure and cumulative heating rate from the effective data. The results and variation of the accurate and reliable to become an effective tool for the future engine combustion analysis.

Acknowledgment

The project is supported by the Jiangsu Province ordinary university graduate student research and innovation program(NO.SJZZ_0134).

References

1. Colin O, Truffin K. A Spark Ignition Model for Large Eddy Simulation Based on an FSD Transport Equation (ISSIM-LES). [J]. Proceedings of the Combustion Institute, 2011, 33(2):3097-3104.
2. Crete F, Matalon M. Propagation of wrinkled turbulent flames in the context of hydrodynamic theory.[J]. Journal of Fluid Mechanics, 2011, 680:225-264.
3. Jiang Feng, Sun Zhiyi. Design of video image acquisition system based on DSP [J]. Industrial control computer, 2012(5):23-26.

4. Huang Zhiliang, Lei Yong. A data acquisition system for the whole process parameter measurement of aero engine based on LABVIEW [J] Measurement and control technology 2012(5):23-26.
5. Li Yu hua, Zhang Li, Kang Jiang, using the crankshaft speed fluctuation to diagnose the internal combustion engine combustion state [J]. Sensor and micro system, 2010,29(10).
6. Bai Jing xiao, Cao Yi peng, Zhang Wen ping. Transmission path and evaluation of combustion noise in internal combustion engine [J]. Journal of internal combustion engines 2010,52(11):123-126.
7. Sun Wei chao, Yin De, Wang Ding guo, Wu Zhong de. The design of the environmental data acquisition system based on C8051F040 [J]. electronic measurement technology, 2011.10.
8. Jenny H Sailer W, Walter T, Huang Bao kang. Kistler instrument company new vehicle combustion analysis system [J]. Foreign internal combustion engine, 2010(3):52-56.
9. Lei Zhen shan, Wei Li. LABVIEW advanced programming and virtual instrument engineering application [M]. Beijing: China Railway Publishing House, 2009, 5.
10. Nada Z, cycle analysis of vehicles powered by a fuel cell and by internal combustion engine for Canada [J]. Journal of Power Souces, 2006, 155(2): 297-310.

Simulation analysis of turbocharged diesel engine performance under exhaust gas recirculation step change conditions on the basis of design of experiments

Ling-Ge Sui

Computer science and technology postdoctoral flow station, Jilin University, Changchun, 130012
Academic affairs division, Changchun Finance College, Changchun, 130022, China
E-mail:suilingge@163.com

Lan Huang

College of Computer Science and Technology, Jilin University, Changchun, 130012

Yong-Qiang Han* and Zhong-Chang Liu*,†

**State Key Laboratory of Automotive Simulation and Control, Jilin University, Changchun, 130025*
†State Key Laboratory of Automotive Simulation and Control, Jilin University, Changchun, 130025,
†E-mail: liuzc@jlu.edu.cn

Improving the exhaust gas recirculation (EGR) response characteristics is essential for optimizing the combustion and emission performance of diesel under transient conditions. This study compares four control strategies of the EGR opening valve characteristics through a simulation analysis. The transient response characteristics were more improved by the upcast and comprehensive control strategies than by the linear and sunken strategies, whereas torque ripple was better controlled by the comprehensive and linear strategies. The comprehensive control strategies were simplified and quantitatively validated by the design of experiments method. Relative to the linear strategy, the comprehensive strategy reduced the NOX emissions by 6% while improving the response time by 19%. Relative to the upcast control strategy, it reduced soot emissions by 11% and delayed generating time by 17%.

Keywords: EGR; DOE; simulation analysis; transient conditions.

†Corresponding author.

1. Introduction

Exhaust gas recirculation (EGR) is the traditional in-engine purification method for diesel engines. It operates to reduce the combustion temperature and oxygen concentration, thereby suppressing NO_X formation under steady-state conditions [1, 2]. By defining transient conditions as a combination of step change conditions [3], we can analyze complex transient conditions [4-7]. Under step change conditions, the combustion boundary conditions are delayed relative to the EGR rate step change, which deteriorates the combustion. The EGR transient response time is related to the combustion deterioration level. In addition, shortening the EGR transient response time can provide adequate backflow exhaust in the least engine cycles, and reduce the nitrogen–oxygen reaction time. Therefore, the performance of a turbocharged diesel engine under transient conditions can be improved by decreasing the EGR transient response time.

Homogeneous charge compression ignition (HCCI) is a new combustion mode characterized by controllable, pre-mixed, and low temperature combustion. Currently, HCCI is extensively researched for its many advantages such as multipoint ignition, high thermal efficiency, and extra-low NO_X and particulate matter (PM) emissions [8]. To control the energy release rate during self-combustion in a large area, HCCI combustion usually requires lean burn and a high EGR rate during engine operation. Therefore, the HCCI combustion mode is more easily realized under low-speed or small-load operation conditions. Under heavy-load operation conditions, the traditional combustion mode is needed. Because both combustion modes demand different EGR rates, the EGR rate undergoes a step change under transient conditions. Therefore, by improving the transient response of the EGR step change, we can achieve rapid combustion mode switchover and a smooth combustion process.

This study analyzes the main factors affecting the EGR response characteristics. Various methods for improving the EGR response characteristics are investigated in a simulation study. The results are important for revealing and controlling the whole engine transient process and improving the combustion and emission performance under transient conditions.

2. Performance Simulation Platform of Turbocharged Diesel Engine Under EGR Step Change Condition

Our EGR step change performance simulation platform is based on a current simulation platform of turbocharged diesel engines under transient conditions [9]. The platform, which combines GT-Power and Matlab/Simulink, is schematized in Figure 1. On one hand, the parameters of the engine working process (such as pressure, temperature and speed) generated by the GT-Power engine module are sent to control unit (the Simulink control module), on the other hand, the control module outputs the control instructions decided by a specific control algorithm. The corresponding control, such as injection mass, EGR valve opening, and constant speed/torque transient condition switching, is implemented by the GT-Power engine module. The results computed by the joint simulation of GT-Power and Simulink are sent to STAR-CD, which provides detailed cylinder data. From the microcosmic aspects of these data, we can infer the main influencing factors and the internal rules of the EGR step change response time under transient conditions. These results are important for controlling the engine transient combustion process and reducing transient emissions.

To verify the accuracy of the EGR loop model of the transient simulation platform of the turbocharged diesel engine, we imposed a step change in the EGR valve opening and compared the simulation results with experimental results at 1650 rpm and 50% load. The comparisons are presented in Figure 2.

The step change in the EGR valve opening alters the intake mass flow, air fuel ratio (AFR), EGR ratio, and turbine speed. The simulated intake air flow and AFR agree with the experiment results. As the EGR valve opens, less exhaust gas enters the turbine. Consequently, the turbine power capability declines, the turbo speed reduces, the intake mass flow drops and the AFR decreases. The turbine speed is not measurable by the experimental devices, but the turbo speed simulation results are indirectly verified through the veracity of the simulated intake mass flow and AFR.

Experimentally, the EGR rate is determined by the flow-rate method and the CO_2 tracing method [10], here labeled Experiment 1 and Experiment 2, respectively. The flow-rate method has good transient responsibility, but the measured data are too fluctuant for the experimental equipment, precluding a quantitative analysis. The data collected by the CO_2 tracing method are stable and provide a criterion for the maximum EGR rate. However, because of the length of the pipeline that gathers the exhaust gas, these data cannot reflect the transient response time. We observe that the simulated response times and maximum EGR rates are consistent with the results of Experiment 1 and Experiment 2, respectively. By comparing the simulation and experimental results, we confirm that the model accurately simulates the varying EGR rate under an EGR step change condition.

It is also seen that the EGR rate peaks at the moment the EGR valve begins the step change. The simulated and experimental data are compared in

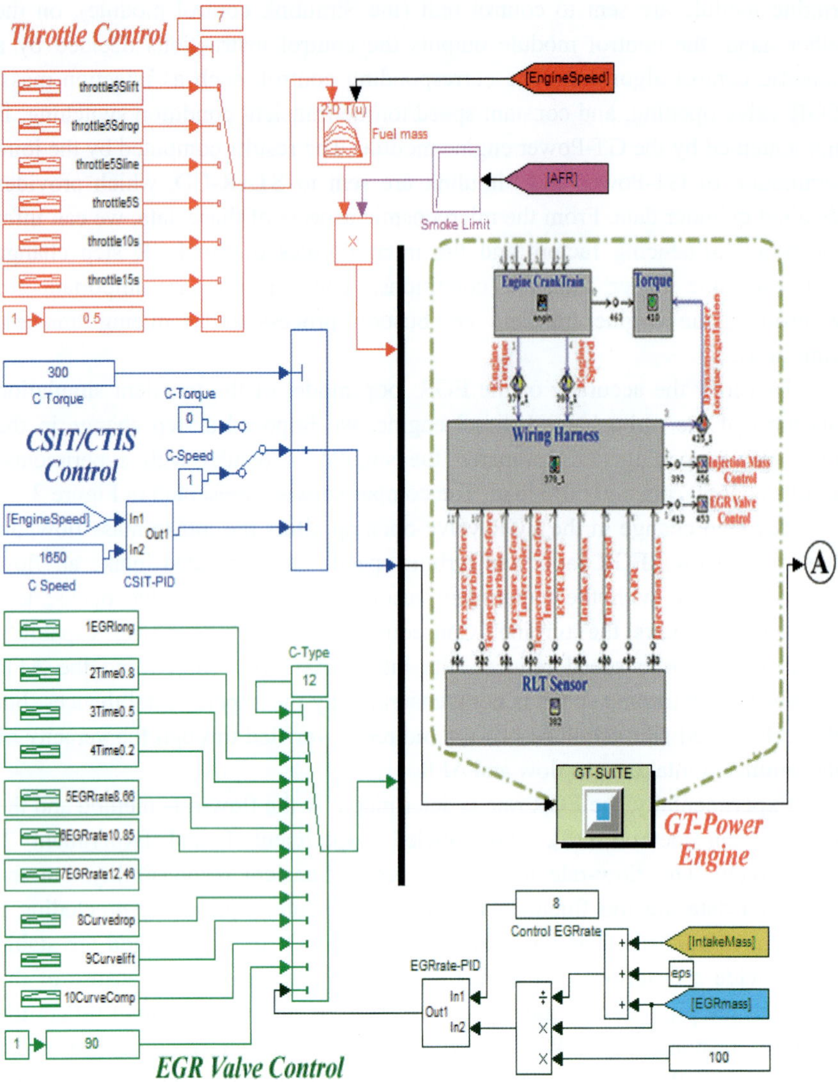

(a) The left part of EGR step change performance simulation platform

Fig. 1. EGR step change performance simulation platform of turbocharged diesel engine.

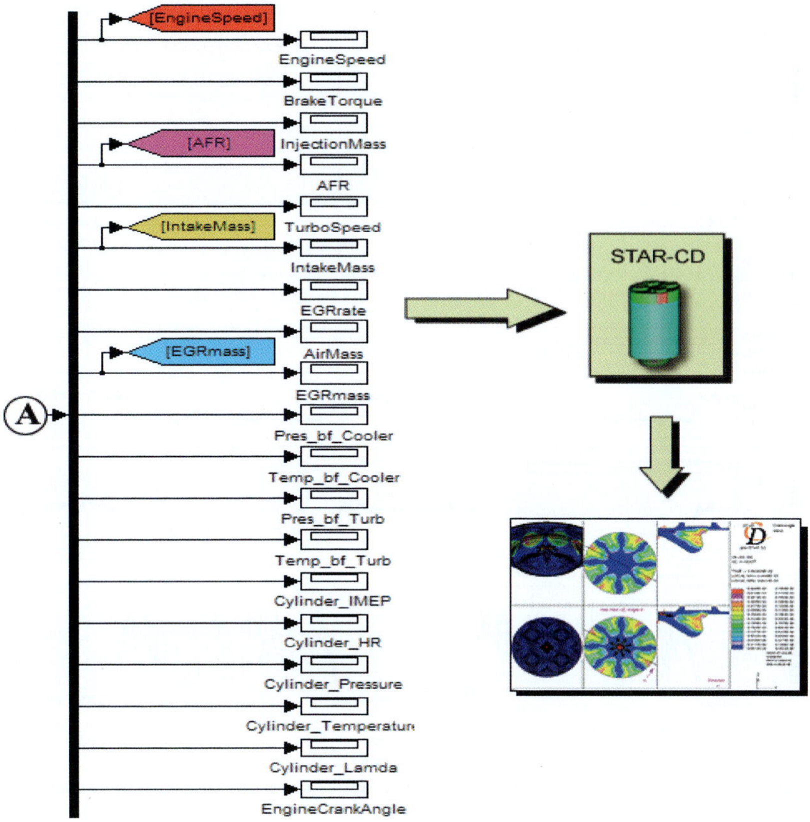

(b) The right part of EGR step change performance simulation platform

Fig. 1. (*Continued*)

Figure 3. The difference between the simulated and experimental results is quantified by the following equation:

$$\dot{P}_{EGR} = \frac{RT_{EGR}}{V_{EGR}} \dot{m}_{EGR} \tag{1}$$

In Eq. (1), R is the universal gas constant, and T_{EGR} and V_{EGR} are the temperature (in K) and the volume (m^3) of the EGR pipeline, respectively.

When the EGR valve opens, the pressure and the rate of change of pressure $\dot{P}_{EGR}$ rise rapidly in the EGR pipeline, while the temperature of the EGR valve changes relatively slowly. Since the EGR flow $\dot{m}_{EGR}$ is proportional to the pressure change rate $\dot{P}_{EGR}$, the EGR flow suddenly increases at the moment of

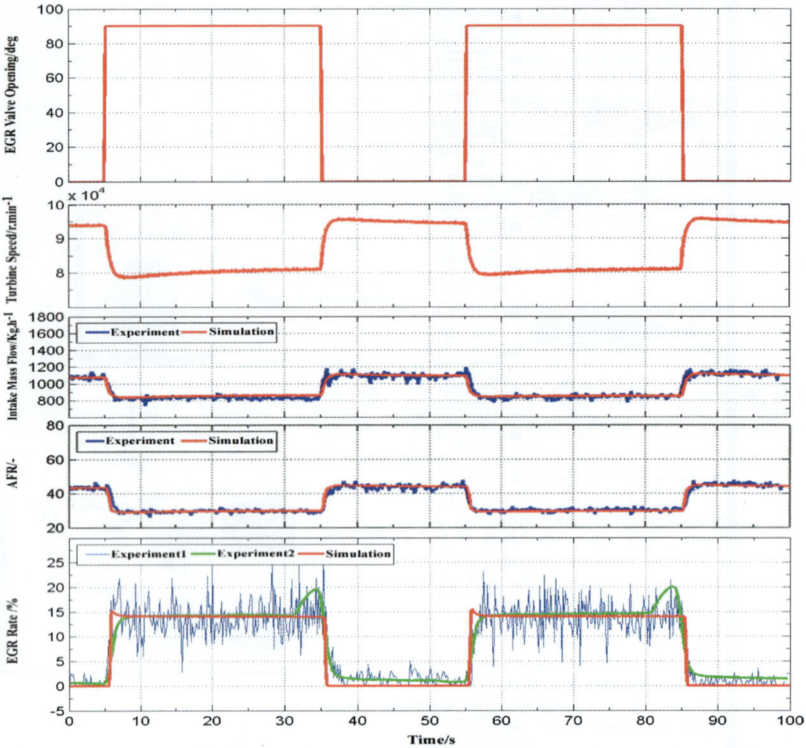

Fig. 2. Comparison of experimental and simulation results under an EGR step change condition.

the step change happens, leading to peak EGR rate at that moment. This behavior was obscured in the experimental data because of the limited device accuracy.

The accuracy of simulations conducted by the transient simulation platform with the EGR loop was validated by comparison with the experimental EGR step change under transient conditions. Therefore, the simulation platform is suitable for studying the transient response characteristics of turbocharged diesel engines under EGR step change operating conditions.

3. Optimization of EGR Valve-Opening Characteristic Curve Based on DOE

To elucidate the factors affecting the transient response characteristics under EGR step change operation conditions, we employed a single-factor method. Reduction in the transient response time and torque fluctuations is investigated

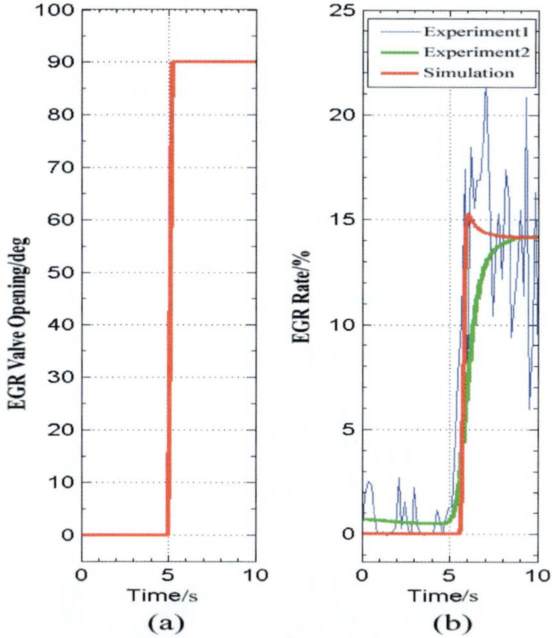

Fig. 3. Simulation analysis of EGR rate under an EGR step change condition.

for constant maximum EGR valve opening (90°) and the EGR valve-opening time (0.2 s).

The original EGR opening-valve characteristic is a linear function of time. Here we simulate the linear control strategy and three additional control strategies, namely, upcast, sunken, and comprehensive. The opening-valve characteristics under the four strategies are shown in Figure 4. Comparing panels (a) and (b) of this figure, we find that under the upcast control strategy, the change in the difference between the pressures before the turbine and before the intercooler are most significant early in the EGR step change, leading to a torque increase at that time. Under the sunken control strategy, the pressure difference changes most significantly at the later stage of the EGR step change; consequently, the torque increases at a later time. Under the comprehensive control strategy, the EGR valve-opening degree is higher in the earlier stage of the EGR step change, improving the EGR transient response. In this strategy, the response rises smoothly and steadily during the middle and later stages of the step change, and the pressure difference remains stable. Consequently, the torque ripple Ttq_2 under the comprehensive control strategy is 18% lower than Ttq_1 in the upcast control strategy.

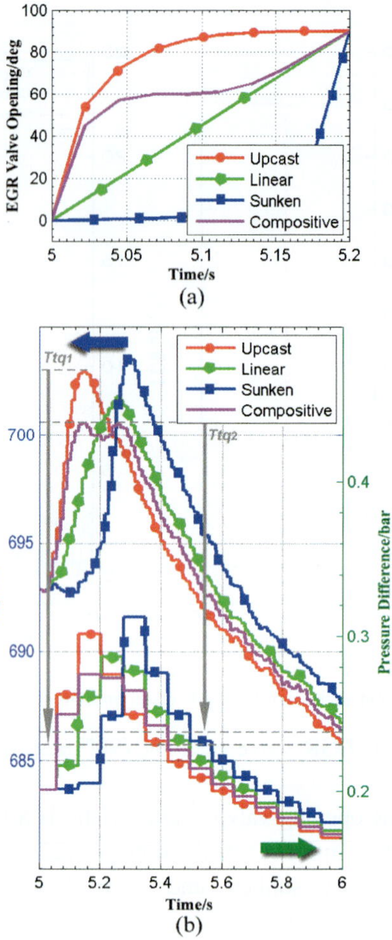

Fig. 4. Comparison among four control strategies of EGR valve-opening characteristics.

By comprehensively comparing these four control strategies, we conclude that the upcast and comprehensive control strategies yield superior transient responses, while the comprehensive and linear better suppress the torque ripple during the transient process. Therefore, the comprehensive control strategy is the most advantageous strategy, because it improves the transient response characteristics while avoiding torque ripple.

Following this qualitative analysis, the control strategies were quantitatively analyzed by a design of experiments (DOE) method [11]. This analysis identifies the optimal characteristic curve that improves the EGR step change transient response characteristic while preventing torque fluctuations.

In the DOE method, the EGR valve opening time (0.2 s) and the total EGR valve opening angle (90°) were divided into four sections with equal intervals, generating different characteristic curves of the EGR valve opening. The above divisions yielded 25 curve points of the EGR valve-opening characteristics, covering the entire space of the characteristic curves (Figure 5). Here, the abscissa and ordinate are the EGR valve opening time and the EGR valve-opening angles, respectively. For example, the coordinate (0, 0) denotes the start point of the EGR valve-opening characteristic (i.e., at time 0, the EGR valve-opening angle is 0°), and (0.2, 90) denotes the end point of the EGR valve-opening characteristic (at 0.2 s, the EGR valve has opened by 90°).

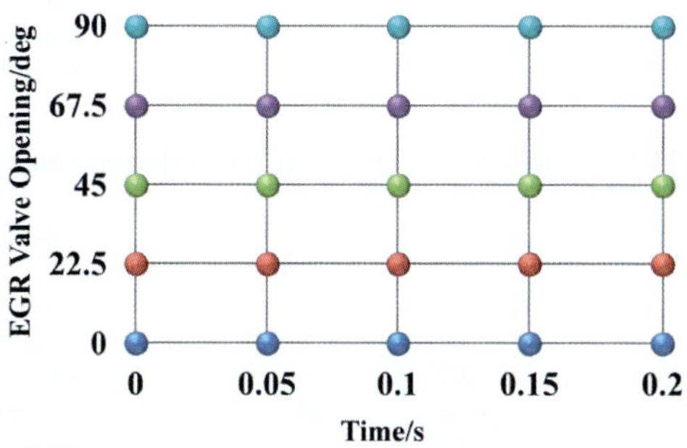

Fig. 5. Coverage of the characteristic curve points under an EGR valve-opening control strategy.

Because the EGR valve opening angle is fixed and predetermined at the start and end times, this parameter can change only at three intermediate time points. Therefore, we can generate $5^3 = 125$ different EGR valve-opening control strategy curves. Such a large number of curves require optimization by the DOE method.

To ensure a practical EGR valve control strategy, the DOE must comply with the basic properties of actual EGR valve-opening characteristics:

1) The opening angle of the EGR valve is 0° at time = 0 s (start point of the curve), and 90° at time = 0.2 s (end point of the curve).

2) The EGR valve opening angle can remain unchanged or can increase, but cannot decrease, within a 0.05 s interval.

3) The EGR valve opening angle cannot suddenly jump to another angle.

These conditions exclude most of the possible control strategies of the EGR valve-opening characteristics. The 35 feasible strategies are shown in Figure 6.

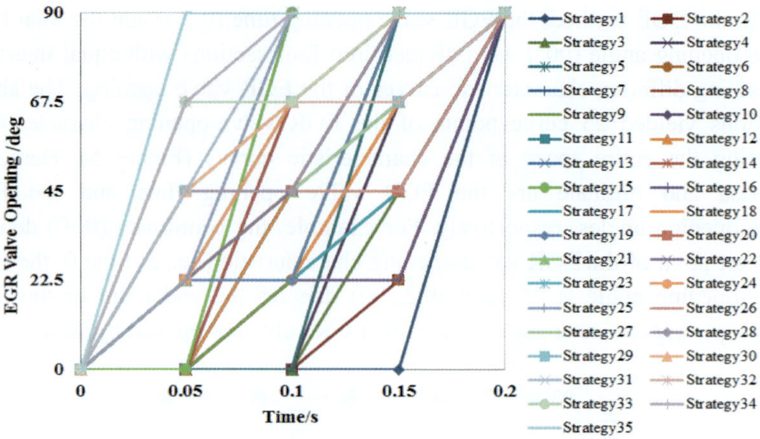

Fig. 6. The 35 feasible strategy curves for EGR valve-opening control.

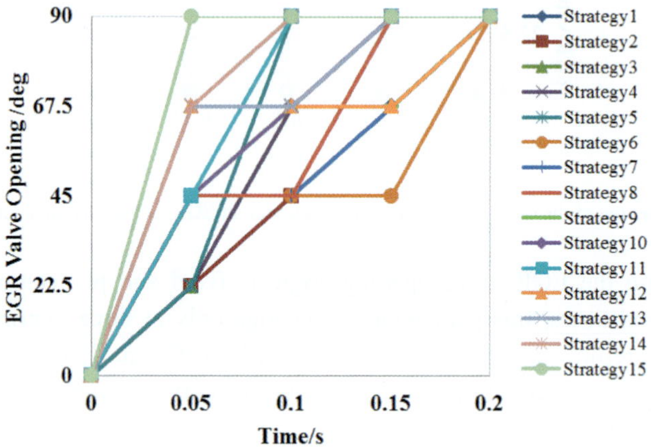

Fig. 7. Simplified set of control strategy curves.

As evident from the above discussion, the sunken control strategy proved inferior to the linear, comprehensive and upcast control strategies for improving the transient response characteristics. Therefore, most of the control strategies whose EGR valve-opening curves lie beneath the diagonal connecting (0, 0) and (0.2, 90); that is, the curves generated by the sunken strategy, were removed from the pending strategies. However, the control strategies covering the points (0, 0), (0.05, 45), (0.1, 45), (0.15, 45), and (0.2, 90) were retained because half of their curves lie above the diagonal, and their principal trend is comprehensive. Thus, 15 control strategies survived the optimization, as shown in Figure 7. Table 1 is the DOE table of the 15 candidate control strategies of the EGR valve-opening characteristics.

The 15 control strategies in Table1 were simulated, and the best control strategy was identified by comparing the transient response characteristics and torque fluctuations among the simulation results.

Table 1. DOE table of control strategies for the EGR valve opening characteristics.

Time/s	Strategy 1/deg	Strategy 2/deg	Strategy 3/deg	Strategy 4/deg	Strategy 5/deg
0	0	0	0	0	0
0.05	22.5	22.5	22.5	22.5	22.5
0.1	45	45	67.5	67.5	90
0.15	67.5	90	67.5	90	90
0.2	90	90	90	90	90
Time/s	Strategy 6/deg	Strategy 7/deg	Strategy 8/deg	Strategy 9/deg	Strategy 10/deg
0	0	0	0	0	0
0.05	45	45	45	45	45
0.1	45	45	45	67.5	67.5
0.15	45	67.5	90	67.5	90
0.2	90	90	90	90	90
Time/s	Strategy 11/deg	Strategy 12/deg	Strategy 13/deg	Strategy 14/deg	Strategy 15/deg
0	0	0	0	0	0
0.05	45	67.5	67.5	67.5	90
0.1	90	67.5	67.5	90	90
0.15	90	67.5	90	90	90
0.2	90	90	90	90	90

The torques generated under the 15 control strategies are plotted in Figure 8. As a measurement standard for torque fluctuations, we computed the torque variation, defined as the difference between the maximum and minimum torque generated in the transient process. Control strategies 12 and 13 yielded the smallest torque fluctuations (Figure 9). The EGR mass flows under the 15 control strategies are plotted in Figure 10. Again, control strategies 12 and 13 yielded the best transient responses. Since EGR valve control is more simply implemented by strategy 12 than by strategy 13, strategy 12 is selected as the ideal strategy for controlling the EGR valve opening characteristics (Figure 11).

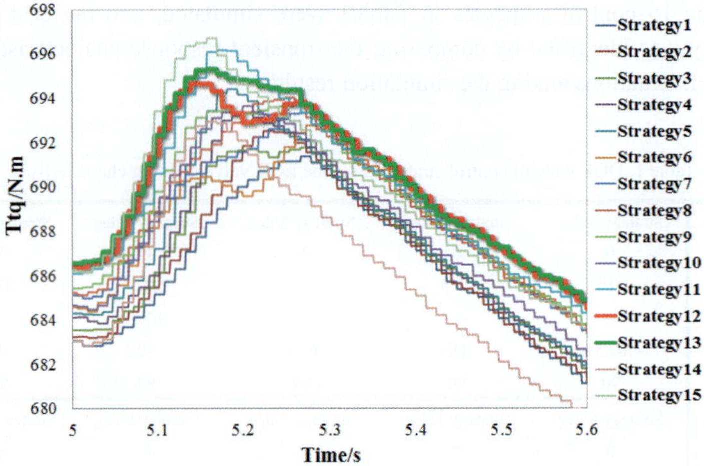

Fig. 8. Torques generated under the 15 candidate control strategies.

Fig. 9. Comparison of torque variation among the 15 candidate control strategies.

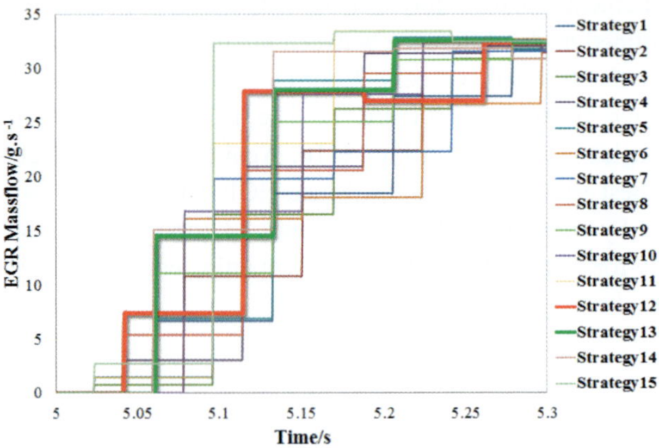

Fig. 10. EGR mass flows under the 15 candidate control strategies.

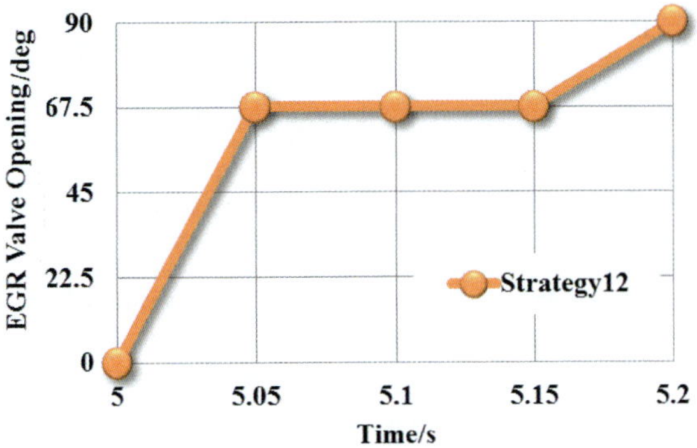

Fig. 11. DOE optimization result of the control strategies for EGR opening characteristics.

The optimal control strategy of the EGR valve-opening characteristics proceeds as follows: During the first quarter of the EGR valve opening time, the EGR valve should be rapidly opened to three-quarters of its opening angle. The opening angle should remain constant throughout the middle half of the EGR valve-opening time, and opened to its full angle during the last quarter of that time.

In summary, the optimal control strategy obtained by a quantitative DOE analysis of the EGR valve-opening characteristics is consistent with the results of the qualitative analysis, in which the comprehensive control strategy curve yielded the best performance.

4. Analysis of Double Objective Optimization of EGR Control Strategy Coupled with Emission Targets

The upcast, comprehensive, and linear control strategies are further analyzed in the cylinder by coupling them to STAR-CD. The sunken control strategy is excluded from this analysis since it improved neither the transient response nor the torque ripples.

As shown in Figure 12, five representative time points were sequentially selected from the EGR rate curves generated by the combine simulation between GT-Power and Simulink. At these time points, the STAR-CD was initialized by the combustion boundary parameters from the combine simulation, such as the total intake mass flow, injection mass, temperatures and pressures after the intercooler and before the turbine, the EGR rate, and pressure in the cylinder. The combustion and emission in the cylinder were then analyzed in detail.

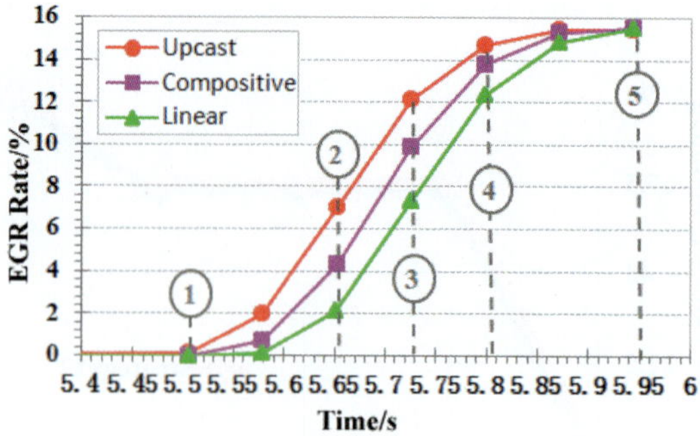

Fig. 12. Emissions under different EGR step change control strategies compute by STAR-CD at various time points.

Figure 13 compares the NO_X emissions in the cylinder computed by STAR-CD under different control strategies. Assuming the NO_X emissions without EGR as 100%, the EGR rates in the cylinder were increased within a shorter time under the upcast and comprehensive control strategies than that under the linear control strategy because their EGR step change transient responses were improved over the linear strategy. Specifically, the NO_X emissions under the upcast and comprehensive control strategies were respectively 11% and 6% less than that under the linear control strategy at any given time. In the upcast and comprehensive control strategies, the response time for reducing the NO_X emissions by 12% was 38% and 19% lower, respectively, than that in the linear control strategy (cf. times T_1, T_2 and T_3 in Figure 13).

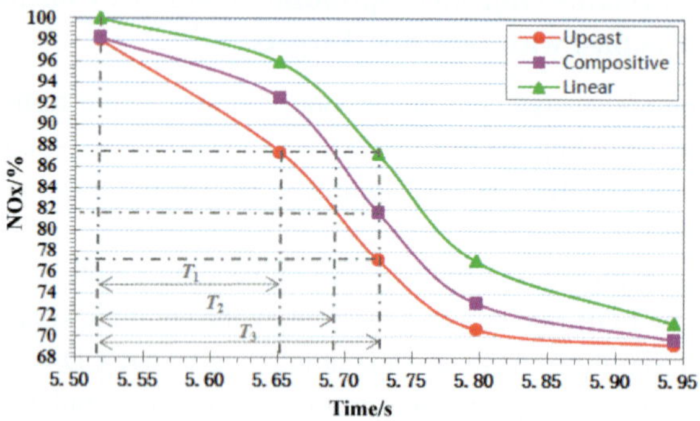

Fig. 13. Comparison of NO_x emissions under different control strategies.

Assuming the soot emissions after the EGR step change as 100%, the linear and comprehensive control strategies reduced the soot emissions by 23% and 11%, respectively, relative to that in the upcast control strategy, at any given time (Figure 14). The generating time for reaching the same soot emissions was 36% and 17% slower in the linear and comprehensive control strategies, respectively, than that in the upcast control strategy (cf. times T_1, T_2, and T_3 in Figure 14).

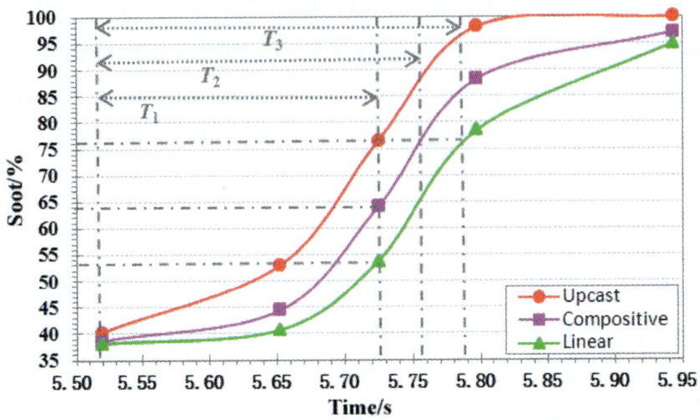

Fig. 14. Comparison of soot emissions under different control strategies.

5. Conclusions

1) Four control strategies of the EGR valve-opening characteristics were qualitatively compared under a constant maximum EGR valve-opening angle and opening time. The upcast and comprehensive control strategies yielded superior transient responses to the linear and sunken strategies. However, the comprehensive and linear control strategies better suppressed the torque ripple.
2) The EGR valve-opening characteristics were optimized by a DOE, enabling the quantitative analysis of the characteristic curves. Finally, the double-objective optimization control strategy (that is, improving the transient response of the EGR step change while avoiding torque fluctuations) was established by comparing the 15 candidate control strategies after the DOE optimization. It was determined that the EGR valve should be rapidly opened to three-quarters of its opening angle during the first quarter of the EGR valve-opening time, retained constant during the middle half, and fully opened during the last quarter of the opening time.
3) Coupled with STAR-CD technology, our platform, which is based on the whole-parameter transient simulation platform of a turbocharged diesel engine, successfully analyzed the emission performance of the double-objective optimization control strategy. Relative to the linear control strategy, the comprehensive control strategy reduced the NO_X emissions by 6% while improving the response time by 19%. It also reduced the soot emissions by 11% and delayed the generating time by 17% relative to the

upcast control strategy. Therefore, the comprehensive control strategy yields superior emission performance to the upcast and linear control strategies.

Acknowledgments

This work is supported by the National Natural Science Fund Project of China (61472159), the Science & Technology Development Projects of Jilin Province (20121805, 20140101180JC), and the education department of Jilin province twelfth five year science and technology research project ([2015]446).

References

1. Yoshizawa K, Mori K, Matayoshi Y, et al. Development of an exhaust gas recirculation distribution prediction method using three-dimensional flow analysis and its application [J]. *Journal of Engineering for Gas Turbines and Power* 2003; 125(4): 1066–1074.
2. K. S. Kalsi and N. Collings. Study of steady state and transient EGR behaviour of a medium duty diesel engine [J]. SAE Paper 2008-01-2438, 2008.
3. Han Yong-qiang, Liu Zhong-chang and Zhu Chang-ji. Response of DI TCA automotive diesel engine under EGR rate step change operations [J]. *Journal of Combustion Science and Technology* 2007; 13(3): 209–213.
4. Yokomura H, Kouketsu S, Kotooka S, Akao Y. Transient EGR control for a turbocharged heavy duty diesel engine [J]. SAE Paper 2004-01-0120, 2004.
5. Liu Z.-C, Wang Z.-S, Li J, Li W.-X. Investigation on NOx emission behavior from CA6DE1-21K diesel engine under transient operating conditions [J]. *Journal of Combustion Science and Technology* 2004-03.
6. Li Z.-C, Wang Z.-S, Li J, Wang Y.-H. Smoke emission behavior of CA6DE1-21K diesel engine under transient operating conditions [J]. *Journal of Combustion Science and Technology* 2004-06.
7. Han Y.-Q, LI Z. -C, Cheng P and Li X.-J. Effect of combustion parameters on diesel engine smoke opacity under constant torque and increasing speed transient operating conditions [J]. *Transactions of CSICE* 2003-05.
8. Dubreuila A and Foucher F. HCCI combustion: Effect of NO in EGR [J]. In *Proc. of the Combustion Institute* 2007; 31(2): 2879–2886.
9. Sui L.-G, Liu Z.-C, Han Y.-Q and Tian J. Simulation of transient emission of turbocharged diesel engine. *Transactions of CSICE* 2012-02.

10. Xu Y, Liu J.-W, Liu Z.-C and Liu X.-J. Control and exhaust particulate measurement of automotive DI diesel engine under transient operating conditions. *Journal of Combustion Science and Technology* 2006-01.
11. Lach R, Kiehn R, Hans C, et al. DOE application within the analytical valve train development[c]//3rd Conference: DOE in Engine Development, Berlin Germany: Expert-verlag, 2005:197–202.

Dynamics analysis and application of the folding mechanism based on the Lagrange method

Sheng Chen, Xiao-Guang Wu†, Xiao-Long Lei and Xia Liang
School of Mechanical and Automation,
Wuhan University of Textile,
Wuhan, Hubei, China
†*E-mail: 12006wist@163.com*

This paper, regarding to a particular folding fixture, has analyzed and studied the dynamic model of the folding mechanism driven by elastic potential energy during its deployment process. Based on that, in light of the Lagrange Method utilized in rigid multi-body dynamics, the dynamic equation of this folding mechanism was established. By analyzing and calculating through the Matlab, the moving properties of the mechanism have been obtained. In addition, through simulation in Solidworks motion and comparison of the results, it can be seen that theoretical analysis basically accords with the simulation outcomes. The researching subject in this paper is mainly used on fully automatic shoe-cover machine, and the analysis method and ideas involved can also be used as the references or lessons for problems of the same sorts.

Keywords: The elastic potential energy; folding fixture; Lagrange method; the shoe covers machine.

1. Introduction

As a structural system, folding structure can be changed from tight contracting gradually to fully expanded state[1]. The structure has a shape of uncertainty. In the whole process, the system is a process with large displacement and little strain. Unlike conventional structure, design and analysis of folding structure need to meet developable conditions[2]. Through the kinematics and dynamics analysis of the process to ensure that the folding structure can along a predetermined orbit to the working condition. Dynamics analysis of the folding structure belongs to the category of multi-rigid-body system dynamics theory[3,4]. This paper analyzed the folding mechanism by the driving of the elastic potential energy, and established the dynamics model of folding mechanism based on the second Lagrange method[5]. Dynamics analysis was completed, and motion simulation was operated by the SolidWorks to obtain reasonable movement characteristics.

2. The Model of Folding Mechanism

2.1. *The application principle of folding mechanism in the fixture*

Fixture is composed of stepper motor, synchronous belt, trapezoidal screw, the upperplate, the underplate, folding mechanisms and self-locking mechanism. Opening and closing of the fixture is controlled by clockwise and counterclockwise of stepper motor. Folding mechanisms is situated in the middle of the fixture. The structure is shown in Fig 1, Support framework 1, Guide shaft 2, Activity board 3, Fixed board 4, slider 5, hinge 6, roller 7, Folding rod 8, stents 9. With the characteristics of stored energy and instantaneous power, the spring driving device solves the power supply problem of the folding mechanism in the narrow space. Under the impetus of the external pole, the folding mechanism completes folding and makes spring store energy. When the fixture opened, the folding mechanism completes unfolding and supports the underplate under the action of the stored energy spring, which avoids the cantilever plate producing the front-end warp under the weight.

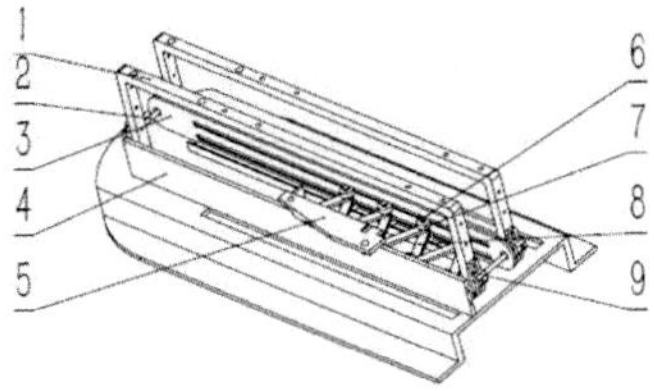

Fig. 1. The structure of folding mechanisms.

2.2. *Dynamics analysis of the system*

The folding mechanism is made up of eight rigid homogeneous rods and a slider.Adjacent two rods are connected with a hinge, and the circles are the hinged points. Simplified mechanical model of the system is shown in Fig. 2. To facilitate the analysis, putting forward the following assumptions and rules on the mechanical model of folding mechanism:The rods have no deformation in the process of the whole movement and their centers of mass are in the midpoint; Ignore the friction of each kinematic pair.

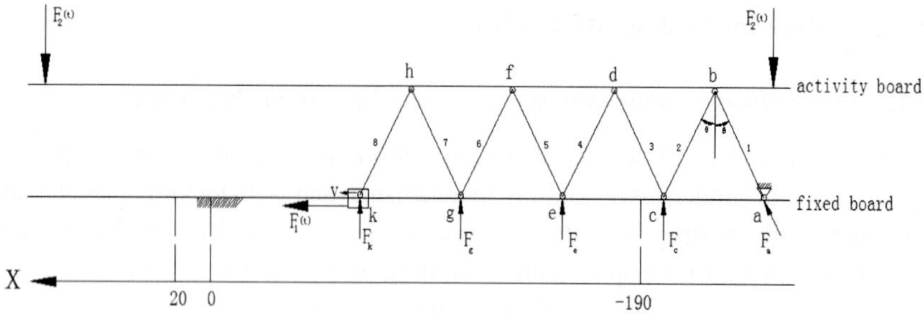

Fig. 2. Simplified mechanical model of the system.

The whole as the research object, the balance equation of the system arelisted.

$$F_1(\mathrm{t}) + F_a sin\theta = m_s a_s(t) + \textstyle\sum_{i=1}^{8} m_i\, a_{ix}(\mathrm{t})\ (i = 1, \dots ,8) \tag{1}$$

$$2F_2(\mathrm{t}) - F_a cos\theta - F_c - F_e - F_g - F_k = m_{ab} a_{ab}(t) + \textstyle\sum_{i=1}^{8} m_i\, a_{iy}(\mathrm{t}) \tag{2}$$

Here: m_i represents the quality of the rod i; m_s represents the quality of the slider; m_{ab} represents the quality of the activity board; $a_s(t)$ represents the acceleration of the slider; $a_{ab}(t)$ represents the acceleration of the activity board; $a_{ix}(t)$ represents the x direction acceleration of the rod i; $a_{iy}(t)$ represents the y direction acceleration of the rod i.

2.3. *The system dynamics model based on the Lagrange method*

In this paper, the folding mechanism is a multi-rigid-body mechanical system, and the active force is the powerful force.The second Lagrange equation of the systemis shownin Eq. (3).

$$\frac{d}{dt}\left(\frac{\partial L}{\partial \dot{q}}\right) - \frac{\partial L}{\partial q} = 0\ , L = T - V \tag{3}$$

Here: L represents the Lagrange function; T represents the kinetic energy of the system; V represents the potential energy of the system; q represents the generalized coordinates.

(1) The kinetic energy of the system:

The slider always makes translation along the x direction in the whole movement;Rod 1 rotates around the fixed hinge a,and Rod 2 ~ 8 do plane motion in the whole movement;Theactivity board always makes translation along the y direction in the whole movement.So the total kinetic energy of folding mechanisms is shown in Eq. (4).

$$T=\frac{1}{2}m_s v_s^2+\frac{1}{2}J_1\omega_1^2+\sum_{j=2}^{8}(\frac{1}{2}m_j v_{jx}^2+\frac{1}{2}m_j v_{jy}^2+\frac{1}{2}J_j\omega_j^2)+\frac{1}{2}m_{ab}v_{ab}^2 \quad (4)$$

Here: v_s represents the velocity of the slider; J_1 represents therotational inertia of the rod 1 relative to its center of mass;ω_1 represents the angular velocity of the rod 1; v_j represents the velocity of the center of mass for the rod j; J_j represents the rotational inertia of the rodj relative to its center of mass; ω_j represents the angular velocity of the rod j; v_{ab} represents the velocity of the activity board.

In the dynamic model of folding mechanisms, select the displacement x of the slider as the generalized coordinates, namely, $q=x;\dot{q}=\dot{x}=v_s;\ \ddot{q}=\ddot{x}=a_s$. The kinetic energy of the system can also be shown as follows.

$$T=\frac{1}{2}M_e v_s^2 \quad (5)$$

The effective mass is shown below:

$$M_e=m_s(\frac{v_s}{v_s})^2+J_1(\frac{\omega_1}{v_s})^2+\sum_{j=2}^{8}(m_j(\frac{v_{jx}}{v_s})^2+m_j(\frac{v_{jy}}{v_s})^2+J_j(\frac{\omega_j}{v_s})^2)+m_{ab}(\frac{v_{ab}}{v_s})^2 \quad (6)$$

(2) The potential energy of the system:

For the rigid system, the potential energy generated by the component deformation can be neglected. Therefore, the potential energy of the system is mainly provided by the main pressure spring and auxiliary thrust spring provides. The potential energy of system is shown in Eq. (7).

$$V=\frac{1}{2}k_1*(\Delta x)^2+2*\frac{1}{2}k_2*(\Delta y)^2 \quad (7)$$

Here: k_1 represents the stiffness of the auxiliary thrust spring; k_2 represents the stiffness of the main pressure spring.

(3) The differential equation of the system:

According to the second Lagrange equation, the derivations of the corresponding Lagrange functions are solved, and the system dynamics differential equation is shown in Eq. (8).

$$M_e\ddot{x}+\frac{1}{2}\frac{\partial M_e}{\partial x}\dot{x}^2-k_1*\Delta x=0 \quad (8)$$

3. The Dynamics Simulation and Solution of the Model

3.1. *The dynamics simulation*

According to the function of the fixture, folding mechanisms needs to satisfy this condition: by the driving of the elastic potential energy, the folding mechanism is unfolded to 210mm within 0.5s. By the Solid Works software, the physical model of the folding mechanisms is built and dynamic simulation analysis of the process is completed. The main parameters as follows: the length of the rod i L_i=40.0mm (i=1,…,8),m_i=1.0g,m_s=46.0g,m_{ab}=47.0g,k_1=0.46N/mm, k_2=0.1N/mm. The simulation time is set to 0.5s,and the simulation results are shown in the figures below.

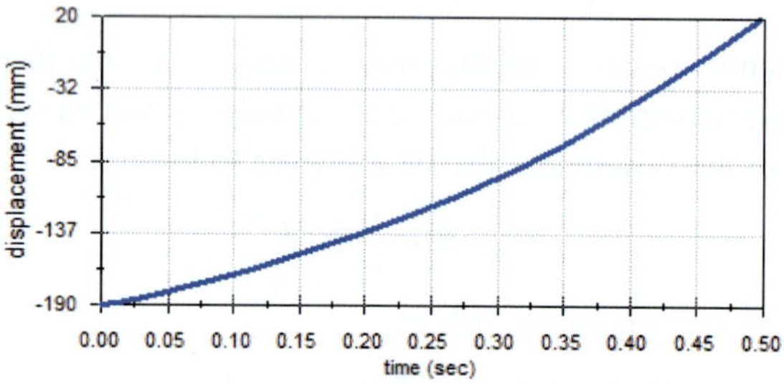

Fig. 3. The displacement curve of slider.

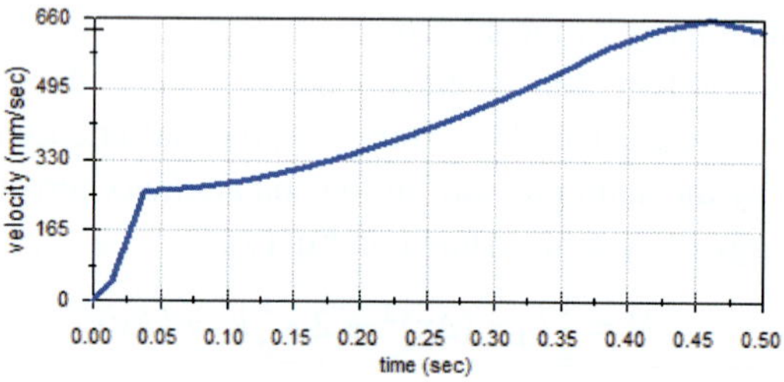

Fig. 4. The velocity curve of slider.

It can be seen from the Fig 3, the slider moves from x= -190 to x=20. From the Fig 4, it can be seen that there is a transient impact process between the auxiliary thrust spring and the slider during 0 ~ 0.03 s, which is generated by the assembly clearance.

3.2. *The solution of the model*

When using numerical method to solve the dynamic equation, the discrete movement parameters of the dynamic simulation need to be used during a single cycle. Expressions of the Me and d Me/dx can be fitted by Matlab and then the dynamics differential equation can be solved[6]. According to the dynamics simulation, the initial value can be obtained as follows: the initial position x=-190; because of the transient impact process between the auxiliary thrust spring and the slider during 0 ~ 0.03 s, the initial velocity is 200 mm/s. The results are shown in the figures below.

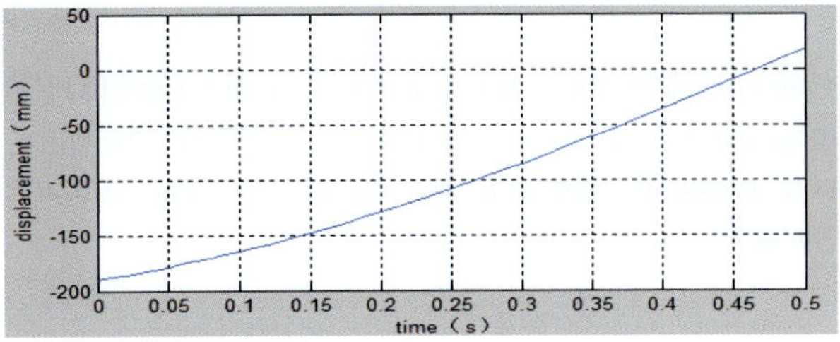

Fig. 5. The displacement curve.

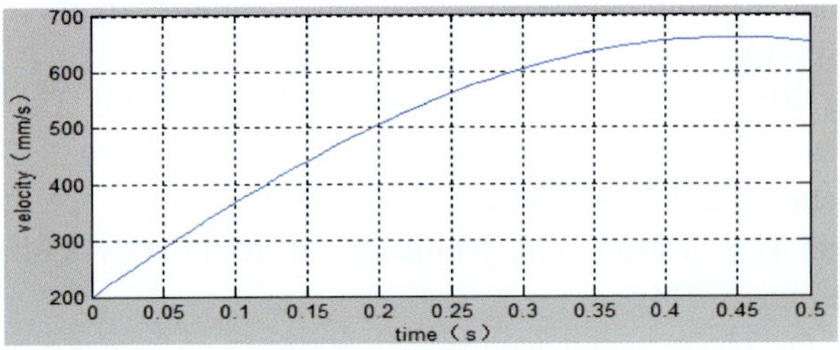

Fig. 6. The velocity curve.

Comparing Figs. 3 and 5, Figs. 4 and 6, it can be seen that the variation tendency of the two displacement curves and velocity curves are similar after 0.03 s. It indicates that the mechanism motion characteristics of the theoretical model and dynamic simulation are basically identical.

4. Conclusion

(1) In this paper, the dynamic theory of model folding mechanism was establishedby the second Lagrange method in the generalized coordinates. And the movement characteristics of the mechanism were obtained by the numerical solution. With SolidWorks software, dynamic simulation of the folding mechanism was completed toobtain displacementand velocity curve of the slider. By comparison, theoretical results and simulation results were basically identical, the correctness of the theory model was verified.

(2) Dynamic performance of folding mechanism basically meet the functional requirements of the fixture, which will provide the reference for the research of new type automatic non-woven shoe covers machine. In this

paper, the basic analysis methods and thoughts will be also suitable for the same and similar problems.

Acknowledgments

This work is supported by the national natural science fund (51175384) and the Hubei province natural science fund (2014 CFA099). Professor Xiao-Guang Wu, the main research direction for digital knitting equipment and key technology research.

References

1. Hongbin Zhao, Zhifeng Wu, Lili Xie et al. Application of SVD method in the folding structure analysis [J]. Journal of engineering mechanics, 2006, 23 (3) : 11-16.
2. Shihong Zhong, Sun Wei, Chonggeng Zhang et al. The design and dynamics research of new deployable folding wing [J]. Aerospace manufacturing technology, 2013, (5): 17-20, 23.
3. Zhenkuan Pan, Yanzhu Liu, Xinsheng Ge. Multi-body system dynamics [M], 2 edition, 2014.
4. Zhaohui Qi. The system dynamics [M], 2008.
5. Jiang Wei, Gengxiang Wang. Dynamic model of five-bar mechanism with clearance joints [J]. Journal of mechanical engineering and automation, 2014, (4) : 91-91. The DOI: 10.3969 / j.i SSN. 1671-5276.2014.04.031.
6. Bin Zhang. The operational reliability simulation and experimental research in deployable structure of the folding wings [D]. Zhejiang university of technology, 2015.

Simulation on the dynamic characteristics of the pulse modulation hydraulic hammer

Gang Wang

The State Key Laboratory of Fluid Power and Mechatronic Systems, Zhejiang University, Hangzhou, Zhejiang 310027, China
School of Mechanical Engineering, Anhui University of Technology, Maanshan, Anhui 243002, China
E-mail: ahut.wg@163.com

Jian-Hua Wei* and Jin-Hui Fang†

The State Key Laboratory of Fluid Power and Mechatronic Systems, Zhejiang University, Hangzhou, Zhejiang 310027, China
*E-mail: *jhwei@zju.edu.cn and †jhfang@zju.edn.cn*

In order to analyze the dynamic characteristics of the pulse modulation hydraulic hammer, the dynamics analysis and simulation model of the hydraulic hammer are carried out in the paper. Firstly, the dynamic equations of the piston and the distributor are established based on the Newton's second law. Secondly, the simulation models of the piston, the distributor and the integrated pulse modulation hydraulic hammer are built on AMESim based on the dynamic equations and the working principle of the hydraulic hammer. It can be concluded from the dynamics analysis and simulation results that for a group fixed system parameters there are the optimal frequency and duty ratio of the pulse control signal. As the optimal frequency and duty ratio are varying with the system parameters nonlinearly, it is a challenging work to find the optimal frequency and duty ratio.

Keywords: Hydraulic hammer; pulse modulation; duty ratio; dynamic characteristics; AMESim.

1. Introduction

The hydraulic hammer, also called hydraulic breaker, is an indispensable equipment for civil engineering, construction, demolition, and even recycling works[1]. The hydraulic hammer turns the hydraulic energy supplied by a hydraulic pump into mechanical energy by means of sequential percussions

---

†Corresponding author.

against a chisel[2]. The mechanical energy is transferred to the chisel by means of a stress wave. The stress wave travels through the chisel until it reaches the opposite end where the chisel is in contact with the rock or concrete. The chisel is appointed to crush a certain material by alternating impacts. Due to its versatility for various applications, the hydraulic hammer is one of the common tools used in an excavator and its performance warranted by the manufacturer has a major effect on the efficiency of an excavator. The crushing theory and practice indicates that for a certain crushing object there is specific optimal impact energy to get optimal working efficiency. When working if the impact energy of a hydraulic hammer can be adjusted according to the physical characteristics of different crushing objects, its efficiency can be improved greatly. Currently, the most popular hydraulic hammer on market is stroke feedback hydraulic hammer, and most of the literature on hydraulic hammer focuses on this type hydraulic hammer[3-5]. Due to theirs structural limitations, its impact energy and impact frequency can be only adjusted with two or three speed ratio. The pressure feedback hydraulic hammer is also proposed by domestic and foreign universities and research institutions [6][7]. Its advantages are that its impact energy and impact frequency can be regulated easily and infinitely, and its structure can be simplified greatly. While, its disadvantages are that its electrical testing equipment is complex, and its failure rate is higher. So the pulse modulation hydraulic hammer is put forward, its impact energy and impact frequency can be regulated directly by the pulse control signal from a computer[8]. The structure of this hydraulic hammer is relatively simple, and its failure rate is also lower because the pressure detection device is omitted. The dynamic characteristics of this hydraulic hammer is the key factor to generate the pulse control signal from a computer. So the dynamics analysis and simulation model on the dynamic characteristics of the pulse modulation hydraulic hammer is carried out in the paper.

2. Working Principle of the Pulse Modulation Hydraulic Hammer

As shown in Fig. 1, the pulse modulation hydraulic hammer is composed of a chisel, a piston, a cylinder body, a distributor, a nitrogen chamber, a pulse control signal from a computer, oil source and so on. The distributor is made up of a main valve and a high speed on/off valve, and whether the hydraulic hammer is in crushing stroke or return stroke is decided by the state of the distributor. The high speed on/off valve is controlled by the pulse control signal form a computer. When the high speed on/off valve is energized, the control chamber of main valve is lined to tank, its spool moves from left to right under the oil pressure from the oil source. The back chamber of the hydraulic hammer is lined to tank by the distributor, so the hydraulic hammer piston begins its return stroke under the high pressure of front chamber and the low pressure of back chamber. When the high speed on-off valve is de-energized, the control chamber of main valve is lined to high pressure chamber from the oil source, its spool quickly switches from right to left under the pressure of the high pressure chamber. The back chamber of the hydraulic hammer is lined to the high pressure chamber by the distributor, so the hydraulic hammer piston ends the its return stroke and begins crushing stroke under the high pressure of nitrogen chamber till the piston strikes the chisel. The effective action area of the front chamber and back chamber is equal.

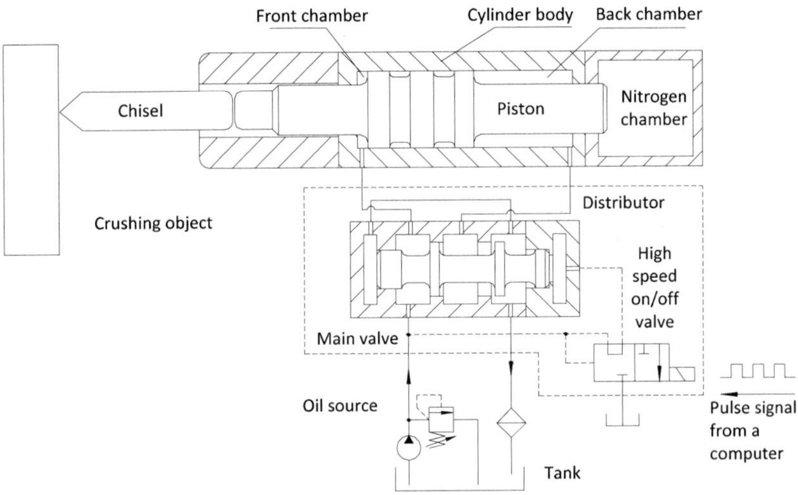

Fig. 1. Schematic diagram of the pulse modulation hydraulic hammer.

Therefore, as mentioned above, the pulse modulation hydraulic hammer is controlled by the pulse control signal generated by a computer. The performance of the pulse modulation hydraulic hammer depends on the two key parameters of pulse control signal: its frequency and duty cycle. The work frequency of the hydraulic hammer is coincident with that of the pulse control signal, and the impact energy of the hydraulic hammer is changing with the frequency of the pulse control signal inversely. The duty cycle of the pulse control signal determines the time distribution of crushing stroke and return stroke of hydraulic hammer in a working cycle, which should be determined by the dynamic characteristics of the hydraulic hammer piston in crushing stroke and return stroke. The following work is to analyze the dynamic characteristics of the hydraulic hammer piston through the dynamics analysis and simulation.

3. Mathematical Model of the Pulse Modulation Hydraulic Hammer

The piston is subjected to pressure, viscous friction, gravimetric force, mass to mass interactions (impact forces) and so on, however, the gravimetric force is omitted because it is negligible if compared to hydraulic forces. Its dynamic equation can be described by the following equation[9]:

$$m\frac{d^2x}{dt^2}=B\frac{dx}{dt}+p_bA_b+p_fA_f+p_nA_n+F_c \tag{3.1}$$

Where m is the mass of piston, B is the damping coefficient, x is the displacement of piston, p_b is the pressure of back chamber, p_f is the pressure of front chamber, p_n is the pressure of nitrogen chamber, A_b is the effective action area of back chamber, A_f is the effective action area of front chamber, A_n is the effective action area of nitrogen chamber, F_c is the impact force between the piston and the chisel, which exists only when the piston strikes the chisel.

Through the dynamic analysis of the piston above, its simulation model can be established using AMESim as Fig. 2.

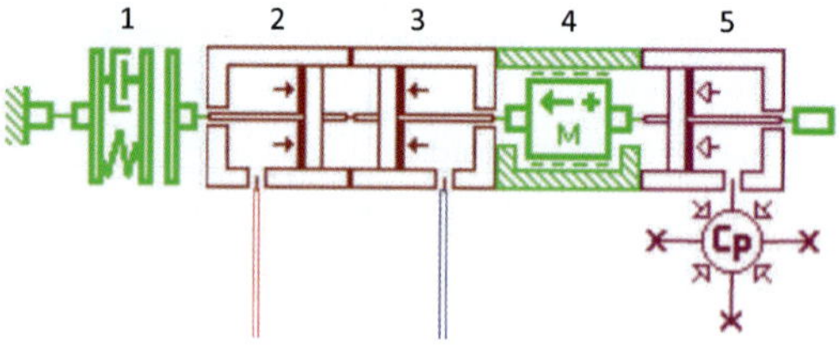

Fig. 2. The simulation model of pulse modulation hydraulic hammer body.

In Fig. 2, part 1 is a sub model of an elastic contact between two bodies capable of linear motion, and is used to simulate the elastic contact between the piston and the chisel. The chisel is supposed to behave like a spring, whereas the anvil below is supposed to be infinitely stiff. Part 2 and 3 are used to model the hydraulic pressure of front and back chamber, part 4 is used to simulate the mass and the viscous resistance of hydraulic hammer piston, part 5 is a variable volume pneumatic chamber with poly tropic law, and is used to model the force of nitrogen chamber. It is modeled as an adiabatic process.

The main valve spool is subjected to pressure, viscous friction, and gravimetric force and so on. In the current stage of modelling the gravimetric force is also omitted, and the end stop of the main valve spool was supposed to be ideal, then no contact force was accounted for when such a mass reaches its limit positions. Its dynamic equation can be described by the following equation[10]:

$$m\frac{d^2y}{dt^2}=B\frac{dy}{dt}+\Delta F \tag{3.2}$$

Where m is the mass of main valve spool, B is the damping coefficient, y is the displacement of main valve spool, and ΔF is the resultant force of pressure, jet reaction, and so on.

Through the force analysis of the main valve spool above, its simulation model can be established on AMESim as Fig. 3.

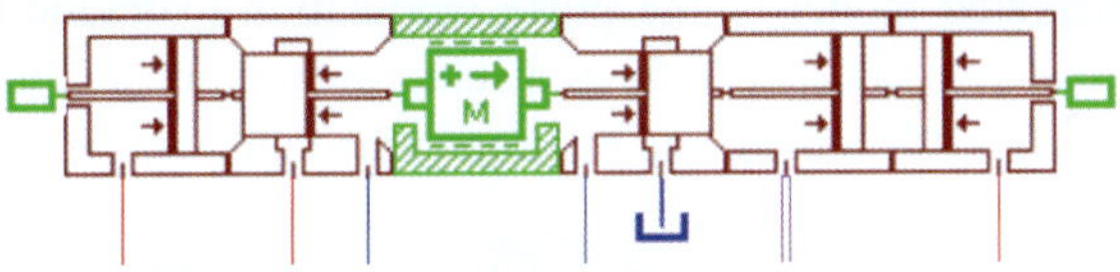

Fig. 3. The simulation model of main valve spool.

The high speed on/off valve used in the pulse modulation hydraulic hammer is a ball valve designed by the Guizhou Honglin and the America BKM company. It srated pressure is 20MPa, rated flow is 1.8L/min, closing time is 2ms, and opening time is 3 ms[11]. The structure of the high speed on/off valve is shown in Fig. 4. When its coil is energized, the control oil is lined to input oil. When its coil is de-energized, the control oil is lined to tank. The high speed on/off valve can be established on AMESim as Fig. 5.

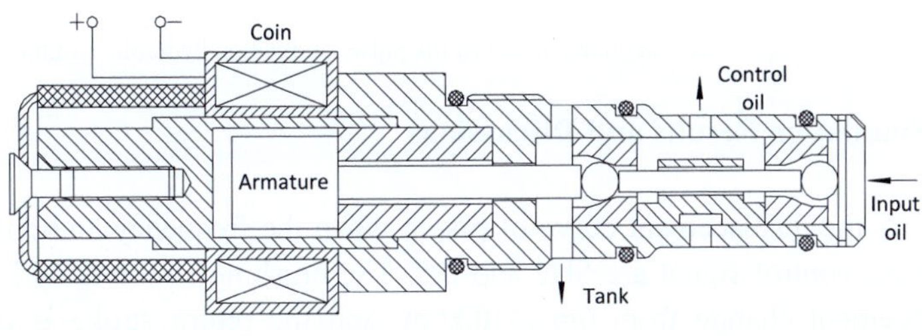

Fig. 4. The structure diagramof high speed on/off valve.

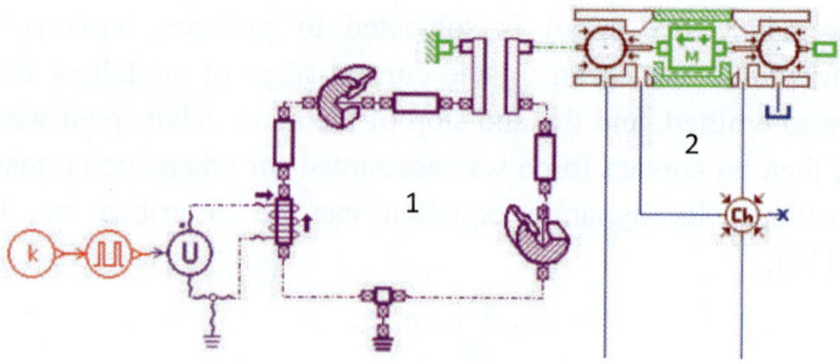

Fig. 5. The simulation model of high speed on/off valve.

In Fig.5, the drive circuit of high speed on/off valve is shown as part 1, which is established based on magnetic circuit analysis. Part 2 is the spool and body of high speed on/off valve.

The integrated model of the pulse modulation hydraulic hammer can be demonstrated as Fig. 6[12]. The fixed displacement pump and the pressure relief valve is used as the oil source, and this conforms to its working condition.

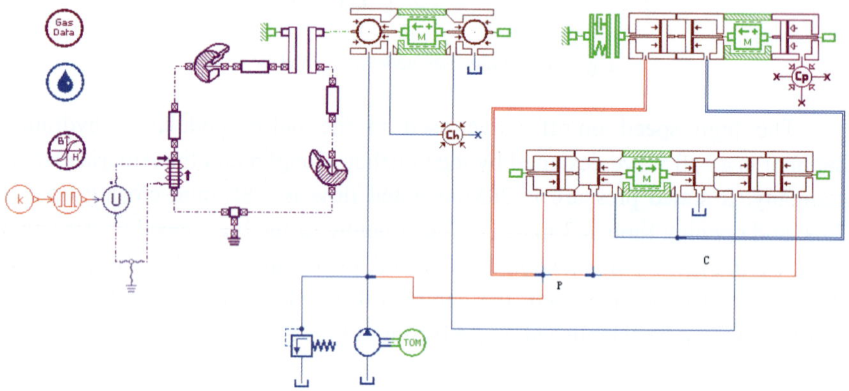

Fig. 6. The integrated model of the pulse modulation hydraulic hammer.

4. Simulation Results and Discussions

Fig. 7 is the displacement curve of piston when the frequency and duty ratio of the pulse control signal are 5Hz and 0.5. The crushing stroke is defined as the displacement change from 0m to 0.85m, and the return stroke is defined as displacement change from 0.85m to 0m. It can be seen from Fig. 7 that there are pause at the end of return stoke and second impact at the end of crushing stroke. When the piston is at the state of the pause, the hydraulic energy from the hydraulic pump is wasted form the pressure relief valve of the oil source. When the piston is at the state of second impact, the hydraulic energy from the hydraulic pump is consumed by the hydraulic hammer, but the impact energy is too small to be practicable. The second impact will arouse the vibration of hydraulic hammer, and the rebound energy of the piston is also wasted. So the frequency and duty ratio of the pulse control signal are required to select proper values to eliminate the phenomenon of pause at the end of return stoke and second impact at the end of crushing stroke.

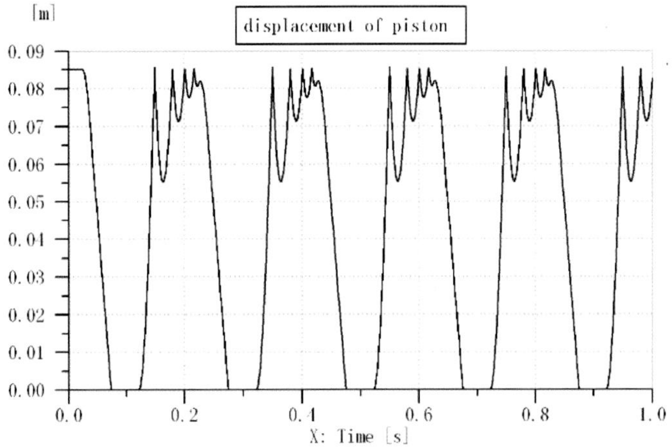

Fig. 7. The piston displacement curve when the frequency and duty ratio are 5Hz and 0.5.

Fig. 8 is the displacement curve of piston when the frequency and duty ratio of the pulse control signal are 13.2Hz and 0.465. As shown in Fig. 8, the displacement of the first return stroke is smaller than that of the other return strokes. This can be explained by the fact that there is no mechanical energy from the rebound of the piston in the first return stroke, that is to say, the mechanical energy from the rebound of the piston can accelerate the return stroke. There are no pause at the end of return stoke and second impact at the end of crushing stroke, in theory the working condition as shown in Fig. 8 is probably the most efficient with the highest impact energy when the system parameters are set as the Table 1. When those parameters are changing, the optimal frequency and duty ratio of the pulse control signal are also required to vary accordingly.

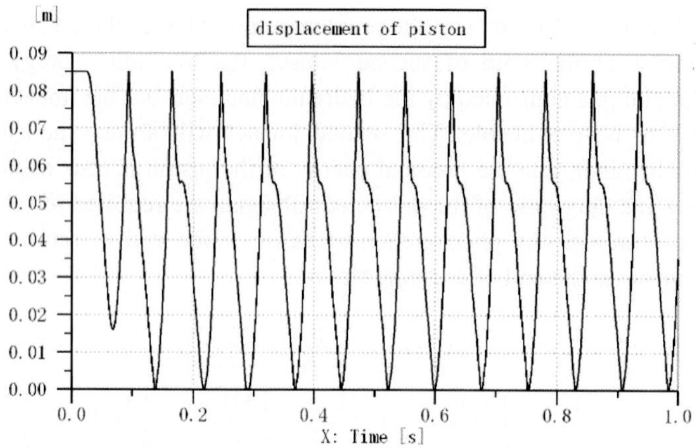

Fig. 8. The piston displacement curve when the frequency and duty ratio are 13.2Hz and 0.465.

Table 1. List of main parameters used in simulation model.

Schedule	Capacity
System flow rate	60 L/min
System pressure	15MPa
Pressure of the nitrogen chamber	1.4MPa
Contact stiffness	$1.2e^9$N/m
Contact damping	8000N/(m/s)

5. Conclusions

(1) For the fixed system parameters, there are optimal frequency and duty ratio of the pulse control signal to obtain the most efficient with the highest impact energy.

(2) Through the dynamics analysis and simulation results, it is found that the optimal frequency and duty ratio of the pulse control signal are varying with the system parameters nonlinearly. It is a challenging work to find the optimal frequency and duty ratio especially when the system parameters and the working condition of the hydraulic hammer also maybe be time-varying.

(3) Combining the advantages of the pulse modulation hydraulic hammer and that of the other hydraulic hammer maybe be a feasible way to improve the pulse modulation hydraulic hammer.

Acknowledgments

This work is sponsored by the Scientific Research Foundation for Young Teachers in Anhui University of Technology (Grant No. QZ201407) and the National High Technology Research and Development Program of China (Grant No. 2012AA041804).

References

1. Antonio Ficarella, Antonio Giuffrida and Domenico Laforgia, Investigation on the impact energy of a hydraulic breaker, SAE Technical Paper 2007–01-4229, 2007. http://dx.doi.org/10.4271/2007–01-4229.
2. JB/T 10844-2008 Hydraulic breaking hammers standard.
3. Z.C. Qing, Z.H. Zhou, Z.R. Zhou, Modeling and simulation of hydraulic rock drillimpact mechanism based on AMESim, Hydraulics Pneumatics & Seals, Vol. 30 (2010), pp. 32-34.
4. Z.J. Yin, Y.X. Hu, Dynamic simulation of hydraulic rock drill impact mechanism basedon AMESim, J. Min. Mach. Vol. 39 (2011), pp. 1-4.
5. Y.Z. Lin, G.Y. Huang, Z.S. Tang, Z. Feng, Overview of research status on hydraulic rock drill, Hydraulics Pneumatics & Seals Vol. 34 (2014), pp.1-4.
6. Xiangbi Yang, Guoping Yang, Xin Zhang, Zhong Liu,A New Type of Pressure Feedback Hydraulic Impactor,China Mechanical Engineering. Vol. 35 (2000), pp. 4-5.
7. Xinqiang Guo, The research of intellective embeded controlling and measuring system of assembled nitrogen inflating hydraulic breaker, Master's thesis, Anhui University of Technology (2007).
8. Gang Wang, Yuwan Cen, Simulation on the Regulating Characteristics of the Pulse Modulation Hydraulic Hammer, Advanced Materials Research, Vol. 422 (2012), pp. 176-183.
9. A. Ficarella, A. Giuffrida, D. Laforgia, Numerical investigations on the working cycle of a hydraulic breaker: off-design performance and influence of design parameters, International Journal of Fluid Power, Vol. 7 (2006), pp. 41–50.
10. Antonio Gluffrida, Domenico Laforgia, Modeling and simulation of a hydraulic breaker,International Journalof Fluid Power, Vol. 2 (2005), pp. 47-56.
11. Information on http://www.gzhonglin.com.cn
12. Imagine, S. A.. *AMESim User Manual*. Version R13.0.

Study of modal test for a large articulated extensible system

Rui-Xiang Li
National University of Defense Technology,
Changsha, Hunan 410073, China
Shanghai Aerospace System Engineering Institute,
Shanghai, Shanghai 201100, China
E-mail: qingbinzhang@sina.com

Long-Long Zhang and Qing-Bin Zhang*
National University of Defense Technology,
Changsha, Hunan 410073, China
**E-mail: longer162@qq.com*

Qi-Feng Cui and Zhi-Long Peng
Shanghai Aerospace System Engineering Institute,
Shanghai, Shanghai 201100, China

For satellites and other spacecraft demand for large area of solar cell wing, this paper proposes a rod framed stretched institutions to increase its developed area. Analysis of the modal test, it is concluded that state in institutions collapsed and the corresponding natural frequency, damping ratio and modal shape and so on. It provides a basis for the following structural optimization and dynamic characteristics analysis. To further model correlation analysis, correction and correction of the model evaluation foundation, make the accurate after correction verification model more truly represent the actual structure, accurate response and load analysis for structure parts technology basis.

Keywords: Modal test; stretching agencies; natural frequency; damping ratio; modal shape.

1. Introduction

Spatial developable mechanism is to realize the spacecraft accessories such as solar cell wings, stretching institutions and payload equipment in orbit or on one of the important device or system. It is not only change the spacecraft spatial position and posture of the important device in the attachment, and provide satellite overall configuration and ensure spacecraft attachment support stiffness and positioning accuracy, load-bearing characteristics and the working stability of the important products.

In the 60's of last century, the National Aeronautics and Space Administration (NASA) put forward the concept of the development of the organization, it is with the development of science and technology and the birth of a new mechanism, and in the continuous development. In the 80's, with the deepening of the research, the technology is gradually mature, has begun to be widely used in many kinds of spacecraft, such as the free space station, Alfa space station, etc. With the increasing demand for energy space station and large spacecraft, area only dozens of square meters of small solar wing have been unable to meet the demand, so the area reached 100 square meters or even higher solar wing needs more and more urgent [1]. In order to solve this problem, application in solar wing extension mechanism is invented, which has repeatedly unfolding and folding ability [2].

China from the 90s of the last century began to deployable mechanism theory research, over the past 30 years, China has yet to carry out major Aerospace Science and technology project, and of space can development agencies caliber and precision demand is not high[3]. In view of the deployable structure truss type, most of the domestic use is very rigid body dynamics and generalized inverse method for the analysis of the process [4, 5]. Harbin Industrial University make an in-depth research to the deployable mechanism and locking and tempered, aerospace dynamics and failure model, and other related technologies, successfully developed a 7.3m articulated arm extension and the modal test[3,6]. Hou Guoyong and Guan Fuling [7] of Zhejiang University have studied the truss deployable mechanism, and have made the corresponding theoretical analysis, simulation analysis and modal test, and successfully developed the 30m truss ring deployable antenna. In order to study the dynamic characteristics (frequency, damping and vibration type) of the large-scale structures, the modal test analysis is carried out [8,9].

The space solar wing of a rod frame type deployable mechanism as the object of study, put forward a rod-shaped truss deployable mechanism to increase the area of expansion. By modal experiment and simulation analysis to ensure that the design of deployable mechanism in the unfolded state and in the folded state, with enough rigidity and strength, achieve the design index requirements, for the subsequent structure optimization and dynamic characteristics of precise analysis provides the basis and further model correlation analysis, and the evaluation of the correction model and modified model to lay the foundation [10].

2. The Oretical Support for Modal Test

Modal analysis is a modern method to study the dynamic characteristics of structures, and it is the application of the system identification method in the engineering field. Its significance is not only to avoid the resonance of the structure in the work, but also the basis of the analysis of the dynamic response and other dynamic characteristics of the structure [11]. Mode is the natural vibration characteristics of the mechanical structure, each mode has a specific natural frequency, damping ratio and modal vibration mode. Modal analysis is divided into two kinds, one is the finite element method to calculate the modal analysis, the other is through the test to the system input and output signals through the parameter identification of the experimental modal analysis [12].

The spread of solar battery wing institutions in the launch and the space environment will be impacted by the strong, so for modal test is necessary. The main purpose of the modal test analysis is the identification and evaluation of existing structure system, to identify the dynamic characteristics of structure system, to verify and modify the structural dynamics analysis model to provide experiment basis [13].

Based on the theory of modal analysis, the differential equation for the vibration of the system with multiple degrees of freedom is

$$[M]\{\ddot{x}(t)\} + [C]\{\dot{x}(t)\} + [K]\{x(t)\} = \{f(t)\} \tag{1}$$

Where

$$\text{Mass matrix } M = \frac{1}{6}\rho A l \begin{bmatrix} 2 & 0 & 0 & 1 & 0 & 0 \\ 0 & 2 & 0 & 0 & 1 & 0 \\ 0 & 0 & 2 & 0 & 0 & 1 \\ 1 & 0 & 0 & 2 & 0 & 0 \\ 0 & 1 & 0 & 0 & 2 & 0 \\ 0 & 0 & 1 & 0 & 0 & 2 \end{bmatrix} \tag{2}$$

ρ is density, *A* is cross-sectional area, *l* is unit length.

$$\text{Stiffness matrix } K = \frac{EA}{l} \begin{bmatrix} 1 & 0 & 0 & -1 & 0 & 0 \\ 0 & 0 & 0 & 0 & 0 & 0 \\ 0 & 0 & 0 & 0 & 0 & 0 \\ -1 & 0 & 0 & 1 & 0 & 0 \\ 0 & 0 & 0 & 0 & 0 & 0 \\ 0 & 0 & 0 & 0 & 0 & 0 \end{bmatrix} \tag{3}$$

E is Young's modulus.

Damping matrix C can be obtained by K and M.

To type on both sides as a Fourier transform, displacement frequency response function matrix $[H(w)]$ as follows

$$[H(w)] = \frac{[X(w)]}{[F(w)]} \tag{4}$$

Assuming that$[\varphi] = |\{\varphi\}_1, \{\varphi\}_2, \{\varphi\}_3, \dots, \{\varphi\}_n|$is the main mode matrix of the structure, and the damping of the structure system is proportional damping, it can be proved that the displacement frequency response function matrix and the modal parameters have the following relations

$$[H(w)] = \frac{\sum_{i=1}^{n}\{\varphi\}\{\varphi\}_i^T}{-w^2M_i + jwC_i + K_i} \tag{5}$$

Where, M_i, C_i, K_i are respectively *i* order modal mass, modal damping and modal stiffness, and $\{\varphi\}_i$ is the *i* order natural mode.

It can be proved that the acceleration frequency response function has the following relations with the displacement frequency response function

$$\left|H^{(a)}\right|(w) = \left|H(w)w^2\right| \tag{6}$$

Traditionally, said the acceleration frequency response function with, $H_{lp}(w)$ so as to achieve the theoretical basis for the modal parameters of the acceleration frequency response function. In order to find the vibration mode, the frequency response function matrix of a row or a column can be determined.

3. Test Object and Scheme

3.1. *Test object*

Stretching mechanism is important stand-alone solar array wing, supporting the solar wing on orbit of unfolding and folding, and solar panels in orbit of stiffness and strength [15]. The full extension of the stretching mechanism is near 24m, the expansion speed is 0.4m/min. Stretch institutions have a large number of component parts (Types of components to reach 100, the number of parts more than 7000, standard parts up to more than 2000), the structure is more complicated. Stretching mechanism can expand and change state from the folded state, is completely closed and fully expanded two stable states [1]. In order to ensure its stiffness and strength to meet the actual needs, the modal analysis of the expanded state is carried out. The physical object of the extension mechanism is shown in Figure 1.

a Collapsed state b Expansion state

Fig. 1. This is the caption for the figure. It is a extensible mechanism. The a is collapsed state, and the b is expansion state.

3.2. *Test scheme*

For structural modal testing analysis research, through the measured structure (input) and response (output) to determine its dynamic characteristics, thus the analysis model of system is established, is a major aspect of the structural dynamic inverse problem, its content can be divided into system identification and parameter identification[13]. According to the excitation and response to a different number, identification of vibration parameters can be divided into single input and single output (SISO), single input / multi output (SMO) and multi input/ multi output (MMO) three, this kind of method is mainly used in the small structure and quality structure.

In the modal analysis of the extended mechanism, it is necessary to exert an exciting force to make it vibrate, and the vibration response is measured by the acceleration sensor. Incentive can be classified as single point excitation and multi support excitation and single point excitation can't will be the complete structure of the modal characteristics of goal motivation, and multi-point excitation can will inspire a complete energy assigned to structures, especially the large structure, thus the structure mode to complete and effective motivation [16]. In this paper, the method of multi point excitation is used to analyze the modal test of the extended mechanism.

The modal shapes of the structure are obtained by the test model vibration test, so the position of the measuring point can be fully and accurately to obtain the various order vibration modes of the structure [16]. The principle of measuring point arrangement is to try to fully the vibration mode of the reaction structure [14].

3.2.1. *Test boundary conditions*

When the modal experiment of an extension mechanism of the solar cell wing is carried out, the constraint conditions are different. When it is horizontal expansion, stretching mechanism has a base fixed in the wall is simulated and hanging device of ultra low frequency modal test gravity balance system with connection to balance the gravity. Vertical expansion, extension mechanism is fixed with cast iron disc, just in front of weight increase mechanism. In order to check the connection rigidity of the fixed support boundary, a number of sensors are placed in the test fixture, which can be used to test the boundary condition of the modal test.

3.2.2. *Expansion state*

Stretching institutions roots fixed in the simulation of the wall, arms spread out with ultra-low frequency modal test hanging device connection of gravity balance system, to balance the gravity of the extended arm. Extended arm and ultra-low frequency modal test connection way as the triangle side of the gravity balance system horizontally, directly by the hanging device hook the extended arm hanging points. The modal experiment, increased weight in the front stretch institutions. Using Dytron3263M9 three-way accelerometer, sensitivity of 100 mv/g. A sensor to protect products, paste 3m adhesive tape on the product first, and the bending, lateral bending and torsion three directions for incentives, incentives position as shown in Figure 2, each direction driving test results validate each other.

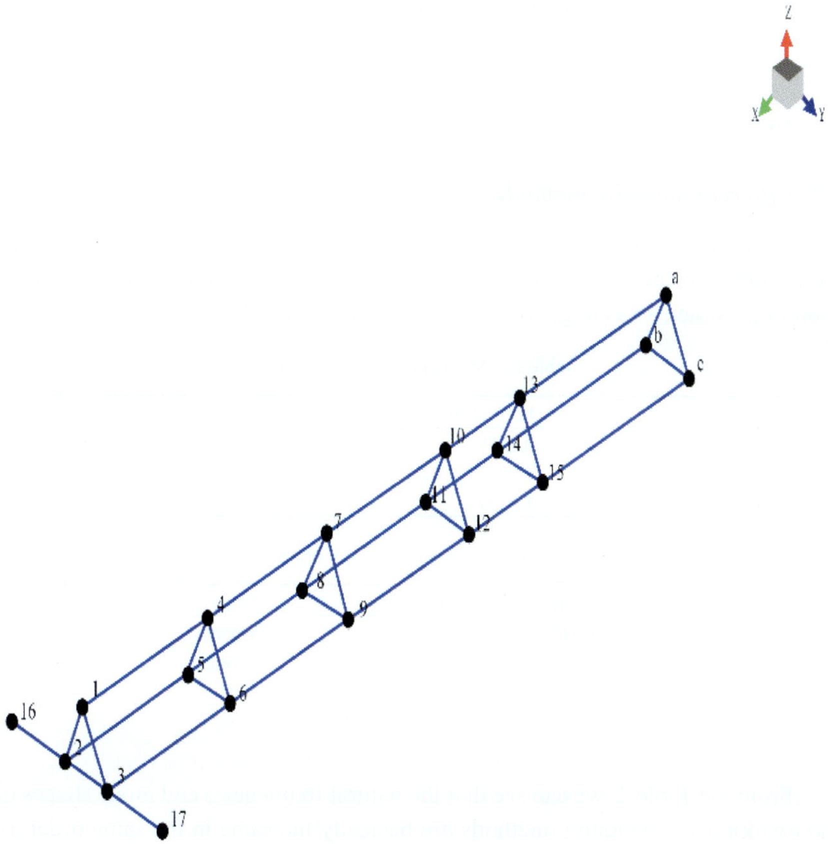

a Measurement point distribution

Fig. 2. This is the caption for the figure. The a is measurement poin distribution, and the b is horizontal expansion modal test.

b Horizontal expansion modal test

Fig. 2. (*Continued*)

4. Comparative Analysis of Test Data

4.1. *Different incentive methods*

The modal test of the extension mechanism is carried out under the horizontal expansion, and the mode is divided into two kinds: the manual release and the vibration exciter, so as to get the results shown in Table 1.

Table 1. Manual release test results.

Order	Natural frequency/Hz	Mode
1	0.8789	Y bend
2	1.0742	X bend
3	1.6797	Twist

Table 2. Excitation test results of vibration exciter.

Order	Natural frequency/Hz	Mode
1	0.8824	Y bend
3	1.1162	X bend
4	1.6733	twist

From the Table 2, we can see that the natural frequencies and mode shapes of the two kinds of excitation methods are basically the same in the same order. In the actual test, manual release incentive design time is short, but the incentive force should not be too large, otherwise will result in structural nonlinear change; exciter set is more complex, but it can obtain more accurate test data through multi-point excitation. Therefore, it can be said that the manual release to obtain rough data, and the vibration exciter to obtain accurate data.

4.2. *Different expansion lengths*

The modal test analysis is carried out on the extension mechanism with different deployment conditions, so as to obtain the influence of the expansion situation on the extension mechanism. As can be seen from the Table 3, in the case of the extension length of the 18 sections, with the increase of modal order, the natural frequency increases gradually, and the damping ratio decreases. As can be seen from the Table 4, in the case of the extension length of the 15 sections, with the increase of the modal order, the natural frequency increases, the damping ratio is also increased. Table 3 and Table 4 show that with the increase of the expansion length, the natural frequency becomes smaller and the damping ratio becomes larger.

Table 3. Expand the length of the 18 sections.

Order	Natural frequency/Hz	Damping ratio/%	Mode
1	0.8824	7.83	Y bend
2	1.1162	6.03	X bend
3	1.6733	4.09	twist
4	4.1085	4.83	Y bend
5	6.4348	1.84	X bend

Table 4. Expand the length of the 15 sections.

Order	Natural frequency/Hz	Damping ratio/%	Mode
1	1.6368	1.91	twist
2	1.8119	3.17	Y bend
3	1.8243	3.15	X bend
4	8.0487	3.36	Y bend
5	9.0155	3.50	X bend

4.3. *Comparison of modal test and finite element simulation results*

Table 5 is the results of finite element simulation for the extension of the length of 18 sections. From Table 3 and Table 5 shows, in considering the boundary conditions in the finite element simulation settings and parameters setting of reason, the natural frequencies and mode shapes are basically the same, show that modal test can be obtain the integrity of the main mode.

Table 5. Finite element simulation results 18 sections).

Order	Natural frequency/Hz	Mode
1	0.9513	Y bend
2	1.3789	X bend
3	1.9754	twist
5	5.0137	Y bend
6	7.2572	X bend

5. Conclusion

This paper tests all test equipment and instrument are in within the period of validity of the measurement, the measurement data has certain rules, measured frequency response function curve is smooth, peak of state obviously and the main vibration mode corresponding to the mode indicator function MIF value far less than 0.1, the measured data quality is better. Based on the results of study and data analysis, we can get the following conclusion:

1. The modal vibration obtained by the modal test is identical with that of the finite element simulation method. The results show that the method can obtain the full main mode.

2. Modal test to obtain the solar wing subsystem extension mechanism is complete and accurate modal parameters, including natural frequencies, modal damping ratios and modal shapes, for the follow-up of structure optimization and dynamic characteristics of precise analysis provides the basis.

3. By using the obtained experimental modal parameters, the model correlation analysis, model updating and the evaluation of the correction model can be further developed in this model. This makes it more realistic to represent the actual structure, and also lay a foundation for the accurate response and load analysis of the structural components.

References

1. Zhao Guowei, Du Shanshan, Wei Wei, Space application and development trend of deployment mechanism, Mechanical technology history and mechanical design (7): the 7th machinery technology history and mechanical design international academic conference papers in 2008.
2. Lin Shangmin, Space deployable structure design and analysis of arm, Master's thesis, Xi'an Electronic and Science University (2012).
3. Huang Hailin, Unit configuration synthesis and large scale networking of articulated deployable mechanism, Master's thesis, Harbin Institute of Technology (2012).

4. He Xuchu, Basic theory and calculation method of generalized inverse matrix, (Shanghai science and Technology Press, 1985).
5. Liu Yanzhu, Multi rigid body system dynamics, (Beijing: Higher Education Press, 1987).
6. Tian dake, Modular space deployable antenna support truss design and experimental study, PhD thesis, Harbin Institute of Technology (2011).
7. Hou Guoyong, Design, analysis and experiment of deployable truss structure, PhD thesis, Zhejiang University (2008).
8. Cao Shuqian, Zhang Dewen and Xiao Longxiang, Modal analysis of vibration structure, (Publishing House of Tianjin University, 2001).
9. Li Debao and Lu Qiuhai, Experimental modal analysis and its application, (Science Press, 2001).
10. Wei Juanfang. 4 ~ 10 meters of space deployable antenna coupling dynamic analysis and experimental research, PhD thesis (2002).
11. Zhang Linbo, Ren Ge and Chen Hongbin, Modal analysis for horizontal shafting of a large-aperture telescope, Vol. 30 (Opto-electronic Engineering, 2003), pp 8-10.
12. Jin Xianding, Lu Xinsen and Cheng Xueming, The modal analysis of a solar array, Vol. 12 (Journal of Astronautics, 1991), pp 58-62.
13. Xu Yougang, Feng Jiaquan and Du Qiang, etc. Modal test method of large structural parts, Vol. 2012 (science and technology and engineering, 2007), pp 3231-3234.
14. Nie Yongjun and Liao Qizheng. Finite element analysis and experimental modal analysis of truss structures, Vol 27 (mechanical design and research, 2011), pp 61-64.
15. Guo Qiwei, Wu Song and Liu Fang, etc. Research on engineering practice of modal analysis-test of spacecraft, Vol. 2014 (Journal of dynamics and control, 2014), pp 274-278.
16. Jiao Anchao and Yao Qi Feng. Method of structural modal test for Tiangong-1target spacecraft, Vol 28 (Spacecraft environment engineering, 2011), pp 593-596.

Research on the characteristic of high-speed grinder debris penetrating into multi-layer shields

Li-Zhong Wang[*], Jin-Feng Shang[*,§], Meng Zhao[†], Hua-Ying Wu[‡] and Lu-Kai Feng[†]

[*]*School of Mechanical Engineering, Xi'an Jiaotong University, Xi'an 710049, P. R. China*

[†]*School of Materials Science and Engineering, Xi'an Jiaotong University, Xi'an 710049, P. R. China*

[‡]*The National Engineering Research Center of Rapid Manufacturing, Xi'an Jiaotong University, Xi'an 710049, P. R. China*

[§]*Emile:jinfeng.shang@foxmail.com*

Aiming at the penetration of the high-speed grinder debris to the external protection board, the numerical simulation research is carried out by using ABAQUS. The influence of stacking sequence on the penetration resistance of the guard plate is discussed, and the corresponding residual velocity of debris and the regularity of the order of the thickness of the protective plate are obtained.

Keywords: High-speed grinder debris; outer shield; penetration resistance; simulation analysis.

1. Introduction

With the rapid development of industry, NC machine tools in automobile, military industry and other manufacturing gained more and more widely used. In order to improve the processing speed, quality, accuracy, high-speed machine tools are more and more get the attention of people.NC machine tools in both China and overseas are developing in the trend of high-speed, complex and environmentally friendly directions. And China has successfully made a lot of advanced high speed machine tools. However, with the development of high speed machine tool, the demands of safety and health are getting stronger and more urgent[1]. It is a great challenge how to protect workers and the machine from chips and knife fragments made in the processing. In general, it belongs to the high-speed grinder when the wheel speed exceeds 45m/s. In this paper, the line speed of high-speed grinder debris is150m/s, machine outer shield material is Q235 and thickness is 3mm.It research the best way to manufacture the multilayer shields.

2. Calculation Model

2.1. *Geometric model*

Debris model is the round bullet model, which head diameter is 15mm and the total length is 32.5mm; machine outer shield model is a square planar plate which length is 300mm. Although the whole high-speed impact model is symmetrical, the damage of the projectile target is not symmetrical, so it cannot be simplified by using the partial model analysis. The model is established as shown in Fig. 1.

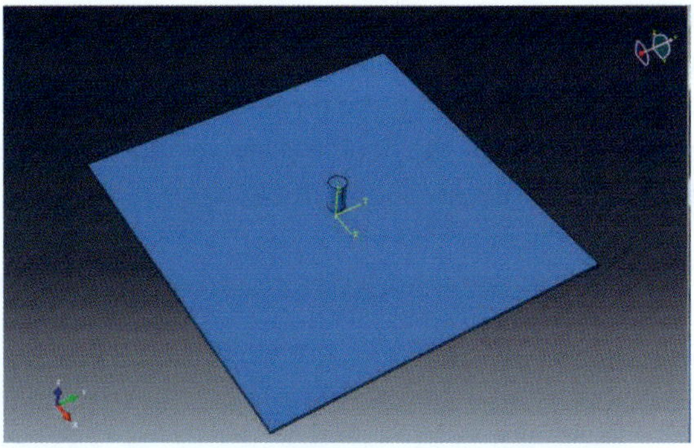

Fig. 1. Solid model.

2.2. *Finite element model*

In this paper, the initial line speed of high speed grinding machine is 150m/s. The deformation of the debris is very small at the speed of 150m/s, so the debris is set as a rigid body. Its elastic modulus input is 210GPa, the Poisson's ratio is 0.3, the density is 8000kg/m^3.

The Johnson-Cook failure model was used to describe the mechanical properties of the steel plate material [2-3], considering the adiabatic temperature rise effect, without considering the heat conduction effect. The J-C failure model of Q235 steel are shown in Table 1 as follows[4].

Table 1. Material parameters of Johnson-Cook model.

Material	$\rho / (kg / m^3)$	E / Gpa	μ	θ_t / K	θ_m / K
Q235	7800	200	0.3	300	1800
Material	A/MPa	B/Mpa	n	C	m
Q235	410	20	0.08	0.100	0.55
Material	$d1$	$d2$	$d3$	$d4$	$d5$
Q235	0.3	0.9	2.8	0	0

The debris and shield board choose C3D8R unit, which is the eight node linear hexahedral element [5]. In order to reduce the computation time and does not affect the calculation results, it is needed to make the mesh of the steel plate to be locally refined. In order to be more precise in the direction of the thickness of the shield board, the unit of the double plate should be divided into smaller and more dense[6]. The results are shown in Fig 2.

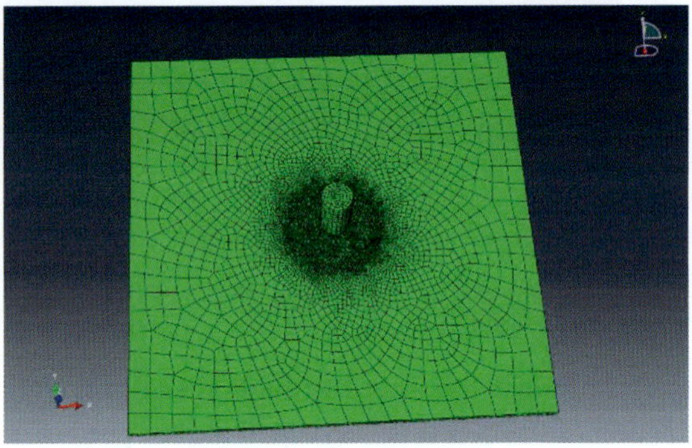

Fig. 2. Finite element mesh generation.

3. Simulation Results and Analysis

In the penetration, the energy loss mainly includes deformation and damage energy absorption of the shield board and debris. Therefore, the thickness of the shield board, the number of layers and the layer spacing are all factors that influence the penetration resistance. The initial speed of the debris is set to

-150m/s because of the Z axis along the negative direction. So if the final speed is negative, it indicates that the protection board is broken down.While the final speed is positive, it indicates that the protective plate is not broken down and the debris occurred rebound. The experimental scheme is to simulate the 3mm thickness of the monolayer shield board first, and then the total thickness is divided into two layers to simulate the different stacking sequence. The results of each model are extracted and analyzed, and the final velocity of debris is shown in Table 2.

Table 2. Experimental results.

Test	A	B	C	D	E	F	G	H
H_1 (mm)	3	1	1.2	1.4	1.5	1.6	1.8	2
H_2 (mm)	0	2	1.8	1.6	1.5	1.4	1.2	1
V_t (m/s)	5.94	-18.44	3.91	3.24	2.27	2.48	3.94	4.31

As shown in Fig. 3, the results show that the penetration performance of Q235 single layer shield with 3mm thickness is weaker than that of the double deck shield which the total thickness is 3mm(except 1mm+2mm). Analysis of its causes, on the one hand, there are interactions between two layered slab which plays the part of the damping effect.On the other hand, the response of the double deck plate to the stress wave is more complex than that of the single plate, and the energy absorbing effect produced by the elastic plastic wave superposition is more excellent.

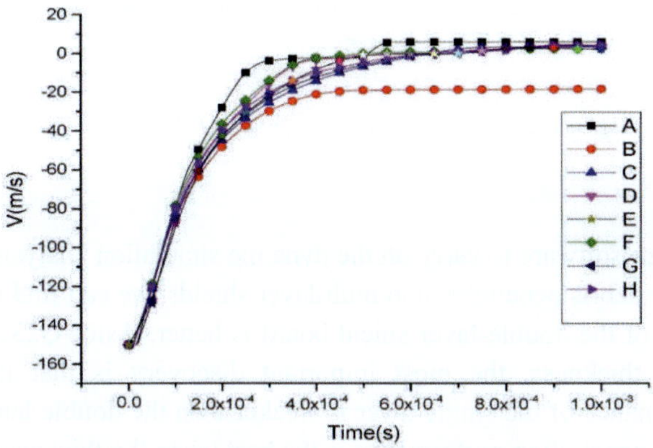

Fig. 3. Speed-time curve of A-H test.

Using origin to fit the data of the program B-H, the curve of the thickness of the front plate of the double layer plate with the total thickness of 3mm and the residual velocity of the fragment is shown in Fig. 4. It can be drawn from the fitting curve thatthe rebound speed is the minimum when the current shield board is about 1.5mm, which is consistent with the simulation results. The stacking sequence has great influence on the penetration resistance of the double layer shield board with the total thickness of 3mm. The anti penetration performance is the best when the thickness of the front and back boards is equal. Then is the order that the front board is much thicker than the back board. The worst stacking sequence is that the front board is thinner than the back board.

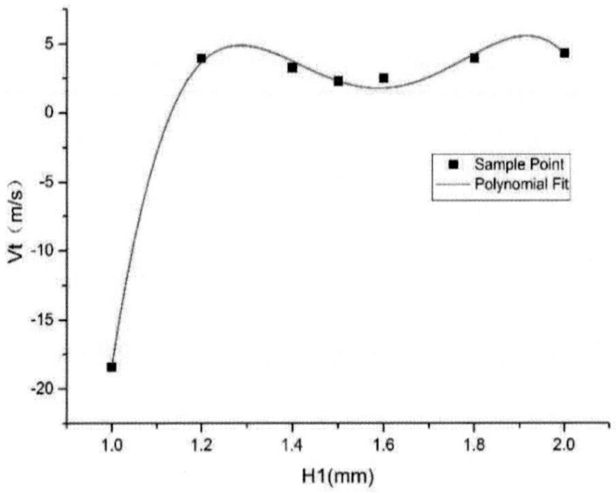

Fig. 4. The data fitting curve of the program B-H.

4. Conclusions

Through the Abaqus software to carry on the dynamic simulation analysis of the high-speed grinder debris penetrates into multilayer shields, we can find that the energy absorption of the double-layer shield board is better. Asfor Q235 shield board with 3mm thickness, the most important discovery is that the anti penetration performance of the single layer is weaker than the double laminated plate. And the anti penetration performance is the best when the thickness of the front and back boards is equal.

Acknowledgments

This work is supported by high-end CNC machine tools and basic manufacturing equipment science and technology major project (No.: 2014ZX 04008032).

References

1. Zhu Xiaochun. Research on the application of new technology to machine tools' appearance design[D]. Shenyang University of Technology, 2011
2. Johnson G R, Cook W H. A constitutive model and data for metals subjected to large strains, high strain rates and high temperatures[J]. Proceedings of the Seventh International Symposium on Ballistics the Hague, 1983.
3. Johnson G R, Cook W H. Fracture characteristics of three metals subjected to various strains, strain rates, temperatures and pressures[J]. Engineering Fracture Mechanics, 1985, 21(85):31–48.
4. Lin Li, Xu Dong Zhi, Fan Feng, Meng Shang-Jiu, Su Junjie. Determination of parameters of Johnson-Cook models of Q235B steel[J]. Journal of vibration and shock, 2014, 09:153-158+172.
5. ABAQUS official document. Abaqus 6.12 Analysis User ABAQUS official document 's Manual[M]. Dassault SIMULIA company, 2010.
6. Zhuang Zhuo. Based on ABAQUS of finite element analysis and application [M]. Beijing: Tsinghua University press, 2009.

Fuzzy sliding mode variable structure control of semi-active air suspension

Ming-Jun Zheng*, Xiao-Lei Zhang and Jing Zhao
School of Mechanical Engineering, Shijiazhuang Tiedao University, Shijiazhuang, 050043, China.
E-mail: zcl1971@163.com

In order to improve semi-active air suspension vehicle ride comfort, Proposing the fuzzy sliding mode control strategy by combining fuzzy control and sliding mode variable structure. Through simulation in the time domain and frequency domain to verify the effectiveness of this control algorithm. Compared to the sliding mode control suspension, the damping effect of fuzzy sliding mode control suspension is obvious, the evaluation indexes are decreased 14.21%, 23.55%, and 17.03%, respectively. It promotes damping performance and ride comfort of the suspension system significantly.

Key words: Air suspension; fuzzy-SMC; ride comfort analysis.

1. Introduction

The fuzzy sliding mode control is particularly suitable for air suspension system that is strongly nonlinear and difficult to establish a precise mathematical model[1,2]. In this paper, a fuzzy sliding mode controller for air suspension vehicle is designed and analyzed in time-domain and frequency-domain.

2. Fuzzy Sliding Mode Controller Design

Figure 1 is 1/4 vehicle model of 2 DOF air suspension, and figure 2 is mechanics model of sky hook system.

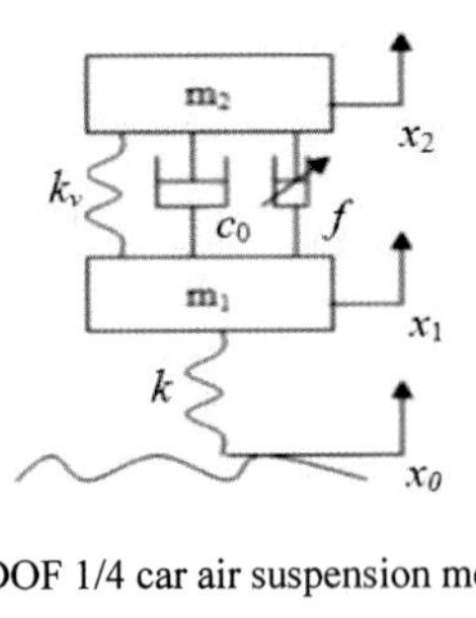

Fig. 1. 2-DOF 1/4 car air suspension model.

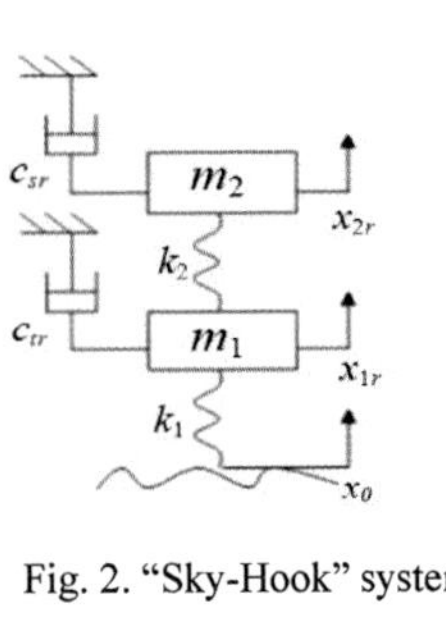

Fig. 2. "Sky-Hook" system.

From figure 1, the dynamic equations of motion for this quarter-vehicle air spring suspension system can be derived as Eq. (1).

$$\begin{aligned} m_2\ddot{x}_2 + c_0(\dot{x}_2 - \dot{x}_1) + k_1(x_2 - x_1) + f = 0 \\ m_1\ddot{x}_1 - c_0(\dot{x}_2 - \dot{x}_1) - k_1(x_2 - x_1) + k_2(x_1 - x_0) - f = 0 \end{aligned} \quad (1)$$

According to Figure 2, the dynamic equations of "Sky-Hook" system can be derived as follows:

$$\begin{aligned} m_2\dot{x}_{2r} + c_{sr}\dot{x}_{2r} + k_1(x_{2r} - x_{1r}) = 0 \\ m_1\dot{x}_{1r} - c_{tr}\dot{x}_{1r} - k_1(x_{2r} - x_{1r}) - k_2(x_0 - x_{1r}) = 0 \end{aligned} \quad (2)$$

where m_2 is the sprung mass; m_1 is the unsprung mass; c_0 is the damping coefficient of the suspension; c_{sr} is the damping coefficient of the reference model tire; c_{tr} is the damping coefficient of the reference model damper; k_1 is the spring stiffness; k_2 is the tire stiffness; $x_2 - x_1$ and $x_{2r} - x_{1r}$ are the suspension stroke; $x_1 - x_0$ and $x_{1r} - x_0$ are and tire deflection. x_0 is the vertical ground displacement caused by the road unevenness; and f is the active force.

Sky hook system state variables are:

$$X_r = [x_{1r} \quad x_{2r} \quad \dot{x}_{1r} \quad \dot{x}_{2r}]^{\mathrm{T}}. \quad (3)$$

The suspension system error e is defined as vector sprung mass velocity error, sprung mass displacement error and the integration of sprung mass displacement error[3], which can be expressed in matrix form as Eq. (4) and Eq. (5).

$$e = [\dot{x}_2 - \dot{x}_{2r} \quad x_2 - x_{2r} \quad \int(x_2 - x_{2r})]^{\mathrm{T}}. \quad (4)$$

$$\dot{e} = [\ddot{x}_2 - \ddot{x}_{2r} \quad \dot{x}_2 - \dot{x}_{2r} \quad x_2 - x_{2r}]^{\mathrm{T}}. \quad (5)$$

From the preceding discussions, the error control system dynamic equation can be expressed as Eq. (6).

$$\begin{cases} \dot{e} = A_e e + B_e X + H_e X_r + G_e U \\ s = Ce \end{cases} \quad (6)$$

$$A_e = \begin{bmatrix} -\frac{c_0}{m_2} & -\frac{k_1}{m_2} & 0 \\ 1 & 0 & 0 \\ 0 & 1 & 0 \end{bmatrix}, \quad B_e = \begin{bmatrix} \frac{k_1}{m_2} & 0 & \frac{c_0}{m_2} & 0 \\ 0 & 0 & 0 & 0 \\ 0 & 0 & 0 & 0 \end{bmatrix}, \quad H_e = \begin{bmatrix} -\frac{k_1}{m_2} & 0 & -\frac{c_0}{m_2} & 0 \\ 0 & 0 & 0 & 0 \\ 0 & 0 & 0 & 0 \end{bmatrix} \quad G_e = \begin{bmatrix} -\frac{1}{m_2} \\ 0 \\ 0 \end{bmatrix}, U = [f]。$$

Sliding movement must be asymptotically stable and all control points lays in the sliding plane. Switching process should have good dynamic quality[4]. Sliding motion characteristic equation is $C(s)=c_2+c_1s+s^2$, and sliding switch function will be determined after getting c_1 and c_2 values:

$$s=[(\dot{x}_2-\dot{x}_{2r})+c_1(x_2-x_{2r})+c_2(\int(x_2-x_{2r}))]=0. \tag{7}$$

$$\dot{s}=C\dot{e}=CA_e e+CB_e X+CH_e X_r+CG_e U. \tag{8}$$

Matrix CA_e, CB_e, CH_e, CH_e all can be reversible, then

$$U=(CG_e)^{-1}[\dot{s}-(CA_e e+CB_e X+CH_e X_r)]. \tag{9}$$

The necessary and sufficient conditions for the establishment of sliding control is $s\cdot\dot{s}<0$ [5]. In the premise of sliding to meet the presence and achieve condition, it can be used to improve the sliding mode control dynamic quality by constant reaching law:

$$\dot{s}=-\varepsilon\cdot\mathrm{sgn}(s). \tag{10}$$

3. Design of Fuzzy Sliding Mode Controller

Resetting membership functions are triangles. In order to make the system moving point reach sliding switching surface within the specified time, the control rules are set as follows: When $s(x)>0$ and $\dot{s}(x)>0$, should take $U>0$ and large absolute amount of control; when $s(x)>0$ and $\dot{s}(x)<0$, the system has been close to the sliding surface movement, then, should take $U<0$ and a small absolute amount of control[6,7]. The fuzzy control rules are shown in Table 1.

Table 1. Rules of fuzzy control.

$s(x)$	$\dot{s}(x)$						
	NB	NM	NS	ZE	PS	PM	PB
NB	NB	NB	NM	NM	NS	NS	ZE
NM	NB	NB	NM	NS	NS	NS	ZE
NS	NM	NM	NS	ZE	ZE	ZE	ZE
ZE	NM	NM	NS	ZE	ZE	PS	PS
PS	NS	NS	ZE	ZE	PS	PS	PM
PM	ZE	PS	PS	PS	PS	PM	PB
PB	PS	PS	PS	PM	PB	PB	PB

For more intuitive representation of fuzzy control rules, fuzzy control rules are given 3-D model, as shown in Figure 3. Sliding control block diagram above analysis, we can get semi-active air suspension fuzzy sliding model control block diagram, as shown in Figure 4.

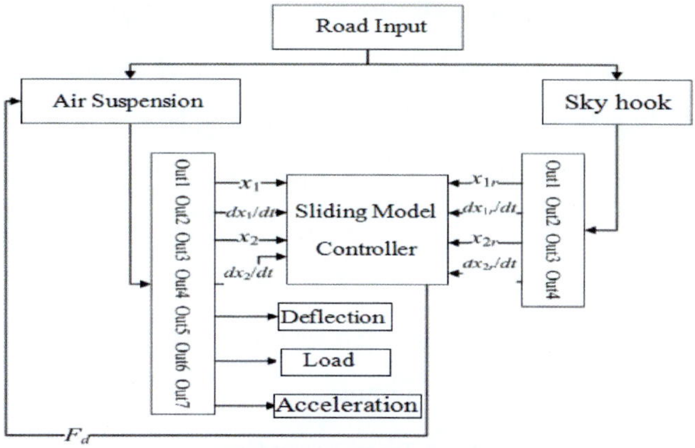

Fig. 3. 3-D figure of fuzzy control rules.

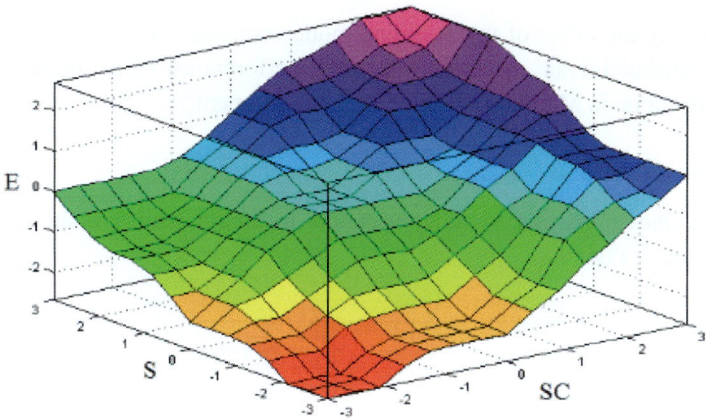

Fig. 4. Fuzzy sliding mode control diagram.

The road excitation power spectral density $G_q(n)$ is:

$$G_q(n) = G_q(n_0)\left(\frac{n}{n_0}\right)^{-\omega} \tag{11}$$

Where reference spatial frequency n_0=0.1m^{-1}, frequency index ω=2, cut off frequency f_0=0.1Hz, road roughness coefficient $G_q(n_0)$=2.56×10^{-4}m^3. *C*-level pavement resulting spectrum is shown in Figure 5.

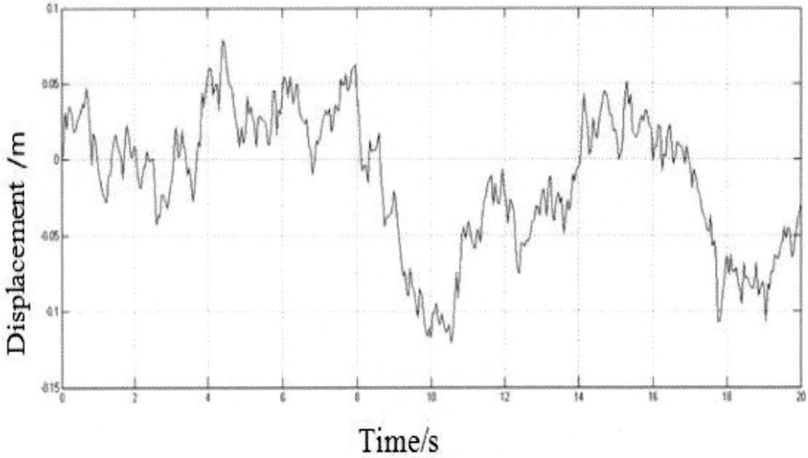

Fig. 5. The random road input.

4. Simulation Analyses

In order to verify the effect of fuzzy sliding mode controller for semi-active air suspension, semi-active air suspension damping performance of the numerical simulation was done in MATLAB/SIMULINK environment.

4.1. *Time domain analysis*

Based on the above analysis, simulation results in time-domain are shown in Figure 6, Figure 7, Figure 8.

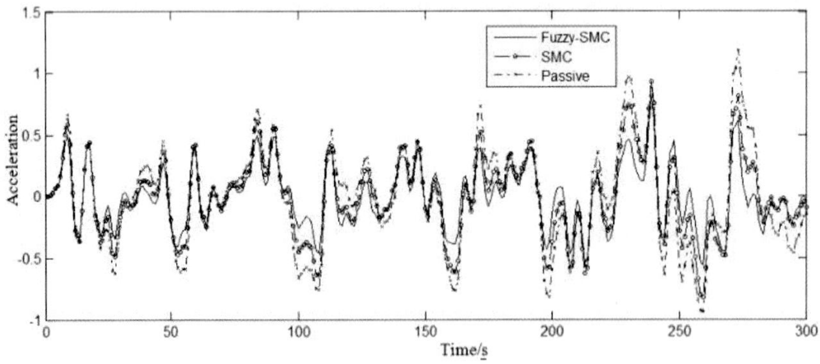

Fig. 6. Body vertical acceleration.

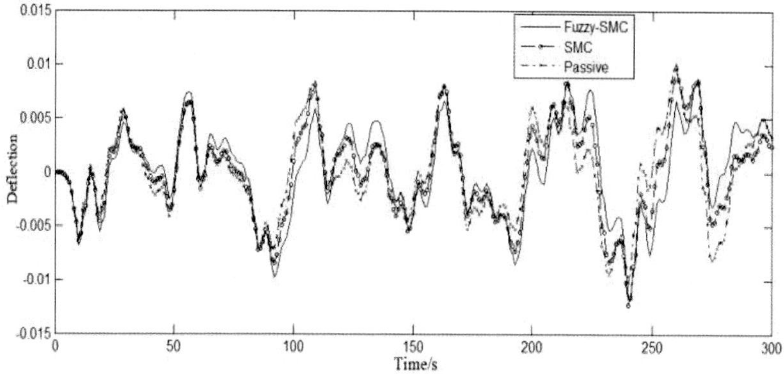

Fig. 7. Suspension deflections.

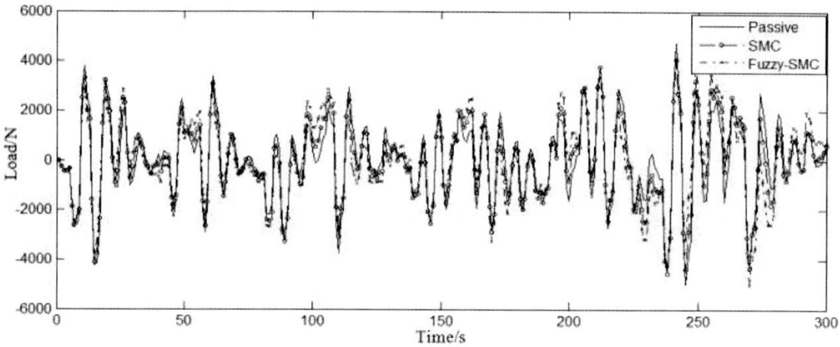

Fig. 8. Dynamic wheel load.

In order to illustrate more intuitively that fuzzy sliding mode control strategy is superiority in the semi-active air suspension system control, Table 2 lists the RMS evaluation simulation statistics.

Table 2. The statistical results of simulation.

Evaluation	Control Type		
	Passive	SMC	Fuzzy-SMC
Body vertical acceleration/($m{\cdot}s^{-2}$)	0.230	0.183	0.157
Suspension dynamic deflection/m	0.00 472	0.00 361	0.00 276
Tire dynamic load/N	2 030	1 732	1 437

Integrated figures and Table 2 shows that fuzzy sliding mode control can suppress the body vertical acceleration more effectively, compared with the sliding mode control. Simulation statistics prove that semi-active air suspension outperforms passive suspension and further enhances the car ride comfort.

4.2. *Frequency domain analysis*

Frequency response characteristics of the suspension reflect the energy distribution in the frequency domain when the vehicle vibrates. Grasping the system frequency characteristics of vibration is of great significance. The paper gives the Power Spectral Density(PSD) in the frequency domain analysis that are shown in Figure 9~ Figure 11.

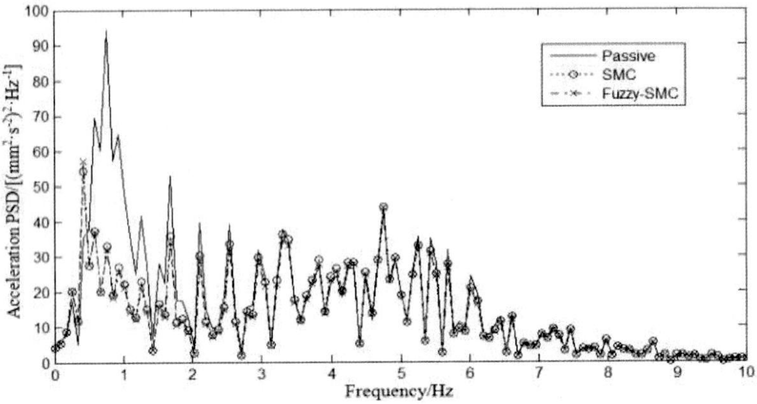

Fig. 9. Body vertical acceleration PSD.

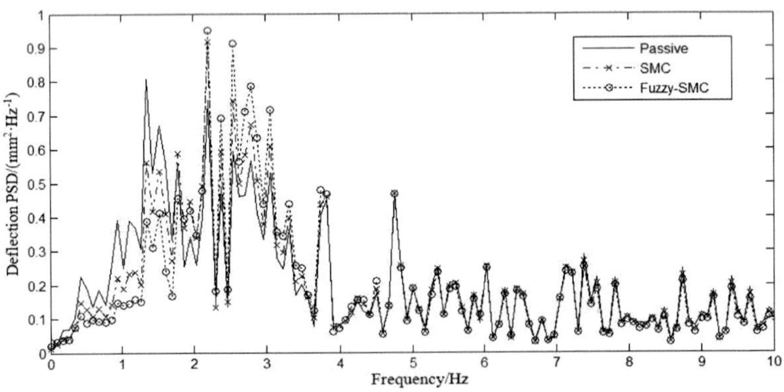

Fig. 10. Suspension deflections PSD.

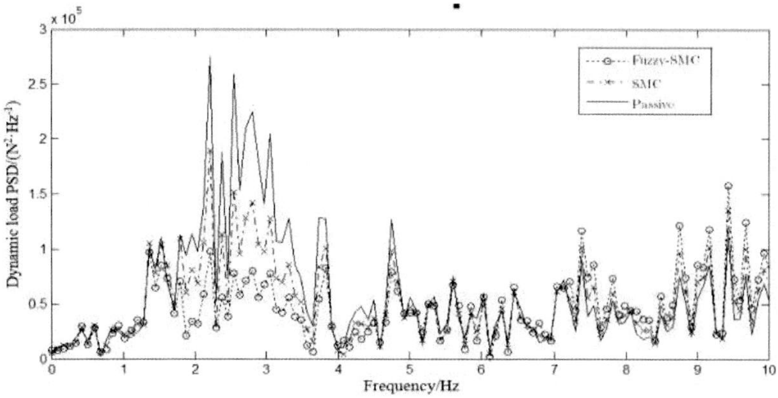

Fig. 11. Dynamic wheel load PSD.

In general, 4.0-12.5 Hz region most likely cause human organs and spine systems resonance. Figure 8 shows that fuzzy-SMC can effectively restrain the resonance of the body and the body of the human body and spine, which helps to improve the comfort of the vehicle. Fuzzy-SMC is better than the other two control strategies.

5. Conclusions

(1) Compared to the "sky-hook" suspension system and sliding mode control, fuzzy sliding mode control is more superior,
(2) Compared to simple slide mode control, the vertical acceleration, suspension dynamic deflection, and dynamic wheel load statistics RMS value of fuzzy sliding mode control semi-active air suspension are decreased 14.21%, 23.55%, and 17.03%, respectively, indicating that the proposed semi-active air suspension fuzzy sliding mode control strategy is practicable.

Acknowledgment

This work is supported by Hebei province science and technology support project (No. E2013210018), all support is gratefully acknowledged.

References

1. Hui Wang, Sihong Zhu. *Neural network adaptive control for semi-active air suspension*. Transactions of the Chinese Society for Agricultural Machinery, 2006, 37(1):28-31.

2. Abidi, Khalid, Xu Jian-xin, Shi Jin-hua. *A discrete-time terminal sliding mode control approach applied to a motion control problem.* IEEE Transactions on Industrial Electronics, 2009, 56(9): 3619-3627.

3. Guangqiang Wu, Huanjun Huang, Guanghu Ye. *Semi-active control of automotive air suspension based on fractional calculus.* Transactions of the Chinese Society for Agricultural Machinery, 2014, 45(7):19-24.

4. Shanyue Wu, Shijian Zhu, Yingyun Huang. *Impact characteristics analysis of rubber air spring with auxiliary chamber.* Journal of Vibration Engineering, 2005, 18(2):248-251.

5. Cheng Zhao, Dayue Chen. *Floating raft isolation system of semi active fuzzy sliding mode control.* Journal of Mechanical Engineering, 2008, 44(2):163-169.

6. Yuzhuang Zhao, Zongcheng Wang, Sizong Chen. *Air suspension vehicle high control system research based on sliding mode control.* Journal of Beijing Institute of Technology (Natural Sciences Edition), 2014, 34(11):1125-1139.

7. Liqin Sun, Zhongxing Li, Xing Xu. *Semi-active air suspension sliding mode variable structure control and test.* Journal of Jiangsu University (Natural Science Edition), 2014, 35(6):621-626.

Vibration simulation of EMU carbody underframe considering rail short wave

Cheng-Qiang Wang
CRRC Changchun Railway Vehicles Co. LTD,
NO.435, Qingyin Road, Changchun, Jilin, China
E-mail: wcq_2004_2004@163.com

Fan-Song Li
Traction Power State Key Laboratory, Southwest Jiaotong University,
NO.111, North 1st Section of Second Ring Road,
Chengdu, Sichuan, China
E-mail: lifansong2013@163.com

Abnormal vibration of Electric Multiple Unit (EMU) carbody underframe was found by the line test, so a three dimensional rigid-flexible coupled vehicle system dynamics model was built to simulate the vibration response of carbody. Due to the excitation frequency was comparatively high, so the rail short wave was considered during the multi-body dynamic simulation. The concerned positions of carbody underframe, which located in center of underframe, was chosen to compare the results from theoretical model with line test. The comparison results illustrate that rigid-flexible coupled model combined with high frequency excitation should be adopted to simulate the high frequency vibration of carbody appropriately.

Keywords: EMU; flexible; vibration; rail; short wave.

1. Introduction

Weight reduction of railway vehicle has greatly contributed to speed-up. However, lightweight of car body often lead to less structural stiffness. From these reasons, the flexible vibration of car body need to be considered during vehicle system dynamics simulation.

In recent years, many experts utilize different vehicle dynamics model to study the vibration characteristics of car body. P.F. CARLBOM gives a method to combine multi-body system (MBS) with FEM for rail vehicle dynamics analysis[1]; P.B. Wu et al evaluate dynamic response of high-speed passenger car considered flexible car body model[2]; M. NAGAI et al use the flexible car body dynamics model to simulate the coupled vibration of passenger and lightweight car body, during the simulation, every passenger is modelled as one degree-of-freedom system[3]; T. TOMIOKA et al provide a new method to calculate car body service weight modal, in the method, a car body is modeled as a box-type structure consisting of plates and beam, and then combine the MBS and FEM model to simulate the vibration of car body[4].

The conventional ride quality and running comfort computing method does not consider the flexible effect of car body, which is acceptable when the concerned positions do not contain local high frequency vibration. However, when the local vibration generated in the underframe of car body, the rigid multi-body dynamics method may produce large error. Among the references mentioned above, no one considers the high frequency (20~40Hz) vibration of car body, a method to simulate high frequency is provided in this paper, which combines the rigid-flexible coupled model and short wave effect of track spectrum. The concerned frequency range of ride quality is shown in Figure 1 [5], so the max frequency considered in this paper is 40 Hz.

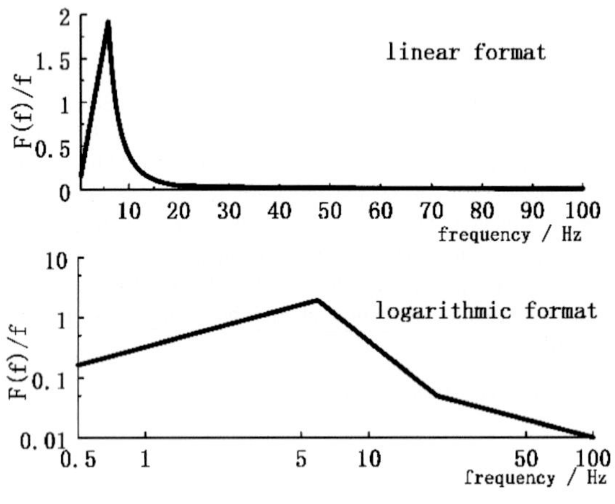

Fig. 1. Concerned frequency range of ride quality.

2. Vibration Analysis of EMU Carbody Frame by Line Test Data

Acceleration of the center of the EMU carbody frame was tested when the EMU was running on the track. The fast Fourier transform (FFT for short) of the acceleration is shown in Fig. 2. We could find that the dominant frequency is about 30Hz, the low frequency (less than 10Hz) do not exit. So, we could guess that resonance may be exit, the flexible vibration should be considered.

3. Rigid-flexible Coupled Modeling Approach

The process of modeling a flexible carbody is shown in Figure 3, which mainly consists of three steps. The first step is to set up carobody FEM model with service weight, and generate the substructure model of carbody by ABAQUS, then produce the intermediate file for SIMPACK; the second step is to translate the intermediate file into SID format file for SIMPACK; the last step is to import the SID file and set the parameters in SIMPACK.

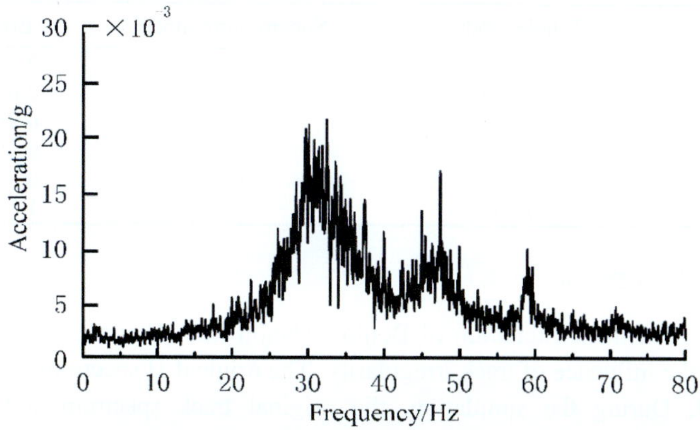

Fig. 2. FFT of the acceleration of the center of underframe.

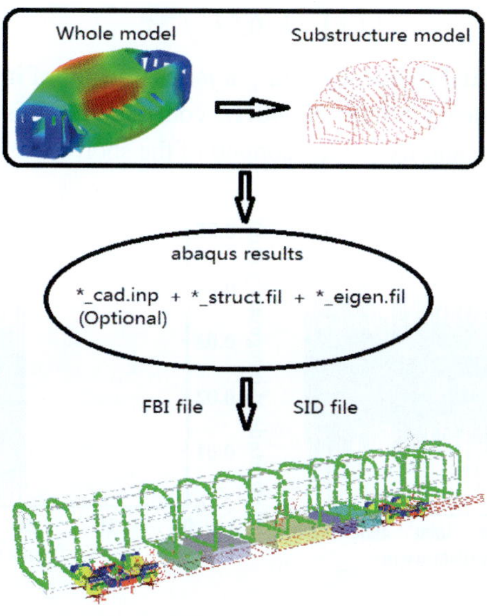

Fig. 3. Process of establishing rigid-flexible multi-body dynamics model.

The use of substructure should be emphasized in the preprocessor. The basic idea of substructure is to consider a substructure separately and eliminate all but the degrees of freedom needed to concern in the whole model. So the calculation errors between substructure model and whole degree of freedom model is given in Table 1. The maximum error is 1.67% in the sixth mode, therefore, the substructure model meets the engineering requirements.

Table 1. Calculation error between whole DOF model and substructure mode.

Mode	Whole model	Substructure model	Error
1	10.187	10.201	0.14%
2	11.479	11.550	0.62%
3	14.644	14.788	0.98%
4	15.008	15.121	0.75%
5	15.327	15.373	0.30%
6	17.337	17.047	1.67%

4. Rail Short Wave

The measured track excitation of Beijing-Tianjin line (JJ spectrum) is used to simulate the influence of track irregularity. The original JJ spectrum is shown in Figure 4. During the simulation, the original track spectrum add a track spectrum with short waves. Then excitation between wheel and rail contains the short wave effects. The power spectral density of track irregularity is [6]

$$S(f) = k \cdot 0.036 f^{-3.15} \tag{1}$$

Where f is the spatial frequency [1/m], k is scale factor of irregularity.

The random frequency ranges 20~40Hz, considered in this paper. Figure 4 shows the time domain and frequency domain of the short wave.

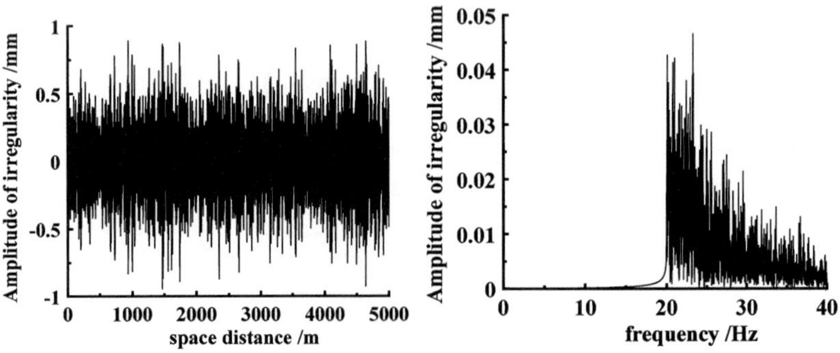

Fig. 4. The track spectrum of short wave.

5. Rigid-flexible Coupled Vehicle System Dynamics Model

A three-dimensional rigid-flexible coupled vehicle system dynamics model established for high-speed coach by ABAQUS and SIMPACK shows in Figure 5. The primary suspension is modeled as parallel spring-damper elements, and the nonlinear wheel-rail contact are considered. Air springs, yaw dampers, lateral stops, anti-roll bars, lateral dampers and traction links are considered in secondary suspension system. In addition, all of the equipment are mounted on the underframe according to the actual parameters in the dynamics model.

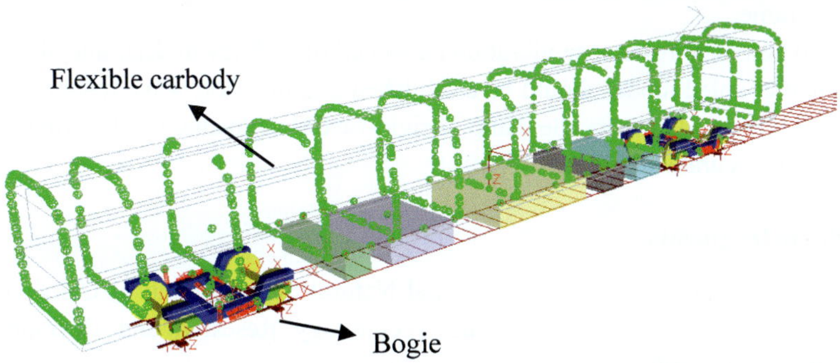

Fig. 5. A rigid-flexible coupled vehicle system dynamics model.

6. Vibration Simulation Results of Carbody Frame

Considering effects of flexible carbody and short wave, vibration simulated result of the center of underframe is shown in Fig. 6.

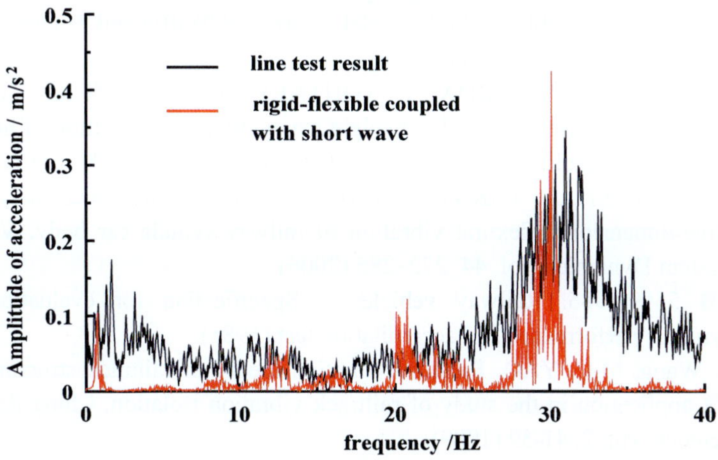

Fig. 6. Comparison between line test result and simulation result.

The rigid-flexible coupled model with track short wave could give a good result, the local resonance frequency fitting better. The amplitude and effect frequency range need more study.

7. Summary

The vibration simulation of the center of EMU carbody underframe was studied in this paper, and the rigid-flexible coupled model was established considering the short wave effect of track spectrum and the flexible mode vibration of underframe.

When local resonance vibration happened of carbody underframe, the rigid-flexible coupled model should be established and the short wave track spectrum should be considered. By the damp of rigid-flexible dynamic needs further study to obtain accurate amplitude.

Acknowledgments

This work is supported by the National Natural Science Fund of China [U133 4206, 51475388] and Science and Technology Research and Development Program [2012G002].

References

1. P.F. CARLBOM. Combining MBS with FEM for Rail Vehicle Dynamics Analysis. Multibody System Dynamics, vol. 6, 291–300 (2001).
2. P.B. Wu, S.H. Xue, C.H. Yang. Dynamic response of high-speed passenger car based on flexible car body model. Journal of Traffic and Transportation Engineering, vol. 5, p. 5-8 (2005).
3. M. NAGAI, H.YOSHIDA, T. TOHTAKE and Y. SUZUKI. Coupled vibration of passenger and lightweight car-body in consideration of human-body biomechanics. Vehicle System Dynamics, vol. 44, p. 601–611 (2006).
4. T. TOMIOKA, T. TAKIGAMI and Y. SUZUKI. Numerical analysis of three-dimensional flexural vibration of railway vehicle car body. Vehicle System Dynamics, vol. 44, 272–285 (2006).
5. GB 5599-1985. Railway vehicles – Specification for evaluation the dynamic performance and accreditation test (1985).
6. L. Wang, M.C. YAO. Random vibration theory of railtrack structure and it's application in the study of railtrack vibration isolation. China Railway Science, vol. 2, 41-59 (1989).

Research of knowledge pushing actively in complex aerospace products MRO process

Zhao-Wei Nie* and Hai-Cheng Yang
School of Mechanical Engineering,
Nanjing University of Science and Technology
Nanjing, 210094, China
**E-mail: ggcua@163.com*

Dan-Dan Xiong
Beijing Shenzhou Aerospace Software Technology Co. Ltd.,
Beijing, 100094, China

In order to overcome the difficulties of inadequate ability of knowledge reuse in the maintenance, repair and overhaul of the aerospace products, this paper focused on the key technology of knowledge actively pushing on repair process management and knowledge reuse, provided data source which suit the repair data acquisition for repair knowledge learning. Furthermore, repair knowledge is mined from repair data with Markov model, formed the knowledge model based on the maintenance data sequence. At last, knowledge actively pushing system with the knowledge demand of the repair node is build, the traditional workflow engine is also rebuild, which realized the extensible mechanism of knowledge files, and pushed knowledge actively according to the repair node requirement.

Keywords: MRO; process management; knowledge pushing.

1. Introduction

The MRO knowledge of the complex aerospace products is the key factor of repairing efficiency and quality, the mining and using methods of maintenance knowledge is of great researching potential and value in MRO process of complex aerospace products[1-4]. There still exists some knowledge management deficiencies in aerospace manufacturing enterprises nowadays, such as lack of non-structural data collecting methods, lack of MRO knowledge model and the ways of MRO knowledge reuse. This paper studied the reuse methods of repair data collecting and repair knowledge pushing that can be used in process of repair task.

2. The Knowledge Model of Repair based on Mro Service Data

In the aerospace products lifecycle process, the MRO service data are generated continually[5]. The data type of complex aerospace products MRO mainly contains:

- repair BOM data.
- data which repair, maintenance process needed.
- repair and maintenance process data.
- the feedback data of the repair and maintenance process.

The knowledge model that established based on the data type and process of the complex aerospace product MRO is shown as Fig. 1.

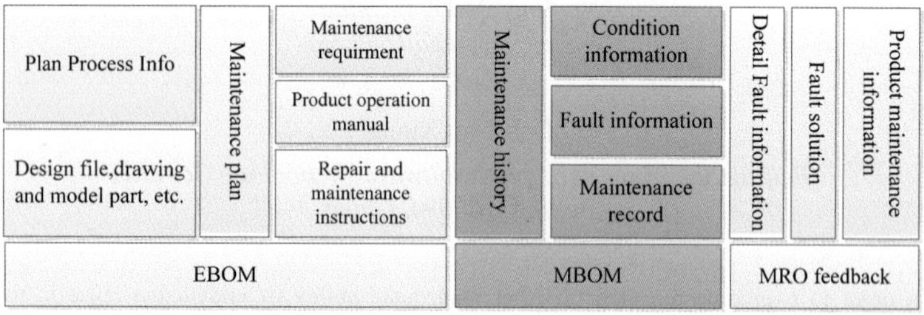

Fig. 1. The data of complex aerospace product MRO.

Because of the need of aerospace product development, The data form is established in the standard text editor. So the knowledge acquire method is established based on the repair knowledge model, which is shown as Fig. 2.

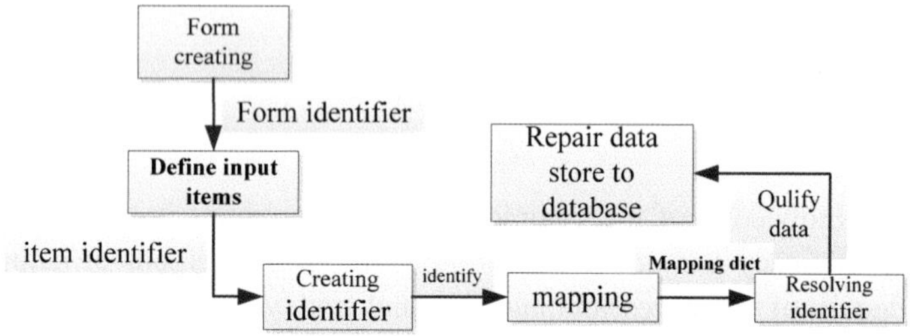

Fig. 2. The knowledge acquire methods based on the repair knowledge model.

Assign unique form identification for each maintenance form, denoted by *TF*, it is used as the unique name when identifying a form. Define the input items in the form and assign the unique data item identification for every input items, denoted by *IF*, which is used to identify the repair information when inputting the data. Create table structure of the corresponding form in the repair database at the same time, denoted by *DT*, create table field, denoted by *FT.TF* is corresponding to the input form in the database, Identification of the data item *IF* is corresponding to a repair data field *FT*. Define the identifier of every business data form, create a mapping relationship between the identifier, the input identification form *TF* and the identification of the data item *IF*, denoted by *BM*. A one-to-one mapping is created between each identifier *BM* and the unique identification of the data item from one form identification.

3. The Repair Knowledge Learning based on Markov Model

The markov model is a kind of statistic model which is wildly used in various natural languages processing in application area[6]. We can mine the knowledge of repair by training the using sequential patterns of repair data since MRO data can be seen as ordered repair data composition. In this paper, we assume that the subject of repair knowledge is k, training markov model individually on the dataset of KLk In the markov model. Jump probability matrix Pand and initial probability vector are two essential parameters that need to be studied. Jump probability $P(K_i|s_i^k)$ represents the given current data learning condition S_i, the next data item that need to be studied is the probability of K_i. Use the jump probability matrixPand, according to the present repair situation of the repair person, data items that are most likely to learn is predictable, establish the sequence of repair data items based on the context. The initial probability represents the probability that each maintenance data item is used as the initial data, the efficiency of knowledge learning is improved by recommending the higher popular initial probability item to the repair person. According to the maximum likelihood estimation, the calculation method for the jump probability is as in Eq. (1). that is the proportion of Ki appears after S_i. Correspondingly, K_i is the probability of the initial repair data item.

$$P(k_i \mid S_j^k) = frequency(S_j^k, k_i) / frequency(S_j^k) \tag{1}$$

In the markov model, with the growing of order k, Add more context information to the model, the accuracy of prediction will go up correspondingly; but at the same time, data sparsity problem will reduce the coverage of forecast. Use the method of linear interpolation of first order model and second order model, strike a balance between accuracy and recall as in Eq. (2).

$$P(k_i \mid S_j) = \lambda * P(k_i \mid S_j^1) + (1 - \lambda) * P(k_i \mid S_j^2) \tag{2}$$

$0 \le \lambda \le 1$ is the weight coefficient. Based on the markov model, the repair data can be organized as a sequence of repair knowledge that can be used in the maintenance process.

4. The Research about the Knowledge Active Pushing Facing De Working Process of Complex Aerospace Product MRO

The repair process is the source of actuate the application of repair knowledge of complex aerospace product MRO[7], visual intelligent workflow engine is the important tool of pushing the process intelligent. In the repair process, the repair crews work according to their respective maintenance working content, the specific work process follows the actuate of workflow engine. In this mode, repair personnel need to adjust the specific repair process according to the specific problems of maintenance, combined with the current working status, tool usage, the environmental factors, involving related post maintenance information and results.

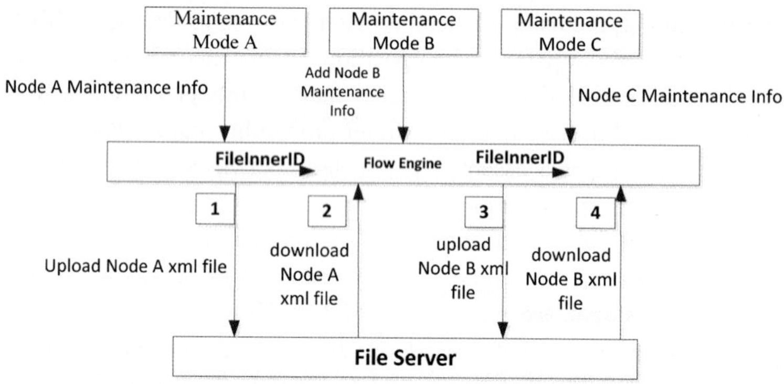

Fig. 3. The working process of workflow engine after the reform.

The structure of repair knowledge is organized by xml fragment. information context is added to xml fragment from different MRO process. Every xml fragment will be transferred by workflow engine. Maintenance knowledge pushing process are as follows:

(i) On the basis of the packaging repair knowledge, establish a packaged repair knowledge document, and stored it in the file server. Repair knowledge documents will be going through a status update by the repair data sequence before every call, to ensure the accuracy of the application information of dynamic knowledge in the repair knowledge. The knowledge file stored in the document server is distributed globally unique identification number by the workflow engine.

(ii) Workflow engine is preset to the repair task box of node b in accordance with the maintenance process template and *FileInnerID*. When the process entered B, the repair node will obtain A.xml which repair node A produced according to *FileInnerID*. And in the repair node B, system will retrieve maintenance knowledge that they need automatically.

(iii) After completion of the current maintenance nodes knowledge push, A.xml that repair node obtained are merged and expand to B.xml. Then upload them to the document server, and assign a globally unique identification number *FileInnerID*. And continually working according to the pushing process.

The detail of workflow engine which supports maintenance knowledge pushing actively is shown as Fig. 3.

5. Conclusion

This article studied established mapping between dynamic form and the database table fields, realized the structured storage of repair data. And then mined knowledge sequence from the structured storage maintenance data based on the markov model, Finally workflow engine is build which supports maintenance knowledge pushing actively.

References

1. X Yang, W Wang, F Xue, et al. MRO knowledge management system based on knowledge flow. Machinery Design & Manufacture. 2012(02): 250-252.
2. Innovating General Aviation MRO's through IT: The Sky Aircraft Management System - SAMS. Procedia - Social and Behavioral Sciences. 2015, 195: 1503-1513.
3. H Li, Y Ji, G Qi, et al. Integration model of complex equipment MRO based on lifecycle management. Computer Integrated Manufacturing Systems. 2010(10): 2064-2072.
4. B Huai, H Yang, N Wan. The technology research on knowledge modeling and reusing for aero-engine MRO. Manufacturing Automation. 2011(21): 64-67.
5. J Wang, S Si. The Integrated Modeling Method for the Maintenance BOM of Complex Equipments. ManufactureInformation Engineering of China. 2011(23): 50-54.
6. J Liu, J Xu, Y Zhang. Summarization Based on Hidden Topic Markov Model with Multi-features. Acta Scientiarum Naturalium Universitatis Pekinensis. 2014(01): 187-193.
7. W Cheng, H Yan. On Behaviour-based Quality Management System. Journal of Chongqing Institute of Technology. 2004(05): 29-31.

Integrated optimization method for trajectory planning in the magnet-assisted reticle stage

Fan Zhi, Ming Zhang* and Yu Zhu

The State Key Laboratory of Tribology, Tsinghua University, Beijing, 100084, China

**E-mail: zm01@mails.tsinghua.edu.cn*

The acceleration and deceleration trajectory of the magnet-assisted reticle stage mainly depends on the force created by the magnet structures which is of the exponential type. This trajectory may not be able to accomplish the collaborative exposing which requires the reticle stage and the wafer stage to derive the specified positon, velocity and acceleration respectively at the same moment for exposure. In this paper, the scanning motor in the reticle stage is introduced as a compensating force to generate the desired trajectory, which aims to achieve the collaborative exposing and minimal residual vibration of the reticle stage. Since the peak force in the motor should be limited to reduce the force ripples which has an adverse effect on the tracking performance, the integrated optimization method embodying both peak force design and trajectory planning is proposed to achieve both the collaborative exposing and tracking precision.

Keywords: Integrated optimization; trajectory planning; magnet-assisted reticle stage.

1. Introduction

Collaborative exposing requires the reticle stage and the wafer stage to derive the specified positon, velocity and acceleration respectively at the moment for exposure. However, the acceleration and deceleration trajectory of the magnet-assisted reticle stage mainly depends on the force created by the magnet structures. This force varies exponentially with the relative position of the magnet opponents, which would yield an acceleration and deceleration trajectory of the exponential type, and this may not satisfy the requirements for collaborative exposure. To deal with this, the scanning motor in the reticle stage is used to produce a compensating force, and a new acceleration and deceleration trajectory of the reticle stage is generated to meet the demands for collaborative exposing. Furthermore, the residual vibration of the reticle stage caused by the trajectory is minimized to reduce the settling time before exposure starts, which increases the efficiency of the reticle stage. However, the peak force of the motor should be limited since the force ripples in the motor will deteriorate the tracking performance. In this paper, the integrated optimization incorporating the design of the peak force of the motor as well as the trajectory planning is proposed to derive the desired trajectory of the reticle stage.

2. Method

In position control of robotic manipulators or other track accessing operations, it is expected to achieve a finite rigid-body displacement with minimum power dissipation in the actuator or minimum excitation of the flexible modes. This is referred to as the point to point control problem [1]. In this paper, we wish the residual vibration of the reticle stage is as small as possible so as to reduce the settling time. The most common technique for reducing residual vibration is to profiler the control input by a low-pass or a notch filter. However, this method could only deal with the specified resonant frequencies [1]. Unfortunately, regardless of the number of flexible modes that is taken into account, there will always be higher frequency resonances that contribute to residual vibration. This problem can be solved by the minimum jerk solution [2]. The criterion can be formulated as Eq. (1).

$$J(u(t)) = \|jerk\|_q \tag{1}$$

while $\|\bullet\|_q$ can be any in R^n norm. $u(t)$ is the driving force. in this paper q is chosen to be 2 for practical purposes. To guarantee that the driving force $u(t)$ is a minimum jerk solution, $u(t)$ must be continuous at the beginning and the end of the trajectory[1], i.e. and, in which t is the execution time. To meet the requirements for collaborative exposing, the acceleration and deceleration should be executed within time t, and the positon, the velocity and the acceleration reference trajectory should satisfy the constraints for exposure. Since the trajectory is strictly subject to an exponential-type magnet force, the compensating force in the motor should be adopted to generate the desired trajectory. The magnet force varies with the position of the magnet opponents exponentially, which implies that the magnet force cannot be reduced to zero in the scanning period, therefore, the compensating force should keep working to maintain the zero acceleration during this time. However, the force ripples in the motor will adversely affect the tracking performance and this effect will grow as the force in the motor grows. Therefore, the peak force in the motor should be designed to achieve both the desired trajectory and tracking performance. To solve this problem, an integrated optimization method is adopted in this paper. The peak force $f_{\max}$ is designed and verified for precision through dynamic analysis in the outer-loop, and the trajectory planning is studied under the constraint of the given peak force in the inner loop. The scheme is as follows.

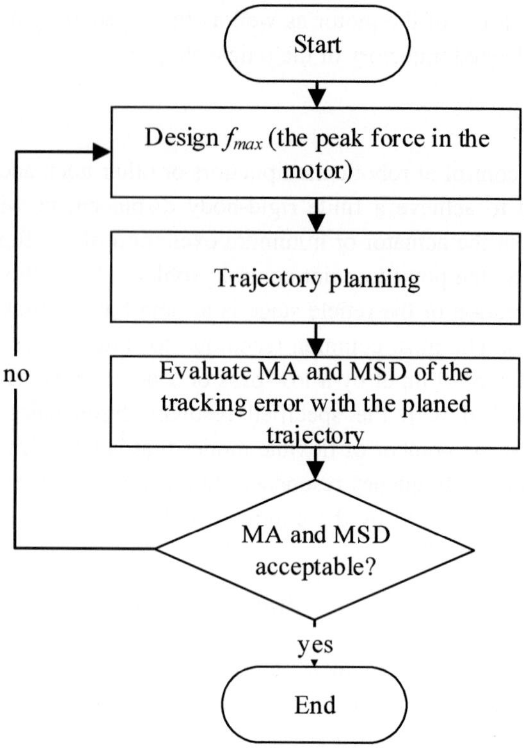

Fig. 1. Scheme of the integrated optimization.

Given the peak force in the motor $f_{\max}$ as one of the constraint, the trajectory planning can be defined as an optimization problem as follows.

$$\begin{aligned} &\min \|jerk\|_2 \\ &s.t. pos(T) = s_1 \ vel(T) = v_1 \ acc(T) = 0 \\ &\delta u(0) = 0 \ \delta u(T) = 0 \\ &\|u - g(pos)\|_\infty \le f_{\max} \end{aligned} \tag{2}$$

In which s_1 and v_1 are the desired positon and velocity for exposure, $g(pos)$ is the magnet force of a function with position. Nonlinear Programming is an effective approach to numerically solve such optimization problem [3]. The numerical solution discretizes the time domain in N steps, which creates N corresponding force inputs and N moving parameters. The nonlinear programing can be formed as

$$
\begin{aligned}
&\min F(x) \\
&s.t.\ \ l_x \le x \le h_x \\
&Ax = 0\ \ b_l \le Bx \le b_u \\
&c_i(x) = 0\ \ c_d(x) \le 0
\end{aligned}
\tag{3}
$$

Solving the optimization problem could yield the desired trajectory and the driving force in the motor. The force ripples in the motor then is verified for precision through the approach of dynamic analysis. This approach considers the effect of various disturbances acting on the closed loop of the reticle stage, and evaluate the performance through transfer functions from the disturbance and the tracking error.

3. Application

In the scanning period, the performance of the reticle stage is evaluated by MA and MSD, which represents the lower frequency and the higher frequency of the tracking error [4]. According to the specifications of the 65nm reticle stage, MA should be less than 3nm and MSD should be less than 8nm. The position of the fixed magnet is zero, the initial position of the stage is 0.01m, the initial velocity and acceleration is zero. The acceleration and deceleration should be executed in 20.8ms, and the positon should reach 0.0243m, the velocity should reach 1m/s and the acceleration should be reduced to zero. The magnet force is expressed as Lists may be laid out with each item marked by a dot:

$$
f = f_0 e^{-\frac{2\pi}{l_m}(pos - pos\_ref)}
\tag{4}
$$

In which f_0 is the coefficient, l_m is the period of the magnet arrays, pos is the position of the moving magnet, and pos_ref is the positon of the fixed magnet, which is set to zero.

The main disturbances that will cause tracking error in the scanning period are the driver noise d_1, the vibration where sensors are located d_2 and the force ripples in the motor d_3. DA quantization noise is small compared to the driver noise and is neglected in this paper. The reticle stage adopts the interferometer as measuring system, and the sensor noise is also neglected. The propagation of the disturbances to the tracking error is illustrated as in Fig. 2.

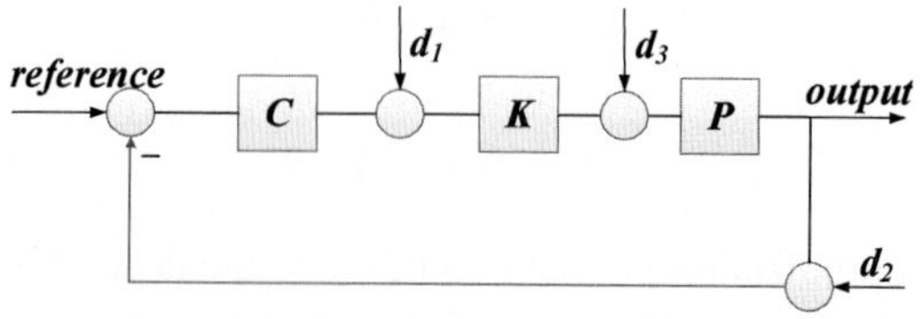

Fig. 2. Propagation of the disturbances to the tracking error.

According to the previous research, dynamic analysis can be used to evaluate the tracking performance. PID compensators with classical parameters setting is adopted as the controller C, the motor is modeled as a force constant, and the plant is treated as a rigid body. The tracking error caused by the driver noise and the vibration where sensors are located is calculated as follows.

$$error = \frac{d_1 KP}{1+CKP} + \frac{d_2}{1+CKP} + \frac{d_3 P}{1+CKP} = \frac{d_1 \frac{K}{m} + d_2 + d_3 \frac{1}{m}}{s^2\left(s^2 + \frac{k_p K}{m}\left(1 + f_i / s + s / f_d\right)\right)} \tag{5}$$

Apply the methods proposed above to solve the integrated optimization problem. It was shown that with the peak force in the motor designed to be 100N, the derived trajectory can satisfy the requirements for collaborative exposing, which is shown as in Fig. 3. The corresponding MA and MSD is less than 2nm and less than 6nm respectively, which also meet the requirements for tracking performance.

4. Conclusion

Collaborative exposing requires the reticle stage and the wafer stage to derive the specified positon, velocity and acceleration respectively at the moment for exposure. However, the magnet force may not satisfy these requirements. To deal with this, the scanning motor in the reticle stage is used as a compensation to generate a new acceleration and deceleration trajectory which could meet the demands for collaborative exposing and minimize the residual vibration of the reticle stage. The peak force of the motor which is a constraint to the trajectory planning is integrated in the optimization to achieve both the desired trajectory and tracking precision. The integrated optimization method is applied to the trajectory planning of the 65nm magnet-assisted reticle stage to derive the desired

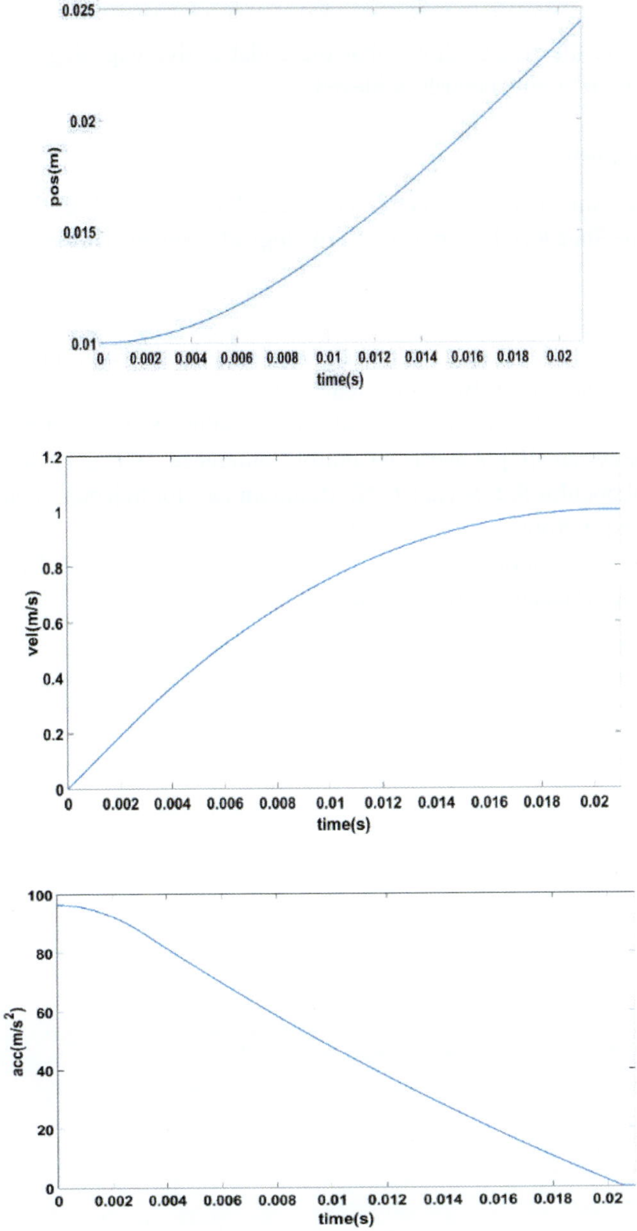

Fig. 3. The derived trajectory (position, velocity and acceleration from top to bottom).

trajectory, and the results shows that the collaborative exposing and tracking precision can be simultaneously achieved.

Acknowledgment

This research was financially supported by the Autonomous Scientific Research Project of the State Key Laboratory of Tribology of Tsinghua University at China.

References

1. Miu D K. Mechatronics: electromechanics and contromechanics[M]. Springer Science & Business Media, 2012.
2. Skaar S B. Closed form optimal control solutions for continuous linear elastic systems[C]//American Control Conference. 1984 (21): 1653-1658.
3. Kyriakopoulos K J, Saridis G N. Minimum jerk for trajectory planning and control [J]. Robotica, 1994, 12(02): 109-113.
4. Butler H. Position control in lithographic equipment [J]. IEEE Control Systems Magazine, 2011, 31(5): 28-47.

Load performance analysis and parameter optimization of gas bearing textured with micro-dimples

Yanjun Lu*, Shasha Yang, Cheng Liu and Sha Li

School of Mechanical and Precision Instrument Engineering, Xi'an University of Technology, Xi'an 710048, China

**E-mail: yanjunlu@xaut.edu.cn*

Micro-dimple texturing is an effective approach to improve the load-carrying performance of mechanical components. The influences of different micro-dimples on the load-carrying capacity of gas bearing were studied based on the compressible Reynolds equation. The finite difference method was used to solve the equation. The load-carrying forces of gas bearing textured by different dimples varied with different geometric parameters are compared. Meanwhile, the parameter optimization of textures was implemented by GA (genetic algorithm). The results indicate that the load-carrying capacity of gas bearing is increased and the friction force is reduced by choosing optimal texture area density, depth and radius. The model and numerical results can provide a theoretical direction to the design of practical gas-lubricated bearing-rotor system.

Keywords: Gas bearing; load-carrying capacity; surface texture; optimization.

1. Introduction

In recent years, gas bearings have been widely used in the machine tools, high speed and precision rotary machinery, due to their high accuracy, low friction. Since gas bearings are usually applied to the light load situations, hence how to improve the load-carrying capacity of gas bearings becomes an important problem in the application and development of gas bearing with higher speed. The surface texturing technology can improve the load-carrying performance of journal bearings by texturing certain shape and distribution of pits and micro dimples [1]. Therefore, the texturing parameters of micro-dimples have to be optimized to obtain maximum load-carrying capacity of journal bearing.

Since the 1960s D.B. Hamilton et al. put forward the form of face texturing, many surface texturing studies have been implemented on the sliding surfaces of bearings and mechanical seals [2]. L.P. Shi et al. compared the effects of micro-grooves and micro-dimples on the load-carrying performance of mechanical gas seals [3]. The results indicated that various load-carrying capacities can be obtained by different texturing parameters. M.F. Qiu et al. used the compressible Reynolds equation to predict the pressure of the gas lubricated bearing with grooves, and simulated the lubrication performance of gas bearings with different texture forms [4].

In order to obtain the maximum load-carrying capacity of the bearing, the texture parameters of micro-dimples have to be optimized. C.B. Ma et al. investigated the influence of the elliptical-shape dimples with various depths, diameters, area ratios on friction coefficient, and optimized the texture parameters of air bearing by the method of "trial and error" [5]. M. Fesanghary et al. studied the effect of surface grooves on the load-carrying capacity with optimization methods. The results showed that optimally-designed grooves can provide up to 36% more load carrying capacity by comparing to the conventional ones [6].In this paper, the influences of circular dimples and square dimples with various depths, radii, area densities on the load-carrying capacity of gas bearing are investigated, meanwhile, the parameters of dimples are optimized by GA method.

2. Analytical Model

2.1. *Model description*

Cylindrical gas bearing with surface texture is modeled. The schematic description of the model is shown in Fig.1. W is the load of the rotor, ω is the angular velocity of the rotor, O_b is the center of gas bearing, O_j is the center of the rotor, R is the radius of the rotor, c is the clearance between the bushing and the journal, e is the eccentricity, θ is the eccentric angle, ϕ is the angle between the negative direction of the y axis and the gas film location.

2.2. *Reynolds equation of gas lubricated bearing with texture*

The steady Reynolds equation describing the flow in gas film journal bearing with texture can be written as Eq. (1). The corresponding assumptions are: (i) the inertia force is smaller than viscous force and can be neglected; (ii) gas is compressible and viscous with a constant viscosity; (iii) gas flow is ideal and laminar.

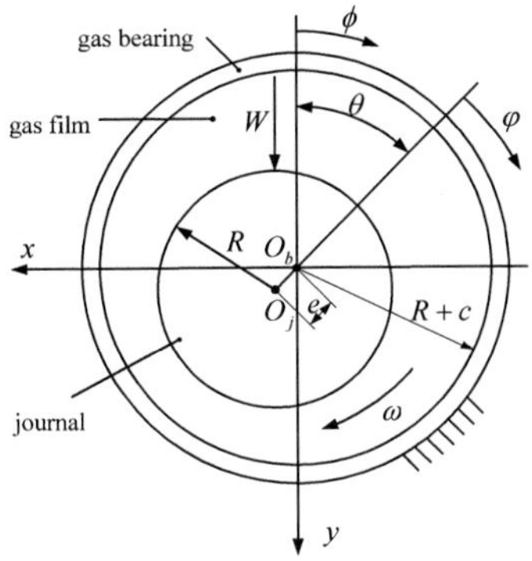

Fig. 1. Schematic diagram of cylindrical gas bearing.

$$\frac{\partial}{\partial x}\left(\rho h^3 \frac{\partial p}{\partial x}\right)+\frac{\partial}{\partial z}\left(\rho h^3 \frac{\partial p}{\partial z}\right)=6\mu U\frac{\partial(\rho h)}{\partial x} \tag{1}$$

where x and z are the coordinates in the circumferential and axial directions, ρ is gas density, p is gas film pressure, μ is the gas viscosity and U is the circumferential velocity of the surface of the bearing journal. h is the gas gap between the journal and the bushing.

Let us define dimensionless variables as in Eq. (2).

$$x = R\phi, z = R\lambda, U = \omega R,\ p = p_a P, h = cH \tag{2}$$

where p_a is the ambient pressure.

The dimensionless compressible Reynolds equation is written as in Eq. (3).

$$\frac{\partial}{\partial \phi}(PH^3\frac{\partial P}{\partial \phi})+\frac{\partial}{\partial \lambda}(PH^3\frac{\partial P}{\partial \lambda})=\Lambda\frac{\partial(PH)}{\partial \phi} \tag{3}$$

where $\Lambda = (6\mu\omega/p_a)(R/c)^2$.

In this study, two different dimples are discussed: circular dimples and square dimples. Micro-dimples are arranged on the surface of the bushing in the circumferential direction. The thickness of gas film h consists of the gas film thickness without dimples h_0 and the depth of the dimples h_d. The parameters of a single dimple are shown in Fig. 2. h_d is the dimple depth, r_d is the circular dimple radius.

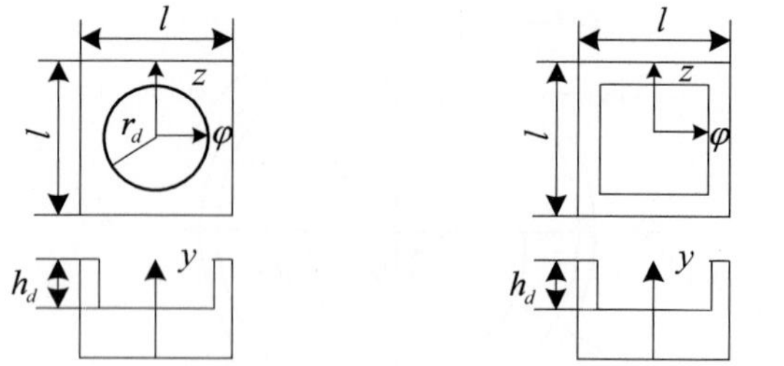

(a)single circular dimple (b)single square dimple

Fig. 2. The geometrical model of single dimples.

To simplify the calculation, the relative depth H_d, relative radius R_d and area density S_p are defined as in Eq. (4).

$$R_d = \frac{r_d}{R},\ H_d = \frac{h_d}{c},\ S_p = T/S \tag{4}$$

where T is the texturing area, S is the bearing surface.

The thickness of gas film is written as in Eq. (5).

$$h(x,z) = \begin{cases} h_0 & (x,z) \notin \Omega \\ h_0 + h_d & (x,z) \in \Omega \end{cases} \tag{5}$$

where Ω is the texturing field.

The pressure boundary conditions of gas bearing with dimples are: gas pressure on the both ends of the bushing equal to the ambient pressure p_a, gas pressure P is a periodic function.

The dimensionless radial gas film force F_n and the tangential gas film force F_t are obtained as in Eq. (6).

$$\begin{cases} W_n = -\int_{-\frac{B}{2R}}^{\frac{B}{2R}} \int_0^{2\pi} (P-1.0)\cos\phi d\phi d\lambda \\ W_t = -\int_{-\frac{B}{2R}}^{\frac{B}{2R}} \int_0^{2\pi} (P-1.0)\sin\phi d\phi d\lambda \end{cases} \tag{6}$$

The dimensionless of gas film force is obtained as in Eq. (7).

$$W = \sqrt{W_n^2 + W_t^2} \tag{7}$$

3. Results and Discussions

The geometric parameters of gas bearing are: the width of the bearing B=110mm, R=40mm, c=0.432mm, e=0.3024. Fig. 3 illustrates the effects of texture depth H_d on the load-carrying force with different dimple shapes. As shown in Fig. 3, the load-carrying force increases and then decreases with the increase of the depth. Fig. 4 depicts the effects of area density S_p on the load-carrying force with different dimple shapes. As shown in Fig. 4, the load-carrying force decrease with the increase of area density. Fig. 5 describes the effects of texture radius R_d on the load-carrying force. The load-carrying force decreases and then increases with the increase of the radius.

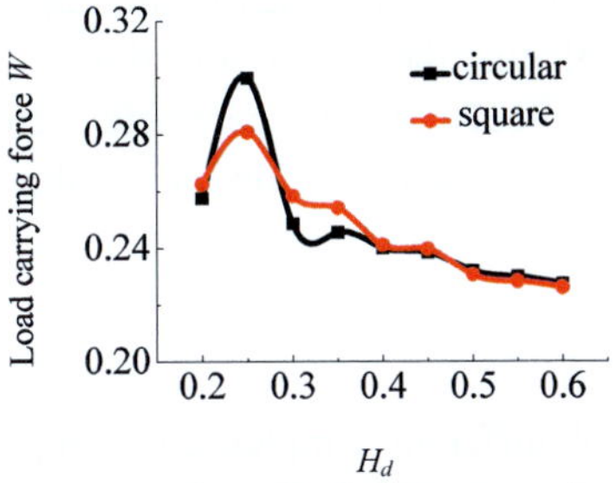

Fig. 3. Load-carrying force W versus the relative depth Hd for the case of Rd=0.2,Sp=0.36.

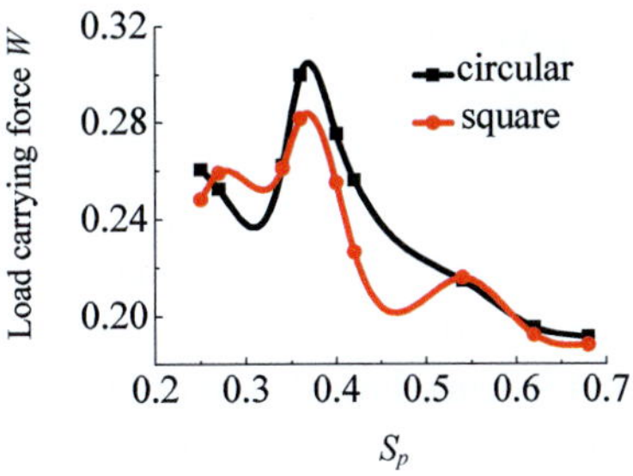

Fig. 4. Load-carrying force W versus to the relative area density S_p for the case of H_d=0.2, R_d=0.2.

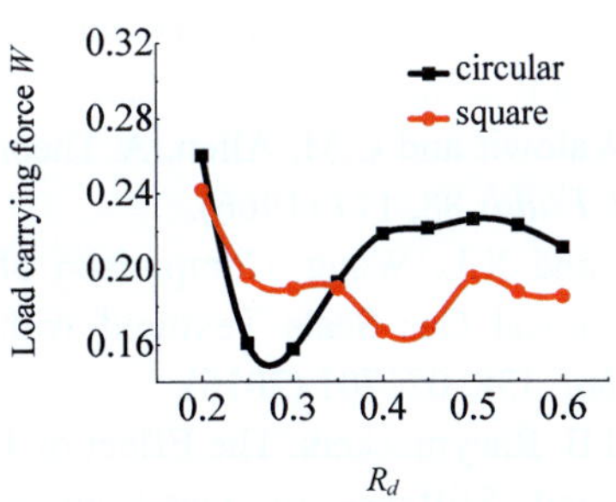

Fig. 5. Load-carrying force W versus the relative radius R_d for the case of H_d=0.2, S_p=0.36.

4. Optimum Texture Parameters

The key of this study is to optimize the texture parameters that can generate the maximum load-carrying force by using genetic algorithm. H_d, R_d, S_p are the population individuals and the load-carrying force of gas bearing with texture is the objective function in genetic algorithm. The process of selection, crossover and mutation are conducted and consequently the optimization of texture parameters is obtained by imitating the natural selection.

The optimal values of circular dimple parameters are obtained when H_d =0.4115, R_d=1.2176, S_p=0.7059, and the optimal values of square dimple parameters are obtained when H_d =0.447, S_p=0.5718.

5. Conclusions

The load-carrying forces of different dimples are compared and the texture parameters, such as area density S_p, depth H_d and R_d radius are optimized. The numerical results show that the texture parameters have a great influence on the load-carrying performance of gas bearing, and micro-dimples can significantly improve the load-carrying capacity of gas bearing.

Acknowledgments

This work is supported by Natural Science Foundation of China (Grant No. 51505375), Natural Science Foundation of Shaanxi Province of China (Grant No. 2014JM2-5082), Natural Science Foundation of Department of Education of Shaanxi Province of China (Grant Nos. 15JS068, 15JK1549).

References

1. D.M. Lei and S.H. Wang, Effect of Surface Texture on Mixed Lubrication Characteristics of Journal Bearings, *Bearing* 138, 36 (2013).
2. D.B. Hamilton, J.A. Walowit and C.M. Allen, A Theory of Lubrication by Micro-Irregularities, *J. Fluids* 88, 177 (1966).
3. L.P. Shi, X.Y. Wang and X.L. Wang, Comparison of the Load-Carrying Performance of Mechanical Gas Seals Textured with Microgrooves and Micro Dimples, *J. Tribol.* 138, 021701 (2015).
4. M.F Qiu, A. Delic and B. Raeymaekers, The Effect of Texture Shape on the Friction Coefficient and Stiffness of Gas-Lubricated Parallel Slider Bearings, *Tribol. Trans.* 55, 210 (2015).

5. C.B. Ma and H. Zhu, An Optimum Design Model for Textured Surface with Elliptical-Shape Dimples under Hydrodynamic Lubrication, *Tribol. Int.* 44, 987 (2011).
6. M. Fesanghary and M.M.Khonsari, On the Optimum Grooves Shapes for Load-Carrying Capacity Enhancement in Parallel Flat Surface Bearings: Theory and Experiment, *Tribol. Int.* 67, 254 (2013).

A newly developed R-Theta-Z motion stage for high-speed wafer inspection system

Pu-Liang Yu, Wei Jiang*, Xiao-Li Wang and Qiu-Cheng Wen

State Key Laboratory of Digital Manufacturing Equipment and Technology, Huazhong University of Science and Technology, Wuhan, Hubei, P. R. China

**E-mail: jiangw@hust.edu.cn*

Wei Jiang

School of Mechanical Science and Engineering, State Key Laboratory of Digital Manufacturing Equipment and Technology, Hongshan District, Wuhan City, Hubei Province Luo Yu Road, No. 1037

Wuhan, Hubei, P. R. China

E-mail: jiangw@hust.edu.cn

A newly developed R-Theta-Z motion stage included a horizontal translational mechanism, a lifting mechanism and a rotational mechanism was designed for high-precision and high-speed wafer inspection. A spiral path which was employed to sample the entire wafer surface was generation of the motion of the horizontal translational mechanism and the rotational mechanism. The positioning performance of the motion stage was evaluated through a series of positioning experiments. The positioning accuracy of the horizontal translational mechanism is less than ±2μm along the X-axis at a speed of about 1.2mm/s. The positioning accuracy of the rotational mechanism is less than ±2arcsec along the Theta-axis at a speed of about 12 revolutions per second. The axial error of the rotational mechanism along Z-axis is less than ±1μm at a speed of about 15mm/s. The experimental results demonstrate that the motion stage has a superior positioning performance.

Keywords: R-Theta-Z motion stage; spiral path; positioning performance.

1. Introduction

As the moving and positioning carrier of the wafer inspection system, precision motion stages are key components of the wafer inspection system. As the feature size continuously reduces and the wafer dimensions ever grows, there has been an increasing need of precision scanning stage to meet the demand requires the development of new design concepts and techniques.

Previously, Linear mechanisms with a lead-screw, linear guides and a rotary motor, have been widely applied in precision motion stages[1,2,3]. The raster scan motion can also be achieved by XY motion stage which are an orthogonal combination of two linear mechanisms. However, such XY stages have some deficiencies in terms of precision positioning. The XY motion stage with large inertia is not suitable for fast movement, and frequently alternates direction reduces the stability and accuracy of positioning. They are also not fit for clean-use due to friction and wear, which also causes positioning accuracy degradation over time.

To get over these disadvantages of conventional XY motion stages, The combination of a linear mechanism and a rotatory mechanism has been developed[4,5]. Such stages which can generate spiral scan motion specifically for wafer inspection system have no reciprocating motion and frequently alternates direction. However, the traditional rotatory mechanism is supported with mechanical bearings in radial direction and axial direction. They may cause a degradation of positioning accuracy over time due to friction and wear. So, aerostatic bearing spindles have been widely applied for precision motion stages because of merits of high speed, high precision, no friction heat and no pollution. Furthermore, aerostatic bearing spindle should be designed for a rotatory mechanism. However, aerostatic bearings can easily reduce the capacity and rigidity and even cause instability due to the compressibility of gas. In the previous study, the dynamic characteristics of aerostatic bearings have been studied in different approaches, such as CFD simulation[6,7], numerical methods[8,9]. Numerical simulations and experiments carried out by Chen et al.[10,11] have proved that vortices existed in the recess which might cause instabilities of aerostatic bearings. Chen et al.[12] studied the effects of recess shapes on the phenomenon of pneumatic hammering in aerostatic bearings, and pointed out that the recess shape significantly affected the vibration status of the pneumatic hammer. Zhu et al.[13] investigated the transient flow field in the aerostatic bearing using the large eddy simulation method, and further discovered vortex shedding phenomenon in the bearing recess. Computational results demonstrate that vortex shedding causes pressure fluctuation in the bearing clearance, which might cause nano-vibration of aerostatic bearings. Yu et al.[14] investigated the influence of perturbations on the dynamic stiffness of aerostatic bearings by numerical simulations and experiments.

In this paper, a novel design of R-Theta-Z motion stages with aerostatic bearing spindle is presented for 300mm wafer inspection system. The design and construction of the precision motion stage are presented in detail, and the results of experiments performed to evaluate the performance of the stage system.

2. Design of R-Theta-Z Motion Stage

2.1. *Scanning strategy*

The R-Theta-Z motion stage should be designed for suppling wafer to, inspecting wafer at, and extracting wafer from the R-Theta-Z motion stage. The motions of the horizontal translational mechanism along the X-axis and the rotational mechanism along the vertical axis make up the wafer inspecting motion. The motion stroke of horizontal translational mechanism is up to 180 mm, and the accuracy should be less than ±2μm. The maximum speed is 100mm/s, but the inspecting speed is less than 10mm/s. The motion of the lifting mechanism is designed for suppling wafer, extracting wafer and focusing control of the optical inspection sensor. The motion stroke of the lifting mechanism is up to 20 mm for coupling to the wafer, and the motion accuracy of the lifting mechanism for coupling to the wafer is not very high. But the lifting mechanism should drive the wafer moves up for quickly accurate focusing, and the position accuracy of the lifting mechanism should be less than ±0.5μm. Therefore, those questions have put forward a challenge for the design of the R-Theta-Z motion stage.

The R-Theta-Z motion stage may follow the same order: 1) a wafer handler extracts a first wafer from a wafer carrier and supplies the wafer to the R-Theta-Z motion stage; 2) the wafer handler return, and the wafer vacuum chuck moves up for coupling to the wafer; 3) the wafer is transferred to the position just below an optical inspection sensor, and the vacuum chuck moves up and down for focus; 4) the inspection device inspects the first wafer, while the wafer revolves around the Z-axis at the same time translates along the X-axis under the optical inspection sensor; 5) the wafer return to the starting position, and the wafer vacuum chuck moves down for decoupling to the wafer; 6) the wafer handler returns the inspected first wafer to the wafer carrier. Thereafter the order repeats and the wafer handler extracts a second wafer from the wafer carrier and transports the second wafer to the R-Theta-Z motion stage, etc.

The motion equations of spiral-scan which is employed to scan the entire wafer surface are defined as follow:

$$x = vt\cos(\omega t) \tag{1}$$

$$y = vt\sin(\omega t) \tag{2}$$

$$v = \omega d / 2N\pi \tag{3}$$

where: x is the coordinate in X-axis. y is the coordinate in Y-axis. v is the translational speed of the vacuum chuck. ω is the rotational speed of the vacuum chuck. d is diameter of the laser spot which can be adjusted by using different apertures. N is number of the optical detection unit.

In order to improve the motion stage accuracy cost-effectively for wafer surface metrology, the motion stage errors have to be considered and predicted for error compensation.

$$x(t) = (v + \Delta v)t\cos\left[(\omega + \Delta\omega)t\right] + \Delta e_x(t) \tag{4}$$

$$y(t) = (v + \Delta v)t\sin\left[(\omega + \Delta\omega)t\right] + \Delta e_y(t) \tag{5}$$

where, Δv is X-axis velocity error; $\Delta\omega$ is Theta-axis velocity error ; $\Delta e_x(t)$ is the radial errors along X-axis; $\Delta e_y(t)$ is the radial errors along Y-axis.

2.2. *Stage design*

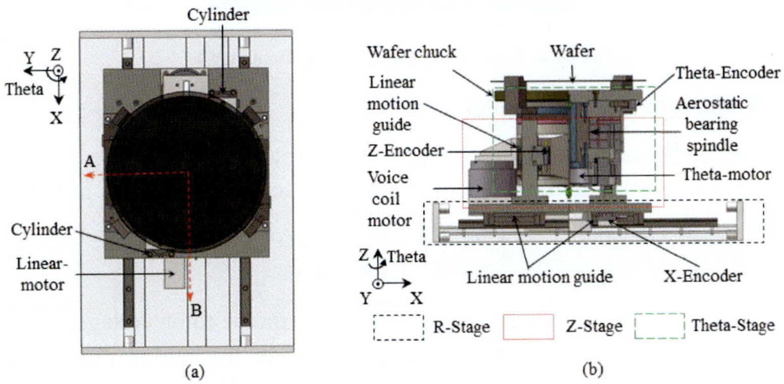

Fig. 1. Schematic view of R-Theta-Z motion stage: (a) Top view; (b) A–B cross-sectional view.

Figure 1 shows the schematic view of the proposed stage which is called R-Theta-Z motion stage, and a photograph of the prototype is shown in Figure 2. Most of the structural modules were made of aircraft-grade aluminium alloy (7075) for weight reduction, high stiffness and low thermal expansion coefficient. The R-Theta-Z motion stage is mainly composed of a horizontal motion mechanism, a vertical lifting mechanism and a rotation mechanism. The horizontal motion mechanism is mainly composed of linear glass scales, linear motor, linear rolling guide, etc. Two guideway and four slider bearings are installed according to H style. The linear motor is installed at the centre of R-stage for mass centre driving. The vertical lifting mechanism is mainly composed of linear glass scales, voice coil linear motors, cylinders and precision pressure reducing valve, etc. Four guideway and four slider bearings are installed at either of four corners respectively. Two air cylinders are installed for gravity balance. Precision pressure reducing valve is applied for precision pressure. The rotation mechanism is mainly composed of a circular glass scale, an aerostatic bearing spindle, a brushless motor, a vacuum chuck, a precision pressure reducing valve, a wafer support and a high-speed rotary joint, etc.

Aerostatic bearing spindle includes two thrust bearing and one journal bearing. The vacuum chuck has a wafer supporting surface for coupling to the wafer, when the vacuum chuck moves up and down.

Fig. 2. Fabricated prototypes (top-front view).

2.3. *Compound control strategy*

Based on the Permanent-Magnet Synchronous Motor (PMSM) vector control, the volt-age equation in the d-q coordinate system can be defined as follow:

$$\begin{cases} u_d = R_s i_d + p\psi_d - \omega_d \psi_q \\ u_q = R_s i_q + p\psi_q + \omega_d \psi_d \end{cases} \tag{6}$$

where, u_d is motor rotor voltage in *d*-axis; u_q is motor rotor voltage in *q*-axis; R_s is the motor rotor winding resistances; ψ_d is the motor rotor winding flux linkage in *d*-axis; ψ_q is the motor rotor winding flux linkage in *q*-axis; ω_d is the motor rotor rotational angular velocity around *d*-axis ; i_d is the motor rotor winding current in *d*-axis ; i_q is the motor rotor winding current in *q*-axis. Flux linkage equation can be written as

$$\begin{cases} \psi_d = L_d i_d + \psi_f \\ \psi_q = L_q i_q \end{cases} \tag{7}$$

where, L_d is motor rotor winding inductance in *d*-axis; L_q is motor rotor winding inductance in *q*-axis; ψ_f is fundamental wave flux linkage of permanent magnet. The electromagnetic thrust equation of permanent-magnet synchronous motor can be given as follows:

$$F_{em} = \frac{3\pi}{2\tau}\left(\psi_f i_q + (L_d - L_q) i_q i_d\right) \tag{8}$$

Based on the basic principle of vector control, the control strategy ($i_d = 0$) is adopted. The equation can be simplified as

$$F_{em} = k_f i_q \tag{9}$$

where, $k_f = 3\pi\psi_f / 2\tau$ is the electromagnetic thrust coefficients.

Dynamic equation for permanent magnet synchronous motor can be obtained:

$$F_{em} = M\ddot{\theta} + F_L + F_f \tag{10}$$

where, M is the load mass; F_L is load resistance; F_f is the friction; $\ddot{\theta}$ is motor acceleration.

A compound controller which combines feedback and feedforward controls are presented based on the vector control. Figure 3 shows a block diagram of the compound controller.

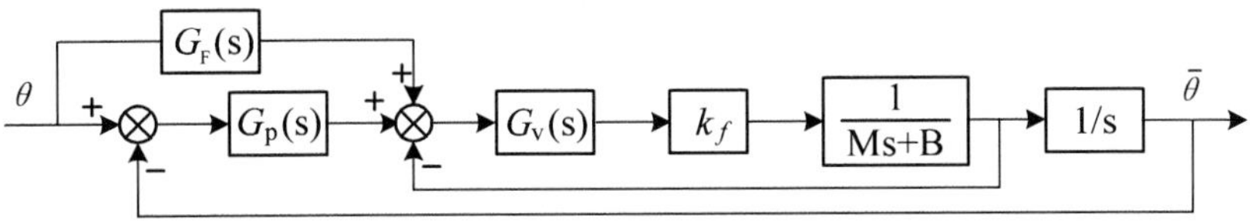

Fig. 3. Block diagram of compound controller.

The closed-loop transfer function of the compound controller can be written as

$$G(s) = \frac{G_F(s)G_V(s)k_f + G_P(s)G_V(s)k_f}{Ms^2 + Bs + G_V(s)k_f s + G_P(s)G_V(s)k_f} \tag{11}$$

where, M is the load mass; $G_F(s)$ is feedforward controller, which is composed of velocity feedforward control and acceleration feedforward control.; $G_V(s)$ is P controller of position loop; $G_P(s)$ is PI controller of velocity loop; B is the friction coefficient, which of aerostatic bearing spindle is equal to zero.

3. Results and Discussion

3.1. *Positioning performances*

The positioning performances of the R-Theta-Z motion stage was performed with the proposed controller. Figure 4 shows positioning performance of R-stage at four different velocities (0.6mm/s, 0.8mm/s, 1.0mm/s and 1.2mm/s). It can be clearly observed that the positioning accuracies are less than ±2μm on the x-axis. Figure 5 shows positioning tracking performance of Theta-stage at four different velocities (6r/s, 8r/s, 10r/s and 12r/s). It shows that the positioning tracking accuracies are less than ±2arcsec on the Theta-axis. Figure 6 show that the positioning errors are less than ±0.4μm along the Z-axis.

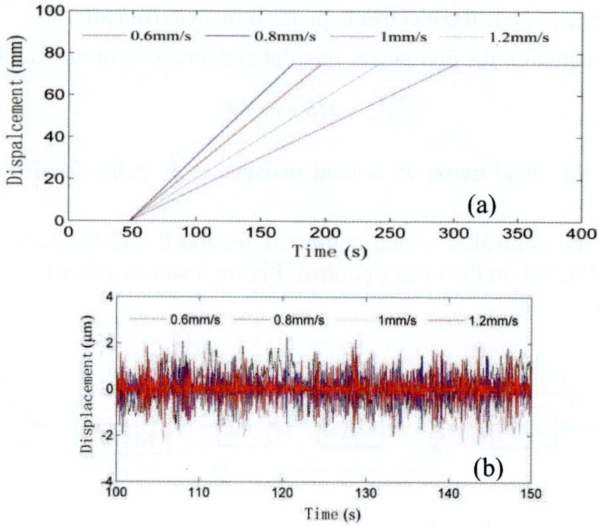

Fig. 4. Positioning performance on the R-stage: (a) Displacement profile; (b) Tracking error.

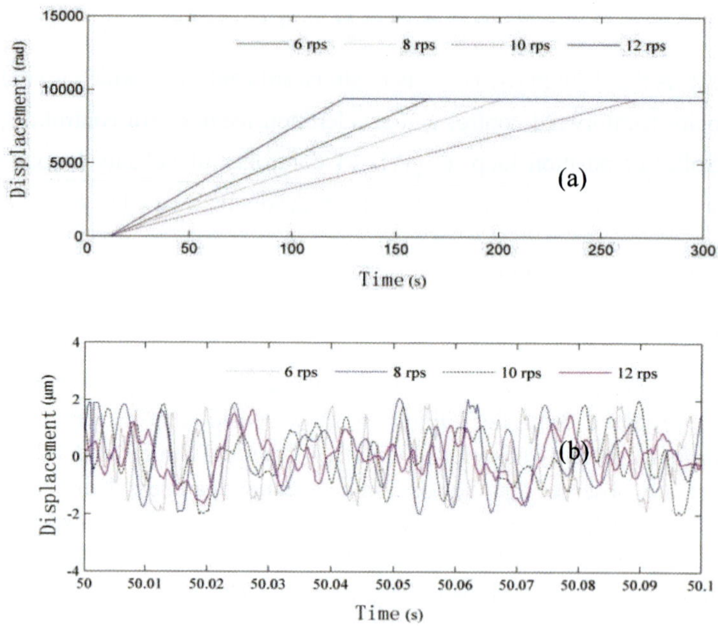

Fig. 5. Positioning performance on the Theta-stage: (a) Displacement profile; (b) Tracking error.

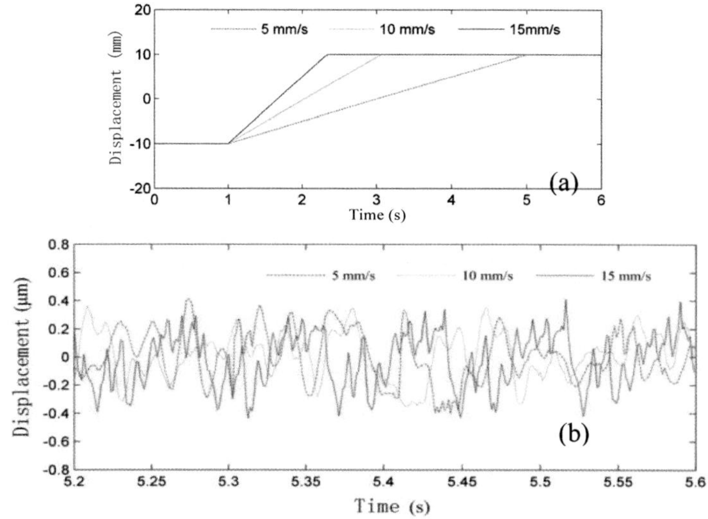

Fig. 6. Positioning performance on the z-stage: (a) Displacement profile; (b) Positioning error.

3.2. *Run out error of aerostatic bearing spindle*

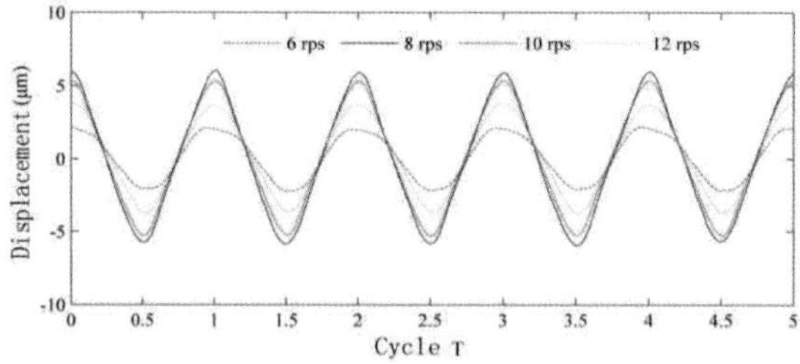

Fig. 7. Radial error polar plot along X-axis.

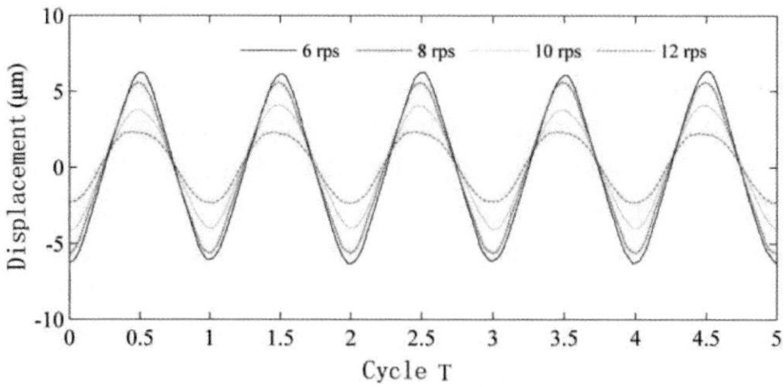

Fig. 8. Radial error polar plot along Y-axis.

The radial errors of the vacuum chuck have effect on the positioning performance of the wafer. The experiments were performed at five different rotate speeds. As shown in Figure 7 and Figure 8. The radial errors along X-axis and Y-axis are less than ±6μm. It can be clearly observed that the radial errors along X-axis and Y-axial decrease gradually with the increase of speeds. It can be infer that the radial motion error will be decrease with increasing rotational speed due to "squeeze effect" of the film. As shown in Figure 9. The axial errors along Z-axis are less than ±1μm. The axial error along Z- axis doesn't increase with the increase of speed.

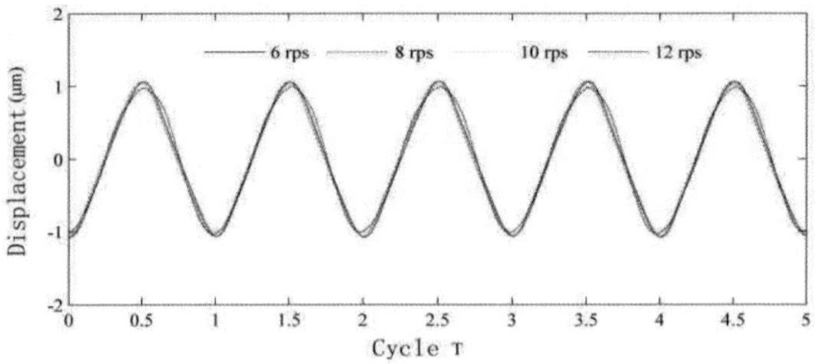

Fig. 9. Axial error polar plot along Z-axis.

4. Conclusion

The R-Theta-Z motion stage with nanometer resolution is developed for 300mm wafer automatic inspection system. The positioning performance of the R-Theta-Z motion stage was evaluated by a series of experiments. As a result, the conclusions could be drawn as follow:

1. A new scanning strategy to generate a spiral path is employed to scan the entire wafer surface, while the wafer is rotated around the Z-axis and translated along the X-axis under the optical system.
2. A position and velocity double closed-up control algorithm with velocity/acceleration feedforward control and feedback control is proposed.
3. The positioning accuracies are less than ±2μm along the X-axis at the speed of about 1.2 mm/s. The positioning accuracies are less than ±2arcsec along the Theta-axis at the speed of about 12 revolutions per second. The positioning accuracies are less than ±0.5μm along the Z-axis at the speed of about 15 mm/s.
4. The radial errors along X-axis and Y-axis are respectively less than ±2μm at the speed of about 12 revolutions per second. The axial error along Z-axis is less than ±1μm at the speed of about 12 revolutions per second.

5. Experiments results confirmed that the R-Theta-Z motion stage has a superior positioning performance for 300mm wafer automatic inspect system.

Acknowledgments

This research was supported by the National Natural Science Foundation of China (No. 51435006, No. 51405174, No. 51235005, and No. 51421062).

References

1. M.C. Lin, S.A. Velinsky, B. Ravani, Design of the ball screw mechanism for optimal efficiency, *J. Mech. Design.* **116**, 856 (1994).
2. K.K. Varanasi and S.A. Nayfeh, The dynamics of lead-screw drives: low-order modeling and experiments, *J. Dyn. Control Syst.* **126**, 388 (2004).
3. M.F. Zaeh, T. Oertli, J. Milberg, Finite element modelling of ball screw feed drive systems, *Cirp Ann-Manuf. Tec.* **53**. 289 (2004).
4. R. Kendall, S. Doran, E. Weissmann, A servo guided X-Y-theta stage for electron beam lithography, *J. Vac. Sci. Technol.B.* **9**, 3019 (1991).
5. A. Joubair, M. Slamani, I. A. Bonev, A novel X-Y-Theta precision table and a geometric procedure for its kinematic calibration, *Robot. Cim-Int. Manuf.* **28**, 57 (2012).
6. Y. Li and H. Ding, Influences of the geometrical parameters of aerostatic thrust bearing with pocketed orifice-type restrictor on its performance, *Tribol. Int.* **40**, 1120 (2007).
7. M.E. Eleshaky, CFD investigation of pressure depressions in aerostatic circular thrust bearings, *Tribol. Int.* **42**, 1108 (2009).
8. T.S. Luong, W. Potze, J.B. Post, et al. Numerical and experimental analysis of aerostatic thrust bearings with porous restrictors, *Tribol. Int.* **37**, 825 (2004).
9. S. Yoshimoto, M. Yamamoto, K. Toda, Numerical calculations of pressure distribution in the bearing clearance of circular aerostatic thrust bearings with a single air supply inlet, *J. Tribol-T. Asme.* **129**, 384 (2007): 390.
10. X.D. Chen and X.M. He, The effect of the recess shape on performance analysis of the gas-lubricated bearing in optical lithography, *Tribol. Int.* **39**, 1336 (2006).
11. X.D. Chen, H. Chen, X. Luo, et al. Air vortices and nano-vibration of aerostatic bearings, Tribol. Lett. **42**, 179 (2011).
12. Y.X. Ye, X.D. Chen, Y. T. Hu, et al. Effects of recess shapes on pneumatic hammering in aerostatic bearings, *P. I. Mech. Eng. J-J. Eng.* **224**, 231 (2010).

13. J.C. Zhu, H. Chen, X.D. Chen, Large eddy simulation of vortex shedding and pressure fluctuation in aerostatic bearings, *J. Fluid. Struct.* **40**, 42 (2013).
14. P.L. Yu, X.D Chen, X.L. Wang, et al. Frequency-dependent nonlinear dynamic stiffness of aerostatic bearings subjected to external perturbations, Int. *J. Precis. Eng. Man.* **16**, 1771 (2015).

Research on Missile-borne Aerial Vehicle Group with a Variant of Bare-Bones particle swarm optimization

Hui Yang
Computer Science College,
Beijing Institute of Technology,
Beijing, China
zlj-1943@163.com

Shou-Jun Zhao
The Second Artillery Equipment Research Academy,
Beijing,China
Shoujun12@263.com

Lin-Lin Ci
The Second Artillery Equipment Research Academy,
Beijing, China
cilinlin@263.com

Ming-Hua Yang
The Second Artillery Equipment Research Academy,
Beijing, China
yang_mh@foxmail.com

In Missile-borne Aerial Vehicle Group (MAVG) field, previous attempts at improving the algorithm have only been able to produce better results in a handful of test cases, which is not suitable for rapid deployment in the real battle field. We introduced the concept of velocity and position into Bare-Bones particle swarm optimization (BBPSO), called Variant of Bare-Bones particle swarm optimization (VBBPSO), which consider the velocity and position term, these terms are additional constraint, but it doesn't damage the structure of simplicity of VBBPSO. In this paper, we present VBBPSO algorithm with mathmatics. Secondly, we have testified our thesis by several benchmark function and simulation in Matlab.

Keywords: MAVG; VBBPSO; conditional constraint.

1. Introduction

As a new style of Missile-borne Aerial Vehicle Group (MAVG) in military domain, MAVG is composed of multiple sensor-node embedded MAV, which can navigate freely in three-dimensionalspace. This kind of micro MAVG has many merits, such as low construction cost, small radar reflecting surface, infrared signature is not obvious, no extra communication infrastructure, high mobility, usually invisible to the human eyes and high-tech detection equipment, and so on. So MAV can carry out missions, for instance, close observation on enemy targets, it also can make up the problem of military reconnaissance satellite and early-warning aircraft's search blind spot in some particular regions that human could hardly reach. The MUAVs can achieve destruction appraisal, communications relay, war situation sensing, precision weapons guide, electronic counter measure and electronic protection in coordination.

Particle swarm optimization was introduced by Kennedy and Eberhart (1995)[1]. It has roots in the simulation of social behaviors using tools and ideas taken from computer graphics and social psychology research.It is a new optimization technique inspired by swarm intelligence. Like other evolutionary algorithms , PSO is also a population-based stochastic search algorithm,but it does not containany cross over or mutation operator.

There are five major problems among all the related researches in MAVG.

- After release MUAVs by missile platform, how to quickly build high-quality communication links within MAV coverage? In order to meet the information needs of the interaction between MUAVs[6].
- Secondly, how to faster and more efficiently to complete MAVG deployment in target field? grasp the initiative in battle, then attack the enemy in electromagnetic interference, and make coordinated plan to cope with.
- Additionally, different combat missions put forward several kinds of requirements on spatial deployment.
- Just like to the common type of 2-dimensional data acquisition, MAVG is also faced with how to save energy and improve anti-interference ability, in order to extend the network life cycle, when performs communication, situational awareness, operational effectiveness assessment in 3-D environment [14].
- These problems are equivalent to design a high-performance protocol to solve the wireless channel resource optimization, conflict avoidance and resolve interference issues in MAVG.

2. Bare-Bone PSO Model

Bare-Bones Particle Swarm Optimization (BBPSO) is a powerful algorithm, which has shown potential to solving multimodal optimization problems. Kennedy's Bare-Bones PSO was not the only method proposed but it is the simplest and very effective. This is the reason we investigate it further. This strategy was developed to escape from local minima using jumps with promising results[13].

Consider a simplified situation shown in Figure 1, which with only two particles .We know that particle A is the best one in the field, it will always stand in the same position a when particle B moves to the right side. If particle B too close to A, the PSO algorithum will become inactive. The particles will move very slowly in future iterations[7].

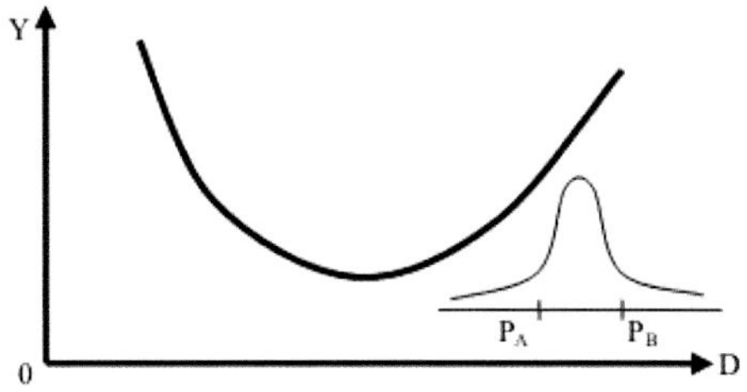

Fig. 1. An example with two paticles.

Fortunately, when the number of particles grows, efficiency of this algorithum becomes decreases. If the problem becomes complicated, determining an appropriate number of particles might be a difficult task because the growing size of swarm also increases the time of computation.

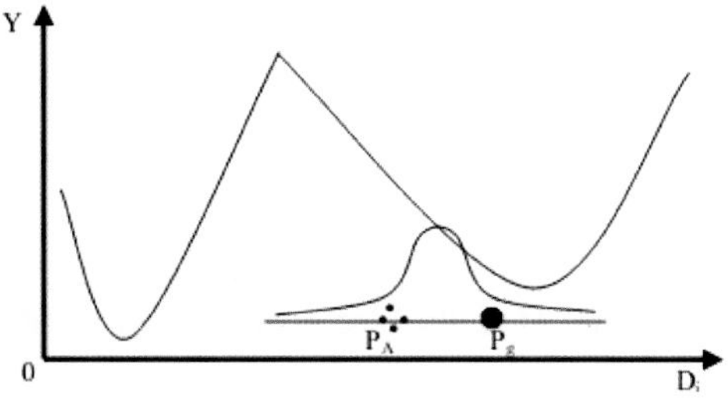

Fig. 2. Eeach paticle moving to the optimal solution.

Figure 2 shows all the particles move fast toward to the inside a local optimum area. When each of the particles fails to reach the region of the global optimum, the whole system might lose exploration capability as the standard deviation cannot be back to large value.

3. The Variant of Bare-Bone PSO

3.1. *Bare-Bone PSO*

Some of the most important developments. For a more detailed description of many of the existing particle swarm optimization variantssuch as Discrete PSO, Constriction Coefficient, Fully informed PSO and Bare-Bones PSO [1,3,8].

$$\begin{aligned} v_{ij}(t+1) &= \omega v_{ij}(t) + c_1 rand_1()[y_{ij}(t) - x_{ij}(t)] \\ &+ c_2 rand_2()[\hat{y}_{ij}(t) - x_{ij}(t)] \end{aligned} \tag{1}$$

Where $v_{ij}(t)$ for the current velocity of the ith particle. for the presently (j iterations) best position that particle i explores after j steps. $j = 1,2,\ldots,n$, ω is called the inertia weight. c_1 and c_2 are two cognitive and social parameters. $rand_1$ and $rand_2$ are two vectors of random numbers gaussian distributed

$$y_i(t+1) = \begin{cases} y_i(t) \text{ , if } f(x_i(t+1)) \geq f(y_i(t)) \\ x_i(t+1) \text{, if } f(x_i(t+1)) < f(y_i(t)) \end{cases} \tag{2}$$

Where the row vector x_{ij} stands for the current position of the ith particle in the system $j = 1,2,\ldots,n$. $y_i(t)$ stand for the particle's positon at time t.

3.2. *VBBPSO with velocity and position constraint*

The VBBPSO eliminates the velocity and instead uses the Gaussian distribution to explore the search space based on the the local best postion and globally best postion,

$$y_i(t) = N\left(\frac{y_{jg} + y_{jl}}{2}, abs(y_{jg} - y_{jl})\right) \tag{3}$$

Where y_{ig} stand for the globally best postion, and y_{j1} stand for the locally best postion (j iterations).

As we already kown, two standard iterative equations (1) and (2) are used to express the updating formulas, a velocity constraint and a position constraint are adopted inpractical applications, or the optimization results may be worse. So we can figre out the new detailed two iterative equations (4) and (5).

$$\bar{v}_{ij}(t+1) = p(v_{ij}(t+1)) \tag{4}$$

Where fumular p() denote a constraint function of velocity in case the value of velocity surpass the range of our algorithum's permission ($v_{min,}$ v_{max}). As we know, battlefield situation is easy to change; faster deployment means that we can grasp enemy's combat deployment ahead. The possibiliy of the MAV crashing will increase if the velocity surpass the v_{max}. However, if the velocity is too small, the efficiency of explorer or deployment of MAVG will be worse than standard PSO.

$$\overline{y_i}(t+1) = q(y_i(t+1)) \tag{5}$$

The utility of the fuction $q()$ is similarly with $p()$ when the current position is larger than $y_{\max}$ the current position is set as $y_{\max}$ or the current position remains unchanged.

3.3. *The equations set of the VBBPSO*

In the standard BBPSO, Kennedy et.al abandoned the velocity term in order to simplify the complexity of the standard PSO[14]. we present the concept of velocity and position constraint into the standard BBPSO in order to improve a potential risk. The equations set of the VBBPSO and the meaning of character in the equation set have been descripted in section B.

$$\begin{cases} \boldsymbol{v}_{ij}(t+1) = \omega \boldsymbol{v}_{ij}(t) + c_1 rand_1()[\boldsymbol{y}_{ij}(t) - x_{ij}(t)] \\ \qquad\qquad + c_2 rand_2()[\hat{\boldsymbol{y}}_{ij}(t) - x_{ij}(t)] \\ \boldsymbol{y}_i(t+1) = \begin{cases} \boldsymbol{y}_i(t) \ , \text{ if } \ \mathrm{f}(\boldsymbol{x}_i(t+1)) \geq \mathrm{f}(\boldsymbol{y}_i(t)) \\ \boldsymbol{x}_i(t+1), \text{ if } \ \mathrm{f}(\boldsymbol{x}_i(t+1)) < \mathrm{f}(\boldsymbol{y}_i(t)) \end{cases} \\ \boldsymbol{y}_i(t) = N(\frac{\mathrm{y}_{jg} + \mathrm{y}_{jl}}{2}, \ abs(\mathrm{y}_{jg} - \mathrm{y}_{jl})) \\ \overline{\boldsymbol{v}_{ij}}(t+1) = p(\boldsymbol{v}_{ij}(t+1)) \\ \overline{\boldsymbol{y}_i}(t+1) = q(\boldsymbol{y}_i(t+1)) \end{cases} \tag{6}$$

4. Simulation and Analysis

In consideration of the authenticity of our program, Initial positions were dispersaled randomly in the search space.

The number of particles was set to be 10， which means we dispersaled 10 MAVs in the battle field. Each experiment was run 20 times and terminated at the maximum number of iterations.Finally, connected topology was adopted, in which all particles can own their adjacent nodes.

4.1. *Several benchmark funtion*

With our equations set (6) figured out in chapter III, the concept of introducing the velocity constraint into VBBPSO was tested on five benchmark functions: sphere, Schaffer's ,Rosenbrock, Rastrigin, Griewank.

Table 1. VBBPSO with considering constraint.

v	sphere	Rosenbrock	Rastrigin	Griewank	Schaffer's f6
0.1	7.86E-250	**3.48E+00**	**3.71E+01**	8.62E-03	2.43E-03
0.2	**9.01E-252**	4.52E+00	4.37E+01	1.40E-02	**1.94E-03**
0.3	1.99E-251	1.18E+01	4.56E+01	1.23E-02	**3.89E-03**
0.4	3.04E-249	4.56E+02	5.11E+01	1.18E-02	2.43E-03
0.5	3.37E-249	1.02E+01	5.02E+01	1.16E-02	2.43E-03
0.6	**2.42E-248**	9.16E+03	5.53E+01	9.84E-03	1.94E-03
0.7	3.64E-250	6.44E+02	7.85E+01	8.99E-03	1.94E-03
0.8	2.09E-250	5.33E+03	7.85E+01	**7.76E-03**	2.43E-03
0.9	1.59E-250	**3.21E+04**	9.06E+01	1.31E-02	2.91E-03
1.0	3.75E-249	1.84E+04	**9.90E+01**	**1.51E-02**	**3.89E-03**
min	9.01E-252	3.48E+00	3.71E+01	7.76E-03	1.94E-03
max	2.42E-248	3.21E+04	9.90E+01	1.51E-02	3.89E-03

As is shown in the table 1, if there is no velocity constraint (such as v=0.1) on Rosenbrock, the function evaluations are high away from the optimum of zero, even from v=0.4 to 1.0. In addition, our proposed velocity constraint on function Rastrigin has the best performance at v=0.1. As for the other three functions, there is no distinct change. So we can infer that v=0.1 is also a practical choice.

4.2. *Simulation on Matlab*

We introduce the velocity factor and simulate VBBPSO algorithm with Matlab ,and the yellow smear area as the target area. Figure 3 is the initial state of the paticles distribution, in the non-target area of the paticles in the cluster to reach the target area smoothly, as shown in Figure 4.

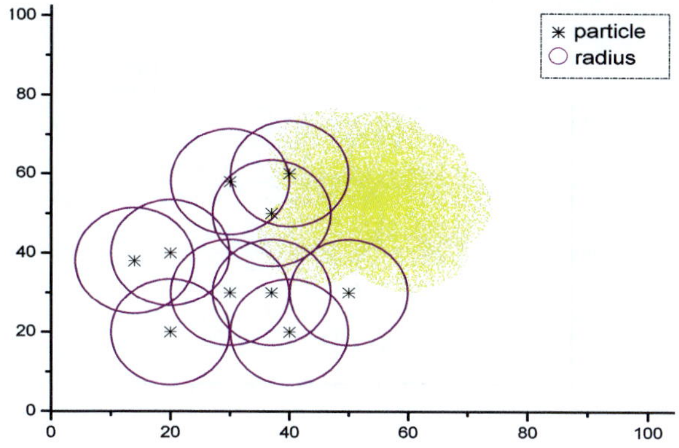

Fig. 3. The paticles deviated the best field.

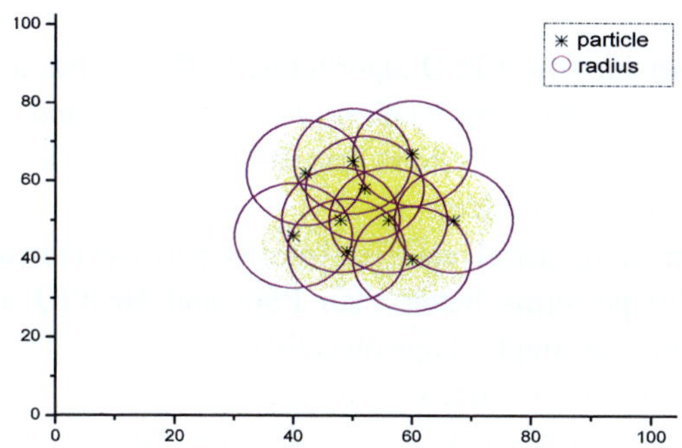

Fig. 4. The paticle find the optimum result.

As we can see in the Figures 3 and 4, the paticles move to the ideal field at the suitable speed according to the equations (1) and (4). as shown in the Figures 5 and 6 , using the renewal position information given by equations (2) and (5), the paticles will continuous adjust its position in order to achive the best position.

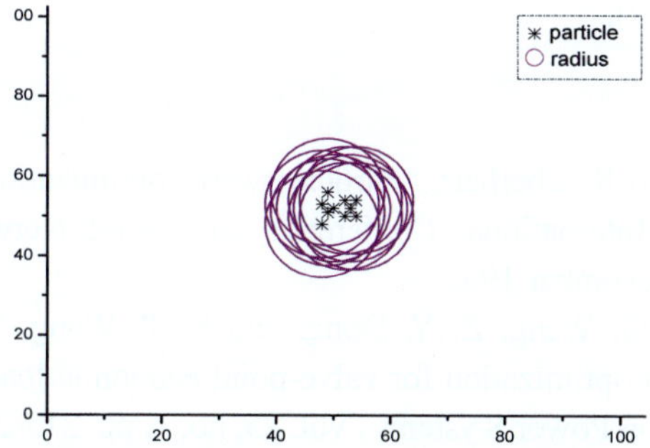

Fig. 5. The paticles in the inappropriate position.

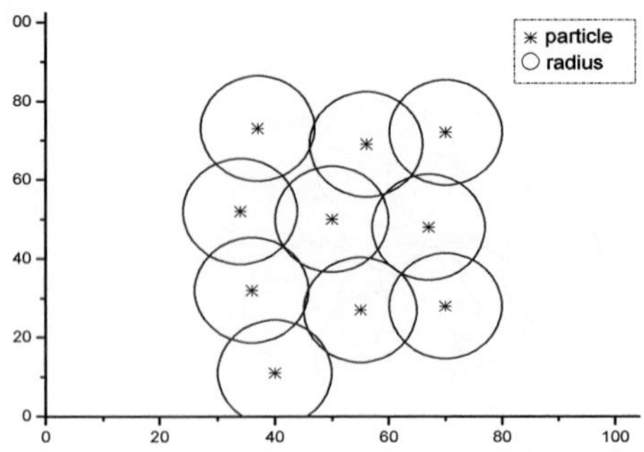

Fig. 6. The paticles disperse uniformly.

5. Conclusion

Compared to other improved PSO algorithms, VBBPSO has achieve a tradeoff between exploration and exploitation with velocity and position, which Although some reported results show that BPSO is better than PSO, it still gets stuck if solving complex problems.

For depolyment of aerial vehicles (means particles) in the the MAVG program,VBBPSO performs better than PSO and BBPSO algorithms when optimizing a single and simple shape objective.

Meanwhile ,we can conclude several facts .

- The best particle in Bare-Bones acts different from the canonical one. In a particle swarm, particles change their positions at each time step.
- The best particle of the adjacent nodehood in VBBPSO simply stands in its best previous position depend on the the function rand() in this paper.

Acknowledgment

This work is supported by the National natural science foundation of China (grant no. 61063042).

References

1. J. Kennedy and R. Eberhart, "Particle swarm optimization," in Proceedings of the IEEE International Conference on Neural Networks. Part 1, pp. 1942–1948, December 1995.
2. K. Meng, H. G. Wang, Z. Y. Dong, and K. P. Wong, "Quantum-inspired particle swarm optimization for valve-point economic load d ispatch," IEEE Tr ansactions on Power S ystems , vol. 25, no. 1, pp. 215-222, 2010.

3. L.D.S. Coelho, "Gaussian quantum-behaved particle swarm optimization approaches for constrained engineering design problems," Expert Systems with Applications, vol.37, no.2, pp. 1676–1683, 2010.
4. C. L. Sun, J. C. Zeng, and J. S. Pan, "A new vector particle swarm optimization for constrained optimization problems," in Proceedings of the International Joint Conference on Computational Sciences and Optimization (CSO '09) , pp. 485–488, April 2009.
5. H. Liu, Z. Cai, and Y. Wang, "Hybridizing particle swarm optimization with differential evolution for constrained numerical and engineering optimization," Applied S oft Computing Journal, vol. 10, no. 2, pp. 629–640, 2010.
6. G. Venter and R. T. Haftka, "Constrained particle swarm optimization using a bi-objective formulation," Structural and Multidisciplinary Optimization, vol. 40, no. 1–6, pp. 65–76, 2010.
7. H. Lu and X. Chen, "A new particle swarm optimization with a dynamic inertia weight for solving constrained optimization problems," Information Technology Journal, vol. 10, no. 8, pp. 1536–1544, 2011.
8. M. Daneshyari and G. G. Yen, "Constrained multiple-swarm particle swarm optimization within acultural framework," IEEE Transactions on Systems, Man, and Cybernetics, Part A, vol. 18, pp. 1–16, 2011.
9. J.Peña, "Simple dynamic particle swarms without velocity," in Ant Colony Optimization and Swarm Intelligence Berlin/Heidelberg, Germany: Springer, 2008,
10. T. M. Blackwell and D. Bratton, "Examination of particle tails," J. Artif. Evol. Applicat., vol. 2008, no. 893237, p. 10, 2008.
11. T. M. Blackwell and D. Bratton, "Origin of bursts," in Proc. Conf. Companion GECCO, 2007, pp. 2613–2620.
12. Evol. Applicat., vol. 2008, no. 654184, p. 10, 2008. R. A. Krohling, "Gaussian Particle swarm with jumps" in Proceedings of the IEEE Congress on Evolutionary Computation, pp.1080-1085, 2006.
13. J. Kennedy, "Bare bones particle swarms," in Proc. IEEE Swarm Intell. Symp., Apr. 2003, pp. 80–87
14. K. Parsopoulos and M. Vrahatis, "UPSO: A unified particle swarm optimization scheme," Lecture Series on Computer and Computational Sciences, vol. 1, pp. 868–873, 2004.
15. R. Mendes, J. Kennedy, and J. Neves, "The fully informed particle swarm: simpler, maybe better," Evolutionary Computation, IEEE Transactions on, vol. 8, no. 3, pp. 204–210, 2004.

Application of simulation technique in spinning

Yu Yang
Changchun University of Science and Technology,
Changchun, 130022, China
yangyu1983@sina.com

Spinning is an advance process integrating forging, extrusion and stretching. Every new type of work-piece requires process experiment before the spinning, which consumes a lot of economic and time costs. As a result, simulation technique is increasingly applied to spinning process.

Keywords: Spinning; simulation; process experiment.

1. Introduction

Spinning is an advance process integrating forging, extrusion and stretching. It is a plastic processing of hollow rotary parts. Blanks are clamped to and rotating along with the mandrel, and spinning tool rotates around the blanks. Blanks are stressed by the relative feed between spinning tool and mandrel stresses to undergo consecutive local plastic deformation. The hollow rotary parts herein are cylindrical work-pieces.

Generally, there are three ways to produce a cylindrical work-piece. The first is to bend the sheet material into a circle before welding. Materials are subject to secondary processing with machine. This method has the advantage of saving materials. However, the welding seams are easy to snap, and welding defects like blowhole and bubble are frequent. The second is to lathe the hollow rod directly. No welding seams are produced by this method, but most materials are cut away by the machine in processing cylindrical work-pieces, wasting a lot of materials. In addition, the two methods above have a common trouble. That is, tooling cuts off the metal fiber in turning. The cylindrical work-pieces are degraded in strength, rigidity, tensile strength, yield limit and other mechanical properties due to lack of continuous and complete fibers. The third method is punching. However, the punch forming requires extremely large punching force. The costs are very high.

As an advanced process, spinning is widely applied to both defence and civil industries. Compared with the aforesaid three methods, spinning has the following features:

(1) Involving little or no cutting, it has a high-usage of materials.

(2) During the power spinning, the blanks maintain continuous and complete fibers, so that the spinning products are improved in strength, rigidity, tensile strength, yield limit and other mechanical properties.

(3) The roller levels off the blanks during spinning, making the spinning products in high precision of dimension and smooth surface.

(4) During the spinning, the contact area between roller and blank approximates to point contact, with unit pressure over 2500~3500N/mm^2. It is suitable for materials hard to deform. Additionally, the power spinning requires less force, consuming less power.

2. Application of Simulation Technique in Spinning

Dynamic simulation technique is a process of solving mathematical equation under fixed solution conditions making use of computer, and inferring the material deformation in virtue of arithmetic solution. As to power spinning of cylindrical work-pieces, dynamic simulation is mainly used for analysis on spin forming process. As mathematic models and assumptions are neatened and improved, finite element has been widely applied to the dynamic spinning simulation. At the beginning of 1950's, it was proposed to apply the FE method to dynamic simulation of spinning process. Hindered by the spinning technique and computer state-of-the-art, this technique has been developing slowly [1~3].

FE method focuses on the dispersion of continuous body. A continuous body is dispersed into several units for the physical performance analysis. The analyzed units are then reconstructed into a body similar to the original one, and that is the approximate result required by engineering. In general, it is a method solving approximate solution of simultaneous differential equations. The advantage of FE method for spinning simulation lies in avoiding the wastes of money and time spent on process experiments, because the spinning process parameters are determined by the computer through simulating the spinning process.

FE method was firstly proposed in 1956, by Clough in his analysis on structural mechanics of airplane. It was also referred to as direct stiffness method at that time, because this method was widespread to elastic plane subjects from matrix displacement method. After that, Clough designated it FE method. Its mathematic foundation is the Ritz method of piece based upon variation principle. As early as 1943, Courant utilized variation principle and piecewise interpolation to solve the Saint-Venant's torsion problem, but did not get enough attention. In 1954, A Chinese scholar, Hu Haichang, proposed the generalized variation principle of three kinds of elastic mechanics under Hu-Washiru variation principle, laying theoretical foundation for multi-variable finite element. Early FE method was based upon principle of virtual work and principle of minimum potential energy. Considering the complexity of parameter matching and the stability and convergence theory, most engineering applications at that time chose displacement methods with displacement as basic variable. By 1970's, FE method developed towards high-precision units, shell elements, iterative solution of non-linear problem, MSFEM, multi-field coupling, plastic forming simulation, etc.

Metal plastic forming is a series of complicated process involving grain growth and evolution, dislocation movement, and etc. In terms of continuum mechanics, metal plastic forming is a non-linear process, including material nonlinearity, which refers to the nonlinearity of strain-stress relation; geometrical nonlinearity, which is the nonlinearity between strain and displacement; and boundary nonlinearity, which is the change of boundary conditions along with the object motion. Elastic-plastic mechanics is the theoretical foundation of metal plastic forming.

The dynamic simulation of plastic forming utilizes non-linear FE method. Argris and Marcal were the earliest to work on elastic-plastic FE analysis. In 1968, Yamada deduced elastic-plastic stress and strain matrix of small deformation. In 1979, Hibbit proposed the elastic-plastic finite element method of large deformation based upon Lagrange descriptive method. Afterwards, Mcmeeking proposed the finite element mothod of large deformation based upon Euler descriptive method. In 1973, Lee and Kobayashi proposed the rigid-plastic finite element method introducing incompressibility by Lagrange method. In 1975, Zienkiewicz proposed the rigid-plastic finite element method, wherein incompressibility was treated with penalty function method. In 1977, Mori proposed the compressible rigid plastic finite element method. In 1981, Park and Kobayashi simulated the flat anvil compression process by 3-D plastic finite elements. At the NUMIFORM'98 Conference in 1989, Honecker stated the numerical simulation results of oil-pan forming process. After 1990, supported by the improved computer processing capacity, non-linear analysis software, like DYNA3D, MARC and ABAQUS, and professional plastic forming software, like DEFORM, DYNAFORM and FORGE, were widely applied.

Dynamic simulation of plastic forming made a great progress in China since 1980's. In1983, Wang Zutang and Xie Shuisheng analyzed the process of hydrostatic extrusion making use of elastic finite element. In 1990, Xiong Huolun simulated the forming process of automotive heater cover by elastic-plastic finite element. In 1994, Guo Gang utilized elastic-plastic finite element of large deformation to analyze the forming process and tearing of stamping parts. In recent years, numerical simulation technique developed rapidly in the field of spinning. Tian Hui, Huang Haiqing and Chen Guoqing adopted numerical simulation for the measurement of forward and backward spinning forces and analysis on filming using TC4 titanium alloy cylindrical work-pieces. In combination with the data obtained from experiments and simulations, they analyzed the effects of main process parameters on the forming quality of cylindrical work-pieces. Xu Yinli, Zhan Mei and Yang He made a detailed analysis on the spindle rotating speed, feed ratio and resilience rules of conical work-pieces during the power spinning. Based upon ABAQUS, they adopted dynamic display method and static implicit method, respectively, to numerically simulate the power spinning of conical work-pieces. In order to accurately analyze the effects of varying three-dimensional loads applied to the roller on service life of die, Wang Miao, Wang Zhongtang and Wang Benxian, in their research on the rules of material changes and work-piece forming for ball spinning forming of thin-walled tubes, shortened the dimensions of blanks to simplify the numerical simulation model, and accordingly improve the analysis efficiency of such model. Xia Qinxiang and Hu Yu simulated the spinning process of rectangular internal gear in respect of effects of roller profile and parameters on the forming quality. The simulation was based upon MSC.MARC. At present, some software for plastic forming have been developed at home, for example, Sheetform developed by Beijing University of Aeronautics and Astronautics for simulation of metal sheet material punch forming , KMAS developed by Jilin University for forming simulation and die design of automotive panels, etc. Dynamic simulation technique has been therefore an important approach for research on stress, strain and metal materials flow trend as well as for accelerated optimization of process parameters.

3. Conclusion

Along with the development of spinning industry, simulation technique will be more and widely applied to this field, assisting spinning industry in China catching up with and surpassing the international advanced level.

References

1. Bao Meiling. The DEFORM Secondary Development for Dynamic Recrystallization Model of 316LN Stainless Steel. D. Shanxi, Taiyuan. (2013).

2. Zhang Lipeng, Liu Zhichong, Zhou Hongyu. The Development and Present Situation Analysis of Power Spinning of Cylinders. J. Journal of Plasticity Engineering, Vol.13 (2006), pp. 43-46.
3. Xie Shuisheng, Li Lei. FE Simulation in Metal Forming and Application. M. Beijing: Science Press. (2008), pp. 107-199.

Simulation and verification of piezoelectric torsional effect of quartz pillar

Chang-Yin Gao* and Li Liu

School of Mechatronics Engineering, Zhengzhou University of Aeronautics, Zhengzhou, Henan 450015, China

**E-mail: changyingao@163.com*

The simulation and experimental verification of torsional effect of a piezoelectric quartz pillar is studied in the article. Based on the theory of elasticity of an anisotropic body, the torsion stress function is obtained. According to the Maxwell electromagnetic theory the Poisson equation and boundary conditions of the electrostatic field in the quartz pillar is deduced, and then by the piezoelectric equation the distribution of bound volume and surface charge densities are computed. By Finite element method, the distributions of the potential and the electric field intensity by torsion are simulated. Based on theoretical analyses of the bound charge densities and simulation, the detection electrodes are effectively disposed on the surface of the piezoelectric pillar. The theoretical and experimental results show that the charge measured is linear with the torque applied. The research on torsional effect of piezoelectric quartz will provide a new method for torque measurement.

Keywords: Piezoelectricity; torsional effect; electric field; finite element method; polarization.

1. Introduction

Since the brothers Curie discovered piezoelectricity in 1880, many scientists have investigated extensively piezoelectric effect, and then the piezoelectric theory of the longitudinal effect, transverse effect and shear effect has been established[1]. In engineering application, it is more convenient to measure mechanical parameter directly by distortion. Therefore, it is meaningful to research the torsional effect of quartz pillar.

In this paper from the application point of view, using the theory of anisotropic elasticity, the electromagnetic theory and piezoelectricity, the torsional effect of a rectangular quartz pillar is investigated, and then by disposing the measuring electrodes on the quartz surfaces the quantitative relationship of torque and charge measured is obtained.

---

*Corresponding author.

2. Theoretical Calculation

Considering the elastic equilibrium of a rectangular quartz pillar with the general rectilinear anisotropy, one of whose ends is fixed and the other is subjected to a twisting moment M_t, whose direction is along the z-axis, as shown in Figure 1. The dimension of the quartz pillar is a=15mm, b=13mm, h=15mm.

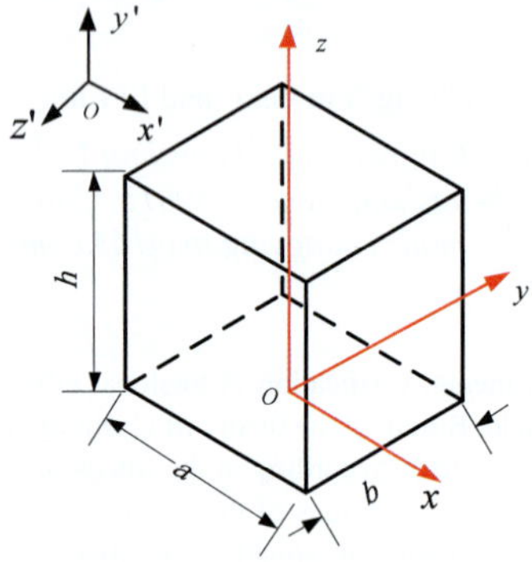

Fig. 1. Quartz beam subjected to a torque.

In order to facilitate numerical calculation, we need rotate the crystallographic coordinate system *O'x'y'z'* to the calculated coordinate system *Oxyz*. During a transformation of coordinate axis, the elastic compliance coefficients and the piezoelectric tensor change accordingly in terms of the transformation rule of tensor [2].

According to the theory of elasticity of an anisotropic body [3], it is obvious that the plane stress function $F(x, y)$ is zero. For rectangular section, the torsional stress function $\phi(x, y)$ can be expressed in Fourier sine series[3]:

$$\phi(x,y) = \sum_{m=1,3,5\cdots}^{\infty} Y_m(y) \sin\frac{m\pi(2x+a)}{2a} \tag{1}$$

Based on the condition on the sides $y = \pm b/2, \phi = 0$, and using the definitions of the plane stress function $F(x, y)$ and the torsion stress function $\phi(x, y)$, the components of state of stress can be calculated as follows:

$$\sigma_x = \frac{\partial^2 F(x,y)}{\partial y^2} = 0, \sigma_y = \frac{\partial^2 F(x,y)}{\partial x^2} = 0, \tau_{xy} = -\frac{\partial^2 F(x,y)}{\partial x \partial y} = 0 \quad \sigma_z = 0 \tag{2}$$

$$\tau_{yz} = -\frac{\partial \phi}{\partial x} = -3.06331\times 10^6 M_t \sum_{m=1,3,5\cdots}^{\infty} \frac{1}{m^2}\left(1 - \frac{ch\frac{2.5m}{a}y}{ch1.08m}\right)\cos\frac{m\pi(2x+a)}{2a} \tag{3}$$

$$\tau_{xz} = -\frac{\partial \phi}{\partial y} = 2{,}4377 \times 10^6 M_t \sum_{m=1,3,5\cdots}^{\infty} \frac{1}{m^2} \frac{sh\frac{2.5m}{a}y}{ch1.08m} \sin\frac{m\pi(2x+a)}{2a} \tag{4}$$

According to the piezoelectric equation of quartz [4] and the torsional stress results of Eqs. (1)-(4), and then using the electrostatic theory the equivalent bound surface charge density can be computed by the following equations[5]:

$$\eta_{bx+} = 4.1270 \times 10^5 M_t \sum_{m=1,3,5\ldots}^{\infty} \frac{1}{m^2}(1 - \frac{ch2.5my/a}{ch1.25m}) + 9.0846 \times 10^{-7} M_t \tag{5}$$

$$\eta_{by+} = 0 \tag{6}$$

$$\eta_{bz+} = 2.8078 \times 10^6 M_t \sum_{m=1,3,5\ldots}^{\infty} \frac{1}{m^2} \frac{ch2.5my/a}{ch1.25m} \sin\frac{m\pi(2x+a)}{2a} \tag{7}$$

3. Experiment Verification

3.1. *Arrangement of detection electrode*

In order to effectively dispose detection electrodes to measure the electric result of torsional effect, from Equation (5) and the simulation of the electric field the distribution of the bound charge density on the two opposite surfaces perpendicular to x coordinate axis depends upon the torque M_t and the x and y coordinate. Therefore, we can arrange two detection electrodes on the surfaces of the quartz pillar, as indicated in Figure 2.

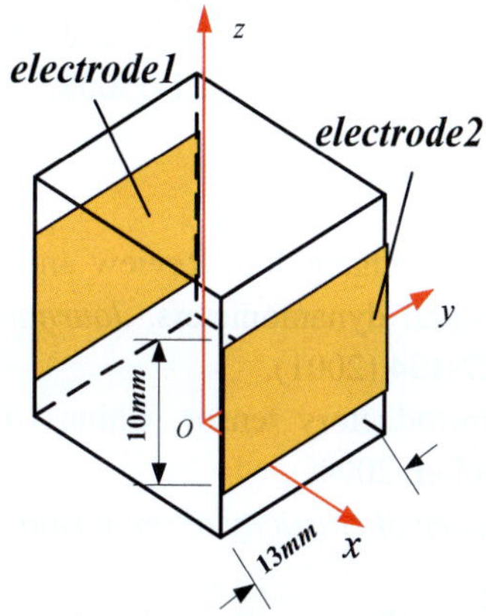

Fig. 2. Arrangement of detection electrodes.

3.2. *Result and analysis*

By experiments we obtain the relationship between the bound charge and the torque load, as indicated in Figure 3. It is evident that the bound charges are proportional to the torque.

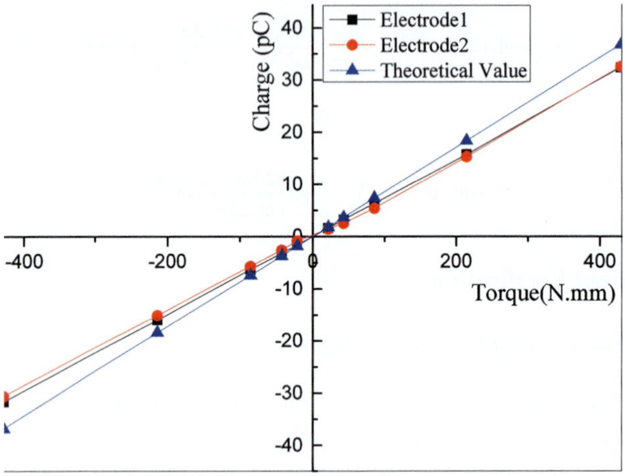

Fig. 3. Relationship between the torque and polarization charge.

Acknowledgments

This work has been partially supported by Foundation of He'nan Scientific Committee, project number 142102210507, and the Aeronautical Science Foundation of China, project number 2013ZD55008.

References

1. Sun Baoyan, Qian Min, Zhang Jun, Review and Prospect on research for piezoelectric sensors and dynamometers, *Journal of Dalian University of Technology* V41, 127-134 (2001).
2. Lu Pangming. An introductory tensor. China University of Science and Technology press, HeFei, 2004.
3. Lekhnitskii, S.G, *Theory of elasticity of an anisotropic body* (Mir Publisher, Paris, 1981).
4. Sun Baoyuan, Zhang Yigong. *Piezoelectric quartz force sensor and dynamic dynamometer* (China Measurement Press, Beijing, 1985).
5. Wang Yulun, Guo Wenyan, *Electromagnetic Field and Electromagnetic Wave* (Harbin University of Industry Press, Harbin, 1985).

Research on failure mode and ultimate bearing characteristics of the single pile foundation in layered soil

Bin Li

Tianjin Port Engineering Institute Ltd. of CCCC First Harbor Engineering Company Ltd.; Key Laboratory of Port Geotechnical Engineering, Ministry of Communications; Key Laboratory of Port Geotechnical Engineering of Tianjin; Tianjin, China,

E-mail:lee_binbin@163.com

Xiao-Qiang Kou

Tianjin Port Engineering Institute Ltd. of CCCC First Harbor Engineering Company Ltd.; Key Laboratory of Port Geotechnical Engineering, Ministry of Communications; Key Laboratory of Port Geotechnical Engineering of Tianjin; Tianjin, China,

E-mail:xiaoqiangKou@163.com

The single pile foundation of offshore wind turbine is currently the most common form, but the huge loads and the overturning moment make characteristics of foundation is different from offshore oil platforms. Using elastic-plastic element of ABAQUS simulation of pile foundation in bearing stress and failure modes in soil layers. The results show that the single pile foundation subjected to vertical ultimate load in soil layers, no plastic strain. Under the horizontal ultimate load, pile shows plastic strain. Under limit bending moment load, the plastic zone in the soil is mainly distributed in the surface soil in front of pile.

Keywords: Pile foundation; ultimate bearing capacity; theoretical analysis; finite element.

1. Introduction

Offshore wind power generation technology has become a hot research and application in recent years, which is rich in wind energy resources and broad flat area [1-3]. China's offshore wind energy resources are expected to reach 750000000 kilowatts, which is 3 times than onshore wind energy resources, offshore wind power will become the focus of future development. At present, the basic structure of offshore wind power has the gravity type, the pile foundation type, the triangle type, the structure of the jacket and the floating structure. Single pile mechanism is a kind of structure which is widely used in offshore wind farms. The horizontal load and overturning moment of the offshore wind turbine foundation are much larger than the offshore oil platform, and the vertical load is less than the offshore oil platform, so the bearing form and characteristics of the foundation are different from the offshore oil platform. In particular, the pile diameter (4~7m) of single pile is far larger than the diameter of offshore oil platform (1~2m), and Bearing mechanism of single pile foundation is also one of the hot spots in the research of offshore wind turbine [4-7].

2. Model Parameters

The finite element software ABAQUS is used to establish the finite element model of different type foundation. The calculation range of soil foundation is larger to eliminate the effect of boundary effect. In order to better simulate the interaction between piles and soil, this paper chooses Mohr-Coulomb models in the model calculation. The foundation inclayed soil foundation, specific parameters see Table 1. In this paper, the failure mode of large diameter piles in different foundation is studied by theoretical analysis, numerical simulation and model experiment [8-12]. In the calculation of the vertical ultimate bearing capacity of single pilefoundation, the effect of soil arch effect on the single pile foundation is considered.

Table 1. Soil parameters.

type	*Soil name*	*H(m)*	*γ(kN/m³)*	*E (MPa)*	*C (kPa)*	*f (°)*
layered soil	silt	8.5	17	2.49	11.4	11.3
	clay	10.5	18.5	3.22	25.0	10.4
	Silty clay	10	20.3	6.03	33.8	10.5
	Clayey silt	5	20.5	6.76	30.7	18.8
	Silty sand	66	20	21	0	38

Table 2. Material parameters of pile body material.

	γ (kN/m³)	*Poisson's ratio*	*E(MPa)*
Steel pipe	68.5	0.3	2.1×10^5
Concrete (C60)	24.5	0.2	3.0×10^4

The elastic model is used for thesimulation of pile foundation structure, and the concrete parameters of the pile body material are shown in Table 2.

Normalcontact and shear contact are respectively friction contact and hard contact between the pile and the soil.

Single pile foundation steel type is Q345 which using the elastic plastic model, specific material parameters are shown in Table 2, steel plastic material parameters as shown in Figure 1.

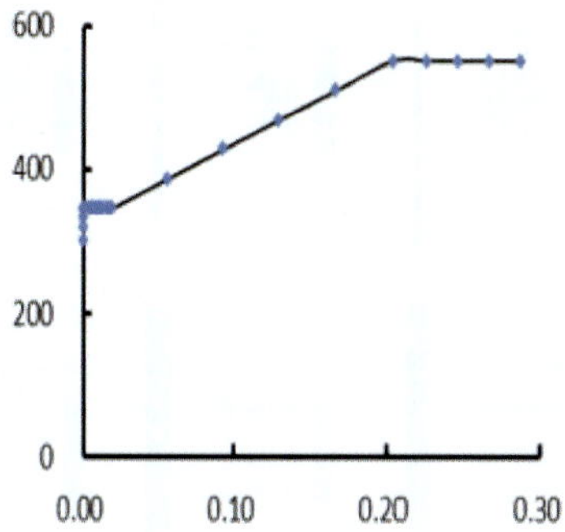

Fig. 1. Q345Relationship of stress and strain.

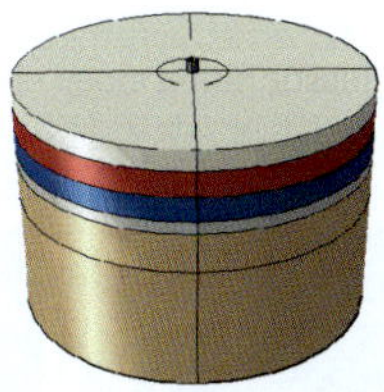
Fig. 2. Finite element model diagrams.

3. Foundation Failure Mode

The failure mode and load bearing characteristics of single pile foundation are analyzed based on the finite element software ABAQUS. The finite element model is shown in Figure 2.

(1) Foundation failure mode under vertical load

When the pile body under the ultimate vertical load in clayed soil, different diameter of pile equivalent plastic strain cloud chartand the plastic region distribution in soil respectively as in Figure 3, Figure 4.

From Figure 3, it is known that the equivalent plastic strain of piles in different diameters is 0, which has no plastic strain. Therefore, The ultimate failure of single pile foundation under vertical load is the bearing failure of the soil.

Fig. 4 show that with the change of the pile diameter, the plastic zone distribution in the soil gradually increases, the fracture surface is not extended to the surface of the foundation; the failure of the soil is manifested by local shear failure. With the longer of the pile diameter, the distribution of plastic zone in the soil increases gradually, but the shape is basically the same. The plastic point of the soil is first appeared below soil mass at the side wall in the bottom of the pile

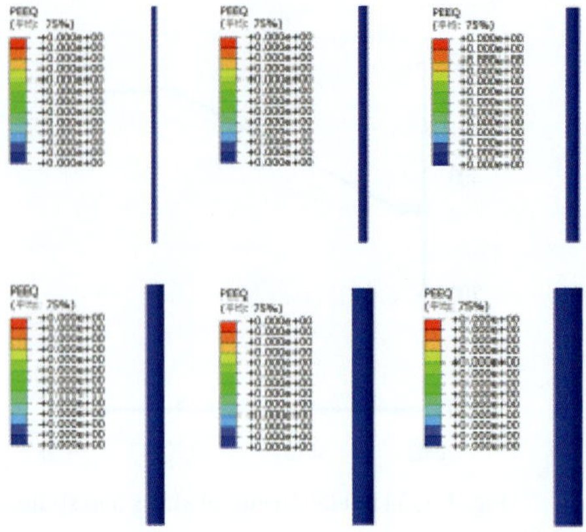

Fig. 3. Equivalent plastic strain cloud chart of pile body.

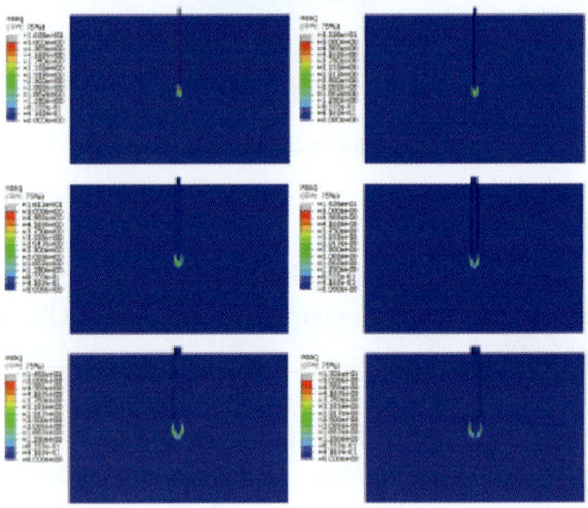

Fig. 4. Distribution of plastic zone in soil.

foundation, then gradually extension to the pile bottom soil and both sides of the pile tip soil, the slip surface is formed at the bottom of the pile in the end, the plastic zone in the soil is different from sand soil, not only the pile tip of the plastic zone, but also the pile side appear plastic zone.

(2) Foundation failure mode under horizontal loading

In clayed soil foundation, when the pile body under horizontal load , different diameter of pile equivalent plastic strain cloud chart and plastic region distribution in soil and the displacement vector diagram in soil are shown as Figure 5 and Figure 6.

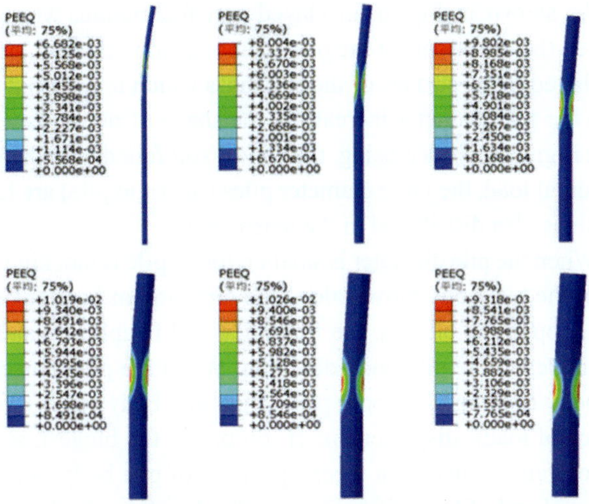

Fig. 5. Equivalent plastic strain contour of pile body.

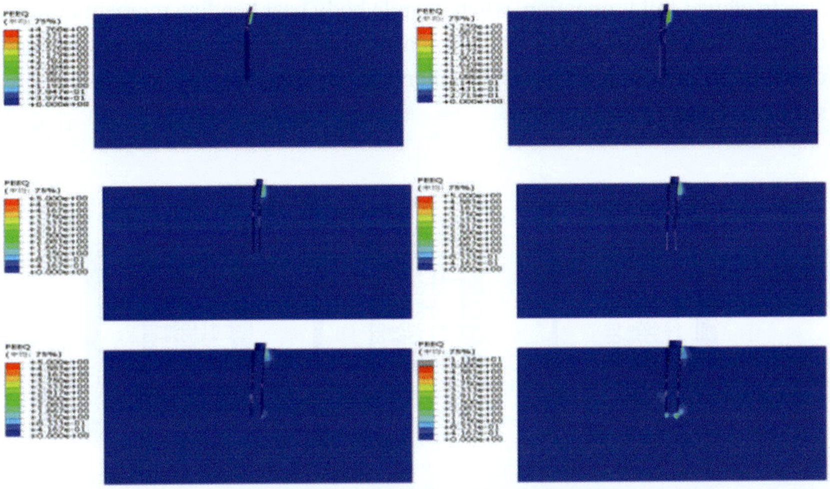

Fig. 6. Distribution of plastic zone in soil.

From Figure 5, plastic deformation of piles appears in different diameter piles under horizontal ultimate load. When the diameter is smaller, the range of the plastic zone is distributed in the upper part of the pile; when the diameter is increased, the plastic zone is distributed in the middle part of the pile. Compared with the sand foundation, the clayed soil foundation has less resistance, and change position of pile body is increased along the load direction when the pile is bearing the horizontal load, and the plastic zone gradually moves down.

As shown in Fig. 6, in clayed soil foundation, when the pile diameter is smaller, the lower part of the pile is embedded in the soil, plastic zone are mainly distributed in the surface of the soil mass which in the front side of the pile body; when the pile diameter increasing, the shape of plastic zone is always same, the range is gradually increasing, the clayed soil foundation has less resistance, under horizontal load, the large diameter piles (the rigid pile) are turned., and the end of the pile is also distributed in the plastic zone..

When the pile diameter is smaller, lower pile is blocking in the soil, The upper part of the pile body moves along the direction of horizontal loads, displacement of pile top is the biggest, the front side soil is uplifting and the soil of the back side moves to the pile body direction. When the pile diameter increases and the pile body rotation, the upper part of the pile body along the moving direction of horizontal loads, displacement of top pile is the biggest, soil produces spherical rotation failure surface at theend position of pile body, soil uplift in front of pile body, and soil of back side moves to the direction of pile body.

(3) Foundation failure mode under the bending moment.

In clayed soil foundation, when the pile body bear bending moment limit load, different diameter of pile equivalent plastic strain cloud chart and plastic region distribution in soil and the displacement vector diagram of pile and the soil as shown in Figure 7 and Figure 8.

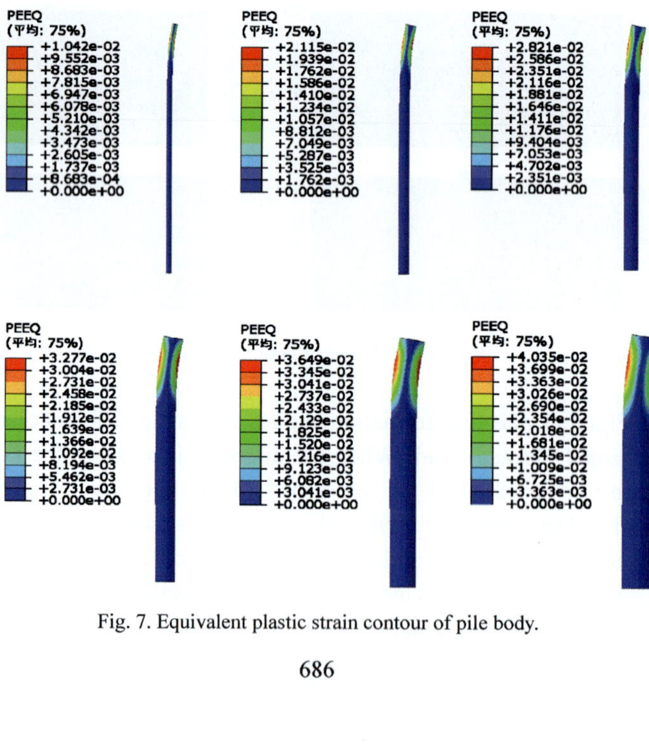

Fig. 7. Equivalent plastic strain contour of pile body.

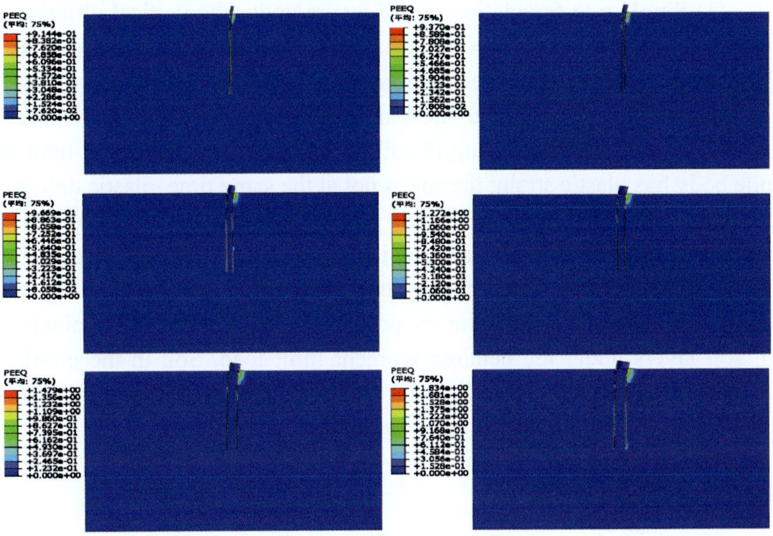

Fig. 8. Distribution of plastic zone in soil.

From figure 7, the plastic deformation of piles appears in different diameters of piles under the ultimate bending moment load, the plastic zone is distributed in the area which near the top of the pile, the plastic deformation of the foundation is the same as sand soil foundation, which is different from the plastic zone distribution to shift down of clay under the horizontal load.

From Fig. 8 it is known that the plastic zone of the soil is mainly distributed in the upper part of the pile body when the pile body bear bending moment load, Plastic zone is relatively smaller than horizontal load condition. The plastic zone in front ofpile body is increased when the diameter is increased; the distribution of plastic zone is almost unchanged compared with the sand foundation.

The lower pile embedded in clay when pile body bear the moment load, upper pile has rotation, soil in front of pile body is heaved. When pile diameter is smaller, a smaller range of soil uplift, the surface soil in back of pile moved towards pile body direction, the situation of the foundation under bending moment condition is similar to the sand soil.

4. Conclusion

In this paper, the bearing characteristics and failure modes of the single pile foundation in soil is analyzed by using the large finite element software ABAQUS.

(1) Single pile foundation in soil layers under the vertical ultimate load, no plastic strain, in the elastic working stage, single pile foundation under the vertical load appears failure is the destruction of the soil. The distribution of plastic zone in the soil is closely related to the soil layer of the single pile foundation.

(2) Single pile foundation in soil layers under the horizontal limit load, the pile body has plastic strain; the soil is not at the same time, plastic deformation of the part has a slight change, single pile foundation bearing layer is clay, soil surface in front of pile side and the pile tip appear plastic zone distribution.

(3) The position of the plastic strain of the pile is always the top of the pile and the soil is not the same but the position is same, when single pile foundation in soil layers under the bending moment limit load. Soil in the plastic zone is mainly distributed in the soil surface of pile body's front side; plastic zone is smaller when comparing with the horizontal load condition.

References

1. Global wind report. Global Wind Energy Council. 2010.
2. European Environment Agency (EEA). Europe's onshore and offshore wind energy potential. 2009.
3. Lian J J, Sun L Q, Zhang J F, et al. Study on bearing characteristics and technology advantages of the composite bucket foundation of offshore wind turbines. Transactions of Tianjin University. 2011, 17(1): 55-60.
4. Lian J J, Wang Haijun, Liu Meimei. The foundation of offshore wind turbine[p], 2010-10.
5. Lian J J, Ding Hongyan, Zhang Puyang. A combination of guiding and negative pressure technology for thick foundation sinking method[p], 2009-10.
6. Lian J J,Wang Haijun. Uses the prestressed concrete cylinder type structure of offshore wind turbine foundation[p], 2010-10.
7. Huan Choi. Offshore pile type fan foundation structure design and research of [D]. Dalian University of Technology, 2009.
8. Chen Guangsi, Liu Run, Liu Yuchen, et al. Analysis of the bending characteristics of the wide and shallow cylindrical foundation with offshore wind power [J]. Journal of Tianjin University: Natural Science and engineering technology [J], 2013, 46 (5): 393-400.
9. Le Cong Huan, Ding Hongyan, Pu Yang Zhang. The deck concrete bucket foundation of offshore wind turbine bearing model. Engineering mechanics [J] 2013, 30 (4): 429-434.

10. Chen Fei, Yang Xu, LianJijian, et al. Numerical analysis and experimental verification of the soil compaction effect of tube type foundation. Rock and soil mechanics [J], 2014, 35 (12): 3587-3591.
11. Hu Caiqing, Ding Hongyan, Zhang Puyang, et al. Analysis of horizontal load bearing capacity of single pile foundation in a single layer of clay. Marine engineering [J], 2014, 32 (2): 30-37.820-826.
12. in fluent, Wang Haijun. Under cyclic loading composite barrel type foundation of pore water pressure changes and liquefaction analysis. Rock and soil mechanics[J], 2014, 35 (3).

Chapter 3

Information Technology

The development of mobile end web system development based on HTML5 + CSS3 + java framework

Gai-Zhen Yang

Department of Computer, Huanggang Normal University, Huanggang, Hubei, China
E-mail:29712527@qq.com

With the popularity of mobile terminals, Web system must adapt to the mobile terminals. In this paper, introduction mobile web systems development, analyze the characteristics of HTML5, CSS3, JavaScript and popular framework technology. On this basis, to explore the development of a more rational structure, information architecture comb. Proposed two JavaScript script code optimization program to make system development more lightweight, improves data transfer speeds.

Keywords: HTML5; CSS3; JavaScript framework; Web; mobile.

1. Introduction

Mobile end Web system development needs more and more, but using HTML4 of B/S mode mobile development client system there are bottlenecks. Web technology appears revolutionary update JavaScript + HTML5 + CSS3 comprehensive technical support.

2. Move the Mobile Terminal end Web Development Needs Analysis

The mobile terminal refers to the computer can be used in mobile devices, including mobile phones, notebooks, tablet PCs, POS machines, in most cases refers to the phone, the user can customize the installation of a variety of APP. Compared to install APP, more convenient Web-based applications and more popular. Mobile terminal operating system platform Android, iOS, webOS, blackberry, WP7, Web browser mainly to webKit core Chrome, Firefox and so on.

Web-based development is more lightweight, including the creation of specific regulatory compliance according to W3C semantic standard pages for basic functionality required to respond to, the completion of loading and the corresponding speed server script, and the establishment of common mobile web component libraries.

3. Key Technology Analysis

3.1. *HTML5 technology*

HTML5 can make the browser without plug-ins to achieve the current functionality that already exists and there is no capability to enable location tracking, local storage, offline applications, video and audio playback and communications. It will bring the Internet to a new era, so that Internet-based applications more convenient, efficient and safe. HTML5 Wireless Application Protocol is an open standard protocol, which allows you to transmit information over the network to a mobile phone or other wireless communication terminals. HTML5 development based on existing Internet standards, but the characteristics of the wireless network is optimized.

3.2. *CSS3 technology*

CSS3 is an upgraded version of CSS 2 technologies, which advocates a high degree of modularity. Using Cascading Style Sheets technology when creating web pages, you can make the page more presentation, beautiful appearance. As long as the page corresponding code to make some simple changes, you can change the appearance and format of the page. Although CSS3 is China Mobile's products, but SMS terminal to access their coverage Mobile, China Unicom, Telecom three networks, do not worry about non-mobile phone users are not able to use the system. Users can rent a server with CSS3 way their companies access to mobile information services business, the secondary development through CSS3 management platform, developed its own data processing platform.

4. System Terminal Adaptation Needs Analysis

In order to provide the different mobile terminals need to be adapted to the best version of the terminal, to establish a project back User Agent adaptation system. UA is a SIP protocol in a logical entity, SIP is a signaling protocol, meaning the user agent handles signaling protocol in place, it simply is for the user to send and receive signaling information. The physical entity may be a SIP soft and hard phone terminals, SIP multimedia terminals, IAD and SIP network access to other network elements and other equipment. In addition to handling signaling there PROXY, RS and other logical entity.

5. System JavaScript Development Framework to Build

Rational use of JavaScript framework is very important, JS Development Framework is divided into three levels. The first layer is a layer API, is relying on the project to build a library of components suitable for mobile web needs of the project, with reusability, for example, a number of predefined functions and components, which allows developers to understand the program from source code or not access to a set of routines at the inner workings of details. JS API layer primary role is to provide a framework for functional components in the structure in the outermost layer, the developer of the most closely. In the development of the outermost layer of the functional components of integration, some common features extracted through the organization and coordination, it becomes a second layer function components, which mainly to facilitate the development of the outermost layer and API written. The third layer components, is for the first layer and the second layer components function together with the service, including the bottom of some of the most basic functions, such as is Empty Object, is Function, is Null. These features and functions very close to native JS,

These three layers using relationship is shown in Fig. 1:

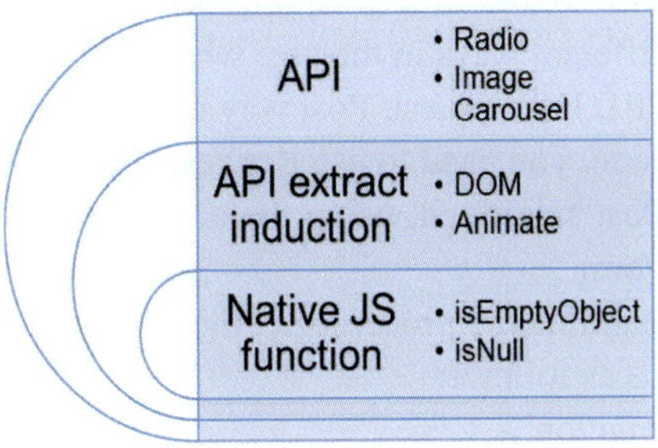

Fig. 1. Three layers using relationship.

The outermost component JS framework, need to use a second layer, but also often used in the core layer. Intermediate core layer rarely call each other, and to the second intermediate layer is often need to call each other to achieve some of the features, such as Animate may use often Dom. To the third floor as much as possible independent of each other.

6. Touch System Events to Write Function

In this paper, the Web system is running in the mobile terminal, so the code will be repeatedly invoked touch trigger event. Dojo 1.7 in the development of the document, the new touch API can simplify the code and to call this part of the added dojox.mobile (lightweight mobile application components), which is part of the technology development of key technology of mobile terminal systems. Dojo reference document code: <script src = "js / dojo-1.7 / dojo.js"></ script>. Touch event consists mainly Touchstart, Touchend, Touchmove, Touchcancel,

Touch event code written in the following format:

```
[javascript]
node.addEventListener("touchmove", function(event){
if(event.touches.length == 1){
var touch = event.touches[0],
style = touch.target.style;
style.position = "absolute";
style.left = touch.pageX + "px";
style.top = touch.pageY + "px";
}}, false);
```

7. Form Data Processing

Form submit the form to the content server, there are two ways, POST and GET, submitted in two different ways in the use: when (1) Get the user's input mode to pass through the URL link request, Post way is through other the way. (2) Get in the way of submission, you need to use Request. Query String to get the value of a variable, while Post submitted, using the access method of content submitted through Request. Form.

In order to streamline the process site, the choice of post handled page form data, the key code is as follows:

```
<div id="information">
<form method="post" action="http://www.thewebaddress.com" on submit="to Post()" target="_blank">
<form action="demo_form.asp" method="get">
Songname:<input type="text" name="fname" /><br />
songer: <input type="text" name="lname" /><br />
<input type="submit" value="Submit" />
</form><input type="submit" class="inputBtn" style="margin:5px 0;" value="提交"/>
</form>
</div>
```

8. System Performance Optimization

A web page is inevitable to introduce a lot of external files: JavaScript, CSS, background images, because no state of the http protocol, each user access to these files will be re-request all files to a remote server, and accumulated a large number of requests, speed is the most important reason for the impact site. Therefore, by streamlining the code to reduce the number of http requests to speed up access. In this paper, the following two optimization schemes:

Merge code and picture files: Web introduced only one css file, a file and a js sheets background. The main method is to merge background images using CSS Sprites, all the images together to form a large map, and then use a combination of CSS "background-color", "background-image", "background-repeat", "background-position" of background positioning, background-position can be digitally can accurately locate the position of the background image to display a different image. For the purposes of the current popular network speed, not more than 200KB of single image loading time required for this group is about the same, so the load time is greatly reduced. When visiting the page at the same time, the pictures are not loading a piece of land slowly display, but all together is loaded, it will give users a better experience.

Reduce the size of the requested file, thereby reducing network bandwidth request data.

The first Volume is compressed .js file: Remove js blank lines, remove unnecessary comments, simply change the variable name, the use of global variables, the code as concise as possible. Using server-side GZIP compressed JS files, shrink the size of the file transfer.

The second Volume is compressed CSS file: Remove unnecessary CSS comments, deletion CSS property is no longer used to, try to use shorthand writing. Improve CSS reuse, to reduce the HTML file sizes. Using server-side CSS GZIP compressed file, transfer the file size shrink.

9. Sum Up

Formal commercial 4G network makes the domestic mobile Internet business gradually developed. Android and iPhone and other smart phones, leaving the field of mobile Internet to be fully exploited. The future of mobile Internet services will be moving in the direction of diversification. Powerful, open platform, user-friendly interface and a better user experience will enable the mobile Web technologies (HTML5, CSS3, JavaScript framework) program has a broader application, and more far-reaching development.

Acknowledgment

I thank reviewers for constructive reviews and suggestions that improved the quality of this manuscript. This essay is the fruit of our several months' consisting work. This work was supported by the School Natural Science Project 2015016403.

References

1. Nicholas C. Zakas, JavaScript Advanced Programming [M], Posts & Telecom Press, 2006.
2. Lu Ling cattle, HTML5 and CSS3 Definitive Guide [M], Machinery Industry Press; 1st edition (2011).
3. David Garlan, Mary Shaw An Introduction to Software Architecture [M]. Carnegie Mellon University:. Technique Report, 2006.
4. Len Bass, Pick Kazman, Paul Clements. Software architecture in practice [M]. Addison.Wesley Educational Publishers Inc, 3rd Revised edition, 2012.
5. Ding-nan, based on HTML5 and CSS3 Netease WAP mobile station touch-screen version of the product [J]. 2012, 29-35.

Based on the CAS of campus network SSO system research

Yu-Rong Guan
Computer Department,Huanggang Normal University,
Huangzhou, Hubei, China
jsjgyr@hgnu.edu.cn

Nowadays, campus network are often constructed in the different period and different technical solutions, different uses of a plurality of independent application system, the user when using these applications must one by one corresponding user name and password to login, time-consuming and laborious. In this paper, we first introduce the SSO system, then expounds the basic advanced technology for the mechanism of CAS implementation of campus network single sign on system principle, finally summarize the SSO system for the relative merits of campus network construction, can significantly improve the time efficiency of the majority of teachers and students.

Keywords: CAS; SSO; campus network.

1. The Overview of the SSO Basic Concepts

With the continuous development of network technology, in the campus website construction step by step in the process, according to the actual needs of the various levels of business and information constructed corresponding application system, e-mail login system such as my school website, educational management system, office automation system unification, financial integrated query system, scientific research management system, library management system, quality course management system, student psychology archives management system, student dormitory management system, of course, with the development of the school, there will be many new application system to generate.

As a matter of fact, we have seen, these application systems and not in the same period the unified design and development of technology to complete and use of the environment of the application of these systems also very different, some systems focus on teacher use. Some systems focus on students, therefore, at present, in the campus network using these application systems are mutually independent of the use of, is the application system formed independently of the user authentication system, in other words, is for users to access without application system to log on the corresponding different username and password. Not say a single in between each application system repeatedly switch login the waste of time and not say a single multiple user name password to remember mixing, remember to influence the work efficiency is a school students and teachers a tiny part of the adjustment will also bring huge workload for system administrators and maintenance personnel. Obviously, the use of these independent application system will undoubtedly make the information system convenient to bring weakened.

Single sign on, referred to as SSO[1], is a popular method to solve the enterprise business integration. SSO is defined in multiple application systems, users only need to log in once to access all the applications of mutual trust, without the need for access to each application system when they enter the password again. IBM is an image of the interpretation of "single sign on, the whole network roaming on SSO". A key problem in Web service security authentication and access control plays an important role in the safety design of Web service system. Single sign on technology can be used for web service requester to implement authentication and authorization. In the design of Web services security, identity authentication and access control is the guarantee to access the web service in a safe and effective way, and single sign on technology is a powerful weapon to resolve identity authentication and access control.

2. The Principle of SSO in Campus Network

Some domestic research in single sign on the development of Web service, of course, also made the corresponding research results, and puts forward some solutions, which can be roughly divided into three types, the first is to propose a Web service security model based on XKMS and SAML; the second is put forward based on the SSO method to realize Web service of SAML; the third is put forward another the SSO system based on SAML.

As can be seen, the SSO's main function is to make the system all applications share an identity authentication system, which is the key to solve the key technology there are many implementation mechanism, generally can be divided into two major categories of cookie and session[2]. Most of the current SSO product is used in the mechanism of Cookie, of course CAS is no exception, CAS of the Central Authentication Service is a set of Yale developed at the University of JAVA based open source SSO service. From the point of view of the structure, the CAS consists of two parts, the CAS server and CAS client, the CAS server need independent deployment, is mainly responsible for the authentication of the user to work; CAS client is responsible for processing to the client by the access requests of the conservation of resources, you need to log in and redirected to the CAS server. CAS advanced design concepts, reasonable structure, simple allocation of resources, a wide range of client support technologies and make it convenient to release in Web online.

The following CAS as an example[3], the use of Cookie mechanism to achieve single sign on system principle view as shown in Figure 1. And on the campus network and its implementation mechanism are as follows.

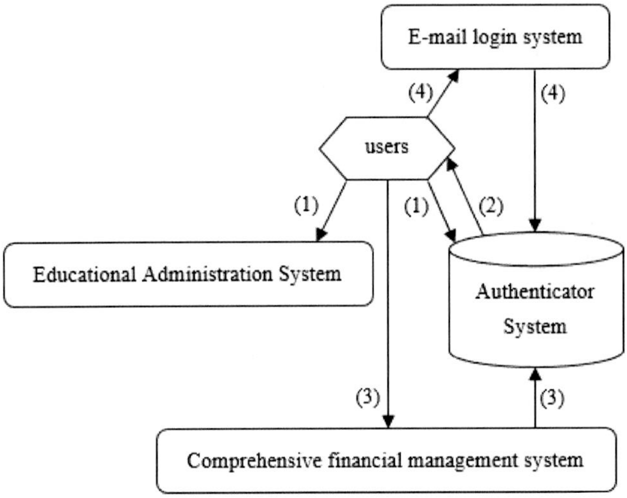

Fig. 1. Campus network single sign on principle view.

(1) First when the users first login educational management system, because this is the first application of the system login, you will be guided to the CAS login authentication system;

(2) Then, in CAS authentication system according to users with the user name and password to login information for identity verification, if the verification is correct passing, educational administration management system login is successful, also returned to the users a certification document, also is the so-called ticket;

(3) And then when the users to access a comprehensive financial management system, you automatically will just get the ticket with credentials as login of the application system, application system will check a ticket redirected to the CAS authentication system, if the check, users can login success, of course, the entry there is no again through the user name and password login information to access the, is to achieve single sign on;

(4) Eventually in accordance with the implementation of the same mechanism, you can also access the E-mail login in the system to experience the SSO system without login again under the condition of convenient.

Of course, you can also use single point login system other management information system in the campus network, such as the student management system, young teachers home information system, Huanggang Normal college scientific research management system and so on, and convenient.

Single sign on system looks simple, easy to operate, as if only to solve the authentication system, that seems to be the only solution to the log on this single issue, in fact in the campus network of each application system integration to deal with a lot of unified authentication, assembly, and system architecture, these problems will relate to the user's usability, system security, if improper handling may will bring the system collapse, is mainly due to the various application systems in the campus network login, identity identification authentication, the client language processing have very different. That is to say, in the processing system may sometimes be according to some small plug or some subsystems of service management program for users is very simple, but increased the single sign on system stability, maintainability and flexibility. Of course, there are many applications of large enterprises have SSO, realized its strong function.

3. The Advantages of SSO

From the point of view of development trend, effective management of the campus network information system, reduce user operation, improve the work efficiency and information safety is the campus networks are facing an important issue, but also the realization of the basic requirements of the construction of digital campus network[4]. Therefore, the use of single sign on system is based on the need of system management, user needs, convenience is based on needs, based on the security of the application is based on Application integration. The main advantages of SSO are as follows:

(1) Substantially reduces the possibility errors of user login in different application system;

(2) Increase the various application management system security and maintainability;

(3) To be effectively avoid redundant authentication information processing and save multiple sets of users of the system;

(4) In essence, decrease the a considerable amount of cost about the management information system and the application of system complexity;

(5) This is this handy for the user provide more humanized operation, no need to remember all kinds of user name and password.

In the real life, we also experience the some and similar single point login application service, for example, to login the large number of electronic mailbox, microblogging, business service platform [5], baidu application system, not only can use traditional username and password to login, can also use QQ account number, mobile phone number and even more popular two-dimensional code scanning, micro letters in a variety of ways. These are the convenient network development to people's lives. It can be said that the campus network using the SSO system will become like a large enterprise network platform management trends.

4. The Conclusion of Campus Network SSO

The most important SSO in the unified identity authentication system, which can identify the user has logged, and can automatically judge whether the current user login, thus completing the single point login function. In process of construction of the campus network, in order to improve the learning and work efficiency of the school staff and students, the SSO system will play a multiplier effect, in order to save a lot of time and energy, the unified, perfect, high-efficient, scientific user information management system is a kind of necessity.

Therefore, this paper combined with the actual situation of our college, we design and implement a web service environment of single sign on system applicable to. The system can save the campus network resources effectively, and the system has high security and good scalability etc. Based on the secondary development of CAS code SSO component in the application of the campus network resources integration process, for subsystems of the project implementation process, also encountered a many problems, through debugging repeatedly, for single sign on also have more in-depth understanding.

Acknowledgment

This work supported by the scientific research project of Huanggang Normal University.

References

1. Qingfeng Zhang, Yun gejing, Yuli yang. Based on the resources of campus network authentication study and concept [J]. Network security technology and application, 2008.06.

2. Chongzhang, Chaowu, et al. The design and implementation of unified identity authentication system of campus network [J]. Zhongyuan University of Technology, the 2008 volume fourth issue nineteenth.
3. http://www.blogjava.net.
4. Andrej Volchkov, A Revisiting Single Sign-on: a pragmatic approach in a new context[J]. IT Professional, Vol. 3: 39-45, 2001.
5. Qinyong yuan, et al. NET Framework programming [M]. Beijing: Tsinghua University press, 2002.

The matching method of web service based on semantic similarity in cloud computing environment

Zuo-Ping Xu
School of Informatics, Linyi University, China
E-mail: lyxzp2008@163.com

In order to solve the problem that the precision and recall of the semantic web services is not ideal in cloud computing environment, the paper researches on the concept similarity algorithm, proposes a web service matching method enhancing the matching accuracy of semantic web services, and then gives the detailed matching flow chart. The experimental results show that the method presented in this paper not only makes full use of the semantic relations of concept, but also cuts a lot of web services having nothing to do with the demand. Both the recall and the accuracy rate are superior to the traditional method, and the efficiency of the algorithm can support the large-scale service application in the dynamic environment.

Keywords: Cloud computing; semantic similarity; web service; matching method.

1. Introduction

Cloud computing is a calculation method based on the Internet. The shared hardware and software resources and information may be provided to end-users on demand by the way. The appearance of semantic web technology provides a good method of increasing semantic level content for web service description, and promotes the web service search automatically and intelligently. Currently, the web service discovery and matching based on semantic has gradually become one of the main research areas of web services.

In recent years, the network enters into people's daily life, and brings great changes to people's lives, study and work. The Internet provides a lot of web services which have rich functionality and variety of species to meet the needs of different groups, such as weather services, travel agency services, bank services, transport services, hotel services and so on. A large number of services meeting those needs exist on the Internet, and are constantly updated. For example, some services may also be used in the last second, but may not be used in the next second. There is an urgent need of a method identified automatically and implemented by computer to meet user's service demand. Moreover, not only to meet the needs of users, but also as fast as possible, because the long waiting time is too difficult to accept by user. The search and automatic assembly technology based on semantic provides a solution to achieve the above needs.

2. Related Research

Since Tim Berners-Lee first proposed the Semantic Web concept in 1998, people have been studying and discussing the Semantic Web. Semantic web services connect web services and the semantic web through OWL-S technology [1]. Boualem analyzed a class of service request to rewrite the web service matching technology, they realize inference based web service automatically matches (found), which proposes a flexible web service matching algorithm [2].

Dekang Lin from University of Manitoba in Canada proposed that not only the amount of information shared of concepts would be considered, but also the amount of different information of concepts, while calculating the semantic similarity. The semantic similarity of two concepts is that the shared information of the most recent ancestor concept divides the sum of the unshared information of two concepts [3].

The professors from Indiana University of the United States, School of Information, Department of Computer Science, introduced graph theory into the calculation of semantic similarity, and gave a calculation method. They established a structure of graph using the relationship of the concept node, regarded the link edge between nodes as a set of edges, and considered the link relationship in the three kinds of concepts in general. But the central idea of the calculation is consistent with the method based on the amount of information, so there is no qualitative change but more rigorous in theory. The inherent shortcomings of information model do not be solved[4].

3. Concept Similarity Algorithm

Ali cloud open storage services (OSS) provides a large cloud storage services for users. Users can call the API interface provided by OSS to upload and download data through the network, and can manage the data through the web console. The storage capacity of OSS supports flexible expansion, and provides the settlement function paid by usage based on the cloud computing model. This provides appropriate technical support for the development of web services.

3.1. *APS calculation method*

APS(Advanced Planning System), also known as advanced planning and scheduling. Running demand plan and capacity plan is a complicated man-machine dialogue process with repeated calculation and adjustment, which is the process spending most of time while running a MRP (Material Requirement Planning). When the customer puts forward the order requirements, such as a long and tedious running speed, the customer can't give a quick response, but also can't respond to the requirements of the rapid changes of the environment.

APS calculation method considers that the similarity between the two concepts can be judged by the number of the information they share, while the shared information of the two concepts is approximately equal to the amount of information of their recent ancestors which contain the two concepts. The calculation method is calculated based on the probability of the concept of a specific text focus. From the information amount calculation formula, we can see that the shallower the level of concept in the concept hierarchy is, the weaker the concept similarity is; the deeper the level of concept in the concept hierarchy is, the stronger the concept similarity is.

The ideal computation method of semantic similarity is that, having a good intuition and a certain theoretical basis, not overly dependent on the structure of the concept hierarchy, fully embodying the influence of distance, depth and density, and not using the text set. It also can distinguish the concept nodes with different depth, the brothers concept nodes with the same depth and different density, and the concept nodes with different depth and different density. It is technically difficult to achieve.

3.2. *Semantic change measure*

The concept similarity computation based on semantic is important to the matching of semantic web service, information retrieval, data mining, information integration, knowledge management and so on. In order to express the degree of matching, this paper introduces the matching degree of the concept, and provides the semantic service query. The concept matching degree in the same domain ontology is mainly measured by the semantic similarity between two concepts, which plays a key role in web service matching field. Intuition tells us that the similarity between objects is related to their similarities and differences. The more common two objects have, then the greater the similarity is, while the more difference the two objects have, the smaller the similarity is. On the contrary, when the two concepts are completely independent, the similarity is the smallest, and the similarity is 0. In the case of a certain relationship between the two concepts, the similarity is between 0 and 1.

3.3. *Semantic distance*

Another important indicator of measuring the semantic relation is the semantic distance. In many researches, semantic distance is regarded as the opposite measure of semantic relevance and semantic similarity, and it is used to express the degree of semantic similarity and semantic relevance. That is to say, the distance is farther, the relevance and the similarity is weaker (monotonic decrease), the difference is bigger; the distance is closer, the relevance and the similarity is stronger (monotonic rise), the difference is smaller. The value of the semantic distance is usually from 0 to ∞.

4. Optimal Matching Algorithm

4.1. *Optimal matching algorithm*

The paper proposes a matching method of semantic web service based on semantic reasoning, improves the accuracy of service matching, reduces the computation times of the semantic similarity between concepts, increases the service resolution performance, and retrieves the service that the user needs. In order to simplify the calculation process of semantic distance, it uses semantic reasoning to calculate the semantic distance, and uses the process of semantic reasoning to guide the calculation process of semantic distance. It only searches the path of semantic reasoning, so as to reduce the computation of semantic distance. It simplifies the calculation complexity and improves the performance of web service matching.

4.2. *Service similarity calculation*

The semantic distance of concepts between two sub-graphs should be calculated first in order to determine whether any service description is related to the current service request, when service query is carried out. In order to avoid a large number of irrelevant concept of similarity computation, this paper puts forward the service matching method of semantic similarity, while calculating the concept similarity based on the structure information of ontology hierarchical. Semantic similarity not only cuts out the service which has nothing to do with the user's needs, but also gets the similarity of concept between the service and the demand, which is the basis of bipartite graph's matching. However, it is necessary to filter the relevant service set to improve the efficiency of the matching calculation before the bipartite graph's matching. First of all, considering the three different concepts contained in the relevant service: matching concept, mismatching concepts and irrelevant concept. For the specific user needs, the matching concept and the mismatching concept of the service reflect the semantic relevance of the demand, and the unrelated concepts are meaningless, which has no influence on the overall similarity of service.

The matching and reasoning of ontology concept is important links in the matching process of semantic web service. The matching relationship between ontology concepts (such as insert matching, exact matching, including matching, etc.) is obtained through the matching and reasoning of ontology concepts, and then the matching degree is obtained. For example, a service needs to enter the account and the password, another service may need the username and the password, but the account is usually equal to the username, how to make the system to judge that they are equivalent. Ontology inference engine carries out corresponding reasoning and similarity calculation in the field of ontology library, regarding the concept reasoning of ontology as a tool.

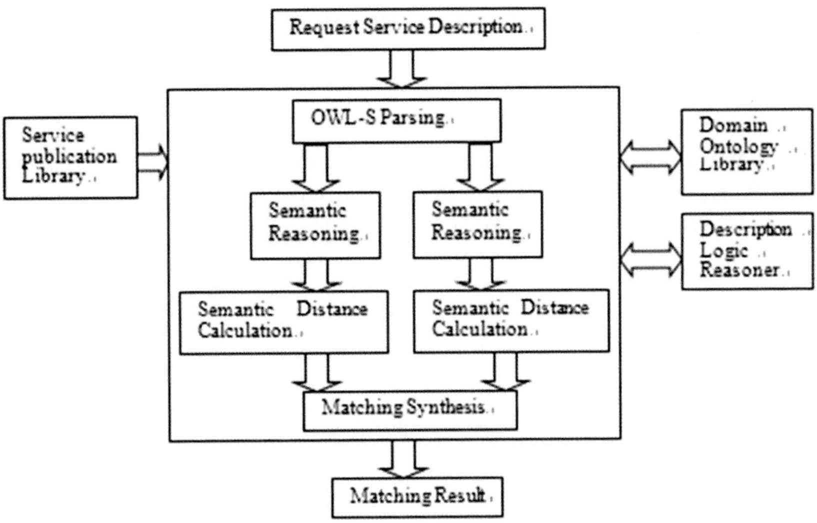

Fig. 1. The structure of the matching method to enhance the semantic accuracy.

4.3. *Service matching process*

The paper proposes a web service matching method which can enhance the semantic precision, the structure of which is shown in Figure 1, and all matching algorithm modules are in the dashed frame. We use OWL-S language to describe the web service. First, we use OWL-S analysis module to resolve the web service that needed by service requester or registered by service publisher, and set the input and output parameters of the semantic reasoning. Then we use the description logic inference engine to carry out the semantic reasoning of the input and output parameters of concepts. Then the result obtained by semantic reasoning may be used to calculate the semantic inference between concepts. Then, we will carry out a comprehensive matching calculation on the result of semantic distance of the input and the output. Finally, the service list of the service matching result obtained by the weight is sent to the service requester. In this system, it also needs the support of service publication library, domain ontology library and description logic inference engine.

5. Web Service Matching Experiment

5.1. *Experimental environment and data*

The performance and scalability of the paper are compared with the traditional service query method. The experimental data uses the standard test set OWL-STC given in the literature[4], which contains 1083 services divided into nine different areas. The similarity service query in this paper is implemented in the on-demand service platform, and the performance of 10 thousand real WSDL services contained by the service library is tested. The main comparison objects and data are shown in Table 1, because the OWLSMX can only handle OWL-S services, so it cannot be tested in real service set. So the experiment is run on the host of the Intel Core i5-760 2.8 GHz processor and 4GB memory, the operating system of which is Windows 7. The test code is written by Java and Eclipse3.6 compiler through.

Table 1. Experimental data set.

	Data Size	Performance Index
Standard Test Set	1083 services, divided into 9 areas	precision ratio recall ratio
Real Service Set	10000 services, divided into 6 groups of different size of the service set	response time Matching efficiency

5.2. *Experimental results and analysis*

In order to measure the scalability of this method, we extract six different size (1000-10000) service sets from the service library to carry out searching, and compare the response time between the method propose in this paper and the traditional method. Through the comparison of Figure 2, we can find that if the maximum response time is 100s, the semantic similarity service query method can only support the query within 1000 services. While caching the concept similarity, although it improves the efficiency of the query, but also it can't be applied to the application scenarios of 4000 services. The execution time of semantic similarity service query method is basically maintained at the millisecond level, even if it is a backtracking method, the query time processing 10 thousand services is less than l0s. In the aspect of scalability, the method proposed in this paper has obvious advantages.

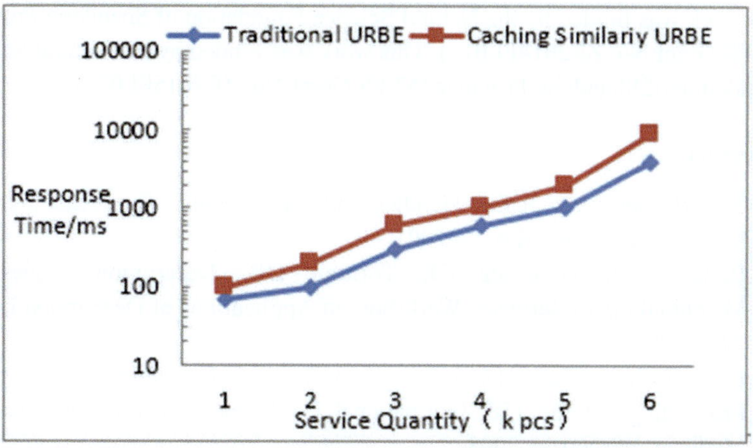

Fig. 2. Comparison of query efficiency with traditional methods.

According to the test data and the experimental results of real service, it is shown that the service query based on semantic similarity proposed in the paper not only improves the recall and the accuracy ratio of service query, through considering the semantic relations among concepts in the ontology level, but also can greatly improve the scalability of similarity service query. It can adapt to the application scenarios that a sharp increase in the number of services in cloud computing environment.

6. Conclusion

Web service is one of the key supporting technologies of cloud computing from the idea to the realization in summary, while how to improve the accuracy and efficiency of web service based on similarity is the key point of this paper. It proves that the method considering different query strategy on semantic similarity is in line with the actual situation by experiment, and the similar query method proposed by the paper is superior to the traditional method not only in the recall but also in the accuracy ratio. Algorithm efficiency can be used to support large scale service applications in dynamic environments. It will be to consider using the Map Re-duce method to further improve the performance of semantic similarity next step, which has practical significance for the application of massive web services in cloud computing.

Acknowledgments

This work was funded by the Natural Science Foundation of Shandong Province (NSFS Grant No. ZR2014FL013). This work was funded by the Natural Science Foundation of Shandong Province (NSFS Grant No. ZR2015FL032).

References

1. Tim Berners-Lee, James Hendler and Ora Lassila, The Semantic Web Service, American, May 17, 2001.
2. Boualem B, Gonzalez CJ, Trastour DB. Description Logics for Matchmaking of Services. Workshop on Applications of Description Logics, 74-85, 2001.
3. Dekang Lin and Jarrardan, Information-Theoretic Definition of Similarity in Proceedings of the 15th International Conference on Machine Learning, Madison, Wisconsin, 2008.
4. Klusch M, Fries B, Sycara K. A Hybrid Semantic Web Service Matchmaker for OWL-S Sservices. Web Semantics Science Services and Agents on the World Wide Web, 7(2):121-133, 2009.
5. Dongwei Bai, Chuan-Chang Liu, Junliang Chen. A Web Service Matching Method of Enhancing Semantic Precision[J]. Journal of Beijing University of Posts and Telecommunications, 21(5):25-30, 2012.
6. Zhongzhi Shi, Yuncheng Jiang, Haijun Zhang. Agent Service Matching Based on Description Logic[J]. Journal of Computer Science, 27(5):625-635, 2014.
7. Research Group of Knowledge Grid, Institute of Computing Technology, Chinese Academy of Sciences. The Improved Semantic Web Service Matching Method[J]. Computer Engineering and Applications. 2011.13.
8. Dasgupta S, Bhat S, Lee Yugyung. Taxonomic Clustering and Query Matching for Efficient Service Discovery. Proceedings of 2011 IEEE International Conference on Web Services (ICWS2011), 2011.
9. Jue Wang, Yuqian Cheng, WenDong Zhao, Chang Tian. The Analysis of Semantic Web Service Matching Method[J]. Military Communication Technology. 2012.03.

10. Xundeng Wang, Shiping Chen, Lei Wang, Research on Service Discovery Architecture in Cloud Computing Environment, Software Guide, 2015.09.
11. Xiaohui Cui, Guisheng Yin, Hongbin Dong, Collaborative Evolutionary Algorithm for Service Matching Problem Journal of Software, 2015, 07.

Research on information encryption and authentication in big data environment

Xia Lin
School of Informatics, Linyi University,
Linyi, 276000, China
E-mail:linx166@163.com

Big data brings convenience to the people, at the same time, it also brings some potential risks during the collection, storage and use. Especially, the information security of big data has become one of the bottlenecks of big data development. For these problems, this paper puts forward a scheme based on AES and ECC, designs a secure model of information encryption and authentication, which ensures the confidentiality, integrity and non-repudiation in big data storage.

Keywords: Security; big data; authentication.

1. Introduction

With the development of big data and cloud computing technology, more and more information is transmitted and exchanged by users. A dramatic increase in the amount of data brings convenience to people, but brings a new challenge to the information security. The information in big data era is large capacity, a wide variety, then the secure management requirements are getting higher and higher. How to verify the reliability of the information and access data efficiently is the most important thing. The information in big data environment should:

(1) Confidentiality: In order to protect the important data, users save the cipher text by using the symmetric algorithm. It is the core technology to ensure the security of users' sensitive data in big data environment. Storage and encryption technology is use key management, data encryption and algorithm parameter management in order to allow users to control their own resources.

(2) Integrity: The data is accessed and modified by authorized users. Finally, availability refers to resources being available to authorized user, includ ing both data and services.

(3) non- repudiation: Users can't deny the information they transfer.

2. Encryption Operations

There are many encryption algorithms, we select Advanced Encryption Standard (AES) [1]and Elliptic Curve Cryptography(ECC) [2,3]in this paper.

AES is based on Rijndael algorithm, which is provided by Joan Daemen and Vincent Rijimen. The length of key is flexible, such as 128bits, 192bits and 256bits, but the length of group is 128bits, and the encryption process is inconsistent with decryption process. If the length of the key is 128bits, the number of iterations is 10; If the length of the key is 192bits, the number of iterations is 12; If the length of the key is 256bits, the number of iterations is 14.Each encryption process is made up of 4 operations: SubBytes, ShiftRows, MixColumns and AddRoundKey.

AES takes a byte as an element in $GF(2^8)$ and use polynomials presents each elements.To a byte b, the binary is $b_7b_6b_5b_4b_3b_2b_1b_0$, now b is a polymials: $b_7x^7 + b_6x^6 + b_5x^5 + b_4x^4 + b_3x^3 + b_2x^2 + b_1x + b_0$.

The encryption process:

```
AESCipher(byte in[16], byte out[16],word w[4*(Nr+1)])
{
byte s [4,4];
s=in;
AddRonudKey(state,w[0,3]);
For(int round=1;round<Nr-1;round++)
{
SubBytes(s);
ShiftRows(s);
MixColumns(s);
AddRoundKey(s,w[4*round,4*round+3]);
}
SubBytes(s);
ShiftRows(s);
AddRoundKey(s,w[4*N,4*Nr+3]);
out=s;
}
```

The decryption process is difference with the encryption process:

```
AESDecipher(byte in[16],byte out[16],word w[4*(Nr+1)])
{
byte s[4,4];
s=in;
AddRoundKey(s,w[4*N,4*Nr+3]);
For(int round=Nr-1;round>0;round--)
```

```
{
InvShiftRows(s);
InvSubBytes(s);
AddRoundKey(s,w[4*round,4*round+3]);
InvMixColumns(s);
}
InvShiftRows(s);
InvSubBytes(s);
AddRoundKey(s,w[0,3]);
}
```

DES algorithm is grouped with 64 bits, the useful key is 56 bits, the output ciphertext is 64bits, and 16 round interative block symmetric algorithm. Table1 shows the time of encryption of DES and AES.In this paper,we select AES as the symmetric algorithm.The speed of encryption of DES and AES comparison as follows:

Table 1. The comparison of the process time of DES and AES.

	1K	10K	100K	1000K	10000K
DES	31	73	155	1471	14684
AES	16	25	37	349	2823

3. Authentication Method

The main attraction of ECC compared to RSA is that the best algorithm known for solving the elliptic curve discrete logarithm problem takes fully exponential time. On the other hand, the best algorithms known for solving the underlying hard mathematical problems in RSA and DSA (the integer factorization problem, and the discrete logarithm problem, respectively) take sub-exponential time. This means that significantly smaller parameters can be used in ECC than in other systems such as RSA and DSA, but with equivalent levels of security.[4]

3.1. *Key generation*

Let F_p be a prime finite field so that p is an odd prime number, and let $a, b \in F_p$, satisfy $4a^3 + 27b^2 \neq 0(mod\ p)$. Then an elliptic curve $E(F_p)$ over F_p defined by the parameters $a, b \in F_p$ consists of the set of solutions or points $P = (x, y)$ for $x, y \in F_p$ to the equation:

$$y^2 = x^3 + ax + b(mod\ p)$$

together with an extra point O called the point at infinity. For a given point $P = (x_{p,}y_p)$, x_p is called the x -coordinate of P, and y_p is called the y-coordinate of P. The addition operation in $E(F_p)$ is specified as follows:

(1) $P + O = O + P = P$ for all $P \in E(F_p)$.

(2) If $P = (x, y) \in E(F_p)$, then $(x, y) + (x, -y) = O$.

(3) Let $P = (x_1, y_1) \in E(F_p)$ and $Q = (x_2, y_2) \in E(F_p)$, where $P \neq \pm Q$ .Then $P + Q = (x_3, y_3)$, where $x_3 = \lambda^2 - x_1 - x_2 (mod\ p)$, $y_3 = \lambda(x_1 - x_3) - y_1 (mod\ p)$ and $\lambda = \frac{y_2 - y_1}{x_2 - x_1}$.

(4) Let $P = (x_1, y_1) \in E(F_p)$, then $P + P = 2P = (x_3, y_3)$,where $x_3 = \lambda^2 - 2x_1 (mod\ p)$, $y_3 = \lambda(x_1 - x_3) - y_1 (mod\ p)$ and $\lambda = \frac{3x_1^2 + a}{2y_1}$.

3.2. *Signing and verifying operations*

Elliptic curve domain parameters over F_p are a sextuple:

(1) $T = (p, a, b, P, n, h)$consisting of an integer p and two elements a, b, a base point $P = (x_p, y_p)$ on $E(F_p)$, a prime n which is the order of P and an integer h which is the cofactor $h = \#E(F_p)/n$.

(2) Select a secret key k which is a random integer in the interval [1,n-1].

(3) Select an elliptic curve public key $Q = kP$.

3.2.1. *Signing operation*

(1)Select a key pair (k,Q) with$Q = (x_{q,}y_q)$ associated with the elliptic curve domain parameters T established during the setup procedure.

(2) Select $d \in [1, n-1]$, $dP = (x_1, y_1)$,and convert x_1 to an integer $\overline{x_1}$ using a proper conversion routine, set $\gamma = \overline{x_1} (mod\ n)$.

(3) Use the hash function selected to compute the hash value:

$e = hash(M)$.

(4) Computer:

$$s = d^{-1}(e + rk) mod n$$

(5) The signature of M is $S = (r, s)$.

3.2.2. *Verifying operation*

(1) Given Q and T to verify r and s, if r and s are not both integers in the interval [1,n-1],the signature is invalid.

(2) Use the hash function to compute the hash value:

e = hash(M).

(3) Compute $u_1 = s^{-1}e \bmod n$ and $u_2 = s^{-1}r \bmod n$

(4) Compute $R = u_1P + u_2Q = (x_{R,}y_R)$

(5) Convert the field element x_R to an integer $\overline{x_R}$ using the conversion routine, and set $v = \overline{x_R} \bmod n$

(6) Accept the signature if and only if $v = r$.

4. Security Model

The information is processed by AES and ECC, then storage; when users access the data, they should be authorized, otherwise, they can't access the correct data. The process is described as Fig. 1.

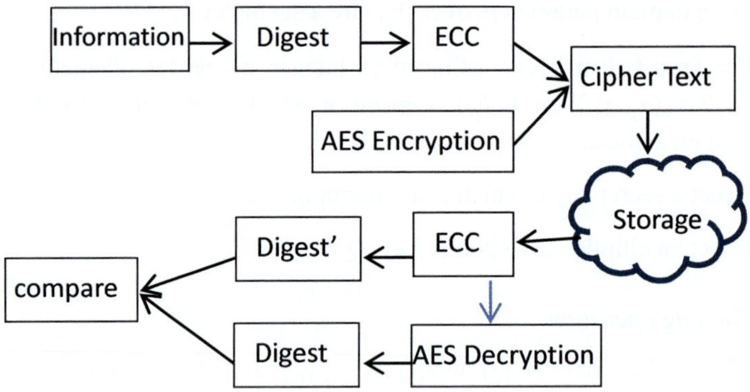

Fig. 1. The security model.

5. Conclusion

Big data brings great value, at the same time, it also increase the risk of information security. The paper uses the method of cryptography to store and protect the data based on the data storage security, ensures the reliability and security of data. If we want to solve the security problem of big data better, only combine technical means and regulations.

References

1. Nan Li, Research of Database Encryption Based on Fast AES Algorithm Implementation,CSEE, (Wuhan, China 2011).

2. Certicom Research, SEC1: Elliptic Curve Cryptography Version 1.0, 2000.
3. Yuanling Hao, Shiwei Ma, Guanghua Chen, Xiaoli Zhang, Hui Chen, and Weimin Zeng, "Optimization algorithm for scalar multiplication in the elliptic curve cryptography over prime field," Lecture Notes in Computer Science, (2008).
4. Xia Lin,The Application of Elliptic Curve Cryptography in Electronic Commerce, (ICCII2012).
5. Sivaganesh E, Velayutham R, Manimegalai, A secure software implementation of non linear AES S-box with the enhancement of biometrics, (ICCEET 2012).
6. R. Velayutham,, E. Siva Ganesh, D. Manimegalai, Enhancing the Security of AES Algorithm in Wireless Network, Power Electronics and Renewable Energy Systems (CPERES 2014).
7. A. Kannammal, S. Subha Rani, DICOM Image Authentication and Encryption Based on RSA and AES Algorithms, Trends in Intelligent Robotics, Automation, and Manufacturing (2012).
8. Demchenko , Canh Ngo, Cees de Laat, Peter Membrey, Daniil Gordijen ko,Big Security for Big Data: Addressing Security Challenges for the Big Data Infrastructure, 10th VLDB Workshop, (SDM 2013).
9. Pushpak Verma, Tej Bahadur Chandra, A. K. Dwivedi, Network Security in Big Data: Tools and Techniques, Proceedings of Third International Conference (INDIA 2016).
10. Sithu D. Sudarsan , Raoul P. Jetley, Srini Ramaswamy, Security and Privacy of Big Data, (2015).
11. Roundup of Big Data Pundits' Predictions for 2013. Blog post by David Pittman. 18 Jan 2013. http://www.ibmbigdatahub.com/blog/roundup-big-data-pundits-predictions-2013.
12. Demchenko, Y., Zhao, Z., Grosso, P., Wibisono, A., de Laat, C.: Addressing big data challenges for scientific data infrastructure. In: The 4th IEEE Conference on Cloud Computing Technologies and Science (CloudCom2012).
13. Demchenko, Y., Membrey, P., Grosso, P., de Laat, C.: Addressing big data issues in scientific data infrastructure. In: First International Symposium on Big Data and Data Analytics in Collaboration (BDDAC 2013).

The communication realization based on VC and PLC for zinc flow two-level computer system

Yong Chen† and Yong-Fang Xie

Mechanical & Electrical Engineering College,
Foshan University, Foshan 528000, China
†E-mail: iicechen@163.com

For two-level computer system on high temperature zinc flow detection and control, this paper introduces the programming method based on VC serial communication between the PC and the PLC, and gives part of the program. Actual operation shows that the system can work reliably, the flow control accuracy of crude zinc is ±1.5%, reaches the desired results. The communication method is simple, stable and reliable, may have some value to test and control system for other similar industrial parameters.

Keywords: Flow control; VC; PLC; communication realization.

1. Introduction

In the modern production process automation, flow is one of the important process parameters and important measure of equipment efficiency and economy, it can indicate the working condition of the production process [1]. In the actual production, the fluid of crude zinc liquid put into Pb column from furnace is manually controlled by adjusting the valve opening. However, the artificial method of zinc fluid flow control with a lot of subjectivity is not accurate, some accidents may be brought for high temperature environment and hard labor. To improve the accuracy and reliability of the crude zinc liquid flow control, soft measurement technology to detect the crude zinc liquid flow is proposed, two-level zinc liquid flow detection and control computer system is established. With the system, crude zinc liquid flow can be monitored, unmanned operation and control of high temperature zinc liquid flow is realized basically [2].

With the rapid development of electromechanical integration technology in recent years, programmable logic controller (PLC) is applied in industrial production more and more widely for its compact design, good scalability, low price, strong instruction and outstanding advantage in the switch quantity control [3]. However, the display function of PLC is not very strong. The combination of computer and PLC has many advantages of powerful man-machine interface, rich application software and low price. Integration of management and control has become a new developing trend [4], the interconnection and communication between the computer and PLC is the key technology to realize the integration of management and control. Many application systems have been designed based on VC + + platform which has strong function in graphics processing and database management, and can achieve the underlying communication control with high real-time requirement [5]. Real-time communication between computer and PLC has become a key point of two-level zinc liquid flow computer system.

2. System Structure

The zinc liquid flow control system is composed of two levels of computer system. The host computer is an industrial computer IPC610, and the volume of the crude zinc liquid is set, monitored and recorded according to the output request. 8 sets of SIEMENS's S7-226 PLC are used as lower computers. The flow of zinc liquid is controlled according to the flow set value given by the host computer, MPI communication interface is used between host computer and lower computer to exchange information exchange efficiency. Figure 1 shows the system framework.

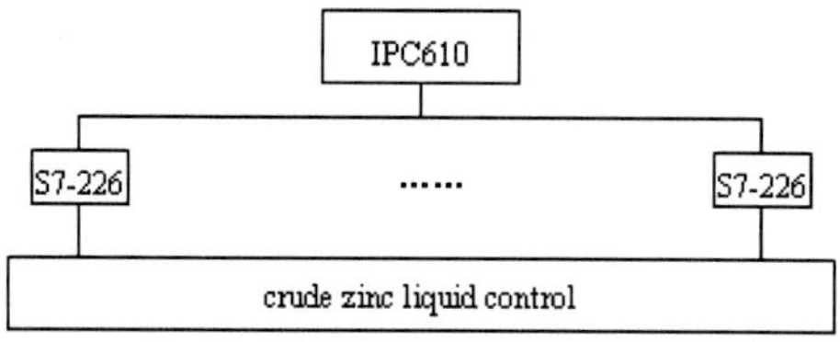

Fig. 1. System framework.

The structure of the control system is shown in figure 2. Flow set value is input from the computer keyboard by operator according to actual expected output. The value is set by the host computer to the PLC via the communication network. Actual flow signal detected is input to A/D module of PLC, the output of A/D module of PLC is flow feedback value. By comparing flow set value and flow feedback value, two output signals are direction signal and pulse signal (CP). Direction signal represents the size relationship between flow set value and flow feedback value, if the flow set value is greater than the flow feedback value, the output signal is 1, otherwise is 0. Pulse signal reflects the differential degree between flow set value and flow feedback value, more number means larger degree. Direction signal and pulse signal drive the operation of stepping motor together, a pulse for an certain angle. Rotation of the stepping motor drives the valve vertical moving through worm gear to make the valve opened or closed. If the set flow value is greater than the flow feedback value, the direction signal on motor is 1, positive revolution, valve opening increases; and the opposite condition had the opposite result. Therefor, the system can control the crude zinc liquid flow from lead weighing launder to lead column.

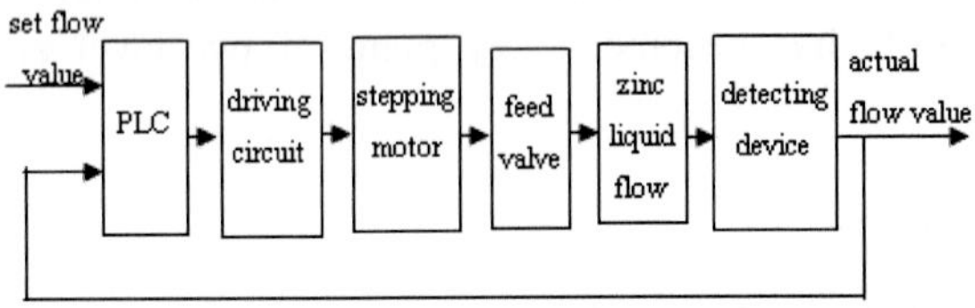

Fig. 2. System structure.

Software design includes three parts: host computer monitoring program, PLC control program, and communication program. Host computer monitor programis designed with Visual C ++ 6.0 platform, software design includes three parts: host computer monitoring program, PLC control program, and communication program. Host computer monitor programis designed with Visual C ++ 6.0 platform, which includes the following features: (1) PLC sents processing zinc liquid flow data, stored in the flow database; (2) Displays zinc liquid flow curves and real values according to operation command; (3) Querys historical flow data and abnormal flow data for production management of distillation process; (4) Inputs flow seting value, and transmits this value to the PLC after processing; (5) Carries out model experiments, the model parameters can be selected according to actual situation.

Communication program is executed only after the system has sent the execution instruction. Communication function can easily achieve by using the VC MSComm communication control. PLC control program mainly receives flow seting value from the host computer, samples real-time flow feedback value to A/D channel, generates direction signal and pulse signal through operation processing, controls the stepping motor running to achieve stable flow output.

3. Host computer communication programming

Host computer can use event-driven mode or inquiry mode to process communication data with lower computers (PLC). Serial communication program requires VC MSComm communication control with a series of attributes and user interface. Current value of the property is given by “Get...” function. The property value can be set by “Set... ” function.

(1) Firstly, establish a dialog-based project with VC++ 6.0 platform, add controls as listed in Table 1 in the dialog box;

(2) Add program code in OnInitDialog () function to initialize the serial port parameters as following:

```
BOOL CSetValueDlg::OnInitDialog()
{    m_Comm.SetCommPort(1); //select COM2
if(!m_Comm.GetPortOpen())
m_Comm.SetPortOpen(true); // Open the serial port, Set 9600 baud rate, 8 //
                                data bits, no parity, 1 stop bit
m_Comm.SetSettings("9600,n,8,1");
m_Comm.SetInputMode(1); // Set mode of  binary input
m_Comm.SetInputLen(0); // Read all the data in the receive buffer with a
                       // reading operation period
                       // Trigger a OnComm event when send a
                       //character
m_Comm.SetSThreshold(1);
}
```

Table 1. Controls list.

Type	ID	Caption	Variable	Function	ID
MSComm	IDC_MSCOMM		m_Comm		IDC_MSCOMM
Static	IDC_STATIC1	Furnace number selection			IDC_STATIC1
Combo Box	IDC_COMBONO		m_cmbNO	OnSelChange ComboNo()	IDC_COMBONO
Static	IDC_STATIC2	Flow value setting			IDC_STATIC2
Edit	IDC_SETVALUE		m_fSetValue		IDC_SETVALUE
Button	IDC_BTNOK	Determination		OnBtnok()	IDC_BTNOK
Button	IDCANCEL	Exit			IDCANCEL

(3) Data frame format sending from Host computer to PLC

Data frame format sending from Host computer to PLC is shown in Figure 3.

#	IP	Data1	Data2	V	*

Fig. 3. Data frame format.

#:Start symbol;
IP: Node identifier of PLC;
Data1,Data2: Two bytes of communication data;
V: Frame check code, 1 byte;
*:Data frame end flag.

The following is the data check function.

```
int CSetValueDlg::verify(int number, int upper, int low)
{
int a[4],b=0;
a[0]='#';
a[1]=number;
 a[2]=upper;
a[3]=low;
for(int i=0;i<4;i++)
b=a[i]^b;
return b;
}
```

Number means node identifier of PLC, upper means Data1, low means Data2.

(4)Sending data program

```
void CSetValueDlg::OnBtnok()
{
UpdateData(TRUE);
int n,digital,uppervalue,lowervalue;
float setvalue;
n=m_cmbNO.GetCurSel()+1;
setvalue=m_fSetValue;
digital=(int)((setvalue-16.0)*4000/(24.0-16.0));
                                        //digital is 12-bit input digital signal
uppervalue=(int)(digital/64);
                                        //uppervalue is the high 6 bits of digital
lowervalue=digital-uppervalue*64;
                                        //lowervalue is the low 6 bits of digital
CByteArray s;
s.RemoveAll();
```

```
s.SetSize(6);
s.SetAt(0,'#');
s.SetAt(1,n);
s.SetAt(2,uppervalue);
s.SetAt(3,lowervalue);
s.SetAt(4,verify(n,uppervalue,lowervalue));
s.SetAt(5,'*');
m_Comm.SetOutput(COleVariant(s));
}
```

4. Communication Program of PLC

There are two kind of communication modes of S7-226 PLC, one is the point to point communication protocol mode (PPI), and the other is free port mode which is completely opened and free to user interface. The latter is more commonly used. Free port mode means serial communication port of the CPU can be controlled by the user program. Free port communication is feasible only when the CPU is in RUN mode. Serial port initialization is absolutely necessary before communication. Port control register SMB30 and SMB130 are used to configure communication port 0 and 1, including baud rate, parity, and data bits as a free port communication mode.

RCV instruction can be used to receive information, which may receive one or more characters, up to 255 characters in the buffer, and the characters are stored in data buffer. If there is an interrupt routine attached to the receiving complete event, it will generate an interrupt (interrupt event 23 for port 0, interrupt event 24 for port 1) when the last character receipt in the buffer.

RCV buffer format is shown in figure 4.

Number of characters	Initial character	M	E	S	S	A	G	E	Ending character

Fig. 4. RCV buffer format.

Part of the communication program is as follows.

```
//port initialization
MOVB  16#09, SMB130// select serial port 1, initialize free port,
                   //Set 9600 baud rate, 8 data bits, no parity
MOVB  16#B0, SMB187 // initialize RCV information
MOVB  42, SMB189//set ending information 42
MOVB  10, SMB194//set maximum number of characters 10
```

```
ATCH   INT_1, 24        // interrupt 1 attached to the receiving
                        //complete event
ENI//enable interrupt
RCV   VB100, 1 // VB100 as receiving buffer of port 1
……
 // program of interrupt 1 for processing received data
LD    SM0.0
MOVW   +0, VW200
MOVW   +0, VW202
BTI   VB103, VW202 //VB103 as high 6 bits of digital
MOVW   +64, VW200
*I    VW202, VW200
BTI   VB104, VW202 //VB104 as low 6 bits of digital
+I    VW202, VW200
MOVW   VW200, VW100  // result in unit VW100
CRETI // interrupt return
```

5. Conclusion

Communication program using VC++6.0 platform is proposed, which designed for two-level computer system on high temperature zinc flow detection and control. The method and the program has been successfully applied in a smelting plant for the process of zinc refining. Actual operation shows that the system can work reliably, the flow control accuracy of crude zinc is ±1.5%, has achieved expected effect. This communication method is simple, stable and reliable.

Acknowledgments

This work is supported by National Natural Science Foundation Project under Grant 61473318; Science and Technology Plan Project of Guangdong Province under Grant 2011B010100015; Higher Education Institutions Characteristic Innovation Project of Guangdong Province under Grant 2015KTSCX152, 2014KTSCX156; Science and Technology Innovation Special Fund Project of Foshan city under Grant 2014AG10018.

References

1. YS. Shao, *Process control and instrumentation,* Shanghai: Shanghai Jiao Tong University Press, 1994: 76–77.

2. Y. Chen, WH. Gui, and CH. Yang, *Instrument to measure fluid of coarse liquid Zinc with high temperature based on soft sensor method*, Instrument Technique and Sensor, 12:10-11,15, 2006.
3. SIEMENS AG, *S7-200 SIMATIC programmable controller system manual*, 2001.
4. M. Song, T. Wang, *Implementation of communication between Siemens PLC and level two computer with Ethernet*, Metallurgical Industry Automation, 35(5):65-68, 2011.
5. BJ. Wang, XJ. Wang, *Centralized control system in washery based on VC++ and PLC*, Coal Mine Machinery, 34(2):229-230, 2013.

Author Index